Springer Geography

The Springer Geography series seeks to publish a broad portfolio of scientific books, aiming at researchers, students, and everyone interested in geographical research.

The series includes peer-reviewed monographs, edited volumes, textbooks, and conference proceedings. It covers the major topics in geography and geographical sciences including, but not limited to; Economic Geography, Landscape and Urban Planning, Urban Geography, Physical Geography and Environmental Geography.

Springer Geography—now indexed in Scopus

Subodh Chandra Pal · Uday Chatterjee
Editors

Surface, Sub-Surface Hydrology and Management

Application of Geospatial and Geostatistical Techniques

Springer

Editors
Subodh Chandra Pal
Department of Geography
The University of Burdwan
Bardhaman, West Bengal, India

Uday Chatterjee
Department of Geography
Bhatter College Dantan
Paschim Medinipur, West Bengal, India

ISSN 2194-315X ISSN 2194-3168 (electronic)
Springer Geography
ISBN 978-3-031-62375-2 ISBN 978-3-031-62376-9 (eBook)
https://doi.org/10.1007/978-3-031-62376-9

This Springer imprint is published by the registered company Springer Nature Switzerland AG
The registered company address is: Gewerbestrasse 11, 6330 Cham, Switzerland

To Young Scholars in the Field of Hydrology, Water Resource Management, Environmental Science, Hazard and Disaster Management, Geospatial Science, Geography, Social Science and Sustainability Science and Policymakers

Foreword

It is indeed a great pleasure to pen down a foreword for the book entitled *Surface, Sub-Surface Hydrology and Management: Application of Geospatial and Geostatistical Techniques* edited by Dr. Subodh Chandra Pal and Dr. Uday Chatterjee, hailing from the discipline of Geography. The book covers comprehensive articles from the fields of surface and sub-surface hydrology, environmental science, water resource management, soil moisture and erosion estimates, floods, geospatial science, geography, social and sustainability science and policy planning. This book is being published in *Springer Geography* Book Series, Springer International Publishing, Springer Nature Switzerland AG (Publisher).

We all understand the value of water and more so in the wake of changing climatic scenario. The availability of freshwater on the globe is under tremendous pressure due to increasing population, urbanization, industrialization and adverse impact of climatic changes which are resulting into major environmental challenges as well at many places. Natural distribution of freshwater resources on the globe is skewed and uneven which makes it more pertinent to conserve, preserve, recycle and reuse the water in a prudent manner. The increased dependency on groundwater in major agriculture dominating parts is resulting into sharp decline of groundwater levels across various regions in the world. These management issues are becoming more complex in the wake of climate change. The climate change is threatening the environmental ecosystem on the earth resulting into rise in temperature, increasing glacier melt, sea level rise, reduction in agricultural production, and erratic rainfall pattern leading to threatening disasters across the globe. The perceptible change in the rainfall intensity and pattern and erratic behaviour of many environmental parameters is conspicuous in recent years. It is therefore necessary to study the surface, sub-surface water resources in various regions and understand its interplay with long-term hydrological and hydrogeological datasets. The basin-wise integrated water resources planning needs to be practiced. Geospatial technology and geostatistical techniques provide a handy tool for addressing these challenges on sustainable basis.

Geospatial technology, an integrated use of remote sensing, geographical information systems, global positioning system and information and communication technology, helps in generating baseline information for the effective and optimal utilization of water resources on one hand and parameterization of many environmental phenomenon on the other as it aids to better decision-making for planners and decision-makers based on sound techno-economic feasible considerations. The increasing availability of geospatial data from various sources in the open domain is encouraging the researchers to address these issues at global, regional and local level with increasing inputs from site-specific measurements. The multi-temporal and multi to hyperspectral earth observation data through optical and radar satellites is continuously generating huge spatial data source for the scientific community which is easing water resources management challenges created across the globe due to natural and anthropogenic influences. Visual image interpretation and digital image processing along with limited field checks create timely baseline information for prudent decisions. Continuous monitoring of glaciers, methane emissions from different land uses, changes in the river flow, forest cover and urban heat islands through earth observation devices are certainly helpful for addressing multifarious challenges. The information from global circulation models is being integrated with site-specific data sets to create local specific plans for addressing water science and environmental challenges. Geospatial technology and geostatistical tools certainly provide standardized solutions for assessing potential and planning processes across geographical regions.

The volume in hand presents a compilation of thirty research articles in the form of chapters under three different parts, namely hydrology, environment and management covering topics ranging from basin hydrological phenomena, hydrological drought, community flood initiatives, flood susceptibility, soil moisture estimation, relationship between surface water and groundwater, water quality, morphometric and morphotectonic indices, water logging, soil erosion risk and vulnerable zone identification and watershed hydrology to natural resources management using various geostatistical tools and geospatial technology.

I congratulate the editors for this nice initiative to produce such a valuable treatise for catering the needs of academicians, researchers, scientists, policy planners and interventionists in the domain of earth system sciences. The book is rich resource of various research articles covering state-of-the-art approaches in the domain. I am quite sure that it will not only add new knowledge to the domain of earth system sciences but will also be helpful in igniting young minds to deal the problems in a multidisciplinary way.

Prof. (Dr.) Bhagwan Singh Chaudhary,
DAAD Fellow
Secretary
Indian National Chapter
of International Association
of Hydrogeologists
Department of Geophysics
Kurukshetra University
Kurukshetra, Haryana, India
bschaudhary@kuk.ac.in

Former Registrar
Kurukshetra University
Kurukshetra, Haryana, India

Chaudhary Bansi Lal University
Bhiwani, India

Preface

Surface and below-surface water systems are essential parts of Earth's water cycle, crucial for sustaining life and shaping the environment. Comprehending how water moves across the earth's surface and underground is crucial for managing water resources, protecting the environment and promoting sustainable development. The world is rapidly running out of freshwater, with many regions around the globe experiencing severe water scarcity in recent years. Worldwide, the progressive rise in groundwater pumping has caused significant drops in groundwater levels in numerous regions as a result of unsustainable water pressure. Groundwater resource exploitation has occurred at a rate that prevents the water table from recovering its losses. As a result, shallower aquifers dried up, river water levels dropped, and wells were dug continuously. This book explores the surface and sub-surface hydrological conditions and their management in a tropical or sub-tropical setting applying geospatial and geostatistical derived predictive methods. The book is a research of the surface and sub-surface hydrological processes as a result of over-exploitation for agricultural production and irrigation, which resulted in a significant drop in water even during pre-monsoon period and a charge up study to control surface and sub-surface water scarcity. Recent and expected trends indicate that future water challenges will become more complex. Over-exploitation of groundwater and surface water scarcity are posing an increasing threat to food security and rural livelihood stability. Understanding the features of groundwater and surface water and how they interact with the environment, population and social aspects is a difficult challenge for researchers. They must provide guidance for sustainable resource management while balancing economic efficiency and ecological health. This book stands out for its focus on using geospatial and geostatistical methods to tackle practical issues. The chapters demonstrate how advanced analytical tools can be used to extract useful insights from complex hydrological data sets, such as mapping surface water flow patterns and delineating groundwater contamination plumes. Additionally, including case studies from various geographic areas offers readers a worldwide view on hydrological challenges and approaches to managing them. This book is catered towards undergraduate, postgraduates, researchers, policymakers, hydrologists, environmentalists, managers of natural hazards and disasters, NGOs, corporate sectors, social scientists and government officials. This book acts as a catalyst for more research and dialogue in the areas of environmental sciences, geography, hydrology, natural hazards, geospatial sciences, remote sensing and GIS, agriculture, crop science, forestry, soil science, agronomy, humanistic and social sciences, etc., promoting organization in the field of surface and sub-surface hydrology. We aim to advance scientific knowledge and promote sustainable water resource management by combining expertise from various disciplines and highlighting the potential of geospatial and geostatistical techniques.

Bardhaman, West Bengal, India — Subodh Chandra Pal
Paschim Medinipur, West Bengal, India — Uday Chatterjee

Acknowledgments

Our heartfelt appreciation goes out to the contributors, colleagues, researchers and reviewers who were essential in the creation and release of the book entitled—"Surface, Sub-Surface Hydrology and Management: Application of Geospatial and Geostatistical Techniques." Their knowledge, assistance, advice and relentless commitment greatly impacted the development of this in-depth resource focusing on a better comprehension of the surface and sub-surface hydrology dynamics and the application of geospatial and geostatistical techniques for sustainable water resource management. Our families deserve special recognition for their continued support, understanding and encouragement throughout the challenging process of completion of the book. We are also thankful to the publishing team at Springer for their exceptional guidance, expertise and dedication to excellence. Their precise editing, design and production work have turned the book into a comprehensive and reader-friendly resource.

Bardhaman, West Bengal, India — Subodh Chandra Pal
Paschim Medinipur, West Bengal, India — Uday Chatterjee

Contents

About the Editors

Dr. Subodh Chandra Pal Assistant Professor has been working in the Postgraduate Department of Geography, The University of Burdwan, Purba Bardhaman, West Bengal, India. He received his Ph.D. in Geography from Visva-Bharati (Central University in India) in 2015. His main fields of expertise are fluvial geomorphology, pedo-geomorphology, land degradation, soil erosion, gully erosion, landslide vulnerability, climate change and water resources monitoring and management and published nearly two hundred international and national research articles in various renowned journals (Springer, Elsevier, Taylor & Francis group, Wiley, MDPI and Nature Publishing group.). He published one author book from Springer Nature publishing house on *Climate Change Impact on Soil Erosion in Sub-tropical Environment: Application of Empirical and Semi-empirical Models*. He is Editor of four running edited books published by Elsevier and Springer Publishing house. His research work has been funded by the University Grants Commission (UGC), Government of India, and Department of Science and Technology (DST), Government of India. Dr. Pal has served as Reviewer for many international reputed journals. He is Editorial Board Member of *MDPI Water* journal and *Frontiers of Earth Science* journal. Since 2022, Stanford University has recognized Dr. Pal as one of the world's top 2% scientists.

Dr. Uday Chatterjee is Assistant Professor at the Department of Geography, Bhatter College, Dantan, Paschim Medinipur, West Bengal, India, and Applied Geographer with a Doctoral Degree in Applied Geography at Ravenshaw University, Cuttack, Odisha, India. His areas of research interest cover urban planning, social and human geography, applied geomorphology, hazards and disasters, environmental issues, disaster governance, community-based disaster risk management, climate change adaptation, urban risk management and disaster. He has delivered 10 invited lectures in University Grants Commissions (UGC) sponsored by national seminars and various academic departments of different colleges in India. In addition, he presented 20 papers in national and international seminars/conferences held in India as well as chaired and co-chaired more than 5 technical sessions. He has successfully guided project dissertations to undergraduate students. He has also conducted (Convener) one Faculty Development Programme on 'Modern methods of teaching and advanced research methods' sponsored by Indian Council of Social Science Research (ICSSR), Government of India. Currently, Dr. Uday Chatterjee has completed the Special Issue (S.I.) of Urbanism, Smart Cities and Modelling, *GeoJournal*, Springer as Lead Editor and Book Series Editor, Development in Environmental Science, Elsevier (https://www.elsevier.com/books-and-journals/book-series/developments-in-environmental-science). His research work has been funded by the West Bengal Pollution Control Board (WBPCB) Government of West Bengal, India. He has served as Reviewer for many International journals. Currently, Dr. Chatterjee is doing the international project, in collaboration with Indonesia, Malaysia and Japan funded by APN (Asia Pacific-Global Change Research), and he is Guest Editor of the special issue *Social Ecology, Human-Well-being and Sustainability, Global Social Welfare* journal, Springer. He has published 30 research papers, 10 edited books, 2 author books, 15 book chapters and 2 conference proceedings.

Hydrology

Assessment of Basin Hydrological Phenomena Using State-of-the-Art Methods and In-situ Field Measurements

Asmit Roy Burman

Abstract Water holds significance not only for society but also for crucial Earth system processes like weather, climate, and biogeochemical cycles. Advancements in remote sensing technology have demonstrated the ability to study water resources and hydrologic processes from a spatial perspective. On a global scale, approximately 30% of the Earth's freshwater resources can be attributed to groundwater, making it the source of 96% of all available liquid freshwater. The storage of groundwater, which is part of terrestrial water storage (TWS), is influenced by both human-induced activities and climate-related factors. Terrestrial water storage (TWS) encompasses all forms of water present both above and below the Earth's surface, including soil moisture, snow, ice, water stored in vegetation canopies, groundwater, and other similar reservoirs. The Gravity Recovery and Climate Experiment (GRACE) missions employ satellite–based gravimetry to precisely estimate the Terrestrial Water Storage Anomaly. The GRCAE TWSA values show very high negative correlation to groundwater storages, almost all of India showing depletion, highest being in Haryana while recharge happening in parts of the Deccan Plateau highest in Madhya Pradesh. Assessing the TWSA enable us to comprehend this intricate system and aid in resolving water resource and environmental management challenges.

Keywords Gravity anomaly and climate experiment (GRACE) · Terrestrial water storage (TWS) · Remote sensing · Climate change

A. R. Burman (✉)
Disaster Management Studies Department, Indian Institute of Remote Sensing-Indian Space Research Organisation, Dehra Dun, India
e-mail: asmit.iirs@gmail.com

S. C. Pal and U. Chatterjee (eds.), *Surface, Sub-Surface Hydrology and Management*, Springer Geography, https://doi.org/10.1007/978-3-031-62376-9_1

1 Introduction

Water holds significance not only for society but also for crucial Earth system processes like weather, climate, and biogeochemical cycles. Consequently, comprehending this intricate system not only aids in resolving water resource and environmental management challenges but also contributes to a better understanding of Earth's system processes. Nevertheless, the scientific community has faced hurdles in terms of insufficient continuous data and monitoring networks. However, advancements in earth observation technology have demonstrated the ability to study water resources and hydrologic processes from a spatial perspective (Rodell and Famiglietti 2002; Andreadis and Lettenmaier 2006; Birkinshaw et al. 2010; Vinukollu et al. 2011; Duan and Bastiaanssen 2013; McCabe et al. 2017; Blöschl et al. 2019). Studies indicate that groundwater provides 96% of the liquid freshwater that is available on Earth, accounting for around 30% of the planet's freshwater resources globally (Niu et al. 2007; Shiklomanov). Approximately 43% of irrigated agriculture worldwide relies on groundwater as its primary source, while around half of the global population depends on groundwater for their drinking water consumption (Siebert et al. 2010; Richard 2015). Groundwater depletion has become a pressing issue in numerous regions across the globe, including India, due to the excessive extraction of groundwater for irrigation purposes in the past few decades (Tiwari et al. 2009; Rodell et al. 2009; Wada et al. 2010; Aeschbach-Hertig and Gleeson 2012; Asoka et al. 2017). India is the nation that uses the most groundwater, an invaluable resource that supplies around 85% of the nation's drinking water and 60% of its water for agriculture (Sundaray et al. 2009; Pahuja et al. 2010; Siebert et al. 2010; Shankar et al. 2011; Bhattarai et al. 2021). Decline in the monsoon precipitation and unsustainable extraction of ground water has come up as the primary cause of the groundwater depletion in India (Tiwari et al. 2009; Rodell et al. 2009; Asoka et al. 2017). The storage of groundwater, which is part of terrestrial water storage (TWS), is influenced by both human-induced activities and climate-related factors. As a result, changes in TWS can be attributed to the combined effects of anthropogenic actions and climatic conditions on groundwater storage among other factors (Taylor et al. 2013). TWS encompasses all forms of water present both above and below the Earth's surface, including soil moisture, snow, ice, water stored in vegetation canopies, groundwater, and other similar reservoirs (Meng et al. 2019). In recent years, the progress made in earth observation technologies and climate system modeling has led to significant advancements (Fig. 1).

Modern numerical weather models and space-borne sensors have been instrumental in producing vast precipitation datasets with almost worldwide coverage and unmatched spatiotemporal resolution (Sunilkumar et al. 2016; Prakash et al. 2018). The integrated multi-satellite retrievals for GPM (IMERG) product, which is now available at a temporal resolution of 30 min and a spatial resolution of 0.1°, is a prime example. This dataset has demonstrated its critical importance in the observation of meteorological and hydrological conditions (Beria et al. 2017; Massari et al. 2019). Several satellite sensors have been employed in water-related studies, each associated

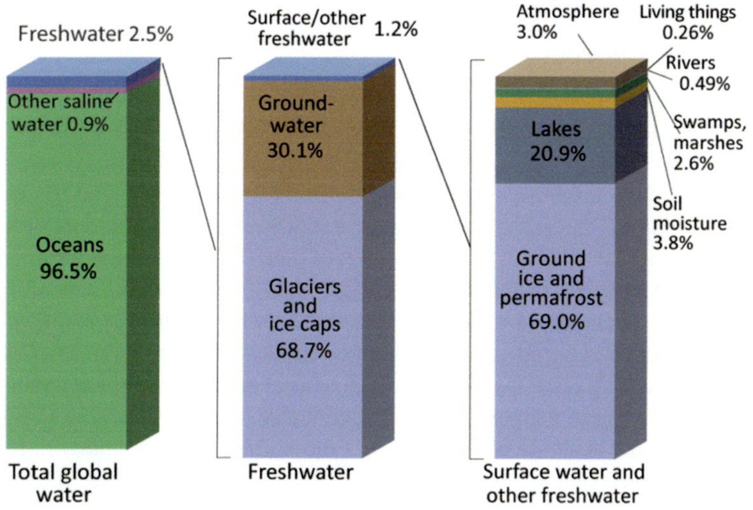

Fig. 1 Earth water system (Mehta et al. 2020)

with specific hydrologic variables. These include the advanced microwave scanning radiometer—Earth observing system (AMSR-E), the global precipitation measurement (GPM) for quantifying precipitation, the surface water and ocean topography (SWOT) for assessing water resources overall, the gravity recovery and climate experiment (GRACE)/gravity recovery and climate experiment-follow on (GRACE-FO) for monitoring terrestrial water storage anomalies (TWSA), the soil moisture active passive (SMAP) for measuring soil moisture, the moderate resolution imaging spectroradiometer (MODIS) for evaluating vegetation and land surface temperature, and the moderate. To comprehend the effects of human and climate change on groundwater storage (GWS) and shifting occupant usage patterns, an assessment of terrestrial water storage is crucial. The land surface models and TWSA obtained from GRACE/GRACE-FO data are widely used to evaluate the variations and oscillations in TWS. These techniques are also applied to estimate regional and worldwide variations in groundwater storage (Rodell et al. 2009; Asoka et al. 2017; Nie et al. 2018; Tapley et al. 2019).

The objectives of this chapter delve into the utilization of GRACE missions to assess and comprehend basin hydrology in India, shedding light on the dynamics of surface and subsurface water resources. It aims to investigate the contrast between depletion and recharge scenarios. Through data retrieval, preprocessing, and analysis, the chapter seeks to unravel spatial and temporal patterns of water storage variation. Furthermore, the findings will undergo validation against groundwater level data and be contextualized within the existing literature.

2 Evolution of GRACE Missions to Assess Earth System Water Resources

TWSA estimation significantly improved in hydrological studies after the introduction of the Gravity Recovery and Climate Experiment (GRACE) satellite in 2002. This satellite mission, a collaborative effort between NASA and the German Aerospace Center (DLR), provided reliable and accurate measurements of TWSA on a global scale. The GRACE satellite uses a precise gravity measurement technique to detect changes in the Earth's gravitational field, enabling the estimation of TWSA. The availability of GRACE data revolutionized hydrological studies by providing comprehensive and consistent measurements of TWSA. Prior to GRACE, estimating changes in terrestrial water storage was challenging and relied on limited ground-based observations and modeling techniques. The introduction of GRACE allowed for a more accurate assessment of TWSA, including variations in groundwater storage, soil moisture, snow, and surface water. The GRACE satellite mission provided a continuous record of TWSA at a spatial resolution of several kilometers and a temporal resolution of approximately one month. This dataset has been widely utilized in hydrological studies to monitor long-term trends, understand hydrological processes, assess water resource availability, and study the impacts of climate change and human activities on terrestrial water storage (Andreadis and Lettenmaier 2006; Taylor et al. 2013; Felfelani et al. 2017; Zhu et al. 2021). Accurately measuring and evaluating the impact of human activities and climate change on terrestrial water resources is crucial for ensuring the sustainable management of these valuable water sources. By quantifying these influences, we can gain a better understanding of water availability, identify trends and patterns, assess ecosystem impacts, manage water demand effectively, and develop strategies to adapt to the challenges posed by climate change. Ultimately, quantification serves as a foundation for informed decision-making and the implementation of sustainable water management practices (Gleeson et al. 2012; Famiglietti 2014; Felfelani et al. 2017). The estimation of Terrestrial Water Storage Anomaly (TWSA) using the Gravity Recovery and Climate Experiment (GRACE) method provides valuable information on changes in vertically integrated surface and subsurface water storage. These changes are influenced by both climate factors and human activities. Climate factors include variations in precipitation patterns, temperature, and evapotranspiration, while anthropogenic activities involve activities such as groundwater pumping, reservoir construction, and land-use changes. By considering both climate and anthropogenic influences, the GRACE-based TWSA helps us understand the complex dynamics of water storage changes and enables the assessment of the combined effects of natural and human-induced factors on terrestrial water resources. This information is essential for sustainable water resource management and decision-making processes (Huang et al. 2015).

2.1 Measurement Principle of GRACE Missions

The GRACE missions which include GRACE and GRACE-FO was first launched on March 17, 2002, and then in 2018 was a collaborative effort between NASA and the German Aerospace Centre, Deutscen Für Luftund Raumfahrt (DLR). It was initiated based on years of research and recommendations (Natural Research Council 1997; Marti 2012). The primary objective of GRACE was to track changes in mass to monitor changes in the hydrosphere, cryosphere, and oceans using monthly combined observations of Earth's gravitational field (Tapley et al. 2004). GRACE uses a constellation of two satellites, as opposed to single satellite techniques, which depend on a single dedicated sensor. Each of these spacecrafts has an array of measurement instruments installed, and they orbit one after the other. The measurements of GRACE and GRACE -FO are based on two identical satellites (GRACE A/B) orbiting the planet in a near-polar orbit plane, one behind the other (Fig. 2). Within a range of 220 ± 50 km, the GRACE satellites maintain their spacing along their trajectory. The satellites undergo both positive and negative along-track accelerations due to differences in the mass distribution beneath them. Because each satellite contacts the local mass at a little different time, this causes a differential acceleration. The mass attraction of the gravitational potential (U) is directly proportional to variations in distance (range) and velocity ($\varDelta$V) caused by differential acceleration. $\varDelta V = V_a - V_b$; where V_a = velocity at initial point *a*, V_b = velocity at another point *b*. A dual-frequency, K-band intersatellite ranging system measures the relative distance between the satellites to within microns. Each satellite is equipped with a three-axis accelerometer, which is used to monitor pressure from solar radiation, pushing, and other nongravitational forces. Position and time synchronization are achieved via GPS receivers, while orientation data are obtained by a dual star camera system. The satellites cover the whole surface of the Earth in around 30 days, enabling monthly estimations of a global gravity model with a high accuracy of 2 cm and a typical surface spatial resolution of 300 km (Vishwakarma et al. 2018). The readings are used to estimate a global spherical harmonic model of Earth's gravity field after a month of data collection. Then, this model is used to evaluate changes in Earth's surface mass. It is crucial to remember that differences in the inferred mass time series acquired from the GRACE data can be introduced by the processing decisions made by the science data system (SDS) centers and the users (Luthcke et al. 2006; Watkins et al. 2015; Save et al. 2016).

2.2 GRACE Data Used

The Centre for Space Research (CSR) provided GRACE and GRACE-FO RL06.2 Mascon solutions for this investigation. The mascon grids for CSR RL06.2 have been adjusted to account for an ellipsoidal Earth. The analysis of the data spanned

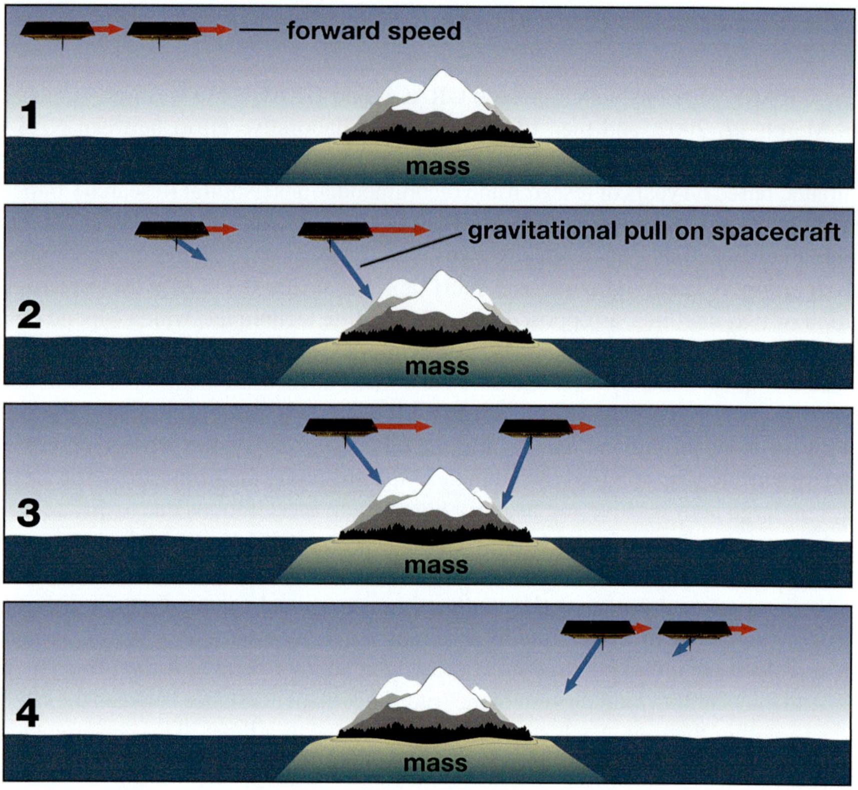

Fig. 2 GRACE working principle (Tapley et al. 2019)

225 months, from April 2002 to September 2023. The dataset has a temporal resolution of one month and a geographical precision of 0.25°, or around 27.22 km. For the GRACE/GRACE-FO time variable gravity, the mass anomaly grids in CSR RL06.2 mascons are represented on an ellipsoidal earth with a semi-major axis of 6378.1363 km. The mass anomaly grid representation ellipsoidal adjustment has been done in accordance with (Ditmar 2018). The necessary corrections (C20 replacement, C30 replacement, Degree 1, GIA, and GAD corrections) have already been applied to the CSR RL06.2 solution (Purcell et al. 2016; Richard Peltier et al. 2018). Every anomaly that has been identified is compared to the mean baseline for 2004–2009. The updated grid used in the mascon solutions RL06 version is different from the one used in the RL05 version. The hexagonal tiles that span the shoreline are split into two tiles in this modified grid along the actual coastline. This separation is made to lessen signal interference or mixing between places that are on land and those that are at sea (Save et al. 2016; Save 2020).

2.3 Terrestrial Water Storage

Measurement errors and noise cause the signal quality used to estimate fluctuations in terrestrial water storage (TWS) to deteriorate. These mistakes fall into two categories: random errors that increase in spectral degrees of spherical harmonics (Wahr et al. 1998), as well as systematic mistakes that show association within particular spectral orders (Swenson and Wahr 2006). Land-hydrology model simulations run by (Landerer and Swenson 2012) created connections between regional time series representing different geographic scales. These correlations were used to extend GRACE TWS estimations to finer spatial scales (about 100 km) from their effective spatial resolution (a few hundred kilometers), irrespective of the actual GRACE data. Three scaling relationships were examined: one gain factor based on regionally averaged time series, spatially dispersed gain factors based on time series at each grid point, and gridded-gain factors estimated as a function of temporal frequency. Their results showed that, in contrast to the widely utilized regional gain factors, scaled time series based on gridded gain factors can yield equivalent accuracies. However, applying gain factors from the previous two techniques could result in a loss in the accuracy of the scaled time series when various temporal modes of TWS variability had significantly different spatial scales. In these situations, distinct gain factors that were calculated using frequency were necessary to provide accurate results. The study's gridded fields of leakage and GRACE measurement errors allowed users to calculate the level of uncertainty surrounding regional TWS. TWS is also defined as liquid water equivalent (LWE) in the mascons data. This basically refers to the equivalent amount of water stored in all forms, i.e. ice, snow, soil moisture, vegetation moisture, and liquid as well. The measurement errors showed a latitudinal relationship, with the highest values (standard deviation up to 35 mm) close to the equator and the lowest values (15 mm) closer to the poles (Landerer and Cooley 2021). In India, the issue of TWSA (Total Water Storage Anomaly) change is a major worry due to the excessive utilization of water, specifically in agriculture, as well as inadequate governance. The results of this research hold potential value in the effective organization and control of water resources, agriculture, and the country's long-term development. Moreover, these findings are crucial for any forthcoming strategic planning. These problems can be addressed through efficient water resource management, implementation of an intelligent water distribution system, and promoting sustainable water usage (Hasan et al. 2023). The GRACE mission has made significant contributions, particularly in revealing the dynamic alterations occurring in Earth's freshwater distribution, which carry significant implications for water security, food production, and human well-being. Global GRACE calculations show that whereas mid-latitudes have seen a decline in water storage (drying), both high and low latitudes have shown a tendency of increasing water storage (wetness) (Reager et al. 2016; Rodell et al. 2018). The finding of significant changes in the global hydrological cycle, even though the GRACE mission's record has only been in place for a short time, has provided crucial preliminary support for the changes projected by climate models for the twenty-first century (Held and Soden 2006;

Stocker et al.). However, reliable predictions of future water availability need a thorough evaluation of soil moisture trends using models such as the Coupled Model Intercomparison Project Phase 5 (CMIP5) in combination with data from GRACE and other measurements. Accurate projections of future water availability remain highly dependent on climate models. There could be changes in the accessibility of freshwater in the future, which would make it difficult to maintain supplies of water for industrial uses, agriculture, human consumption, and food security. These changes are indicated by the observed drying in mid-latitudes, wetting in high and low latitudes, and declining water tables in mid-latitude aquifers. Notably, 13 of the 37 major aquifer systems in the globe were found to be critically depleted over the GRACE observing period (Richey et al. 2015).

3 Materials and Methods

3.1 Study Area

The study area is India. It comprises of 28 states and 8 union territories, namely Andhra Pradesh, Arunachal Pradesh, Assam, Bihar, Chhattisgarh, Goa, Gujarat, Haryana, Himachal Pradesh, Jharkhand, Karnataka, Kerala, Madhya Pradesh, Maharashtra, Manipur, Meghalaya, Mizoram, Nagaland, Odisha, Punjab, Rajasthan, Sikkim, Tamil Nadu, Telangana, Tripura, Uttar Pradesh, Uttarakhand, West Bengal, Andaman and Nicobar Islands, Chandigarh, Dadra and Nagar Haveli and Daman and Diu, Lakshadweep, The government of NCT of Delhi, Puducherry, Ladakh, and Jammu and Kashmir. Approximately 24% of the global population resides in the Indian subcontinent which occupies only 3.61% of the global land area (FAO 2013) out of which the major portion belongs to India. India's terrain is a patchwork of interesting and varied landscapes that showcase a wide variety of topographical elements and ecosystems. The Himalayas are a dominating mountain range in the northern, distinguished by snow-capped peaks, vast valleys, and unspoiled river systems (Fig. 3).

These mountains provide an amazing geological profile, including the well-known Mount Everest. As one moves south, one finds that the rich plains of the Indo-Gangetic basin sustain a sizable population and make a major contribution to agricultural sustenance. The coast-to-coast Western and Eastern Ghats are home to a multitude of endemic species, lush forests, and abundant biodiversity. The arid western regions of India, particularly the Rajasthani deserts, provide distinctive landscapes featuring golden sand dunes. India is drained by numerous rivers which make through numerous basins, which contain some of the world's top yielding aquifers. The aquifers associated with these river basins span international borders within the contiguous countries that make up the Indian subcontinent. These aquifers are known as transboundary aquifers, and they include the aquifers of the tributaries to the Ganges (between Nepal and India), the aquifers of the tributaries to the Brahmaputra

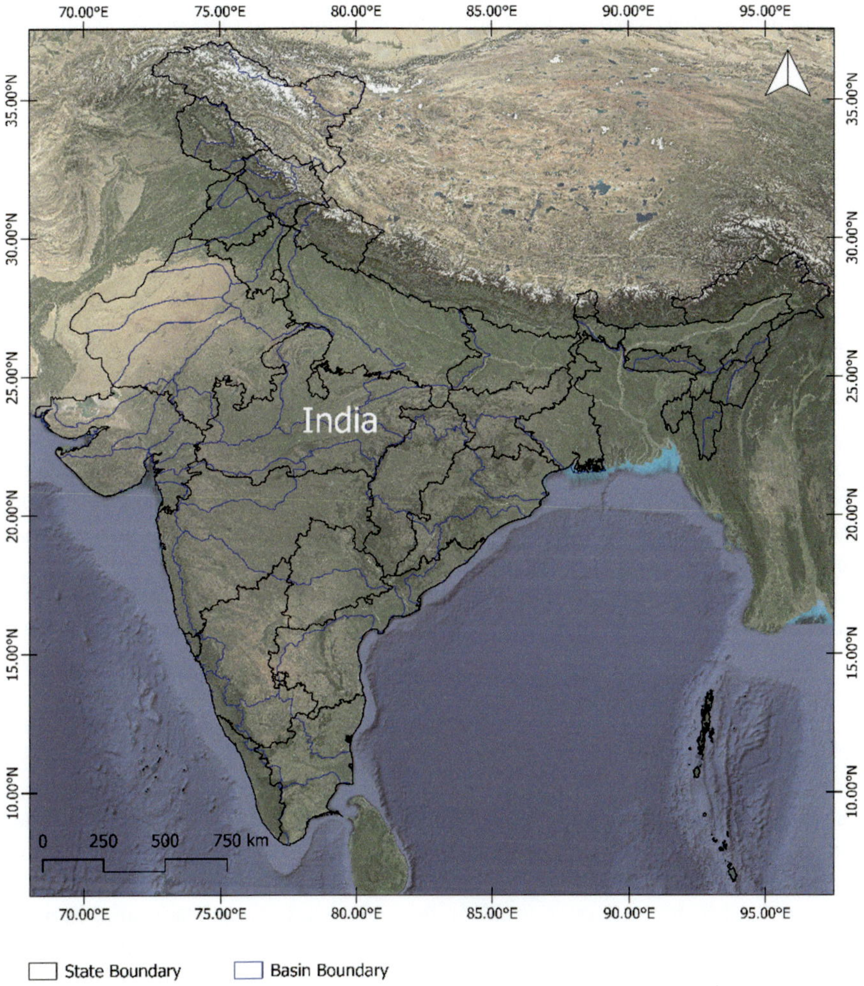

Fig. 3 Study area (India)

(between Bhutan and India and between India and Bangladesh), and the aquifers associated with the Indus basins (between India and Pakistan), the Ganges and Brahmaputra basin aquifers (between Bangladesh and India), and so on (Mukherjee et al. 2015). Groundwater withdrawals in the India are projected to surpass 350 billion cubic meters (bcm) annually, making it the largest consumer of groundwater globally (Siebert et al. 2010). The region is currently experiencing a severe scarcity of potable water and other usable water sources in numerous areas. This situation is exacerbated by the escalating demand for water due to rapid urbanization, shifting societal water consumption patterns, and changes in lifestyle (Scanlon et al. 2010). In many urban and rural areas of the India, surface waters have been historically used as receptacles of sewage and industrial waste, rendering them unfit for domestic use, prompting a switch to groundwater and rainwater sources to meet drinking and agricultural water needs (Mukherjee et al. 2011). At present, groundwater sources fulfill approximately 60–80% of the domestic water requirements in the Indian subcontinent, including countries such as Bangladesh, India, and Pakistan. Most groundwater withdrawals, over 85%, are allocated to irrigation purposes in the region. This extensive use of groundwater for irrigation is recognized as the primary factor contributing to the depletion of terrestrial water resources (Tiwari et al. 2009; Rodell et al. 2009; Famiglietti 2014). The terrestrial water storage conditions in the India may become more complex due to the anticipated consequences of impending climate change. The headwaters of the India's river basins receive their water supply from the melting of glaciers at high altitudes. In recent years, there has been an increase in glacial lake outburst floods in the Himalayas, specifically in Bhutan, India, and Nepal, with the occurrence rising from 0.38 events per year in the 1950s to 0.54 events per year in the 1990s. Global warming is expected to lead to higher surface runoff in the period from 2090 to 2099 compared to 1980 to 1999, thereby increasing the potential for flood risks downstream (Bates et al. 2008). The eastern side of the Indian subcontinent has been observed to have high recharge rates, exceeding 300 mm per year. This is attributed to the relatively higher levels of current precipitation in that region. However, the western parts of the India receive limited recharge, indicating lower levels of replenishment of groundwater resources in those areas (Scanlon et al. 2010).

3.2 Methodology

The GRACE and GRACE-FO CSR RL06.2 Mascons data have been downloaded and thoroughly analyzed for terrestrial water storage changes for India and further narrowed down to Central-Northwestern regional basins where the areas of maximum depletion and recharge happen to fall in the entire study area. The spatial average, maximum, minimum, and standard deviation have been extracted and plotted for analysis. The onset of monsoon happens in July according to IMD, consequently the seasonal averages have been calculated and the seasonal trends during pre–monsoon, monsoon, and post-monsoon seasons have been analyzed. The reasons behind these trends have also been investigated and validated with in situ measurements of ground data (Fig. 4).

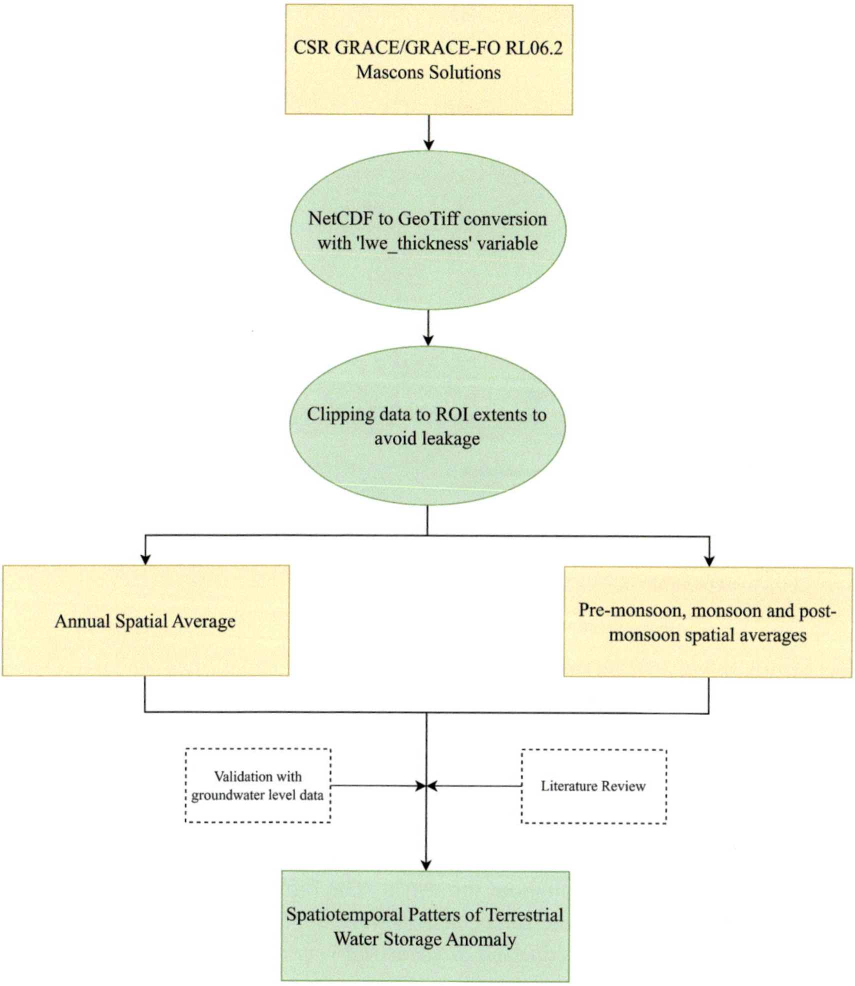

Fig. 4 Methodology

4 Results and Discussion

4.1 TWSA in India

From Fig. 5, it is well evident that the terrestrial water storage of which groundwater storage is a major component of Mehta et al. (2020) has been alarmingly decreasing in the Indian subcontinent over the past two decades. Not only that the total storage is decreasing but the rate of increase of the depletion has increased in the recent years as is observable from the plots. It is also a matter of grave concern that the amount of recharge happening in the monsoon season through the years has also been

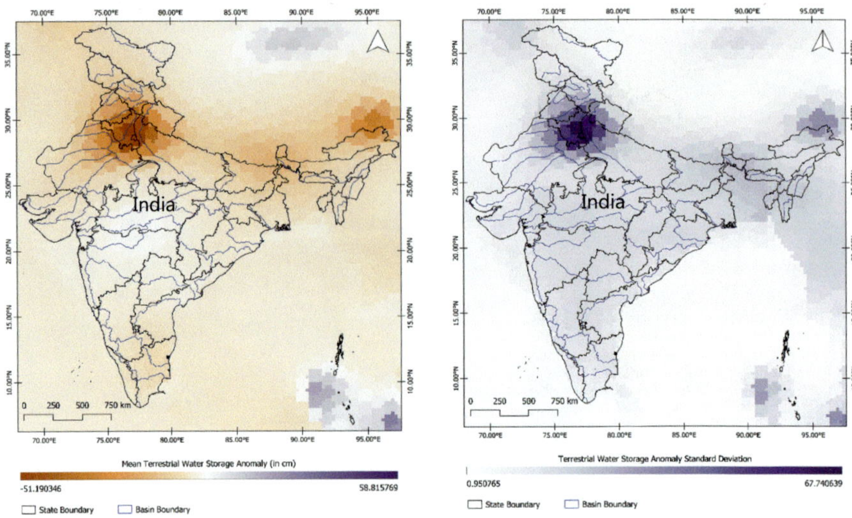

Fig. 5 Mean (left) terrestrial water storage anomaly and standard deviation (right) over India during April 2002 to September 2023

decreasing. It is also observable that the zones of highest depletion as well as recharge are in India. The depletion can be attributed to India using the most groundwater for different purposes (Rodell et al. 2009; Wada et al. 2010; Asoka et al. 2017; Asoka and Mishra 2020; Chatterjee et al. 2020). The zones of relatively high recharge, even though the net change in terrestrial water storage is negative, are mostly due to the hydrogeological setup of the regions (Chatterjee et al. 2020; Pranjal et al. 2023).

As per IPCC (2021) global warming and the greenhouse effect has triggered the rapid melting of glacier throughout the world. The rapid melting has disturbed the glacier mass balance and in turn has cause severe depletion in the terrestrial water storage of these areas. The reduction of terrestrial water storage in the northwestern Indian region and in the Indo-Gangetic plain is not a contribution of the glacier ice melting but extraction of groundwater for irrigation. In the Upper Indus Basin (UJB), there is a noteworthy decline in total water storage (TWS) during winter and on an annual scale, with statistical significance at either the 0.1 or 0.05 level. The decrease observed in the winter season can be attributed to a decline in winter precipitation and the loss of glacier mass caused by global warming in the Western Himalaya. On the other hand, the decrease in annual TWS, despite an increase in precipitation in low-altitude areas, is primarily due to the rapid generation of runoff, which leaves less time for groundwater recharge (Ansari et al. 2022; Kulkarni et al. 2023). TWSA can also provide important insights on the hydrological condition of a particular basin. The LWE can tell us about the deficiency or abundancy of water resources in a basin and assist us on different aspects of water resources management such as the National River Linking Project of the Government of India (GoI) (Fig. 6).

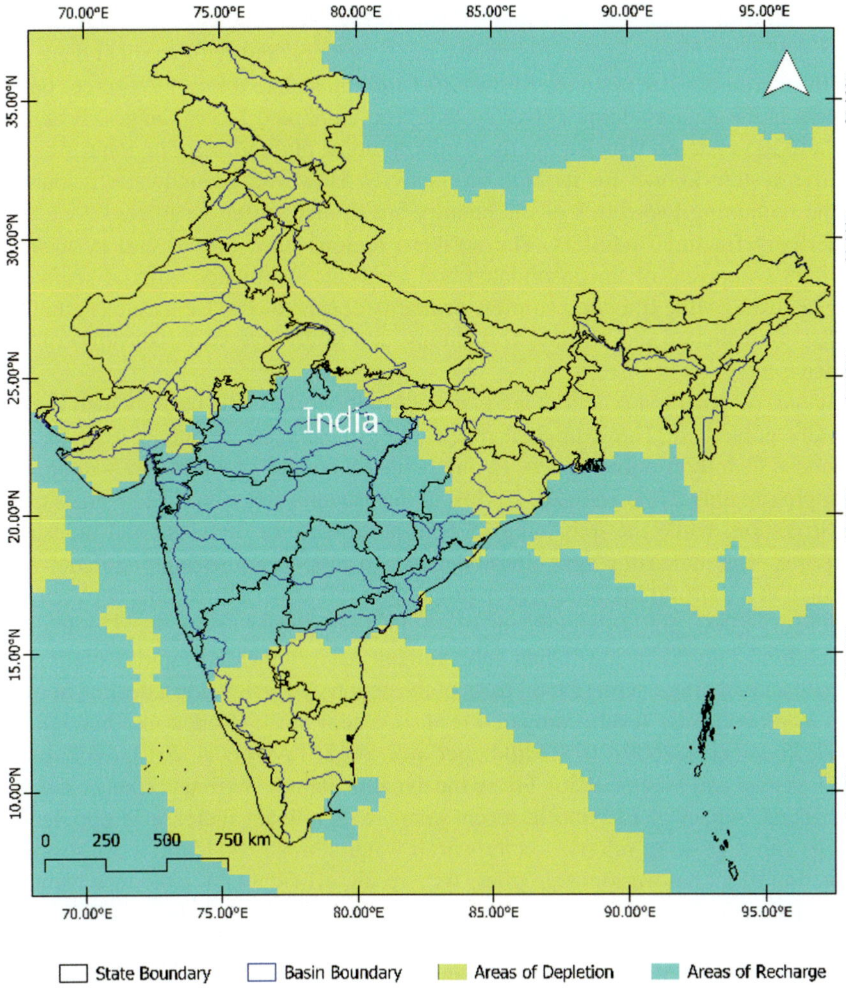

Fig. 6 Areas of net depletion and recharge of TWS over India during April 2002 to September 2023

India, as the largest country within the Indian subcontinent, holds the highest land area and population. The country receives between 750 and 1500 mm/year of precipitation annually, with the western regions receiving somewhat less rain (2500 mm/year). The majority of precipitation falls between June and September, during the southwest monsoon season. The main river basins that drain India are connected to the nation's major aquifers. Based on their origin and flow pattern, these basins can be divided into four groups: (i) Himalayan rivers, like the Ganges, Brahmaputra, and Indus, which originate from high-altitude glaciers and snow and flow continuously; (ii) Indian craton rivers, like the Mahanadi, Godavari, Krishna, Pennar, Cauvery, Narmada, and Tapi, which are mostly fed by rain and depend on baseflow; (iii) coastal rivers, most of which are nonperennial; and (iv) rivers in the western desert, which are small-scale fluvio-aeolian basins that are ephemeral and fed by rain and are cut off from groundwater systems. At over 86.1 million square miles, the Ganges basin system is the nation's largest river system. The combined Indus-Ganges–Brahmaputra (IGB) systems, which drain the northern Indian plains, form a regional alluvial aquifer system known for its high productivity. However, groundwater resources in the remaining two-thirds of the country are limited to the weathered zone and underlying fractured aquifers. The northern regions have porous and permeable aquifers composed of unconsolidated and semi-unconsolidated alluvial sedimentary deposits, while fractured aquifers primarily consist of Pre-Cenozoic crystalline rocks, consolidated sedimentary formations, and multi-layered basalt flows of the Indian craton.

Intensive irrigation practices are prevalent in the fertile IGB basin, which is also densely populated. The estimated renewable groundwater resources in India are around 433 billion cubic meters (bcm), with an annual groundwater withdrawal of approximately 245 bcm in 2011. Most of the groundwater, about 223 bcm, is used for irrigation, while the remaining 23 bcm is utilized for domestic and industrial purposes. The increasing agricultural demand to support a growing population has resulted in a fourfold increase in crop production (50–204 million tons) between the 1950s and 2000, putting significant strain on groundwater resources. The intense agricultural regions in the Indian subcontinent have witnessed rapid depletion of groundwater storage, with several parts of the country experiencing a decline of over 4 m in groundwater levels compared to the decadal mean (Rodell and Famiglietti 2002; Rodell et al. 2009, 2018; Mukherjee et al. 2011; Taylor et al. 2013; Asoka et al. 2017; Asoka and Mishra 2020). Given the dynamic nature of climate change and the increasing impact of El Nino in recent years, a significant increase in population along with a growing reliance on terrestrial water storage has led to a catastrophic depletion of global warming supplies. As a result, efficient GW table management and monitoring become essential and ought to be given top priority (Gautam et al. 2024). As established by previous studies (Rodell et al. 2009; Taylor et al. 2013; Asoka et al. 2017), the problem of terrestrial water storage depletion in India is quite alarming as it can be observed from the plots that the terrestrial water storage change follows a clear path of decline over entire India. The maximum terrestrial water storage has dipped significantly, and the rate of decrease is seeming to be rising. The deviation of ΔTWS from the mean is following a similar trend as that of the Indian subcontinent, i.e., increasing at a rapid scale. The increase in value of standard deviation has accelerated around 2013–14 and is seeming to be increasing still. The recovery rate of depleted water storage during the monsoon seasons has decreased over the years till the point where the contribution of monsoon rainfall does not show positive ΔTWS but remain in subzero levels. This also seems to have increased post 2013–14 (Figs. 7 and 8).

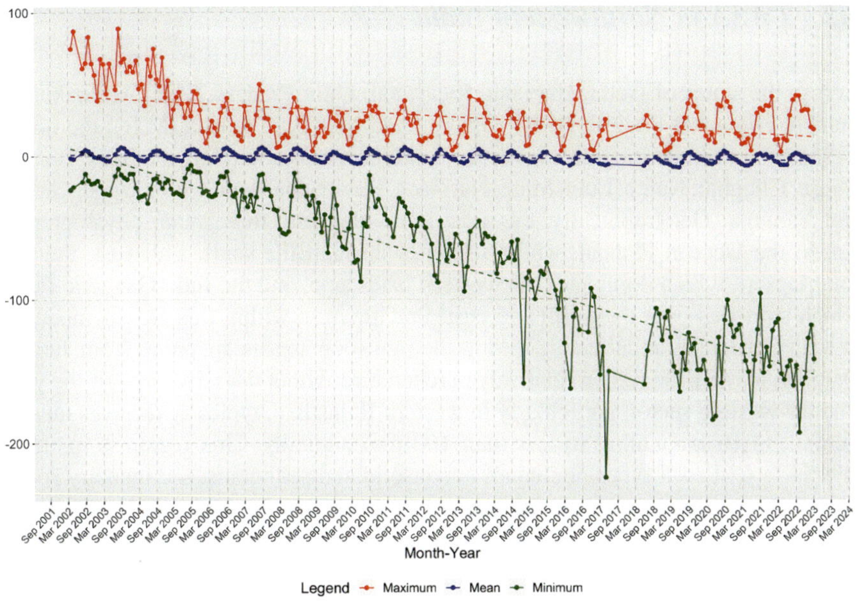

Fig. 7 Area-averaged TWS changes over India (in cm)

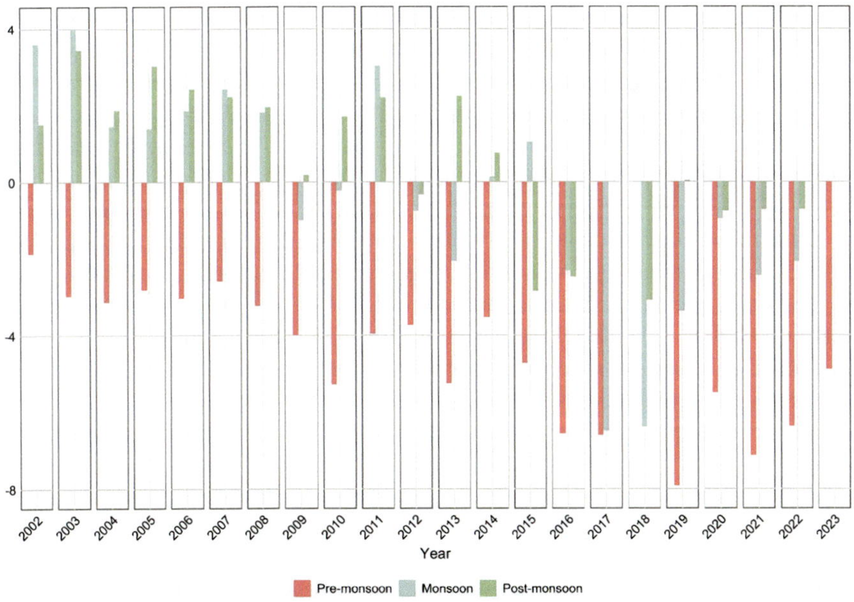

Fig. 8 Area-averaged premonsoon, monsoon, and postmonsoon terrestrial water storage change over India

4.2 TWSA in Northwestern India

Across the states of Delhi, Haryana, Rajasthan, Uttar Pradesh (U.P.), Uttarakhand, Punjab, Himachal Pradesh (H.P.), Gujarat, and Madhya Pradesh (M.P.), the northwest region of India is home to a number of river basins, including the Beas, Bhadar, Churu, Ghaghar, Kutch, Luni, Mahi, Narmada, Ravi, Sabarmati, Sutlej, Upper Ganga, and Yamuna. Geographically, this area is divided into three main physiographic zones: the Deccan Plateau, which stretches through the south and west, the arid Thar Desert, which dominates the western landscape, and the Indo-Gangetic Plain, which stretches over the north and east. Precipitation within this area is moderate to scant, primarily manifesting during the monsoon season spanning from June to October, leaving the region relatively parched throughout the remainder of the year. The far-western sector, particularly in western Rajasthan, typically receives meager rainfall, often amounting to less than 100 mm annually. This region is made up of a variety of rock types, including unconsolidated sediments, sedimentary rocks, igneous formations, and a combination of igneous, metamorphic, and sedimentary rocks that are classified as part of the Deccan trap, Vindhyan formations, and Indo-Gangetic alluvium geological groups. The Indo-Gangetic alluvium is made up of aeolian sediments, older and more recent alluvium, and other Quaternary sediments. Together, these sediments form alluvial aquifers, which are distinguished by high transmissivity, storage capacity, and infiltration rates. Contrarily, the Deccan trap consists of basaltic lava flows exhibiting limited infiltration, storage, and transmissivity, save for select areas with highly weathered, fractured basalts, showcasing moderate hydrological characteristics. Because primary and secondary porosities are present in Proterozoic sandstones and conglomerates, the Vindhyan rock formations exhibit a variety of infiltration and storage capacities. These rock formations are composed of consolidated clastic and nonclastic sedimentary and meta-sedimentary rocks. This geological diversity contributes to the region's wide spectrum of soil types, including alluvial, desert, sandy, black, and red soils, shaped by various geological and climatic influences. A multitude of geomorphological processes has sculpted the landscape, giving rise to an assortment of landforms. Fluvial landforms shaped by river processes, aeolian formations crafted by wind action, volcanic features, and structural configurations collectively manifest diverse topographies across the region. A variety of vegetation types are supported by this variable topography, including littoral and swamp vegetation, tropical semi-evergreen forest, tropical thorn forest, tropical dry deciduous forest, and subtropical pine forest. The intricate interplay between climate patterns and physical terrain engenders this rich tapestry of ecological diversity within the northwestern domain of India (Chatterjee et al. 2020). These areas include the areas of most intensive agriculture practices (Haryana, Punjab, Uttar Pradesh) and the most intensive irrigation practices, drawing groundwater leading to a major depletion of the terrestrial water resources (Fig. 9).

Punjab and Haryana, both known for intensive agriculture, heavily rely on groundwater extraction for irrigation throughout the crop season. Unfortunately, this practice has led to severe depletion of groundwater in both states. Changes in terrestrial water

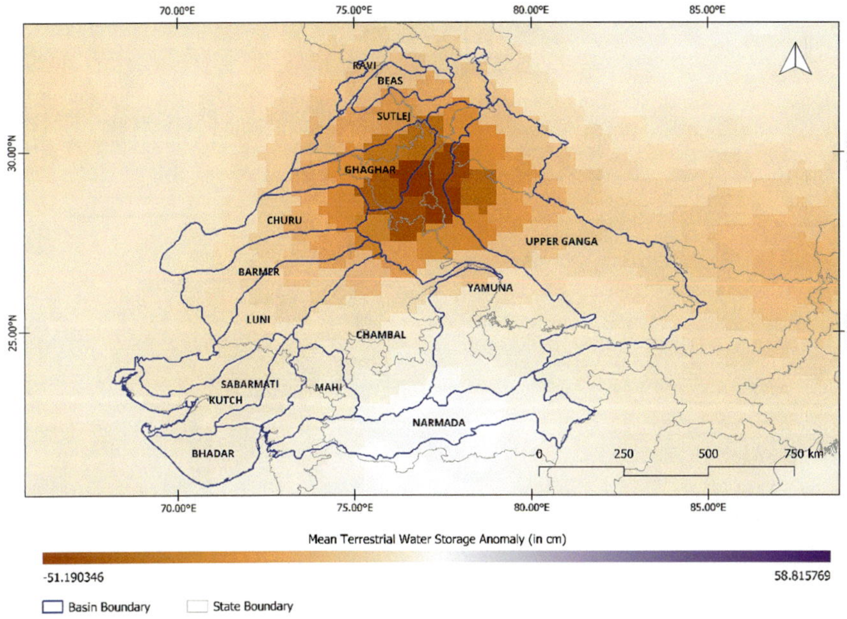

Fig. 9 Mean TWSA over northwestern river basins of India during

storage have been more pronounced in Haryana compared to Punjab, with similar trends observed in minimum, mean, and maximum storage changes. The deviation of these storage values from the mean has been increasing over time. While the rate of change in standard deviation has been consistent in Haryana, Punjab has recently experienced a significant increase in deviation, particularly in subzero levels since 2021. The plots for premonsoon, monsoon, and postmonsoon periods indicate that groundwater depletion has become so severe that even monsoonal precipitation is insufficient to recharge the groundwater. Notably, the depletion during the monsoon and postmonsoon seasons is even higher than during the premonsoon season. Climate change and other factors have contributed to altered rainfall patterns (Stocker et al.) And hindered groundwater recharge during the monsoon, exacerbating the situation (Figs. 10 and 11).

In contrast to Haryana and Punjab, the zones of highest recharge also lie in the northwestern India. Gujarat and Madhya Pradesh show positive trends of ΔTWS over the period of last two decades as assessed from GRACE/GRACE-FO satellites. The trend of increase in the mean terrestrial water storage change seems to be stable and increase over the years. The increase in ΔTWS is more in Madhya Pradesh as compared to Gujarat. But the rate of depletion in terrestrial water storage changes in Haryana, Punjab, and Delhi is much more than the rate of recharge in Gujarat, Madhya Pradesh, and Maharashtra. Which is an indicator of overall loss of terrestrial water resources. The differencing mean values between Haryana-Delhi and Madhya Pradesh reveal no significant changes in the trends of the ΔTWS. It is also to be noted

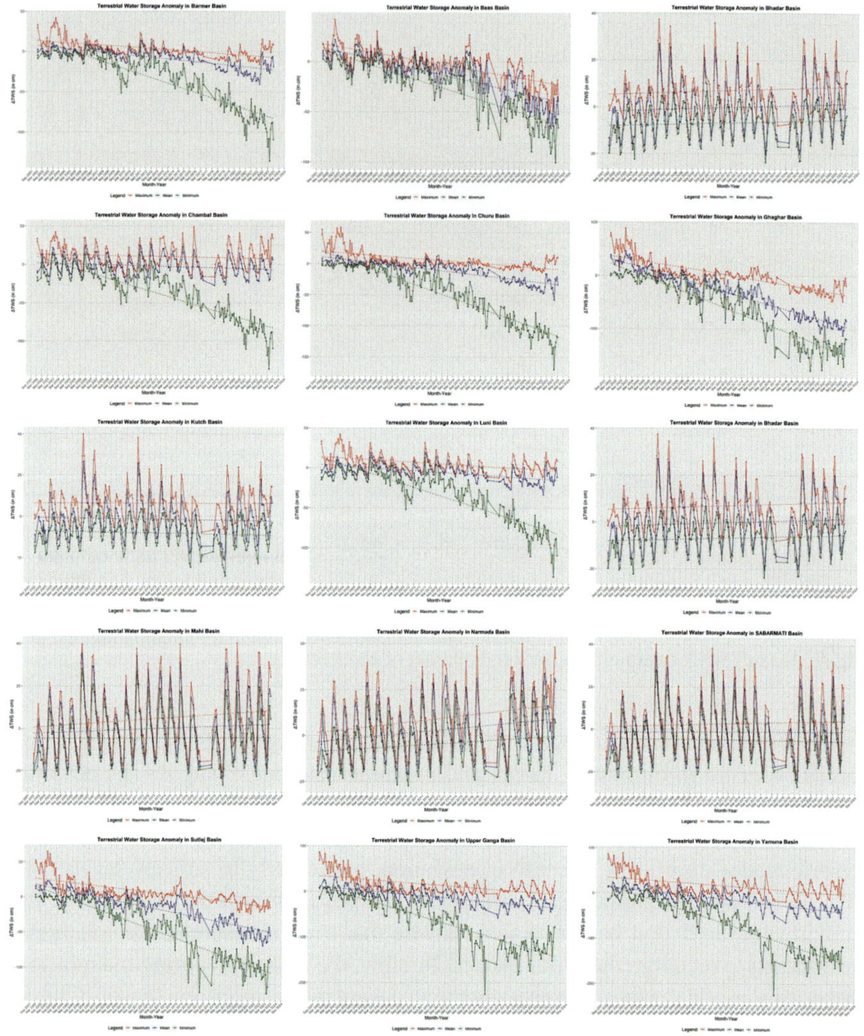

Fig. 10 Maximum, minimum, and mean TWSA plotted basin wise in the northwestern India. The maximum and minimum correspond to the max. and min. pixel values for every timestamp anywhere in a particular basin. The mean is the spatial average

that the availability of terrestrial water storage is quite high in the Haryana–Delhi–Punjab regions and so is the depletion.

Districts of Jind in Haryana and East Nimar in Madhya Pradesh have been identified as the places having highest depletion and recharge, respectively, in India and as a matter-of-fact entire Indian subcontinent. The seasonal variability is similar as compared to that of the entire respective states. But the values are much higher than the spatial averages of the particular basins.

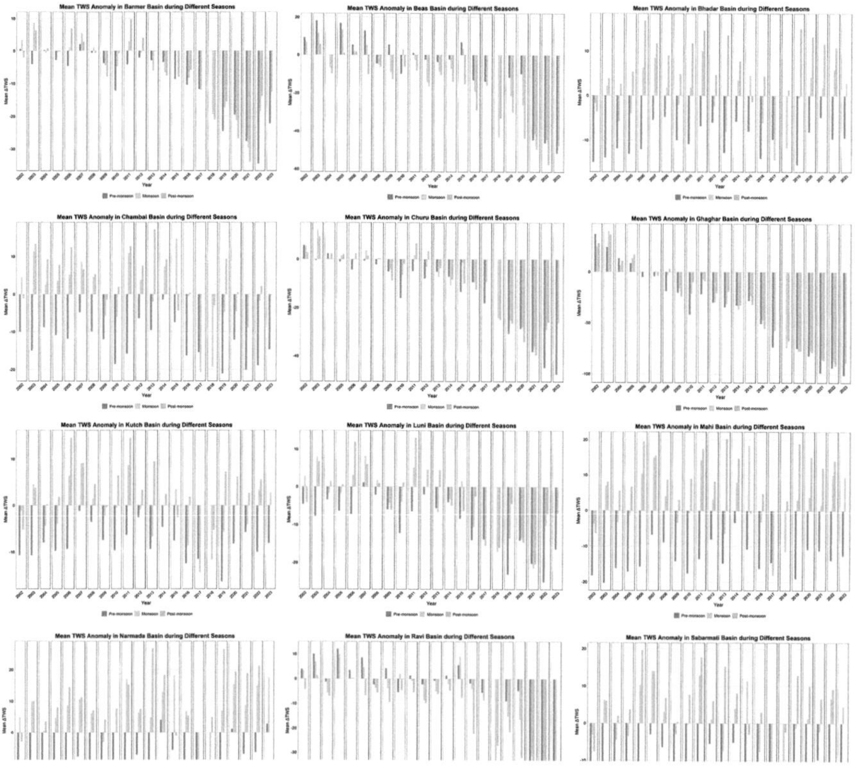

Fig. 11 Mean TWSA (in cm) premonsoon, monsoon, and postmonsoon through different northwestern river basins

4.3 *Validation*

In the northwestern region of India, the major component of TWS is groundwater storage and surface water bodies. But the volume of groundwater storage is much larger as compared to surface water bodies. Since there is absence of ice and snow, we can rule out the influence of the same on TWS. Nandi and Biswas (2024) discovered that after removing the other water budget components of the total water storage anomaly (TWSA), such as the soil moisture anomaly (SMA), the snow water equivalent anomaly (SWEA), and the plant canopy surface water storage anomaly (CSWA), from the TWSA, the GRACE groundwater storage anomalies (GWSA), estimated using the mass balance approach, followed a similar pattern to the TWSA but with a different amplitude. Hence, the most convenient way to validate the measurement of TWS with the groundwater data available with Central Ground Water Board (CGWB) sourced from the India-WRIS website.

CGWB has many measurement wells spread out throughout India, and the values of the groundwater level are measured in meters below ground level (m bgl). That

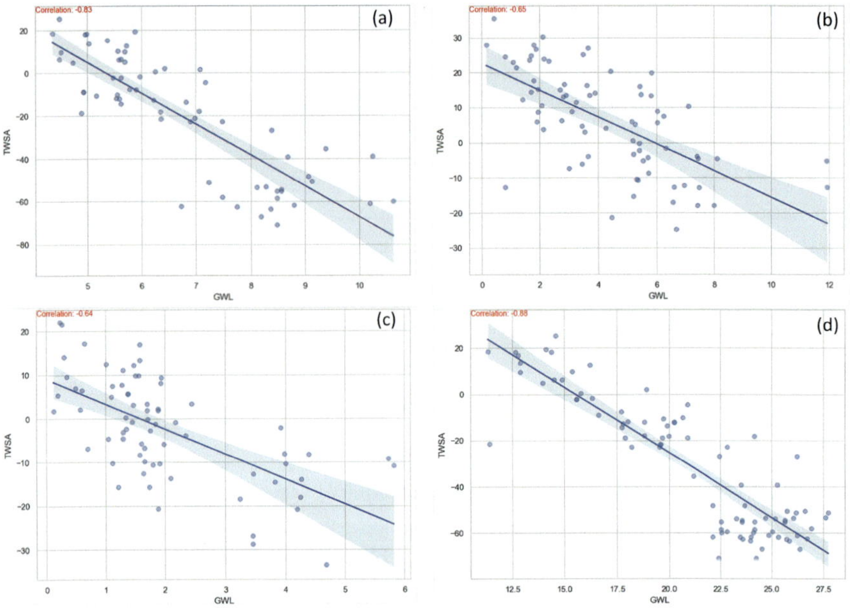

Fig. 12 Linear regression of TWSA with groundwater level data at randomly chosen locations in 4 basins **a** Ghaghar, **b** Narmada, **c** Upper ganga and **d** Yamuna with the maximum depletions and recharge

means, a reduction in groundwater resources will show an increase in the values of the measurements. In context of TWSA from space, an increase in liquid water equivalent (LWE) thickness will correspond to reducing values of meters below ground level. If the available groundwater is of bigger volume, it will be found at a lesser depth as compared to scarce resources. In this chapter, LWE has been linearly regressed over groundwater levels from CGWB and has shown moderate to high negative correlation, hence validating the values of TWSA (Fig. 12).

5 Limitations of the Study

The interpretation of GRACE-derived Terrestrial Water Storage Anomaly (TWSA) data for basin hydrology encounters significant challenges. The spatial resolution, averaging at 0.25° (~27.22 km), may not sufficiently capture localized variations, presenting difficulties in detecting rapid hydrological changes within the basin. Additionally, the presence of data gaps within the GRACE dataset impedes a comprehensive representation of temporal water storage changes, limiting a thorough understanding of basin dynamics during specific periods. Utilizing CGWB groundwater level data for validation, although beneficial, brings about uncertainties in confirming the accuracy of GRACE-derived TWSA measurements due to indirect correlation.

Furthermore, the baseline of measurement for TWS, set between 2004 and 2009 in the CSR mascons data, remains fixed and cannot be adjusted. This outdated baseline introduces a bias, wherein changes diverging from this period might be exaggerated to a certain degree, potentially affecting the accuracy of interpreting basin hydrology solely based on GRACE-derived TWSA data.

6 Conclusions

The analysis of GRACE and GRACE-FO CSR RL06.2 mascons data for terrestrial water storage changes in India, specifically focusing on river basin wise study within India, has revealed alarming trends of depletion in the terrestrial water resources over the past two decades. The study shows that the rate of depletion has increased in recent years, indicating a worsening situation. Moreover, the recharge during the monsoon season has been decreasing, with no positive changes observed since 2016.

India, being a largest country, faces significant challenges in managing its terrestrial water resources. The depletion of groundwater, which is a major component of terrestrial water storage, is primarily attributed to excessive extraction for various purposes, particularly irrigation. The Indus-Gangetic plain, the largest river system in India, has witnessed a rapid decline in total water storage, both during the winter season due to declining winter precipitation and glacier mass loss caused by global warming, and on an annual scale due to increased runoff and reduced time for groundwater recharge. The deviation of change in terrestrial water storage from the mean has been increasing at a rapid scale in India. The recovery rate of depleted water storage during the monsoon season has decreased, and the contribution of monsoon rainfall to positive terrestrial water storage has become negligible. In the northwestern region of India, which includes states like Haryana, Punjab, and Delhi, intensive agricultural practices and excessive groundwater extraction for irrigation have led to severe depletion of groundwater resources. The plots show a consistent trend of declining terrestrial water storage and increasing deviation from the mean. Even during the monsoon and postmonsoon seasons, groundwater depletion exceeds the premonsoon levels, indicating the insufficiency of rainfall for recharge.

On the other hand, Gujarat and Madhya Pradesh exhibit positive trends in terrestrial water storage changes, indicating recharge. However, the rate of depletion in Haryana, Punjab, and Delhi is much higher than the rate of recharge in Gujarat, Madhya Pradesh, and Maharashtra. This highlights the overall loss of terrestrial water resources in the region.

The analysis also identified districts of Jind in Haryana and East Nimar in Madhya Pradesh as areas of maximum depletion and recharge, respectively. The seasonal variability in these districts follows similar patterns as observed at the state level, but the magnitudes are significantly higher.

In conclusion, the study underscores the urgent need for sustainable management and conservation of terrestrial water resources in the Indian subcontinent, particularly in regions facing severe depletion. Effective measures to regulate groundwater extraction for irrigation, improve water use efficiency in agriculture, and promote alternative water sources are crucial to mitigate the ongoing depletion and secure water availability for future generations. Additionally, addressing climate change impacts and implementing policies for efficient rainfall and runoff management are essential to ensure adequate recharge and sustainable water resource management in the region.

Acknowledgements I am deeply grateful to Dr. R.S. Chatterjee, Scientist 'G' & Group Head of the Geosciences and Disaster Management Group at the Indian Institute of Remote Sensing-ISRO, Dehradun, whose guidance and support were invaluable in completing this study. My sincere appreciation goes to NASA's Center for Space Research, The University of Texas at Austin, for providing the GRACE and GRACE-FO RL06.2 Mascon solutions (URL: https://www2.csr.utexas.edu/grace/), and to the Central Ground Water Board (CGWB), India, for granting access to groundwater level data through the India-WRIS portal (URL: https://indiawris.gov.in/wris/#/groundWater). I am also thankful to the Survey of India for the India state boundary shapefiles.

References

Aeschbach-Hertig W, Gleeson T (2012) Regional strategies for the accelerating global problem of groundwater depletion. Nat Geosci 5:853–861. https://doi.org/10.1038/ngeo1617

Andreadis KM, Lettenmaier DP (2006) Assimilating remotely sensed snow observations into a macroscale hydrology model. Adv Water Resour 29:872–886. https://doi.org/10.1016/j.advwatres.2005.08.004

Ansari R, Usman Liaqat M, Grossi G (2022) Evaluation of gridded datasets for terrestrial water budget assessment in the Upper Jhelum River Basin-South Asia. J Hydrol 613:128294. https://doi.org/10.1016/j.jhydrol.2022.128294

Asoka A, Gleeson T, Wada Y, Mishra V (2017) Relative contribution of monsoon precipitation and pumping to changes in groundwater storage in India. Nat Geosci 10:109–117. https://doi.org/10.1038/ngeo2869

Asoka A, Mishra V (2020) Anthropogenic and climate contributions on the changes in terrestrial water storage in India. J Geophys Res Atmos 125:e2020JD032470. https://doi.org/10.1029/2020JD032470

Bates B, Kundzewicz ZW, IPCC (eds) (2008) Climate change and water

Beria H, Nanda T, Singh Bisht D, Chatterjee C (2017) Does the GPM mission improve the systematic error component in satellite rainfall estimates over TRMM? An evaluation at a pan-India scale. Hydrol Earth Syst Sci 21:6117–6134. https://doi.org/10.5194/hess-21-6117-2017

Bhattarai N, Pollack A, Lobell DB et al (2021) The impact of groundwater depletion on agricultural production in India. Environ Res Lett 16:085003. https://doi.org/10.1088/1748-9326/ac10de

Birkinshaw SJ, O'Donnell GM, Moore P et al (2010) Using satellite altimetry data to augment flow estimation techniques on the Mekong River. Hydrol Process 24:3811–3825. https://doi.org/10.1002/hyp.7811

Blöschl G, Bierkens MFP, Chambel A et al (2019) Twenty-three unsolved problems in hydrology (UPH)—a community perspective. Hydrol Sci J 64:1141–1158. https://doi.org/10.1080/02626667.2019.1620507

Chatterjee RS, Pranjal P, Jally S et al (2020) Potential groundwater recharge in north-western India vs spaceborne GRACE gravity anomaly based monsoonal groundwater storage change for evaluation of groundwater potential and sustainability. Groundw Sustain Dev 10:100307. https://doi.org/10.1016/j.gsd.2019.100307

Ditmar P (2018) Conversion of time-varying stokes coefficients into mass anomalies at the Earth's surface considering the Earth's oblateness. J Geod 92:1401–1412. https://doi.org/10.1007/s00190-018-1128-0

Duan Z, Bastiaanssen WGM (2013) Estimating water volume variations in lakes and reservoirs from four operational satellite altimetry databases and satellite imagery data. Remote Sens Environ 134:403–416. https://doi.org/10.1016/j.rse.2013.03.010

Famiglietti JS (2014) The global groundwater crisis. Nat Clim Change 4:945–948. https://doi.org/10.1038/nclimate2425

Felfelani F, Wada Y, Longuevergne L, Pokhrel YN (2017) Natural and human-induced terrestrial water storage change: a global analysis using hydrological models and GRACE. J Hydrol 553:105–118. https://doi.org/10.1016/j.jhydrol.2017.07.048

Food and Agriculture Organization of the United Nations (FAO) (2013) Statistical Yearbook 2013. United Nations

Gautam PK, Chandra S, Henry PK (2024) Monitoring of the groundwater level using GRACE with GLDAS satellite data in Ganga plain, India to understand the challenges of groundwater, depletion, problems, and strategies for mitigation. Environ Chall 15:100874. https://doi.org/10.1016/j.envc.2024.100874

Gleeson T, Wada Y, Bierkens MFP, van Beek LPH (2012) Water balance of global aquifers revealed by groundwater footprint. Nature 488:197–200. https://doi.org/10.1038/nature11295

Hasan MSU, Saif MM, Ahmad N et al (2023) Spatiotemporal analysis of future trends in terrestrial water storage anomalies at different climatic zones of India using GRACE/GRACE-FO. Sustainability 15:1572. https://doi.org/10.3390/su15021572

Held IM, Soden BJ (2006) Robust responses of the hydrological cycle to global warming. J Clim 19:5686–5699. https://doi.org/10.1175/JCLI3990.1

Huang Y, MhdS S, Krol MS et al (2015) Estimation of human-induced changes in terrestrial water storage through integration of GRACE satellite detection and hydrological modeling: a case study of the Yangtze River basin. Water Resour Res 51:8494–8516. https://doi.org/10.1002/2015WR016923

IPCC (2021) Climate change 2021: the physical science basis. In: Contribution of working group I to the sixth assessment report of the intergovernmental panel on climate change. Cambridge University Press

Kulkarni A, Prasad V, Arya AR et al (2023) Differential loss of glacier stored water in the Indus River basin. Curr Sci 124

Landerer FW, Cooley SS (2021) GRACE-FO Level-3 data product user handbook. NASA Jet Propuls Lab

Landerer FW, Swenson SC (2012) Accuracy of scaled GRACE terrestrial water storage estimates: ACCURACY OF GRACE-TWS. Water Resour Res 48. https://doi.org/10.1029/2011WR011453

Luthcke SB, Zwally HJ, Abdalati W et al (2006) Recent Greenland ice mass loss by drainage system from satellite gravity observations. Science 314:1286–1289. https://doi.org/10.1126/science.1130776

Marti U (2012) Gravity, geoid and height systems: In: Proceedings of the IAG symposium GGHS2012. Springer, Venice

Massari C, Maggioni V, Barbetta S et al (2019) Complementing near-real time satellite rainfall products with satellite soil moisture-derived rainfall through a Bayesian Inversion approach. J Hydrol 573. https://doi.org/10.1016/j.jhydrol.2019.03.038

McCabe MF, Rodell M, Alsdorf DE et al (2017) The future of Earth observation in hydrology. Hydrol Earth Syst Sci 21:3879–3914. https://doi.org/10.5194/hess-21-3879-2017

Mehta A, Podest E, McCartney S (2020) Groundwater monitoring using observations from NASA's gravity recovery and climate experiment (GRACE) missions. NASA Appl Remote Sens Train Program ARSET

Meng F, Su F, Li Y, Tong K (2019) Changes in terrestrial water storage during 2003–2014 and possible causes in Tibetan Plateau. J Geophys Res Atmos 124:2909–2931. https://doi.org/10.1029/2018JD029552
Mukherjee A, Fryar AE, Scanlon BR et al (2011) Elevated arsenic in deeper groundwater of the western Bengal basin, India: extent and controls from regional to local scale. Appl Geochem 26:600–613. https://doi.org/10.1016/j.apgeochem.2011.01.017
Mukherjee A, Saha D, Harvey CF et al (2015) Groundwater systems of the Indian sub-continent. J Hydrol Reg Stud 4:1–14. https://doi.org/10.1016/j.ejrh.2015.03.005
Nandi S, Biswas S (2024) Spatiotemporal distribution of groundwater drought using GRACE-based satellite estimates: a case study of Lower Gangetic Basin, India. Environ Monit Assess 196:151. https://doi.org/10.1007/s10661-024-12309-7
Natural Research Council (1997) Satellite gravity and the geosphere: contributions to the study of the solid Earth and fluid envelopes. National Academia Press
Nie W, Zaitchik BF, Rodell M et al (2018) Groundwater withdrawals under drought: reconciling GRACE and land surface models in the United States high plains aquifer. Water Resour Res 54:5282–5299. https://doi.org/10.1029/2017WR022178
Niu G-Y, Yang Z-L, Dickinson RE et al (2007) Development of a simple groundwater model for use in climate models and evaluation with gravity recovery and climate experiment data. J Geophys Res Atmos 112. https://doi.org/10.1029/2006JD007522
Pahuja S, Tovey C, Foster S, Garduño H (2010) Deep wells and prudence: towards pragmatic action for addressing groundwater overexploitation in India
Prakash S, Mitra AK, AghaKouchak A et al (2018) A preliminary assessment of GPM-based multi-satellite precipitation estimates over a monsoon dominated region. J Hydrol 556:865–876. https://doi.org/10.1016/j.jhydrol.2016.01.029
Pranjal P, Chatterjee RS, Kumar D et al (2023) Satellite gravity observation and hydrological modelling-based integrated groundwater storage change in Northwestern India. J Hydroinform 25:226–242. https://doi.org/10.2166/hydro.2023.072
Purcell A, Tregoning P, Dehecq A (2016) An assessment of the ICE6G_C(VM5a) glacial isostatic adjustment model. J Geophys Res Solid Earth 121:3939–3950. https://doi.org/10.1002/2015JB012742
Reager JT, Gardner AS, Famiglietti JS et al (2016) A decade of sea level rise slowed by climate-driven hydrology. Science 351:699–703. https://doi.org/10.1126/science.aad8386
Richard C (2015) The United Nations world water development report 2015: water for a sustainable world. UNESCO Publishing
Richard Peltier W, Argus DF, Drummond R (2018) Comment on "An assessment of the ICE-6G_C (VM5a) glacial isostatic adjustment model" by Purcell et al. J Geophys Res Solid Earth 123:2019–2028. https://doi.org/10.1002/2016JB013844
Richey AS, Thomas BF, Lo M-H et al (2015) Uncertainty in global groundwater storage estimates in a total groundwater stress framework. Water Resour Res 51:5198–5216. https://doi.org/10.1002/2015WR017351
Rodell M, Famiglietti JS (2002) The potential for satellite-based monitoring of groundwater storage changes using GRACE: the high plains aquifer, Central US. J Hydrol 263:245–256. https://doi.org/10.1016/S0022-1694(02)00060-4
Rodell M, Famiglietti JS, Wiese DN et al (2018) Emerging trends in global freshwater availability. Nature 557:651–659. https://doi.org/10.1038/s41586-018-0123-1
Rodell M, Velicogna I, Famiglietti JS (2009) Satellite-based estimates of groundwater depletion in India. Nature 460:999–1002. https://doi.org/10.1038/nature08238
Save H (2020) CSR GRACE and GRACE-FO RL06 Mascon solutions v02. Mascon Solut 12:24. https://doi.org/10.15781/cgq9-nh24
Save H, Bettadpur S, Tapley BD (2016) High-resolution CSR GRACE RL05 mascons. J Geophys Res Solid Earth 121:7547–7569. https://doi.org/10.1002/2016JB013007
Scanlon B, Mukherjee A, Gates J et al (2010) Groundwater recharge in natural dune systems and agricultural ecosystems in the Thar Desert Region, Rajasthan, India. Pap Earth Atmos Sci

Shankar PSV, Kulkarni H, Krishnan S (2011) India's groundwater challenge and the way forward. Econ Polit Wkly 46:37–45

Shiklomanov IA (1998) World water resources: a new appraisal and assessment for the 21st century

Siebert S, Burke J, Faures JM et al (2010) Groundwater use for irrigation—a global inventory. Hydrol Earth Syst Sci 14:1863–1880. https://doi.org/10.5194/hess-14-1863-2010

Stocker TF, Qin D, Plattner G-K et al IPCC climate change 2013: the physical science basis. Cambridge University Press

Sundaray SK, Nayak BB, Bhatta D (2009) Environmental studies on river water quality with reference to suitability for agricultural purposes: Mahanadi river estuarine system, India—a case study. Environ Monit Assess 155:227–243. https://doi.org/10.1007/s10661-008-0431-2

Sunilkumar K, Narayana Rao T, Satheeshkumar S (2016) Assessment of small-scale variability of rainfall and multi-satellite precipitation estimates using measurements from a dense rain gauge network in Southeast India. Hydrol Earth Syst Sci 20:1719–1735. https://doi.org/10.5194/hess-20-1719-2016

Swenson S, Wahr J (2006) Post-processing removal of correlated errors in GRACE data. Geophys Res Lett 33:L08402. https://doi.org/10.1029/2005GL025285

Tapley BD, Bettadpur S, Ries JC et al (2004) GRACE measurements of mass variability in the Earth system. Science 305:503–505. https://doi.org/10.1126/science.1099192

Tapley BD, Watkins MM, Flechtner F et al (2019) Contributions of GRACE to understanding climate change. Nat Clim Change 9:358–369. https://doi.org/10.1038/s41558-019-0456-2

Taylor RG, Scanlon B, Döll P et al (2013) Ground water and climate change. Nat Clim Change 3:322–329. https://doi.org/10.1038/nclimate1744

Tiwari VM, Wahr J, Swenson S (2009) Dwindling groundwater resources in northern India, from satellite gravity observations. Geophys Res Lett 36. https://doi.org/10.1029/2009GL039401

Vinukollu RK, Wood EF, Ferguson CR, Fisher JB (2011) Global estimates of evapotranspiration for climate studies using multi-sensor remote sensing data: evaluation of three process-based approaches. Remote Sens Environ 115:801–823. https://doi.org/10.1016/j.rse.2010.11.006

Vishwakarma B, Devaraju B, Sneeuw N (2018) What Is the spatial resolution of grace satellite products for hydrology? Remote Sens 10:852. https://doi.org/10.3390/rs10060852

Wada Y, van Beek LPH, van Kempen CM et al (2010) Global depletion of groundwater resources. Geophys Res Lett 37. https://doi.org/10.1029/2010GL044571

Wahr J, Molenaar M, Bryan F (1998) Time variability of the Earth's gravity field: hydrological and oceanic effects and their possible detection using GRACE. J Geophys Res Solid Earth 103:30205–30229. https://doi.org/10.1029/98JB02844

Watkins MM, Wiese DN, Yuan D-N et al (2015) Improved methods for observing Earth's time variable mass distribution with GRACE using spherical cap mascons. J Geophys Res Solid Earth 120:2648–2671. https://doi.org/10.1002/2014JB011547

Zhu Y, Liu S, Yi Y et al (2021) Overview of terrestrial water storage changes over the Indus River Basin based on GRACE/GRACE-FO solutions. Sci Total Environ 799:149366. https://doi.org/10.1016/j.scitotenv.2021.149366

A Geographical Appraisal of Hydrological Drought—A Case Study

Samira Bayati, Akbar Norouzi-Shokrlu, Sara Mardanian, and Khodayar Abdollahi

Abstract Drought arises from periods of insufficient precipitation, leading to deficits in both surface and subsurface water resources. In this chapter, various factors affecting hydrological drought have been evaluated. Also, hydrological drought has been investigated using the Standardized Streamflow Index in a case study (Karun 4 basin in Iran). For this purpose, the monthly discharge data of Armand station during the years 1987–2020 have been used. First, these data were examined in terms of homogeneity, outliers, and trends. Then the 3-month SSI was calculated. Our case study reveals a trend of increasing drought severity from 2000 to 2018, with extreme droughts in 2008 and 2018, indicating an urgent need for intervention. The use of GIS and spatial data has improved real-time monitoring and assessment of droughts, providing valuable data on their severity and duration. Understanding the spatial distribution and dynamics of droughts, influenced by local catchment characteristics, is essential for effective management, especially in regions with variable drought conditions. The study emphasizes the threat of climate change and land use changes, which can alter and disrupt the hydrological cycle, increasing vulnerability to drought. To address these challenges, the study suggests a focus on emerging technologies like remote sensing, the development of early warning systems, drought management plans, and water conservation strategies. It advocates for transboundary cooperation and continuous research to enhance resilience against droughts, ensuring sustainable water management. This serves as a call to action for coordinated efforts to mitigate the impacts of hydrological droughts.

Keywords Hydrological drought · Land use · Climate change · Drought indices · SSI

S. Bayati (✉) · S. Mardanian · K. Abdollahi
Department of Nature Engineering, Faculty of Natural Resources and Earth Science, Shahrekord University, Shahrekord, Iran
e-mail: bayatisamira36@yahoo.com

A. Norouzi-Shokrlu
Water Resources Planning and Allocation at Regional Water Company of Ardabil, Ardabil, Iran

S. C. Pal and U. Chatterjee (eds.), *Surface, Sub-Surface Hydrology and Management*, Springer Geography, https://doi.org/10.1007/978-3-031-62376-9_2

1 Introduction

The history of life on Earth reveals that mankind has always faced various natural hazards, with drought being a particularly significant challenge (Merzdorf 2019). Understanding the causes and consequences of hydrological drought is crucial for developing effective strategies to manage water resources and mitigate its impacts on both the environment and human societies.

A drought refers to a lack of rainfall which has an impact on humans, animal life cycles, and crops (Achite et al. 2023). Drought is a significant concern and one of the oldest natural phenomena that humans have been challenged with for a long time. It is a natural phenomenon that poses numerous challenges in the realms of environment, society, and economy (Zhang et al. 2017). Drought effect on Water resources, soil degradation, desertification, and food security (Danandeh Mehr et al. 2023).

The gradual emergence of drought is the most significant difference between droughts and other natural events (Khan et al. 2020). Drought is commonly forecasted based on various meteorological states and drought indices over monthly or seasonal periods (Hao et al. 2018). Drought arises due to natural elements like shifting weather patterns, fluctuating ocean temperatures, air circulation, and diminishing soil moisture (Ndayiragije and Li 2022). Over 150 instances of drought have been identified in the literature. Drought is categorized into four primary types: meteorological [a decrease in precipitation), hydrological (decrease in stream-flows and runoffs (Nalbantis and Tsakiris 2009)], agricultural (lack of moisture in the soil), and socio-economic droughts (reduction in human water use and ecological water shortage in the environment) (Keyantash and Dracup 2004; Yihdego et al. 2017).

1.1 Importance of Studying the Hydrological Drought

There have been numerous extreme droughts over the last few years. The occurrence of these extreme drought events, which result in extremely low water levels in reservoirs and groundwaters and restrictions on water use for irrigation and domestic use (Aghakouchak et al. 2014). For example, in Scandinavia, severe wildfires were caused by a winter drought in 2014. Drought disaster relief was required in Namibia and Angola in 2013, as well as in Brazil, central Europe, and New Zealand. In the year 2012, a concurrent drought in the central and southern regions of the United States and Russia resulted in a rise in food prices. During the spring of 2011, western Europe experienced a scarcity of water and severe water shortage. Similarly, in 2011, the Horn of Africa endured a prolonged drought that led to famine, widespread migration, and loss of human lives (Dettinger and Cayan 2014). Therefore, the assessment of hydrological drought represents a crucial role in water resources management (Mishra and Singh 2010). The significance of managing and assessing hydrological drought arises from the fact that human activities heavily rely on both surface water and groundwater resources (Shahid and Behrawan 2008). In order to delineate and signify the characteristics of drought, it is imperative to utilize variables and drought indicators due to their pivotal function in determining the extent of drought response and the timing, quantifying, and examination of drought conditions (Temam et al. 2019). Sutanto et al. (2024) emphasized the importance of catchment characteristics in hydrological drought prediction and that their results provide an opportunity to develop a global-scale hydrological drought early warning system to achieve the target of the Sendai framework for disaster risk reduction by 2030 and support disaster risk reduction initiatives.

Drought indicators have generally been applied as practical tools for evaluation and monitoring of drought situations over the past years (Yang et al. 2021; Chen et al. 2022). One of the indicators used in the assessment of hydrological drought is Standardized Streamflow Index (SSI). This index is used to evaluate both the duration and intensity of hydrological drought within a specific river basin. SSI is crucial in assessing the impact and extent of drought conditions on water resources within a particular geographical area, providing valuable insights into the intensity and duration of drought events that can significantly affect water availability and ecosystem health in the region. According to the contents mentioned about the importance of investigating hydrological drought, this chapter has investigated hydrological drought using the SSI in a case study.

1.2 *Regional Variability in Hydrological Drought*

In recent years, droughts have been observed in various regions across the globe, including North America (Cook et al. 2007; Kauffman and Vonck 2011). Among all continents, Africa is particularly susceptible to drought occurrences, especially in sub-Saharan countries (Shiru et al. 2019; Kamali et al. 2022), Asia (Guo et al. 2018), Australia (Kirono et al. 2020), and Europe (Hänsel et al. 2019). The increase in global warming is believed to have an impact on the hydrological cycle, which is a key factor influencing extreme climate events such as droughts, floods, and storms (Masson-Delmotte et al. 2022). Drought indices have been designed based on regional and local drought definitions in term of time and space including:

- A consecutive period of 15 days without any rainfall
- A duration of more than 21 days with precipitation levels below one-third of the expected average
- Annual precipitation less than 75% of normal expectancy
- Monthly precipitation lower than 60% of normal expectancy
- Lastly, any rainfall less than 85% of normal expectancy

These definitions reflect the differences in regions, needs, and approaches to drought assessment and management (Heim 2002). All forms of drought commence with a deficiency in precipitation either in terms of time or space.

Investigations reporting drought effects are more localized, with emphasizing the specific geographical regions, particularly in northern mediterranean countries. Drought researches tend to concentrate on particular regions due to the complex interactions between physical and human processes occurring in those areas, which are influenced by local contexts. Drought impact studies are frequently found in more prevalent locations such as sub-Saharan Africa and Australia-Oceania for impacts related to food and water security, respectively. The distribution of studies on food and water safety during droughts varies across different geographical regions, indicating that the impact of droughts is influenced by both the natural and human factors present in the specific area, commonly referred to as the local context. The indicator category reveals that the impact of drought on food security in sub-Saharan Africa is considerably greater than in any other region, while Australia-Oceania faces significant challenges in terms of water security. Food depending on drought is commonly documented in sub-Saharan Africa, while water security related to drought is frequently reported in Australia, the USA, Mexico, Brazil, the Middle East, and South Africa. One of the most factors influencing regional differences in hydrological drought is climatic changes. Climatic changes will turn fundamental aspects of the hydrological cycle such as groundwater availability, soil moisture, and the magnitude and timing of runoff. Depending on the spatial distribution and magnitude of the climatic changes in combination with the hydrological characteristics of the study area above effects will differ from region to region (Servat 2003). In regions where catchment conditions are relatively uniform and convective thunderstorms are the dominant precipitation type might reverse the spatial drought patterns, with spatially more coherent hydrological droughts and more patchy meteorological droughts (Grayson et al. 2001). For instance, Tao et al. (2024) emphasized that the drought situation in China was severe in 1950s and 1960s, gradually decreased in the 1970s, 1980s, and 1990s, and the 1990s were the mildest. At the beginning of the twenty-first century, the drought situation began to worsen significantly.

In summary, the regional variability in hydrological drought is influenced by the complex interactions between physical and human processes occurring in specific geographical areas, which are influenced by local contexts. Drought impact studies are more commonly located in regions such as sub-Saharan Africa and Australia-Oceania for impacts related to food and water security, respectively (Kchouk et al. 2022).

1.3 Spatial Analysis of Drought Severity and Duration

The characteristics and outcomes of drought can vary across various climatic regions worldwide. In arid areas characterized by high vulnerability, a deficiency in rainfall significantly impacts water resources, often leading to consecutive occurrences of meteorological droughts are often succeeded by hydrological and environmental droughts. Drought is predominantly experienced in arid and semiarid areas, like Iran, resulting in significant economic and environmental repercussions. Arid and semiarid regions experience higher drought frequency, severity, and deficit compared to subhumid regions (Pandey and Ramasastri 2001). Extraction of drought severity–duration–frequency (SDF) curves, or alliteratively magnitude–duration–frequency (MDF) curves, and iso-severity maps are crucial priority in both environmental and agricultural planning (Dalezias et al. 2000). It was proposed that the long duration of drought rather than its severity should be given priority in some dry regions such as Iran, Australia, and Pakistan (Wu et al. 2007). Nevertheless, even brief periods of drought frequently coincide with heatwaves lasting a few days, which are typical of arid regions. These occurrences have the potential to devastate an entire crop, resulting in food insecurity (Hatfield and Prueger 2015). In summary, spatial analysis of drought severity and duration is crucial for environmental and agricultural planning, particularly in arid and semiarid are as like Iran. Remote sensing-based indices, such as the SPI and SSMDI, can be applied to assess drought duration and severity provide valuable information for monitoring drought conditions and assessing its impacts. However, drought definitions can vary widely, reflecting differences in regions, needs, and disciplinary approaches (Saghafian et al. 2003).

1.4 Geographical Patterns of Hydrological Drought Occurrence

The hydrological drought's spatial pattern relies heavily on the unique characteristics of the local catchment area and the alteration of the drought signal as it progresses through the hydrological cycle (Hannaford et al. 2011). Catchment-scale drought propagation displays several typical characteristics, such as the sequential occurrence of drought in different domains of the hydrometeorological cycle and the corresponding time lag. Drought propagation on a regional level may not occur evenly, as the propagating the drought signal is influenced by meteorological variability and additionally altered by variations in surface and subsurface geological features. Numerous studies have examined the intricate and geographically varied occurrences of hydrological droughts throughout Europe in the last sixty years. These studies have highlighted the fact that hydrological droughts exhibit intricate spatial patterns across the European continent (Peña-Angulo et al. 2022). In summary, the spatial pattern of hydrological drought occurrence is influenced by local catchment characteristics and the complex propagation dynamics as the drought signal moves through the hydrological cycle. Catchment-scale drought propagation displays several typical features, and drought propagation may not happen uniformly on a regional scale. Understanding these propagation dynamics is crucial for effective drought monitoring and management.

1.5 Identification of Hotspots and Cold Spots of Hydrological Drought

In the past decade, numerous instances of severe drought have taken place. Looking on recent droughts indicate the recurring and global occurrence of such events despite common misconceptions that drought only affects semi-arid regions, it is crucial to emphasize that droughts in wet and cold areas can also lead to significant damage (Van Loon 2015).

According to Van Loon and Van Lanen (2012), the process of recovering from hydrological drought can be impeded due to the buildup snow in cold seasonal climates and the loss of water through evapotranspiration in warm seasonal climates. While there are uncertainties and ongoing debates surrounding the most effective methodology for studying droughts in the future. There are specific areas where models and approaches align in their projections of changes in hydrological drought. Hotspots of higher drought occurrence are anticipated in South Africa and Central America, while areas experiencing intensified drought severity include the Middle East, the southeastern United States, Chile, and southwestern Australia (Prudhomme et al. 2014). Contrary to expectation, droughts can have severe consequences in wet and cold regions, despite the common misconception that they only affect semiarid areas For instance, Scandinavia has experienced issues with electricity production and drinking water supply, as highlighted by Cattiaux et al. (2010). Similarly, regions like Mongolia have faced challenges such as livestock mortality and economic losses, as mentioned by Davi et al. (2010) and Sternberg (2010). The frozen precipitation in the winter months is used to cover future drainage; so, a dry winter can cause hydrologic drought over the coming months (Yihdego et al 2019). Therefore, identification of hotspots and cold spots of hydrological drought is crucial for understanding and managing water resources in various regions.

Various research studies have delved into the analysis and prediction of drought patterns using diverse methodologies and data sources. These investigations have yielded significant findings, including a screening procedure proposed by Galletti et al. (2023) to identify drought hotspots by evaluating aggregated severity and frequency indices between different time periods. This methodology offers insights into hydrologic behavior for subcatchment analysis. Additionally, Islam et al. (2022) utilized the local indicators of spatial autocorrelation (LISA) method to pinpoint extreme drought hotspots on a combined map. Moreover, Brunner et al. (2023) discovered that the significance of drought generation processes and severity is evolving in the Alps, particularly in higher elevation catchments. The collective results emphasize the importance of advancing methodologies and data sources to enhance the identification of hydrological drought hotspots and cold spots, thereby contributing to the development of more effective drought management strategies.

1.6 Mapping Hydrological Drought Using Remote Sensing and GIS

Environmental and meteorological added values of remote sensing (RS) data are the most important viewpoint of remote sensing in drought studies. RS products provide good data sources for land change analysis, hence, recently they have paved a precious way to monitor droughts, especially through identifying the spatial–temporal landscape changes remotely. The information is gathered through different platforms including ground-based (such as surface radars or ground-based research instruments), aerospace (such as balloons, helicopters, and aircraft), and space (satellites and shuttles) bases that use sensors to collect information. Due to recent progress, a variety of sensors are available for better decision-making and acquiring enhanced information on a range of different scales (West et al. 2019). Recent technological progress in collecting satellite and RS data has given a good opportunity to use RS data for drought assessment aims (Thenkabail and Gamage 2004; Senay et al. 2015; West et al. 2019). Looking at recent advances exposes that remotely sensed data have provided a better prospect for monitoring of drought-related spatiotemporal changes and hydrological components, such as rainfall, soil moisture, groundwater, and evapotranspiration (AghaKouchak et al. 2015). In addition, this progress provides a good opportunity to quantify the effects of drought, vegetation health, and observational changes in land cover. The products based on RS can provide real-time global information since high spatial and temporal coverage. This can provide a special opportunity to manage water resources (Sheffield et al. 2018; AghaKouchak et al. 2015). In the drought management cycle, RS could provide a good range of data that could be found useful for preparedness, mitigation plans, risk reduction, and crisis management. The most important advantage of RS data is that RS data reflect the dynamics of phenomena, such as plant phenology, canopy stress, biomass burning, and other hydrological dynamics in acceptable time and space scales (Asner and Alencar 2010).

Geographic information system (GIS) and remote sensing have a significant impact on the near real-time monitoring of drought conditions across extensive regions. The utilization of remote sensing and GIS enables the spatial depiction of an area, making it a widely studied technology for analyzing agricultural drought and monitoring large areas. The utilization of the drought index has expanded over the years. Abdollahi and Eslami (2023) provided a comprehensive outline of the various dimensions of hydrological drought, its impacts, and the measures needed to address its challenges. They discusses the significance of remote sensing and GIS in monitoring and assessing drought conditions, as well as the utilization of drought indices as practical tools to evaluate and monitor drought conditions. Additionally, this work addresses the identification of drought-prone regions and the vulnerability of specific areas to drought conditions, emphasizing the need for tailored measures to diminish the impacts of drought and guarantee the sustainable management of water resource of the sustainable water resource management. It also includes recommendations for hydrological drought manage management, highlighting the importance of ongoing research, the development of advanced methodologies, and the utilization of data sources to enhance the identification of hotspots and cold spots of hydrological drought, ultimately contributing to better drought management strategies.

2 Materials and Methods

2.1 *Study Area*

Here, hydrological drought is explored for a case study. The study area is the Karun 4 basin that located in the southwest of Iran country and between 49°32′59″–51°45′00″ E longitude and 31°19′47″–32°40′12″ N latitude with 14,553 km^2 area (Fig. 1). This basin is a relatively mountainous area (about 76% of the area is mountainous) and has a high altitude. Its elevation range is between 785 and 4148 m, and its average elevation is 2313 m. Within this basin, the long-term annual averages reveal an annual rainfall of approximately 603 mm, an average temperature of 13 °C, potential evaporation of 1878 mm, and a mean discharge rate of 70 m^3/s.

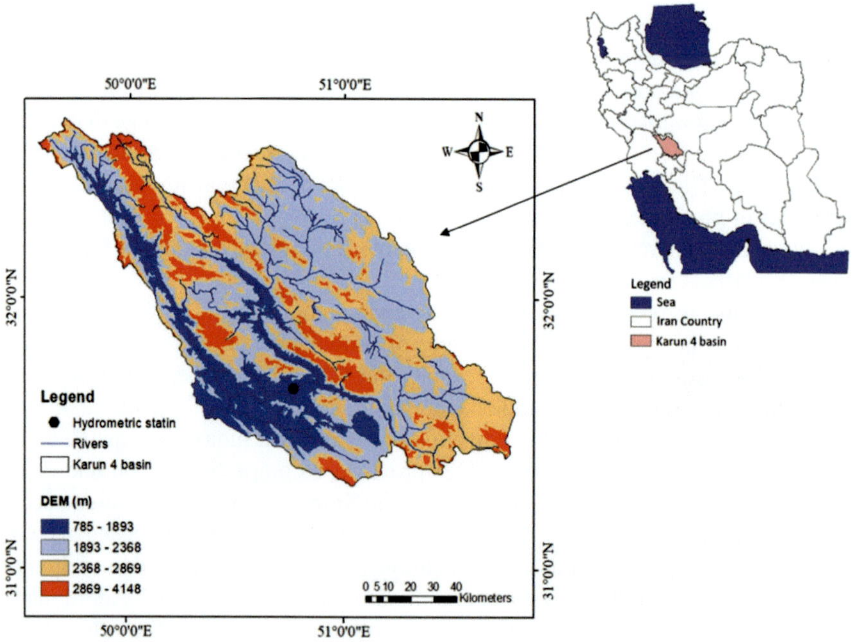

Fig. 1 Study area location: elevation, river system, and hydrometric station

2.2 Research Methodology

Figure 2 depicts the methodology followed in this research. The monthly discharge data of Armand hydrometric station were used in the study area in order to calculate the SSI hydrological drought index. The data were obtained from the Chaharmahal-va-Bakhtiari's Regional Water Authority in daily scale. The first step before the statistical analysis of the data is to check the quantity and quality of the data and ensure the completeness of the statistical data series. Therefore, first, the data were checked in terms of homogeneity, the presence of outliers, and the presence of patterns in the time series of the data. The discharge data of Armand station had no missing data in the statistical period from 1987 to 2020. The histogram of the data was drawn in SPSS software in order to assess the homogeneity of the data. According to the single exponential of the data histogram, the homogeneity of the discharge data was confirmed (Fig. 3).

In order to detect outlier data among the discharge data of Armand station, Grubs-Beck (1971) (G-B) test was used. To use this test, the following two values are calculated.

$$X_H = \text{EXP}\left(\overline{X} + K_N \times S\right) \tag{1}$$

$$X_L = \text{EXP}\left(\overline{X} - K_N \times S\right) \tag{2}$$

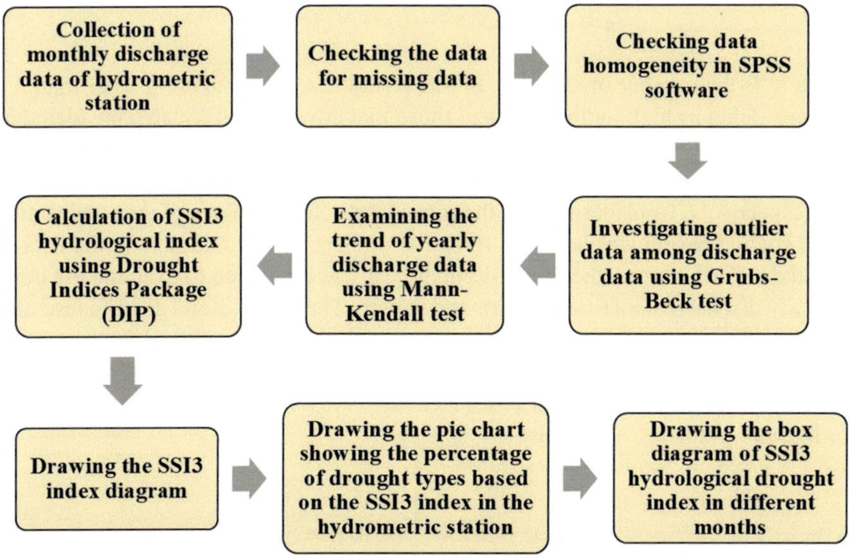

Fig. 2 The flowchart of research methodology

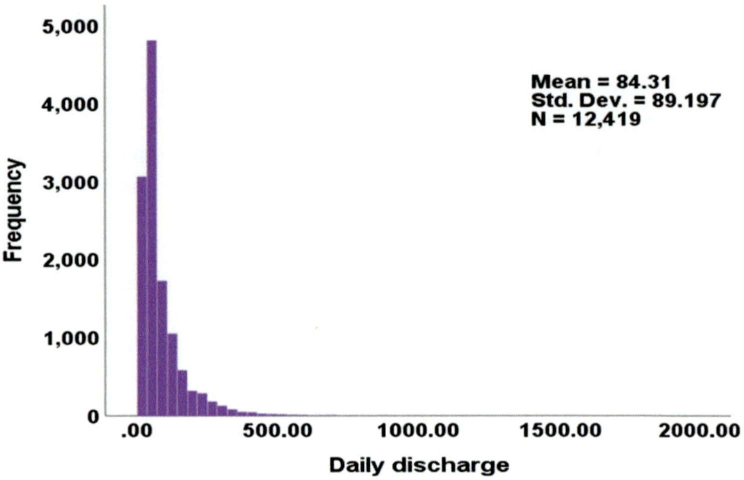

Fig. 3 Daily discharge histogram of Armand station

where $\overline{X}$ and S are, respectively, the mean and standard deviation of the natural logarithm of the sample and the value of K_N at the 10% confidence level is obtained from the following approximate polynomial.

$$K_N = -3.32201 + 6.28446\mathrm{N}^{0.25} - 2.49835\mathrm{N}^{0.5} + 0.491436\mathrm{N}^{0.75} - 0.037911\mathrm{N} \quad (3)$$

where N is the number of samples. In this (G-B) test, data that are greater than X_H are considered as high outlier data and those that are less than X_L are considered as low outlier data. The values of parameters X_L, $X_{H,}$ and K_N calculated by Groves and Beck test related to the monthly discharge data of Armand station were 6, 681.6, and 3.3, respectively. Considering that the range of discharge values of Armand station in the studied period was between 16 and 426, there were no outliers in the data of this study. Furthermore, the Mann–Kendall test was employed to examine the trend of yearly discharge data from the Armand station. The Mann–Kendall trend test, also known as the M–K test, is a nonparametric statistical test. It is utilized for examining data gathered over a period of time to identify consistent upward or downward trends (monotonic) in Y values (Mann 1945; Kendall 1975). Mann–Kendall's test is in the form of the following equations:

$$\mathrm{Sgn}\left(x_j - x_i\right) = \begin{Bmatrix} 1;\ \text{if } x_j > x_i \\ 0;\ \text{if } x_j = x_i \\ 1;\ \text{if } x_j < x_i \end{Bmatrix} \quad (4)$$

$$S = \sum_{i=1}^{n-1} \sum_{j=i+1}^{n} \text{sgn}\left(x_j - x_i\right) \tag{5}$$

where x_i and x_j are the data values at times i and j, and n is the length of the data set. Positive and negative values of S demonstrate increasing and decreasing trends, respectively. The given expression is utilized as an assumption for the series in which n is greater than 10 and the data are approximately normally distributed, with a variance ($\sigma^2 = 1$) and a mean ($\mu = 0$) value (Kisi and Ay 2014).

$$\text{Var}(S) = \frac{n(n-1)(2n+5) - \sum_{i=1}^{P} t_i(t_i - 1)(2t_i + 5)}{18} \tag{6}$$

where P is the number of tied groups, t_i is the number of data values in the Pth group. This summary process maynot be taken into account if there are not tied groups. The standard Z value is then computed using the Equation provided below:

$$Z = \left\{ \begin{array}{ll} \frac{S-1}{\sqrt{\text{Var}(S)}}; & \text{if } S > 0 \\ 0; & \text{if } S = 0 \\ \frac{S+1}{\sqrt{\text{Var}(S)}}; & \text{if } S > 0 \end{array} \right\} \tag{7}$$

After comparing the calculated standard Z value with the standard normal distribution table with two-tailed confidence levels ($\alpha = 10\%$, $\alpha = 5\%$ and $\alpha = 1\%$), if the calculated Z is greater than $|Z| > |Z_1\text{-}\alpha/2\,|$, the null hypothesis ($H_0$) is invalid. So, the trend is statistically significant. On the other hand, the H_0 hypothesisi can be accepted that there is no trend in the time series and the trend is not statistically significant (Kisi and Ay 2014). In this data study, the annual discharge of Armand station had a decreasing trend at a significance level of 0.05. After checking the amount of annual precipitation in this basin, it was found that the decrease in discharge occurred in this basin following the decrease in precipitation.

2.3 SSI

In order to assess hydrological drought state on a monthly basis, SSI (Vicente-Serrano et al. 2012) was calculated, which was developed through the theoretical framework of the SPI (McKee et al. 1993). The flow deficit is characterized by the SSI in relation to the long-term mean at a specific location and time scale, using standard deviations (Rivera et al. 2018). The SSI encompasses various time scales, ranging from 1 (SSI1) to 48 (SSI48) months, for the collection of streamflow data. For this particular study, a time scale of three months (SSI3) was utilized to effectively eliminate high-frequency fluctuations (Rivera et al. 2017, 2018), and the Drought Indices Package (DIP) was used to calculate it. The following equation is used to calculate this index.

Table 1 Drought severity classification of the SSI (*Source* Adapted from WMO World Meteorological Organization 2012)

SSI values	Classification
≥ 2.00	Extremely wet (EW)
1.50 to 1.99	Very wet (VW)
1.00 to 1.49	Moderately wet (MW)
0.99 to – 0.99	Near normal (NN)
– 1.00 to – 1.49	Moderately dry (MD)
– 1.50 to – 1.99	Severely dry (SD)
≤ – 2.00	Extremely dry (ED)

$$\text{SSI3} = W - \frac{C_0 + C_1 W + C_2 W^2}{1 + d_1 W + d_2 W^2 + d_3 W^3} \tag{8}$$

$$W = \sqrt{-2\ln(P)} \quad \text{for } P \leq 0.5 \tag{9}$$

The probability of exceeding a given value Q, denoted as P, is calculated as $P = 1 - F(Q)$ ($F(Q)$ is probability distribution function). If P is greater than 0.5, it is replaced by $1 - P$, and the sign of the resulting SSI3 is reversed. The constants used in this calculation are $C_0 = 2.515517$, $C_1 = 0.802853$, $C_2 = 0.010328$, $d_1 = 1.432788$, $d_2 = 0.189269$, and $d_3 = 0.001308$ (Rivera et al. 2021) [further details for computing the SSI may be research in Vicente-Serrano et al. (2012)]. Furthermore, Table 1 displays the SSI's drought severity classification, which was derived from McKee's (1993) methodology.

3 Results and Discussion

Figure 4 illustrates three-month SSI of the Armand hydrometric station during the period from January 1987 to December 2020. The year 2000 was the beginning of a moderate drought that intensified in 2001, and the SSI reached a value below – 1.5. Then in the years 2008 to 2018 moderate droughts are observed in some months. Also, severe drought are occurred in some months of 2008 to 2012 and 2018. The extreme droughts are occurred in 2008 and 2018 with a value below – 2 (Fig. 4). In addition, it can be said that since 2008, droughts have been observed in the study area, and with the continuation of this trend, necessary measures must be taken to deal with droughts.

Figure 5 presents the results of the percentage of drought types based on the SSI3 index in Armand station. As shown in this figure, in this study area during the 34-year period from 1987 to 2020, 17% of moderate droughts, 5% of severe droughts, and 2% of very extreme droughts have occurred (Fig. 5).

Figure 6 presents the results of hydrological drought characteristics (duration and amount) in Armand station in different months during the years 1987–2020. Based

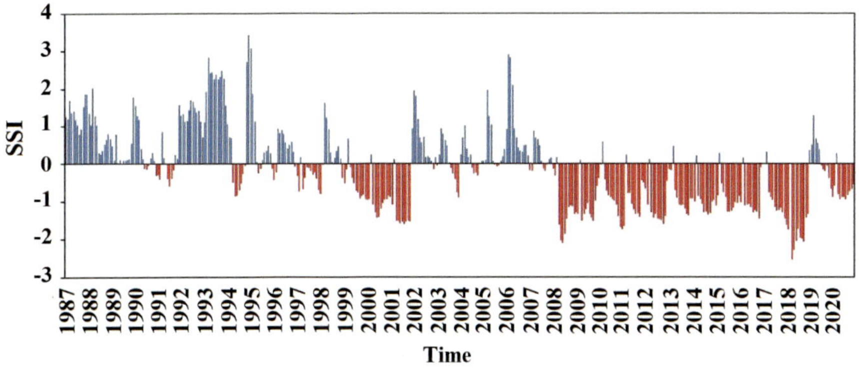

Fig. 4 The SSI3 of the Armand hydrometric station (1987–2020)

Fig. 5 Pie chart showing the percentage of drought types based on the SSI3 in Armand station

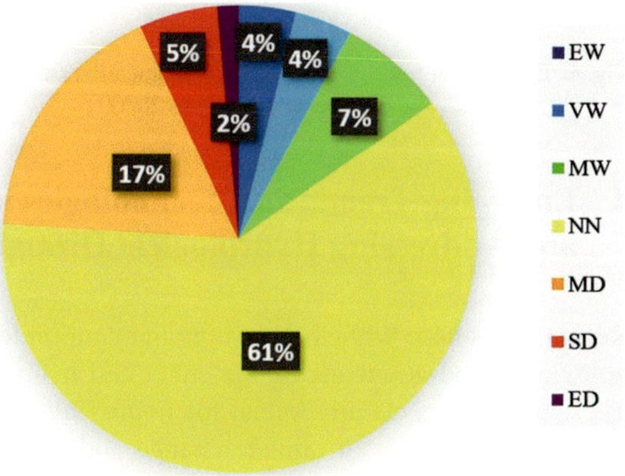

on the box diagram of SSI3 in different months, the most severe droughts occurred in May (Fig. 6).

These findings are in consistent with Ghorbani et al. (2019), who state as although the Armand River has experienced consistent flow, dry periods of varying severity have occurred in recent years. The drought's impact has worsened due to a significant increase in the river's volume deficit in the past few years.

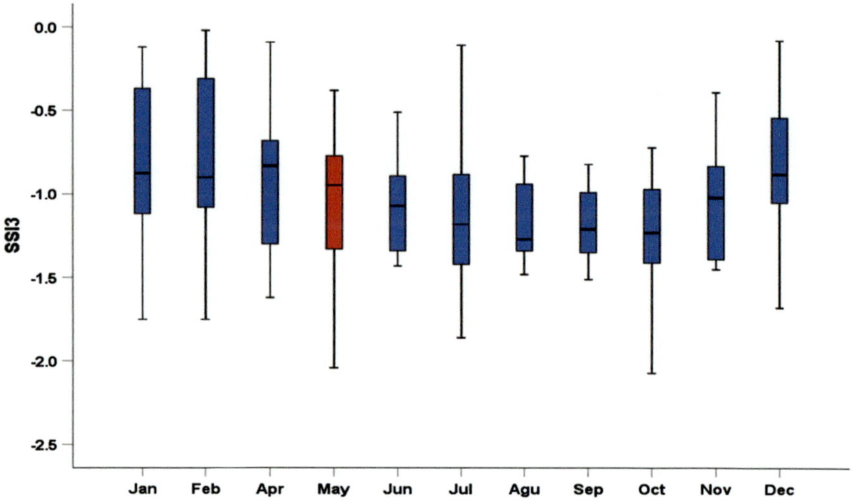

Fig. 6 Box plot of hydrological drought characteristics (duration and amount) in Armand station in different months during the years 1987 to 2020

4 Emerging Trends and Technologies in Studying and Addressing Hydrological Drought

Recent years have witnessed a surge in attention toward emerging trends and technologies aimed at enhancing the study and management of hydrological drought. These advancements are crucial for improving water resource management, especially in the context of climate change and the escalating water demands across various sectors. Key aspects of these trends and technologies include the increased utilization of drought indices for monitoring changes in meteorological, hydrological, agricultural, and economic conditions, as highlighted by Diani et al. (2019). Climate change plays a significant role in influencing hydrological droughts, impacting the water cycle and exacerbating the severity and frequency of drought events, as emphasized by Bănăduc et al. (2021). Researchers are actively engaged in understanding the intricate relationship between drought and climate change while developing adaptation and mitigation strategies. Moreover, the integration of remote sensing technologies like satellite imagery and GIS has become more prevalent in monitoring drought conditions, assessing water resources, and aiding decision-making processes, as noted by Wohlgemuth (2021). Machine learning principles are also being explored to address challenges in probabilistic hydrological postprocessing and forecasting, offering potential improvements in the accuracy and efficiency of drought predictions and forecasting, as discussed by Papacharalampous and Tyralis (2022). Additionally, the impact of climate change and human activities on freshwater fish populations and their habitats during drought periods is a growing concern, prompting research into the role of aquatic refuge habitats in preserving fish diversity and abundance, along with the development of management strategies to safeguard these habitats, as highlighted by Bănăduc et al. (2021). Furthermore, studies analyzing regional and global trends in meteorological, agricultural, ecological, and hydrological droughts are underway to identify regions experiencing worsening drought conditions and to inform the formulation of effective adaptation and mitigation strategies, as indicated by Vicente-Serrano et al. (2022).

5 Lessons Learned from Successful Hydrological Drought Assessment Practices

Various successful methods have been utilized globally for hydrological drought assessment, playing a crucial role in the prediction and management of droughts while offering valuable insights for water resource management and ecological considerations. Noteworthy methodologies include the Regional Multi-Component Gaussian Hydrological Drought Assessment (RMcGHDA), which employs a two-phase statistical procedure to enhance precipitation data characteristics for accurate standardized drought monitoring (Ali et al. 2021). Additionally, the Standardized Precipitation Index (SPI) and Standardized Precipitation Evapotranspiration Index (SPEI) are widely used indices that compare observed precipitation with long-term averages, considering the balance between precipitation and evapotranspiration (Brleković and Tadić 2022). Another approach involves analyzing the hydrological footprint of drought events, which assesses the severity and spatial extent of droughts in comparison with meteorological footprints to draw insights for drought management, aiding in understanding the dynamic evolution of drought over time and space (Laaha et al. 2017). Furthermore, assessing drought by incorporating enhanced precipitation records and auxiliary information, such as land use changes and climate conditions, helps characterize regional hydrological drought (Ali et al. 2021). These methodologies have been applied in diverse regions to evaluate hydrological drought, with examples like the analysis of the European 2015 drought and comparison to the extreme drought of 2003, showcasing their utility in understanding the complexities of hydrological drought and guiding effective management strategies.

6 Recommendations

The future of hydrological drought management involves developing effective strategies and techniques to mitigate the impacts of drought on water resources and society. Nauditt et al. (2017) emphasize the importance of implementing drought early warning systems to detect drought conditions early, enabling timely preparation and response. Yordanova et al. (2022) highlight the significance of creating and updating drought management plans to guide decision-making and ensure a coordinated approach to addressing drought risks. Promoting water conservation measures and encouraging efficient use of water resources can help reduce water demand during drought periods. Yuan et al. (2016) stress the need for developing and implementing drought adaptation strategies to enhance community resilience to drought impacts. Addressing the root causes of drought, particularly climate change, is crucial for reducing the frequency and severity of drought events, as advocated by Satoh et al. (2022). Encouraging transboundary cooperation between countries and stakeholders, as suggested by Katipoğlu and Acar (2022), can facilitate sharing information, resources, and best practices for effective drought management in regions where droughts cross national borders. Lastly, continuous research and monitoring of hydrological drought conditions are essential for understanding spatial and temporal patterns of drought hazards and risks, as well as for developing effective mitigation strategies.

7 Conclusion

7.1 Summary of Key Findings

Underscoring the importance of managing and evaluating hydrological droughts is particularly vital as human activities are heavily dependent on both surface and groundwater resources. The use of drought indices, such as the Standardized Drought Index (SDI), Streamflow Drought Index (SRI), Standardized Precipitation Index (SPI), and others, has become a standard practice for assessing and monitoring drought conditions. These indices are critical tools that reflect regional and local definitions of drought, taking into account the specific needs and approaches to drought management in different areas.

The integration of geographic information system (GIS) and remote sensing (RS) technologies has significantly enhanced the ability to monitor drought conditions across vast regions in near real time. These technologies provide a spatial representation of affected areas, allowing for more effective monitoring and assessment. Remote sensing-based indices, in particular, are instrumental in evaluating the severity and duration of droughts, offering valuable insights for monitoring and impact assessment.

The spatial distribution of hydrological drought occurrences is not uniform and is influenced by local catchment characteristics and the dynamics of drought propagation through the hydrological cycle. Understanding these dynamics is essential for effective drought monitoring and management. Despite this, regions such as semi-arid, subhumid, mountainous, and coastal areas may experience regional variability in drought conditions.

Identifying hotspots of hydrological drought—areas where droughts occur more frequently and intensely—is crucial for managing water resources effectively. Numerous studies have been conducted to analyze and predict drought patterns, employing various methodologies and data sources. These efforts are vital for understanding the complex interplay between climate change, land use changes, and hydrological processes.

Climate change, in particular, poses a significant threat to hydrological drought management by potentially altering precipitation patterns and increasing evapotranspiration rates. Human-induced land use changes, such as deforestation, urbanization, and wetland loss, can further disrupt the hydrological cycle, exacerbating water scarcity and vulnerability to drought.

To address these challenges, emerging trends and technologies in the study of hydrological drought are focusing on enhancing our understanding and management of water resources. These include advancements in remote sensing, machine learning, and the study of aquatic refuge habitats, which are informing decision-making processes related to climate change and drought assessment.

Looking to the future, the sustainable management of hydrological drought will involve the development of early warning systems, implementation of drought management plans, promotion of water conservation and efficient use, adoption of drought adaptation strategies, and mitigation of climate change effects. Additionally, fostering transboundary cooperation and conducting ongoing research and monitoring are imperative for reducing the impacts of drought on water resources and society. Together, these measures will contribute to a more resilient future, safeguarding both human communities and the environment against the adverse effects of drought.

7.2 Future Research Directions

Future research directions in hydrological drought are centered on managing and understanding this phenomenon in the context of changing climate and hydrological conditions. Hasan et al. (2019) emphasize the importance of developing and enhancing drought assessment and management strategies to mitigate the impact on water resources and related sectors. They also highlight the use of nonstationary joint drought management indices to better comprehend and manage drought risks in the future, considering the evolving characteristics of drought events and climate change impacts. Additionally, Satoh et al. (2022) stress the significance of investigating the timing of unparalleled hydrological drought under climate change scenarios and implementing mitigation measures to cope with intensified drought conditions. Furthermore, Brunner et al. (2023) advocate for the development of hydrological drought forecasting and monitoring systems using artificial neural networks to improve accuracy and timeliness in forecasting, aiding decision-making and resource management. Yu et al. (2019) also underscore the need to improve drought assessment and management strategies in light of changing climate and hydrological conditions. Prioritizing the enhancement of drought preparedness and adaptation strategies within specific timeframes is crucial to effectively manage the escalating intensity of the hydrological cycle due to climate change (Satoh et al. 2022).

7.3 *Call for Action to Address Hydrological Drought*

Hydrological drought refers to a prolonged period of below-average water availability, which can have significant impacts on water resources, ecosystems, and society. Addressing hydrological drought requires a comprehensive approach involving water management, conservation, and policy interventions. The call for action to address hydrological drought emphasizes the need for proactive measures to mitigate its impacts and ensure sustainable water management. Hydrological drought can lead to reduced water availability for drinking, agriculture, and industry, as well as negative impacts on aquatic ecosystems and biodiversity. Hydrological droughts not only challenge water availability but can also exacerbate water quality issues and lead to conflicts over water resources. To mitigate these impacts, several strategies are essential:

Implementing water-saving measures in agriculture, industry, and households can help mitigate the impacts of hydrological drought. Developing integrated water management plans that consider the needs of various sectors and prioritize water allocation during drought periods is crucial. Investing in water storage and distribution infrastructure, such as reservoirs and pipelines, can enhance water availability during dry periods. Implementing policies and regulations to promote sustainable water use, reduce water waste, and protect ecosystems can help address hydrological drought. Implementing nature-based Solutions, such as invasive alien tree clearing, has been shown to reduce the severity of hydrological drought caused by human-induced climate change, as demonstrated in South Africa (Seka et al. 2022). Integrating co-creation processes in hydrological modeling can aid in developing multivariate drought indicators for better risk assessment and intervention. This approach considers variables like precipitation, soil moisture, and streamflow to enhance the understanding of drought dynamics and improve response strategies (Steyaert and Condon 2022). Mathematical Modeling for Trend Analysis: Analyzing hydrological drought trends through mathematical models can provide insights into predicting future trends and offering practical solutions for drought-affected regions globally. Studies have shown that declines in reservoir storage have been occurring faster in more arid regions, emphasizing the importance of trend analysis for effective water resource management (Seka et al. 2022). Integrated Drought Management Program: Developing global coordination of efforts to strengthen drought monitoring. Drought forecasting and early warning services: These are crucial for preparing for and mitigating the effects of drought. By embracing these measures, we can work toward a more sustainable and secure water future, even in the face of the increasing challenges posed by hydrological droughts.

Addressing hydrological drought requires a coordinated effort involving government agencies, water utilities, stakeholders, and the public. By implementing proactive measures and adopting sustainable water management practices, the impacts of hydrological drought can be mitigated, ensuring the availability of water resources for current and future generations. Therefore, it is crucial to ensure that emergency assistance is delivered in a way that safeguards the most vulnerable members of society and encourages self-sufficiency. This should also align with a national drought policy focused on reducing risks.

References

Abdollahi K, Eslami Z (2023) Remote sensing capabilities for observational drought assessment. In: Integrated drought management, vol 2. CRC Press, pp 361–378

Achite M, Katipoğlu OM, Jehanzaib M, Elshaboury N, Kartal V, Ali S (2023) Hydrological drought prediction based on hybrid extreme learning machine: Wadi Mina Basin case study, Algeria. Atmosphere 14(9):1447

Aghakouchak A, Feldman D, Stewardson MJ, Saphores JD, Grant S, Sanders B (2014) Australia's drought: lessons for California. Science 343(6178):1430–1431

Aghakouchak A, Farahmand A, Melton FS, Teixeira J, Anderson MC, Wardlow BD, Hain CR (2015) Reviews of geophysics remote sensing of drought: progress, challenges. Rev Geophys 53:1–29

Ali Z, Ellahi A, Hussain I, Nazeer A, Qamar S, Ni G, Faisal M (2021) Reduction of errors in hydrological drought monitoring—a novel statistical framework for spatio-temporal assessment of drought. Water Resour Manage 35(13):4363–4380

Asner GP, Alencar A (2010) Drought impacts on the Amazon forest: the remote sensing perspective. New Phytol 187(3):569–578

Bănăduc D, Sas A, Cianfaglione K, Barinova S, Curtean-Bănăduc A (2021) The role of aquatic refuge habitats for fish, and threats in the context of climate change and human impact, during seasonal hydrological drought in the Saxon Villages area (Transylvania, Romania). Atmosphere 12(9):1209

Brleković T, Tadić L (2022) Hydrological drought assessment in a small lowland catchment in Croatia. Hydrology 9(5):79

Brunner MI, Götte J, Schlemper C, Van Loon AF (2023) Hydrological drought generation processes and severity are changing in the Alps. Geophys Res Lett 50(2):e2022GL101776

Cattiaux J, Vautard R, Cassou C, Yiou P, Masson-Delmotte V, Codron F (2010) Winter 2010 in Europe: a cold extreme in a warming climate. Geophys Res Lett 37(20)

Chen S, Li Q, Zhong W, Wang R, Chen D, Pan S (2022) Improved monitoring and assessment of meteorological drought based on multi-source fused precipitation data. Int J Environ Res Public Health 19(3):1542

Cook ER, Seager R, Cane MA, Stahle DW (2007) North American drought: reconstructions, causes, and consequences. Earth Sci Rev 81(1–2):93–134

Dalezios NR, Loukas A, Vasiliades L, Liakopoulos E (2000) Severity-duration-frequency analysis of droughts and wet periods in Greece. Hydrol Sci J 45(5):751–769

Danandeh Mehr A, RikhtehgarGhiasi A, Yaseen ZM, Sorman AU, Abualigah L (2023) A novel intelligent deep learning predictive model for meteorological drought forecasting. J Ambient Intell Humaniz Comput 14(8):10441–10455

Davi N, Jacoby G, Fang K, Li J, D'Arrigo R, Baatarbileg N, Robinson D (2010) Reconstructing drought variability for Mongolia based on a large-scale tree ring network: 1520–1993. J Geophys Res Atmos 115(D22)

Dettinger M, Cayan DR (2014) Drought and the California delta—a matter of extremes. San Francisco Estuary Watersh Sci 12(2)

Diani K, Kacimi I, Zemzami M, Tabyaoui H, Haghighi AT (2019) Evaluation of meteorological drought using the Standardized Precipitation Index (SPI) in the High Ziz River basin, Morocco. Limnol Rev 19(3):125–135

Galletti A, Formetta G, Majone B (2023) A screening procedure for identifying drought hot-spots in a changing climate. Water 15(9):1731

Ghorbani M, Mozayyan M, Zarei H (2019) Hydrological drought investigation of Armand River using low flows analysis. J Water Soil Conserv 26(3):247–263

Grayson R, Blöschl G (eds) (2001) Spatial patterns in catchment hydrology: observations and modelling. CUP Archive

Guo H, Bao A, Ndayisaba F, Liu T, Jiapaer G, El-Tantawi AM, De Maeyer P (2018) Space-time characterization of drought events and their impacts on vegetation in Central Asia. J Hydrol 564:1165–1178

Hannaford J, Lloyd-Hughes B, Keef C, Parry S, Prudhomme C (2011) Examining the large-scale spatial coherence of European drought using regional indicators of precipitation and streamflow deficit. Hydrol Process 25(7):1146–1162

Hänsel S, Ustrnul Z, Łupikasza E, Skalak P (2019) Assessing seasonal drought variations and trends over Central Europe. Adv Water Resour 127:53–75

Hao Z, Singh VP, Xia Y (2018) Seasonal drought prediction: advances, challenges, and future prospects. Rev Geophys 56(1):108–141

Hasan HH, Mohd Razali SF, Muhammad NS, Ahmad A (2019) Research trends of hydrological drought: a systematic review. Water 11(11):2252

Hatfield JL, Prueger JH (2015) Temperature extremes: Effect on plant growth and development. Weather Clim Extrem 10:4–10

Heim RR Jr (2002) A review of twentieth-century drought indices used in the United States. Bull Am Meteor Soc 83(8):1149–1166

Islam SS, Islam KA, Mullick MRA (2022) Drought hot spot analysis using local indicators of spatial autocorrelation: an experience from Bangladesh. Environ Challeng 6:100410

Kamali B, Jahanbakhshi F, Dogaru D, Dietrich J, Nendel C, AghaKouchak A (2022) Probabilistic modeling of crop-yield loss risk under drought: A spatial showcase for sub-Saharan Africa. Environ Res Lett 17(2):024028

Katipoğlu OM, Acar R (2022) Space-time variations of hydrological drought severities and trends in the semi-arid Euphrates Basin, Turkey. Stoch Env Res Risk Assess 36(12):4017–4040

Kauffman GJ, Vonck KJ (2011) Frequency and intensity of extreme drought in the Delaware Basin, 1600–2002. Water Resour Res 47(5)

Kchouk S, Melsen LA, Walker DW, Van Oel PR (2022) A geography of drought indices: mismatch between indicators of drought and its impacts on water and food securities. Nat Hazard 22(2):323–344

Kendall MG (1975) Rank correlation methods, 4th edn. Charles Griffin, London

Keyantash JA, Dracup JA (2004) An aggregate drought index: assessing drought severity based on fluctuations in the hydrologic cycle and surface water storage. Water Resour Res 40(9):1–13

Khan N, Sachindra DA, Shahid S, Ahmed K, Shiru MS, Nawaz N (2020) Prediction of droughts over Pakistan using machine learning algorithms. Adv Water Resour 139:103562

Kirono DG, Round V, Heady C, Chiew FH, Osbrough S (2020) Drought projections for Australia: updated results and analysis of model simulations. Weather Clim Extrem 30:100280

Kisi O, Ay M (2014) Comparison of Mann-Kendall and innovative trend method for water quality parameters of the Kizilirmak River, Turkey. J Hydrol 513:362–375

Laaha G, Gauster T, Tallaksen LM, Vidal JP, Stahl K, Prudhomme C, Heudorfer B, Vlnas R, Ionita M, Van Lanen HA, Adler MJ (2017) The European 2015 drought from a hydrological perspective. Hydrol Earth Syst Sci 21(6):3001–3024

Mann HB (1945) Nonparametric tests against trend. Econometrica 13:245–259

Masson-Delmotte V, Zhai P, Pörtner HO, Roberts D, Skea J, Shukla PR (2022) Global warming of 1.5 C: IPCC special report on impacts of global warming of 1.5 C above pre-industrial levels in context of strengthening response to climate change, sustainable development, and efforts to eradicate poverty. Cambridge University Press

McKee TB, Doesken NJ, Kleist J (1993) The relationship of drought frequency and duration to time scales. Proc 8th Conf Appl Climatol 17(22):179–183

Merzdorf J (2019) NASA study: human influence on global droughts goes back 100 years. Goddard Institute for Space Studies GISS

Mishra AK, Singh VP (2010) A review of drought concepts. J Hydrol 391(1–2):202–216

Nalbantis I, Tsakiris G (2009) Assessment of hydrological drought revisited. Water Resour Manage 23:881–897

Nauditt A, Firoz ABM, Trinh VQ, Fink M, Stolpe H, Ribbe L (2017) Hydrological drought risk assessment in an anthropogenically impacted tropical catchment, Central Vietnam. In: Land use and climate change interactions in central Vietnam: LUCCi, pp 223–239

Ndayiragije JM, Li F (2022) Effectiveness of drought indices in the assessment of different types of droughts, managing and mitigating their effects. Climate 10(9):125

Pandey RP, Ramasastri KS (2001) Relationship between the common climatic parameters and average drought frequency. Hydrol Process 15(6):1019–1032

Papacharalampous G, Tyralis H (2022) A review of machine learning concepts and methods for addressing challenges in probabilistic hydrological post-processing and forecasting. Front Water 4:961954

Peña-Angulo D, Vicente-Serrano SM, Domínguez-Castro F, Lorenzo-Lacruz J, Murphy HJ, Allan RP, Tramblay Y, Reig-Gracia F, El Kenawy A (2022) The complex and spatially diverse patterns of hydrological droughts across Europe. Water Resour Res 58(4):e2022WR031976

Prudhomme C, Giuntoli I, Robinson EL, Clark DB, Arnell NW, Dankers R, Fekete BM, Franssen W, Gerten D, Gosling SN, Hagemann S (2014) Hydrological droughts in the 21st century, hotspots and uncertainties from a global multimodel ensemble experiment. Proc Natl Acad Sci 111(9):3262–3267

Rivera JA, Penalba OC, Villalba R, Araneo DC (2017) Spatio-temporal patterns of the 2010–2015 extreme hydrological drought across the Central Andes, Argentina. Water 9(9):652

Rivera JA, Otta S, Lauro C, Zazulie N (2021) A decade of hydrological drought in Central-Western Argentina. Front Water 3:640544

Rivera JA, Araneo DC, Penalba OC, Villalba R (2018) Regional aspects of streamflow droughts in the Andean rivers of Patagonia, Argentina. Links with large-scale climatic oscillations. Hydrol Res 49(1):134–149

Saghafian BAHRAM, Shokoohi ALIREZA, Raziei TAYEB (2003) Drought spatial analysis and development of severity-duration-frequency curves for an arid region. Int Assoc Hydrol Sci Public 278:305–311

Satoh Y, Yoshimura K, Pokhrel Y, Kim H, Shiogama H, Yokohata T, Hanasaki N, Wada Y, Burek P, Byers E, Schmied HM (2022) The timing of unprecedented hydrological drought under climate change. Nat Commun 13(1):3287

Seka AM, Zhang J, Prodhan FA, Ayele GT, Finsa MM, Sharma TPP, Melesse AM (2022) Hydrological drought impacts on water storage variations: a focus on the role of vegetation changes in the East Africa region. A systematic review. Environ Sci Pollut Res 29(53):80237–80256

Senay GB, Velpuri NM, Bohms S, Budde M, Young C, Rowland J, Verdin JP (2015) Drought monitoring and assessment: Remote sensing and modeling approaches for the famine early warning systems network. In: Hydro-meteorological hazards, risks and disasters. Elsevier, pp 233–262

Servat E (ed) (2003) Hydrology of Mediterranean and semiarid regions: papers selected for the international conference on hydrology of the Mediterranean and semi-arid regions, held in Montpellier, France from 1 to 4 April 2003 (No. 278). International Association of Hydrological Sciences

Shahid S, Behrawan H (2008) Drought risk assessment in the western part of Bangladesh. Nat Hazards 46:391–413

Sheffield J, Wood EF, Pan M, Beck H, Coccia G, Serrat-Capdevila A, Verbist K (2018) Satellite remote sensing for water resources management: potential for supporting sustainable development in data-poor regions. Water Resour Res 54(12):9724–9758

Shiru MS, Shahid S, Chung ES, Alias N (2019) Changing characteristics of meteorological droughts in Nigeria during 1901–2010. Atmos Res 223:60–73

Sternberg T (2010) Unravelling Mongolia's extreme winter disaster of 2010. Nomadic Peoples 14(1):72–86

Steyaert JC, Condon LE (2022) Historical analysis of reservoir storage trends and resilience across contiguous US from 1980–2019. Egusphere 2022:1–29

Sutanto SJ, Syaehuddin WA, de Graaf I (2024) Hydrological drought forecasts using precipitation data depend on catchment properties and human activities. Commun Earth Environ 5(1):118

Tao Y, Meng E, Huang Q (2024) Spatiotemporal changes and hazard assessment of hydrological drought in china using big data. Water 16(1):106

Temam D, Uddameri V, Mohammadi G, Hernandez EA, Ekwaro-Osire S (2019) Long-term drought trends in Ethiopia with implications for dryland agriculture. Water 11(12):2571

Thenkabail PS, Gamage MSDN (2004) The use of remote sensing data for drought assessment and monitoring in Southwest Asia, vol 85. Iwmi

Van Loon AF (2015) Hydrological drought explained. Wiley Interdiscip Rev Water 2(4):359–392

Van Loon AF, Van Lanen HA (2012) A process-based typology of hydrological drought. Hydrol Earth Syst Sci 16(7):1915–1946

Vicente-Serrano SM, López-Moreno JI, Beguería S, Lorenzo-Lacruz J, Azorin-Molina C, Morán-Tejeda E (2012) Accurate computation of a streamflow drought index. J Hydrol Eng 17(2):318–332

Vicente-Serrano SM, Peña-Angulo D, Beguería S, Domínguez-Castro F, Tomás-Burguera M, Noguera I, Gimeno-Sotelo L, El Kenawy A (2022) Global drought trends and future projections. Phil Trans R Soc A 380(2238):20210285

West H, Quinn N, Horswell M (2019) Remote sensing for drought monitoring & impact assessment: progress, past challenges and future opportunities. Remote Sens Environ 232(June):111291

Wohlgemuth PM (2021) The San Dimas experimental forest: Hydrologic research in semiarid steepland terrain. Hydrol Process 35(5):e14148

World Meteorological Organization (2012) Standardized precipitation index user guide. World Meteorological Organization, Geneva

Wu H, Svoboda MD, Hayes MJ, Wilhite DA, Wen F (2007) Appropriate application of the standardized precipitation index in arid locations and dry seasons. Int J Climatol J R Meteorol Soc 27(1):65–79

Yang X, Lu M, Wang Y, Wang Y, Liu Z, Chen S (2021) Response mechanism of plants to drought stress. Horticulturae 7(3):50

Yihdego Y, Webb JA, Vaheddoost B (2017) Highlighting the role of groundwater in lake–aquifer interaction to reduce vulnerability and enhance resilience to climate change. Hydrology 4(1):10

Yihdego Y, Vaheddoost B, Al-Weshah RA (2019) Drought indices and indicators revisited. Arab J Geosci 12:1–12

Yordanova A, Ilcheva I, Bocheva L, Malcheva K, Lubenova K (2022) Analysis of hydrological drought indices and their relationship with meteorological factors and river basin specifics. Int Multidiscip Sci GeoConfer SGEM 22(3.1):41–48

Yu J, Kim TW, Park DH (2019) Future hydrological drought risk assessment based on nonstationary joint drought management index. Water 11(3):532

Yuan F, Ma M, Ren L, Shen H, Li Y, Jiang S, Yang X, Zhao C, Kong H (2016) Possible future climate change impacts on the hydrological drought events in the Weihe River Basin, China. Adv Meteorol 2016:1–14

Zhang L, Jiao W, Zhang H, Huang C, Tong Q (2017) Studying drought phenomena in the Continental United States in 2011 and 2012 using various drought indices. Remote Sens Environ 190:96–106

Technologies and Approaches to Support Community Flood Initiatives—A Bibliometric Analysis Around the Theme

Ahmed Karmaoui

Abstract New technologies and decision tools have been widely used to reduce flood losses. These techniques are focused on collecting and analyzing information and predicting flood disaster risks. This chapter aims to provide insights into the research structure in the context of using technologies and approaches to support community flood management, utilizing bibliometric and networking analyses. It provides a broad overview about publications, authors, countries, subjects, citations, years, journals, and affiliations and highlights the current and trending dimensions in the field. In this work, the current state of using these techniques in the period 1996 to April 2023 was outlined. Current hot spots highlighted that climate change and risk assessment are the most visible concepts while research trends are turning toward topics such as resilience, green infrastructure, ecosystem services, sponge city, livelihood, strategic approach, nature-based solutions, and urbanization. The analysis reveals a growing number of published documents, with environmental science subject exerting significant influence. The International Journal of Disaster Risk Reduction stands out as the primary source in the field, and funding predominantly comes from the National Science Foundation and the National Natural Science Foundation of China. Articles and reviews constitute the main types of published documents, and the USA emerges as the most influential country. Thaler, T., and the University of Leeds are identified as the leading author and affiliation, respectively. Many other aspects were detailed such as assessment and management techniques, vulnerability dimensions, types of resources, floods, and associated phenomena, flood indicators, and solutions.

Keywords Floods · Climate change · Adaptation · Decision-making support tools · Bibliometric review · Scopus

---

A. Karmaoui (✉)
Faculty of Sciences and Techniques (Health and Environment Research Team), Errachidia, Moulay Ismail University, Meknes, Morocco
e-mail: ah.karmaoui@gmail.com

S. C. Pal and U. Chatterjee (eds.), *Surface, Sub-Surface Hydrology and Management*, Springer Geography, https://doi.org/10.1007/978-3-031-62376-9_3

1 Introduction

Floods represent one of the most prevalent natural hazards experienced globally. The combination of rapid urbanization and heightened rainfall intensity, attributed to climate change, has resulted in escalated flooding occurrences across vast regions worldwide (Bibi and Kara 2023). Consequently, more people worldwide are impacted by floods than by any other natural hazard. The world bank presented a working paper about global flood risk exposure and poverty in which estimated the number of exposed people that is about 1.47 billion people (19% of the global population) where 89% live in low- and middle-income countries (Rentschler and Salhab 2020). According to the last report of IPCC, it was reported that between 2010 and 2020, in comparison with locations with very low vulnerability, human mortality from floods, droughts, and storms was 15 times greater in highly vulnerable regions (Pörtner et al. 2022). As a result of rising flood risks brought on by climate change as well as other variables like urbanization and land use and land cover change, flood control is becoming more difficult to implement in real-world situations (Earl et al. 2023). Floods are affecting many special scales, namely catchment, rural areas, river basins, neighborhood, and primarily, urban and coastal areas (Karmaoui 2023). In a case study of Haikou, a coastal city in China, the authors indicate that future changes in rainfall patterns and sea level rise will both increase the amount, depth, and duration of flooding (Xu et al. 2023). Urban flood protection may face increasing difficulties in the future as a result of the increased frequency of extreme storms and accompanying uncertainty in flood risk brought on by climate change (Liu et al. 2023). In urbanization and climate change scenarios, the low-impact developments that have been put in place may decrease projected peak runoff, the intensity of floods, and flooded intersections (Bibi and Kara 2023). Otherwise, inequity and marginalization associated with gender, ethnicity, low income, or combinations of these factors worsen vulnerability at several spatial levels, particularly for many indigenous peoples and local communities (Pörtner et al. 2022). As several studies have documented that the climate change induced sea level increase (Roy et al. 2023; Pariartha et al. 2023; Pörtner et al. 2022), governments and residents in vulnerable regions are searching for novel ways to support community flood initiatives (CFI). Technological advancements play a pivotal role in supporting CFI, facilitating effective surface hydrology and management practices. Indeed, there is a pressing need to explore and evaluate novel technologies and approaches that can support these initiatives. However, while there is a growing interest in leveraging technology for flood disaster risk reduction, there is limited understanding of the current landscape of these technologies, their effectiveness, and the methods used to implement them. Therefore, the research problem is to conduct a comprehensive bibliometric analysis to review the current and trending dimensions of technologies and approaches utilized in community flood initiatives, including assessment and management techniques, vulnerability dimensions, resource types, flood indicators, and solutions. The utilization of bibliometric analysis provides valuable insights into the knowledge dissemination process and holds significant importance across various fields, including artificial intelligence

(Ivanova et al. 2024; Mesa Fernández et al. 2022), medicine (Geng et al. 2024), geology (Karmaoui 2022), geographic information systems (Shkundalov and Vilutienė 2021), and climate change (Ajibade et al. 2023; Karmaoui et al. 2023a). This chapter aims to offer insights into the research framework regarding the utilization of technologies and methodologies to aid community flood management. It employs bibliometric and networking analyses to present a comprehensive overview of publications, authors, countries, subjects, citations, years, journals, and affiliations. Additionally, it sheds light on current and emerging trends within the field and explore many other aspects including management techniques, dimensions, types of resources, flood indicators, and solutions.

2 Material and Methods

Data collection and research structure

Web of Science was not used because the number of publications recorded was 0. Using the Scopus database, a search was conducted for "floods" together with "technologies to support community initiatives". Using the key terms [TITLE-ABS KEY (floods)] AND (technologies AND to AND support AND community AND initiatives), 637 document results were retrieved from the Scopus database in the period 1996–April 22, 2023 (Fig. 1).

Methodology

A bibliometric review was conducted to analyze the structure, evolution of research, and the international collaboration in the flood field. This technique also provides a broad overview and a mine of information into its publications, authors and co-authors, countries, subjects, and citations based on network and occurrence analyses. The technique allows us to identify knowledge gaps and trending topics. Using the keywords "floods" and "technologies to support community initiatives", the extracted data was processed via the VOSviewer software conceived by Van Eck and Waltman (2010). This software allowed displaying key term networks, linkages, and trends as colored circles connected by proportionally thick colored lines depending on how near they are (Karmaoui 2023). This tool includes three types of visualizations, namely network analysis, density visualization, and trends. "Co-occurrence" for "all keywords" is the counting method performed. The minimum number of key terms (MNK) used as the threshold was 5. Of the 4600 keywords highlighted terms, 437 met the threshold. For each of the 437 keywords, the total strength of co-occurrence links was estimated. In terms of document quality, the MNK used as the threshold was 5. Of the 95 countries, 43 met the threshold. For each of the 43 countries, the total strength of citation links was estimated.

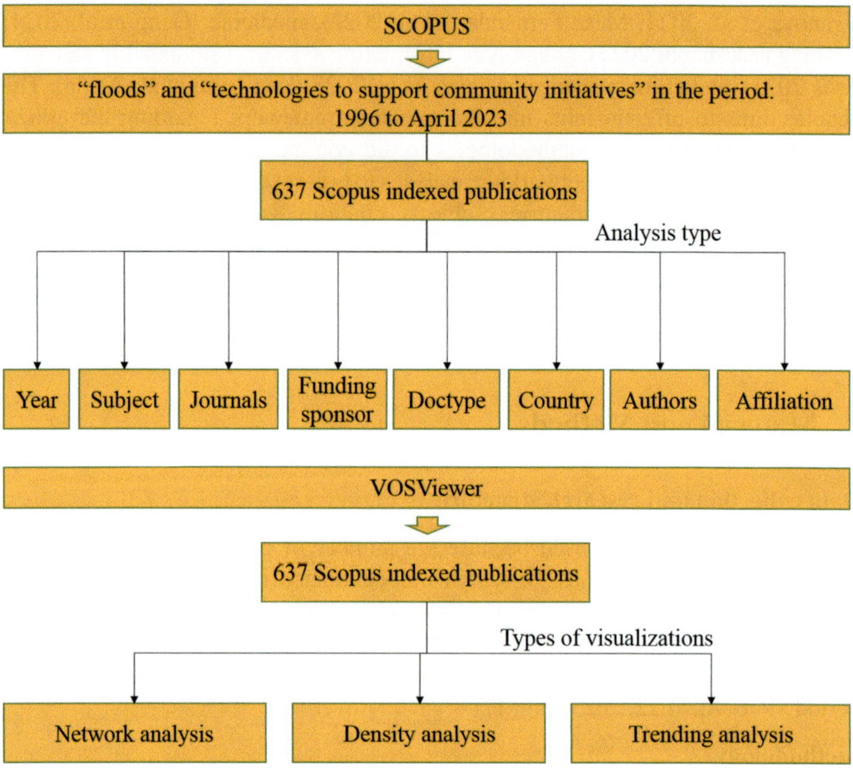

Fig. 1 Extracted data and type of analyses and visualizations

3 Results and Discussion

3.1 Research Structure

Figure 2a represents the evolution of the number of documents from 1996 to April 2023 that show an increasing trend. The first record of the retrieved paper published in 1996 was about freshwater biodiversity. This diversity can be influenced by projects related to dams, flood control, irrigation, bridges, water supply, domestic waste, and waterway modification reporting the importance of World Bank policies to support actions for freshwater biodiversity conservation (Kottelat and Whitten 1996). One of the most recent works deals with natural flood management application and spatial techniques in tropical watersheds. Among the most frequent NFM types are pond, riparian buffers, and reforestation/revegetation (Earl et al. 2023). With a total number of citations of 901 until the date of the search, a work hydrologic services is the most cited publication in the retrieved documents. This paper examined policy and valuation techniques for preserving ecosystem services and provided various instances of land management that used these strategies (Brauman et al. 2007). The second most cited work was entitled "Urban Hazard Mitigation: Creating resilient cities" (Godschalk 2003) with 884 citations (see also Appendix 1 for more information about the other publications using VOSviewer). Based on the number of publications, the subjects the most influencing are environmental science, social science, and earth and planetary sciences (Fig. 2b). The International Journal of Disaster Risk Reduction, Water Switzerland, Sustainability Switzerland, and Natural Hazards are the leading sources in the field (Fig. 2c). However, the National Science Foundation, the National Natural Science Foundation of China, and the European Commission are the main funding sponsors (Fig. 2d) while articles, reviews, and book chapters are the principal types of published documents (Fig. 2e). For the most influencing countries and as in many fields, the USA emerges as the country where most technologies to support community initiatives are studied for "floods", followed by the United Kingdom, Australia, and China (Fig. 2f). The leading position of the USA in terms of scientific productivity has been confirmed across multiple fields by several studies (Geng et al. 2024; Ajibade et al. 2023; Karmaoui et al. 2023b). In terms of leading authors (Fig. 2g) and affiliations (Fig. 2h), respectively, Thaler, T., Demir, I., and Mulligan, J., and University of Leeds, Chinese Academy of Sciences, and Delft University of Technology were found.

3.2 *Network, Density, and Trending Analyses*

The network analysis showed seven group or clusters (Fig. 3a). Cluster 1 highlighted 72 red items including climate change, vulnerability, flooding, adaptation, disaster management, perception, knowledge, government, livelihood, education, traditional knowledge, environmental policy, and poverty… Cluster 2 explored 69 green items, namely water quality, hydrology, biodiversity, ecology, runoff, rivers, ecosystem, land use, catchment, watershed, environmental protection, environmental impact, river basin, and land management… (Fig. 3a). However, Cluster 3 encompassed 67 blue items such as risk assessment, risk management, disasters, remote sensing, disaster risk reduction, resilience, natural hazards, disaster prevention, vulnerability assessment, hazard assessment, GIS, information management, information systems, and planning… The fourth cluster grouped and linked 62 yellow items including the key terms floods, hydrological modeling, forecasting, rain, extreme events, uncertainty analysis, digital elevation model, mapping, GIS, assessment methods, spatiotemporal analysis, and weather forecasting… The fifth cluster included 61 purple items like floods, sustainable development, drought, ecosystems, restoration, ecosystem services, sea level change, coastal flooding, coastal zone management, wetlands, forestry, sea level, hurricanes, greenhouse gases, and satellite imagery… Otherwise, the cluster showed 59 blue clear items primarily decision-making, numerical model, artificial intelligence, decision support systems, green infrastructure, flood damages, urban area, urbanization, urban planning, nature-based solutions, and willingness to pay… The cluster 7 (47 orange clear items) highlighted mainly the terms flood control, flood risk management, water management, stakeholders, early

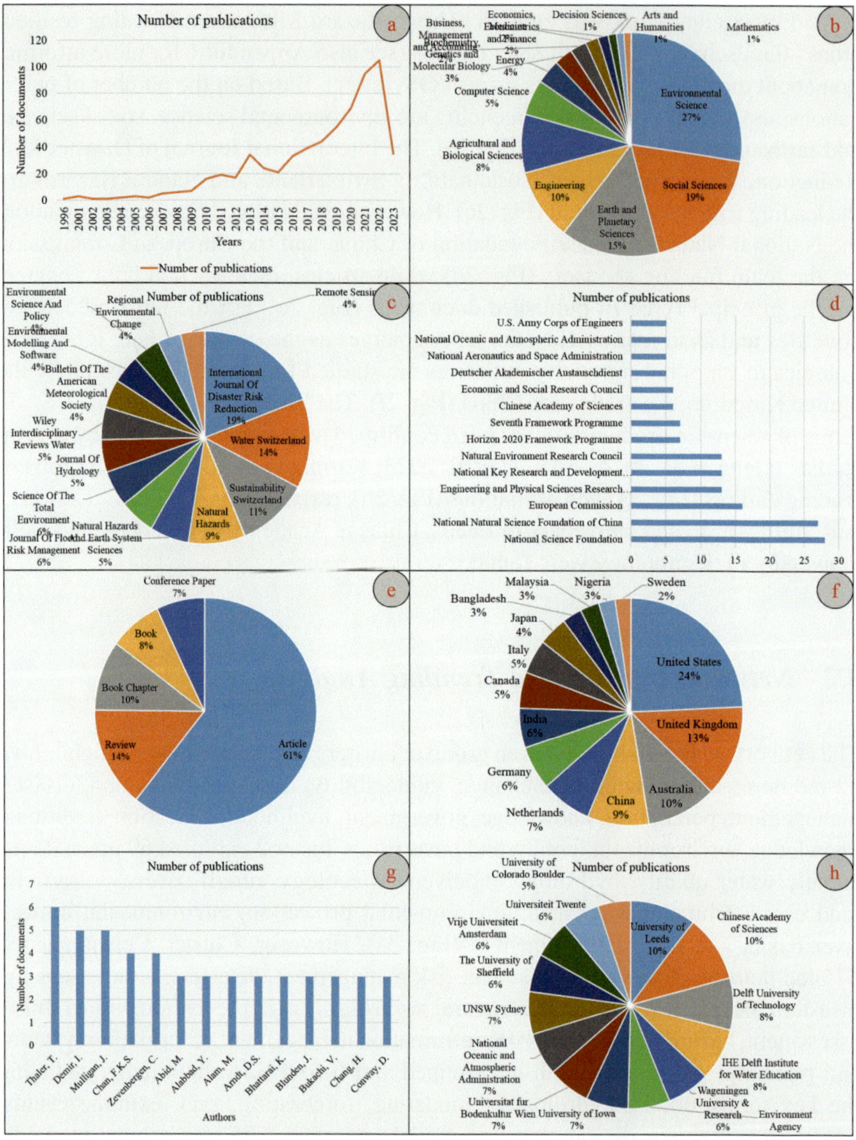

Fig. 2 Research structure using the key terms "floods" and "technologies to support community initiatives" and Scopus database. **a** years; **b** subject; **c** journals; **d** funding sponsor; **e** document type; **f** countries; **g** authors and **h** affiliations

warning system, economic and social effects, governance, infrastructure, water, and water resources (Fig. 3a).

Regarding the hot spot search topics linked to "floods" and "technologies to support community initiatives, climate change and risk assessment are the most visible key terms (Fig. 3b). The impact of climate change at different scales on many sectors including ecosystems and socioeconomic (Karmaoui et al. 2022) makes it a pressing international concern, which leads it to attract growing scientific attention (Fan et al. 2022; Sun et al 2022). However, the trending analysis recorded that research is turning toward topics such as resilience, green infrastructure, ecosystem services, sponge city, livelihood, strategic approach, nature-based solutions, and urbanization (Fig. 3c). The rise of resilience as a trending topic in the field is due to its importance in terms of mitigating the impacts and protecting communities and properties. Green infrastructure (green roofs and permeable pavements) has emerged as well as a trending topic in the field because it offers multiple benefits in mitigating flood impacts including natural flood control and enhances ecosystem services such as floodplain restoration and wetland preservation.

In terms of quality, the United States followed by the United Kingdom and the Netherlands are the most influencing countries (Fig. 4 and Table 1). As mentioned above, the USA has emerged as a leading country in many domains, including the field of floods. For USA and England, this trend has been highlighted highlighted in many relevant studies including those of Lu et al. (2023), Karmaoui (2023), and Gao and Ruan 2018).

The classification of the exported keywords into components such as:

- Flood aspects;
- Type of assessments and management;
- Disciplines, tools, and approaches;
- Ecologic, socioeconomic, physical resources, and ecosystems and areas;
- Types of floods, associated phenomena, and impacts–consequences;
- Indicators, variables, and solutions.

Flood extreme event in research is studied under many aspects such as vulnerability (Fig. 5a), risk (Fig. 5b), resilience (Fig. 5c), disaster (Fig. 5d), exposure (Fig. 5e), adaptation (Fig. 5f), and hazard (Fig. 5g). According to the findings, the concept of disaster followed by vulnerability, adaptation or coping strategies, and resilience are the most occurring keywords considered in terms of flood aspects. Some of these notions such as vulnerability, resilience, and exposure are used in flood vulnerability indices (composite indicators) (Karmaoui et al. 2019a) developed by Balica et al. (2012) in the context of the coastal cities, Karmaoui and Balica (2021) in case of rural and urban spatial scale, and Karmaoui et al. (2021) at mountain scale. Also, the concept "adaptation" was used in the CRiSTAL tool while hazards and adaptation at the pre-Saharan scale in Karmaoui et al. (2019b).

Reviewing the type of assessments and management explored in the retrieved research efforts, several concepts were found. Risk assessment is the most explored in the selected documents (Fig. 6a). Balica et al. 2013 compared Parametric and physically based modeling techniques for flood risk and vulnerability assessment; Rosenzweig et al. (2018) investigated pluvial flood risk and opportunities for resilience.

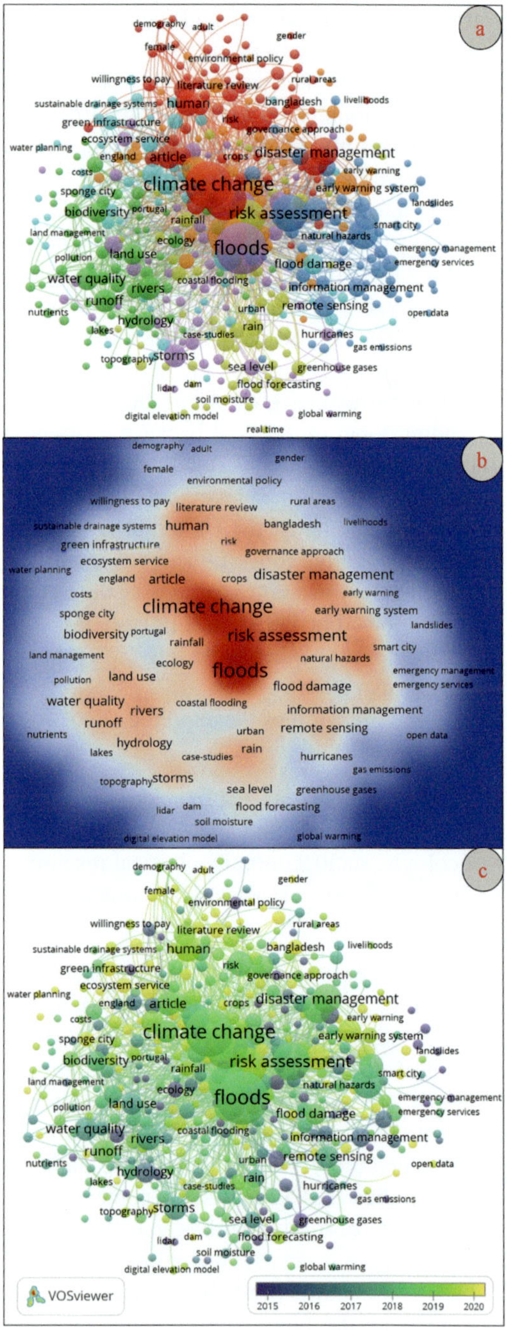

Fig. 3 Bibliometric analysis. **a** Network analysis; **b** density visualization; and **c** trending analysis using the search keywords "floods" and "technologies to support community initiatives" and VOSviewer tool

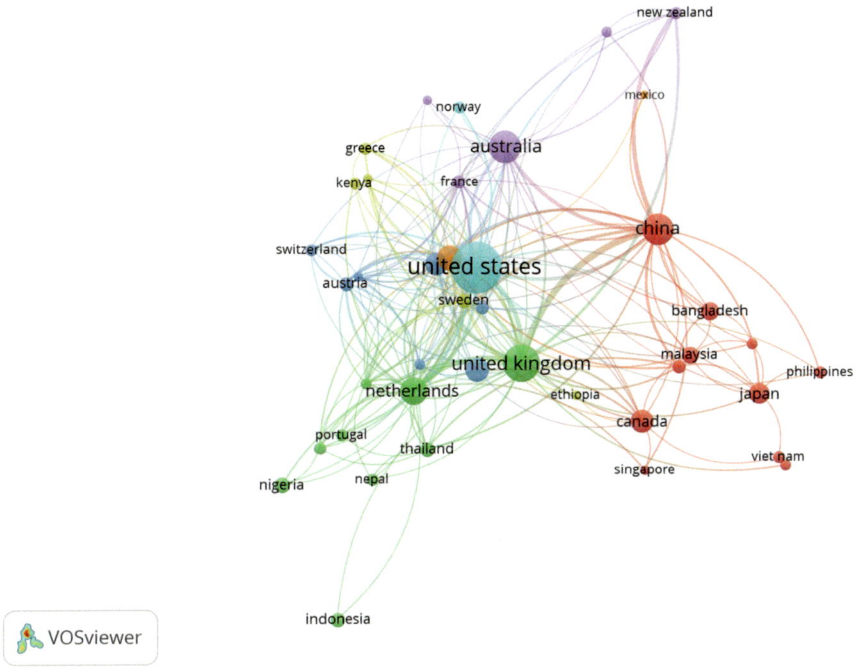

Fig. 4 Number of documents and citation by country using VOSviewer software

Smith (2013) edited a book about "Environmental Hazards: assessing risk and reducing disaster", and Wisner et al. (2004) edited a book entitled "At Risk: natural hazards, people's Vulnerability and Disasters".

The management aspect also well occurred including, primarily, water management, disaster management, decision-making, risk management, and adaptive management (Fig. 6b). This aspect highlighted water management as the most occurring concept. Undoubtedly, managing water and wastewater may help decrease flooding risks. In the context of floods, Ashley et al. (2020) described the old method of managing water and floods as engineer-driven, viewing water as a "problem," while the new perspective sees all types of water as potentially advantageous, offering a variety of benefits. The occurrence of floods in the future is uncertain, which suggests preparing for pessimistic scenarios. The best preparation is to propose the best way to cope with flood risks. Adaptable flood management is required at all special scales in rural urban, coastal, and mountain areas. One of the recorded responses to flood events was reported by Klijn et al. (2015) stating that the Netherlands invests heavily in future planning and research related to adaptation advancing that adaptive management planning should be based on a sound ex-ante policy analysis that considers the future outlook, determines whether a policy transition is necessary, and evaluates alternative flood risk management strategies. Flood preparedness and prevention (one of the most occurring key terms) actions are crucial steps to cope with floods and climate change. Governments and research centers and institutions are working

Table 1 Number of documents and citation by country using VOSviewer software

Country	Documents	Citations	Total link strength
United States	176	8416	108
United Kingdom	94	4120	131
Netherlands	52	2961	82
China	63	2448	123
Australia	72	1728	52
Germany	44	1378	51
Italy	34	1377	56
France	13	1328	31
Austria	14	1304	39
Switzerland	10	954	14
Ireland	7	899	20
Belgium	9	884	27
Canada	36	863	19
Norway	10	860	5
Spain	12	762	16
Finland	6	701	9
New Zealand	10	657	30
India	40	647	9
Bangladesh	21	470	11
Japan	29	426	14
Sweden	15	396	38
Thailand	15	358	18
Pakistan	13	321	21
South Africa	10	311	13
Portugal	10	268	10
Nepal	10	217	4
Czech Republic	8	201	19
Kenya	11	184	9
Ethiopia	7	180	4
Nigeria	16	174	5
Iran	7	152	5
Greece	10	151	11
Brazil	5	133	0
Ghana	8	107	3
Viet Nam	10	85	3
Indonesia	14	81	2
Mexico	5	73	3
South Korea	8	67	8

(continued)

Table 1 (continued)

Country	Documents	Citations	Total link strength
Singapore	6	65	5
Malaysia	20	61	9
Philippines	10	50	5
Saudi Arabia	8	47	6
Uganda	5	22	0

together on public education actions to prepare the population and students against flood risks. To facilitate learning, one reported initiative involves the development of a serious game based on knowledge and real-world flood prevention experiences (Tsai et al. 2020).

Figure 7 shows the disciplines, tools, and approaches explored in the extracted data. Concerning the most occurring terms indicating flood fields, the findings highlighted that hydrology, ecology, and urban planning are the main fields dealing with the topic investigated current review (Fig. 7a). However, geographic information systems, uncertainty analysis, Internet–Web sites–Web services, satellites and satellite imagery data, and the numerical model–theoretical model (s) are the primary tools explored in the retrieved indexed documents (Fig. 7b). Regarding the approaches, remote sensing, hydrological modeling, questionnaire surveys and surveys, mapping methods, and forecasting are the main recorded methods (Fig. 7c). Earl et al. (2023) compared some spatial prioritization methodologies distinguishing three model types, namely the conceptual models (such as multi-criteria analysis and multi-criteria decision analysis—geographic information systems), empirical models (agent-based models), and numerical models such as flood source analysis—unit flood response. For the current study, remote sensing was reported as the main method used to study floods. To forecast and comprehend flooding, remote sensing imagery and flood models are frequently used (Shastry et al. 2023). Currently, in addition to remote sensing techniques, researchers are using some new technologies such as artificial intelligence including deep learning methods. To map floods, Shastry et al. (2023) conceived a deep learning model using Maxar WorldView imagery.

Globally, floods are affecting several types of resources including ecologic, socioeconomic, and physical resources and many ecosystems and areas. The catchment/watershed (s), river (s), and coastal zones are the main ecosystems while ecosystem service (s), biodiversity, and water resource (s) are the most occurring resources (Fig. 8a). However, human (s), agriculture, livelihood (s), public health, and crops are the most impacted socioeconomic resources and sectors (Fig. 8b). In terms of physical resources and areas, urban areas (Cities), green infrastructure, rural area (s), infrastructure, and hydroelectric power are the most occurring in the retrieved research efforts (Fig. 8c).

In research, each extreme event is associated with other phenomena. Figure 9 recorded some categories of floods highlighted several associated phenomena and reported impacts–consequences linked to flood events. Climate change is the most occurring keyword after the primary search keyword "flood (s)", which indicates

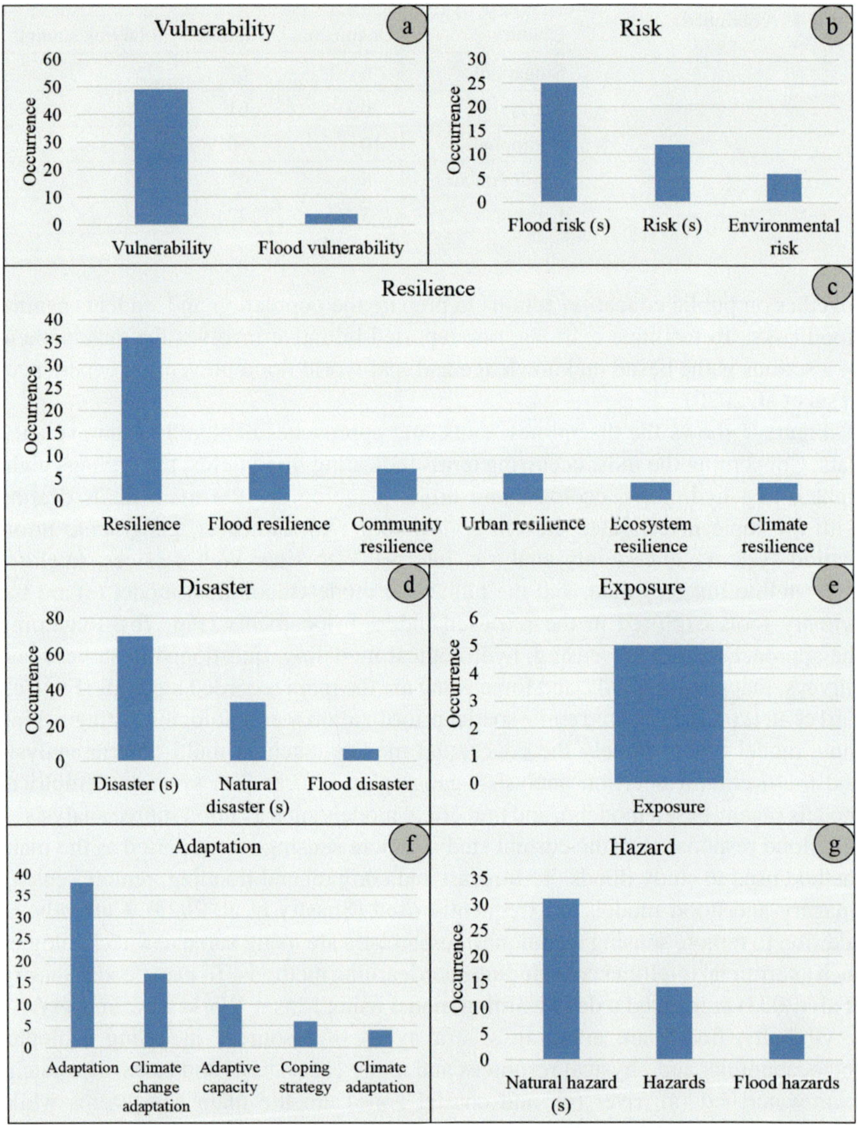

Fig. 5 Flood aspects. **a** vulnerability; **b** risk; **c** resilience; **d** disaster; **e** exposure; **f** adaptation; and **g** hazard

the strong link between them (Fig. 9a). This association was reported by Bibi and Kara (2023) advancing that climate change led to increasing flooding in most of the world. In the last report of the IPCC, Pörtner et al. (2022) reported that in comparison with 1.5 °C global warming without adaptation, projected increases in direct flood damages are 1.4–2 times higher at 2 °C and 2.5–3.9 times higher at 3 °C. The same

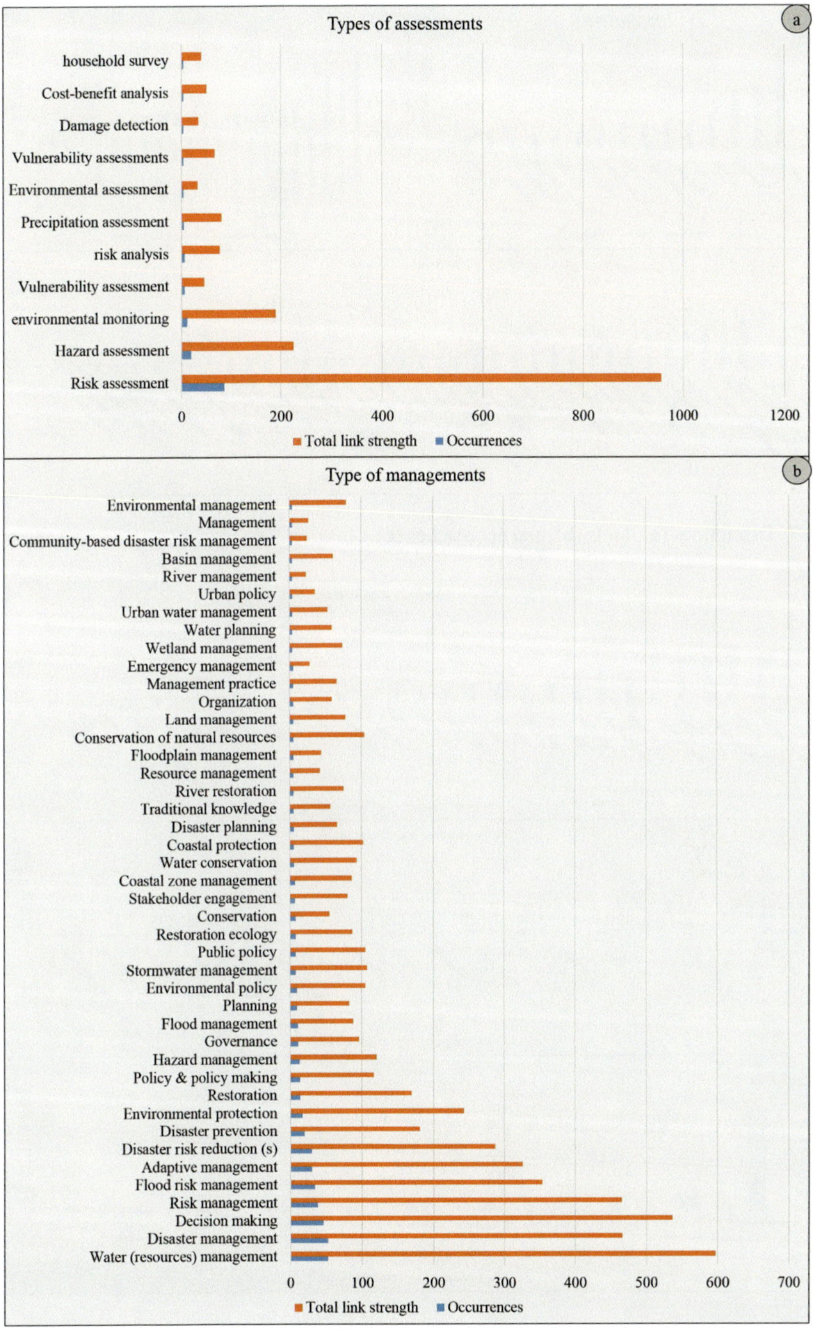

Fig. 6 Type of assessments (**a**) and management (**b**)

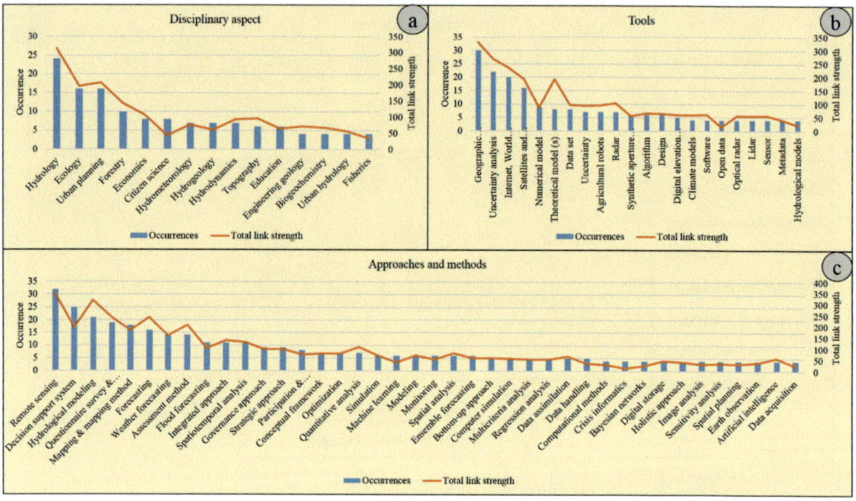

Fig. 7 Disciplines (**a**), tools (**b**), and approaches (**c**)

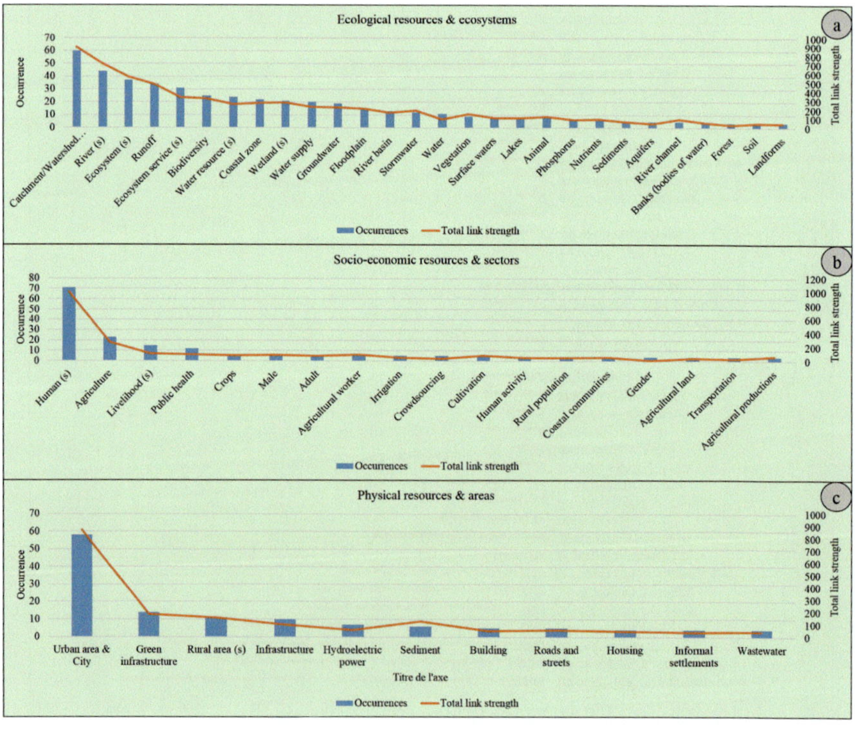

Fig. 8 Resources. **a** ecologic; **b** socioeconomic and ecosystems; and **c** physical resources and areas

report emphasizes that with planning implications for all water use sectors, it is anticipated that at 4 °C of global warming, 10% of the world's land area will see increases in extreme high and low river flows at the same site. Pörtner et al. (2022) added that depending on the extent, speed, and regional specifics of future climate change, challenges for water management will be made more difficult in the short, medium, and long terms. It was also reported that flood vents are linked frequently with storms, drought (s), hurricane (s), and urban floods and are linked sometimes with other events such as volcanic eruptions (Fig. 9a). The flood events induced several impacts or consequences including water pollution, environmental impact, economic and social effects, food security, poverty, costs, and climate change impacts (Fig. 9b).

Research in the field of floods often emphasizes the use of indicators or variables associated with the concepts of vulnerability, exposure, susceptibility, and resilience. Among the most occurring variables used in the retrieved research efforts are water quality, land use, rain–rainfall, sea level change, urbanization, risk perception, drainage, population statistics, soil moisture, local participation, urban growth, flood frequency, losses, organization and management, residence characteristics, sediment transport, etc. (Fig. 10a). The water quality is one of the most threatened elements due to frequent floods. Floods are linked to water quality degradation through the increased pollution loadings. This aspect was reviewed by Fabian et al.

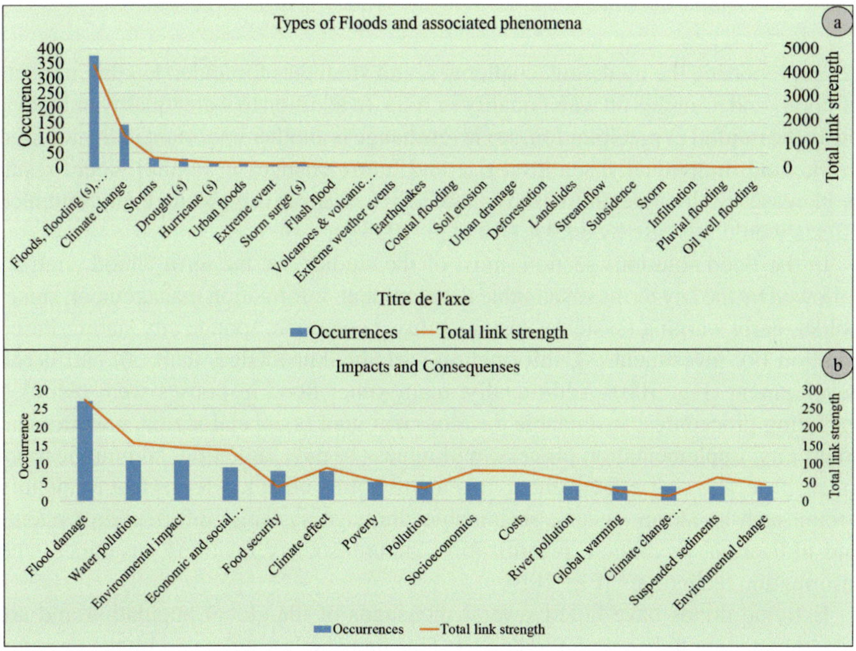

Fig. 9 Flood types and consequences. **a** Types of floods and associated phenomena; **b** impacts–consequences

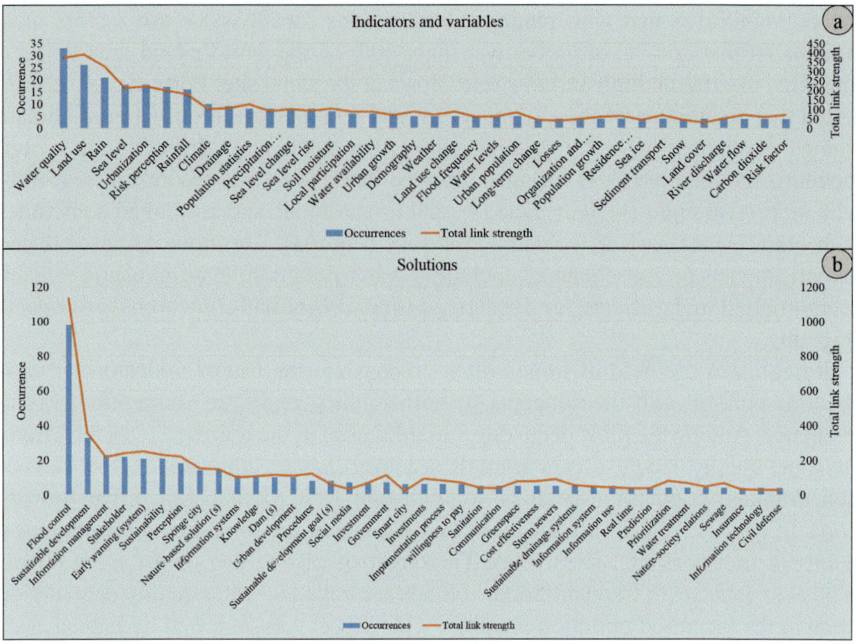

Fig. 10 Indicators and responses. **a** Indicators and variables; **b** solutions

(2023) reporting the modeling, challenges, and strategies for understanding impacts of floods and droughts on water quality in Asia. In addition to water quality and obviously the rainfall or precipitation, sea level change is another important indicator used in the field. Individually, sea level rise and future changes in rainfall would result in increases in flood depth of 10.6% and 32.1%, respectively, while their combined effects would increase by 34.9% (Xu et al. 2023).

In the flood solutions section, most of the studies use the term "flood control" followed by the key terms sustainable development, information management, stakeholder, early warning (system), sustainability, perception, sponge city, nature-based solution (s), investment (s), information systems, knowledge, dam (s), and urban development (Fig. 10b). Additionally, many other flood responses were recorded including procedures, sustainable development goal (s), social media, government, smart city, implementation process, willingness to pay, sanitation, communication, greenspace, and cost-effectiveness. It is worth noting other key terms that frequently occur, such as storm sewers, sustainable drainage systems, information systems and utilization, prediction, prioritization, nature–society relations, insurance, and information technology (Fig. 10b).

Extreme floods have killed several thousands of the global population and are now threatening to trigger a food crisis. This damage is accelerated by the growing trend of climatic and anthropogenic changes, the vulnerability, and the exposure of a large part of the international population, especially in coastal regions. To address the effects of this phenomenon, governments must find effective solutions. To reduce and avoid possible effects of floods, structural measures including physical construction, engineering techniques, or non-structural measures such as education, knowledge, practice, training, and laws have been proven to be effective in terms of flood response and recovery. Currently, many new technologies have been proposed and applied in many aspects of floods before, during, or after the occurrence of the flood. GIS, remote sensing, Internet, numerical model, theoretical model, artificial intelligence, and ex-ante technologies are used to collect and analyze data and assess vulnerability, risk, exposure, and resilience.

To summarize, risk-hazard assessment, environmental monitoring, vulnerability assessment, and risk analysis are the main types of flood assessment. However, water (resource) management, disaster risk management, adaptive management, and disaster prevention are the most occurring management vocabulary. The reduction of floods needs flood control and implementation of sustainable development, information management, early warning (system), and the engagement of stakeholders. The use of cited technologies in addition to traditional measures requires an effective engagement of all linked sectors and stakeholders in the education field, economy, politics, and research. The new technologies have considerable potential analyses and reduce flood risks. To further explore the current topic the following areas for future research can be suggested:

- Identify emerging technologies utilized in community flood initiatives, such as remote sensing, the Internet of things (IoT), artificial intelligence, and big data analytics.
- Compare and contrast different approaches employed in community flood initiatives, such as structural (e.g., flood barriers, levees) versus non-structural (e.g., land-use planning, ecosystem-based adaptation) measures,

Efforts to develop more artificial intelligence and geospatial technologies are promising frontiers to be investigated and developed. The findings suggest a set of management methods and tools for decision-makers to evaluate the flooding impacts and propose effective solutions to reduce the risk and effects.

4 Conclusion

Flooding is one of the most frequent, pervasive, and destructive natural disasters (Shastry et al. 2023). New technologies have been widely employed to cope with floods due to the necessity of analyzing data and reducing vulnerability and risks. In this context, this chapter reviewed the scientific literature to highlight the current and trending dimensions of the technologies and approaches to support community flood Initiatives. In this regards, this chapter reviewed the scientific literature to highlight the current and emerging dimensions of technologies and approaches supporting community flood initiatives. The analysis indicates an increasing volume of published documents, with a notable influence from the field of environmental science. The

International Journal of Disaster Risk Reduction emerges as a key source while funding primarily originates from the National Science Foundation and the National Natural Science Foundation of China. Thaler, T., the University of Leeds, and the United States are recognized as the leading author, affiliation, and country, respectively. The findings also showed that hot spot search topics linked to "floods" and "technologies to support community initiatives are climate change and risk assessment are the most visible key terms. However, the trending analysis depicted that research efforts are turning toward topics such as resilience, green infrastructure, ecosystem services, sponge city, livelihood, strategic approach, nature-based solutions, and urbanization. This research effort provides a set of management techniques and tools for decision-makers to assess the effects of proposed flood mitigation measures.

Appendix 1

Quality of research using the key terms "floods" and "technologies to support community initiatives". Documents with the associated number of citation.

Document	Citations	Links	Document	Citations	Links
Brauman et al. (2007)	901	1	Bhattacharyya (2015)	229	0
Godschalk (2003)	884	5	Abid (2016)	220	3
Dorigo (2017)	640	4	Glenn (2005)	206	0
Fletcher (2013)	573	4	Babaeian (2019)	194	1
Jacka (2008)	531	0	Berne (2013)	184	1
Blann (2009)	450	1	Alfieri (2012)	177	2
Wisner (2014)	412	0	Van Beijma (2014)	174	0
Zhang (2011)	353	3	Balica (2013)	170	1
Blöschl (2011)	327	0	Renöfält (2010)	170	0
Norris (2009)	306	0	Mei (2018)	160	8
Buijse (2002)	294	0	Tilcsik (2013)	160	0
Tompkins (2010)	277	1	Islam (2014)	158	0
Chan (2018)	266	10	Levin (2009)	152	0
Jiang (2018)	244	7	Huntjens (2011)	150	3

References

Ajibade SSM, Zaidi A, Bekun FV, Adediran AO, Bassey MA (2023) A research landscape bibliometric analysis on climate change for last decades: evidence from applications of machine learning. Heliyon 9(10). https://doi.org/10.1016/j.heliyon.2023.e20297

Ashley R, Gersonius B, Horton B (2020) Managing flooding: from a problem to an opportunity. Phil Trans R Soc A 378(2168):20190214. https://doi.org/10.1098/rsta.2019.0214

Balica SF, Popescu I, Beevers L, Wright NG (2013) Parametric and physically based modelling techniques for flood risk and vulnerability assessment: a comparison. Environ Model Softw 41:84–92. https://doi.org/10.1016/j.envsoft.2012.11.002
Balica SF, Wright NG, Van der Meulen F (2012) A flood vulnerability index for coastal cities and its use in assessing climate change impacts. Nat Hazards 64(1):73–105. https://doi.org/10.1007/s11069-012-0234-1
Bibi TS, Kara KG (2023) Evaluation of climate change, urbanization, and low-impact development practices on urban flooding. Heliyon e12955. https://doi.org/10.1016/j.heliyon.2023.e12955
Brauman KA, Daily GC, Duarte TKE, Mooney HA (2007) The nature and value of ecosystem services: an overview highlighting hydrologic services. Ann Rev Environ Resour 32:67–98. https://doi.org/10.1146/annurev.energy.32.031306.102758
Earl E, Johnson F, Marshall L, Sanderson D (2023) A critical review of natural flood management application and spatial prioritisation approaches in tropical island catchments. Sci Total Environ 878:162776. https://doi.org/10.1016/j.scitotenv.2023.162776
Fabian PS, Kwon HH, Vithanage M, Lee JH (2023) Modeling, challenges, and strategies for understanding impacts of climate extremes (droughts and floods) on water quality in Asia: a review. Environ Res 115617. https://doi.org/10.1016/j.envres.2023.115617
Fan J, Liu G, Xia Z, Cai S (2022) A bibliometric analysis of climate change risk perception: hot spots, trends and improvements. Front Environ Sci 10:917469. https://doi.org/10.3389/fenvs.2022.917469
Gao C, Ruan T (2018) Bibliometric analysis of global research progress on coastal flooding 1995–2016. Chin Geogra Sci 28:998–1008. https://doi.org/10.1007/s11769-018-0996-9
Geng Z, Wang J, Liu J, Miao J (2024) Bibliometric analysis of the development, current status, and trends in adult degenerative scoliosis research: a systematic review from 1998 to 2023. J Pain Res 153–169. https://doi.org/10.2147/JPR.S437575
Godschalk DR (2003) Urban hazard mitigation: creating resilient cities. Nat Hazard Rev 4(3):136–143. https://doi.org/10.1061/(ASCE)1527-6988(2003)4:3(136)
Ivanova M, Grosseck G, Holotescu C (2024) Unveiling insights: a bibliometric analysis of artificial intelligence in teaching. Informatics 11(1):10. https://doi.org/10.3390/informatics11010010
Karmaoui A (2023) Flood vulnerability in developing countries, international collaboration, and network visualization: a bibliometric analysis. In: Handbook of flood risk management in developing countries. Routledge, pp 183–196. https://doi.org/10.1201/9781003160823-15
Karmaoui A, Yoganandan G, Sereno D, Shaukat K, El Jaafari S, Hajji L (2023a) Global network analysis of links between business, climate change, and sustainability and setting up the interconnections framework. Environ Dev Sustain 1–25. https://doi.org/10.1007/s10668-023-03883-w
Karmaoui A, El Jaafari S, Chaachouay H, Hajji L (2023b) A bibliometric review of geospatial analyses and artificial intelligence literature in agriculture. GeoJournal 88(Suppl 1):343–360. https://doi.org/10.1007/s10708-023-10859-w
Karmaoui A (2022) Ordovician-Cambrian palaeontological heritage of Zagora Province: a bibliometric analysis from 1984 to 2020 (Anti-Atlas, Morocco). Geoheritage 14(2):55. https://doi.org/10.1007/s12371-022-00695-8
Karmaoui A, Ben Salem A, El Jaafari S, Chaachouay H, Moumane A, Hajji L (2022) Exploring the land use and land cover change in the period 2005–2020 in the province of Errachidia, the pre-sahara of Morocco. Front Earth Sci 10:962097. https://doi.org/10.3389/feart.2022.962097
Karmaoui A, Balica S (2021) A new flood vulnerability index adapted for the pre-Saharan region. Int J River Basin Manage. https://doi.org/10.1080/15715124.2019.1583668
Karmaoui A, Zerouali S, Ayt Ougougdal H, Shah Ashfaq A (2021) A new mountain flood vulnerability index (MFVI) for the assessment of flood vulnerability. Sustain Water Resour Manage 7(6):1–13. https://doi.org/10.1007/s40899-021-00575-z
Karmaoui A, Minucci G, Ben Salem A (2019a) Composite indicators as decision support method for flood analysis: flood vulnerability index category. In: Decision support methods for assessing flood risk and vulnerability. IGI Global. https://doi.org/10.4018/978-1-5225-9771-1.ch002

Karmaoui A, Moumane A, Akchbab J (2019b) Environmental hazards assessment at pre-Saharan local scale: case study from the Draa Valley, Morocco. In: Decision support methods for assessing flood risk and vulnerability. IGI Global. https://doi.org/10.4018/978-1-5225-9771-1.ch012
Klijn F, Kreibich H, De Moel H, Penning-Rowsell E (2015) Adaptive flood risk management planning based on a comprehensive flood risk conceptualisation. Mitig Adapt Strat Glob Change 20:845–864. https://doi.org/10.1007/s11027-015-9638-z
Kottelat M, Whitten T (1996) Freshwater biodiversity in Asia: with special reference to fish, vol 343. World Bank Publications, pp XI–55. ISSN 02537494
Liu W, Feng Q, Engel BA, Yu T, Zhang X, Qian Y (2023) A probabilistic assessment of urban flood risk and impacts of future climate change. J Hydrol 618:129267. https://doi.org/10.1016/j.jhydrol.2023.12926
Lu S, Huang J, Wu J (2023) Knowledge domain and development trend of urban flood vulnerability research: a bibliometric analysis. Water 15(10):1865. https://doi.org/10.3390/w15101865
Mesa Fernández JM, González Moreno JJ, Vergara-González EP, Alonso Iglesias G (2022) Bibliometric analysis of the application of artificial intelligence techniques to the management of innovation projects. Appl Sci 12(22):11743. https://doi.org/10.3390/app122211743
Pariartha IGS, Aggarwal S, Srinivas R, Egodawatta P, McGree J, Goonetilleke A (2023) Compounding effects of urbanization, climate change and sea-level rise on monetary projections of flood damage. J Hydrol 129535. https://doi.org/10.1016/j.jhydrol.2023.129535
Pörtner HO, Roberts DC, Adams H, Adler C, Aldunce P, Ali E et al (2022) Climate change 2022: impacts, adaptation and vulnerability. IPCC, Geneva, Switzerland, p 3056
Rentschler J, Salhab M (2020) People in harm's way: flood exposure and poverty in 189 countries. Policy Research Working Paper;No. 9447. © World Bank, Washington, DC. http://hdl.handle.net/10986/34655 (License: CC BY 3.0 IGO)
Rosenzweig BR, McPhillips L, Chang H, Cheng C, Welty C, Matsler M et al (2018) Pluvial flood risk and opportunities for resilience. Wiley Interdiscip Rev Water 5(6):e1302. https://doi.org/10.1002/wat2.1302
Roy P, Pal SC, Chakrabortty R, Chowdhuri I, Saha A, Shit M (2023) Effects of climate change and sea-level rise on coastal habitat: Vulnerability assessment, adaptation strategies and policy recommendations. J Environ Manage 330:117187. https://doi.org/10.1016/j.jenvman.2022.117187
Shastry A, Carter E, Coltin B, Sleeter R, McMichael S, Eggleston J (2023) Mapping floods from remote sensing data and quantifying the effects of surface obstruction by clouds and vegetation. Remote Sens Environ 291:113556. https://doi.org/10.1016/j.rse.2023.113556
Shkundalov D, Vilutienė T (2021) Bibliometric analysis of building information modeling, geographic information systems and web environment integration. Autom Constr 128:103757. https://doi.org/10.1016/j.autcon.2021.103757
Smith K (2013) Environmental hazards: assessing risk and reducing disaster. Routledge, p 504. eBook ISBN9780203805305. https://doi.org/10.4324/9780203805305
Sun YL, Zhang CH, Lian YJ, Zhao JM (2022) Exploring the global research trends of cities and climate change based on a bibliometric analysis. Sustainability 14(19):12302. https://doi.org/10.3390/su141912302
Tsai MH, Chang YL, Shiau JS, Wang SM (2020) Exploring the effects of a serious game-based learning package for disaster prevention education: the case of battle of flooding protection. Int J Disast Risk Reduct 43:101393. https://doi.org/10.1016/j.ijdrr.2019.101393
Van Eck NJ, Waltman L (2010) Software survey: VOSviewer, a computer program for bibliometric mapping. Scientometrics 84(2):523–538. https://doi.org/10.1007/s11192-009-0146-3
Wisner B, Blaikie PM, Blaikie P, Cannon T, Davis I (2004) At risk: natural hazards, people's vulnerability and disasters. Psychology Press, 471 pp. ISBN 0415252156, 9780415252157
Xu K, Zhuang Y, Bin L, Wang C, Tian F (2023) Impact assessment of climate change on compound flooding in a coastal city. J Hydrol 617:129166. https://doi.org/10.1016/j.jhydrol.2023.129166

Assessing Flood Susceptibility Using GIS and AHP Method: Case of Mandovi River Basin, Goa

R. Rajeev, Saishree Naik, and N. Nivedha

Abstract This research focuses on performing a comprehensive flood susceptibility analysis in the Mandovi river basin in Goa, in order to increase flood resilience. The study employs the analytical hierarchy process (AHP) model and geographic information system (GIS) techniques to prioritize and map various factors influencing flood susceptibility. Beginning with an extensive background review and preliminary analysis, the research involves the generation of parameter maps, including distance from rivers, topographic wetness index (TWI), drainage density, rainfall, elevation, slope, land use and land cover (LULC), soil type, normalized difference vegetation index (NDVI), and distance from roads. Applying a weighted overlay method, these parameters are integrated to develop a flood susceptibility map, classifying areas within the basin as very high, high, moderate, or low and very low susceptible to flooding. Each parameter has been reclassified based on chances of the class being very high susceptible to very low susceptible to flooding. The conclusion from the susceptibility map was that higher elevation and slope are associated with lower flood susceptibility, while distance to the river, topographic wetness index (TWI), drainage density, and high vegetation areas are factors that increase susceptibility to flooding. It provides a comprehensive classification of potential flood risk in various areas within the basin underscoring the importance of flood risk assessment in comprehending the intricate interplay of factors contributing to flooding. It proposes basin-level interventions as strategic measures to mitigate the impact of floods, contributing valuable insights for formulating measures to reduce the vulnerability of the Mandovi river basin to flooding. The findings hold significance for informed land-use planning, zoning regulations, and targeted mitigation measures. In summary, this research offers a comprehensive understanding of flood susceptibility in the Mandovi river

R. Rajeev (✉)
Department of Planning, School of Planning and Architecture, Vijayawada 520008, India
e-mail: rajeevnair@spav.ac.in

S. Naik
Goa State Urban Development Agency, Panaji, Goa, India

N. Nivedha
National Institute of Technology, Surathkal, Karnataka 575025, India

S. C. Pal and U. Chatterjee (eds.), *Surface, Sub-Surface Hydrology and Management*, Springer Geography, https://doi.org/10.1007/978-3-031-62376-9_4

basin, providing a valuable foundation for proactive flood resilience strategies in the region.

Keywords Susceptibility · Flood · Mitigation · Mandovi · River Basin

1 Introduction

Sever natural calamities like floods cause huge and irreversible damage to life and property. These natural calamities can be monitored and predicted using the latest technology and information systems (Swain et al. 2020). This natural disaster is caused by climate change because global warming affects the intensity and rate of rainfall which causes heavy rainfall leading to over overflow of seas and streams resulting in flooding (Souissi et al. 2019). The effect of flood in coastal cities is aggravated as the location is prone to rise in sea level, tropical cyclones, and storm surges which intensify because of climate change. The coastal cities of India face several significant risks concerning food security, many of which are tied to environmental factors and urban development practices. The primary factors are the effect of cyclonic storms, tidal inundation and tsunamis, riverine flooding, and failure in unsustainable land-use practices and urban planning (Dhiman et al. 2018). Riverine flooding occurs when the water handling capacity of a high volume of water from upper catchments, caused by intense rainfall, is reduced by the drainage system. The flood risk is exacerbated by factors like river banks erosion and sediment accumulation which minimize the water-holding capacity rivers (Kumar et al. 2023). It can concluded that the flood is the combined effect of both hydrological conditions and meteorological events. Since many years, effects have been made to reduce the impact and damage caused by flood, and most of the solution has been structural measures such as the construction of drainage channels, reservoirs, embankments, and more. Achieving a permanent solution is proven to be infeasible and uneconomical which has forced the researchers to come up with non-structural measures for flood-affected regions. Key components of non-structural flood mitigation in India include flood plain zoning (FPZ) and the implementation of flood forecasting and warning systems. However, the lack of comprehensive flood data along with ongoing urban expansion in flood-prone regions has led to a rise in estimated annual losses due to flood disasters. Consequently, there is a pressing need to assess flood-prone regions by developing flood susceptibility maps, which can help identify and prioritize areas at risk of flooding across different scales. This prioritization plays a crucial role in directing interventions to the areas which are more vulnerable to flood in the region (Kumar et al. 2023). Hence, identifying the flood-prone areas using flood susceptibility mapping to assess and rank the various scales of flood vulnerability is decisive.

Riverine flood susceptibility mapping of India using fuzzy analytical hierarchy process (FAHP) and geographic information system (GIS) considering geomorphological, meteorological, hydro climatic, and anthropogenic data showed that the identified high-risk areas consist of a significant portion of India's agricultural land, potentially impacting food security. The study emphasizes that the primary factor influencing riverine floods is the distance from the floodplain, and it identifies cities with high proximity to rivers which are Guwahati, Sri Ganganagar, Jagdalpur, Madurai, Burhanpur, Mathura, Bhiwani, Nellore, Panjim, and 43 other cities (Kumar et al. 2023). The researchers state choosing factors for flood susceptibility mapping as there are various dependent and independent factors. The paper on Markham riverine flooding using GIS and frequency ratio model emphasizes the significance of ten conditional factors for mapping out of which soil texture and soil drainage are the most important factors as high to very high vulnerable areas are categorized as having very poor soil properties (Samanta et al. 2018). Mohana–Khutiya River in Nepal is highly vulnerable to flood due to the presence of steep slopes and varying geological structure associated with flat and vulnerable flood zone. The study reveals that the proximity of the river exerted a greater influence on the incidence of flooding and the level of annual precipitation throughout the year showed no evident effect on the occurrence of floods in the study area. The study highlights the importance of framing disaster management policy by local government for the study area and the contribution of identifying flood-vulnerable areas in preparing disaster plan to reduce its impacts (Maharjan et al. 2024). Flood susceptibility mapping is useful for watershed management, and it can be used for flood hazard and risk management (Moazzam et al. 2020).

India is more prone to riverine flooding, and it is necessary to prioritize this method of flood mitigation. Floods in coastal area rivers are influenced by a convergence of factors, including tidal effects, heavy rainfall, and the overarching threat of sea level rise. These dynamic elements contribute to the vulnerability of rivers as well as the coastal region, such as Goa, to flooding incidents. Coastal regions, including the riverbanks, are susceptible to tidal effects due to their proximity to the sea. When tidal effects coincide with heavy rainfall, the risk of flooding intensifies, impacting areas adjacent to the riverbanks. The state of Goa experiences heavy rainfall during monsoon seasons, and the river systems are intricately connected to the region's topography. Intense rainfall can result in rapid runoff and overwhelming river capacities. The combination of heavy rainfall and tidal effects can amplify the flood risk, affecting both the interior and peripheral areas of the river basin. Sea level rise due to climate change poses an increasing threat to coastal areas, including the riverbanks of Goa. As sea levels gradually increase, the risk of riverbanks being submerged rises. This becomes particularly pronounced during extreme weather events, including the impact of tidal effects and heavy rainfall on river systems.

Hence, the interplay of tidal effects, heavy rainfall, and sea level rise compounds the vulnerability of coastal area rivers to flooding, and this is the reason why the paper focuses on flood susceptibility mapping in coastal areas. The proximity of rivers to the coast makes them susceptible to tidal surges, and when coupled with heavy rainfall, the riverbanks may overflow while accommodating the increased water volume. Sea level rise further elevates the baseline flood risk. This necessitates a comprehensive understanding and mapping of regions having flood risk.

To address the multifaceted nature of flood risk in coastal areas, analysing various parameters, including topography, land use, and hydrological conditions, helps identify areas prone to flooding. This approach aids in the development of mitigation strategies and adaptive measures to protect vulnerable communities and infrastructure. The Sahyadris, situated on the eastern periphery, form watersheds crucial for river sources. The river Mandovi rises at an elevation of 600 m in the Sahyadri and flows through Karnataka and Goa. Focusing on the Mandovi river basin in North Goa, the research employs risk analysis, incorporating various parameters to develop a flood susceptibility map and identify areas which are at risk of flooding.

Conducting such assessments is important, as it enables communities to prepare for unexpected severe flood events, especially in the face of climate change uncertainties (Dhiman et al. 2018). The basin-level flood susceptibility assessments help in disaster recovery planning, community awareness, and contributing to overall resilience. Additionally, if we consider proposing flood risk mitigation strategies at the micro-level focusing on a particular village, zone, or river stretch that are highly at risk of flooding based on the flood susceptibility map, then these assessments can help in informed land-use planning, guide responsible urbanization and infrastructure development, and protect ecosystems.

This study aims to assess flood susceptibility in the Mandovi river basin, identifying and mapping potential flood risk areas. The objectives include identifying influential parameters and mapping them, assessing their impact on flood susceptibility, and proposing macro-level flood mitigation measures. As per the first objectives, influential parameters were identified, that are distance from rivers, drainage density, TWI, slope rainfall, land use/land cover, elevation, soil type, distance from roads, and NDVI. For the second objective, these parameter maps are generated, and using weightages gained through AHP method, a weighted overlay analysis in GIS is conducted to create a flood susceptibility map. This map shows high, medium, and low flood susceptible zones. The third objective focuses on recommending strategies at the basin level for the susceptible zones identified in the map.

2 Rationale of the Study

Aim

To assess the flood susceptibility in the Mandovi river basin in order to identify and map potential areas having risk of flooding issue.

Objectives

1. To identify parameters having impact on flood susceptibility and map them.
2. To assess the flood susceptibility of the Mandovi river basin using identified influential parameters.
3. To propose flood mitigation measures at the macro-level.

Need

Conducting flood vulnerability and susceptibility assessments in a river basin, even in regions with historically low flood risk, is imperative for various reasons. Firstly, it allows communities and authorities to prepare for rare but potentially severe flood events that can occur unexpectedly, safeguarding lives and property. With the uncertainty brought about by climate change, such assessments help anticipate how shifting weather patterns might increase the flood risk due to sea level rise, increasing rainfall, and tidal influence in the future. Additionally, they inform land-use planning and zoning regulations, guiding responsible urbanization and infrastructure development that could otherwise exacerbate flood hazards. These assessments are also vital for safeguarding critical infrastructure and economic investments in the regions having potential flood risk and for protecting sensitive ecosystems that could be impacted by flooding. Flood susceptibility provides valuable information for insurance, risk management, and disaster recovery planning and increases community awareness about potential flood risks. Moreover, assessing susceptibility can enable more effective resource allocation for immediate response and recovery efforts in case of a flood, contributing to overall community resilience.

3 Methods and Materials

3.1 Study Area

Goa state is known in India as one of the famous tourist destinations. It covers an area of approximately 3702 km^2 has with Western Ghats towards east and the Arabian Sea to the west. The state is divided into two districts, North and South, with administrative centres in Panaji (the capital) and Margao, respectively, as shown in Fig. 1. Goa has eleven rivers, including Terekhol, Mandovi, Baga, Zuari, Colval, Saleri, Mandre, Harmal, Sal, Talpona, and Galjibag. These rivers, along with their 42 tributaries, play a crucial role by serving as a source of water, supporting agriculture through irrigation, contributing to the region's biotic and mineral resources, used to facilitate ore transportation from mining areas to the port, and acting as transportation routes for people and goods across the state. These rivers predominantly flowing from the Western Ghats to the Arabian Sea create intricate ecosystems of wetlands, tidal marshes, cultivated paddy fields, canals, lakes, lagoons, salt pans, khazan land and creeks influenced by regular tides.

The Sahyadris, situated along the eastern periphery of Goa, create numerous watersheds that serve as the primary water source for rivers to flow down. These rivers, characterized by highly erosive flow towards west, shape 'V'-shaped valleys in the Western Ghats region, transitioning to 'U'-shaped valleys in the central midlands and coastal plains. Occupying about 600 km^2 with an average elevation of 800 m above mean sea level, the Sahyadri feature plateaus range from 100 to 30 m above MSL in the central portion of Goa. The state's 105-km coastline includes large estuaries of

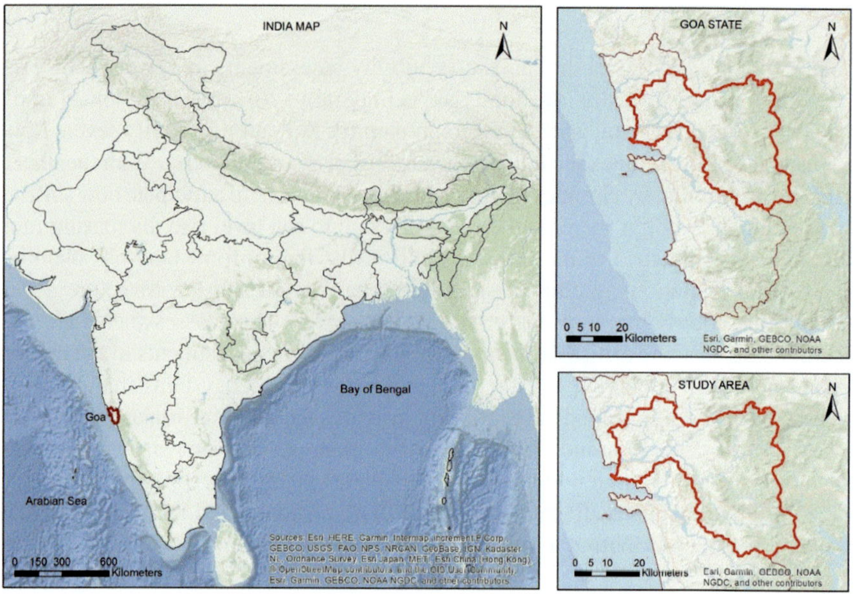

Fig. 1 Key map of the study area

Mandovi and Zuari rivers. The estuaries, including the Mandovi–Zuari confluence at Arabian Sea, cover approximately 12,000 ha, with mangrove forests over 2000 ha (Kerkar et al., SANDRA) (Fig. 2).

The River Mandovi rises at an elevation of 600 m in the Sahyadri, flows through Karnataka and Goa, and drains an area of 1580 km^2 within Goa alone. Referred to as Mhadei in its upper reaches, it holds ecological significance as 85% of its watershed

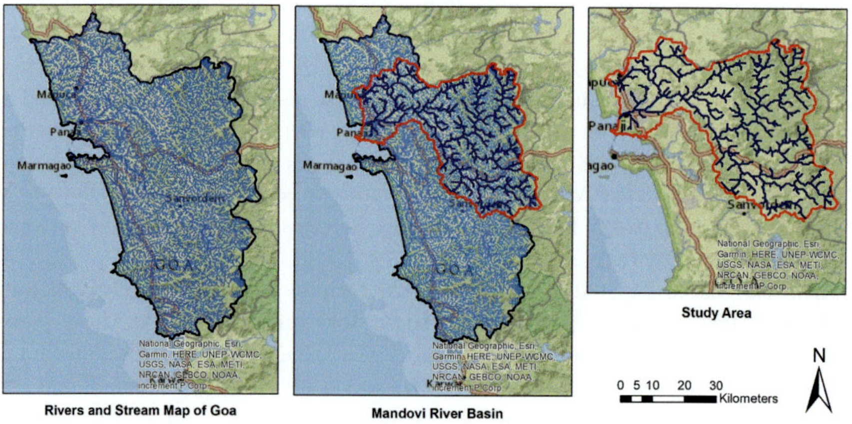

Fig. 2 Study area delineation

area covered by forests and forms the Mhadei Wildlife Sanctuary. Mandovi River has the largest basin in Goa as shown in Fig. 4. The length of the river Mandovi within the state is 52 km with basin area of 1580 km^2, and the study area taken is majorly falling into North Goa with some part of it in South Goa as shown in Fig. 3. The Mandovi river basin is delineated and considered as study area as shown in Fig. 2. Goa being coastal area is under threat of sea level rise. The river in Goa is often subjected to tide levels. Hence, whenever there is tidal level rise along with heavy rainfall, there is considerable rise in the water level. The interior parts of Mandovi River do not face much effects of this phenomena. Having said this, during heavy rainfall, certain portions of the river face flooding issue for certain period.

As we observe sea level rise due to climate change, Goa also faces heavy rainfall issue. And there are everyday tidal effects experienced by the rivers as they are very near to the coast. Due to these three aspects, there are chances of the areas adjacent to the river getting flooded. There are definitely other factors that may contribute to flooding issue. To know which areas are susceptible to flooding, this research is conducted. The risk analysis involved using of different parameters and finally coming up with flood susceptibility map (Fig. 4).

Figure 5 represents a comprehensive methodology for flood susceptibility mapping, encompassing various parameters and analytical techniques to generate a flood susceptibility map for the case area.

It begins with a background study on these flooding phenomena and the selection and understanding of the case area. A literature review was conducted to gather

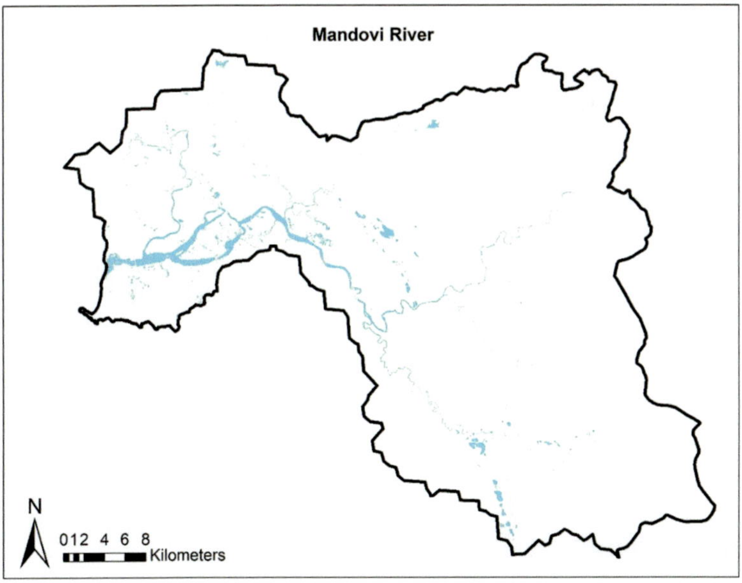

Fig. 3 Map showing Mandovi River

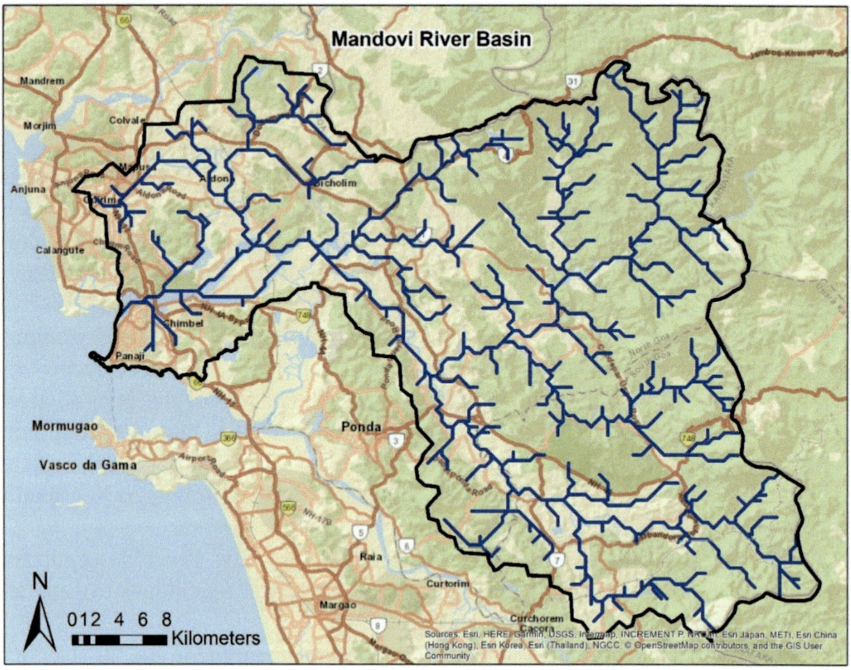

Fig. 4 Maps showing Mandovi river basin and Mandovi River

relevant information. This was then followed by a preliminary analysis involving the identification of stream order, basin mapping, and delineation of the case area.

The flood susceptibility analysis is carried out using various parameters such as distance from rivers, drainage density, TWI, slope rainfall, land use/land cover, elevation, soil type, distance from roads, and NDVI. The maps were reclassified as per the values based on the chances of susceptibility. The AHP method is then applied to prioritize these factors, and a weighted overlay analysis in GIS is conducted to create a flood susceptibility map. The culmination of these analyses results in the creation of a flood susceptibility map. The application of GIS and AHP methods demonstrates a systematic approach to the study, and the conclusions and recommendations emphasize the importance of the resilient planning and mitigation efforts.

3.2 *Materials*

Parameters Influencing Flood

The parameters considered for mapping susceptibility include distance from the river, topographic wetness index (TWI), drainage density, rainfall, elevation, slope, land use land cover (LULC), soil type, normalized difference vegetation index (NDVI),

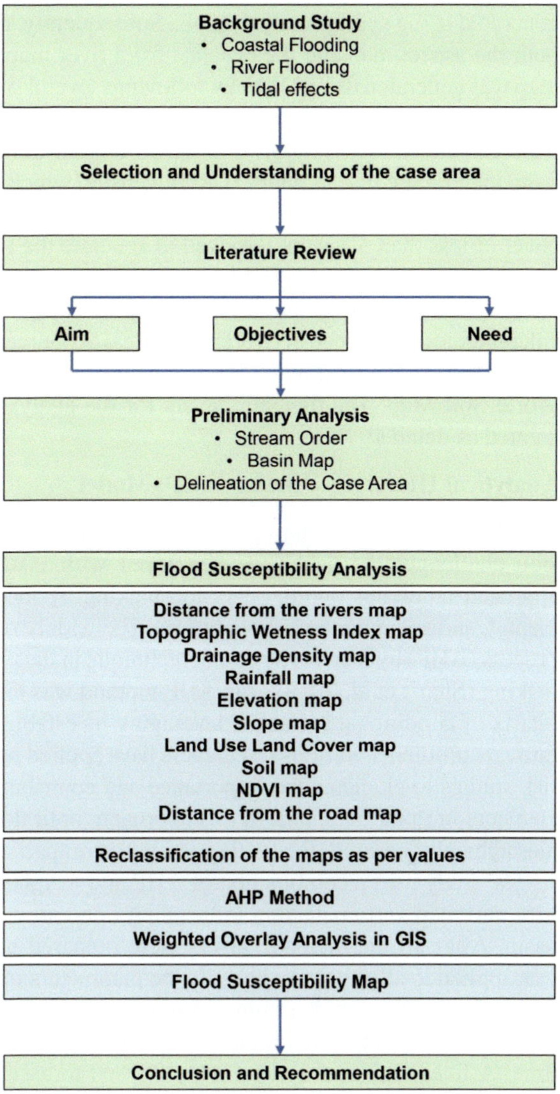

Fig. 5 Methodology diagram for flood susceptibility mapping

and distance from the road. These parameters are examined to have major impact on flooding on the study area. Open-source spatial data and secondary information from various sources were collected for the study.

The Shuttle Radar Topography Mission (SRTM) global (30 m) Digital Elevation Model (DEM) data was obtained from the USGS Earth Explorer, specifically using a polygon feature to identify the study region, with less than 5% of cloud cover. The unprocessed DEM data was imported into ArcGIS 10.3, where an elevation map of the designated research area was generated. Slope percentage results were generated using spatial analysis tools. Subsequently, the DEM was utilized to create both the aspect map and the distance from river map for the specified region. TWI map was generated using DEM by following several steps. The stream order map was generated from the DEM from which the distance from river and drainage distance were calculated. The rainfall time series dataset utilized in this study was sourced from the climate research unit (CRU) website, which provides open-source data.

The latest Landsat 8–9 satellite images from USGS Earth Explore with cloud cover of less than 5% were retrieved to generate the LULC 2023 map. The research area was then extracted using the mask command. Bands 4 and 5 among satellite images were used to generate NDVI map by applying the formula using the raster calculator tool. The road network shapefile was sourced from open source to generate the distance to the road map. The soil map was developed from the FAO-UNESCO World Soil Map. The thematic layers for the analysis of flood susceptibility were created as detailed.

Analytical Hierarchy Process (AHP) Model

Multi-criteria decision analysis serves as a crucial tool for examining intricate decision challenges, particularly when faced with insufficient data or criteria. This approach facilitates optimal decision-making by incorporating technical, environmental, and socio-economic objectives. The widely recognized analytical hierarchy process (AHP) by Saaty is a popular technique in the realm of multi-criteria decision-making (Sharir et al. 2019). The AHP method was employed to produce a pairwise matrix of conditional variables, leveraging its established reputation for addressing intricate problems. Various researchers have applied pairwise rating methods in flood risk studies to elucidate the importance and contributions of individual conditional elements in flood mapping. In this approach, both flood susceptibility factors were hierarchically ordered, facilitating pairwise comparisons (Faizan et al. 2023).

The study uses a combination of AHP and weighted overlay techniques to prioritize and map various factors influencing flood susceptibility in the Mandovi river basin. After all the thematic layers were prepared using GIS, the AHP technique was applied to allocate weightage to the parameters impacting flooding. In this technique, the weights are determined through pairwise comparison matrix. The process involves contrasting criteria and assessing the significance of two parameters at a time. A linear scale of 1–5 on Satty is used to assign the value to build pairwise comparison matrix.

1. Very Low Flood Susceptibility
2. Low Flood Susceptibility
3. Moderate Flood Susceptibility
4. High Flood Susceptibility
5. Very High Flood Susceptibility

The pairwise comparison matrix for criteria is used to input the pairwise comparisons and generate weights as output. The consistency ratio indicates the probability that the matrix ratings were randomly generated. Upon completing the pairwise comparison matrix, the parameters undergo normalization to generate the normalized matrix. This involves calculating normalized values. By normalizing each element within a column through division by the total sum of the column values, subsequent determination of criterion weights involves averaging all individual components within each row (Das 2020). Following this, an ideal pairwise comparison matrix should exhibit a consistency ratio of less than 10%. In our research, we have achieved a consistency ratio of 8.9%, confirming the validity of employing the pairwise comparison matrix. The analysis involved a total of 45 pairwise comparisons. The resulting weights and ranks of the parameters are provided in the Table 1. The susceptibility class ranges and ratings are given based on which the chances of the class being very high susceptible to very low susceptible to flooding.

The AHP model's outcome indicates that distance from the river is the most significant parameter, followed by TWI and drainage density. This suggests that these factors have the most influence on the flood susceptibility as the high prominence of these parameters increases the risk of having flooding issue in the region.

4 Results and Discussion

4.1 Approach and Intention of the Study

The study employs a combination of the analytical hierarchy process (AHP) model and geographic information system (GIS) techniques. These methodologies are utilized for conducting a comprehensive flood susceptibility analysis in the Mandovi river basin in Goa.

Flood susceptibility mapping is done to identify the potential flood-prone areas based on certain parameters that cause flooding. Flood susceptibility maps can help in planning mitigation measures and help in decision-making during such events (Swain et al. 2020). Various techniques such as multi-criteria decision analysis (MCDA), hydrological and hydrodynamic models statistical analysis, fuzzy logic, hybrid methods, and machine learning models integrated with geographical information system (GIS) software are employed to perform flood susceptibility analysis (Swain et al. 2020). MCDA is widely used by researchers due to their accuracy, and among these, the most suitable model is the analytic hierarchy process (AHP) as it is suitable for performing complex decision-making (Pradhana et al. 2023). Researchers have used a maximum of up to 21 parameters in such studies among which certain parameters are frequently used. The basic parameters used are slope, elevation, rainfall, land use/ land cover, distance to river, distance from the drainage network, topographic wetness index (TWI), soil type, normalized difference vegetation index (NDVI), and river network density/drainage density (Kaya and Derin 2023).

Presently, remote sensing and geographic information systems (GIS) techniques have become pivotal components in multi-criteria decision-making processes, particularly in the realm of water resource management. Multi-criteria analysis (MCA) is

Table 1 Resulting weights and ranks for the parameters based on pairwise comparison

Flood susceptibility criteria and sub-criteria for flood susceptibility assessment					
Flood causative criterion	Unit	Class	Susceptibility class ranges and ratings	Susceptibility class ratings	Weight (%)
Distance from river	m	0–200	Very high	5	28
		200–400	High	4	
		400–600	Moderate	3	
		600–800	Low	2	
		> 800	Very low	1	
TWI	Level	− 1 to 10	Very low	1	21
		− 3 to 2	Low	2	
		− 5 to − 4	Moderate	3	
		− 9 to − 6	High	4	
		− 19 to − 10	Very high	5	
Drainage density	m/km	0–74	Very low	1	15
		75–149	Low	2	
		150–223	Moderate	3	
		224–298	High	4	
		299–372	Very high	5	
Rainfall	mm/year	2280–2308	Very low	1	11
		2308–2344	Low	2	
		2344–2391	Moderate	3	
		2391–2453	High	4	
		2453–2500	Very high	5	
Elevation	m	0–22	Very high	6	
		23–49	High	5	8
		50–73	Moderate	4	
		74–111	Low	3	
		112–220	Very low	2	
		221–255	Extremely low	1	
Slope	%	74–90	Very high	5	7
		50–73	High	4	
		27–49	Moderate	3	
		11–26	Low	2	
		0–10	Very low	1	
LULC	Level	Water body	Very high	5	4
		Agriculture Land	High	4	
		Settlement	Moderate	3	

(continued)

Table 1 (continued)

Flood susceptibility criteria and sub-criteria for flood susceptibility assessment					
Flood causative criterion	Unit	Class	Susceptibility class ranges and ratings	Susceptibility class ratings	Weight (%)
		Barren Land	Low	2	
		Vegetation	Very low	1	
Soil type	Level	Alluvial	Very high	5	3
		Red	Moderate	3	
		Laterite	Low	2	
NDVI	Level	– 0.052 to 0.13	Very high	5	2
		0.131–0.23	High	4	
		0.231–0.297	Moderate	3	
		0.298–0.349	Low	2	
		0.350–0.519	Very low	1	
Distance from road	m	0–500	Very high	5	1
		500–1000	High	4	
		1000–1500	Moderate	3	
		1500–2000	Low	2	
		2000–7000	Very low	1	
Total					100.0

a method that systematically assigns weights to physical parameters within GIS studies. The integration of MCA, specifically using the AHP method, with GIS is extensively applied in diverse research areas related to water management, particularly in flood susceptibility mapping (Anon 2022). The combined use of AHP and GIS methods proves instrumental in addressing various water management challenges, including identifying, assessing, and simulating flood-prone areas and their impacts. This integrated approach aims to minimize the costs incurred in terms of lives and property, particularly in the context of developing countries' urban areas (Sakmongkoljit et al. 2021). In this study, it is imperative to align an AHP method that assigns weights to the relative importance of parameters with the physical parameters under consideration.

For data collection and analysis, the research involves various parameters, including distance from rivers, topographic wetness index (TWI), drainage density, rainfall, elevation, slope, land use and land cover (LULC), soil type, normalized difference vegetation index (NDVI), and distance from roads. Subsequently, the parameters are analysed using GIS techniques to generate maps.

The collected parameters are integrated using a weighted overlay method to develop a flood susceptibility map. This map classifies areas within the basin into different susceptibility categories, ranging from very high to very low.

The study aims to provide a comprehensive understanding of flood susceptibility in the river basin. By analysing various parameters and developing a susceptibility map, the research seeks to identify areas at risk of flooding.

The study, through conducting basin-level flood susceptibility assessments, not only contributes to disaster recovery planning and overall resilience-building efforts but also assists in preparing communities for severe flood events. The findings enable the proposal of micro-level flood risk mitigation strategies based on the susceptibility map, enhancing the proactive management of flood risks at local levels.

4.2 Flood Susceptibility Analysis

The flood susceptibility map depicted in Fig. 6 was generated using the pertinent factors outlined in Table 1 for effective flood susceptibility and the criteria weights derived from the analytic hierarchy process (AHP), represented as percentages. More detailed information on the outcomes is provided in the subsequent sections.

Distance from River

Increased river network density in a locale indicates heightened surface runoff, amplifying the risk of flooding. An additional key criterion for assessing flood risk is the 'proximity to rivers,' as excessive runoff from these water bodies often leads to recurrent flooding (Khoirunisa et al. 2021). The severity of flooding was most pronounced

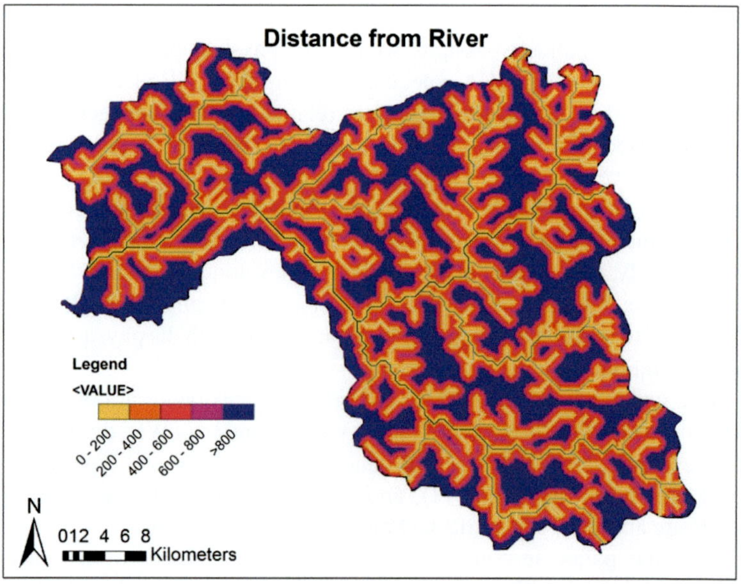

Fig. 6 Map showing distance from the river

in areas near the catchment outlet. Consequently, the distance from river proves to be a critical parameter in flood mapping, categorized into ranges of 0–200 m, 200–400 m, 400–600 m, 600–800 m, and > 800 (Fig. 6).

TWI

TWI effectively captures the combined influences of slope and water accumulation aiding in the identification of vulnerable areas (Swain et al. 2020). Due to the low slope, a high index value signifies a greater potential for water accumulation, and conversely, a low index value suggests less water accumulation. Consequently, areas with a higher topographic wetness index (TWI) are more susceptible to flooding (Saputra et al. 2023) (Fig. 7). The equation gives the TWI:

$$\mathrm{TWI} = a = \ln\left(\alpha / \tan\beta\right)$$

where

α upslope contributing area (flow accumulation raster map), and
$\tan\beta$ slope angle (the slope raster map in degrees).

In this study, TWI values exceeding 10 are categorized as high risk, while those below 3 are associated with a lower risk (Table 1).

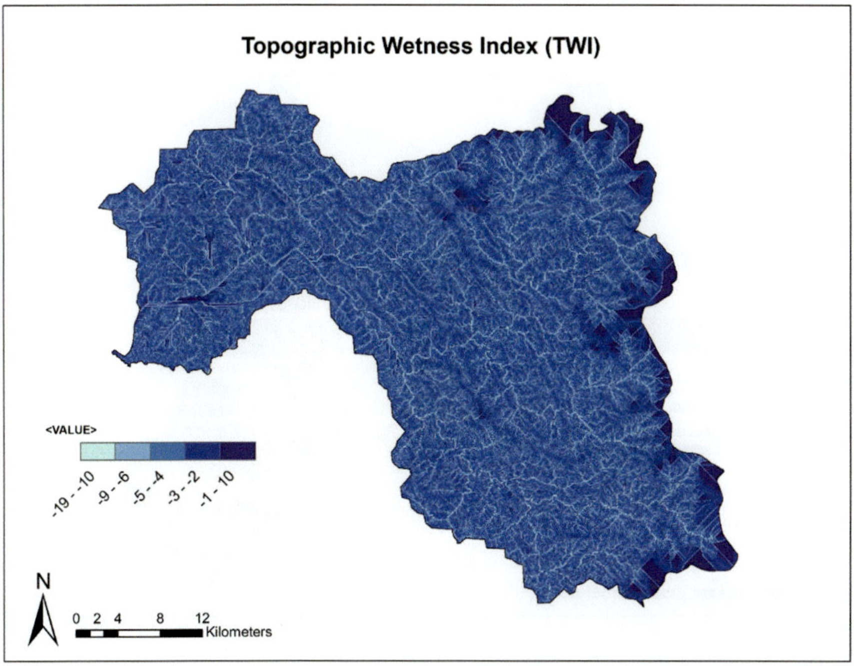

Fig. 7 Map showing TWI

Drainage Density

The extent of flooding in the basin is linked to the proximity of a location to the drainage system, with areas farther away experiencing reduced instances of flooding compared to those in closer proximity (Kaya and Derin 2023). Regions with higher drainage density are likely to be more susceptible to flood when compared to areas with lower drainage density. Elevated drainage density implies a larger catchment area prone to erosion, leading to sedimentation in lower-lying areas. Additionally, drainage density serves as a critical indicator influencing hazards by revealing information about soil nature and properties.

The reclassification of the drainage density layer was performed utilizing the natural break method. Regions characterized by exceptionally low density were designated as < 150 km/km^2, whereas areas with notably high drainage density were categorized as > 224 km/km^2 (Fig. 8).

Rainfall

Rainfall data provides critical information on precipitation patterns, enabling the assessment of potential flood risks in a given area (Tariq et al. 2023). Elevated precipitation rates significantly raise the likelihood of floods in flood-prone areas, compounded by other contributing factors (Danumah et al. 2016). Intense rainfall and extreme weather have most direct impact on amplifying the intensity of other flood-causing factors, contributing to severe flooding and elevated flood risk. In the

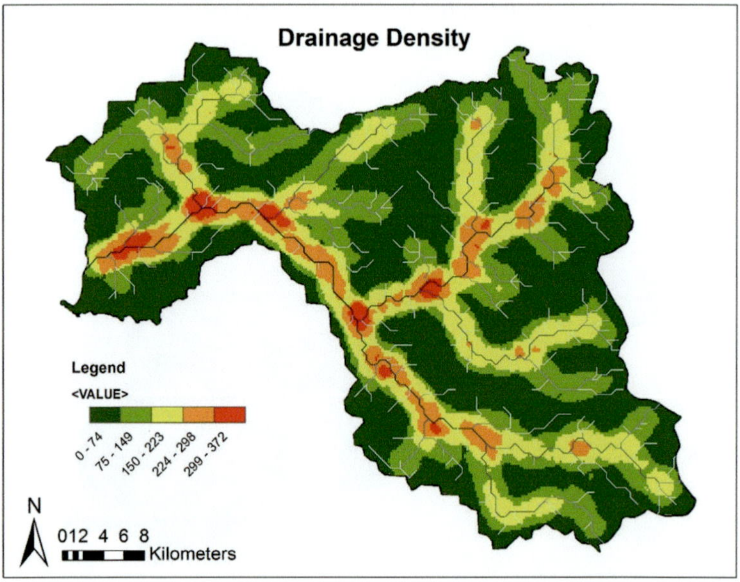

Fig. 8 Map showing drainage density

study area, rainfall ranges from 2280 to 2500 mm (Fig. 9). Higher precipitation levels (2500 mm) are associated with severe flooding.

Elevation

Elevation significantly influences the mapping of areas prone to flooding. It represents the topography of a given location. A higher elevation signifies an elevated area with a lower likelihood of flooding, while a lower elevation indicates a lower-lying region with higher susceptibility to waterlogging. Hence, there exists an inverse relationship between elevation and flood susceptibility (Hissain and Adhikary 2022). Within the domain of flood mapping, specialists underscore that elevation stands out as a fundamental determinant affecting the risks associated with floods. The inherent movement of water from elevated to lower terrains implies that flat surfaces positioned at lower elevations are inherently more predisposed to experiencing floods. The elevation layer was spatially distributed into five classes: 0–22 mts, 23–73 mts, 74–111 mts, 112–220 mts, and 221–255 mts (Fig. 10). Most of the study zone is located in lowland or floodplain areas. In this investigation, elevated importance was attributed to lower elevations compared to their higher counterparts, as delineated in Table 1.

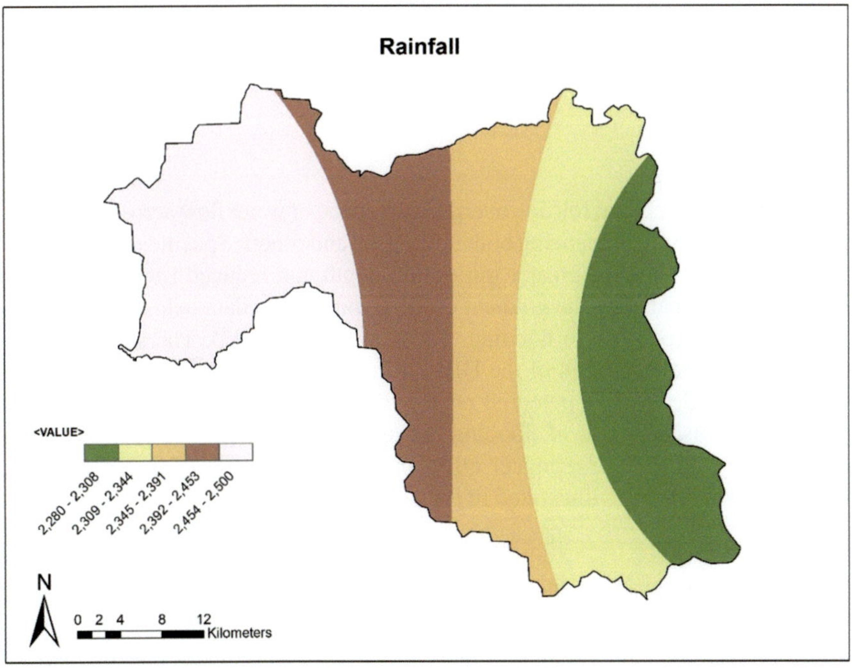

Fig. 9 Map showing rainfall

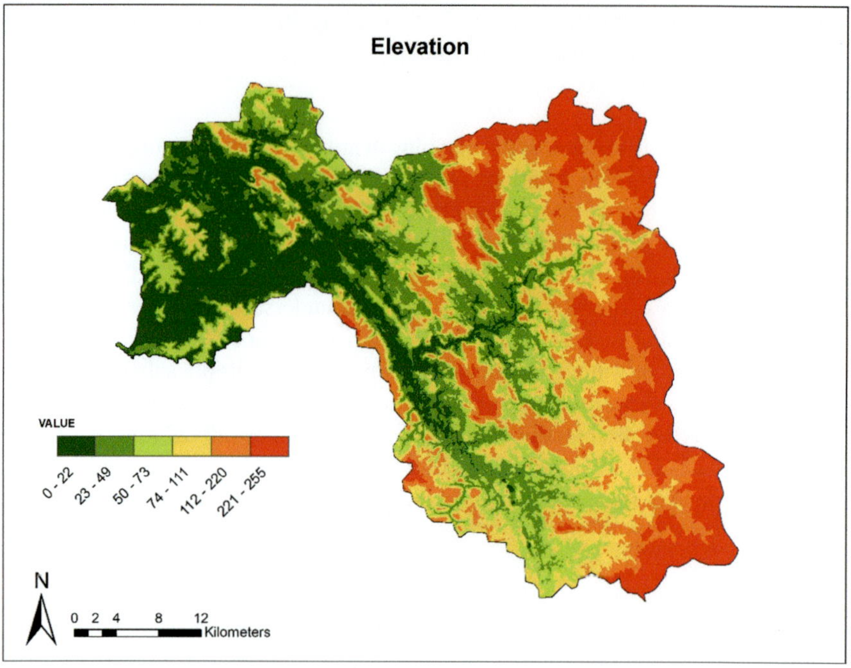

Fig. 10 Map showing elevation

Slope

The incline plays a crucial role in governing the speed of water flow across the Earth's surface, consequently influencing both infiltration and runoff. Specifically, areas with minimal slope experience greater infiltration depth and reduced runoff (Mehravar et al. 2023). By analysing the gradient of the terrain, slope data helps identify areas with higher susceptibility to flooding (Kaya and Derin 2023). The slope factor is determined by the resolution of the DEM grid and the range of elevations. As the slope rises, the destructive potential of flooding intensifies. In essence, flat terrain maintains a consistent risk of flooding because of its negligible slope. Slope and elevation are said to have a greater effect on flooding in most of the studies. The slope in the study area is illustrated in Fig. 11.

LULC

The way land is utilized has a noticeable effect, either directly or indirectly, on how quickly water is absorbed, the process of evapotranspiration, and the creation of surface runoff. The configuration of land use in a region provides insights into the kind of activities carried out by its residents and the associated natural processes. Recognizing alterations in land usage over time and comprehending the impact of both natural forces and human actions is vital, as these practices can serve as catalysts for severe natural disasters (Penki et al. 2022). We organized satellite data into six

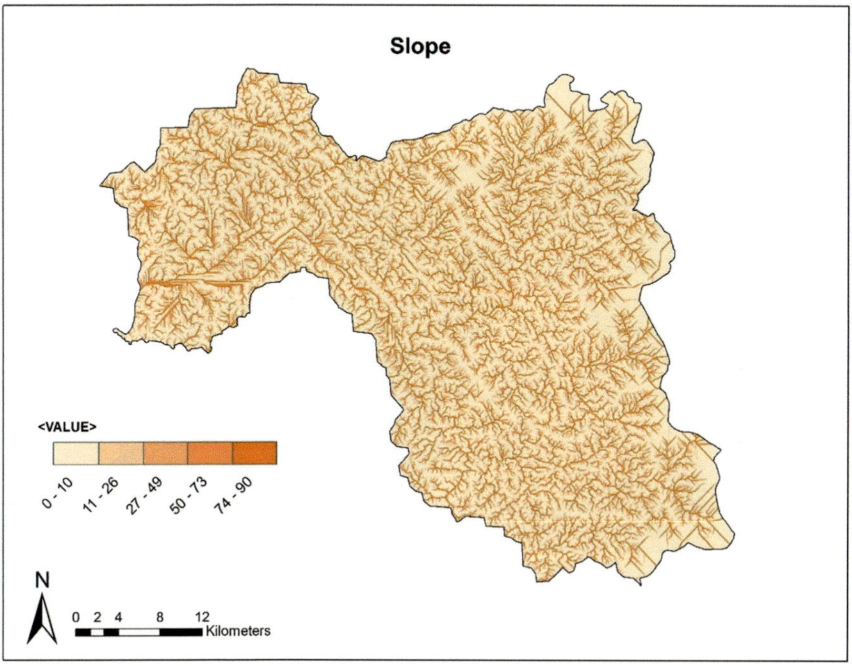

Fig. 11 Map showing slope

different land-use types: cultivated areas, commercial zones, barren lands, forests, public infrastructure spaces, and residential areas. In the study area, the most common land cover is vegetation, followed by water bodies, agricultural land, settlements, and barren lands as shown in (Fig. 12).

Soil

The soil type within the basin governs the extent of water infiltration into the soil, influencing the precipitation threshold that triggers flooding. Soil type and texture play a crucial role in infiltration, influencing flood susceptibility. Clay soils typically result in more rapid and extensive surface runoff compared to sandy soils. Impermeable clay soils are particularly prone to flooding, while permeable sandy soils facilitate relatively easy infiltration, absorbing significant surface water. In essence, soil type regulates the amount of water that can infiltrate the ground (Lappas and Kallioras 2019). The weighted soil map was created by assigning weights to various soil classes (Fig. 14). In the case study, lithological units were categorized into three broad classes based on infiltration capacity: highly infiltrated for alluvial soil moderately infiltrated for red soil and less infiltrated for laterite soil.

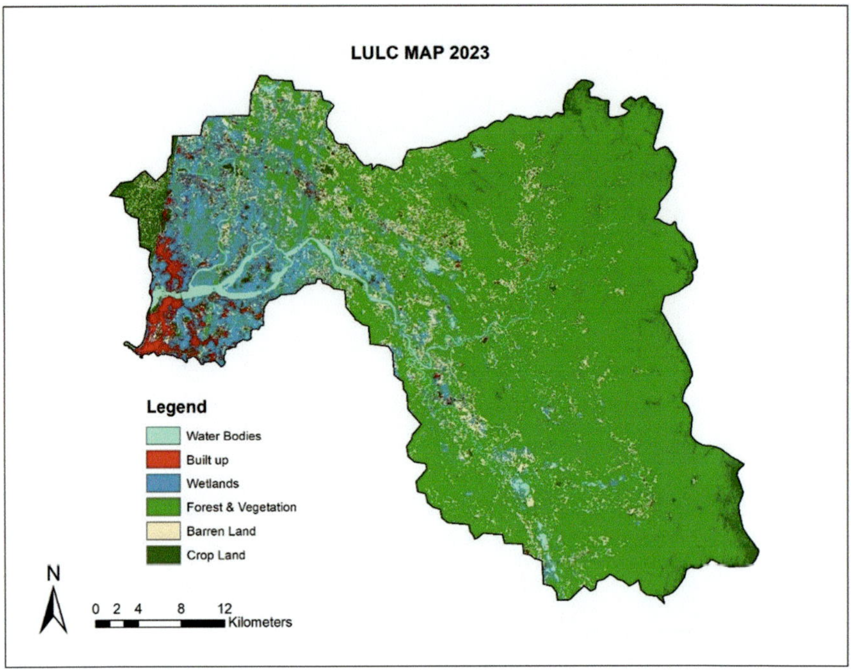

Fig. 12 Map showing LULC

NDVI

The crucial vegetation index is commonly employed in global-scale studies addressing climatic and environmental changes. The sensitivity of vegetation to its surroundings serves as a robust defence against natural disasters, exerting influence not only on ecological equilibrium but also on climate dynamics (Hissain and Adhikary 2022). The presence of forest and vegetation cover enhances infiltration rates and reduces runoff depth, thereby lowering the potential risk of flooding (Kaya and Derin 2023). According to the NDVI, areas with vegetation levels below 0.23 were identified as having a higher risk of flooding (Fig. 13). This suggests that places with less greenery are more susceptible to flooding, and the opposite is true for areas with higher vegetation levels (Fig. 14).

Distance from Road

An anthropogenic interference criterion, i.e. distance to road network was considered much like the influence of proximity to rivers, and flooding can also occur in areas adjacent to roads, especially in sloping terrains. The construction of roads contributes to a higher proportion of impervious surfaces, leading to diminished groundwater recharge and alterations in topography. These changes, in turn, impact flow accumulation and result in increased runoff (Swain et al. 2020). Flood susceptibility can be influenced by the distance from the road map, highlighting the role of roads in acting

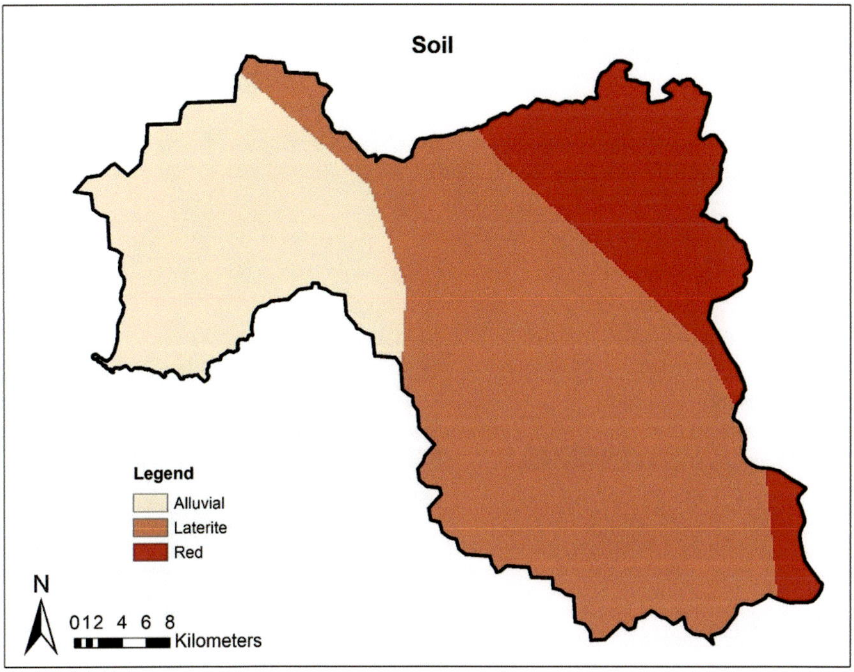

Fig. 13 Map showing soil type

as barriers to water flow. This factor shows an inverse relationship with flood susceptibility, ranking tenth among the factors according to the AHP result. The index is created based on the road map, and Fig. 15 illustrates the categorized distance to the road map for the research area.

The influencing factor mentioned above, with its weightage obtained through AHP analysis, was incorporated through weighted overlay analysis using the ArcGIS platform to generate the flood susceptibility map. The flood susceptibility map for the Mandovi river basin categorizes the area into five levels of susceptibility. The levels range from very low to very high flood susceptibility. The resulting map is depicted in (Fig. 16).

The analysis reveals that higher elevation and slope lead to lower flood susceptibility, while greater distance to the river, TWI, drainage density, and high vegetation areas increase the susceptibility to flooding. This classification provides a comprehensive understanding of the potential risk of flooding in different areas within the basin. The categorization allows for a targeted approach to managing the impact of potential floods, thereby enhancing the overall resilience of the region.

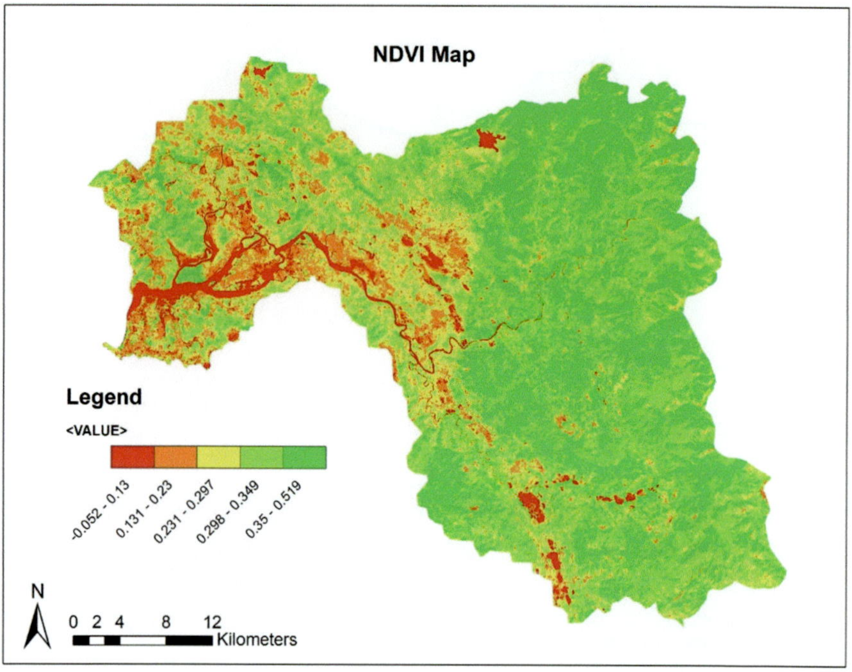

Fig. 14 Map showing NDVI

5 Challenges/Limitations of the Study

Due to the unavailability of micro-level data, this study has focused only on the parameters that can be derived using Geographic Information Systems (GIS). Additionally, due to time constraints, the analysis has been restricted to the macro-level, specifically at the basin level, rather than delving into the micro-level intricacies, where detailed basin mapping would be required.

6 Recommendations at the Basin Level

A comprehensive approach to flood mitigation in the Mandovi river basin is imperative for safeguarding communities and ecosystems amidst escalating flood risks. Basin-level strategies must encompass diverse factors contributing to flooding. Integrated watershed management practices, such as afforestation and soil conservation, should be implemented to maintain the natural balance of water flow. Floodplain zoning regulations and land-use planning, considering the natural flow of water, are essential to restrict construction in high flood susceptibility areas.

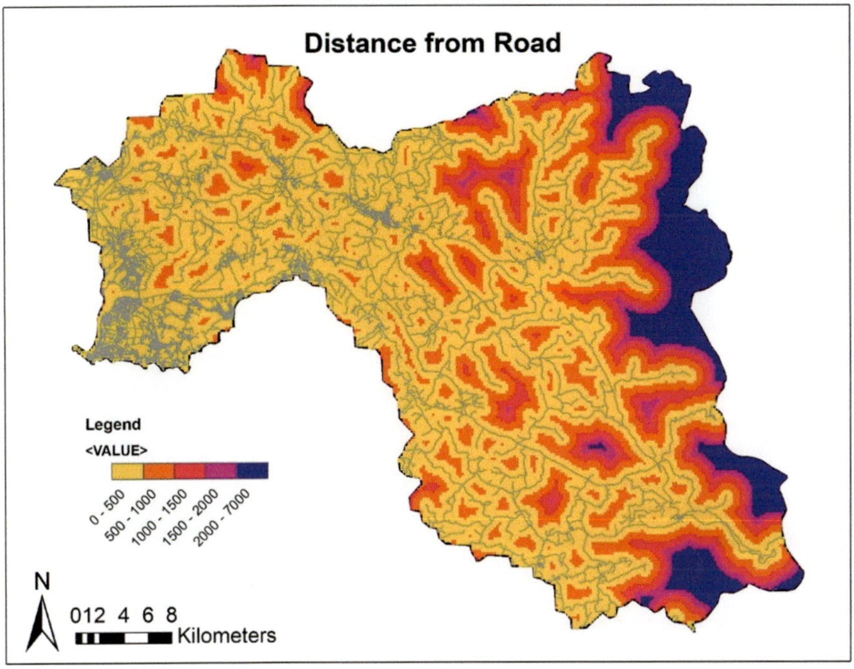

Fig. 15 Map distance from road

The flood susceptibility map, delineating areas into high, medium, and low flood-prone zones, can be used for strategic planning. High susceptibility zones necessitate immediate focus on risk mitigation, including infrastructure improvements and establishment of early warning systems during high rainfall. Medium susceptibility areas require careful land-use planning to mitigate risk and resources allocation to maintain resilience in low susceptibility zones. Strategic planning based on flood susceptibility mapping facilitates proactive decision-making to minimize flood impacts.

Enhancing early warning systems, particularly in very high and high flood susceptible zones, is critical for timely alerts, utilizing technologies like river gauges, weather monitoring, and satellite data. Effective reservoir management strategies, incorporating technologies like IoT or artificial intelligence, should be adopted to control releases during heavy rainfall and prevent water accumulation (Kiedrzyńska et al. 2014). Natural flood control measures, including the restoration of wetlands and floodplains, need to be context-specific and strategically implemented in identified flood susceptibility zones. Community awareness initiatives are crucial, involving periodic campaigns to educate communities about flood risks, evacuation procedures, and sustainable land use. Local disaster management departments can develop apps providing real-time updates on extreme weather events for high flood risk zones. Investment in research and technology, including advanced tools like remote sensing and modelling, is vital for accurate flood forecasting. The establishment of financial

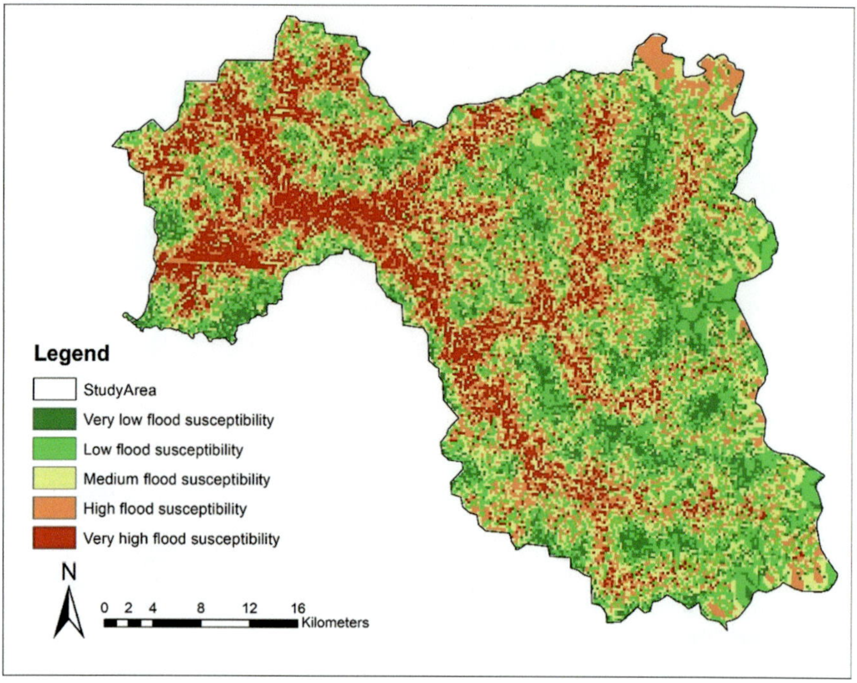

Fig. 16 Flood susceptibility map for the Mandovi river basin

mechanisms, such as flood insurance and risk-sharing arrangements, can provide compensation for losses in high flood risk zones. These proposals, tailored to the Mandovi river basin's specific characteristics, can significantly contribute to effective flood mitigation and management. The involvement of local communities, governments, and stakeholders in planning and implementation processes is crucial for the sustainable and resilient control of floods (Wang et al. 2022).

This paper highlighted several key factors influencing flood susceptibility in the Mandovi river basin and proposed basin-level recommendations for flood mitigation. To ensure actionable outcomes and effective implementation of mitigation measures, practical implications of the findings are necessary. Despite many of such researches being presented, it often remains to the status of a report, with administrative authorities failing to accord it the necessary importance in disaster management implementation. Implementing basin-level strategies may encounter challenges such as regulatory hurdles, limited funding, and coordination among stakeholders. Recognizing these challenges and proposing context-specific strategies to address them is crucial. Engaging local communities in flood mitigation efforts is paramount. Making the communities understand the susceptibility map and identified high-risk zone advocate for participatory approaches that empower communities to contribute to decision-making processes, raise awareness about flood risks, and implement adaptation measures. Integrating innovations such as artificial intelligence and Internet of things for improving flood forecasting and early warning systems into existing frameworks can enhance prediction accuracy and facilitate timely response actions. Policy reforms and institutional strengthening are essential for supporting flood mitigation efforts. Streamlining regulatory frameworks and promoting inter-agency coordination can create an enabling environment for effective flood management. By addressing implementation challenges, engaging local communities, harnessing technology and innovation, and enhancing policy and governance support, we can foster sustainable and resilient flood management practices that benefit both present and future generations.

7 Conclusion

In conclusion, the flood susceptibility mapping for the Mandovi river basin has been carried out by considering a comprehensive set of parameters. Notably, distance from the river emerged as the most influential factor, followed by TWI, drainage density, rainfall, elevation, slope, LULC, soil type, NDVI, and distance from the road. The application of the analytical hierarchy process (AHP) method facilitated the prioritization of these factors, laying the groundwork for a comprehensive flood susceptibility analysis. Subsequent utilization of the weighted overlay analysis in GIS resulted in the generation of a detailed flood susceptibility map for the Mandovi river basin. This map classifies susceptibility into five distinct levels: very low, low, moderate, high, and very high. This can further help in clear delineation of high-risk zones at micro-level, to offer crucial insights for the implementation of targeted mitigation measures. Key findings from the analysis indicate that higher elevation and slope contribute to lower flood susceptibility, while factors such as proximity to the river, TWI, drainage density, and high vegetation areas elevate the risk of flooding. The clear categorization of susceptibility levels contributes significantly to enhancing resilience and preparedness in the face of potential flood events. In conclusion, the flood susceptibility analysis conducted for the Mandovi river basin highlights the importance of natural features that influence the area's susceptibility to flooding. By focusing on physical and environmental factors such as topography, land use, rainfall, hydrology, and soil characteristics, etc., the analysis provides valuable insights into mapping of these parameters and generating map that identified areas with high, medium, and low flood risk. This understanding serves as a foundation for informed land-use planning, zoning regulations, and targeted mitigation measures. It is crucial to note that susceptibility assessments, while instrumental in identifying flood-prone areas, do not consider how well the population and infrastructure in those areas can resist to the flooding. Therefore, the results presented herein offer a basis for further vulnerability assessments and the implementation of specific flood risk reduction measures to enhance the overall resilience of the Mandovi river basin and its communities.

References

Anon (2022) Urban flood susceptibility modelling using AHP and GIS approach: case of the Mfoundi watershed at Yaoundé in the South-Cameroon plateau. Sci Afr 15
Danumah JH et al (2016) Flood risk assessment and mapping in Abidjan district using multi-criteria analysis (AHP) model and geoinformation techniques, (cote d'ivoire). Geoenviron Disast
Das S (2020) Flood susceptibility mapping of the Western Ghat coastal belt using multi-source geospatial data and analytical hierarchy process (AHP). Remote Sens Appl Soc Environ 20
Dhiman R, VishnuRadhan R, Eldho T, Inamdar A (2018) Flood risk and adaptation in Indian coastal cities: recent scenarios. Appl Water Sci 9
Faizan MO, Hudait M, Sengupta K, Chowdhury BR (2023) Flood susceptibility mapping using MCDM-AHP approach and geospatial techniques—a study of Hyderabad District, Telangana, India. Int J Hydrol Sci Technol 16:2
Hissain M, Adhikary S (2022) Flood susceptibility assessment in southwest coastal region of Bangladesh using an AHP-GIS based approach. s.l., s.n
Kaya CM, Derin L (2023) Parameters and methods used in flood susceptibility mapping: a review. J Water Clim Change 16
Kerkar SRP, Pradhan A, Dandekar P, SANDRA (2023) https://sandrp.in/2016/11/25/goa-river-profile/. Accessed 29 Dec 2023
Khoirunisa N, Ku C-Y, Liu C-Y (2021) A GIS-based artificial neural network model for flood susceptibility assessment, vol 17
Kiedrzyńska E, Kiedrzyński M, Zalewski M (2014) Sustainable floodplain management for flood prevention and water quality improvement. Nat Hazards 76:955–977
Kumar R et al (2023) Assessment and mapping of riverine flood susceptibility (RFS) in India through coupled multicriteria decision making models and geospatial techniques. Water 15(22)
Lappas I, Kallioras A (2019) Flood susceptibility assessment through GIS-based multi-criteria approach and analytical hierarchy process (AHP) in a River Basin in central Greece. Int Res J Eng Technol 6(3)
Maharjan M et al (2024) Flood susceptibility assessment using machine learning approach in the Mohana-Khutiya River of Nepal. Nat Hazards Res
Mehravar S et al (2023) Flood susceptibility mapping using multi-temporal SAR imagery and novel integration of nature-inspired algorithms into support vector regression. J Hydrol 617
Moazzam FM et al. (2020) Spatio-statistical analysis of flood susceptibility assessment using bivariate model in the floodplain of River Swat, District Charsadda, Pakistan. J Geosci Environ Protect 8
Penki R, Basina SS, Tanniru SR (2022) Application of geographical information system based analytical hierarchy process modeling for flood susceptibility mapping of Krishna District in Andhra Pradesh. Environ Sci Pollut Res
Pradhana B, Lee S, Dikshit A, Kim H (2023) Spatial flood susceptibility mapping using an explainable artificial intelligence (XAI) model. Geosci Front 14(6)
Sakmongkoljit K, Homtong N, Surakotra N, Silaratana N (2021) Multi-criteria flood susceptibility assessment using remote sensing and GIS-based approaches in Chi River Basin. s.l., IOP Publishing, Thailand
Samanta S, Pal DK, Palsamanta B (2018) Flood susceptibility analysis through remote sensing, GIS and frequency ratio model. Appl Water Sci 8
Saputra HD, Feronika I, Aisah NS, Hadibasyir ZH (2023) Flood susceptibility assessment using topographic wetness index in a part of brantas watershed. s.l., AIP Publishing
Sharir K et al (2019) Assessment of flood susceptibility analysis using analytical hierarchy process (AHP) in Kota Belud Area, Sabah, Malaysia. s.l., IOP Publishing
Souissi D et al (2019) GIS-based MCDM—AHP modeling for flood susceptibility mapping of arid areas, southeastern Tunisia. Geocarto Int
Swain KC, Singha C, Nayak L (2020) Flood susceptibility mapping through the GIS-AHP technique using the cloud. Int J Geo-Inf

Tariq A et al (2023) An effective geospatial-based flash flood susceptibility assessment with hydrogeomorphic responses on groundwater recharge
Wang L et al (2022) A review of the flood management: from flood control to flood resilience 8(11)

Assessing the Seasonal Dynamics of Surface Water Resources Using Satellite-Based Remote Sensing: A Case Study in Indian Sundarbans

Argha Ghosh, Manoj Kumar Nanda, Debolina Sarkar, Sukamal Sarkar, Koushik Brahmachari, and Mohammed Mainuddin

Abstract Indian Sundarbans, a part of the largest delta on planet Earth, is one of the most stressed regions in the world from crop production point of view. Dry season crop production is limited by the non-availability of suitable irrigation water, which can only be supplied from surface water resources. The present research aimed at assessing the dynamics of surface water resources acreage during the dry (post-monsoon and pre-monsoon) seasons of 2017–18, 2018–19, and 2019–2020 cropping years in the Gosaba CD block of Indian Sundarbans. Normalized Difference Water Index (NDWI) computed using multi-temporal Sentinel-2 data was used to estimate the changes in the surface water area during the entire dry seasons. Binary classification of NDWI images was used to separate water bodies from the non-water classes. The results revealed that the surface water resource area was higher just after the monsoon (end of October). About 2.83, 1.63, and 3.74% area of the block was estimated as surface water at the end of monsoon season of 2017–18, 2018–19, and 2019–2020 cropping year, respectively. Surface water area notably decreased

A. Ghosh (✉)
Department of Agricultural Meteorology, Odisha University of Agriculture and Technology, Bhubaneswar, Odisha 751003, India
e-mail: ghoshargha4@gmail.com

A. Ghosh · M. K. Nanda · D. Sarkar
Department of Agricultural Meteorology and Physics, Bidhan Chandra Krishi Viswavidyalaya, Mohanpur, West Bengal 741252, India
e-mail: mknanda@bckv.edu.in

S. Sarkar
School of Agriculture and Rural Development, Ramakrishna Mission Vivekananda Educational and Research Institute, Kolkata, West Bengal 700103, India

K. Brahmachari
Department of Agronomy, Bidhan Chandra Krishi Viswavidyalaya, Mohanpur, West Bengal 741252, India

M. Mainuddin
CSIRO Land and Water, Canberra, ACT, Australia
e-mail: Mohammed.Mainuddin@csiro.au

S. C. Pal and U. Chatterjee (eds.), *Surface, Sub-Surface Hydrology and Management*, Springer Geography, https://doi.org/10.1007/978-3-031-62376-9_5

in December. Higher acreage of surface water in February might be attributed to the presence of standing water in the transplanted *boro* rice field. As the season progressed, the areas under surface water resources decreased and became lowest in March. Intensity and distribution of rainfall influenced the availability of surface water resources. The findings of the present study may contribute to sustainable crop planning in the Indian Sundarbans region.

Keywords Indian Sundarbans · Surface water · Remote sensing · NDWI

1 Introduction: Importance of Surface Water Resource Monitoring

In recent decades, environmental issues have drawn more and more attention. Water resource management, biodiversity protection, overfishing, pollution control, and climate change are major public concerns now on a local, state, and even global level. Crises, environmental degradation, hazards to human health, and the persistence of poverty have all contributed to public mistrust of the advancement of technology and economic progress as well as skepticism about the capacity of public policies to address these issues in a timely manner (Ahmadov 2020).

The biggest issue facing the globe right now is how to deal with the environmental effects of economic expansion. In the late 1980s, sustainable development became popular as a unifying strategy to address issues with the environment, economic growth, and quality of life. Management of water resources is a key component of sustainable development, as acknowledged in the United Nations' (UN) recently adopted development agenda. The Sustainable Developmental Goals (SDGs) of the UN closely mirror the growing global concerns about water supplies. Nearly all of the SDGs, especially those that address food, energy, and the environment, include water. Since water is a basic human right, a key enabler of sustainable development and economic progress, and it is necessary for food security, health, and sanitation, achieving SDG 6 ("ensure availability and sustainable management of water and sanitation for all") will have a significant impact on other 16 SDGs (Cole et al. 2018). The Global Water Partnership (GWP) has consistently highlighted the importance of considering the interdependencies between the sustainable management of water resources and addressing the demands for food production, energy generation, water and sanitation services, and ecosystem support (Sophocleous 2000). Our planet is currently being impacted by a changing climate, rapidly expanding populations, social and economic development, globalization, and urbanization. Water-related problems on a global, regional, and local level are getting more complicated, mostly due to the growing significance of the relationships between water, energy, food, and ecosystems (Ait-Kadi 2016). Population expansion, water-intensive economic growth, pollution, and climate change are all putting strain on water resources (Cole et al. 2018).

Water is an essential element for life. Both human activity and the existence of natural ecosystems depend on it. The demand for water will grow exponentially due to the combined effects of urbanization, growing incomes, and population growth, while the supply of resources will become more unpredictable and erratic (Faye and Dieye 2018). Due to its finite supply and rising global demand, water is under stress. Globally, river flow is extremely erratic and unpredictable. Overuse has made groundwater susceptible, and rainfall is particularly erratic in terms of timing and location (Bhadra et al. 2018). In places where water is currently abundant, it will become scarce if prompt action is not taken. Water scarcity will have an effect on all socioeconomic activities, such as income, energy, agriculture, and health. An essential first step in addressing them and creating a logical management system for water resources is determining the availability of water.

The rapid population growth in India has sparked the "Anthropocene" period of environmental degradation, social deprivation, and inefficient economic growth. Since the population is increasing but the number of renewable water resources accessible is not, a decrease in both the total internal renewable water resource and the overall renewable water resource is anticipated. Future population increase is anticipated to cause the per capita amount of renewable water resources to decline even more. If appropriate mitigating measures are not taken, water scarcity will worsen as the population grows and the rate of per capita total water withdrawal increases. Roy and Pramanick (2019) reported that there has been a 3.07% increase in total water withdrawals per person. A gradual rise in India's total yearly water need has been observed. The amount of water needed in 2050 is expected to nearly triple that of 1990. Additionally, water stress has also been noticed to increase. The water stress level that India is experiencing currently has been estimated as about four times higher than the global average water stress level. A great deal of work needs to go into finding sustainable ways to use water resources as well as increasing the water use efficiencies of processes that use water resources if Indian communities are to achieve sustainable development in the areas of both the water resources and its utilizations.

In India, a major share of the total freshwater resources is utilized in agriculture and allied sectors. When population growth is combined with a very little drop in agricultural water withdrawal, the overall amount of agricultural water withdrawal is likely to rise. It indicates that India's agricultural water productivity has not increased appreciably. In India, surface and groundwater security are essential for maintaining strong public health systems, continuing agricultural output, and ecosystem services (Yuan and Lo 2020; Ram and Irfan 2021). However, India's water systems are severely degrading and becoming scarcer as a result of pollution, over-extraction, unsustainable diversions, and consumption habits combined with a changing climate (Biswas et al. 2022; Dangar et al. 2021; Vanapalli et al. 2021).

Surface water scarcity for irrigation is a significant concern in certain regions of India, impacting agricultural productivity and livelihoods. Several factors contribute to surface water scarcity in the context of agricultural production in India. India experiences uneven distribution of rainfall, both spatially and temporally. Indian agriculture is mostly monsoon-dependent and monsoon-rainfall pattern is changing in many regions in the county (Vinnarasi and Dhanya 2016; Maharana et al. 2021). This ultimately increases the moisture stress of rain-fed crops (Manivasagam and Nagarajan 2018). Some regions receive abundant rainfall, while others face prolonged dry periods. This variability contributes to water scarcity in certain areas, affecting agriculture (Bhat et al. 2014; Dhawan 2017). Excessive reliance on groundwater for irrigation has led to over-extraction in many regions. While groundwater is an essential resource, unsustainable pumping practices contribute to declining water tables and long-term scarcity (Narayanamoorthy 2022). Inefficient water management practices, such as outdated irrigation techniques and infrastructure, lead to wastage of water. Improving the efficiency of irrigation systems is crucial for optimizing water use and reducing scarcity (Priyan 2021). Climate change has altered precipitation patterns and increased the frequency of extreme weather events, leading to uncertainties in water availability for agriculture. Changes in temperature and rainfall patterns can exacerbate water scarcity challenges (Mahato et al. 2022). Rapid population growth in India puts additional pressure on water resources for various purposes, including agriculture. The increasing demand for food necessitates more extensive irrigation, contributing to surface water scarcity (Jain 2011). Pollution of surface water bodies, such as rivers and lakes, further exacerbates water scarcity issues. Contaminated water sources are unsuitable for irrigation, affecting agricultural productivity and the quality of crops (Brindha et al. 2014; Kumar et al. 2015). Inconsistent water management policies, inadequate infrastructure, and challenges in governance contribute to water mismanagement. Clear policies, effective implementation, and participatory approaches are essential for addressing water scarcity.

Keeping these in view, the present research work was conducted in the Indian Sundarbans region to assess the seasonal changes of surface water resources during the dry periods (post- and pre-monsoon periods) using multi-temporal satellite data.

2 Application of Remote Sensing and GIS in Water Resource Monitoring

Remote sensing is a powerful tool used for monitoring and assessing surface water resources (Kaplan et al. 2019). It plays a crucial role in water resource monitoring by providing valuable information about various aspects of water bodies, including quantity, quality, and distribution in both spatial and temporal scales. Remote sensing provides valuable information about various water parameters, such as the physical properties of water, organic properties, and microbiological properties (Riadi et al. 2022). This information is crucial for understanding water quality, tracking changes in surface water resources, and managing water-related issues such as flood mapping and water demand modeling. By using optical remote sensing technology, researchers and scientists can gather data on soil moisture, water usage, and water quality at regional and large scales (Pari et al. 2020). This information can then be used for effective water resource management, planning, and decision-making. Additionally, remote sensing technology allows for long-term monitoring over large areas, making it ideal for applications in precision farming, land and water resource management, and water-related research, and operational applications (Singh and Kathpalia 2007). The application of remote sensing in different areas has gained momentum in the past three decades, largely due to the availability of high-quality satellite data, development of computation facilities, digital imaging techniques as well as improved communication systems and space technology. Remote sensing and GIS are presently playing a vital role in mapping and quantification of water resources as well as their conservation and sustainable utilization. Remote sensing technology has revolutionized surface water resource monitoring in recent years.

The spatio-temporal dynamics of surface water resources can be effectively monitored with the use of remote sensing tools and technology. Remote sensing technology can continuously monitor the Earth's surface at local, regional, and global scales, which makes this technique far more efficient than traditional in-situ observations. In order to map the extent of water resources at a different scale and to regularly and frequently monitor their dynamics, remotely sensed data sets offer an unprecedented opportunity to get observation at fine spatial and frequent temporal resolution (Reshmidevi et al. 2013). The traditional approach to monitoring the parameters of water quality involves taking measurements in-situ and analyzing the results in a laboratory, which is quite complex and time-consuming. The technique typically performs worse when it comes to delivering the temporal and geographical coverage required for an appropriate assessment in big bodies of water (Dube et al. 2022). One of the easiest and most straightforward uses of remote sensing in water resource research has been the identification and mapping of surface water borders (Liu et al. 2023). The surface water resources can be monitored by both the optical and microwave remote sensing techniques (Cazenave et al. 2016; Sheffield et al. 2018).

The use of remote sensing techniques is becoming more important in the assessment of water quality because of its potential to provide greater spatial and temporal sampling frequencies. Water quality parameters are determined via remote sensing by observing changes in the water's optical characteristics brought on by the presence of pollutants (Lillesand et al. 2004). Deeper understanding of the dynamics or variability of our water quality over time, whether it is seasonal or long-term, can be gained by performing a multi-temporal analysis of these remote sensing-based characteristics.

The variation in spectral reflectance of land surfaces with respect to water, vegetation, and dry soil at different wavelengths provides the basis for optical remote sensing of water resources (Kumar and Reshmidevi 2013). The material being measured determines the ideal wavelength to employ to measure the water quality parameter. The best bands for tracking suspended sediments in water were determined to be the VIS and NIR regions of the EMR spectrum, with wavelengths between 0.7 and 0.8 μm, based on many in-situ analyses (Ritchie and Schiebe 2000).

Before, high-resolution optical pictures from satellites like Landsat TM were used for remote sensing of the water quality metric (Brezonik et al. 2005). The advancement of new satellites and sensors has led to a notable improvement in spatial, temporal, and radiometric resolutions. Over wide areas, improved accuracy in the estimation of water quality parameters has been achieved using sensors like MODIS (with 36 spectral bands) and MERIS (with 15 spectral bands) (Wu et al. 2009; Gons et al. 2008).

Microwave signal has longer wavelength, which can easily penetrate clouds and some types of plants. Moreover, active microwave sensors like Synthetic Aperture Radar (SAR) can operate day and night in every weather condition, regardless of solar radiation. Synthetic aperture radar is a valuable tool for surface water monitoring because of its ability to penetrate clouds and collect data in all weather conditions, providing continuous and reliable information (Shi et al. 2022). SAR can capture the backscatter signal, which is influenced by surface roughness, and changes in water content, providing information about water characteristics. It is particularly useful for detecting changes in water levels, identifying flood extent, and mapping variations in water bodies over time. Radar waves are transmitted and received by SAR systems in various polarizations, such as circular (e.g., right- and left-handed) and linear (e.g., horizontal and vertical). The way SAR interacts with surface characteristics, such as water, is influenced by polarization. Furthermore, the polarization of microwave signals used in SAR can provide additional insights into the surface water properties. For example, in flood monitoring, SAR data with HH polarization and an incidence angle of 20°–49° is preferred over VV polarization (Surwase et al. 2019). This is because HH polarization allows for better discrimination of smooth surfaces, which is important for accurately detecting and mapping flooded areas. The launch of the Sentinel-1 mission (by the European Space Agency as a part of their Copernicus Programme), which utilizes SAR technology, has opened up new possibilities for mapping wetlands and monitoring changes. Several researchers demonstrated the efficiency of optical and microwave remote sensing for surface water resource monitoring.

Surface water bodies, such as lakes, rivers, reservoirs, and inland water bodies, have been extensively mapped and monitored with high extraction precision (Jiang et al. 2014; Li and Gong 2016). The overestimation of water resources must be addressed while mapping inland surface water because of the diverse backgrounds (Zhou et al. 2014). A number of researches have proposed the application of water index to identify surface water bodies using remotely sensed data. By applying the algebraic combination of various bands' reflectance, the water and non-water areas can be distinguished more accurately (Ali et al. 2019). It is necessary to identify the ideal band combination in order to reliably and precisely distinguish water from other types of land cover (Yang et al. 2017). Normalized Difference Water Index (NDWI), computed using green and near infrared bands' reflectance, is a remote sensing index commonly used for monitoring surface water. It is particularly effective in distinguishing between water and non-water features in satellite or aerial imagery. According to the research, NDWI performs poorly in shallower water and better in deeper water (Xu 2006). It was reported earlier that NDWI was one of the best indices for detection of surface water in terms of the accuracy of the estimation and the stability of threshold values (Li et al. 2013; Ogilvie et al. 2018).

3 Case Study: Seasonal Dynamics of Surface Water Resources in Gosaba CD Block of Indian Sundarbans Using Satellite-Derived Normalized Difference Water Index (NDWI)

3.1 Rationales of the Study

The main supply of freshwater in the Sundarbans is rainfall. The South-West monsoon provides the majority of the yearly precipitation received by the Sundarbans. Usually, it begins in mid-June and disappears in the second week of October. The average annual rainfall is 1625 mm, however in years with heavy rainfall, it can reach 2000 mm, and in other years, it can even go as low as 1300 mm (Dasgupta 2008). Frequent occurrence of depression and tropical cyclones contributes to the surface and sub-surface water resources during the pre-monsoon and post-monsoon season.

Nestled in the tidal plain of southern West Bengal, the Indian Sundarbans are home to 4.43 million people (Hazra et al. 2016). The prevailing freshwater flow into the Sundarbans depends on the hydrological conditions of its rivers (Mirza 2006). In the Indian Sundarbans Delta, the rivers are primarily salty. With a few limited aquifers that are a few meters thick, the groundwater is likewise salty. Although there are a large number of creeks and rivulets in the Sundarbans Island, the freshwater is not available in plenty. Simultaneously, the Sundarbans get an enormous amount of precipitation, but this precipitation is not adequately exploited to meet the water needs of the 4.43 million people living in 19 administrative blocks within the Biosphere Reserve (Hazra et al. 2016). In the Indian Sundarbans Delta, the rivers are primarily salty. With a few limited aquifers at a few meters below the surface, the groundwater is likewise salty. Although there is source of fresh groundwater at a depth of more than 250 m, the same is not used for agricultural purpose considering the priority for domestic use in the island. A common way for saltwater intrusion into shallow groundwater occurs through river influent discharge (Bhadra et al. 2018). The existing climate change and rising sea level could make this issue even worse. Agriculture is made more difficult and drinking water sources are contaminated when soil and aquifers become salinized (Hazra et al. 2016). It is exceedingly difficult for the people of the Sundarbans to guarantee their livelihood through conventional farming due to factors such as increased siltation, high soil and water salinity, drainage congestion, and a lack of surface water or groundwater during the dry season (Hazra et al. 2016).

Out of the total area of the Sundarbans of about 7908 square km, the water area covers approximately 2400 km^2 that includes both sweet water area and brackish water area (Shodhganga). Out of this vast water resource, most of the water is saline. The water salinity has been a serious constraint to crop production and hence food security of this region. Water depth, tidal state, and season vary in the Sundarbans based on salinity, pH, and temperature. Previous researchers have described the distributions of mean temperature, pH, and salinity at several experimental sites of the three main river systems during the post-monsoon, dry winter, and monsoon seasons (Rahaman et al. 2015).

In Sundarbans, salinity is largely influenced by the amount of freshwater entering from upstream. The variance depends on the local tide conditions. The Sundarbans' yearly patterns of salinity variations are correlated with variations in the freshwater input from rivers upstream (Sundarbans Biodiversity Conservation Project 2003). Reduced productivity of forest sites, delayed growth of forest kinds, withering crowns of Sundari trees, and retrogression of forest types are all signs of the detrimental consequences of rising salt on the Sundarbans ecosystem (Mirza et al. 1998).

The Sundarbans' economy is based primarily on agriculture, although due to limited landholdings and inadequate irrigation systems, it still operates at a subsistence level. The Sundarbans' net sown area for the 2010–11 census years was 285,632 ha, or 62.44% of the total land area. Of the 376,150 ha of gross cultivated area, 80.53% were not irrigated. In the Sundarbans, the average cropping intensity is 129.97%, whereas the average irrigation intensity is 18.25%. It is discovered that the area experiences a shortage of water during the rabi crops as a result of inadequate irrigation facilities (Sinha Ray 2010).

During the post-monsoon seasons, very less proportion of land is being cropped in the Sundarbans. It was reported earlier that salinity and moisture stress are the limiting factors for the winter season crop production. To ensure the most efficient utilization of the residual soil moisture in the paddy field, relay cropping of grass pea is the best possible way in the study area. Potato can also be grown successfully under zero tillage condition with paddy straw mulch. Irrigation from the surface water bodies, such as ponds and canals, is also recommended to provide lifesaving irrigation to the crops. Water quality analysis is to be done for the water bodies distributed all over the region to identify the water bodies that are safe for irrigating crops. Furthermore, a proper contingent crop planning must be done to mitigate the crop failure after cyclonic storms, heavy rainfall, etc.

The people of Sundarbans region are forced to encroach into the reserve forest for illegal wood cutting, honey collection, and unscientific shrimp cultivation just to sustain their livelihood during the lean period (post- and pre-monsoon periods). These cause loss of biodiversity in the Sundarbans reserve forest. Necessary actions are to be initiated which involve the local community leading to income generation. Integrated mangrove shrimp farming, bee keeping, aquaculture, poultry farming, etc. are the probable ways for improving people's livelihood status while conserving the biodiversity, maintaining the ecological balance, and ultimately ensuring the sustainable development of the entire Sundarbans region.

Keeping the above-mentioned facts in view, the present study was conducted in the Gosaba CD block of Indian Sundarbans region. The study aimed to estimate the seasonal changes in the availability of surface water resources in terms of area. Normalized Difference Water Index (NDWI) derived from satellite-based remote sensing was applied to estimate the areas under surface water resources.

3.2 Study Area

The case study was conducted in the Gosaba community development (CD) block of Indian Sundarbans region. Gosaba CD block serves as an administrative entity within the Canning subdivision of the South 24 Parganas district in the state of West Bengal. The study block, extending from 22° 3′ N to 22° 15′ N and from 88° 42′ E to 88° 60′ E (Fig. 1), is bordered to the north by the Sandeshkhali II CD block in the North 24 Parganas district, to the west by the Basanti CD block, and to the east and south by the Sundarbans mangrove forests. The average elevation of the block is six meters. The Gosaba CD block belongs to the marine delta zone and coastal saline agro-climatic zone of West Bengal. Gosaba CD block occupies a geographical area of around 297 km^2. There are fourteen Gram Panchayats and fifty inhabited villages in the Gosaba CD block (Human Development Report 2009). Majority of the total population of the study block are engaged in agriculture to sustain their livelihood. Agriculture is mainly rain-fed and predominantly *kharif* rice-based mono-cropping. Crop production during the dry (winter and summer) seasons is adversely affected by seasonal soil and water salinity built-up, soil moisture scarcity, unavailability of suitable irrigation water, etc. (Ghosh et al. 2019, 2023). Surface water resources like ponds and canals are the only sources of irrigation during winter and summer seasons. So it is essential to monitor the availability of surface water resources during the dry seasons.

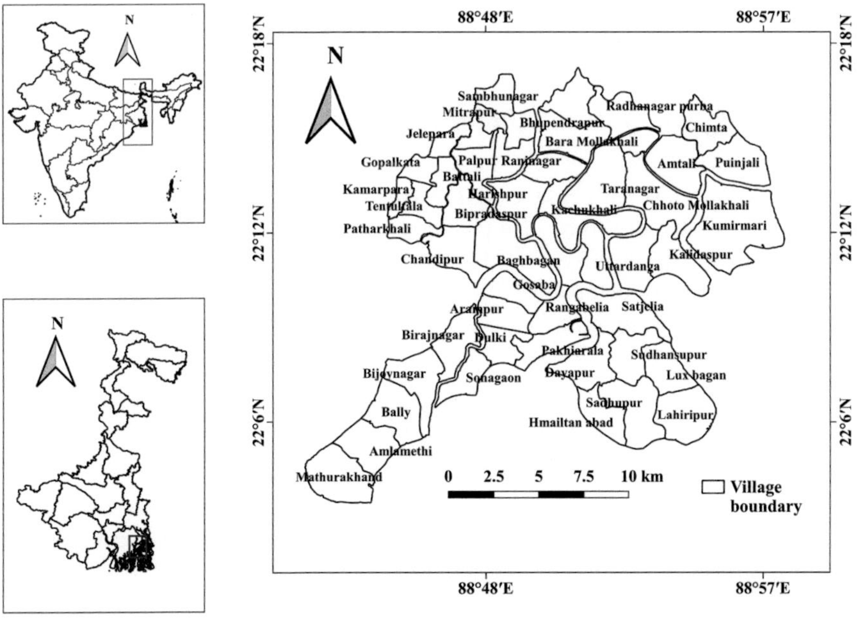

Fig. 1 Geographical location of Gosaba CD block of Indian Sundarbans

Table 1 Acquisition dates of the Sentinel-2 data used for computing NDWI

2017–18	2018–19	2019–20
22 Oct 17	27 Oct 18	1 Nov 19
11 Nov17	16 Nov 18	11 Nov 19
16 Dec 17	11 Dec 18	11 Dec 19
4 Feb 18	4 Feb 19	4 Feb 20
1 Mar 18	24 Feb 19	10 Mar 20
21 Mar 18	26 Mar 19	30 Mar 20
15 Apr 18	15 Apr 19	9 Apr 20

3.3 Study Period

The present study was carried out for the dry seasons (October–April) of three consecutive cropping years, i.e., 2017–18, 2018–19, and 2019–20.

3.4 Data

In the present study, multi-temporal Sentinel-2 MSI (Multi-Spectral Instrument) data was used for surface water resource monitoring. The acquisition dates of the satellite data have been given in Table 1.

3.5 Estimation of Areas Under Surface Water Resources

Normalized Difference Water Index (NDWI) was computed for the Sentinel-2 images acquired during the dry (post- and pre-monsoon) periods. Presence of water body was indicated by the positive NDWI values. Binary classification of NDWI images was used to separate water bodies from the non-water classes. NDWI was calculated in the present study using the formula of McFeeters (1996).

$$\mathrm{NDWI} = \frac{R_G - R_{\mathrm{NIR}}}{R_G + R_{\mathrm{NIR}}}$$

where, RG = reflectance in green band and RNIR = reflectance in near infrared (NIR) band. The value of NDWI ranges from − 1 to + 1. The positive values of NDWI represent water bodies while the negative values represent vegetation and bare soil surface. In the present study, Sentinel-2 MSI band 3 (Green: 0.56 μm) and band 8 (NIR: 0.84 μm) were used to compute the NDWI.

The pixels exhibiting positive NDWI values were extracted using binary classification in the "Raster calculator" of QGIS 3.4.5. After that, the total area under

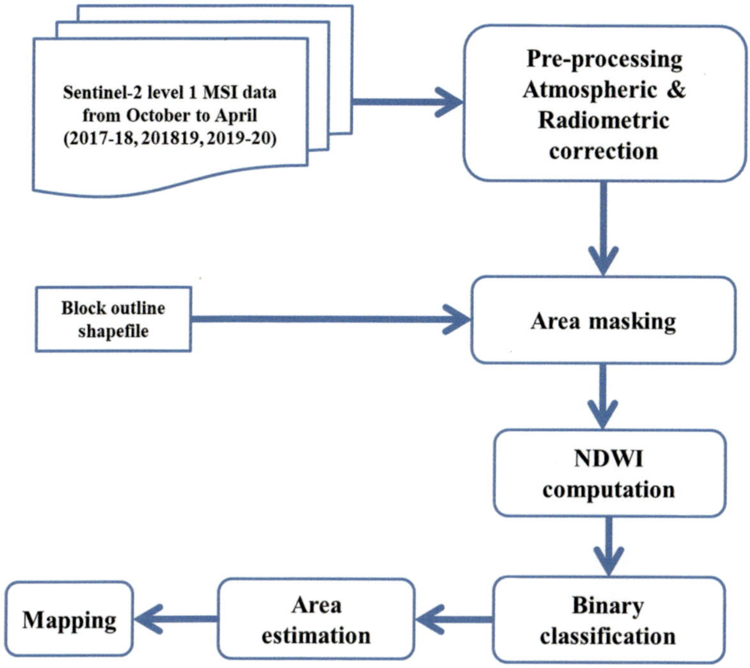

Fig. 2 Model flowchart depicting the process of area estimation of surface water resources

the pixels with positive NDWI was calculated. The variation in the surface water areas among the villages of the study block was mapped in QGIS 3.4.5. The data processing steps are represented in Fig. 2.

3.6 Results

Seasonal changes of surface water bodies in terms of area were determined for the three years viz. 2017–2018, 2018–2019, and 2019–2020 (Figs. 3, 4, and 5).

The results revealed that the surface water area was higher just after the monsoon, i.e., during the end of October. However, the acreage of the surface water resources varied among the villages in Gosaba. On the one hand, it was observed that about 2.83% (835.56 ha) area of Gosaba CD block was estimated as surface water bodies during the end of monsoon season (22nd October) of 2017–2018 cropping year. On the other hand, 1.63% (479.62 ha) and 3.74% (1101.73 ha) of the total block area was estimated as surface water bodies during the end of monsoon season (27th October and 11th November) of 2018–2019 and 2019–2020 cropping years, respectively. In November 2019, surface water area was the highest which was caused by the heavy rainfall occurred during the tropical cyclone Bulbul occurred in the November 2019.

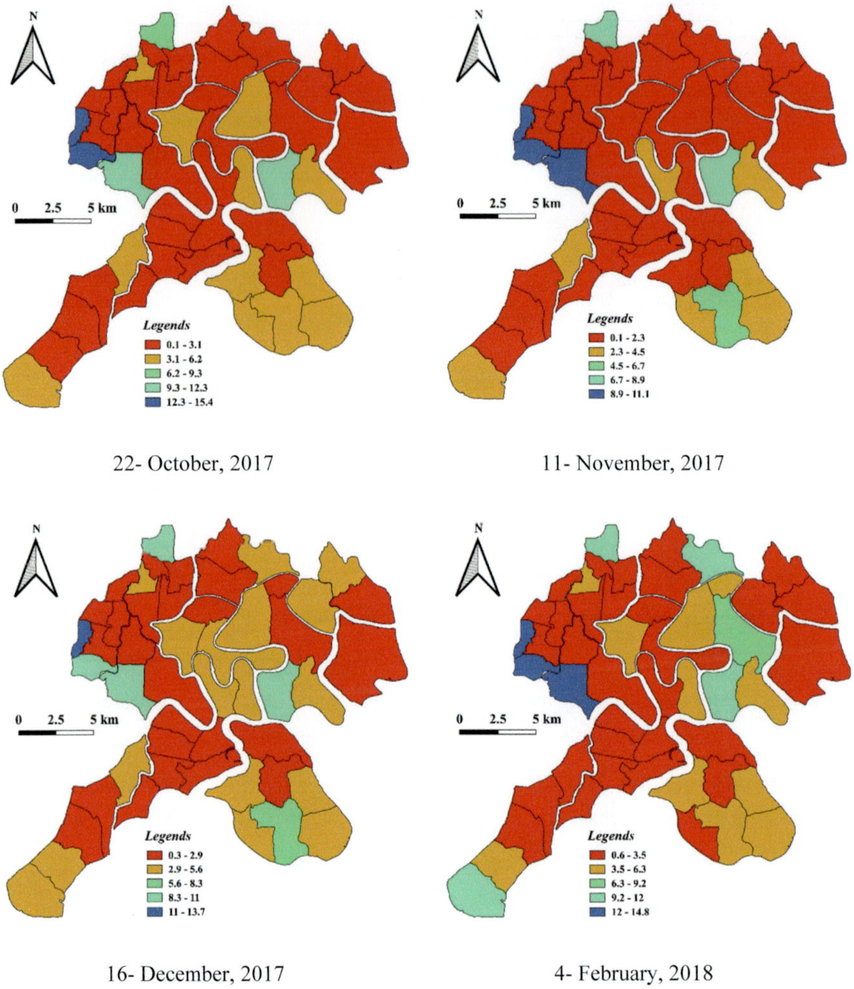

Fig. 3 Seasonal change in areas under surface water resources (as % of total village area) in different villages from October 2017 to February 2018. Seasonal change in areas under surface water resources (as % of total village area) in different villages from March 2018 to April 2018

Heavy rainfall was recorded during the third week of October 2017 and this resulted in higher areas under surface water resources. On the contrary, very less rainfall was experienced during the month of October 2018, and this was reflected in the lowest area under the surface water resources.

Surface water area notably decreased during the month of December in all the years. Water resource area was estimated as only 1.03% (302.22 ha) of the block on 1st December 2019 which was the lowest. In the cropping year, 2017–2018, the surface water area was 2.10% (619.91 ha) of the block on 11th November while in the cropping year, 2018–2019, the surface water area was estimated as 1.57% of

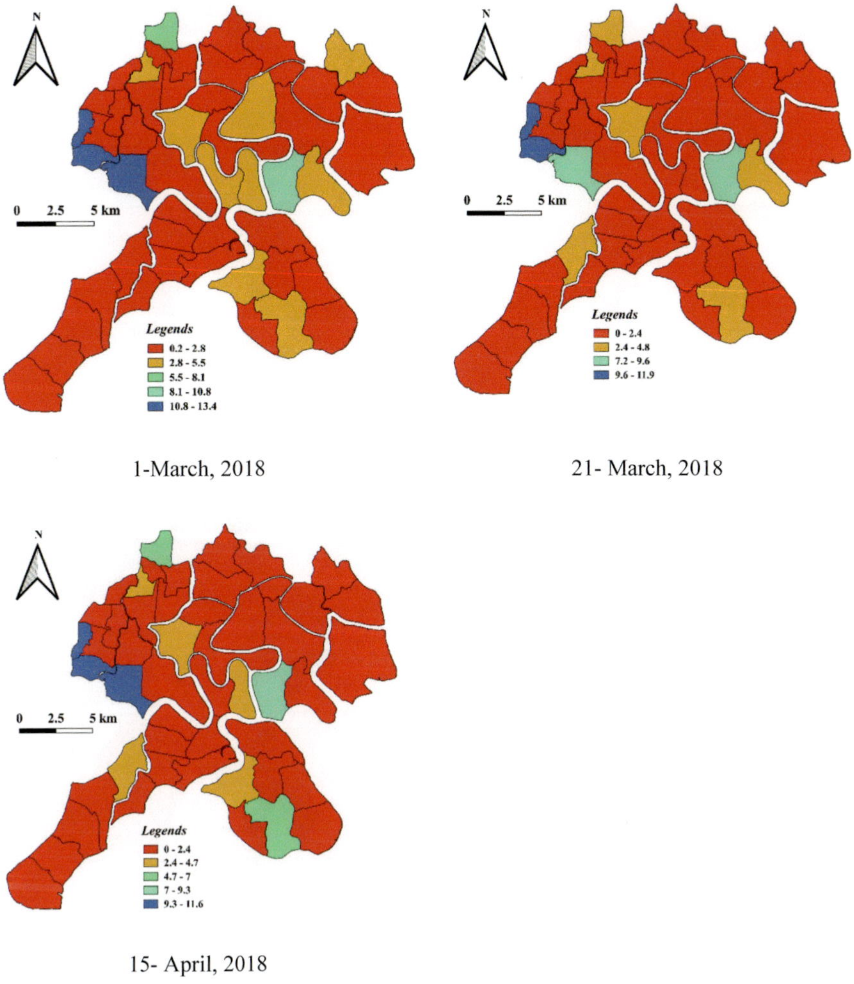

Fig. 3 (continued)

the block (462.14 ha) on 16th November. Gradual increase in the area under surface water resources was found in all the cropping years from December to February. Surface water surface area was estimated as the highest on 4th February in all the three cropping years. This might be attributed to the presence of standing water in the transplanted boro rice field. It was found that surface water area was 4.08% (1201.6 ha), 3.57% (1053.52 ha), and 4.07% (1200.62 ha) of the study block on 4th February 2018, 4th February 2019, and 4th February 2020, respectively. As the season progressed, the areas under surface water resources decreased and became lowest in the month of March.

Surface water resources area gradually decreased from February and this trend continued till the month of April as shown in Table 2. Very less variation among the

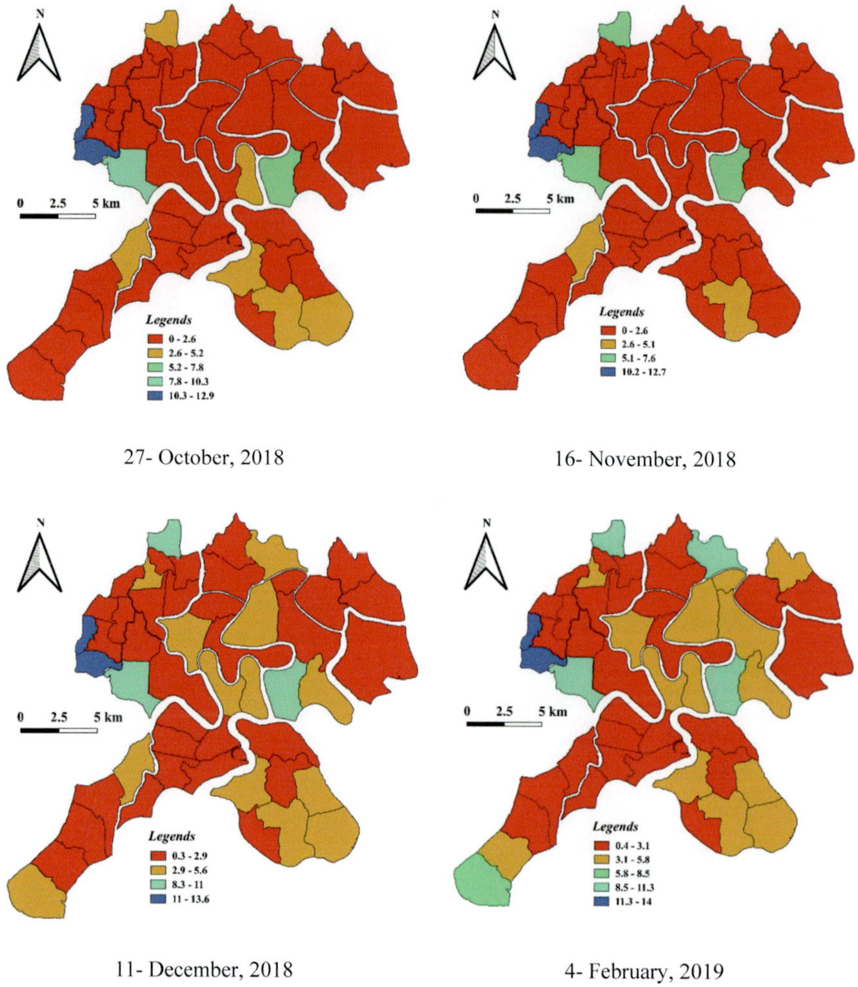

Fig. 4 Seasonal change in areas under surface water resources (as % of total village area) in different villages from October 2018 to April 2029. Seasonal change in areas under surface water resources (as % of total village area) in different villages from February 2019 to April 2029

water area at the end of post-monsoon season and starting of pre-monsoon season during the three study years was found. Results showed that about 2.56% (755.81 ha), 2.56% (754.61 ha), and 2.67% (787.16 ha) of the block was occupied by ponds and canals on 1st March 2018, 24th February 2019, and 10th March 2020, respectively. Further decrease in the water area was observed during the end of March. About 1.69% (499.29 ha), 1.84% (542.22 ha), and 1.43% (420.33 ha) of the block area constituted surface water resources on 21st March 2018, 26th March 2019, and 30th March 2020, respectively. Thus, it was evident from the results that availability of surface water resources decreased with the progress of post- and pre-monsoon

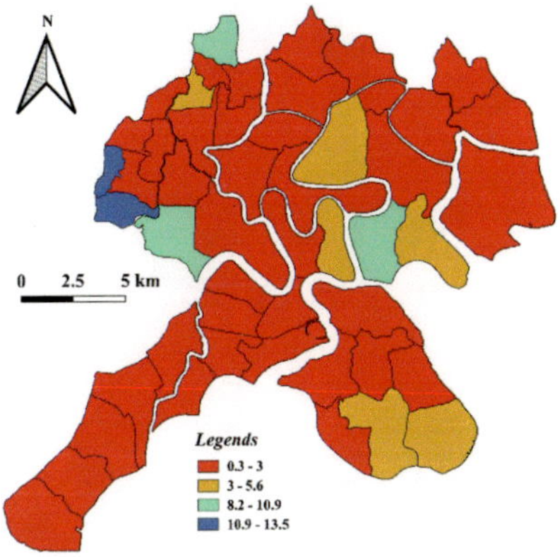

24- February, 2019

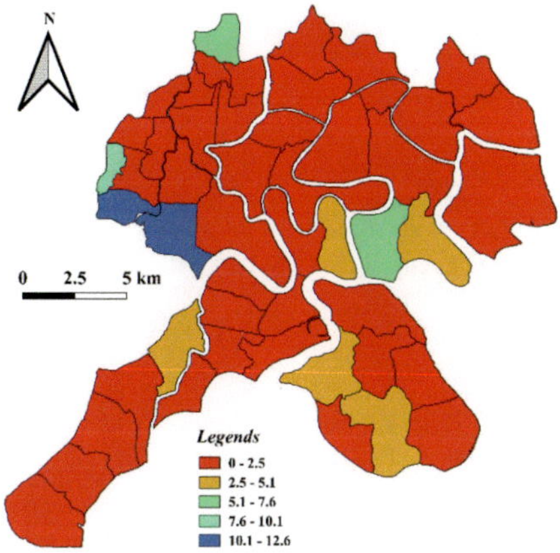

26- March, 2019

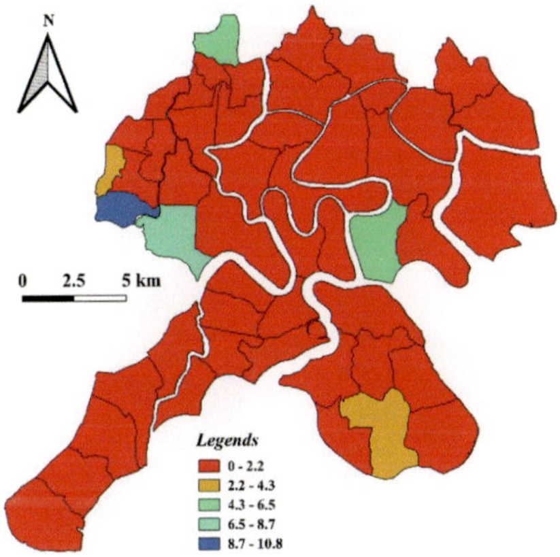

15- April, 2019

Fig. 4 (continued)

seasons. Intensity and distribution of rainfall influenced the availability of surface water resources.

Temporal changes in the village-wise distribution of the surface water bodies were estimated and presented in Figs. 3, 4, and 5. It was observed that the village Patharkhali constituted the highest proportion of water bodies throughout the entire winter and summer season during all the three years. Surface water bodies occupied about more than 10% of the total area of Patharkhali village during the whole study periods. The villages Kamarpara, Chandipur, and Hental bari constituted higher proportion of surface water bodies. On the other hand, the villages namely,

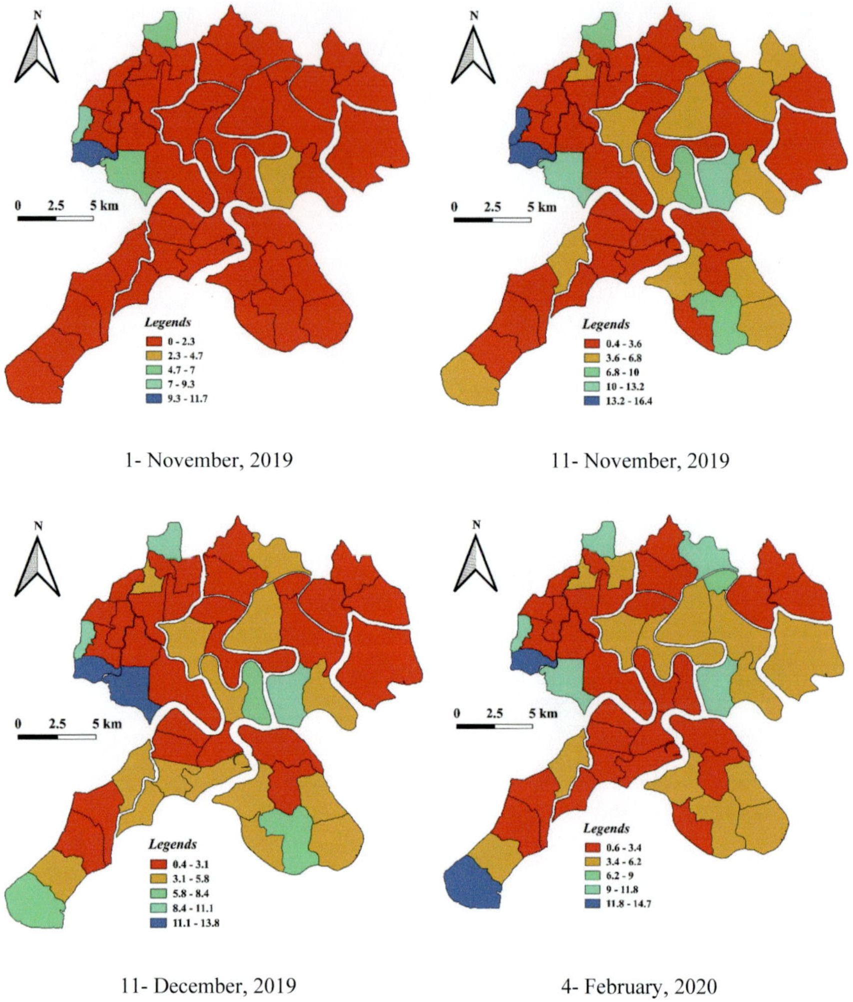

Fig. 5 Seasonal change in areas under surface water resources (as % of total village area) in different villages from November 2019 to February 2020. Seasonal change in areas under surface water resources (as % of total village area) in different villages from March 2020 to April 2020

Bipradaspur and Radhanagar paschim accounted for the lowest proportion of surface water bodies during the study periods.

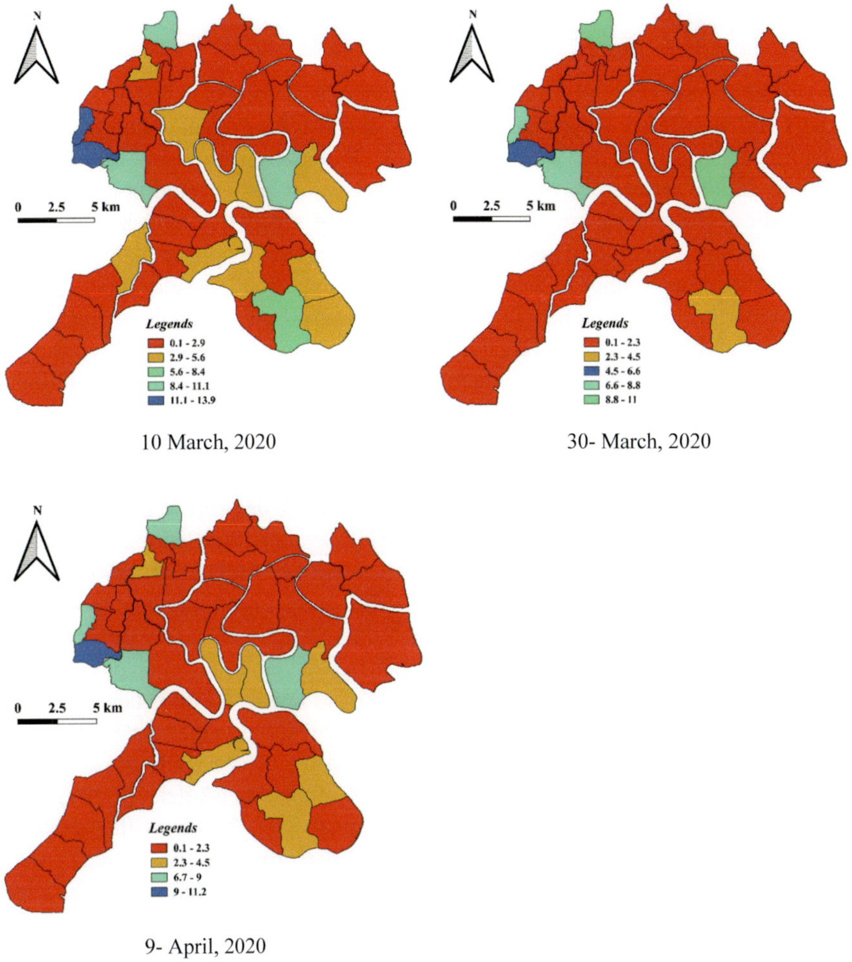

Fig. 5 (continued)

Table 2 Seasonal changes in the area under surface water resources during different cropping years in Gosaba CD block

Cropping years	22 Oct 2017	11 Nov 2017	16 Dec 2017	4 Feb 2018	1 Mar 2018	21 Mar 2018	15 Apr 2018
2017–2018	835.56	619.91	965.43	1201.6	755.81	499.29	526.76
	27 Oct 2018	16 Nov 2018	11 Dec 2018	4 Feb 2019	24 Feb 2019	26 Mar 2019	15 Apr 2019
2018–2019	479.62	462.14	815.19	1053.52	754.61	542.22	336.86
	1 Nov 2019	11 Nov 2019	11 Dec 2019	4 Feb 2020	10 Mar 2020	30 Mar 2020	9 Apr 2020
2019–2020	1101.73	302.22	1084.27	1200.62	787.16	420.33	489.03

3.7 Discussion

The variation in spectral reflectance of land surfaces with respect to water, vegetation, and dry soil at different wavelengths provides the basis for optical remote sensing of water resources (Kumar and Reshmidevi 2013). Water absorbs most of the energy in near and middle infrared wavelengths. In contrast, soil and vegetation reflect more light in these wavelength ranges. Thus, water can be easily distinguished from soil and vegetation elements in a multi-spectral image because it appears in the IR bands with a darker tone. Studies on watershed management, flood monitoring, and water resource monitoring can all benefit from mapping surface water bodies using remote sensing techniques (Kumar and Reshmidevi 2013). An excellent example of the use of satellite remote sensing is the mapping of surface water resources in Jodhpur District, India, where water bodies up to 0.9 ha in surface area have been mapped using Landsat TM pictures with a 30 m spatial resolution (Sharma et al. 1989).

Seasonal changes in the surface water bodies were estimated by following the binary classification of Normalized Difference Water Index (NDWI) derived from multi-temporal Sentinel-2 data. The green and NIR channels of Sentinel-2 data with 10 m spatial resolution were used to compute NDWI (McFeeters 1996). NDWI can effectively differentiate between water bodies and surrounding land features by exploiting the unique spectral characteristics of water. Open water surface could be detected by NDWI because this approach increased the reflectance properties of water by low reflectance in near Infrared (NIR) and enhanced reflectance in green wavelength (Li et al. 2013; Ali et al. 2019). The threshold values for NDWI classification were set to zero. Water bodies have positive values while the non-water body features have negative values. Application of Sentinel-2-derived NDWI for water body monitoring was reported earlier (Yang et al. 2017). With the use of NDWI, remote sensing and GIS technologies have proved to be powerful tools for surface water resource monitoring. They provide a cost-effective and efficient means of acquiring and analyzing data, allowing for timely decision-making in water resource management. The application of NDWI for surface water resource monitoring using remote sensing and GIS allows for accurate and reliable assessment of water availability, detection of drought conditions, and identification of flooding events (Sepuru and Dube 2018). By comparing NDWI values from different time points, it's possible to identify changes such as seasonal fluctuations, long-term trends, or sudden variations due to factors like rainfall, drought, or human activities.

Shortage of the surface water resources during post- and pre-monsoon seasons was observed from the NDWI classification of multi-temporal Sentinel-2 data. The water holding capacity of the existing ponds as indicated by the depth gradually decreased during the winter (rabi) and summer seasons as confirmed by the findings of Hazra et al. (2016) who revealed that throughout the production of the summer and rabi crops, there was a water deficit in the central and southern blocks.

3.8 Limitations

- Atmospheric conditions such as clouds, haze, and aerosols might affect NDWI results; these effects are particularly noticeable in regions with high humidity or air pollution. These atmospheric factors have the potential to contaminate NDWI imaging with noise and artifacts, making it less accurate in identifying and characterizing water bodies.
- Since NDWI makes advantage of the variation in reflectance between NIR and green bands, it is sensitive to both water and vegetation. Dense vegetation can occasionally confuse NDWI analysis, causing water bodies to be overestimated or incorrectly classified, especially in regions with a lot of floating or covered vegetation.
- Depending on the spectral and spatial resolution of the remote sensing data used, NDWI's efficacy can change. The spectral bands and spatial resolutions of various sensors vary, which may have an impact on NDWI's capacity to precisely identify and describe water bodies, particularly tiny or thin water features.
- NDWI's reliance on satellite imaging, with a low temporal resolution, may prevent it from capturing quick or transient changes in surface water dynamics. Furthermore, seasonal variations in water body features and vegetation phenology might impact NDWI data, necessitating cautious interpretation and analysis over extended time periods.
- NDWI alone may not always provide sufficient discrimination between different water body types or water conditions. For instance, it might find it difficult to distinguish between shallow water bodies with sedimentation or submerged vegetation and deep water bodies, which could result in an incorrect assessment of the depth or quality of the water.
- For accuracy assessment and calibration, NDWI results, like those from other remote sensing techniques, should be verified against ground truth data. For example, in remote or inaccessible places, limited availability of ground truth data or difficulties with field validation might bring uncertainties and biases into NDWI analysis.
- Changes in land cover, such as urbanization, deforestation, or land use conversion, can affect NDWI values and complicate the interpretation of water bodies. Urban areas, in particular, may exhibit spectral properties similar to water bodies, leading to potential confusion in NDWI analysis.
- Urbanization, deforestation, and other changes in land use can have an impact on NDWI values and make it more difficult to interpret water bodies. For instance, urban environments may have spectral characteristics that are comparable to those of bodies of water, which could cause confusion in NDWI analyses.

4 Conclusions

Surface water resource was successfully analyzed in terms of total surface water body areas including ponds and canals of Gosaba CD block using satellite-based remote sensing. This study confirms the applicability of multi-temporal optical satellite data for surface water dynamics monitoring in the coastal areas. Surface water area notably decreased during the month of December in all the years. Higher surface water area in the first of February was due to the presence of standing water in the transplanted boro rice field. Surface water resources area gradually decreased from February to April. Intensity and distribution of rainfall influenced the inter-annual variability of available surface water resources. Pond water level depleted from the middle of December due to increasing rate of evaporation. Sharp rise in the water level was recorded after heavy rain storms. The findings of this chapter will supplement potentially in the choice of the crops and cropping systems for the dry seasons. Based on the availability of the surface water, farmers can select the crops which can suit in the Indian Sundarbans region.

References

Ahmadov E (2020) Water resources management to achieve sustainable development in Azerbaijan. Sustain Futur 2:100030

Ait-Kadi M (2016) Water for development and development for water: realizing the sustainable development goals (SDGs) vision. Aquatic Proc 6:106–110

Ali MI, Dirawan GD, Hasim AH, Abidin MR (2019) Detection of changes in surface water bodies urban area with NDWI and MNDWI methods. Int J Adv Sci Eng Inf Technol 9(3):946–951

Anonymous (2003) Annual report-2003. Sundarban Biodiv Conserv Proj 41:5

Bhadra T, Das S, Hazra S, Barman BC (2018) Assessing the demand, availability and accessibility of potable water in Indian Sundarban biosphere reserve area. Int J Recent Sci Res 9(3):25437–25525

Bhat TA (2014) An analysis of demand and supply of water in India. J Environ Earth Sci 4(11):67–72

Biswas JK, Mondal B, Priyadarshini P, Abhilash PC, Biswas S, Bhatnagar A (2022) Formulation of Water Sustainability Index for India as a performance gauge for realizing the United Nations Sustainable Development Goal 6. Ambio 51(6):1569–1587

Brezonik P, Menken KD, Bauer M (2005) Landsat based remote sensing of lake water quality characteristics, including chlorophyll and colored dissolved organic matter (CDOM). Lake Reserv Manage 22(4):373–382

Brindha K, Neena Vaman KV, Srinivasan K, Sathis Babu M, Elango L (2014) Identification of surface water-groundwater interaction by hydrogeochemical indicators and assessing its suitability for drinking and irrigational purposes in Chennai, Southern India. Appl Water Sci 4:159–174

Cazenave A, Champollion N, Benveniste J, Chen J (eds) (2016) Remote sensing and water resources, vol 55. Springer

Cole MJ, Bailey RM, Cullis JD, New MG (2018) Water for sustainable development in the Berg Water Management Area, South Africa. S Afr J Sci 114(3–4):1–10

Dangar S, Asoka A, Mishra V (2021) Causes and implications of groundwater depletion in India: a review. J Hydrol 596:126103

Dasgupta R (2008) Assessment of vulnerability & adaptation strategies for the Sundarbans island system, West Bengal, India, in the perspective of climate. Unpublished Ph. D. thesis, Jadavpur University

Dhawan V (2017) Water and agriculture in India. In: Background paper for the South Asia expert panel during the global forum for food and agriculture, vol 28, pp 80–85

Dube T, Shekede MD, Massari C (2022) Remote sensing for water resources and environmental management. Remote Sens 15(1):18

Faye C, Dieye S (2018) Valorization of water resources in Senegal for economic, social and sustainable development. Afr J Environ Sci Technol 12(12):449–460

Gons HJ, Auer MT, Effler SW (2008) MERIS satellite chlorophyll mapping of oligotrophic and eutrophic waters in the Laurentian Gt Lakes. Remote Sens Environ 112:4098–4106

Hazra S, Bhadra T, Sinha Roy SP (2016) Sustainable water resource management in the Indian Sundarban Delta. In: Proceeding of international seminar on challenges to ground water management: vision, vol 2050, pp 324–334

Human Development Report (2009) United Nations Development Programme 1 UN Plaza, New York, NY 10017, USA

Jain SK (2011) Population rise and growing water scarcity in India–revised estimates and required initiatives. Curr Sci 101(3):271–276

Jiang H, Feng M, Zhu Y, Lu N, Huang J, Xiao T (2014) An automated method for extracting rivers and lakes from Landsat imagery. Remote Sens 6(6):5067–5089

Kaplan G, Avdan ZY, Goncu S, Avdan U (2019) Evaluation of RapidEye-3 satellite data for assessing water turbidity of Lake Borabey. In: Proceedings, vol 48, no 1. MDPI, p 14

Kumar AS, Reddy AM, Srinivas L, Reddy PM (2015) Assessment of surface water quality in Hyderabad lakes by using multivariate statistical techniques, Hyderabad-India. Environ Pollut 4(2):14

Kumar DN, Reshmidevi TV (2013) Remote sensing applications in water resources. J Indian Inst Sci 93(2):163–188

Li W, Gong P (2016) Continuous monitoring of coastline dynamics in western Florida with a 30-year time series of Landsat imagery. Remote Sens Environ 179:196–209

Li W, Du Z, Ling F, Zhou D, Wang H, Gui Y et al (2013) A comparison of land surface water mapping using the normalized difference water index from TM, ETM+ and ALI. Remote Sens 5(11):5530–5549

Lillesand TM, Kiefer RW, Chipman JW (2004) Remote sensing and image interpretation, Wiley India (P.) Ltd., New Delhi

Liu Z, Xu J, Liu M, Yin Z, Liu X, Yin L, Zheng W (2023) Remote sensing and geostatistics in urban water-resource monitoring: a review. Mar Freshw Res

Maharana P, Agnihotri R, Dimri AP (2021) Changing Indian monsoon rainfall patterns under the recent warming period 2001–2018. Clim Dyn 57(9–10):2581–2593

Mahato A, Upadhyay S, Sharma D (2022) Global water scarcity due to climate change and its conservation strategies with special reference to india: a review. Plant Arch 22(1):09725210

Manivasagam VS, Nagarajan R (2018) Rainfall and crop modeling-based water stress assessment for rainfed maize cultivation in peninsular India. Theoret Appl Climatol 132:529–542

McFeeters SK (1996) The use of the Normalized Difference Water Index (NDWI) in the delineation of open water features. Int J Remote Sens 17(7):1425–1432

Mirza MMQ (2006) The Ganges water diversion: environmental effects and implications, volume 49 of water science and technology library. Springer Science & Business Media, p 368

Mirza MQ, Warrick RA, Ericksen NJ, Kenny GJ (1998) Trends and persistence in precipitation in the Ganges, Brahmaputra and Meghna river basins. Hydrol Sci J 43(6):845–858

Narayanamoorthy A (2022) Groundwater irrigation in India: development, overexploitation and sustainability. In: The irrigation future of india: development, resource and policy. Springer International Publishing, Cham, pp 73–95

Ogilvie A, Belaud G, Massuel S, Mulligan M, Le Goulven P, Calvez R (2018) Surface water monitoring in small water bodies: potential and limits of multi-sensor Landsat time series. Hydrol Earth Syst Sci 22(8):4349–4380

Pari P, Thirumaraiselvan P, Ramalingam M, Jayalakshmi S (2020) A statistical model for estimation of soil moisture in paddy field using microwave satellite data. Prog Electromagn Res M 94:155–166

Priyan K (2021) Issues and challenges of groundwater and surface water management in semi-arid regions. In: Groundwater resources development and planning in the semi-arid region, pp 1–17

Rahaman SB, Rahaman MS, Ghosh AK, Gain D, Biswas SK, Sarder L, Islam SS, Sayeed AB (2015) A spatial and seasonal pattern of water quality in the Sundarbans river systems of Bangladesh. J Coastal Res 31(2):390–397

Ram SA, Irfan ZB (2021) Application of system thinking causal loop modelling in understanding water crisis in India: a case for sustainable integrated water resources management across sectors. HydroResearch 4:1–10

Riadi A, Triatmadja R, Yuwono N (2022) Study of total suspended solids (TSS) distribution and salinity of coastal area using satellite imagery for pond development in pond irrigation areas (DIT) Sei Teras. In: International conference on sustainable environment, agriculture and tourism (ICOSEAT 2022). Atlantis Press, pp 172–178

Ritchie JC, Schiebe FR (2000) Water quality. In: Schultz GA, Engman ET (eds) Remote sensing in hydrology and water management. Springer-Verlag, Berlin, Germany

Roy A, Pramanick K (2019) Analysing progress of sustainable development goal 6 in India: past, present, and future. J Environ Manage 232:1049–1065

Sepuru TK, Dube T (2018) An appraisal on the progress of remote sensing applications in soil erosion mapping and monitoring. Remote Sens Appl Soc Environ 9:1–9

Sharma KD, Singh S, Singh N, Kalla AK (1989) Role of satellite remote sensing for monitoring of surface water resources in an arid environment. Hydrol Sci J 34(5):531–537

Sheffield J, Wood EF, Pan M, Beck H, Coccia G, Serrat-Capdevila A, Verbist K (2018) Satellite remote sensing for water resources management: potential for supporting sustainable development in data-poor regions. Water Resour Res 54(12):9724–9758

Shi W, Hu Z, Liu H, Cen S, Huang J, Chen X (2022) Ship detection in SAR images based on adjacent context guide fusion module and dense weighted skip connection. IEEE Access 10:134263–134276

Singh D, Kathpalia A (2007) An efficient modeling with GA approach to retrieve soil texture, moisture and roughness from ERS-2 SAR data. Prog Electromagn Res 77:121–136

Sinha Ray SP (2010) Status of ground water condition in Sundarban area, North and South 24 Parganas District, West Bengal. WWF-India

Sophocleous M (2000) From safe yield to sustainable development of water resources—the Kansas experience. J Hydrol 235(1–2):27–43

Surwase T, Manjusree P, Nagamani PV, Jaisankar G (2019) Novel technique for developing flood hazard map by using AHP: a study on part of Mahanadi River in Odisha. SN Appl Sci 1(10):1196

Reshmidevi T (2013) Remote sensing applications in water resources. J Indian Inst Sci 93(2):163–188

Vanapalli KR, Dubey BK, Sarmah AK, Bhattacharya J (2021) Assessment of microplastic pollution in the aquatic ecosystems—an Indian perspective. Case Stud Chem Environ Eng 3:100071

Vinnarasi R, Dhanya CT (2016) Changing characteristics of extreme wet and dry spells of Indian monsoon rainfall. J Geophys Res Atmos 121(5):2146–2160

Wu G, de Leeuw J, Liu Y (2009) Understanding seasonal water clarity dynamics of lake dahuchi from in situ and remote sensing data. Water Resour Manag 23:1849–1861

Xu H (2006) Modification of normalised difference water index (NDWI) to enhance open water features in remotely sensed imagery. Int J Remote Sens 27(14):3025–3033

Yang X, Zhao S, Qin X, Zhao N, Liang L (2017) Mapping of urban surface water bodies from Sentinel-2 MSI imagery at 10 m resolution via NDWI-based image sharpening. Remote Sens 9(6):596

Yuan MH, Lo SL (2020) Ecosystem services and sustainable development: perspectives from the food-energy-water Nexus. Ecosyst Serv 46:101217

Zhou YN, Luo J, Shen Z, Hu X, Yang H (2014) Multiscale water body extraction in urban environments from satellite images. IEEE J Select Top Appl Earth Observ Remote Sens 7(10):4301–4312

Soil Moisture Evaluation in an Ungauged Basin in the Southwestern Nigeria

Adedayo Oreoluwa Adewole and Adebayo Oluwole Eludoyin

Abstract Soil moisture, despite its significant role in hydro-geomorphological and biogeochemical processes, is not well understood in many environments, particularly in developing nations with limited monitoring basins. This chapter describes a mode of soil moisture investigation in an ungauged catchment in southwestern Nigeria. The study first explained the rationale for soil moisture investigation in the study area which is typical of a data-dearth environment, due to poor monitoring technology and commitments. Consequently, examples of well-established stations around the world were first presented before the attempt to determine the spatial variability of soil moisture was reported. Reported here are the results of an assessment of soil moisture using open-source satellite data. The study also reports an assessment of the correlation of specific vegetal change indices (such as land cover/use and the Normalized Difference Vegetation Index, or NDVI) with land surface temperature (LST) in the area. ArcGIS (version 10.3) was used to analyze the multidate Landsat 8 OLI (Operational Land Imager) data. The results showed that throughout the study period, average soil moisture (SMI) increased from 6.8 to 9.3%, demonstrating spatial homogeneity. The SMI correlated spatially with land cover and use, classifying forested areas as wetter than built-up areas and reducing wetness in proportion. Land surface temperature (LST) and the Normalized Difference Vegetation Index (NDVI) varied spatially in proportion to variations in land use/cover, indicating that these variations had a substantial effect on LST, SMI, and NDVI. The results of multiple regression analysis point to a complicated interaction among the three variables.

Present Address:
A. O. Adewole (✉) · A. O. Eludoyin
Department of Geography, Obafemi Awolowo University, Ile-Ife, Nigeria
e-mail: adedayoadewole@ymail.com

A. O. Eludoyin
e-mail: oaeludoyin@yahoo.com

A. O. Adewole
Department of Geography, Olabisi Onabanjo University, Ago-Iwoye, Nigeria

S. C. Pal and U. Chatterjee (eds.), *Surface, Sub-Surface Hydrology and Management*, Springer Geography, https://doi.org/10.1007/978-3-031-62376-9_6

Keywords Open-source data · Landsat ungauged basins · Soil moisture and hydrological parameters · Opa River basin in Southwest Nigeria

1 Introduction

1.1 Soil Moisture Variability

Soil moisture, 'water held within soil pores' (Western et al. 2002) or water contained in the unsaturated soil zone or vadose zone, is often classified into Surface Soil Moisture (SSM; water in 0–10 cm profile) and Root-zone Soil Moisture (RSM; water under 2 m). The soil moisture, especially the SSM, has been described as important in the modeling of physical and chemical structure in drainage basins. Studies (e.g., Williamson and Turner 1980; Brocca et al. 2008) noted that soil moisture is crucial for accurate modeling of hydrological behavior, and a significant factor in the understanding of global water systems, land–atmosphere spatial dynamics, and land surface-hydrology. Rondinelli et al. (2015) argued that prior knowledge of soil moisture content is essential for forecasting the degree, timing, and location of flood events while Yang et al. (2004) noted that soil moisture is required to modify latent and sensible heat flux, consequently influencing surface-air temperature, humidity, and atmospheric circulation. Similarly, studies (including Breen et al. 2020) revealed that soil moisture essentially regulates soils' physicochemical properties and activities of microbial biomass, such that knowledge of soil moisture is important in soil, climate, and landscape studies.

Soil moisture has been of continual interest across decades because it is a major factor determining plant growth, crop production, and runoff processes (Smith and Warrick 2007; Zhang et al. 2019; Yang et al. 2021). The debate on the existence of soil moisture probably began with the identification of soil water as significant for hydrogeochemical and physical processes. For example, before soil moisture was identified as part of the end-member contributor to water chemistry through the mixing member analysis (Ogunkoya and Jenkins 1993; Jenkins et al. 1994; Knighton et al. 2020), studies believed that soil moisture was only made up of the event water, precipitation or surface flow and groundwater, and had deciphered the components through graphical hydrograph separation method (Singh 1976). Hewlett and Hibbert (1967) called the graphical hydrograph method capricious and subjective. The introduction of isotopes and chemical tracers made studies (Dewalle et al. 1988; Lee and Krothe 2001) recognize that soil water is a distinguishable end-member of a three- or more component of the mixing model. Water is separated into various sources by the mixing model, viz; soil water, groundwater, and channel precipitation (Eludoyin 2013; Eludoyin et al. 2017).

Soil moisture is a major part of the subsurface flow (SSF), and groundwater is formed as part of rainwater that seeps into the soil and moves below the surface into a river channel. When soil moisture reaches a certain soil depth and comes into contact with a less permeable layer, or when the near-stream water table rises and groundwater levels rise during a rainfall event, it flows quickly as lateral flow (Church et al. 1997; Hagedorn and Bundt 2002). Interflow or throughflow are common terms used to describe SSF that takes place in the soil layer. Rapid subsurface water entry into a river channel is known as interflow (Hewlett and Hibbert 1967). Water tends to move preferentially or as a matrix through the subsurface layer quickly. Solutes move through saturated or unsaturated matrix pores in matrix flow, while water moves slowly and steadily. Over a range of scales, soil moisture content fluctuates both spatially and temporally (Li et al. 2016). A deep knowledge of spatial differentiation of soil moisture is highly significant (Wilson et al. 2003) because determinants of SM content change at various scales. For example, climate is a major influence on soil moisture at larger scales while topographic elements and land use influence soil moisture content at smaller scales (Guo et al. 2020). Climatic factors at larger scales conceal the influence of soil characteristics, land use, and topography, on soil moisture, whereas climatic phenomena are quite uniform at smaller scales, and the influence of other factors is accentuated. Catchment or river basin scale has both the universality of the large scale and the exactness of the small scale.

Irrespective of soil moisture dynamics, recurrent surveys often spot specific locations of temporal stability that can act as representative sites within an environment. Precise changes of soil moisture with its temporal stability can be investigated by conscientiously segmenting the soil profile, which increases during the wet season and reduces in the dry season of the following year (Zhang et al. 2020). In wet seasons, the mean soil moisture content in the soil profile increases and then decreases with depth for tall plants that need strong transpiration to ensure the movement of water and supply of nutrients to their top leaves (e.g., forest and maize cultivated land). As a result of the high water consumption by root absorption and inadequate supply of water at deep (40–60 cm) soil layers, the moisture content is far lesser than in the surface layer. Studies (e.g., Li et al. 2018; Zhang et al. 2020) have buttressed the fact that shallow layers display higher temporal dynamics than deep because precipitation, infiltration, and evaporation may have greater effects on shallow soil layers. Determining soil moisture dynamics and their predicting factors in a deep profile is the premise for understanding the procedures of root–water interchange in deep soil layers, water flow in a thick unsaturated zone, and the regimes of precipitation supplying groundwater.

2 Soil Moisture and Its Relationship with Soil Physical Characteristics

Soil and water are two essential natural resources that influence plant growth. Soil provides the natural substructure and food reserve for plant growth. On the other hand, water is significant for the operations of plant life (Rogers et al. 2014). An understanding of soil moisture and its relationship with soil properties is crucial for effective management of resources. Soil texture, bulk density, soil structure, and soil porosity are determinants of soil physical properties. Sand, silt, and clay are constituents of the mineral solid fraction of the soil, which are determinants of the soil texture. Sand particles vary from 0.05 to 2.00 mm, silt particles vary from 0.002 to 0.050 mm, the clay particles are less than 0.002 mm in diameter. Particles greater than 2 mm (rock fragments) are ignored in soil texture analysis, but they are determinants of soil structure and soil moisture interactions. Soil textural class is determined using a soil textural triangle after obtaining sand, silt, and clay fractions (Vos et al. 2016; Easton 2021). It is important to determine the soil texture because of its significant influence on soil properties. Soil texture determines drainage, susceptibility to erosion, water retention ability, pH buffering capacity, aeration, soil tilth, and cation exchange capacity.

The arrangement of soil particles into units of aggregates, or soil peds, is known as soil structure. Six classes—platy, prismatic, columnar, blocky, single-grained, and granular—are used to categorize the soil structure. The movement of air and water through the soil, root penetration, and the uptake of nutrients by plants are all influenced by the structure of the soil. It's common to refer to sandy (single-grained), loamy, and clayey (massive) soils as structureless. In sandy soils, water quickly percolates through the soil, whereas in loamy and clayey soils, water seeps into the soil slowly. Platy structure hinders the downward flow of water attributable to the parallelly spread-out plates of soil. Soils with granular, prismatic, or blocky structures have beneficial water flow peculiarities for crop production. A well-structured soil is favorable for agricultural practices because of its ability to hold up water and gasses and accommodate load-bearing actions, like field traffic. The soil structure can be destroyed through irrigation with high salt content and tillage of some fine-textured soils occurring mostly when they are moist. Soil structure also changes through the biological activities of earthworms, mammals, ants, and termites, when they create large pores through burrowing and deposition of organic matter. The mass of dry soil divided by the total volume occupied by the soil defines the bulk density (Adediji 2005).

$$\text{Bulk Density} = \frac{\text{mass of soil}(g)}{\text{volume of soil}\left(\text{cm}^3\right)} \tag{1}$$

Particle density is usually represented as 2.65 gm^{-3}. Soil porosity is often applicable in hydrological models to acclimate the rate of water movement dynamics through soils; hence exemplifying soil organic carbon, soil bulk density, and water infiltration capacity (Manns and Berg 2014). Compacted soils have low porosity and high bulk density, whereas loose soils have high porosity and low bulk density. Bulk density makes information available on the leaching potential of agrochemicals, erosion, and crop productivity. High bulk density promotes erosion, runoff, and removal of nutrients as a result of the restriction of water movement through the soil leading to overland flow, erosion, and loss of nutrients. Vertical leaching is encouraged when the bulk density is low, especially in sands (Hartemink 2008). Soil management practices like tillage can also influence the soil bulk density. Furthermore, soil porosity can be determined by soil peds, soil texture, root penetration, and biological activities. Sandy soils are less porous than clayey soils even though the individual pores in sandy soils are larger than in clay soils. The smaller pores in the clayey soils are more numerous, making them more porous than the sandy soils. An increase in the organic matter content leads to an increased level of soil porosity.

Reflected electromagnetic radiation is used in remote sensing to gather data on the land and water surfaces of the planet. It is a technology that uses target locations within the sensor instrument's field of view to detect and record electromagnetic energy (Reddy 2018; Abd El-Ghany et al. 2020). The characteristics and instruments used in the management and organization of geospatial data are described by geographic information systems. Computerized systems for storing, processing, and retrieving data that are equipped with hardware and software that can handle spatial data that is geographically referenced are utilized for remote sensing (Ali et al. 2020; Bhatt et al. 2022). Soil erosion, land degradation, and land use/cover classification are all significantly impacted by the technologies (Sarmah et al. 2012; Beyene 2019). Previous research has shown that some of the models of soil, land use, and land cover were created using modeling and data from remote sensing that was examined in a GIS environment (Rai et al. 2014; Fenta et al. 2016).

A strong, useful administrative and management tool in decision-making, remote sensing, and GIS platforms support the integration of analysis of natural resource inventory, management, and planning strategy for sustainable development (Kala and Kumar 2021). It helps in identifying drainage channels that have been impacted by both natural and man-made forces by providing accurate and up-to-date information about distinctive geological formations and landforms. Furthermore, it is effective in solving problems associated with land and water resources planning and management instead of conventional techniques of data processing (Rai et al. 2014). Remote sensing platforms allow for the synoptic view of large and impenetrable areas (e.g., Amazon Basin).

The interpretation of remotely sensed data coupled with adequate real ground information allows for the recognition of different ground features like geological structures and geomorphic features with their hydraulic attributes. Multispectral satellite sensor data also provides an easy way to analyze drainage and unique landforms at different scales (Bagyaraj et al. 2019). Geo-coded resource databases generated on drainage, landforms, and soil parameters in GIS have successfully stored, retrieved, and analyzed data at the catchment scale to determine their association (Reddy 2018). It is a powerful mechanism for upgrading and monitoring parameters and it allows for the spatial analysis of other associated resource databases. It is also useful for building a suitability model for water conservation structures.

In estimating soil moisture content, two approaches can be considered while using the remote sensing method. They are (i) Active way—the use of radar/synthetic aperture radar (SAR), and (ii) passive ways—the use of radiometers with coarse resolution. Radar or SAR extends from ten to hundred meters and has a long revisiting period of about one month while the passive way extends in kilometers with daily temporal coverage (Dwivedi 2018). In some parts of the African continent, retrieval algorithms have been used in estimating soil moisture values from Special Sensor Microwave imager (SSM/I) observation. However, this approach has failed in some regions like Central Africa. This failure is a result of the dense vegetation in the region; hence, it prevents the microwave signals from penetrating the canopy layers as well as the soil layers. Regions with dense vegetation require longer wavelength channel observation to detect information on the land surface (Li et al. 2021). Major sources of data used for soil estimation vary in space and time and consequently, efforts are geared toward discriminating the sources based on data availability, resolution (temporal and spatial), and accuracy.

3 Examples of Soil Moisture Monitoring at Experimental Sites

3.1 North Wyke Farm Platform, Rothamsted Research

In order to achieve sustainability in the food production process, field experiments are essential. These experiments typically call for increased monitoring and the constant presence of technicians to operate the field instruments. At the catchment scale, the North Wyke Farm platform promotes careful observation, assessment, and management of various land uses. According to Murray et al. (2020), it offers scientific data on livestock and agronomic data, water chemistry, greenhouse gas emissions from soils, water flow rates, and records on farm management. The UK Biotechnology and Biological Sciences Research Council funds the North Wyke Farm Platform (NWFP), which combines modern technology with established research techniques to promote studies on the sustainability of grassland management systems and their impact on soil moisture levels, runoff, and related water quality dynamics (Orr et al. 2016; Murray et al. 2020). The meteorological station on the farm has two sets of meteorological equipment: the NWFP experiment site and the UK Meteorological Office site. Meteorological data, including solar radiation, wind direction, speed, and direction of precipitation, are recorded every 15 min at the NWFP experimental site. The amount of runoff is measured using H-flumes and bubbler flow-meter devices, and each catchment has a soil moisture station where measurements are taken at depths of 10, 20 and 30 cm. There is also a tipping bucket rain gauge at the location.

Furthermore, the soil moisture station (SMS) at NWFP includes soil moisture and soil temperature sensors, as well as a rain gauge for precipitation. Data is collected at a fine resolution of 15-min intervals, providing detailed insights into soil moisture dynamics. The data collected from the SMS has been used to study the effects of different grassland farming systems on soil moisture levels, runoff, and water quality dynamics. For example, research has shown that planned reseeding with grasses bred for enhanced sugar content or deep-rooting traits can improve soil moisture retention and reduce runoff. Similarly, sowing grass and legume mixtures have been found to reduce nitrogen fertilizer inputs and improve soil health. Recent studies at NWFP have focused on the potential unintended consequences of agricultural land use change driven by dietary transitions. By analyzing soil moisture data, researchers have assessed the impact of different farming practices on soil health and water quality. These findings contribute to developing sustainable, resilient, and net-zero land management strategies. Soil moisture monitoring at the North Wyke Farm Platform is a vital component of research to improve agricultural productivity and sustainability. The detailed data collected through the SMS provides valuable insights into soil moisture dynamics, helping researchers and farmers make informed decisions about land management practices. As the world faces increasing challenges related to food security and environmental sustainability, the work being done at NWFP is more important than ever (Eludoyin 2013; Orr et al. 2016).

4 Bruntland Burn (BB) Experimental Catchment

The Bruntland Burn (BB) Experimental Catchment, located in the Scottish Highlands, is a long-term research site dedicated to studying hydrological processes and their interactions with the environment (Sprenger et al. 2017). Soil moisture monitoring at this site is a critical component of understanding water dynamics, ecosystem responses, and the impacts of climate change. The BB Experimental Catchment employs a range of methods to monitor soil moisture, including in-situ sensors, soil water stable isotopes, and remote sensing techniques. In-situ sensors, such as time-domain reflectometry (TDR) probes and capacitance probes, are installed at various depths to measure soil moisture continuously. These sensors provide high-resolution data on soil moisture dynamics, allowing researchers to track changes over time and across different soil layers. Soil water stable isotopes, such as deuterium (^{2}H) and oxygen-18 (^{18}O), are used to trace the movement of water through the soil. By analyzing the isotopic composition of soil water, researchers can determine the sources of water, the pathways it takes, and the processes it undergoes. This information is valuable for understanding evaporation, transpiration, and groundwater recharge. Remote sensing techniques, such as satellite imagery and aerial photography, are also used to monitor soil moisture at larger scales. These methods provide spatially extensive data that can complement in-situ measurements and help identify patterns and trends across the catchment. The data collected from soil moisture monitoring at the BB Experimental Catchment has been used to study various hydrological processes and their responses to environmental changes. For example, research has shown that soil moisture storage is highly sensitive to precipitation patterns and can be significantly affected by droughts. During the summer drought of 2018, soil moisture levels in the catchment were found to be much lower than average, leading to reduced groundwater recharge and streamflow. Another study used soil water stable isotopes to investigate evaporation dynamics at the soil-plant-atmosphere interface. The results showed that evaporation rates varied with vegetation type and soil depth, with soils beneath Scots pine exhibiting higher evaporation losses compared to soils beneath heather. This information is crucial for understanding the water use efficiency of different vegetation types and their impact on the hydrological cycle. Recent advancements in soil moisture monitoring technology have improved the accuracy and resolution of data collected at the BB Experimental Catchment. The integration of multiple methods, such as in-situ sensors, isotopic analysis, and remote sensing, provides a comprehensive view of soil moisture dynamics and their interactions with other environmental factors. These advancements have enhanced the ability to model hydrological processes and predict future changes in water availability and ecosystem health (Birkel et al. 2014).

5 The Murrumbidgee Soil Moisture Monitoring Network (MSMMN)

The Murrumbidgee Soil Moisture Monitoring Network (MSMMN) is a comprehensive research initiative aimed at understanding soil moisture dynamics across the Murrumbidgee River Basin in southern New South Wales, Australia. Established to support hydrological research, agricultural planning, and climate modeling, the network provides valuable long-term soil moisture data. The MSMMN employs a network of soil moisture sensors distributed across the catchment. These sensors measure soil moisture at various depths, typically at 0–5 cm, 0–8 cm, 0–30 cm, and 30–60 cm. The data collected includes soil moisture, soil temperature, and rainfall measurements. The network also integrates meteorological data from nearby weather stations to provide a comprehensive view of the environmental conditions affecting soil moisture. The data collected by the MSMMN has been used in various research studies to understand soil moisture dynamics and their implications for hydrology and agriculture. For example, studies have shown that soil moisture levels in the Murrumbidgee catchment are highly variable, influenced by factors such as precipitation patterns, evapotranspiration, and land cover. This variability has significant implications for water resource management and agricultural productivity. Recent advancements in sensor technology and data analysis have enhanced the capabilities of the MSMMN. The integration of new sensors and improved data processing techniques has allowed for more accurate and detailed soil moisture monitoring. These developments have facilitated the use of MSMMN data in land surface modeling studies, contributing to a better understanding of soil moisture dynamics and their interactions with other environmental factors (Young et al. 2008; Smith et al. 2012).

6 TERENO Experimental Field

The TERENO (Terrestrial Environmental Observatories) network is a comprehensive research initiative in Germany that aims to study the long-term effects of climate change on terrestrial ecosystems. One key component of this initiative is measuring soil moisture at various experimental fields across Germany. These measurements are crucial for understanding soil moisture dynamics, water availability, and ecosystem responses to environmental changes (Bogena et al. 2018). The TERENO network employs various methods to measure soil moisture, including in-situ sensors, soil water stable isotopes, and remote sensing techniques. In-situ sensors, such as time-domain reflectometry (TDR) probes and capacitance probes, are installed at various depths to measure soil moisture continuously. These sensors provide high-resolution data on soil moisture dynamics, allowing researchers to track changes over time and across different soil layers (Wiekenkamp et al. 2014, 2016). Soil water stable isotopes, such as deuterium (^2H) and oxygen-18 (^{18}O), are used to trace the movement of water through the soil. By analyzing the isotopic composition of soil water, researchers can determine the sources of water, the pathways it takes, and the processes it undergoes. This information is valuable for understanding evaporation, transpiration, and groundwater recharge. Remote sensing techniques, such as satellite imagery and aerial photography, are also used to monitor soil moisture at larger scales. These methods provide spatially extensive data that can complement in-situ measurements and help identify patterns and trends across the catchment. The data collected from soil moisture measurement at TERENO has been used to study various hydrological processes and their responses to environmental changes (Robinson et al. 2022). For example, research has shown that soil moisture storage is highly sensitive to precipitation patterns and can be significantly affected by droughts. During the summer drought of 2018, soil moisture levels in the catchment were found to be much lower than average, leading to reduced groundwater recharge and streamflow. Another study used soil water stable isotopes to investigate evaporation dynamics at the soil-plant-atmosphere interface. The results showed that evaporation rates varied with vegetation type and soil depth, with soils beneath Scots pine exhibiting higher evaporation losses compared to soils beneath heather. This information is crucial for understanding the water use efficiency of different vegetation types and their impact on the hydrological cycle. Recent advancements in soil moisture monitoring technology have improved the accuracy and resolution of data collected at TERENO. The integration of multiple methods, such as in-situ sensors, isotopic analysis, and remote sensing, provides a comprehensive view of soil moisture dynamics and their interactions with other environmental factors. These advancements have enhanced the ability to model hydrological processes and predict future changes in water availability and ecosystem health. Soil moisture monitoring at the TERENO Experimental Field is a vital component of hydrological research. By combining various methods and technologies, researchers can gain a deeper understanding of soil moisture dynamics and their implications for water management and ecosystem sustainability. The data collected at this site (https://www.tereno.net/?form=MG0AV3) contributes to the development of more accurate hydrological models and informs strategies for mitigating the impacts of climate change and land use change.

7 Rationale for the Study

Techniques are many that exist for the determination or estimation of soil moisture, as well as its temporal and spatial variability. The techniques, which vary in approaches, capacity, and cost, are characterized by different levels of strengths and weaknesses (Wang 2009). Common methods are laboratory methods, in situ measurement using different versions or models of sensors (including the neutron scattering technique, gamma attenuation technique, and dielectric models) as well as remote sensing methods and modeling methods, which involve applying parameters that were generated using the other methods. In this study, the remote sensing technique is examined, based on the assumptions that data are freely available (open access); soil moisture data can be generated for periods when satellite data are available and fieldwork is limited; this is of particular interest in a large drainage basin. According to Brocca et al. (2017), remote sensing also allows for measurements at different spatial and temporal scales. In Nigeria and many parts of Sub-Saharan Africa, basins are largely ungauged; some gauged stations have also been abandoned, largely due to low technology infrastructure in the region. Whereas instrumented observatories are common in many parts of the world including Asia, the African continent has been poorly studied due to largely unavailable observatories or monitoring sites for soil moisture. Existing studies have focused on plot-scale collection majorly within the context of cropping systems and the datasets used are typically 'occasional-once-for-all' sampling (e.g., Shittu et al. 2017; Adoga and Uvirkaa 2018), which have been largely described as unreliable for parameterization into hydrological models. Professional and research groups, including the International Association of Hydrological Sciences (IAHS) and the International Soil Modeling Consortium (ISMC), have expressed concerns about poorly gauged basins coupled with the impact of soil moisture as well as the need for parameterization and prediction in ungauged basins (Hrachowitz et al. 2013; Choi et al. 2021). Consequently, this study is aimed at providing a background for a planned extensive soil moisture campaign in 2022 using a state-of-the-art Delta SM150T soil moisture and HH2 Handheld meter. The study intends to contribute to existing research on the Prediction in Ungauged Basins (PUB) initiative and provide decision-support information for parameterization in soil moisture modeling and predictions in southwest Nigeria. Specific objectives are to examine variability in soil moisture and its relationship with land use/land cover, LST, and NDVI.

8 Materials and Methods

8.1 Study Area

The Opa River basin in southwest Nigeria (7.43°N–7.58°N, 4.40°E–4.65°E) is the study area (Fig. 1). Studies on morphometric traits have previously been written about (Adewole et al. 2020; Eludoyin and Adewole 2020). The basin spans four local government areas and was dammed (as Opa Dam and Reservoir) at Obafemi Awolowo University in Ile-Ife in 1978. The water is treated for human consumption in the university community and used for research, laboratory, agricultural, and recreational purposes (see Akinbuwa and Adeniyi 1996; Adediji 2005; Babatimehin et al. 2020). Koppen's Af climatic class, which is characterized by heavy (1400 mm) annual rainfall and tropical rainforest vegetation, is the dominant climate (Adediji 2005). The region experiences two distinct seasons each year: the wet season and the dry season. Typically, the dry season starts in November and ends in March, while the wet season starts in April and lasts until October. According to Adejuwon and Jeje (1975), the mean minimum and mean maximum temperatures range from 23.0 to 25.8 °C and 28.8 to 33.8 °C, respectively. The Alfisols soil series, with a ferruginous tropical overlay typically in league with basement complex rocks, make up the majority of the soil in the region. The soil is derived from parent materials of granite and gneiss and is classified by the United States Department of Agriculture (USDA) system (1999) as belonging to the Egbeda and Iwo association and the *Oxic Tropudalf* series (Orewole et al. 2016).

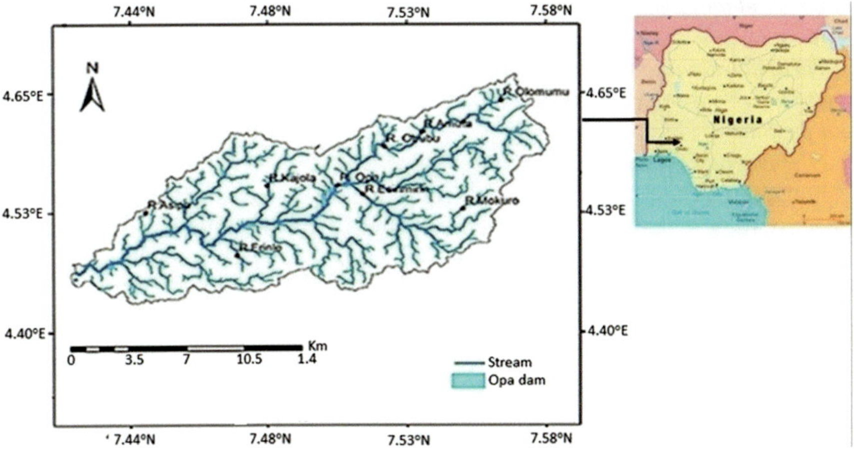

Fig. 1 The study area, Opa River basin in southwest Nigeria

8.2 Data

The Landsat data used in this analysis was a three-year (2017–2019) period that was downloaded from the USGS website (https://landsat.usgs.gov). To mitigate the impact of significant seasonal variations, all downloaded images were taken during a single month in several different years. According to Saha et al. (2018), the images were analyzed using ArcGIS (10.3 version). The topographical map was used to georeference the images by the protocol outlined in the literature (Vaidya et al. 2013; Khare et al. 2014). Inherent geometric and radiometric errors in Landsat images were corrected (Iqbal and Sajjad 2014; Ibitoye et al. 2016). The accuracy of the study's classification operation was ascertained through the application of confusion matrix analysis. The digital number (DN) and the Top-Of-Atmosphere (TOA) values were first obtained in order to retrieve the land surface temperature data from the image bands. The LST values were then obtained using the radiance values and the relevant equations (Adejuwon and Jeje 1975; Adewole et al. 2020; Adoga et al. 2018; Akinbiola et al. 2016).

$$L_\lambda = \frac{\mathit{L}\mathrm{MAX}_\lambda - \mathit{L}\mathrm{MIN}_\lambda}{Q_{\mathrm{calmax}} - Q_{\mathrm{calmin}}} \times (Q_{\mathrm{cal}} - Q_{\mathrm{calmin}}) + \mathit{L}\mathrm{MIN}_\lambda \tag{2}$$

$$P_\lambda = \frac{\left(\pi \times L_\lambda \times d^2\right)}{(ESUN_\lambda \times \sin(\Theta)} \tag{3}$$

Q_{cal} is a quantized calibrated digital number; Q_{calmax} and Q_{calmin} are the maximum and minimum (respectively) quantized calibrated pixel values obtained from the image metadata file; L_λ is the spectral radiance at the sensor aperture in W/(m^2 sr μm).

The spectral radiance scales for q_{calmax} and q_{calmin} are $LMAX_\lambda$ and $LMIN_\lambda$, respectively; $ESUN_\lambda$ is solar irradiance, and d2 is the astronomical distance between earth and the Sun.

Θ = Degrees of Sun Elevation.

Equations (4 and 5) were used to convert the DNs of the OLI image for the reflective (2, 3, 4, 5, 6, and 7) and thermal (band 10) bands for the Landsat OLI data into TOA radiance and surface reflectance.

$$L_\lambda = \mathrm{ML}Q_{\mathrm{cal}} + \mathrm{Al} \tag{4}$$

$$P\lambda = \frac{P\lambda}{\sin(\theta_{SE})} \tag{5}$$

L_λ represents the TOA spectral radiance, and MLB and denotes a particular multiplicative rescaling factor.

A_l is the band-specific additive rescaling factor derived from the metadata, and Q_{cal} = quantized and calibrated standard product pixel values.

P_λ is the corrected TOA reflectance for the angle of the sun; θ_{SE} is the degree of the sun's elevation angle, obtained from the metadata file linked to the OLI image.

9 Results and Discussion

9.1 Soil Moisture Index and Land Cover/Use Variability

Analysis of SMI showed a relatively spatially variable and annually declining pattern (Fig. 2ai–ci). Average SMI decreased from 0.7 through 0.6 to 0.5 between 2017 and 2019 while the coefficient of variability increased by 22.9, 29.7 and 38%, respectively, suggesting increased spatial variability. The pattern of SMI was compared relatively to the built-up areas, as the built-up areas exhibited less moisture than the other places (Fig. 2ai–cii). In general, estimates from land use/cover analysis showed that the river basin has witnessed an increase in built-up areas while the water body declined (Table 1). The rate of change showed an increase in vegetated areas by about 9.1%, and this comprises natural, artificial, and ornamental plants. Areas previously used by farmland were also probably lost to built-up growth, and this loss was obvious when compared to the SMI, which revealed a loss of moisture around the areas (see Fig. 2). Although the change in built-up area was not unexpected, as an increase in population pressure and urbanization has caused drainage basins all over the world to be taken over by human activities, the relative spatially lower indices of soil moisture in the northeastern part of the basin require further study. Also, effects of localized issues, such as reported issues of farm abandonment due to violent inversion from herdsmen (Oli et al. 2018), as well as that of increased gold mining activities which were expected to reveal more water-logged areas (e.g., Akinbiola et al. 2016; Eludoyin et al. 2017), may not have been adequately instructed by the images, probably due to spatial resolution of the images. The uncertainty in the potential of the images to reveal the effects of local activities may however reveal the need for intensive field-based data that the projected Delta T's field measurement is aimed at. Based on the results of the imagery-based analysis, the projected study will hypothesize the impact of land use and soil particle size distribution.

NDVI and LST Patterns

The analysis of the Normalized Difference Vegetation Index (NDVI) showed that the vegetation covering the area consisted of shrubs and grasslands rather than forests (Fig. 3ai–ci). The range of decrease in vegetation greenness over the period also corroborated the possible negative influence of both construction and extractive activities in the area. The mean NDVI ranged from 0.24 to 0.19 to 0.17 in 2017, 2018, and 2019, respectively, suggesting low photosynthetic indicators and consequently, scanty distribution of healthy vegetation in the area. Variability was not significantly different over the years (24–25%). Unlike land cover/land use change, NDVI variation is not unexpectedly high due to the short span of this preliminary study.

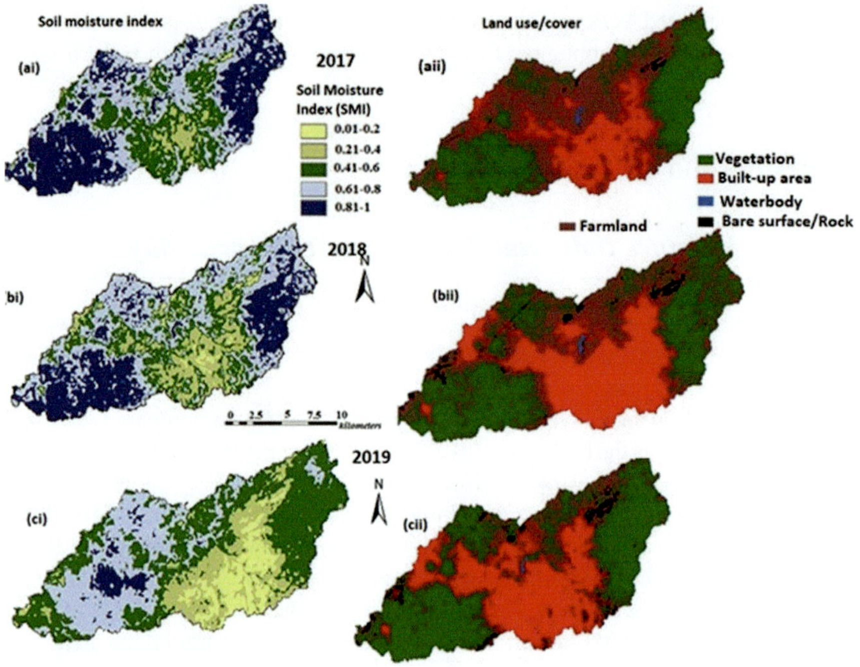

Fig. 2 Annual patterns of soil moisture index and land use/cover over the Opa River basin in southwestern Nigeria

Table 1 Variation in the area (in square kilometers) occupied by a given land cover type

Year	Vegetation	Built-up	Waterbody	Bare surface/ rock	Farmland
2017	96.25	39.95	0.31	1.01	97.38
2018	110.49	62.33	0.35	2.32	59.4
2019	114.47	79.34	0.27	5.95	34.88
Rate of change (annual) Coefficient of determination, R^2	88.9 + 9.1 × 0.904	21.2 + 19.7 × 0.99	0.35–0.02 × 0.25	− 1.82 + 47 × 0.93	126.39–31.3 × 0.99

Furthermore, results from the analysis in LST revealed small variability in the degree of warmth across the study area. Mean LSTs were 28.2 °C, 25.1 °C, and 23.5 °C, with 7.3, 6.3, and 11.2% coefficient of variation in 2017, 2018, and 2019 respectively. Figure 3ai–cii shows that the built-up areas expectedly exhibit warmer temperatures than the vegetated as well as the riparian zones. This was also the same in the study of Chi et al. (2020). The built-up areas have generally been reported to exhibit warmer air and land surface temperature than the vegetated areas as explained

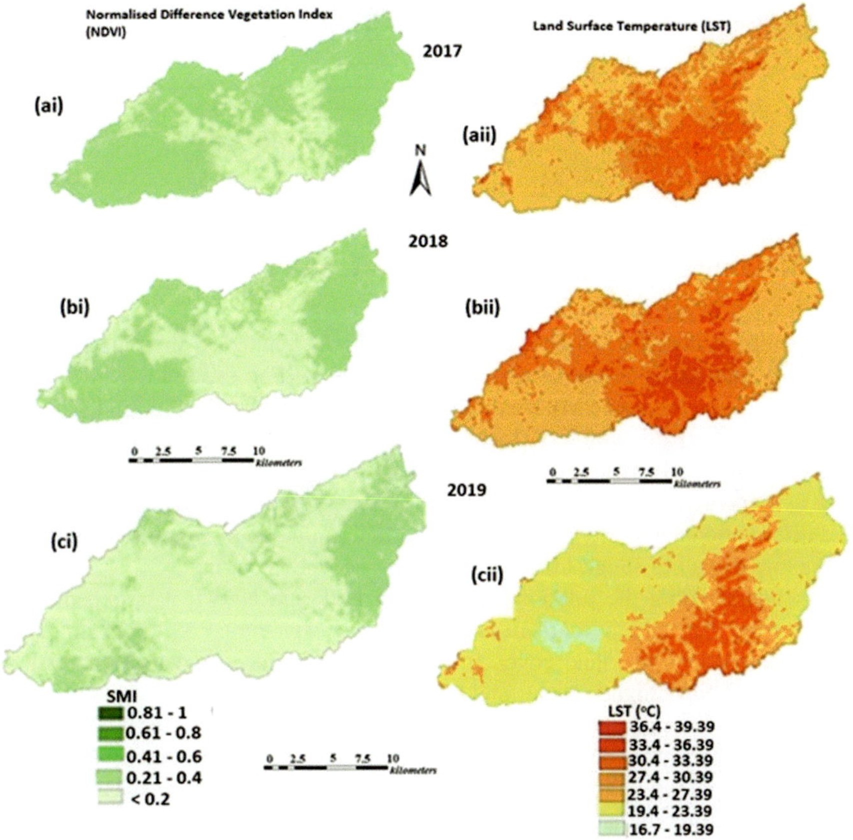

Fig. 3 Spatial and annual variability of SMI and LST

by the concept of urban heat island in urban climate. The river basin has also exhibited such a phenomenon in this study probably because of its poor commitment to river basin protection. In general, the river basin under study is not protected by any policy at either the local or national level and this is typical of river basins in the country. Consequently, many drainage basins have been reported to have increased build-up (Obiahu and Elias 2020). Studies have equally reported that many of the rivers in the disturbed drainage basins may eventually dry up or that the downstream ecosystem may become vulnerable to flooding (Sobowale and Oyedepo 2013; Samela et al. 2016).

Table 2 Relationships among the variables

	SMI	LST	NDVI
	NDVI		LST
Intercept (a)	0.089	1.19	26.5
Slope (b)	0.16	− 0.02	− 3.3
R^2	0.12	0.15	0.006

10 Relationship Between SMI, NDVI, and LST

The results of the simple linear regression that was applied to the data (Table 2) revealed ambiguous relationships among variables. For example, while NDVI appeared to increase with SMI, it exhibited an inverse relationship with LST. The LST also varied negatively with NDVI while the exact link may not be established due to the relatively small amount of data required for more advanced inferential statistics, such as stepwise regression (Yue et al. 2021). The present study has provided baseline information about the low coefficient of determination (R^2) in the relationship, as the low R^2 values suggest the need for more dependent variables.

In understanding the relationship between LST and NDVI, it is expedient to consider the thermal environment and green space signature, i.e., the mean values over different land use land cover. In Fig. 4a, bare surfaces and rocks as well as built-up areas have higher means and lower means for waterbodies, vegetation, and farmland while farmlands and vegetation possess higher means for the NDVI values. Apart from vegetation, the spatial pattern of NDVI is also influenced by solar radiation, slope, and topography. A dense vegetal cover positively influences the rate of evapotranspiration and stimulates latent heat flux over sensible heat. It is difficult to describe the relationship that exists between LST and NDVI because NDVI is influenced by some other factors like the visible and near-infrared reflectance from the plant canopy, reflectance of the same spectra from the soil as well as the atmospheric reflectance which are prone to the impact of observational errors (Marzban et al. 2018). Figure 4b models the interaction among NDVI, SMI, and LST; the result suggests a nonlinear relationship that stresses the importance of SMI in hydroclimatic processes.

11 Limitation of the Study, Conclusion, and Recommendations

This chapter presents ongoing research, and more results will be published with increasing fieldwork details. Results of the preliminary analysis of the characteristics of SMI, NDVI, and LST in the Opa River basin have suggested the nature of the basin and probable impact of human activities, especially when no known protection policy obtains in the area, nor for many basins in the country. In conclusion,

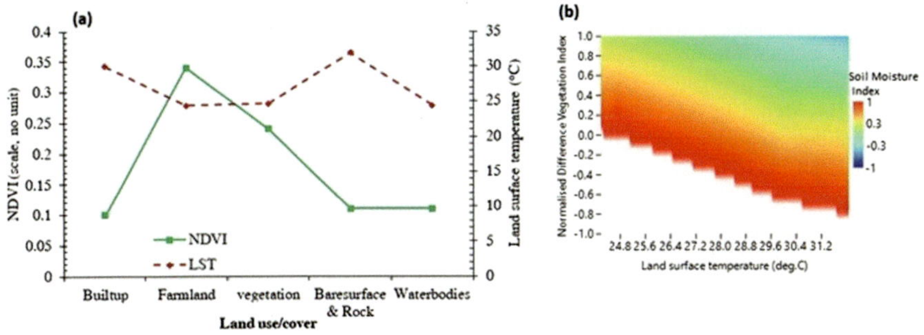

Fig. 4 Relationship between LST and NDVI based on land use land cover types

this study provides baseline information on soil moisture variability in the largely unguaged basin from satellite data. The study recommends more variables beyond the relationship between soil moisture and either NDVI or LST over a longer time scale, in explaining soil moisture variability. Such variables may include longer-term rainfall data, fine-resolution soil moisture sensor probes as well as evaluation of soil particle size distribution for different land uses in the study area. Further investigation may also likely examine the relationship between seasons and time of day on air temperature across different land uses. In all, it recommends awareness of catchment instrumentation for in-depth studies and policy framework for river basin protection in Nigeria.

References

Abd El-Ghany NM, Abd El-Aziz SE, Marei SS (2020) A review: application of remote sensing as a promising strategy for insect pests and diseases management. Environ Sci Pollut Res 27:33503–33515

Adediji A (2005) Reservoir sedimentation: the case of the Opa reservoir catchment, Southwestern Nigeria. South African Geogr J 89:123–128

Adejuwon JO, Jeje LK (1975) Land element of the environmental system of Ife area. In: Ojo A (ed) Environmental resources base project. No. 2, Department of Geography, University of Ife, Ile-Ife

Adewole AO, Ike F, Eludoyin OA (2020) A multi-sensor-based evaluation of the morphometric characteristics of Opa river basin in Southwest Nigeria. Int J Image Data Fusion 11(2):185–200

Adoga PA, Uvirkaa A (2018) Assessment of factors affecting soil moisture for three agricultural landuse management using regression approach. J Earth and Atmospheric Res 1(1):163–184

Akinbiola S, Awotoye O, Adepoju K, Salami A (2016) Floristic indicators of tropical landuse systems: evidence of mining areas in Southwestern Nigeria. Global Ecol Conserv 7:141–147

Akinbuwa O, Adeniyi IF (1996) Seasonal variation, distribution and interrelationships of rotifers in Opa Reservoir, Nigeria. Afri J Ecol 34(4):351–363

Ali MA, Dong L, Dhau J, Khosla A, Kaushik A (2020) Perspective—electrochemical sensors for soil quality assessment. J Electrochem Soc 167(3):037550

Babatimehin OI, Eludoyin AO, Ekundayo OO, Ekwere NJ (2020) Water quality and perception on use in Obafemi Awolowo University, Iile-Ife, Nigeria. Interdisciplinary Environ Rev 20(2):100–117

Bagyaraj M, Mengistie AT, Gnanachandrasamy G, Gemechu B (2019) Data of remote sensing and GIS-to demarcate the potential sector of groundwater in Debre Berhan, Amhara region, Ethiopia. Data Brief 26(104542):1–11

Beyene AA (2019) Soil erosion risk assessment in Nashe Dam Reservoir using remote sensing, GIS, and RUSLE model techniques in Horro Guduru Wollega Zone, Oromia Region, Ethiopia. J Civ, Constr Environ Eng 4(1):1–18

Bhatt CM, Thakur PK, Singh D, Chauhan P, Pandey A, Roy A (2022) Application of active space-borne microwave remote sensing in flood hazard management. Geospatial Technol Land and Water Resour Manage 457–482

Birkel C, Soulsby C, Tetzlaff D (2014) Developing a consistent process-based conceptualization of catchment functioning using measurements of internal state variables. Water Resour Res 50(4):3481–3501

Bogena HR, Montzka C, Huisman JA, Graf A, Schmidt M, Stockinger M, Von Hebel C, Hendricks-Franssen HJ, Van der Kruk J, Tappe W, Lücke A (2018) The TERENO-Rur hydrological observatory: a multiscale multi-compartment research platform for the advancement of hydrological science. Vadose Zone J 17(1):1–22

Breen KH, James SC, White JD, Allen PM, Arnold JG (2020) A hybrid artificial neural network to estimate soil moisture using SWAT+ and SMAP data. Machine Learn Knowledge Extraction 2(3):283–306

Brocca L, Melone F, Moramarco T (2008) On the estimation of antecedent wetness condition in rainfall-runoff modelling. Hydrol Process 22(5):629–642

Brocca L, Ciabatta L, Massari C, Camici S, Tarpanelli A (2017) Soil moisture for hydrological applications: open questions and new opportunities. Water 9(140):1–20

Chi Y, Sun J, Sun Y, Liu S, Fu Z (2020) Multi-temporal characterization of land surface temperature and its relationships with normalized difference vegetation index and soil moisture content in the Yellow River Delta, China. Global Ecol Conserv 23:e01092

Choi J, Won J, Lee O, Kim S (2021) Usefulness of global root zone soil moisture product for streamflow prediction of ungauged basins. Remote Sensing 13(4):756–794

Church CD, Isabelle LM, Pankow JF, Rose DL, Tratnyek PG (1997) Method for determination of methyl tert-butyl ether and its degradation products in water. Environ Sci Technol 31(12):3723–3726

Dewalle DR, Swistock BR, Sharpe WE (1988) Three-component tracer model for stormflow on a small Appalachian forested catchment. J Hydrol 104(1–4):301–310

Dwivedi RS (2018) Active remote sensing. In: Geospatial technologies for land degradation assessment and management. CRC Press, pp 67–96

Easton ZM (2021) Soil and soil water relationships. College of Agriculture and Life Sciences, Virginia Tech, pp 1–9

Eludoyin AO (2013) Modelling the hysteretic patterns of solute concentration-discharge relationships and their significance for hydrological pathways at the farm-scale. University of Exeter (United Kingdom)

Eludoyin AO, Adewole AO (2020) A remote sensing-based evaluation of an ungauged drainage basin in Southwestern Nigeria. Int J River Basin Managem 18(3):307–319

Eludoyin AO, Ojo AT, Ojo TO, Awotoye OO (2017) Effects of artisanal gold mining activities on soil properties in a part of southwestern Nigeria. Cogent Environ Sci 3(1):1305650

Fenta AA, Yasuda H, Shimizu K, Haregeweyn N, Negussie A (2016) Dynamics of soil erosion as influenced by watershed management practices: a case study of the Agula watershed in the semi-arid highlands of northern Ethiopia. Environ Manage 58:889–905

Guo X, Fu Q, Hang Y, Lu H, Gao F, Si J (2020) Spatial variability of soil moisture in relation to land use types and topographic features on hillslopes in the black soil (Mollisols) area of northeast China. Sustainability 12(9)3552:1–17

Hagedorn F, Bundt M (2002) The age of preferential flow paths. Geoderma 108(1–2):119–132
Hartemink AE (2008) Sugarcane for bioethanol: soil and environmental issues. Adv Agron 99:125–182
Hewlett JD, Hibbert AR (1967) Factors affecting the response of small watersheds to precipitation in humid areas. For Hydrol 1:275–290
Hrachowitz M, Savenije HHG, Blöschl G, McDonnell JJ, Sivapalan M, Pomeroy JW, Arheimer B, Blume T, Clark MP, Ehret U, Fenicia F (2013) A decade of predictions in ungauged basins (PUB)—a review. Hydrol Sci J 58(6):1198–1255
Ibitoye MO, Aboyeji OS, Adekemi SOA (2016) Spatio-temporal analysis of landuse dynamics in Upper Opa Catchment, Southwestern Nigeria. Ethiopian J Environ Stud Managem 9(6):780–792
Iqbal M, Sajjad H (2014) Watershed prioritization using morphometric and land use/land cover parameters of Dudhganga Catchment Kashmir Valley India using spatial technology. J Geophys Remote Sens 3:115–127
Jenkins A, Ferrier RC, Harriman R, Ogunkoya YO (1994) A case study in catchment hydrochemistry: conflicting interpretations from hydrological and chemical observations. Hydrol Proces 8(4):335–349
Kala AK, Kumar M (2021) Role of geospatial technologies in natural resource management. Clim Impacts Sustain Nat Resour Manage 19–34
Khare D, Mondal A, Mishra PK, Kundu S, Meena PK (2014) Morphometric analysis for prioritization using remote sensing and GIS techniques in a hilly catchment in the State of Uttarakhand, India. Indian J Sci Technol 7(10):1650–1662
Knighton J, Singh K, Evaristo J (2020) Understanding catchment-scale forest root water uptake strategies across the continental united states through inverse ecohydrological modeling. Geophys Res Lett 47(1):e2019GL085937
Lee ES, Krothe NC (2001) A four-component mixing model for water in a karst terrain in south-central Indiana, USA. Using solute concentration and stable isotopes as tracers. Chem Geol 179(1–4):129–143
Li Y, Luo ML, Li K (2016) A multiwavelet-based time-varying model identification approach for time–frequency analysis of EEG signals. Neurocomput 193:106–114
Li T, Shao MA, Jia Y, Jia X, Huang L (2018) Profile distribution of soil moisture in the gully on the northern Loess Plateau, China. Catena 171:460–468
Manns HR, Berg AA (2014) Importance of soil organic carbon on surface soil water content variability among agricultural fields. J Hydrol 516:297–303
Marzban F, Sodoudi S, Preusker R (2018) The influence of land-cover type on the relationship between NDVI-LST and LST-T air. Int J Remote Sens 39(5):1377–1398
Murray PJ, Griffith BA, Orr RJ, Blackwell MS, Hawkins J, Peukert S (2020) The North Wyke farm platform: a new UK national capability for research into sustainability of agricultural temperate grassland management. In: Proceedings of the 22nd international grassland congress, 1931–1933
Obiahu OH, Elias E (2020) Effect of land use land cover changes on the rate of soil erosion in the Upper Eyiohia river catchment of Afikpo North Area. Nigeria Environ Challenges 1:100002
Ogunkoya OO, Jenkins A (1993) Analysis of storm hydrograph and flow pathways using a three-component hydrograph separation model. J Hydrol 142(1–4):71–88
Oli NP, Ibekwe CC, Nwankwo IU (2018) Prevalence of herdsmen and farmers conflict in Nigeria. Bangladesh e-J Sociol 15(2):171–185
Orewole M, Kappo A, Isa B, Fatile S, Lukman I, Alaga A (2016) Application of geospatial technology in watershed delineation and extraction of hydrologic characteristics in Opa catchment, Southwestern Nigeria. J Geography, Environ Earth Sci Int 8(2):1–9
Orr RJ, Murray PJ, Eyles CJ, Blackwell MSA, Cardenas LM, Collins AL, Dungait JAJ, Goulding KWT, Griffitha BA, Gurr SJ, Harrisa P, Hawkins JMB, Misselbrooka TH, Rawlings C, Shepherd A, Sint H, Takahashi T, Tozer KN, Whitmore AP, Wua L, Lee MRF (2016) The North Wyke Farm Platform: effect of temperate grassland farming systems on soil moisture contents, runoff and associated water quality dynamics. Eur J Soil Sci 67(4):374–385

Rai PK, Mohan K, Kumra VK (2014) Landslide hazard and its mapping using remote sensing and GIS. J Sci Res 58:1–13

Reddy GO (2018) Satellite remote sensing sensors: principles and applications. In: Geospatial technologies in land resources mapping, monitoring and management, pp 21–43

Robinson KL, Bogena HR, Wang Q, Cammeraat E, Bol R (2022) Effects of deforestation on dissolved organic carbon and nitrate in catchment stream water revealed by wavelet analysis. Front Water 4:1003693

Rogers D, Aguilar J, Kisekka I, Barnes P, Lamm F (2014) Soil water plant relationships–an overview. Irrigation Management Series, L904 rev, Kansas State University Research and Extension, p 4

Rondinelli WJ, Hornbuckle BK, Patton JC, Cosh MH, Walker VA, Carr BD, Logsdon SD (2015) Different rates of soil drying after rainfall are observed by the SMOS satellite and the South Fork in situ soil moisture network. J Hydrometeorol 16(2):889–903

Saha A, Patil M, Goyal VC, Rathore DS (2018) Assessment and impact of soil moisture index in agricultural drought estimation using remote sensing and GIS techniques. Proceedings 7(2):1–8

Samela C, Manfreda S, Paola FD, Giugni M, Sole A, Fiorentino M (2016) DEM-based approaches for the delineation of flood-prone areas in an ungauged basin in Africa. J Hydrol Eng 21(2):06015010

Sarmah K, Jha LK, Tiwari BK (2012) Morphometric analysis of a highland micro-watershed in East Khasi Hills District of Meghalaya, India: using remote sensing and geographic information system (GIS) techniques. J Geogr Reg Plann 5(5):142

Shittu KA, Oyedele DJ, Babatunde KM (2017) The effects of moisture content at tillage on soil strength in maize production. Egypt J Basic and Appl Sci 4:139–142

Singh VP (1976) Derivation of time of concentration. J Hydrol 30(1–2):147–165

Smith RE, Warrick AW (2007) Soil water relationships. In: Design and operation of farm irrigation systems, 2nd edn. American Society of Agricultural and Biological Engineers, pp 120–159

Smith AB, Walker JP, Western AW, Young RI, Ellett KM, Pipunic RC, Grayson RB, Siriwardena L, Chiew FH, Richter H (2012) The Murrumbidgee soil moisture monitoring network data set. Water Resour Res 48(7):1–6

Sobowale A, Oyedepo JA (2013) Status of flood vulnerability area in an ungauged basin, Southwest Nigeria. Int J Agricul Biol Eng 6(2):28–36

Sprenger M, Tetzlaff D, Soulsby C (2017) Soil water stable isotopes reveal evaporation dynamics at the soil–plant–atmosphere interface of the critical zone. Hydrol Earth Syst Sci 21(7):3839–3858

Su SL, Singh DN, Baghini MS (2014) A critical review of soil moisture measurement. Measurement 54:92–105

Vaidya N, Kuniyal JC, Chauhan R (2013) Morphometric analysis using geographic information system (GIS) for sustainable development of hydropower projects in the lower Satluj river catchment in Himachal Pradesh, India. Int J Geomat Geosci 3(3):464–473

Vos C, Don A, Prietz R, Heidkamp A, Freibauer A (2016) Field-based soil-texture estimates could replace laboratory analysis. Geoderma 267:215–219

Wang L, Qu J (2009) Satellite remote sensing applications for surface soil moisture monitoring: a review. Earth Sci J 3(2):237–247

Western A, Grayson RB, Bloschl B (2002) Scaling of soil moisture: a hydrologic perspective. Annu Rev Earth Planet Sci 8(30):149–180

Wiekenkamp I, Huisman JA, Bogena HR, Graf A, Lin HS, Drüe C, Vereecken H (2016) Changes in measured spatiotemporal patterns of hydrological response after partial deforestation in a headwater catchment. J Hydrol 542:648–661

Wiekenkamp I, Vereecken H, Huisman JA, Bogena H (2014) Spatial and temporal patterns of preferential flow at the catchment scale. In: TERENO international conference 2014 (No FZJ-2014-05628). Agrosphäre

Williamson RJ, Turner AK (1980) Calibration of a neutron moisture meter for catchment hydrology. Soil Res 18(1):1–11

Wilson DJ, Western AW, Grayson RB, Berg AA, Lear MS, Rodell M, Famiglietti JS, Woods RA, McMahon TA (2003) Spatial distribution of soil moisture over 6 and 30 cm depth, Mahurangi river catchment, New Zealand. J Hydrol 276(1–4):254–274

Yang F, Kumar A, Lau KM (2004) Potential predictability of U.S. summer climate with "Perfect" soil moisture. J Hydrometeorol 5:883–895

Yang M, Wang G, Lazin R, Shen X, Anagnostou E (2021) Impact of planting time soil moisture on cereal crop yield in the Upper Blue Nile Basin: a novel insight towards agricultural water management. Agric Water Manage 243(106430):1–8

Young R, Walker J, Yeoh N, Smith A, Ellett K, Merlin O, Western A (2008) Soil moisture and meteorological observations from the Murrumbidgee catchment. Department of Civil and Environmental Engineering, The University of Melbourne, p 610

Yue W, Xu J, Tan W, Xu L (2021) The relationship between land surface temperature and NDVI with remote sensing: application to Shanghai Landsat 7 ETM+ data. Int J Remote Sens 28(15):3205–3226

Zhang K, Wang Q, Chao L, Ye J, Li Z, Yu Z, Yang T, Ju Q (2019) Ground observation-based analysis of soil moisture spatiotemporal variability across a humid to semi-humid transitional zone in China. J Hydrol 574:903–914

Zhang P, Xiao P, Yao W, Liu G, Sun W (2020) Profile distribution of soil moisture response to precipitation on the Pisha sandstone hillslopes of China. Sci Rep 10(1)9136:1–10

Evaluation of Meandering Characteristics of the Jhelum River Using Geo- Spatial Techniques

Wajahat Annayat, Mohammad Aasif Khaja, and Mohamed Abioui

Abstract River Jhelum is the westernmost river of the five rivers of Punjab region. River Jhelum ends up meeting with Chenab River after flowing for nearly 725 kms. River Jhelum originates from Verinag spring, district Anantnag and flows through Kashmir valley to Punjab Province of Pakistan. River Jhelum after its origin flows through Srinagar and into Wular Lake. This stretch is primary zone of interest of the present work which is subjected to meandering. Due the changing nature of this river, it creates a lot of problems to the people who are residing nearby this river, so an attempt was made to find out meandering parameters of River Jhelum using Digital Elevation Models (DEM) of a study area and Geographical information software (GIS). Reach of River Jhelum which flows through Srinagar district possesses extremely complex meandering parameters. River bends between two points one in Reshipora, Awantipora (latitude 33°44′26.88"N longitude 75° 8′0.25"E) and other in Gund Prang, Near Wular Lake (latitude 34°20′20.68"N longitude 74°38′0.14"E) has been analyzed for meandering morphology possessed by the reach of river at present. Thirty nine (39) bends have been chosen over the reach with width varying as low as sixty six (66) meters to above one hundred and fifty (150) meters at some sections. Sinuosity of the river varies from 1.05 to 1.83 over different reaches of the river and ratio of R_C/w (radius of curvature to width) varies from 1.05 to 8.83 over different reaches of the river. We believe that the outcome of this study will help in understanding the future meandering pattern of Jehlem River.

Keywords Meandering River · Landsat images · GIS · Jhelum River

W. Annayat (✉) · M. A. Khaja
Department of Civil Engineering, National Institute of Technology, Srinagar, Jammu and Kashmir, India
e-mail: Wajahatannayat123@gmail.com

M. A. Khaja
e-mail: aasif_12phd18@nitsri.net

M. Abioui
Geosciences, Environment and Geomatics Laboratory (GEG), Department of Earth Sciences, Faculty of Sciences, Ibnou Zohr University, Agadir, Morocco
e-mail: m.abioui@uiz.ac.ma

S. C. Pal and U. Chatterjee (eds.), *Surface, Sub-Surface Hydrology and Management*, Springer Geography, https://doi.org/10.1007/978-3-031-62376-9_7

1 Introduction

From the earlier times rivers have played significant role in establishing human settlements Annayat et al (2018). Area around rivers used to have highest population density as compared to other locations and those residential still exist in the same way Kunz et al. (2023). Rivers though often called as lifeline of cities can sometimes be called sorrow of states as it does not remain static with time Annayat et al (2021). River behaves dynamic in terms of shape, size, slope, sediment load, sediment size and discharge Annayat et al. (2020a). The change occurring in river does not depend on characteristics of river only but is largely dependent on characteristics of bed materials also Rodrigues et al (2023). In terms of length, river keeps on increasing its length as it meanders by eroding sediments at outer bend. Stream meanders are one of the most prominent features of all the flowing water bodies on earth Annayat et al (2022a, b, 2024). Usually, river flows in series of bends or loops as it passes through plains rather than flowing in a straight path Annayat et al (2021). Meandering is the process by which river changes its course from straight to sinuous form usually in alluvial plains Annayat et al. (2020b); Debnath et al. (2023). The main cause behind the meander in plains is erosion and subsequent deposition of sediments forming the river bed and river bank Annayat et al (2022a, b). Erosion takes place at an outside concave bank whereas sediments are deposited in inner convex bank Sil et al. (2022). From the earlier times meandering has been known to humans as the name "meander" is itself derived from a "Maiandros" who according to Greek mythology was a Greek God known for changing boundaries of countries on its banks and thus often faced law suits. However, due to dynamic nature of meandering rivers they pose several challenges and thus meandering has been the topic of scientific research since the end of nineteenth century. While researchers (e.g. Engels 1926, and Friedkin 1945) provided some landmark insights but the situation changed in the second half of twentieth century with the study of Rozovkii (1957). The design of alluvial channel has been illustrated in experimental studies of common and laboratory channels in an endeavour to get their behaviours Chakravarty (2024). While the early attempts to study meandering parameters was dates back to 1930s, but it was only in 1960s when formulation of empirical equations and interrelations between meandering parameters began to appear with the works of researchers like Leopold and Wolman, 1957 and 1960; Leopold and Langbein (1966). Evaluating meandering process with respect to short time scales began in 1970s with the remarkable works that made a way for whole new dimensions of research (Hooke 1983; Lewin and Brindle 1977; Daniel 1971; Ferguson (1975, 1984). In 1990's models for simulating process were developed (Howard 1992). Lewin and brindle (1977) used simple technique of direct superimposition of photographs taken in different temporal scales and analysed the course for meandering changes with visual inspection. Hooke (1983) identifies five types of approaches that are used for meander analysis. Out of which there are three techniques that use comparison of river reaches of different dates to analyse meandering morphology. It has been shown by both field and lab experiments, meanders in alluvium gradually migrate in the downstream direction (e.g. Tiffany and Nelson

1939; Friedkin 1945; Ackers and Charlton 1970). Hickin (1977) suggests that meanders in unconfined conditions are free to migrate in any direction are more likely to meander more severely even through minute distortions than meanders that are developed in more resistant materials. Meanders are distorted in many different ways yet regularities among the same channel can be observed. Klaassen and Masselink (1992) superimposed sine curve on meanders and five ways in which meander can progress including expansion, rotation and translation and their combinations. There are many theories as to explain the meandering behavior like Excess Energy concept (Inglis 1947; Leopold and Wolman 1960), Earth rotation theory (Gilbert 1884; Neu 1967), Disturbance theory (Friedkin 1945) and Helicoidal motion theory (Onishi et al. 1976). Yet, causes still remain poorly understood and no single theory could explain the reason.

Reach of River Jhelum which flows through Srinagar District possesses extremely complex meandering parameters. The increase in length is due to tendency of fluid particles to erode bed and bank material present at the concave end of bend while depositing the same at subsequent inner convex bends. Due to the reason stated above the concave bend of river goes on increasing and thus the length of river changes. Despite all these evident changes in the river authors found that till date no study has been conducted on the Jhelem River which is based on purely meandering pattern. Therefore, considering the importance of this river and its catchment area our main objective was to find the meandering pattern of the river. In order to achieve this goal we have divided the main objective into sub objectives as follows:

- Calculation of sinuosity
- Calculation of radius of curvature
- Calculation of average width
- Calculation of meander length
- Calculation of meander width.

We believe this study will provide benchmark for meandering parameters and the same will be used for future studies as comparison. Meandering parameters will be compared to previous years for finding changes that has been occurred over a period of time in our study area.

2 Study Area

River Jhelum is the river that originates from Verinag spring, district Anantnag and flows through Kashmir valley to Punjab Province of Pakistan. River Jhelum is the westernmost river of the five rivers of Punjab region. River Jhelum ends up meeting with Chenab River after flowing for nearly 725 kms. River Jhelum after its origin flows through Srinagar and into Wular Lake. However its course with the Kashmir Valley is only 215 kms which after existing ceasefire line is reduced to 165 kms only with catchment area of 17,620 square kilometres and lies 32°-58'-42" to 35°-08'- 02" north latitude and 73°-23'-32" to 75°-35'-57" east longitude. On the upper and lower course of river braided bars are found and the River also possesses meandering pattern for a course of about 106 kms. The Jhelum River is also navigable from Khandanyar to Khanbal (Baramulla), for a distance of about 170 kms. The river throughout its course is joined by number of tributaries. The Jhelum river has 24 main tributaries and its left bank tributaries drain from the slope of the Pir Panjal range while others flow form the Inner Himalayan range and join the river on the right bank. Its right bank tributaries are Bringi, Liddar, Sind Nala, Erin, Madumati and Pohru. While as, its chief left bank tributaries are Vishav, Rambiara, Dundganga and Sukhnag. After Jhelum passes through Anantnag, Pulwama and Srinagar Districts it flows into Wular Lake, which is sometimes regarded its delta in Kashmir and then passing through Baramulla and Uri, it flows through Pakistan. This stretch is primary zone of interest of the present work which is subjected to meandering. Figure 1 shows study area of river reach in Kashmir valley. The river reach was selected because of extreme meandering pattern it shows when viewed through remote sensed images. The river reach consist of series of bends at different locations throughout the valley.

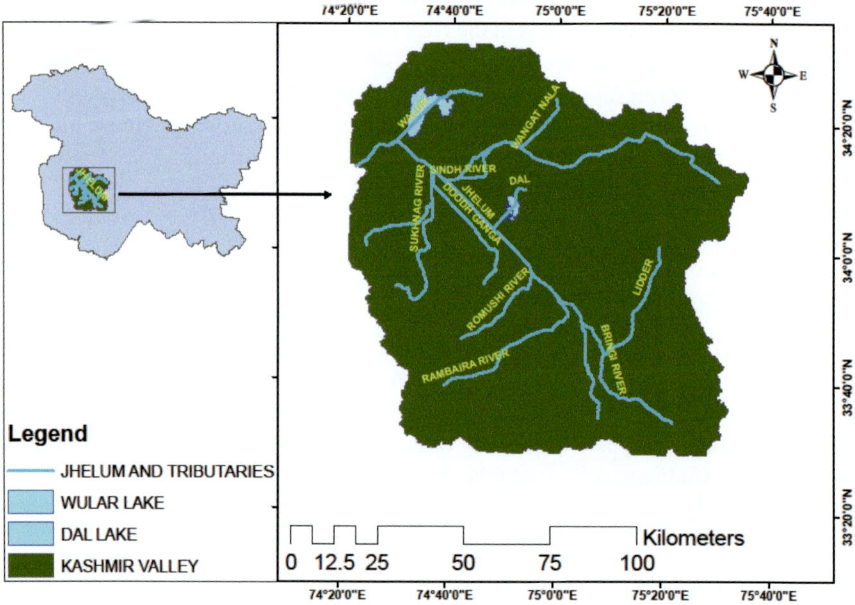

Fig. 1 Map showing the study area

Table 1 Information of data used in the study

Platform	Resolution	Data availability
Cartosat-1:DEM - Version-1	1 arc sec (~32 m)	2006–2008
Cartosat-1:DEM - Version 1.1R1	1 arc sec (~32 m)	2006–2012
Cartosat-1:DEM - Version-2R1	1 arc sec (~32 m)	2005–2014
Cartosat-1:DEM - Version-3R1	1 arc sec (~32 m)	2005–2014
IMS-1:Hyper spectral Imager	Spectral Binned Data (17 bands)	2011–2012
Resourcesat- 1/Resourcesat- 2:AWiFS Ortho	56 m	2009–2016
Resourcesat-1/Resourcesat-2:LISS III Ortho	24 m	2009–2019
Landsat images	30 m	1992 and 2020

3 Data Used

For most of the part, the data was collected from Bhuvan portal that is Indian Geo platform of Indian space research organization (ISRO) of various years such as 1992, 2011, and 2020. Complete information of images for the present study is shown in Table 1. The Shuttle Radar Topography Mission (SRTM) provides the river network of the Jehlem River obtained from digital elevation model (DEM), resolution of 30 m which was downloaded from http://earthexplorer.usgs.gov.

4 Methodology

4.1 Sinuosity

Reach of river Jhelum from Reshipora, Awantipora (latitude 33°44′26.88"N longitude 75° 8′0.25"E) and other in Gund Prang near Wular Lake (latitude 34°20′20.68"N longitude 74°38′0.14"E) measuring about 106 kms was digitalized using Arc GIS version 10.2. After digitalizing the Digital Elevation Models from Resource satellite and Landsat Data, both were transformed to same projection system that is Universal Transverse Mercator (UTM) before the measurements could be made. The UTM coordinate system partitions the world into sixty north–south zones, each 6 degrees of longitude wide. UTM zones are numbered consecutively beginning with Zone 1, which includes the westernmost point of Alaska and progress eastward.

Sinuosity parameter of the river reach was computed by the method discussed by Friend and Sinha (1993) and the Landsat Images of 2020 were used for the analysis. Since the sinuosity of curve is a concept of one line thick curve but in case of rivers we have two lines (banks) running parallel to each other, hence the measurements are to be taken along a line that traverses mid-way between the two channel banks.

The procedure outlined by Friend and Sinha (1993) has made it possible to take measurements readily from topographical maps and aerial photography. Another advantage is that the length of river along a bank may change due to rise in water level. Modified sinuosity parameter given by Friend and Sinha (1993) is given by:

$$P = L_{c\,\max}/Lr \tag{1}$$

L_{cmax} = is the length of channel through mid-line in case of single stream river or length of the widest stream in case of multi Stream River

L_r = is the length of the reach

4.2 Radius of Curvature (R) and Average Width

Before measuring radius of curvature it is most important to locate three points, one apex point and two departure points. Apex point is simply the outermost points of concave or convex banks present in the course of river. For each curve the outer most point where the curve changes its direction is called an apex point. Departure point is the point along the curve where the curvature of curve starts changing towards a more linear trend (Fig. 2). Once these three points are derived from the existing shape file we can arrive at two parameters, one being length of the arc and other being the shortest distance between two departure points. Figure 3 shows different arc parameters namely arc length, chord height, chord length radius of curvature and arc angle. All the parameters are corresponding to each other. If any two parameters are known, all of the rest can be derived.

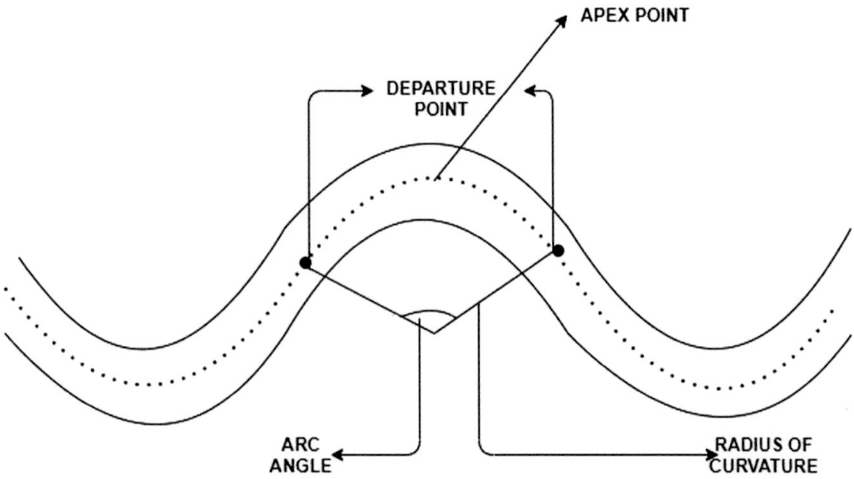

Fig. 2 Location of important points for computing radius of curvature

Fig. 3 Arc parameters

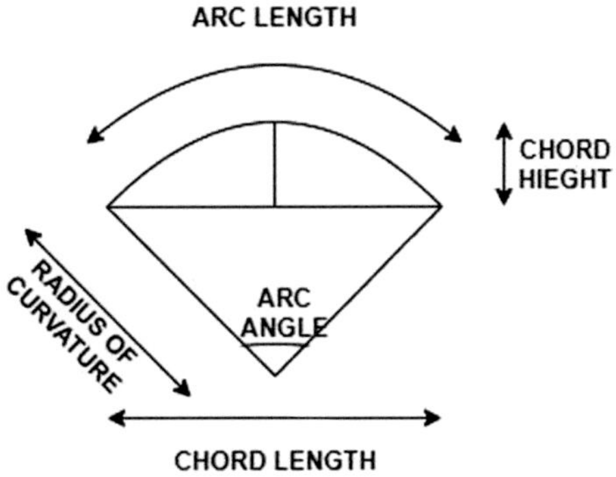

4.3 Average River Width

River width (w) is a basic parameter used in a wide variety of hydraulic applications. It is the measured as the shortest distance between two banks of water surface. In spite of the fact that it is theoretically conceivable to calculate such width values physically, it becomes illogical if hundreds or thousands of stream width values must be obtained. As a result, use of remote sensed images and their measurement may provide valuable information for many hydrologic processes.

Another problem in measurement of river width for a particular reach is the unsymmetrical nature of river, especially in case of river width. Width of river reach keeps on changing as river flows from upstream to downstream. Width of river can increase or decrease even within small reaches. To counter this problem there is a concept of average width used in hydrologic processes. In this study, the average width of a river reach is computed by first finding out water surface area of that reach and then dividing the obtained value by length of flow.

$$\text{Mathematically } W_{\text{avg}} = A/L \tag{2}$$

where,

W_{avg} = Average width of river reach
A = area of water surface within the reach
L = length of flow

4.4 Computation of Stream Meander Length and Meander Width

Computing stream meander length is fairly easy step as the points of inflection have already been derived. In this step, the distance along the river between two alternative inflection points is to be measured. To find out stream meander length of the bend,

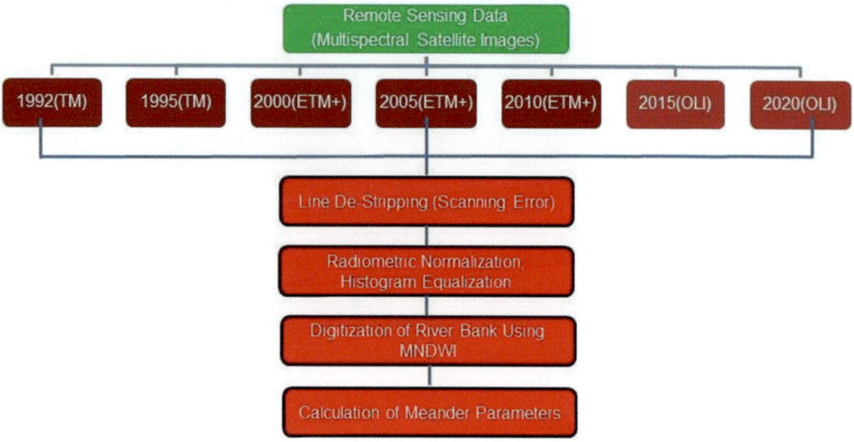

Fig. 4 Flowchart of methodology

distance along the river between two neighboring inflection points, one in upstream and one in downstream are to be taken. Meander width also known as belt width is the lateral distance (perpendicular to centerline) between inflection points of two meanders that occur in opposite direction. For any particular bend, meander width is the distance between inflection point of that bend and inflection point of next opposite bend in the downstream direction. After computing meander length and belt width, one more ratio that is meander ratio can be calculated by dividing belt width with meander length. A methodological flowchart is shown in Fig. 4.

5 Results

5.1 Sinuosity

From Table 2, throughout first segment, the river flows in a single stream and can be regarded as straight because of low sinuosity index (1.2580). As the river enters in second segment, it tends to increase its sinuosity and further in the same segment braid bars appear near Awantipora and Awantiswami Temple. Sinuosity index remained almost same in segment 3 but the value is still appreciable with the index value of 1.3736. Segment 4 and segment 5 possess highest sinuosity throughout the entire river length (Fig. 5). Out of 13 km of segment 5 most of the part flows in Srinagar District, here also river flows in a single stream. After Noorbagh, Srinagar the river sinuosity goes on decreasing with the increase in braid channel ratio. Minimum value of sinuosity was observed at segment 8 with a value of 1.0548. Average value of the sinuosity index is 1.4 and it is quite evident from the numbers that in our study area river Jhelum assumes more meandering pattern. We compared the values of

Table 2 Sinuosity parameter of the Jehlem River for year 2020

Segment	Main channel (L)	Overall length (l)	Sinuosity (P)
1	8736.709	6944.92	1.258
2	9012.698	7183.45	1.3547
3	10,765.33	7837.31	1.3736
4	13,177.05	7464.06	1.7654
5	13,735.02	7482.58	1.8356
6	10,297.59	7318.83	1.407
7	9291.958	6981.71	1.3309
8	7359.561	6977.21	1.0548
9	8056.325	6954.1	1.1585
10	13,030.29	10,125.33	1.2869

sinuosity with other published works such as Annayat et al. (2020a, 2020b) (S.R ~ 1.18–4.25), Bag et al. (2019) (S.R ~ 1.70–2.14). It was also observed that channel reaches developing braided arms are having lower sinuosity values.

From Fig. 6 it is quite clear that the sinuosity of the segment 5 was higher than any other segments and varied from 1.75354 to 1.8356. On the other hand, sinuosity of segment 8, was found lower than others segments and varied from 1.025407to 1.0548. At segment 2 the sinuosity value was decreased over a period. Segment 4 that is in between Lasjan and Rajbagh experienced the highest change in sinuosity throughout the study period. Channel pattern of Jhelum River remained the same despite change in sinuosity parameters. At 1992 only two segments out of five were having sinuosity ($P > 1.5$). The same pattern continued in 2011 and 2020 despite increase in sinuosity parameters at most of segments. Sinuosity values of the year 2020 were compared with the values of year 1992 and 2011 in order to trace changes in sinuosity that have occurred in the different segments of river reach. On comparing the values it was found that the pattern was carried forward all the years. By the looks of the aerial photography and satellite data it is almost evident that the rate of meandering is quite slow and almost uniform. There is no doubt that the river Jhelum has achieved meandering in a young stage as compared to older rivers of world but the process is largely confined now. There may be many reasons for this unexpected check on migration of banks but two of the major reasons that are evident from the satellite data are the extreme urbanization/construction especially over the bends and development of braided bars especially in the region where the river reaches are wider. Most of the river reaches show increase in sinuosity over the period of 28 years but at the segment 2 (Kadlabal to Manwaji) the reverse trend was observed. Presence of geological faults and tectonic activities also add to the abnormalities in sinuosity patterns.

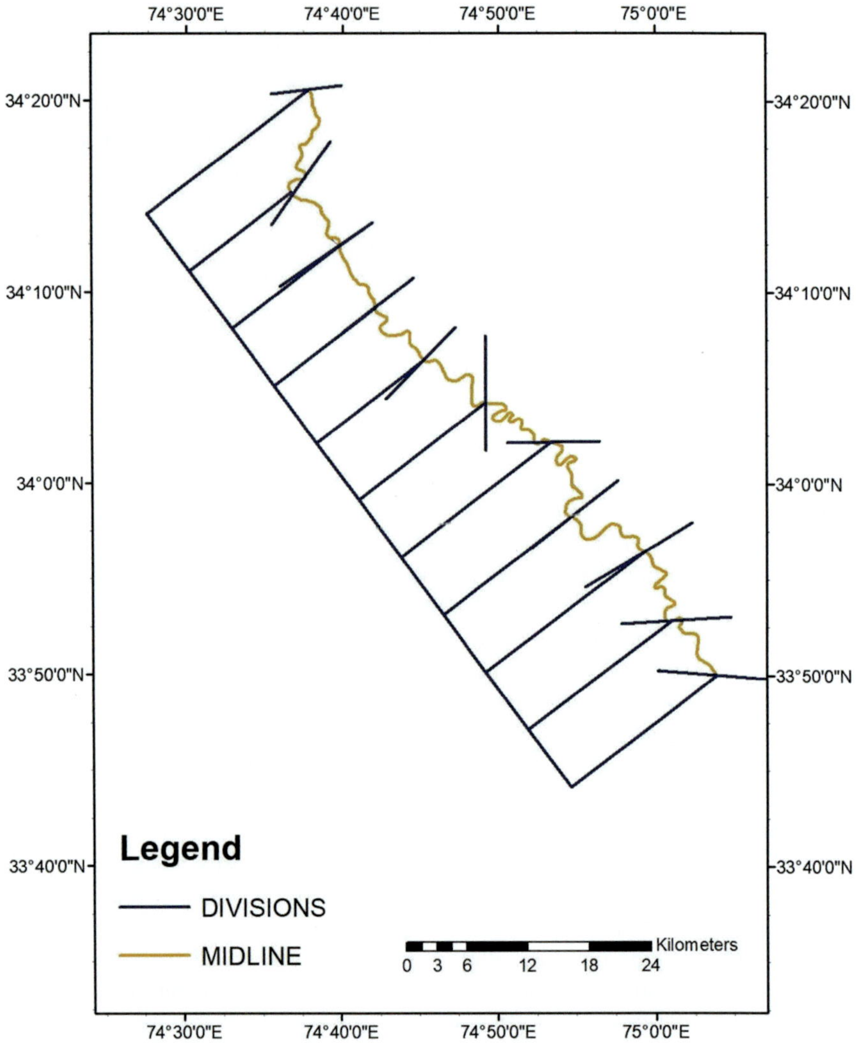

Fig. 5 River Jhelum divided into ten segments for calculation of sinuosity

5.2 *Radius of Curvature and Width of River*

From Table 3, it was observed that the radius of curvature is not constant in the reach of Jhelum River, but ranges from a maximum value of 923.81 m (extremely smooth) to a minimum curvature of 123.2 m around the bend apex (Fig. 7). The average radius of curvature throughout the river reach was found out to be 470 m with a standard deviation of 232 m. Frequency distribution table is centered on the radius of 200 m to 300 m (Fig. 8). These values agree well with the distribution given

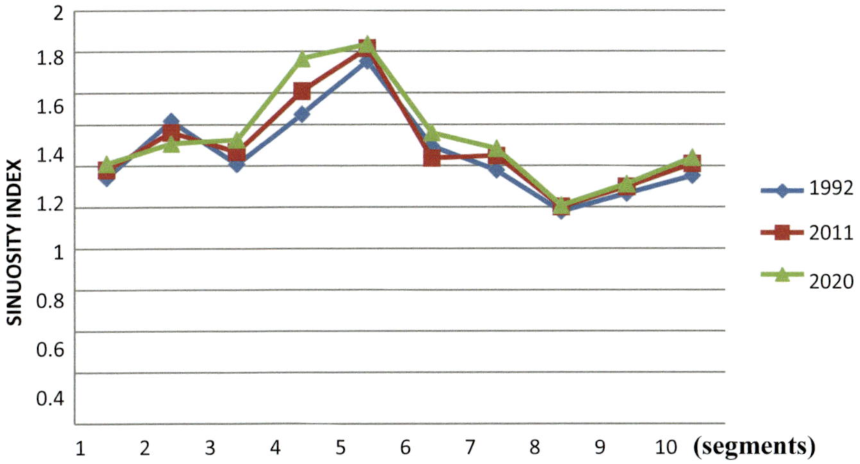

Fig. 6 Variation in sinuosity parameter with time

by the Langbein and Leopold (1966) for 79 naturally occurring R_c/W values. Width profile of Jhelum River shows an interesting pattern. In most of the bends decrease in width is immediately followed by the increase in average width throughout the next bend (Fig. 7). The smallest measured average width was found out to be 66.37 m near Budshah Tomb and the highest measured average width was found out to be 150.30 m at Kharbagh (Fig. 9). Studies suggest that most loops of meandering rivers have radius of curvature to width ratios in between 1.5 and 4.5. Of 39 sites used to evaluate the radius of curvature to width ratio 65 per cent radius of curvature to width ratios lies in the range of 1.5 to 4.5 Langbein and Leopold (1966)**;** Nanson and Hickin (1986)**.** It is interesting to find out that only one data point lies below 1.5 (Figs. 7, 8 and 9).

5.3 Meander Ratio

From Table 4 it was observed that the stream meander length and the belt width varied tremendously in the reach of Jhelum River. Meander stream length ranges from a maximum value of 3120.37 m at Kresbal to minimum of 1199.67 m at Awantipora. The average meander stream length throughout the river reach was found out to 2012.82 m. Belt width ranges from the maximum value of 2261.34 m at Markundal to a minimum value of 190.2276 at Noorbagh. The average value of belt width was found out to be 1053.36 m. We compared our results with the already published works such as Mithun et al. (2012) in Manu River Bangladesh and Bhagirathi River in West Bengal by Islam and Guchhait (2017) and found the almost similar range.

Table 3 Radius of curvature and width computation over Jhelum River (2020)

S. No.	Location name	Arc length (m)	Chord length (m)	Radius of curvature (R_C) (m)	Water surface area (m^2)	Average width of section (w)	R_C/w
1	Reshi pora	1293	1049	589.54	121,319.6	93.82803	6.283197
2	Cheerso	712	542.13	286.23	74,182.03	104.1882	2.747239
3	Beighpora	1342	1144.38	697.50	130,107.9	96.95071	7.194377
4	Kadlabal	556	475.07	290.76	51,048.99	91.81473	3.166812
5	Awantipora	931	785.39	468.82	96,908.67	104.0909	4.503946
6	Padgampora	339	241.74	123.20	39,935.81	117.8048	1.045798
7	Awantiswami	518	450.51	287.04	58,049.88	112.0654	2.561361
8	Manwaji	836	754.78	539.31	95,722.01	114.5	4.710131
9	Owdipora	653	559.85	345.11	61,688.67	94.46964	3.653131
10	Banderpora	1003	931.70	759.56	99,648.81	99.35076	7.645236
11	Lethpora	1242	1119.39	794.58	153,134.9	123.297	6.44444
12	Kakapora	1403	1246.46	842.65	184,465.4	131.4793	6.408994
13	Marhwal	1447	1246.68	776.81	157,607.4	108.9201	7.131924
14	Kandizal	645	506.47	274.43	71,838.47	111.3775	2.463963
15	Tanchabagh	734	507.16	255.98	75,208.23	102.4635	2.498255
16	Bagh Shokar	779	608.94	328.56	71,121.03	91.29785	3.59877
17	Sempora	695	590.51	357.31	62,006.14	89.21747	4.004933
18	Pantha chowk	810	749.47	597.98	95,196.16	117.5261	5.08806
19	Golpora	518	445.60	276.71	48,779.3	94.16853	2.938455
20	Pandrethan	936	853.83	636.16	83,565.03	89.27888	7.125537
21	KIE School	511	445.95	286.61	48,095.75	94.12084	3.045128
22	Pani Mandir	757	628.11	364.52	67,366.96	88.99202	4.096098
23	Soiteng	514	348.69	175.30	44,145.94	85.88704	2.041053
24	Batwara	489	408.04	238.98	47,963.06	98.08396	2.436484
25	KP Bagh	830	692.64	405.70	87,049.82	104.8793	3.868256
26	CNI Church	671	564.17	334.76	63,885.72	95.20972	3.516028
27	Lal Mandi	715	577.65	322.92	56,681.24	79.27447	4.073443
28	Budshah Tomb	1198	954.52	525.04	79,510.35	66.36924	7.910893
29	Gagarzoo	1671	1456.35	932.81	176,694.5	105.7418	8.821585
30	Noorbagh	1117	1009.97	725.78	140,186.9	125.5031	5.782966
31	Parimpora	1054	915.26	580.91	111,058.9	105.369	5.513104
32	Kresbal	773	665.15	413.28	82,492	106.7167	3.872684
33	Mujgund	498	419.75	250.16	59,233.07	118.9419	2.103211
34	Takanwanipora	676	580.05	358.20	87,326.81	129.1817	2.772839
35	Kharbagh	1044	937.05	655.36	156,915.9	150.3026	4.360272

(continued)

Table 3 (continued)

S. No.	Location name	Arc length (m)	Chord length (m)	Radius of curvature (R_C) (m)	Water surface area (m^2)	Average width of section (w)	R_C/w
36	Sumbal	864	762.29	504.71	103,090.2	119.3174	4.229979
37	Markundal	1064	910.96	559.87	150,439.8	141.3908	3.959734
38	Jhelum colony	861	753.29	487.31	114,658.4	133.1689	3.659338
39	Hajin	680	569.28	335.24	79,336.02	116.6706	2.873389

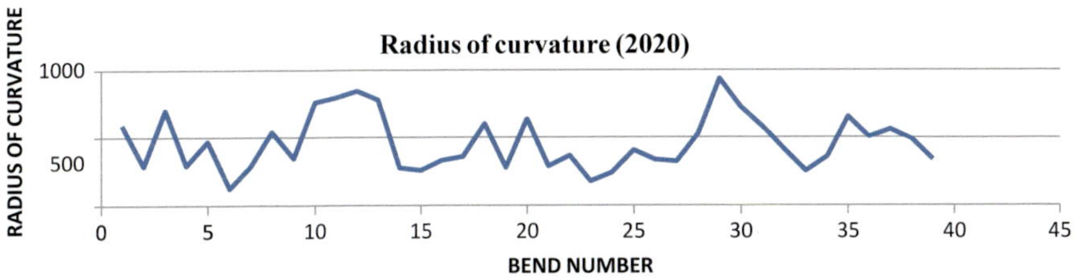

Fig. 7 Graph showing variation in Radius of curvature for the year 2020

Fig. 8 Frequency distribution graph showing frequency of radius of curvature over river reach

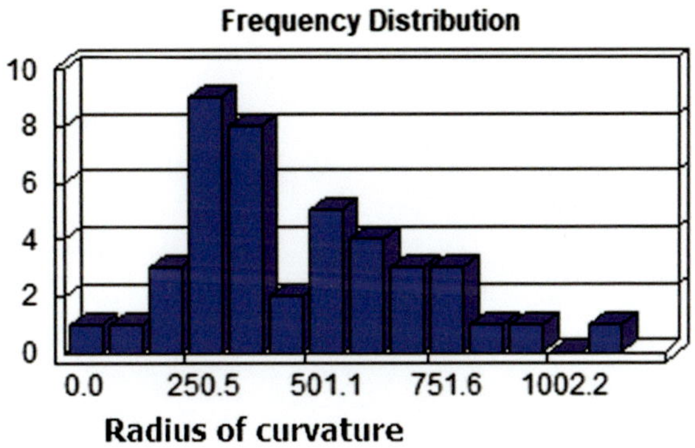

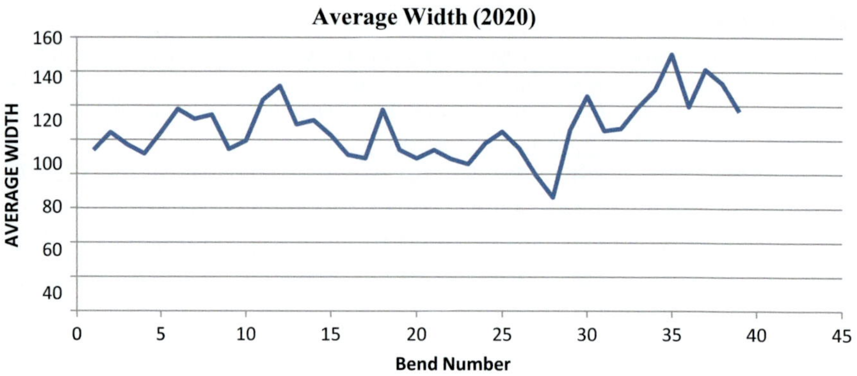

Fig. 9 Average widths of River reach derived from 39 sites

Table 4 Computation of meander ratio for river reaches Jhelum (2020)

Bend number	Meander length	Belt width	Meander ratio
Bend 1	1892.93	756.857	0.399834
Bend 2	2013.206	1204.467	0.598283
Bend 3	2504.577	1676.965	0.66956
Bend 4	1915.55	217.4931	0.113541
Bend 5	1199.695	670.7019	0.55906
Bend 6	1541.586	1733.189	1.12429
Bend 7	2344.446	439.514	0.18747
Bend 8	1353.658	575.895	0.425436
Bend 9	1472.383	946.604	0.642906
Bend 10	1995.632	258.2359	0.129401
Bend 11	2233.197	2012.27	0.901071
Bend 12	2346.714	667.6599	0.284508
Bend 13	2708.394	2170.064	0.801236
Bend 14	1940.955	1017.142	0.524042
Bend 15	2345.648	1756.432	0.748805
Bend 16	1835.831	1439.829	0.784293
Bend 17	1836.772	607.7034	0.330854
Bend 18	1209.345	1013.83	0.83833
Bend 19	1488.859	429.64	0.28857
Bend 20	1483.028	568.5372	0.383362
Bend 21	1219.699	542.5834	0.44485
Bend 22	1434.898	1000.265	0.697098
Bend 23	1448.841	712.007	0.491432
Bend 24	2060.897	1819.468	0.882852
Bend 25	2838.191	1773.858	0.624996
Bend 26	2337.501	1194.471	0.511003
Bend 27	2090.509	1700.601	0.813487
Bend 28	2155.425	1932.154	0.896414
Bend 29	3020.722	389.917	0.129081
Bend 30	2299.39	1071.235	0.465878
Bend 31	1824.675	190.2276	0.104253
Bend 32	3120.377	1994.767	0.639271
Bend 33	2899.705	1220.543	0.42092
Bend 34	2127.256	490.4636	0.230562
Bend 35	1514.003	502.3002	0.33177
Bend 36	1674.021	260.6188	0.155684
Bend 37	2784.332	2261.34	0.812166
Bend 38	2381.303	1171.709	0.492045
Bend 39	1606.088	689.7305	0.429448

6 Conclusion

From the study it was observed that the river Jhelum within the study reach exhibits irregular meandering geometry. The river stream was found to be less sinuous in the upstream followed by more sinuous reach. Furthermore pattern formed by the bends is neither identical nor symmetrical. Each bend possess different meandering parameters. In general over a long time period the sinuosity values showed a slight increase but no major change in pattern was observed throughout the study period. There was absolutely no formation of ox-box lakes during the study. For most of the part River Jhelum can be considered straight except for those parts where sinuosity values are greater than 1.5. River Jhelum throughout its span has meandered in a very unique way. While most of the rivers of world tend to meander at section with larger width, the condition reverses in case of River Jhelum. Srinagar reach of River Jhelum despite having sections of narrow width had shown enormous sinuosity as compared to wider sections of the reach. The frequency distribution of 39 naturally occurring R_c/W values has a geometric mean of 3.9 with nearly three-fifth of the values less than 4.5.

References

Ackers P, Charlton FG (1970) Meander geometry arising from varying flows. J Hydrol 11:230–252

Annayat W, Sil BS (2020b) Changes in morphometric meander parameters and prediction of meander channel migration for the alluvial part of the Barak River. J Geol Soc India 96(3):279–291

Annayat W, Sil BS, Ajay G (2018) Development of rainfall-runoff model using FFBPNN and LRNN for Silchar city-a case study. Dis Adv 11:19–23

Annayat W, Ashwini K, Sil BS (2022a) Channel migration detection and landuse/landcover changes for Barak River in Northeast India. J Geol Soc India 98:981–990

Annayat W, Sil BS (2020a) Assessing channel morphology and prediction of centerline channel migration of the Barak River using geospatial techniques. Bulletin of Eng Geol the Environ 1–23

Annayat W, Sil BS (2021) Predicting meander migration of the Barak river by empirical and time sequence methods. In: Water security and sustainability: proceedings of down to earth 2019. Springer, Singapore, pp 113–129

Annayat W, Gupta A, Prakash KR, Sil BS (2021) Application of artificial neural networks and multiple linear regression for rainfall–runoff modeling. In: Communication software and networks: proceedings of India. Springer, Singapore, pp 719–727

Annayat W, Ashwini K, Sil BS (2022) Monitoring land use and land cover analysis of the barak basin using geospatial techniques. In: Anthropogeomorphology. Springer, Cham, Switzerland, pp 427–441

Annayat, W., Ashwini, K., & Sil, B. S. (2024). Relative prioritizing of sub-watersheds of Barak River built on, morphometric parameters, LULC and SCS-CN model. *International Journal of Energy and Water Resources*, 1–16. J Geogr Sci 29(4):613–634

Chakravarty S, Patel HK, Mohanty B, Kumar B (2024) Review on different shapes of spurs and their effects on channel morphology. Water Practice and Technol 19(1):241–262

Daniel JF (1971) Channel movement of meandering Indiana streams. US Geological Survey Professional Paper 732-A

Jatan D, Sahariah D, Saikia A, Meraj G, Nath N, Lahon D, Annayat W, Kumar P, Chand K, Singh SK, et al. (2023) Shifting sands: assessing bankline shift using an automated approach in the Jia Bharali River, India. Land 12(3):703
Engels H (1926) Movement of sedimentary materials in river bends. In: Freeman JR (ed) Hydraulic laboratory practice. New York, ASME
Ferguson RI (1975) Meander irregularity and wavelength estimation. J Hydrol 26:315–333
Fergusun RI (1984) The threshold between Meandering and braiding. In: Smith KVH (ed) Proceedings of the 1st international conference on water resources engineering. Springer, Berlin, pp 6.15–6.29
Friedkin JF (1945) A laboratory study of the meandering of alluvial rivers. Vicksburg Mississippi, US Waterways Experiment Station
Friend and Sinha (1993) Braiding and meandering parameters. Geological Society London Special Publications, January. https://doi.org/10.1144/GSL.SP.1993.075.01.05
Gilbert GK (1884) ART. XLIX.--The sufficiency of terrestrial rotation for the deflection of streams. Amer J Sci (1880–1910) 27(162):427
Howard AD (1992) Modelling channel migration and floodplain sedimentation in meandering streams. In: PA Carling, GE Petts (eds) Lowland floodplain rivers: geomorphological perspectives. Wiley: Chichester, pp 1– 41
IslamA GSK (2017) Analysing the influence of Farakka BarrageProject on channel dynamics and meander geometry of Bhagirathi river of West Bengal, India. Arab J Geosci 10(11):245
Klaassen GJ, Masselink G (1993) Planformi (changes of a ibraide) rivier with fine sani) as beid) and bank material
Kunz A, Schneider F, Anthony N, Lin HT (2023) Microplastics in rivers along an urban-rural gradient in an urban agglomeration: correlation with land use, potential sources and pathways. Environ Pollut 321:121096
Langbein WB, Leopold LB (1966) River meanders—theory of minimum variance. U.S. Geol Survey, Prof. Page. 422-H, pp 15
Langbein WB (1964) Geometry of river channels. J Hydraulics of the Amer Soc Civil Eng 90(HY2):301–312
Leopold LB, Wolman MG (1960) River meanders. Bull Geol Soc Am 71:769–794
Lewin J, Brindle BJ (1977) Confined meanders. In: River channel changes, pp 221–233
Mithun D, Dabojani D, Misbah U (2012) Evaluation ofmeandering characteristics using RS & GIS of Manu River. J Water Resour Prot
Nanson GC, Hickin EJ (1986) A statistical analysis of bank erosion and channelmigration in western Canada. Geol Soc Am Bull 97(4):497–504
Neu HA (1967) Transverse flow in a river due to earth's rotation. J Hydraul Div 93(5):149–166
Onishi Y, Kennedy JF, Jain SC (1976) Effects of meandering in alluvial streams. J Hydraul Div 102(7):899–917
Rodrigues JM, de Souza JOP, Xavier RA, Santos CAG, da Silva RM (2023) Geomorphic changes in river styles in a typical catchment of the Brazilian semiarid region. CATENA 232:107423
Rozovskii IL (1957) Flow of water in bends of open channels. Academy of Sciences of the Ukrainian SSR, Kiev, Israel Program for Scientific Translations, Jerusalem, 1961
Sil BS, Ashwini K, Annayat W, Debnath J, Farooq M, Meraj G (2023) Observing spatiotemporal inconsistency of erosion and accretion in the Barak river using remote sensing and GIS techniques. Conservation 3(1):14–31
Tiffany JB Jr, Nelson GA (1939) Studies of meandering of model-streams. EOS Trans Am Geophys Union 20(4):644–649
Wajahat A, Sundar SB (2018) Estimation and analysis of possible flood for the Silchar city-a case study. Disaster Adv v.11(1)

Analyzing the Relationship Between Surface Water and Groundwater: Evaluating Integration Techniques in Elaborate Domain Models

Anuj Kumar Dwivedi, Umesh Kumar Singh, and C. S. P. Ojha

Abstract The intricate relationship between groundwater and surface water has garnered significant attention due to the formidable challenges in model validation and calibration. These challenges mainly stem from the dynamic exchange at the soil–water interface, which becomes more intricate at smaller scales, posing significant obstacles for researchers. However, as investigations extend to larger scales, the complexity of these interactions intensifies, reaching unprecedented levels. To gain a deeper understanding of this critical coupling, it is essential to develop robust numerical models capable of accurate simulations. This comprehensive review explores two prominent coupling strategies within detailed domain models: the "fully-coupled" and "loosely-coupled" techniques, which are cornerstones of contemporary groundwater modeling efforts. Through careful analysis, we highlight the unique advantages and inherent limitations of each modeling approach. This review provides invaluable guidance for both researchers and practitioners alike, offering strategic insights for the development of groundwater flow models. It serves as a comprehensive resource, aiding in the formulation of effective strategies and considerations crucial for model crafting. By synthesizing key findings and methodologies, it empowers professionals to make informed decisions in their modeling endeavors. The ultimate goal is to develop models that accurately capture the complexities of real-world conditions governing surface and groundwater interactions, particularly at regional scales. By exploring the intricate landscape of coupling strategies, we pave the way for improved insights and more precise simulations in the field of groundwater dynamics.

A. K. Dwivedi (✉)
Surface Water Hydrology Division, National Institute of Hydrology, Roorkee, Uttarakhand 247667, India
e-mail: akdwivedi2009@gmail.com

U. K. Singh
Water Resources System Division, Nation Institute of Hydrology, Roorkee, Uttarakhand 247667, India

C. S. P. Ojha
Department of Civil Engineering, Indian Institute of Technology, Roorkee, Uttarakhand 247667, India

S. C. Pal and U. Chatterjee (eds.), *Surface, Sub-Surface Hydrology and Management*, Springer Geography, https://doi.org/10.1007/978-3-031-62376-9_8

Keywords Subsurface water · Integrated model · Unified approach · Comprehensive coupling · Semi-independent coupling · Reciprocal effect

1 Introduction

Water security has emerged as a pressing global concern, exacerbated by climate change, variability, and the burgeoning demands of a growing population and expanding economies (Vaux et al. 2011). Both surface water and groundwater are essential resources that are crucial for meeting future water demands. Groundwater, which is extensively utilized across national economies, performs a multitude of functions, including providing potable and industrial water, irrigation, and the production of mineral water. In particular, fresh groundwater serves as a cornerstone for domestic and drinking water systems. Currently, it acts as the primary source of potable water in many European nations and the United States, accounting for 75% of municipal water supplies (Zektser et al. 2004). The realm of irrigation has witnessed rapid expansion over the past five decades, with approximately 40% of irrigated regions worldwide relying predominantly on groundwater resources (Lejars et al. 2017). However, the overexploitation of groundwater has led to its alarming depletion, which has emerged as a critical concern in recent years. A deeper comprehension is essential to effectively manage finite water resources, particularly in regions prone to drought. Comprehending the nuanced interplay between groundwater and surface water, both integral components within the hydrologic cycle, holds paramount importance for fostering sustainable water resource management. Groundwater and surface water dynamics intricately influence each other, impacting availability, quality, and distribution of water resources. This understanding is fundamental for devising effective strategies to ensure the longevity and resilience of water supplies, as well as for mitigating risks associated with water scarcity, contamination, and ecosystem degradation. Therefore, thorough exploration and analysis of these interactions are essential for guiding informed decision-making processes in water management practices. The relationship between surface and groundwater can be impacted by both natural processes and human activities (Toran et al. 2019). Recent research has stressed the importance of using models that integrate both surface and groundwater interactions within a unified framework. However, this approach faces several challenges. The intricate dynamics of surface and groundwater interactions at the soil–water interface can be substantially influenced by the heterogeneous characteristics of porous media and the impacts of climate change. Variations in soil composition, permeability, and porosity create complex pathways for water movement, shaping the intricate interplay between surface and groundwater systems. Moreover, the effects of climate change, including alterations in precipitation patterns, temperature extremes, and shifts in hydrological cycles, further compound these interactions, leading to changes in water availability, recharge rates, and aquifer storage capacities. Understanding and quantifying these complexities are essential for developing robust models and management strategies to address the challenges posed by dynamic environmental conditions and ensure the sustainable utilization and protection of water resources. Furthermore, the diverse array of aquifer types, recharge rates, groundwater flow conditions, and variations in river flow add to the complexity of the system, requiring robust modeling and solution methodologies (Larocque et al. 2016).

The difficulties in validating and calibrating models stem from the complex interactions at the soil–water interface, which become increasingly intricate at smaller scales and more pronounced at larger scales. The complexity of these interactions emphasizes the importance of developing robust numerical models that can accurately capture these dynamics. The advantages and limitations of "fully-coupled" and "loosely-coupled" techniques highlight crucial factors, such as the significant parameter requirements for fully-coupled models and the challenges of nonlinear convergence in loosely-coupled approaches. Addressing these fundamental issues is critical for improving model accuracy and dependability, particularly at regional scales.

Surface water models are essential for understanding surface water systems, as they consider deviations caused by natural phenomena and human activities. These models were first developed by pioneering researchers who recognized the interconnectedness and coupling among the various subsystems of the hydrological cycle, as described in Freeze et al. (1969). The System Hydrology European (SHE) model is the first physically-based and coupled model that focuses on surface water flow fundamentals. The development of hydrological modeling has led to the incorporation of advanced tools, including the precipitation-runoff modeling system (PRMS) of the United States Geological Survey (USGS) and models developed by the National Oceanic and Atmospheric Administration's (NOAA) National Weather Service. These cutting-edge models integrate intricate algorithms and data assimilation techniques to simulate the complex processes governing precipitation, runoff, and hydrological responses at various spatial and temporal scales. The USGS PRMS, for instance, incorporates detailed representations of watershed characteristics, land use, soil properties, and vegetation dynamics to provide accurate assessments of water resource availability and hydrological behavior. Similarly, NOAA's National Weather Service models leverage advanced atmospheric physics and computational methods to forecast weather patterns, precipitation events, and associated hydrological phenomena with high precision. These state-of-the-art models not only enhance our understanding of hydrological processes but also empower decision-makers with valuable insights for effective water resources management and climate resilience planning. Groundwater flow models represent powerful tools capable of providing detailed insights into various aspects of aquifer behavior. These models can anticipate the consequences of overexploitation by simulating the effects of pumping rates on groundwater levels and flow patterns. Additionally, they enable the simulation of contaminant transport, allowing for the prediction of the dispersion and evolution of contaminant plumes within aquifers over time. Furthermore, groundwater flow models are instrumental in forecasting changes in groundwater levels induced by climate change, considering factors such as altered precipitation patterns, temperature variations, and shifts in recharge rates. By incorporating intricate hydrogeological parameters and spatial variations in aquifer properties, these models offer valuable guidance for sustainable groundwater management practices, contamination remediation strategies, and adaptation measures in the face of evolving environmental conditions. However, traditional groundwater flow models generally do not consider surface water fluxes in the subsurface environment. To overcome this constraint, hydrological models such as MikeSHE and SHETRAN have been meticulously crafted and implemented. These models stand out for their comprehensive approach, seamlessly integrating both surface and groundwater flow dynamics. They draw upon the pioneering works outlined by Abbott et al. (1986a, b) and Ewen et al. (2000), which laid the foundation for understanding coupled hydrological processes. By harmonizing surface runoff, infiltration, and subsurface flow mechanisms, these models offer a holistic depiction of water movement within catchment areas. Their advanced algorithms enable the simulation of complex interactions between surface water bodies, soils, and aquifers, facilitating accurate predictions of hydrological responses to various natural and anthropogenic influences. As a result, MikeSHE and SHETRAN have become indispensable tools for hydrologists and water resource managers seeking comprehensive insights into the intricate dynamics of integrated surface and groundwater systems.

Numerous scholars have dedicated their research efforts to exploring the intricate relationship between groundwater and surface water. This relationship is influenced by a multitude of factors, including hydro-climatic variables, physiographic characteristics, disparities in hydraulic head, and the geometric configurations of surface and groundwater flow systems. Insights gleaned from the literature, as documented by researchers such as Scibek et al. (2007), Stefania et al. (2018), Pitz et al. (2016), and Rassam et al. (2012), shed light on the complex interplay between these elements. Through empirical observations and theoretical frameworks, these studies contribute to a deeper understanding of the dynamics governing the interaction between groundwater and surface water, thereby informing effective management strategies and sustainable water resource practices. Although simpler approaches have effectively modeled flows and transport phenomena in discrete groundwater systems (Dahl et al. 2007), the inclusion of surface water and groundwater interaction complicates the modeling process significantly. The unceasing exchange of groundwater with rivers emphasizes the intricate and long-lasting relationships between these two bodies of water, presenting a significant challenge, as highlighted by Larocque et al. (2016). This ongoing interaction is complex and can be divided into several categories, as demonstrated in Fig. 1, which serves as the foundation for this review. The main objectives of this review are twofold: first, to conduct a thorough examination of fully-coupled and loosely-coupled schemes, highlighting their respective advantages and drawbacks; and second, to provide a concise evaluation of domain integrated modeling in comparison to detailed domain models. Through an exploration of these differences, the review aims to elucidate the nuances of integrating groundwater and surface water dynamics, contributing to a deeper understanding and more effective water resource management strategies.

One potential research gap highlighted in this study pertains to the scalability of numerical models for groundwater and surface water interactions. While acknowledging challenges at smaller scales and the escalating complexity at larger scales,

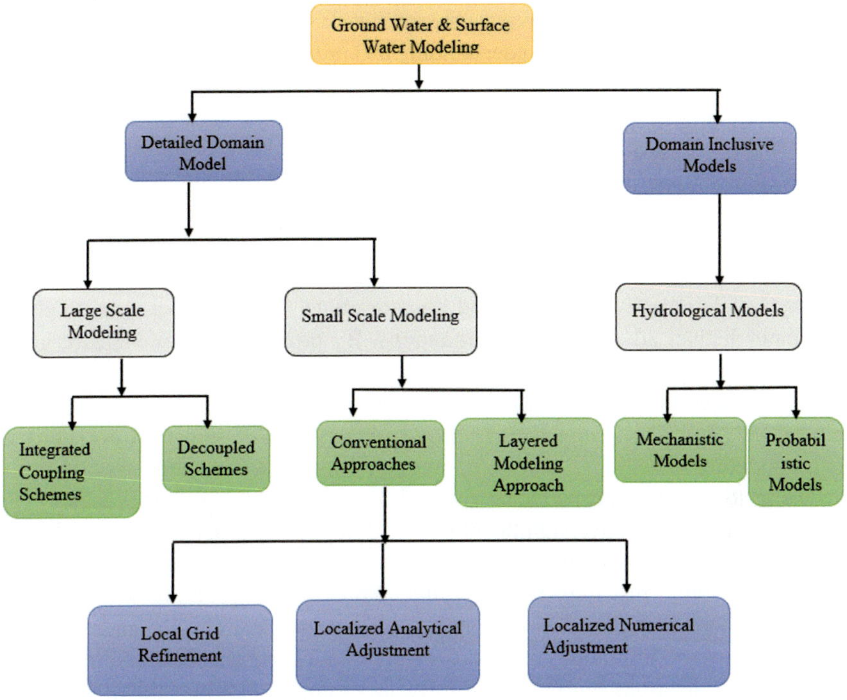

Fig. 1 Surface–groundwater interaction modeling

there appears to be a lack of discussion on specific methodologies or approaches required to effectively bridge these scales. Further research is warranted to explore how modeling techniques can be adapted or refined to maintain accuracy across various spatial scales. Additionally, there is a gap in addressing the integration of new data sources and emerging technologies into existing modeling frameworks, potentially enhancing the accuracy and reliability of simulations, particularly at regional scales.

The novelty of this study lies in its comprehensive exploration of the complex interplay between groundwater and surface water dynamics, emphasizing the critical need for robust numerical models to accurately represent these interactions. By delving into the challenges posed by dynamic exchanges at varying spatial scales, the abstract provides a fresh perspective on the escalating complexity of these relationships. Furthermore, the systematic examination of "fully-coupled" and "loosely-coupled" coupling strategies within detailed domain models offers novel insights into their respective advantages and limitations. This review serves as a pioneering resource, offering strategic guidance for researchers and practitioners in the development of groundwater flow models. Through its synthesis of key findings and methodologies, it empowers professionals to navigate the intricate landscape of model crafting, ultimately aiming to enhance our understanding and predictive capabilities in the realm of groundwater dynamics, particularly at regional scales.

The objective of this study is to underscore the significance of understanding the intricate relationship between groundwater and surface water dynamics, driven by the challenges of model validation and calibration. It highlights the increasing complexity of these interactions across different spatial scales and emphasizes the necessity of developing robust numerical models for accurate simulations. Through an examination of two primary coupling strategies—"fully-coupled" and "loosely-coupled" techniques—within detailed domain models, the abstract aims to elucidate their distinct advantages and limitations. By providing comprehensive guidance for both researchers and practitioners, it seeks to facilitate the formulation of effective strategies crucial for the development of groundwater flow models. Ultimately, the objective is to empower professionals to make informed decisions in their modeling endeavors, leading to improved insights and more precise simulations in the field of groundwater dynamics, particularly at regional scales. The findings of this study have significant practical implications across multiple levels. First and foremost, the identification of terminology gaps highlights the importance of clarity in future research. Secondly, the comparison of coupling strategies aids researchers and practitioners in selecting appropriate models. Thirdly, understanding the advantages and limitations informs decision-making, especially in calibration and uncertainty evaluations. Furthermore, software examples facilitate the implementation of models, while emphasizing the categorization of models enhances efficiency. Overall, addressing uncertainties and improving assumptions can enhance predictive capabilities, ultimately benefiting stakeholders in water resources management.

2 Materials and Method

Detailed domain models refer to comprehensive representations of specific domains, often used in software engineering, system analysis, or other fields where understanding complex systems is necessary. These models typically include intricate descriptions of the entities, relationships, attributes, behaviors, and constraints within a particular domain. (Please refer in Fig. 1).

2.1 Regional-Scale Modeling Approaches

At the regional level, which typically encompasses catchment areas ranging from 1000 to 10,000 km^2, a multitude of factors come into play. These include climate, geology, geomorphology, landscape categories, as well as biological and human influences within the same geographical region, as highlighted in studies by Dahl et al. (2007), and Sophocleous et al. (2002). Within this expansive domain, the interaction between surface and groundwater involves the comprehensive integration of the entire terrestrial hydrological cycle, leveraging observations at larger spatial distances and simplified interaction processes. Regional-scale modeling commonly employs two prominent techniques: fully-coupled and loosely-coupled schemes. These approaches offer distinct methodologies for elucidating the complex dynamics governing water movement and distribution across diverse landscapes, providing valuable insights for effective water resource management and environmental planning initiatives.

2.1.1 Interconnected Scheme

Fully-coupled schemes, grounded in physical principles, seamlessly integrate both surface and groundwater components of the hydrological cycle into a unified software framework. This comprehensive approach enables the simultaneous simulation of surface water processes, such as rainfall–runoff generation, infiltration, surface flow, and evapotranspiration, alongside groundwater dynamics, including recharge, storage, and discharge mechanisms. By accounting for the intricate interactions between surface and subsurface water flows, these schemes offer a holistic representation of hydrological processes across diverse landscapes. Leveraging advanced numerical algorithms and spatially distributed parameters, fully-coupled schemes provide valuable insights into the complex feedback mechanisms shaping water availability, quality, and distribution within watersheds and catchment areas. As such, they serve as indispensable tools for hydrologists, water resource managers, and environmental scientists striving to understand and effectively manage water resources in a changing climate These approaches facilitate the simultaneous resolution of governing equations for both surface and groundwater flow, enabling investigation across various spatial and temporal scales. These models produce stable results, which in turn reduce the need for multiple auxiliary software packages to address different subsystems (Loague et al. 2006). In fully-coupled systems, the groundwater flow equations are typically derived from the modified Richards' equation, which considers water movement in both saturated and unsaturated zones. This equation captures the complex interactions between soil moisture, hydraulic conductivity, and gravitational forces, providing a comprehensive representation of groundwater dynamics. On the other hand, surface water equations stem from the depth-averaged Saint–Venant equation, which characterizes the flow of water in open channels. By accounting for variables such as channel geometry, flow velocity, and surface slope, this equation offers a robust framework for simulating surface water movement and interactions within river networks and floodplains. Together, these equations form the foundation of fully-coupled systems, enabling the integrated modeling of surface and groundwater processes across diverse hydrological environments.

Fundamental Equations

- Richards' equation serves as a fundamental tool for describing the conservation of mass of the water phase within the void spaces of porous media. It accounts for various factors influencing water movement, including soil properties, hydraulic conductivity, and gravitational forces. By integrating these parameters, Richards' equation offers a detailed understanding of how water redistributes within the soil profile under different moisture conditions. Additionally, Darcy's equation complements Richards' equation by providing insights into the volumetric flux of water through porous media. Derived from Darcy's law, this equation quantifies the flow rate of water per unit area and is instrumental in characterizing subsurface flow dynamics. Together, Richards' and Darcy's equations form the cornerstone of hydrological modeling, facilitating the analysis of water movement in porous media and aiding in the development of effective water resource management strategies.

- Richards' equation in three-dimensional porous medium:

$$w_m \frac{\partial \theta_s S_w}{\partial t} + \nabla . w_m \vec{q} \pm \sum \Gamma_{ex} = \pm Q \quad (1)$$

- Darcy–Buckingham equation:

$$\vec{q} = \underset{-K}{\Rightarrow} . k_r \nabla (\psi + z) \quad (2)$$

The following text describes the various parameters that are essential for understanding fluid flow within a porous medium. These parameters include w_m, which represents the volumetric fraction of the total porosity of the medium; Q, which denotes the source/sink rate and indicates the volumetric fluid flux per unit volume, signifying either a source (positive) or a sink (negative) to the system; Γ_{ex}, which accounts for exchange fluxes; S_w, which signifies the degree of water saturation; and θ_s, which reflects the volumetric void fraction of the porous medium. Additionally, ψ represents the pressure head, and z denotes the elevation head. The permeability tensor, k, which governs the flow characteristics, and the relative permeability, kr, which influences flow as a function of saturation, play pivotal roles. The hydraulic conductivity tensor, K, is determined by the water viscosity (μ), acceleration due to gravity (g), and the permeability tensor of the porous medium (k), adjusted for water density (ρ), which is contingent upon solute concentration (C). Moreover, water saturation is linked to moisture content (θ). The parameter Γ_{ex} is closely connected to the volume of fluid transfer between the subsurface domain and other components of the model, such as surface areas, drainage tiles, wells, fractures, and dual-porosity systems. The value and direction of Γ_{ex} are determined by the theoretical framework of the model, which governs fluid exchange between these different domains.

The Saint–Venant equation for surface water flow, which accounts for depth, can be written as:

$$\frac{\partial \Phi_0 h_0}{\partial t} + \nabla . d_0 \vec{q}_0 + d_0 \Gamma_0 = \pm Q_0 \quad (3)$$

Within this context, several parameters play critical roles in understanding the dynamics of fluid flow. Firstly, Q_0 represents the source/sink rate, indicating the volumetric flux per unit area. Next, d_0 signifies the water depth, providing insight into the quantity of water present within the system. Additionally, Γ_0 denotes exchange fluxes, which are pivotal for understanding the exchange of fluid between different components of the system. Lastly, h_0 represents the surface hydrostatic pressure, reflecting the pressure exerted by the water column at the surface. By incorporating the Manning equation, which establishes a relationship between flux and the gradient of the surface water system, the dynamics of fluid flow can be further elucidated. This equation provides valuable insights into how surface water moves in response to variations in topography and hydraulic roughness. Through a comprehensive understanding of these parameters and equations, researchers can gain deeper insights into the behavior of fluid systems and make informed decisions regarding water resource management and engineering applications:

$$\vec{q}_0 = -\frac{d_0^{\frac{2}{3}}}{\eta \Phi^{\frac{1}{2}}} k_{r0} \nabla (d_0 + z_0) \tag{4}$$

In this context, key parameters are instrumental in understanding the coupling of surface and subsurface flows. η represents roughness, influencing the resistance to flow within the system, while φ denotes the surface water gradient, reflecting the change in water level over a specified distance. Additionally, k_{r0} signifies the relative permeability, dictating the flow characteristics through the porous medium. These terms enable the exchange of water between surface and subsurface domains, allowing for a comprehensive representation of hydrological processes within integrated systems. By considering these parameters and interactions, researchers can gain deeper insights into the complex dynamics of coupled surface and subsurface flows, informing effective water resource management strategies and environmental planning initiatives.

$$-\nabla \cdot w_m \vec{q}_0 + \Gamma_0 \pm Q = w_m \frac{\partial \theta_s}{\partial t} \tag{5}$$

$$-\nabla \cdot d_0 \vec{q}_0 - d_0 \Gamma_0 \pm Q = \frac{\partial \Phi_0 h_0}{\partial t} \tag{6}$$

It has been suggested that the calculation of the first-order exchange term may be achieved through the utilization of

$$d_0 \Gamma_0 = k_{rso} K_{so} (h - h_0) \tag{7}$$

Several critical parameters serve to elucidate the dynamics at the surface/subsurface interface in this context. The parameter h signifies the subsurface pressure head, which provides insight into the pressure exerted by groundwater within the subsurface environment. K_{so} denotes the conductance at the surface/subsurface interface, regulating the rate of fluid exchange between these two domains. Furthermore, k_{rso} represents the coupling-relative permeability or rill storage, which influences the flow characteristics and storage capacity at the interface. By examining these parameters, researchers can gain a more profound understanding of the complex interactions between surface and subsurface hydrological processes, ultimately aiding in the formulation of effective water resource management strategies and environmental planning initiatives.

Fully-integrated software packages employ various coupling methods to combine hydrostatic surface water (Saint–Venant) and subsurface flow, each presenting unique benefits and applications. In software packages such as Hydro Geosphere (HGS) and InHM, the first-order exchange formulation, or conductance, is used to couple surface and groundwater flow equations, ensuring flux and pressure continuity across surface and subsurface domains (Maxwell et al. 2015). By incorporating these coupling strategies, fully-coupled software packages offer comprehensive simulations of complex hydrological systems, facilitating well-informed decision-making in water resource management and environmental planning.

The process of transforming hypothetical atmospheric exchanges into actual exchanges across terrestrial surfaces has a substantial influence on the dynamics of surface storage (Camporese et al. 2010; Furman et al. 2008). This technique is rooted in the principle of preserving pressure and flux consistency at the surface–subsurface interface, although it neglects the consistency of momentum and forces.

Few studies have been conducted on the regional applications of these models. There is an absence of small-scale examples, as indicated by Barthel et al. (2016). One of these models is the Penn State Integrated Hydrologic Model (PIHM), which comprises a reservoir simulation component and has been adapted to work within a simulation–optimization (S–O) modeling framework, as outlined by Seo et al. (2018).

The benefits of the fully-coupled approach are apparent, as its simulation outcomes align well with both experimental and observed data from field studies. These models are versatile and can simulate the dynamics of both coastal and lake systems, and they have been extensively validated through a variety of software packages that have been employed in previous research. Nevertheless, it is important to exercise caution when applying the fully-coupled approach in unsaturated zones where lateral flow is dominant (Zhu et al. 2012). To enhance the accuracy of fully-coupled models, it is recommended to use a quasi-adaptive meshing configuration, focusing on correcting local discretization errors resulting from topographic elevation variations (Takeuchi et al. 2010). In conclusion, fully-coupled models serve as dependable tools for examining watershed hydrological characteristics with the desired level of accuracy.

Although fully-coupled models offer numerous advantages, there are also notable drawbacks to consider. Testing fully-coupled approaches on larger scales has proven challenging due to the intricate interactions between surface water and groundwater, as noted by Barthel et al. (2006). Additionally, fully-coupled models are associated with lengthy computational times, posing challenges for model calibration due to the large number of estimated parameters. The uncertainty introduced by boundary and initial conditions further complicates calibration efforts, resulting in increased computational efforts and costs. Therefore, proper scaling for simulations using fully-coupled models is crucial, as emphasized by Fatichi et al. (2016).

2.1.2 Disintegrated Scheme

Surface and groundwater components can be tackled independently through a decoupled scheme, either in a sequential manner without iteration or iteratively within each time step. This method often entails employing two or more separate models to handle surface water and groundwater processes separately. Over the years, numerous decoupled models have been devised based on these methodologies, as documented by Maxwell et al. (2014).

Fundamental Equations

In the field of subsurface hydrology, the fundamental principles are rooted in Darcy's law and the continuity equation, as expounded upon by Bear et al. (2013). The mass conservation equation for a confined aquifer, as delineated by Mehdinejadiani et al. (2013), is represented by the following:

$$\frac{\partial}{\partial x}\left(K_x\frac{\partial h}{\partial x}\right)+\frac{\partial}{\partial y}\left(K_y\frac{\partial h}{\partial y}\right)+\frac{\partial}{\partial z}\left(K_z\frac{\partial h}{\partial z}\right)=S_s\frac{\partial h}{\partial t} \tag{8}$$

In the context of an unconfined aquifer, the governing equations include parameters such as the piezometric head (h), time (t), and principal components of the hydraulic conductivity tensor (K_x, K_y, and K_z), along with specific storage (Ss), as delineated by Bear et al. (2013).

$$\frac{\partial}{\partial x}\left(K_x\frac{\partial h}{\partial x}\right)+\frac{\partial}{\partial y}\left(K_y\frac{\partial h}{\partial y}\right)+\frac{\partial}{\partial z}\left(K_z\frac{\partial h}{\partial z}\right)=S_s\frac{\partial h}{\partial t}=0 \tag{9}$$

The process of solving equations for unconfined groundwater flow is highly intricate, primarily due to the fluctuations in aquifer thickness and the variations resulting from groundwater withdrawal (Furman et al. 2008). To tackle this issue, the Dupuit principle and Boussinesq equation are commonly utilized (Nimmo 2006). In loosely-coupled models, the coupling technique must effectively transfer data both spatially and temporally (Bronstert et al. 2005). It should be noted that many of these models are tailored to specific catchments.

A loosely-coupled approach has been successfully implemented, connecting hydrogeological data to hydrogeological numerical models within GIS platforms like ArC-GIS, as demonstrated by Alcaraz et al. (2017). Notably, unstructured meshes have emerged as a unique characteristic of these loosely-coupled models, offering a streamlined approach to managing boundary conditions.

This approach enables the reuse of hydrogeological conceptual models in GIS, providing alternative software options (Alcaraz et al. 2017). Similarly, it offers flexibility in selecting the platform for each specific stage (Huang et al. 2009).

Although various loosely-coupled frameworks have been developed for regional scales, they come with significant limitations. Deriving concrete decisions from evaluations based on prior studies is difficult, as each framework is generally designed for a particular catchment. Moreover, there is a scarcity of a proper methodology to assess groundwater models, particularly for transient models that have fluctuating recharge rates. Addressing the transition zone requires the construction of a fully-integrated dynamic model with bidirectional linkages between surface and groundwater. Simplifying the geological properties of regional aquifers as heterogeneous for estimating spatial distribution of groundwater recharge poses challenges, particularly in regional-scale modeling (Bhattarai et al. 2013).

Additionally, challenges arise when using loosely-coupled models due to the limited amount of statistical data and information available for interpolating river water levels between gauges. Estimating exchange fluxes based on pressure variations requires accurately determining river stages through interpolation between gauges, which is a task that often exceeds the capabilities of loosely-coupled models to calculate river water levels along the reach of a river.

Currently, MIKE SHE and FEFLOW coupled with MIKE11 are the most frequently utilized loosely-coupled models for regional studies.

2.2 *Microscale Modeling*

Microscale modeling is employed to investigate the interactions between surface and groundwater in smaller spatial areas. This technique involves incorporating a cross-section of a river into a specific region, with the necessary simplifications made for processes occurring outside the area of interest. However, the influence of external processes may vary and could significantly impact the interaction dynamics. Consequently, it is crucial to carefully define the boundary conditions, which should include factors such as groundwater recharge, runoff formation, and regional groundwater flow, as emphasized by Barthel et al. (2016).

2.2.1 Conventional Methods

Local-scale modeling comprises various techniques that involve fine-tuning grids in particular areas, applying targeted adjustments and implementing precise numerical corrections. These methods will be briefly explored. According to a study conducted by Li et al. (2004), a locally developed analytical method that utilizes a finite difference approach is more accurate in predicting drawdown in pumping wells than exact solutions. Instead of directly solving large and complex matrix systems, smaller-scale local models are used to overcome their limitations. By dividing a large matrix system into smaller, more manageable matrices, this approach allows for more efficient analysis. However, it can be time-consuming, especially when new boundary conditions or changes in scale are required.

2.2.2 Systematic Modeling Approach

The hierarchical approach is a methodological technique that is based on the nested grid method, as demonstrated by Emara et al. (2008). This approach is interactive and iterative, and it has been found to be effective in addressing some of the limitations of conventional methodologies by utilizing dynamically coupled and fully-integrated settings, as noted by Li et al. (2004). A hierarchical patch dynamic paradigm (HPDP) has been developed for groundwater modeling, which involves fewer assumptions, simplified data input, and streamlined post-processing, leading to significant improvements in computational efficiency. The simulation results obtained for regional models and groundwater systems using this approach have been found to be satisfactory, without requiring the resolution of large matrix structures, as demonstrated by Wolf et al. (2008).

Moreover, this approach instills confidence in estimating exchange fluxes (Anibas et al. 2012). Real-time integration and subscale modeling are made feasible through the dynamic fusion policy within the model. Additionally, computation steering applications within the approach enable manipulation of the simulation process.

A significant advantage of this modeling strategy is the ability to focus indefinitely on a particular zone, provided that the governing equations can maintain a certain scale. This hierarchical approach presents a realistic and time-efficient method for managing complex hydrological field settings. However, it requires extensive data in a well-organized structure, potentially necessitating a physical representation of the database for users. Hierarchical structures may adapt slowly to changing requirements and involve frequent storage of data in various entities.

3 Results and Discussion

Integrated Domain Modeling

Domain integrated modeling serves to simplify the water cycle and facilitates forecasting of both water quantity and quality. This approach encompasses two main types of models: stochastic-based models and processed models.

3.1 Algorithmic Models

These models encompass streamflow, surface runoff, subsurface flow, and a range of other relevant phenomena and can be applied in both individual-event and continuous simulation models. In the following section, we will discuss the significant findings that have been revealed by previous research studies that have utilized these models.

Saha et al. (2017) utilized a physically-based, distributed interaction model to study the consequences of climate change on greenhouse gas emissions, with a specific focus on changes in surface and groundwater interactions over time. Their simulation results highlighted the significant role of groundwater in stream flow in these scenarios. To enhance the representation of surface and groundwater interactions, substantial resources have been invested in the development of the "CLASS" groundwater model, which merges water table dynamics with land surface parameters to provide a comprehensive understanding of these interactions. Furthermore, the CLASS model has been successfully integrated with other simulation models, such as CK, CK-GW, and CK-GWL, providing insights into regional hydrological variations. CK-GWL, in particular, has been refined to better account for subsurface lateral flow, recharge, and discharge processes. Several studies have employed the CK-GWL model to simulate surface runoff, stream flows, soil moisture changes, and high water tables. Ganji et al. (2018) noted that many studies have used the CK-GWL model. Nevertheless, employing statistical methods suggests the need for additional investigation using regional climate models to enhance simulations of soil moisture and stream flows by incorporating the groundwater component in land surface modeling. Further research has focused on identifying key factors driving groundwater fluctuations, recognizing the impact of parameters such as recharge, extraction rates, and lateral groundwater flow on groundwater level variability, as indicated by Abiye et al. (2018).

3.2 Probabilistic Models

Stochastic models are typically constructed as black boxes, utilizing mathematical and statistical theories to connect input data with the model's output. Commonly employed methodologies in stochastic modeling include regression, neural networks, transfer functions, and system identification. The subsequent section will provide a comprehensive review of the application of these techniques.

Time series modeling has emerged as a prominent approach for estimating hydrogeological properties, such as flow and storage dynamics, due to its high accuracy and cost-effectiveness, as noted by Adhikary et al. (2012). Utilizing extensive data on groundwater over an extended period allows for the assessment of aquifer characteristics. However, the influence of delayed gravitational effects may lead to inaccuracies in groundwater hydrographs when conducting time series modeling. These discrepancies, particularly pronounced during the early stages of pumping activities, can be mitigated by incorporating this effect into a transfer function model. Future investigations should focus on evaluating the efficacy of this modeling approach, as proposed by Shapoorji et al. (2015).

The regression method is frequently used in groundwater studies, especially for examining groundwater table dynamics and management approaches (Mogaji et al. 2015). For instance, researchers have employed a triple linear regression method to simulate short-term fluctuations in groundwater table levels. Through regression and sensitivity analyses, researchers can uncover the interrelationships among key parameters affecting groundwater fluctuations. Notably, precipitation, evaporation, and river stages have emerged as the most influential factors in these analyses (Yan et al. 2018). To enhance the accuracy of groundwater table variations determined by this method, it is advisable to consider geographical locations and hydraulic conductivity.

4 Results and Discussion

This study highlights the pressing need to address the lack of standardized terminology in the integration of groundwater and surface water models. Without a unified framework, evaluating diverse coupling approaches for future modeling applications becomes challenging. Fully-coupled models emerge as powerful tools capable of simultaneously resolving the Richards' and shallow water equations, providing robust explanations for complex physical scenarios and facilitating parameter estimation for regional-scale models. Despite their advantages, they encounter hurdles such as convergence difficulties for nonlinear models and discrepancies among individual models during independent calibrations. To assist researchers in selecting appropriate models, examples of software developed based on both fully-coupled and loosely-coupled models are provided, particularly valuable when interaction flux needs consideration. Moreover, the categorization of models based on interface boundaries between groundwater, the vadose zone, and surface water further refines model selection, enhancing applicability. It is evident that reducing uncertainties and addressing the variable spatial and temporal features of groundwater and surface water flows are imperative for improving model assumptions. Consequently, the development of a suitable and distinct framework is essential to guide future research and modeling endeavors toward enhanced accuracy and reliability.

The intricate interaction between groundwater and surface water underscores the need for a comprehensive framework that integrates various climatic, geological, and hydrological factors. This framework must account for the dynamic interplay between these elements to accurately represent the complex relationships and processes occurring within aquifer systems and surface water bodies. This interaction significantly influences recharge–discharge processes, making it crucial for groundwater research (Huang et al. 2009). Improving the modeling of groundwater and surface water requires a thorough understanding of this interaction. However, from the current review, it is evident that challenges related to scale and uncertainty persist in modeling groundwater and surface water interactions. Clear delineation of temporal and spatial scales, particularly concerning quantity and quality, remains elusive (Kornelsen et al. 2014). Furthermore, precise measurement of local and regional scale alterations and exchanges is lacking, leading to considerable variability in simulation results.

The present analysis evaluates and differentiates diverse coupling strategies employed in hydrological modeling. Prominent among these are fully-coupled models, which offer immediate insights into groundwater and surface water interactions via the integration of process-based equations with 3D subsurface representations. Exceptional in capturing all hydrological processes, these models are well-suited for regional-scale applications where such interactions are crucial. They surpass other integrated systems and effectively manage the intricate complexities inherent in regional-scale modeling. However, the scarcity of field data in large-scale modeling poses a significant challenge to their implementation, as noted by Semenova et al. (2015). Furthermore, the simulation of fully-coupled models necessitates substantial computational resources due to the numerous estimated parameters involved, rendering calibration intricate, particularly given the high uncertainty associated with boundary and initial conditions. Consequently, in the past, a loosely-coupled approach with simpler coupling schemes has been favored.

The research at hand emphasizes the intricate nature of groundwater and surface water interactions, which evolve across various spatial scales. This intricacy underlines the need for modeling efforts to account for real-world complexities effectively.

The study underlines the critical role of numerical models in advancing our understanding of groundwater and surface water dynamics, despite the challenges in model validation and calibration. Further advancements in modeling techniques are essential for addressing these challenges effectively.

The research explores fully-coupled and loosely-coupled coupling strategies, providing valuable insights into model selection and development. It offers guidance on the strengths and limitations of different approaches, aiding researchers and practitioners in crafting effective groundwater flow models.

The study aims to empower professionals by synthesizing key findings and methodologies, enabling them to make informed decisions in their modeling endeavors. This aspect suggests that the research not only contributes to scientific knowledge but also has practical implications for professionals working in the field.

Ultimately, the research aims to contribute to the development of models that accurately capture the complexities of real-world conditions governing surface and groundwater interactions, particularly at regional scales. This long-term goal underscores the study's broader impact on enhancing our understanding and predictive capabilities in groundwater dynamics.

In summary, the study emphasizes the importance of addressing challenges in modeling groundwater and surface water interactions, offering guidance for model development, and ultimately advancing our ability to simulate these dynamics accurately.

5 Conclusions

The integration of groundwater and surface water models lacks standardized terminology in the literature, necessitating a comparison of diverse coupling approaches for future modeling applications. Fully-coupled models, capable of simultaneously resolving Richards' and shallow water equations, offer robust explanations of complex physical scenarios and parameter estimation for regional-scale models. The complexity of the system increases due to the significant number of required parameters and limited understanding of boundary conditions. Loosely-coupled methodologies, which rely on empirical relationships, combine process-based models or algorithms, allowing outputs from individual models to serve as inputs for others. The key advantages of loosely-coupled models include simplified calibration and uncertainty evaluations. However, these models have drawbacks such as convergence difficulties for nonlinear models and discrepancies among individual models during independent calibrations. There are software options that are either tightly integrated or loosely connected, which aid researchers in selecting the most suitable models when accounting for interaction fluxes. These models can also be categorized based on the boundaries at the interface between groundwater, the unsaturated zone, and surface water. The goal is to minimize uncertainties and address the varying spatial and temporal characteristics of groundwater and surface water flows. By creating a robust and distinct framework, it is possible to manage uncertainties and support future advancements.

References

Abbott MB, Bathurst JC, Cunge JA, O'Connell PE, Rasmussen J (1986a) An introduction to the European hydrological system—Systeme Hydrologique Europeen, "SHE", 1: history and philosophy of a physically-based, distributed modelling system. J Hydrol 87(1–2):45–59

Abbott MB, Bathurst JC, Cunge JA, O'connell PE, Rasmussen J (1986b) An introduction to the European hydrological system—Systeme Hydrologique Europeen, "SHE", 2: Structure of a physically-based, distributed modelling system. J Hydrol 87(1–2):61–77

Abiye T, Masindi K, Mengistu H, Demlie M (2018) Understanding the groundwater-level fluctuations for better management of groundwater resource: a case in the Johannesburg region. Groundw Sustain Dev 7:1–7

Adhikary SK, Rahman M, Gupta AD (2012) A stochastic modelling technique for predicting groundwater table fluctuations with time series analysis. Int J Appl Sci Eng Res 1(2):238–249

Alcaraz M, Vázquez-Suñé E, Velasco V, Criollo R (2017) A loosely coupled GIS and hydrogeological modeling framework. Environ Earth Sci 76:1–13

Anibas C, Verbeiren B, Buis K, Chormański J, De Doncker L, Okruszko T, Batelaan O (2012) A hierarchical approach on groundwater-surface water interaction in wetlands along the upper Biebrza River, Poland. Hydrol Earth Syst Sci 16(7):2329–2346

Barthel R (2006) Common problematic aspects of coupling hydrological models with groundwater flow models on the river catchment scale. Adv Geosci 9:63–71

Barthel R, Banzhaf S (2016) Groundwater and surface water interaction at the regional-scale–a review with focus on regional integrated models. Water Resour Manage 30(1):1–32

Bhattarai MP (2013) Impact of climate change on surface and subsurface water interaction and riparian vegetation: linkage between hydrology and invasive Tamarisk in a Semi-Arid Basin. University of Nevada, Reno
Bronstert A, Carrera J, Kabat P, Lütkemeier S (Eds) (2005) In: Coupled models for the hydrological cycle: integrating atmosphere, biosphere, and pedosphere. Berlin, Heidelberg, Springer
Camporese M, Paniconi C, Putti M, Orlandini S (2010) Surface-subsurface flow modeling with path-based runoff routing, boundary condition-based coupling, and assimilation of multisource observation data. Water Resour Res 46(2)
Dahl M, Nilsson B, Langhoff JH, Refsgaard JC (2007) Review of classification systems and new multi-scale typology of groundwater–surface water interaction. J Hydrol 344(1–2):1–16
Emara NA, Hussein AS, Khalifa ME, Abdel-Wahab MS (2008) A hierarchical approach for groundwater modeling. In: Proceedings of the twelfth international water technology conference, IWTC12, January, Alexandria, Egypt, pp 1483–1498
Ewen J, Parkin G, O'Connell PE (2000) SHETRAN: distributed river basin flow and transport modeling system. J Hydrol Eng 5(3):250–258
Fatichi S, Vivoni ER, Ogden FL, Ivanov VY, Mirus B, Gochis D, Tarboton D (2016) An overview of current applications, challenges, and future trends in distributed process-based models in hydrology. J Hydrol 537:45–60
Freeze RA, Harlan RL (1969) Blueprint for a physically-based, digitally-simulated hydrologic response model. J Hydrol 9(3):237–258
Furman A (2008) Modeling coupled surface–subsurface flow processes: a review. Vadose Zone J 7(2):741–756
Ganji A, Sushama L (2018) Improved representation of surface-groundwater interaction in the Canadian land surface scheme. Int J Climatol 38(14):5077–5094
Huang G, Yeh GT (2009) Comparative study of coupling approaches for surface water and subsurface interactions. J Hydrol Eng 14(5):453–462
Kornelsen KC, Coulibaly P (2014) Synthesis review on groundwater discharge to surface water in the Great Lakes Basin. J Great Lakes Res 40(2):247–256
Larocque M, Broda S (2016) Groundwater–surface water interactions in Canada. Canadian Water Resour J/revue Canadienne des Ressources Hydriques 41(4):451–454
Lejars C, Daoudi A, Amichi H (2017) The key role of supply chain actors in groundwater irrigation development in North Africa. Hydrogeol J 25(6):1593
Li SG, Liu Q, Afshari S (2004) Modeling complex groundwater systems across multiple scales: Part 1—a hierarchical patch dynamic paradigm. Michigan State University, East Lansing, MI, USA
Loague K, Heppner CS, Mirus BB, Ebel BA, Ran Q, Carr AE, VanderKwaak JE (2006) Physics-based hydrologic-response simulation: foundation for hydroecology and hydrogeomorphology. Hydrol Processes: an Int J 20(5):1231–1237
Maxwell RM, Putti M, Meyerhoff S, Delfs JO, Ferguson IM, Ivanov V, Sulis M (2014) Surface-subsurface model intercomparison: a first set of benchmark results to diagnose integrated hydrology and feedbacks. Water Resour Res 50(2):1531–1549
Maxwell RM, Condon LE, Kollet SJ (2015) A high-resolution simulation of groundwater and surface water over most of the continental US with the integrated hydrologic model ParFlow v3. Geoscientific Model Developm 8(3):923–937
Mehdinejadiani B, Jafari H, Baleanu D (2013) Derivation of a fractional Boussinesq equation for modelling unconfined groundwater. The Europ Phys J Special Topics 222(8):1805–1812
Mogaji KA, Lim HS, Abdullah K (2015) Modeling of groundwater recharge using a multiple linear regression (MLR) recharge model developed from geophysical parameters: a case of groundwater resources management. Environ Earth Sci 73:1217–1230
Nimmo JR (2006) Unsaturated zone flow processes. In: Encyclopedia of hydrological sciences
Pitz CF (2016) Predicted impacts of climate change on groundwater resources of Washington State. Environmental Assessment Program, Washington State Department of Ecology

Rassam D, Jolly I, Pickett T (2012) Guidelines for modelling groundwater-surface water interactions in eWater source: towards best practice model application

Saha GC, Li J, Thring RW, Hirshfield F, Paul SS (2017) Temporal dynamics of groundwater-surface water interaction under the effects of climate change: a case study in the Kiskatinaw River Watershed, Canada. J Hydrol 551:440–452

Scibek J, Allen DM, Cannon AJ, Whitfield PH (2007) Groundwater–surface water interaction under scenarios of climate change using a high-resolution transient groundwater model. J Hydrol 333(2–4):165–181

Semenova O, Beven K (2015) Barriers to progress in distributed hydrological modelling. Hydrol Process 29(8):2074–2078

Seo SB, Mahinthakumar G, Sankarasubramanian A, Kumar M (2018) Conjunctive management of surface water and groundwater resources under drought conditions using a fully coupled hydrological model. J Water Resour Plan Manag 144(9):04018060

Shapoori V, Peterson TJ, Western AW, Costelloe JF (2015) Estimating aquifer properties using groundwater hydrograph modelling. Hydrol Process 29(26):5424–5437

Sophocleous M (2002) Interactions between groundwater and surface water: the state of the science. Hydrogeol J 10:52–67

Stefania G, Rotiroti M, Fumagalli M, Simonetto F, Capodaglio P, Zanotti C, Bonomi T (2018) Modeling groundwater/surface-water interactions in an Alpine valley (the Aosta Plain, NW Italy): the effect of groundwater abstraction on surface-water resources. Hydrogeol J 26(1):147–162

Takeuchi J, Kawachi T, Imagawa C, Buma N, Unami K, Maeda S (2010) A physically based FVM watershed model fully coupling surface and subsurface water flows. Paddy Water Environ 8:145–156

Vaux H (2011) Groundwater under stress: the importance of management. Environ Earth Sci 62:19–23

Wolf J, Barthel R, Braun J (2008) Modeling ground water flow in alluvial mountainous catchments on a watershed scale. Groundwater 46(5):695–705

Yan SF, Yu SE, Wu YB, Pan DF, Dong JG (2018) Understanding groundwater table using a statistical model. Water Sci Eng 11(1):1–7

Zhu Y, Shi L, Lin L, Yang J, Ye M (2012) A fully coupled numerical modeling for regional unsaturated–saturated water flow. J Hydrol 475:188–203

Environment

Assessment of Surface Water Quality Modelling for Drinking Purposes in Brahmani River Basin (BRB), Odisha, Using Geographical Information System (GIS), IDOCRIW (I) and COPRAS (U_i) Techniques

Abhijeet Das

Abstract Surface water is an important source for drinking water supply in Brahmani River, Odisha, India. However, this basin plays an important role in local economy and ecosystem. This resource is extensively utilized for drinking, irrigation, and industrial activities globally. Thus, rapid use of water quality leads to serious water pollution. An attempt has been made in this work to understand the suitability for human consumption, considering 15 water quality (WQ) parameters collected yearly from seven water sources from Brahmani River Basin, Odisha. Time frame taken into consideration is 10 years (2013–2023). Integrated Determination of Objective Criteria Weights (IDOCRIW/I)-based water quality index (WQI) and MCDMs like Complex Proportional Assessment (COPRAS) method were applied to compute the results. The concept COPRAS was introduced in the current work, to apply weights and rough set theory to generate an impartial and trustworthy depiction of each sample location's overall pollution levels. The interpolation method such as Inverted Distance Weighted (IDW) was applied in ArcGIS software to produce the spatial map. The results showed that the river water is somewhat alkaline and the DO seems to be in good health when compared to the ranges or values provided by standard organizations as addressed by WHO, for evaluating WQ. Most of the parameters evaluated were found to fall within the allowable limits of the WHO standards. The results of integrated (I) weights indicated that WQ status belonged to good-poor quality of water. It is noted that principal stations such as J-(1), (2), (3) and (7) found to be high values, indicating poor H_2O and having more impact on WQ. It was also come into fact that the location J-(7) had the second-highest integrated weight. This culminates primarily with point-based pollution such as wastewater from household and industrial sources, non-point sources such as agricultural activities and natural phenomena such as soil erosion and flooding. The visualization of

A. Das (✉)
Research Scholar, Department of Civil Engineering, C.V. Raman Global University (C.G.U), Bhubaneswar, Odisha 752054, India
e-mail: das.abhijeetlaltu1999@gmail.com; abhijeetlaltu1994@gmail.com

S. C. Pal and U. Chatterjee (eds.), *Surface, Sub-Surface Hydrology and Management*, Springer Geography, https://doi.org/10.1007/978-3-031-62376-9_9

the COPRAS method suggests that J-(1) was the most polluted station, based on its rank and its performance score ($U_i = 100$). The study recommended continuous pollution monitoring and local regulations to reduce the entrance of both diffuse and point-source pollution into the river. Hence, special attention should be given to maintain ecological sustainability. Moreover, the water quality methods, which are discussed extensively in the paper, can also provide an overall idea about indexing for further research.

Keywords Brahmani River · IDOCRIW · COPRAS · Pollution · Rank · Monitoring · Regulations

1 Introduction

Water is essential to the genesis, growth, and survival of all life on Earth, which explains why the majority of human civilizations have flourished in areas near bodies of water (Ravi et al. 2020; Das 2022, 2023). Therefore, to ensure sustainable development, water resources must be managed in an efficient and profitable manner. Once more, without sufficient amounts and qualities of freshwater, sustainable growth is not feasible. Surface water is a valuable resource that India must responsibly use to satisfy rising demands in its industrial, agricultural, and home sectors (Mahadevan et al. 2020). Likewise, surface water quality is also taken into account by considering its biological, chemical, and physical properties. So, the evaluation of the water's quality could provide detailed information on the environments that the water is being taken by the local people, residing thereby (Long-Ling et al. 2021). For any society to flourish socioeconomically, water's quantity and quality are equally important (Su et al. 2020; El-Baba et al. 2020). It's possible that both natural and artificial factors, such as precipitation, rock mineralogy, rock-water interaction, aquifer type, terrain variation, climate, and industrial and domestic use, have affected the quality of surface water. In addition to degrading the quality of water sources, widespread urbanization, community settlement, and relocation brought on by fast industrialization have currently resulted in water scarcity in various locations. On the other hand, because of variations in size, time, and duration, climate change has a significant impact on surface water quantity in arid and semi-arid regions (Karunanidhi et al. 2019; Das 2022). A paradigm shift in wastewater treatment has been required in recent decades due to the wastewater produced from various sources, including mining activities, food and petrochemical businesses, healthcare facilities, home sewage, and others. This wastewater is rich in stubborn harmful contaminants (Wang et al. 2020; Das et al. 2023). Today, as a result of the growth of numerous industries including industry and agriculture, there is an enormous demand for water. Thus, careful concern has been given to the conservation and protection of the water sources through thorough and accurate assessments. Furthermore, surface water's nature matters for a variety of purposes. It is crucial for the advancement of the economy and the enhancement of human health (Zhou et al. 2016; Li et al. 2021).

Surface water resource management is a concern in arid and semi-arid areas because of the scarcity of surface water, the erratic monsoon, and the high rates of evaporation (Li et al. 2017; Das 2023). The water resources are impaired by human-induced activities and geogenic activities (Narsimha 2020a, 2020b). However, it is extremely difficult to clean up the contaminated resource once surface water has been affected. The demand for drinking water has skyrocketed as a result of fast population growth and industrialization, which has a direct impact on crop growth and human health (Li and Qian 2016). As a result, ongoing monitoring is required for the execution of restoration policies and such assessments. Later on, rising irrigational operations and a growing population's constant need for food have contributed significantly to the rate at which irrigation goods are produced (Alemu et al. 2017). In the meanwhile, emerging nations lack regulations on the pollution caused by industrial, agricultural, and residential activities (Muthusamy et al. 2022; Karuppannan et al. 2022). Several studies (Hasan et al. 2019; Sarkar et al. 2019; Parvin et al. 2022) have been carried out to determine the nation's surface water quality. Hence, relying on these past studies, these investigations demonstrated how intensive industrialization, expanding agrochemical-based agriculture, and municipal wastes are continuing to damage this water resource's quality (Parvin and Tareq 2021). Reusing wastewater that has been partially or completely cleaned and discharging it onto agricultural fields for irrigation purposes will have detrimental impacts on the environment, including contaminated soils, contaminated water, and contaminating agricultural products. On the other hand, effective planning, and the repurposing of household wastewater for use in agriculture and other industries mitigates water pollution, preserves freshwater resources, boosts food production, and lowers fertilizer expenses. Numerous techniques and parts have been developed for the evaluation of water contamination and quality. In today's culture, developing a novel technique for determining pollution load is considered a wide topic. In order to evaluate the water contaminants, a variety of techniques and mathematical instruments are used and evaluated to see which works best. Thus, assessing its quality and figuring out how to use it for what it was designed to do are urgent tasks. With the use of software and mathematical tools, a thorough study of the water quality evaluation is simple to complete. As a result, one of the main areas of water analysis study now focused on water quality evaluation. Henceforth, Water Quality Indices (WQIs) are commonly employed as a technique for assessing water quality (WQ), as it condenses several water quality metrics into a single, significant number. (Singh et al. 2018). It is observed that many research efforts have also demonstrated how well WQI works to assess water quality in various parts of the world. (Machiwal et al. 2018). Through the integration of extensive, complicated data and the creation of a score that helps to prioritize the management plan and fund, this seeks to advance understanding of water quality challenges (Zeunert et al. 2017). Subsequently, weights were assigned to certain geochemical criteria, taking into account their relative significance in the overall assessments of surface water quality for drinking purposes. Hence, thematic maps with spatial variation based on vulnerability and potential were created using WQI. The elements' evaluated impact on primary human health was used to provide a weight value between 1 and 5. The parameters that offer a considerable risk to human health and, depending on their frequency above certain key considerations, could limit the surface water's suitability for drinking were assigned the maximum weight. Thus, it can be developed as a trustworthy method for evaluating and classifying the surface water quality for human consumption.

However, this method's weight estimation is fairly subjective, which could have an impact on the water quality results (Amiri et al. 2014). An information entropy-based weight determination technique was suggested in the assessment as a solution to this problem. However, this approach is frequently utilized as a mathematical tool for assessing the quality of surface water (Marghade 2020). Meanwhile, an innovative technique namely, Integrated Determination of Objective Criteria Weights (IDOCRIW/I), is somewhat acknowledged as the original and most likely used technique (Diakoulaki et al. 1995). This technique is a straightforward technique used in the survey of the overall quality of the water. It uses a set of factors to simplify the process and reduce the vast volumes of data to a single value that is typically dimensionless. To compute the value in order to establish the weights, is the most straightforward method of determining the weights of each water quality indicator based on how each evaluated indicator value's fluctuation degree is interpreted (Das et al. 2023). Thus, it provides crucial information outlining the overall state of the water quality, which can be quite helpful in selecting an appropriate water-treatment method to deal with the pollution issue. These uses often resulted in waste water and rural seepage water returning to the waterway as inflows. Any threat, man-made or natural, has the potential to significantly impact the socioeconomic features of the region. Hence, many scholars have developed various techniques for figuring out a criteria's weights, with the utilization of Multi-Criteria Decision-Making (MCDM) problem (Górski et al. 2019). To get rid of the subjectivity in weight allocations, to compute WQI in earlier methods, as instructed by (Mukate et al. 2019; Ukah et al. 2019), this approach was employed by (Gorgij et al. 2019; Singh et al. 2020) that reflected the parameters' innate unpredictability. However, because our environment is mostly made up of random processes, weights derived from the critical method, have discovered a multitude of uses in the domains of sustainability and hydraulic technology. To reduce the chance of accepting unsuitable substitutes in WQ, it presents a new Multi-Criteria Decision-Making (MCDM) approach, that signified to be the COPRAS (Complex Proportional Assessment) method that seeks to develop a better technique for providing a numerical expression that is cumulatively derived and describes a certain water quality level based on critic weights (Peng et al. 2020). However, comparing this procedure to other techniques, it is easier, faster, and simpler (Dash and Kalamdhad 2021). As a result, this application is used to choose among several options by calculating the Euclidean distances between an ideal best that is wanted and an ideal worst that is undesirable (Tahera et al. 2016). The efficiency of such methods is largely dependent on the conceptual framework of assessment processes and the common language used to identify and address complicated water challenges. In later case, very few studies on WQ have been claimed to date, despite the fact that a huge number of research works have been detailed for the optimization of classic WQI utilizing the COPRAS approach.

Among the most well-known of these is making use of the resources based on Geographical Information System (GIS) in the current research investigations on spatial analysis. These days, this approach is used for a variety of tasks, such as continuous monitoring, implementation, and planning at both the macro and micro levels (Srivastava et al. 2012; Azhdarpoor et al. 2019; Judeh et al. 2021). Over the past ten or so years, this method has been effectively applied to various surface water quality evaluation objectives. Additionally, decision-makers can develop or implement solutions connected to water pollution and scarcity with the help of thorough, fast, and trustworthy information provided by the integration of entropy WQIs with GIS (Soujanya et al. 2020). Many of these applications employ GIS to map surface water quality together with certain land use and land cover features, in accordance with criteria from the public and private sectors. Therefore, it has made integrating different databases into a simple manner. Hence, GIS can be a useful tool for managing water resources locally or regionally, preventing flooding, assessing water quality, calculating water availability, and understanding the environment. However, Inverted Distance Weighted (IDW) interpolation technique was applied in order to create the WQIs theme maps. It adheres to an algorithm that is used to estimate values in between observations or interpolate data geographically. Here, weights are calculated by taking the inverse of the distance between the location of an observation and the estimated point (Burrough and McDonnell 1998). When sampling is dense enough to represent the local variation that has to be stimulated, IDW produces its best results. Thus, IDW technique is perfect for analysis of data on water quality from numerous, widely dispersed sample sites. The outputs might not accurately depict the intended surface independently if the input point sample is sparse or highly uneven (Yadav et al. 2018). The Brahmani River is a vital conduit in the eastern Indian state of Odisha. This is also recognized as Odisha's vital artery. Together with industrial, agricultural, and other uses, the water is principally used for drinking by the people who live along the riverbank (Jain et al. 2004; Gupta et al. 2009; Shikazono et al. 2012; Devi et al. 2013; Bhadra et al. 2014; Mishra et al. 2014; Zakir et al. 2017). However, changing landing use patterns, growing populations, deforestation in the riverbanks and watershed, and shifting farming along the river have all put the habitats under jeopardy. Consequently, a great deal of work has gone into preserving and safeguarding the water quality (Solangi et al. 2019).

In the current work, an attempt has been made to assess the water quality of Brahmani River, Odisha in seven key areas, via 15 physicochemical factors. In the next stage, the chosen time frame of 10 years (2013–2023) is being investigated. Since the quality of the drinking water in this river basin has not yet been the subject of substantial research, this study has made an effort to determine the most effective method for investigating the application of IDOCRIW mechanism and COPRAS in conjunction with GIS for WQ assessment. Therefore, it relates to the process of identifying the study area's point and non-point sources, which is the first step in enhancing or reaching WQ and lowering pollution. Moreover, it offers a broad overview of its situation in the areas chosen for drinking.

2 Study Area

The river basin's overall drainage area is roughly 39,268 Km^2, from which 22,516 Km^2 belongs to Odisha State, 15,405 Km^2 in Jharkhand State and the rest 1347 Km^2 falls under Chhattisgarh State. The river signified as Brahmani River is near the point of confluence near Vedvyas in Odisha, at a height of 200 m above sea level (Anand et al. 2023). Here, in the study, the mean annual rainfall of the watershed area is around 1395.36 mm. The latitudinal and longitudinal extensions of the river Brahmani are 22°14′45″ N and 84°47′02″ E. The Panposh gauging station can be located at Rourkela in Odisha. The river basin's estimated yearly renewable water resources are approximately 21,920 million cubic meters. Rice, cotton, and pulses are the main commercial crops grown in the research area (Vashist et al. 2023). Intensive farming requires long-term experience, and the most popular way to increase crop yield is to use chemical fertilizers indefinitely. The Rourkela integrated steel mill in the Sundergarh district, along with the nearby NALCO Smelter plant, the Captive Power mill, the Mahanadi Coal Field Limited, the chromite mines, and the ancillary byproduct and downstream product plants, have all contributed to the region's extensive industrialization. Here, seven gauging stations are chosen based on the neighbouring mining excavations and industrial operations. In addition, rich in materials and labour at a low cost, the river basin often offers the perfect setting for the expansion of industry (Swain et al. 2023). Also, three districts namely, Angul, Dhenkanal, and Sundergarh, possess a significant portion of these sectors. Figure 1 depicts the study region with the riverside sampling locations highlighted.

3 Data Collection and Analysis

Water samples were collected in polyethylene bottles, transported to the laboratory and analysed within 48 h. Here, a stratified sampling technique has been applied due to heterogeneous pollution status throughout the river stretch. It was designed to understand the widespread pollution caused at important sites which characterize the river water quality by considering contributions from various sources that influence the quality of water. In the present research, 15 physicochemical parameters namely Temperature (Temp), pH, Dissolved Oxygen (DO), Electrical Conductivity (EC), Total Dissolved Solids (TDS), Alkalinity, Sodium (Na^+), Calcium (Ca^{2+}), Magnesium (Mg^{2+}), Potassium (K^+), Chloride (Cl^-), Fluoride (F^-), Sulphate ($SO_4{}^{2-}$), Phosphate ($PO_4{}^{3-}$) and Nitrate ($NO_3{}^-$). The parameters' units were stated in milligrams per litre (mg L^-) except pH (non-dimensional) and EC, represented in µS/cm at 25 °C. Here, the sampling protocol was carried out throughout the 2020–2023 study period. Using GPS, a global reference was acquired for every place. Every sample was collected downstream between 10:00 am and 12:00 pm. By sealing the sample vial with air inside water, it becomes watertight. All sampling sites display the water quality in real time at that specific moment. The water quality parameters

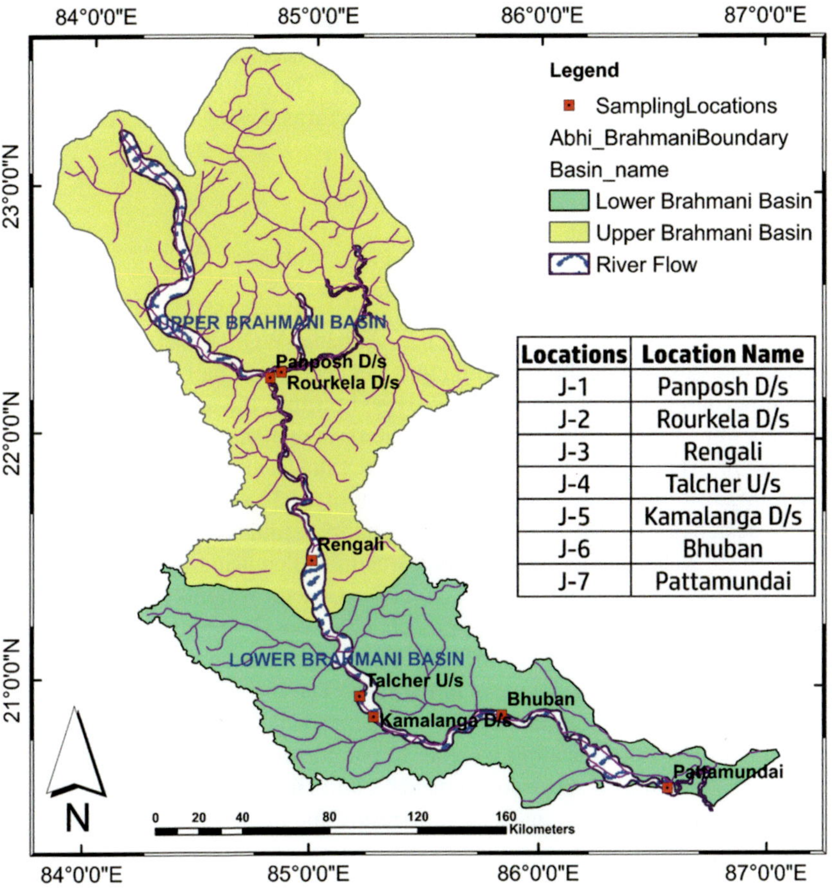

Fig. 1 Location of Brahmani River, Odisha

were evaluated in accordance with the standard methods, based on the American Public Health Association (APHA 2017). The parameters of Temp, pH, DO and EC were established in situ using a portable water analysis kit (Das 2023). After, TDS was performed by evaporation and calculation methods. Cations namely the value of Na^+ and K^+ were established by the use of dual channel flame photometer (model-2655–10). Alkalinity, Ca^{2+} and Mg^{2+} were calculated by the process of titrimetric technique, by using standard EDTA solution. Cl^- was computed by $AgNO_3$ titration. Turbidimetric technique was used for the analysis of $SO_4{}^{2-}$. $NO_3{}^-$ and $PO_4{}^{3-}$ were analysed using the UV–visible spectrophotometer. F^- is measured in ion specific electrode. Additionally, toposheets of the concerned area are gathered from the Survey of India (SOI), that were used to help with the field verification of maps and the well inventory survey in the Brahmani River Basin. The concentrations of cations and anions expressed in mgL^{-1} of each surface water sample were calculated to verify the veracity of the outcomes of the analysis, using ionic balance error (IBE). The calculated value is below the permitted upper bound of plus minus 10%, an additional demonstration of the data's accuracy, whereas, error (E) > 10% was removed from the analysis that followed. The relative standard deviation values were less than 5%. All calculations were performed using MATLAB software (MathWorks 2017).

4 Methodology

Water quality assessment was completed utilizing WQI, which is frequently employed to assess the quality of drinking water. The flow chart description of the study's procedures is displayed in Fig. 2. Based on this, the surface water quality parameters were measured during this investigation at each sampling point. The resultant values of every single parameter were then calculated using the IDW method in Arc GIS software, resulting in the conversion of all 15 variables into grids. It is therefore designated for use in surface water quality testing as well.

The surface water quality for this study was modelled using two effective techniques, i.e. Integrated Determination of Objective Criteria Weights (I) and COPRAS (Complex Proportional Assessment) approach. However, (I) method is a widely used MCDM technique that determines the objective weight for the criteria. It measures the degree of chaos and randomness in the data. The main benefit of the entropy technique is that it produces results that are dependable because human involvement in the process of allocating weights to parameters is eliminated. After that, this phenomenon removes the decision maker's influence from the decision-making process by using correlation analysis to measure the value of each criterion and

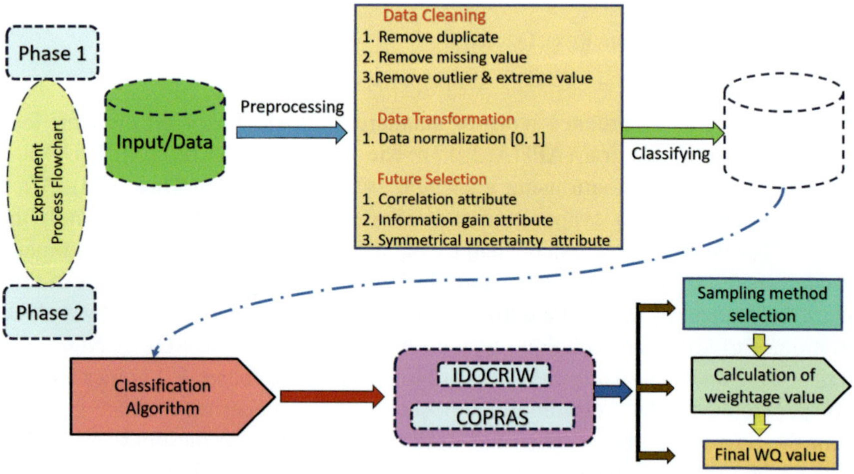

Fig. 2 Flow chart showing the steps involved in the identification of water quality points

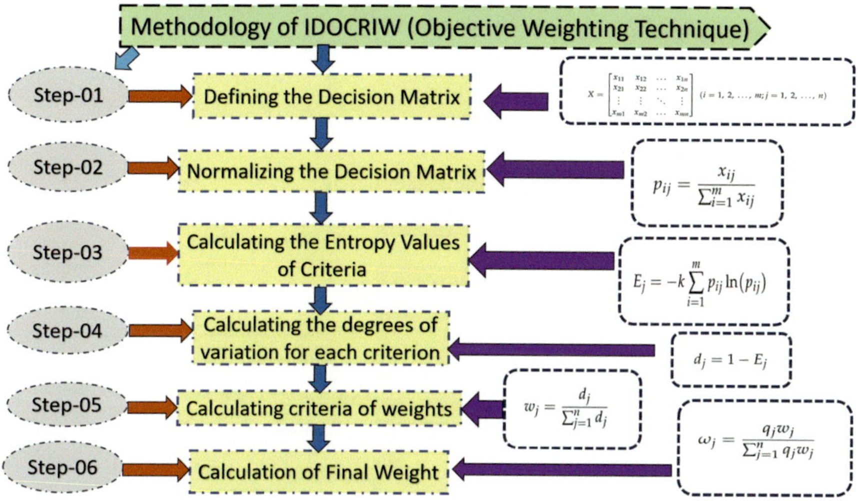

Fig. 3 Framework to calculate IDOCRIW weights in the specified study area under investigation

assist in computing the objective weights of each criterion. Thus, it is a productive method of presenting probability and uncertainty. The stages involved in estimating integrated weights $(w)_j$, are explained in accordance with Diakoulaki et al. (1995). The six-step approach was used to calculate the surface water weights in the research area, as shown in Fig. 3.

Another method popularly named as the COPRAS method, first introduced by Brauer's (2006, 2008), is an important multi-objective optimization method that has the potential to be usefully used to deal with a variety of difficult decision-making situations (Das 2023). It is composed of two elements, namely the reference point method and the ratio (Khan et al. 2016). Similarly, the water quality parameter's weight was assessed using (I) technique. A ratio system will be created by standardizing the decision matrix's data, that can be explained in step 1. Further, $(r)_{ij}$ is a number without dimensions that falls between o and 1. These normalized performance measures are combined when it is advantageous, i.e. 'larger-is-better' attributes and deducted if a feature is not advantageous such as 'lower-is-better'. This gives the row sum's overall assessment value as well as its performance metrics that can be seen in the following step 3. It is common knowledge that in some contexts, some traits are more important than others. Therefore, in order to give these features more weight, they are multiplied by the corresponding weight. Following the weight consideration, the equation will be as stated in step 4. Overall, the weight of each parameter used to determine the performance score value for each sample location is more precise and thoughtful with this ranking. Meanwhile, IDOCRIW data and its weights are crucial factors in this process that represent how different sources of pollution affect the overall quality of the water suitable for drinking; also, it condenses a bigger amount of data into each sample location's complete and illuminating value.

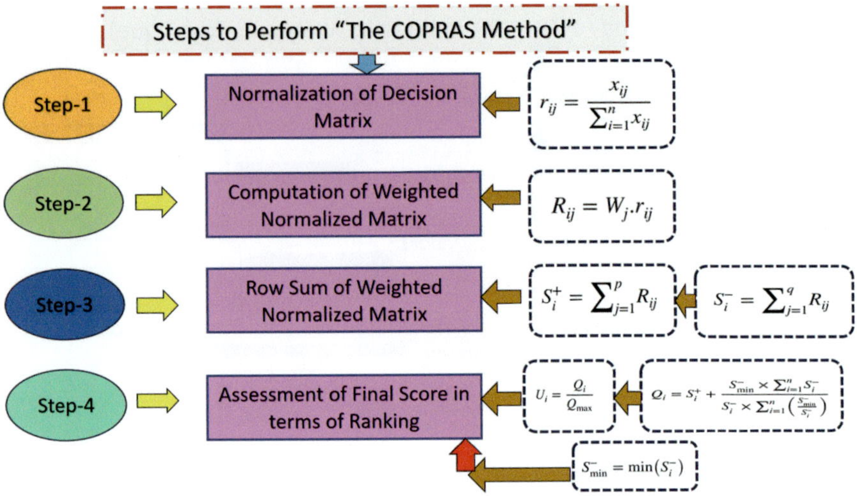

Fig. 4 Model to calculate COPRAS approach, across all sampling sites

The four-step procedure was used to calculate ranking (U_i) of surface water in the current investigation, as demonstrated in Fig. 4.

5 Results and Discussion

The 15 surface water samples' physicochemical properties were statistically assessed from upstream to downstream, and the results, including minimum and maximum values and the associated allowable limits, were displayed in Table 1. Further, it is recommended to compare the parameter's concentrated values with the World Health Organization's (2017) drinking water standards (Laltu 2023). This article discusses the creation and spatial distributional pattern of all the indices of the relevant sites and it is represented below for a brief analysis.

Physical Parameters of Surface Water Quality

The temperature data were recorded as 25.72–30.36 °C. The small range of latitude along the river may be the cause of this outcome. Temperature fluctuations are caused by the action of impervious surfaces (Yang et al. 2015). There isn't a temperature threshold that has been established by the WHO standards (WHO 2017).

pH value is one of the most important indicators for surface water's acidity or alkalinity and hydrogen ion concentration (Ramesh et al. 2010). The value found to be ranged from 7.51–8.07, in the present work. The pH of the river water samples varied little overall, and the influx of sewage or home effluents with high carbonate content from surrounding rural areas, as well as agricultural activities like crop harvesting, may be the cause of a slightly higher pH. Its alkaline nature is due to the presence of hard water minerals, and variables influencing alkalinity can include the outflow of agricultural waste water. Otherwise, changes in salinity and temperature, the breakdown of organic matter, the dilution of bicarbonate, and the mixing or dilution of effluents by aquatic streams can all cause variations in the pH of the water system. When levels are outside or below the advised range, though, it can be a sign of possible problems, including too much bicarbonate dissolved in drinking water. This implies that every sample in the research region is safe to drink. Lower reaches eventually collect and exhibit a progressive rise in pH levels.

DO is an essential parameter for assessment of surface water quality because it influences the organisms living within the water body. From the study, DO was noticed as 6.78–7.67 mgL^{-1}. The projected values of DO exceeded the upper bound of the standard score of 5.0 mgL^{-1}. However, optimum amount of DO is still reported from the river water. As a result, temperature and DO have an inverse relationship because the solubility of oxygen depends on temperature and it decreases as the water temperature increases. In certain cases, low DO can occasionally be seen, and this is brought on by inorganic compounds such as Fe^{2+} and Mn^{2+}. Such inorganic chemicals are thought to be undesirable in excess because they negatively impact sanitation and water distribution systems.

TDS is crucial in figuring out if the water is clear or murky. A higher value in the water will lessen its flavour, odour, and aesthetic appeal. It will also prevent aquatic organisms from seeing light. Also, it measures the quantity of water and serves as a gauge for the overall quality of the water. In the current studied area, the TDS readings were found to be within the range at each sampling station, i.e. WHO, 500 mgL^{-1}. From the calculated results, the value is situated between 86.39–182.72 mgL^{-1}. However, the flavour of water with very low TDS is tasteless (Yadav et al. 2018). Comparatively higher TDS values at J-(1), (2) are the result of home wastewater, industrial wastewater, and agricultural runoff being dumped into the river (WHO 2017).

Pollution may be indicated by a sudden increase or decrease in the water body's conductivity. However, because anions like phosphate, nitrate, and chloride are added, sewage discharge and agricultural runoff will increase conductivity. As a result of the engagement of numerous chemical reactions, it reflects a great deal of variance in the salinity of surface water. EC suggests that is linked to the total amount of soluble salts and plays a significant role in the mineralization of surface water, as stated by Kosal-Sahin et al. (2019). The concentration of dissolved materials, mostly salts, has a considerable impact on it. Additionally, it is observed that an increased EC value indicates the TDS and that this indication is employed to measure total ionic concentration (Song et al. 2020). Water-containing salts have a negative impact on crop productivity because they make the soil more salinized, which makes it unfit for farming and drinking. After this, Bruce and Hardeep (2016) suggested a report, that portrays about crops absorb nutrients in the form of salts, which when they dissolve, create ions, both positive and negative. These ions combine to create an aqueous solution with electrical conductivity (EC). Consequently, elevated EC values at J-(1), (2), and (7) signify an overabundance of nutrients in water, which can cause problems such as nutritional imbalances, osmotic pressure, and ion toxicity. Conversely, very low EC at J-(3), could be a sign of low nitrogen levels in the water, which would hinder plant growth. The observed value in the current work was found to be 151.44–311 μS/cm. It is taken as 250 μS/cm in the current work, which is also pointed towards as the optimum limit for EC in drinking water (WHO 2017). It is noteworthy that every sample falls within the safe category. In turn, the pH, TDS, and EC are important physical and chemical factors that determine water quality indices depending on the hydrology and hydrogeography of the nearby river environment. The usage of river banks and beds for agricultural purposes, together with the application of agrochemicals and vertical erosion of the riverbed, raises the values of the aforementioned parameters, even though they were within norms in the majority of the samples.

Afterwards, Alkalinity is a measurement that gives an idea of the typical salts in water and how well water reacts with soap to create a lather. Higher alkaline water, on the other hand, prevents pH shifts that are harmful to amphibian survival. Desirable limit suggested by WHO (2017) is 200 mgL^{-1}. Value was found in the range of 51.67–72.89 mgL^{-1}. Water at J-(1) was contrasted with alternative stations and revealed a higher alkaline, which could be brought on by the extra presence of these salts, the disposal of sewage, or the release of industrial effluents into the river (Lumb et al. 2011; Goel et al. 2018). All locations, though, are confined within the ideal range. Occasionally, the lithology of the regions may be the defining source of the excess alkalinity at J-(2) as the carbonate minerals dissolve into the rock and soils. Additionally, the increased alkalinity at J-(7) might serve as a buffer to lower the water samples' pH and hardness. Industrial effluents, particularly from the textile and related industries, which use salts and surfactants to improve the colour's binding to fabrics, maybe the anthropogenic source.

Major Cations of Surface Water Quality

Based on Schiavon and Moore (2021), Na^+, Ca^{2+} and Mg^{2+} perform important functions in stabilizing the soil's structure. On the other hand, overapplication in water would cause the ions to be dispersed from soil colloids, resulting in soil deflocculation and ultimately permeability issues.

Na^+ affects agricultural productivity in addition to the physical characteristics of the soil, i.e. permeability (Ali et al. 2018). Here, high salt concentrations can negatively impact soil permeability and structure, resulting in alkaline soils (Muniraj et al. 2021). This restricts the quantity of water and air that may seep into the soil. Based on Ndoye et al. (2018), there exists a conversion process between soil and water. As a result, when clay particles absorb the calcium and magnesium ions that give soil structure, sodium ions take their place. However, Na^+ can leak into naturally occurring water (Islam et al. 2021f), and can alter the structure of the soil, making it harder for plants to grow. Therefore, hypertension and cardio-metabolic disorders may arise from consuming an excessive amount of Na^+, in addition to regular dietary Na^+ usage (Nwankwo et al. 2020). The suggested threshold of Na^+ in drinkable water is taken as 200 mgL^{-1} (WHO 2017). In the current study, the recorded values fall within a spectrum of 5.39–15.38 mgL^{-1}. All locations remained below permissible level.

K^+ is a vital nutrient that, when taken in excess, may have laxative effects (Alam et al. 2019). The main sources of K^+ in surface water are orthoclase feldspars and potassium fertilizer. This indicator is crucial for raising crop yields by improving the quality of irrigation water. Here, without the use of K^+, long-term agriculture will cause soil fertility to decline. This makes it a crucial component for better irrigation and drinking. In this area, the K^+ value was recorded from 1.76–7.24 mgL^{-1}. It is recommended that the most amount that can be present in drinking water is approximated to be 12 mgL^{-1} (WHO 2017). However, all sites in the chosen time duration it showed comparatively lower K^+ concentrations.

The main causes of water hardness are anions like carbonate, bicarbonate, chloride, and sulphate, as well as cations like calcium and magnesium. Thus, Ca^{2+} serves as a crucial dietary supplement for healthy bone formation (Chakraborty et al. 2020a). The studied area's elevated Ca^{2+} content is attributed to the presence of shell pieces mixed with medium- to fine-grained sand. It is plentiful in the direction of the sea, with calcareous bacteria and broken shells predominating. Because of the regular turbulence in the water, the presence of $Ca\ CO_3$, appears to be an increased inflow on account of calcareous biogenic material, while the intermediate and lower levels of metal contamination are building up in the sediments. Numerous variables contribute to this, including the higher concentration of carbonate in carbonate deposits and deposits connected to floods seen in tropical regions and the higher migration and mixing of the terrigenous fraction from continental sediments over the sample period. Industrial waste, agricultural waste, and geological sources could all be utilized to produce Ca^{2+} in river water. Regarding potable water, the suggested threshold is estimated as 75 mgL^{-1} (WHO 2017). The research's observed values varied within the range of 29.6–66.44 mgL^{-1}. Likewise, every sampling place that was investigated, complying beautifully inside the suggested bounds.

On the other hand, biotite and hornblende are the primary resources of Mg^{2+}, within the surface water. Here, long-term use of this indicator may lead to growth retardation, reproductive problems, diarrhoea, cardiovascular disorders, and water hardness. (Fatoba et al. 2017). Furthermore, the increased concentration of Mg2 + in water facilitates the uptake of Ca^{2+} or K^+ by plants, resulting in their deficiency in plant tissue. Additionally, under some soil texture conditions, crop yields are decreased by the use of irrigation water containing Mg^{2+}. It results in surface crusting, pore clogging, and swelling and dispersion of soil clays, which prevents recharge water from penetrating the soil. This circumstance makes it harder to find enough water. Therefore, its values were discovered in the current investigation to be within the range of 18.78–29.22 mgL^{-1}.

In drinking water, the necessary permissible limit is 30 mgL^{-1}. The bulk of surface water is below the allowable Mg^{2+} limit. The river water system's overall cation trend was as follows: $Ca^{2+} > Mg^{2+} > Na^{2+} > K^+$. Numerous natural and related human influences, including adjacent industrial discharges, urban runoff, home effluents, and numerous diffuse or non-point sources as well as atmospheric deposition, are probably to blame for this tendency.

Major Anions of the Surface Water Quality

Despite the fact that the main anions, F^-, Cl^-, $SO_4{}^{2-}$ and $PO_4{}^{3-}$ are especially significant in light of the seasonal variations brought about by human inputs.

Thus, drinking water that has been infused with F^-, contains a suggested value of 1 mgL^{-1}, that is very advantageous to human health (WHO 2017). Also, F^- at this range promotes the remineralization of dental enamel, preventing tooth decay (Abbas et al. 2020). In the area, dental fluorosis is linked to high F^- levels. The value of F^- was shown to be between 0.3–1.16 mgL^{-1}. As a result, every surface water sample that was analysed revealed that it was safe to drink and did not pose any health risks except J-1 (1.16 mgL^{-1}). Because of its low quantities, it might be utilized as a substitute for surface water sources, which in the northeastern region have a high concentration of F^-. It is exhibited that higher concentration of F^- at J-(1) may also be related to agricultural operations, drinking habits, and the discharge of household and livestock excrement, poultry waste, and fertilizer packets that are rinsed with return flow drainage and absorption, before reaching surface water (Pius et al. 2012).

From the results, it is valid that Cl^- and SO_4^{2-} were the most enriched anions. They were heavily loaded, suggesting that anthropogenic activities, rock-water interaction, ion exchange, natural weathering, non-point sources, and chemical dissolution all had an impact on the quality of surface water. Likewise, Cl^- enhances the corrosiveness of water, gives it a salty flavour, and when exposed to high Cl^- levels, it may have a laxative impact on people. (Sadat-Noori et al. 2014). This could be caused by the area's geology, agricultural runoff, industrial effluent, and domestic sources. Since salts contain Cl^-, an important micronutrient for life, it has no negative effects on human health. The quantify of Cl^- concentration in this investigation, was discovered in a span of 8.21–23.59 mgL^{-1}. It is considerably under the maximum allowed for Cl^- in drinking water, i.e. 250 $mgL\text{-}^1$, as stated by WHO (2017). Out of seven locations, J-1 holds a value of 23.59, which surpasses all other sites. Here, elevated Cl^- content is largely brought on by fertilizers and wastewater, which has an impact on ecosystem services including fishing and lake ecology.

Afterwards, NO_3^- is present in surface water as a result of fertilizer runoff from agricultural land and sewage (Gao et al. 2018). This pollutant characteristic is brought on by specific industries improperly disposing of wastewater as well as by the misuse and seeping of chemical fertilizers and manures. Nevertheless, this parameter poses a risk to human health, particularly in newborns, as it can lower blood's ability to carry oxygen, which can result in methaemoglobin and gastrointestinal malignancies. According to the findings, nitrate was within the range of 0.65–3.53 mgL^{-1}. Increased levels result from direct effluent discharge, overuse of surface water, efficient ion leaching, and agricultural effects. Meanwhile, the desirable limit of nitrate is considered 45 mgL^{-1} as instructed by WHO (2017) regulations. Therefore, in the current area, the concentration of NO_3^- recorded is seen to be quite low everywhere.

In the next stage, SO_4^{2-} may get into water streams as a result of surface runoff from contaminated areas and sewage discharge (Bruce et al. 2021). Excessive concentration causes gastrointestinal distress and has a purgative effect on people (CPCB 2008). In this discussion, the increased values at J-(1) are a result of sulphur gas emissions from industry, which oxidize and infiltrate into the water following precipitation. However, runoff from fertilizer, sewage disposal, and industrial effluents are some examples of anthropogenic sources found in water bodies. The value of SO_4^{2-}, in this work, was noticed in a range of 7.89–40.12 mgL^{-1}. When its concentration surpasses the WHO (2017) guideline threshold, i.e. (200 mgL^{-1}), it may also become unstable and can have a laxative impact on people's health. Thus, lowest value was noted at J-(3) and the highest at J-(1). The claimed higher readings because of this parameter, are a result of sewage overflow from factories. Moreover, given the research area's high level of agriculture-driven activity, this might be connected to runoff from agriculture. (Zango et al. 2019; Das (2023). At J-(2), it has a marginally greater value, alters taste, and when combined with sodium and magnesium, has a laxative effect (Narsimha et al. 2017b).

Moreover, the usage of detergents could increase the PO_4^{3-} concentration in water to a considerable degree. The amounts were within the range of 0.07–0.29. In fine, elevated scores of this parameter at J-(3) are noticed. Thus, higher concentrations typically exceeding the allowable limit, can result from a variety of factors, including residential sewage, septic tank leaks, agricultural activities, leaching from the negligent dumping of animal waste, and untreated drainage and sewerage pipes. (Le et al. 2022).

Regarding the aforementioned context, the following order of anions concentration follows the sequence of $SO_4^{2-} > Cl^- > NO_3^- > F^- > PO_4^{3-}$ at J-(1), (2), (4) and (5) whereas $Cl^- > SO_4^{2-}$ at J-(3), (6) and (7). As previously mentioned, the study region includes agricultural areas. The main problems found there were septic tanks, dumping yard leachates, and inappropriate municipal waste disposal management. Additionally, as was seen during the sample collection and preliminary assessment of the research region, the main sources of anthropogenic contamination were the use of synthetic fertilizers and pesticides, trash disposal in open land, municipal waste disposal, and runoff from agricultural waste. Figure 5a-o reflects the above-illustrated spatial map of all water quality parameters. It makes evident how water moves from an upstream (U/s) to a downstream (D/s) direction.

Regarding the two primary cations' concentration ranges such as Ca^{2+} and Mg^{2+}, it displays notable variations in the first two U/s locations (J-(1) and (2)). On the other hand, D/s portion varied greatly, even though the trend weakened as it approached the finish line. Thus, seasonal variations in runoff quantities and natural geochemical processes that regulate catchment weathering are ultimately in charge of causing the downstream to become loaded with dissolved metals (Das 2023). These findings provide important new insight into the potential impacts of surrounding roadside dumps and agricultural fields' leaching and runoff. In an additional instance, Surface water chemistry varies, and the river's upstream is more degraded than mid-stream and also, very poor water is being showed at the downstream site, i.e. J-(7). This is generally brought on by industrial pollutants that have mixed into the river through its tributaries. This demonstrates how the introduction of catchment runoff into this riverine system might be responsible for notable alterations and deterioration of water quality.

The extent to which the river water quality deviates from the drinking standard cannot be determined solely by looking at the values of these factors. Consequently, this study was conducted to assess the river water quality using the IDOCRIW mechanism, and the sources of pollution were identified using COPRAS, helping to ensure

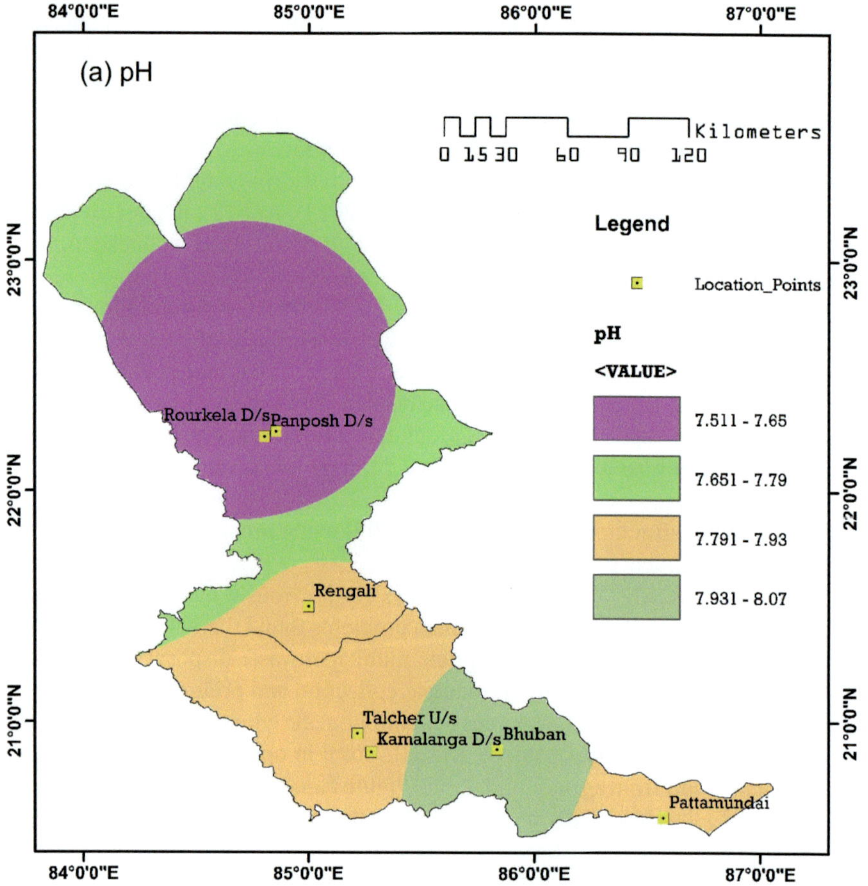

Fig. 5 The annual values for all sampling locations during the period of observation **a** pH, **b** DO, **c** TDS and **d** EC **e** Alkalinity, **f** Na^+, **g** K^+ and **h** Ca^{2+} **i** Mg^{2+}, **j** F^-, **k** Cl^- and **l** NO_3^- **m** SO_4^{2-}, **n** PO_4^{3-}, and **o** Temperature

that the water is safe to drink. Afterwards, the data set is subjected to both of these methods in order to determine whether the water quality is suitable for drinking (Niyoyitungiye et al. 2020). This method makes it possible to evaluate the physiochemical characteristics of water quality more precisely (Singh et al. 2020). The determined amount of I-WQI is situated between 67–175 (Fig. 6), representing "good to poor" class of water. A higher weight score was assessed at J-(1), which included elevated levels of 11 variables namely, "TDS, EC, Na^+, K^+, Ca^{2+}, Mg^{2+}, F^-, Cl^-, NO_3^-, SO_4^{2-}, PO_4^{3-}" and has the minimal DO level. Here, reduced DO encourages organic pollutants. The water quality indicator at J-(4) site has the lowest weight value, indicating good quality. The results indicate that the river's water quality was superior at J-(4), (5), and (6) as a result of the impacts of dilution and higher water depth. So, there was no very bad or inappropriate water quality at any of the sampling

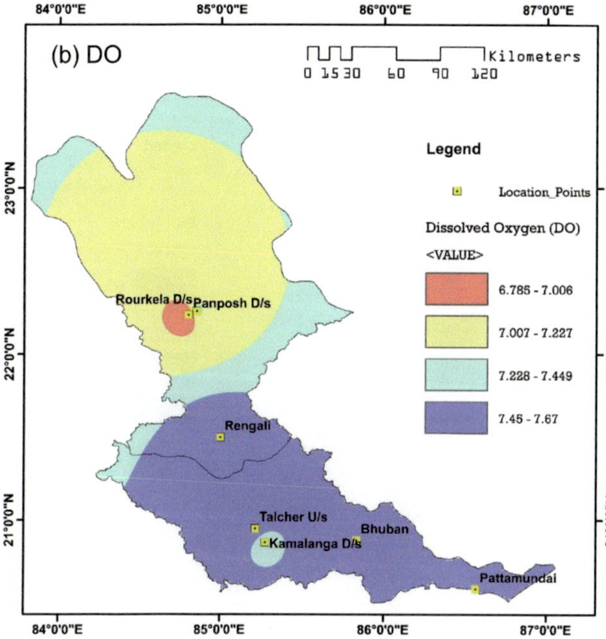

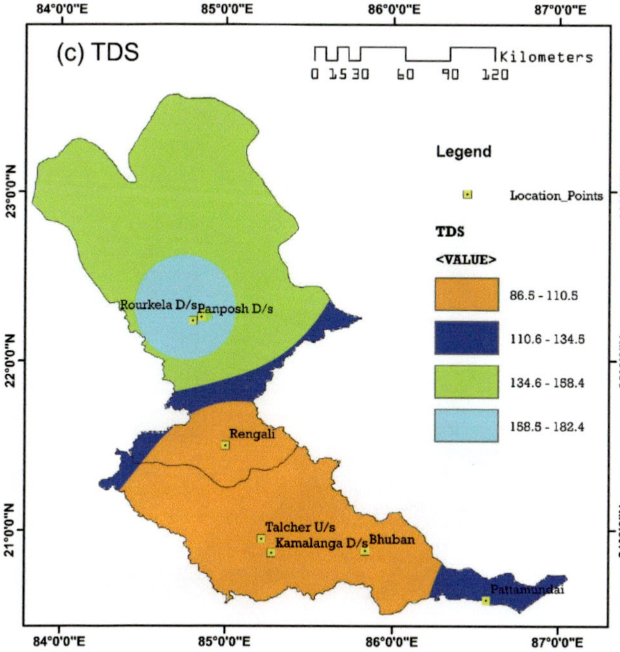

Fig. 5 (continued)

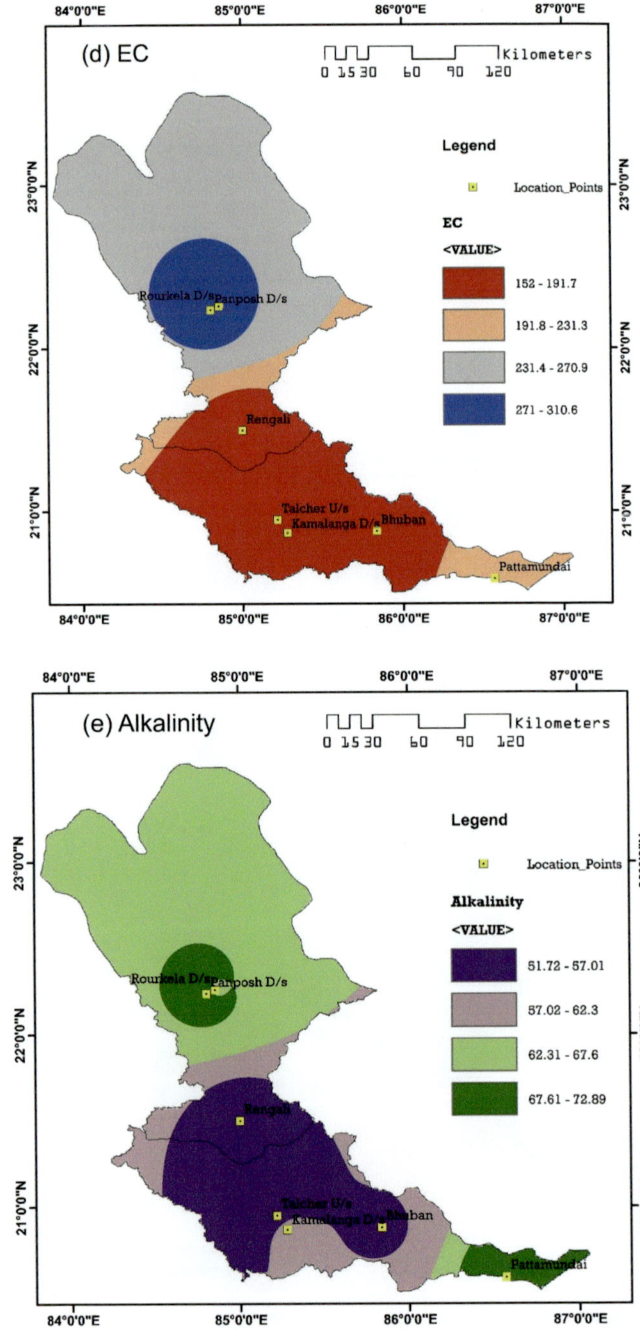

Fig. 5 (continued)

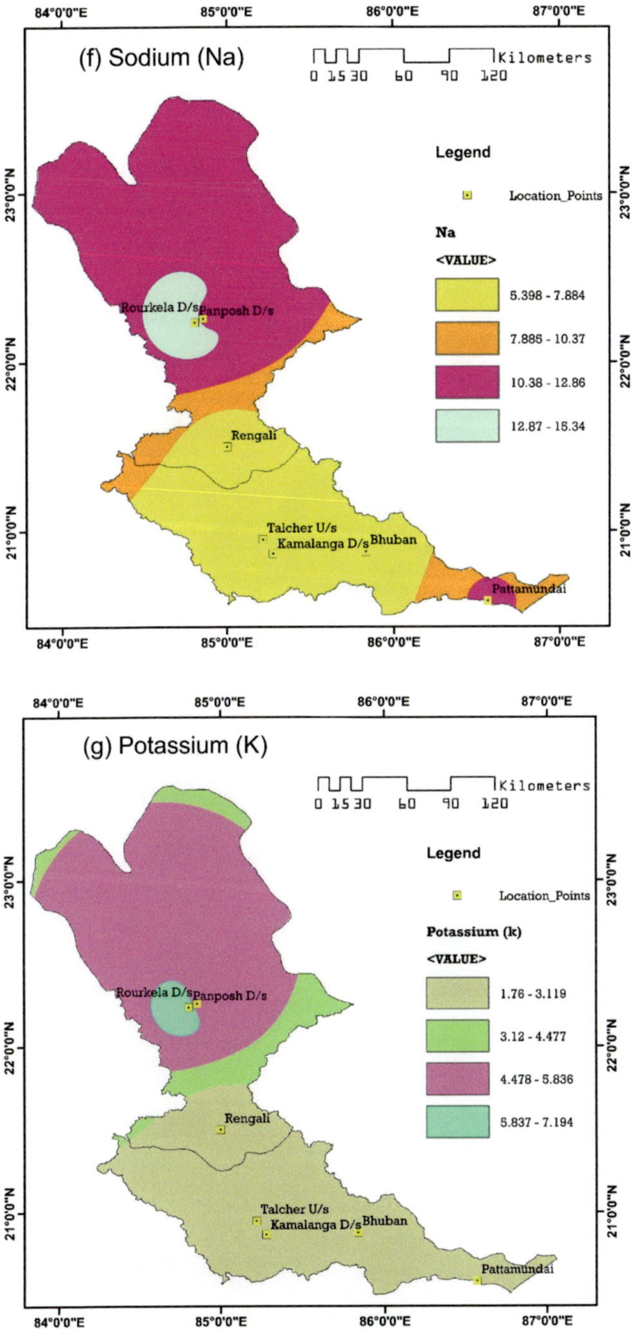

Fig. 5 (continued)

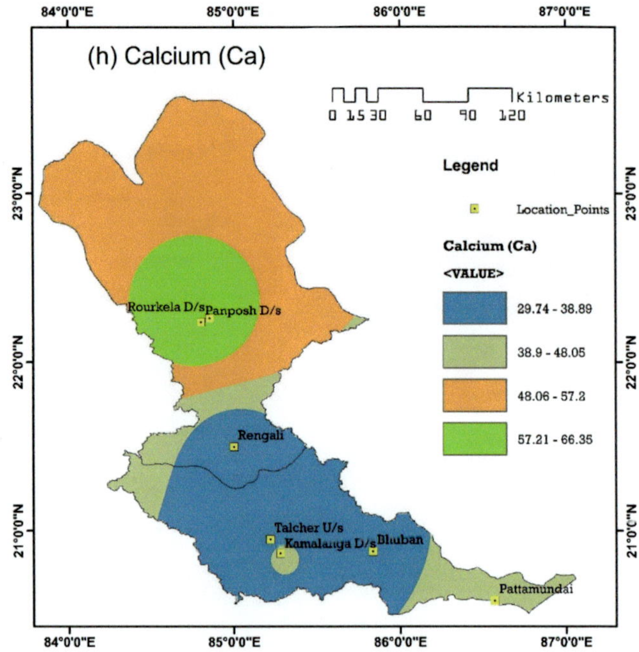

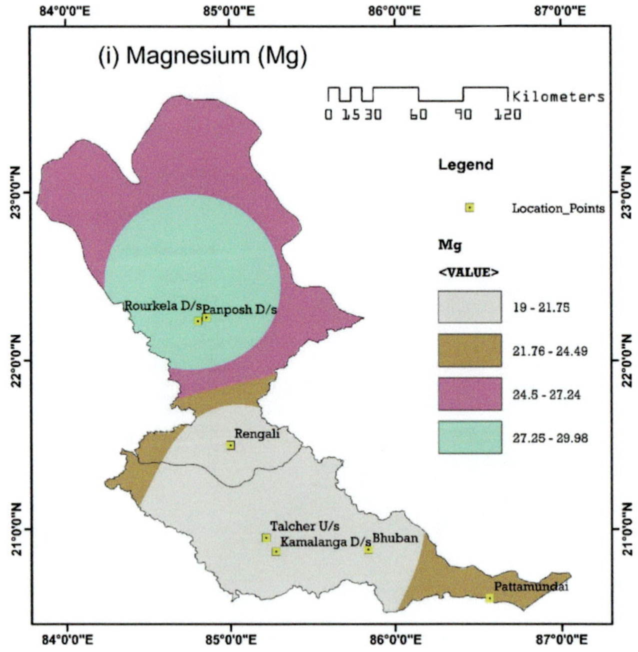

Fig. 5 (continued)

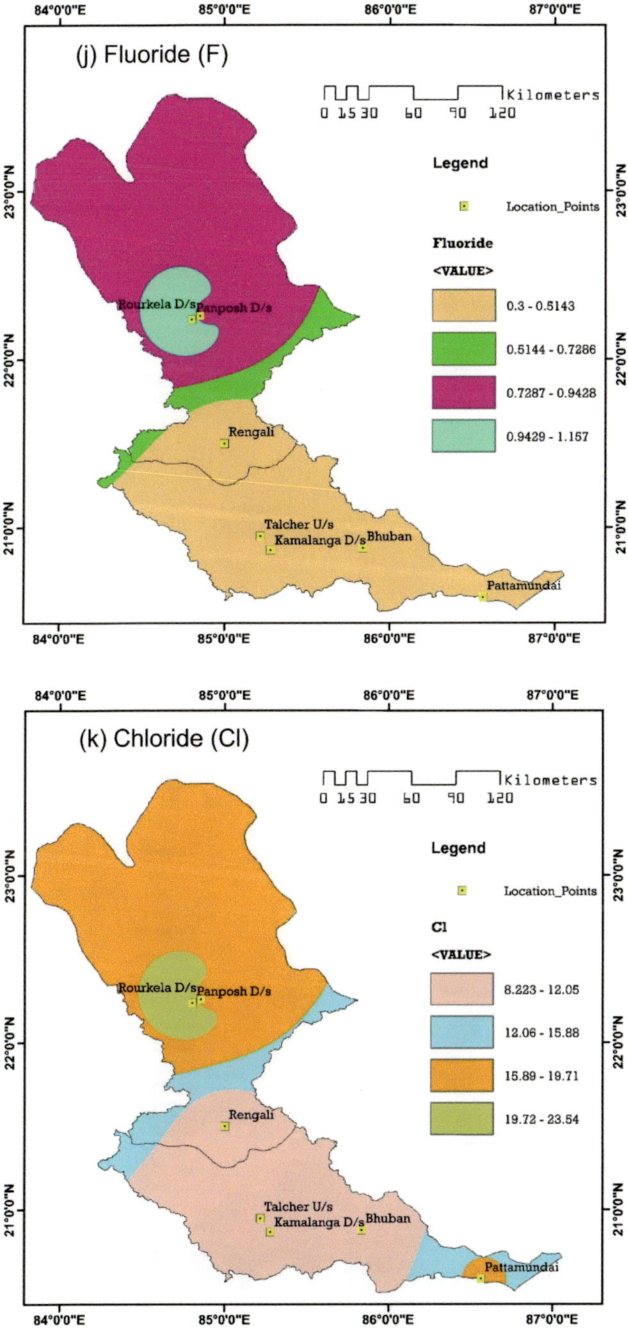

Fig. 5 (continued)

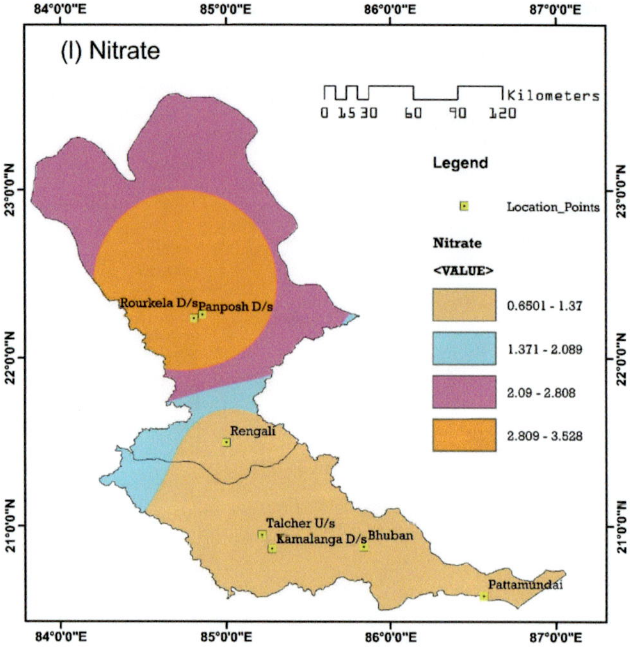

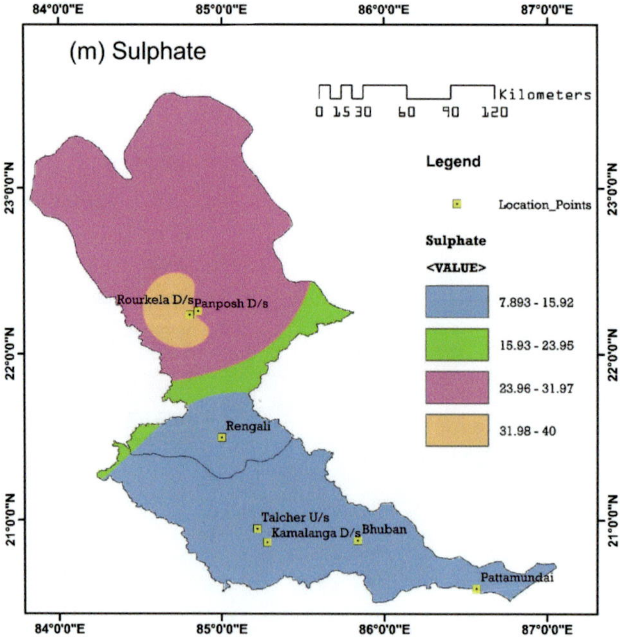

Fig. 5 (continued)

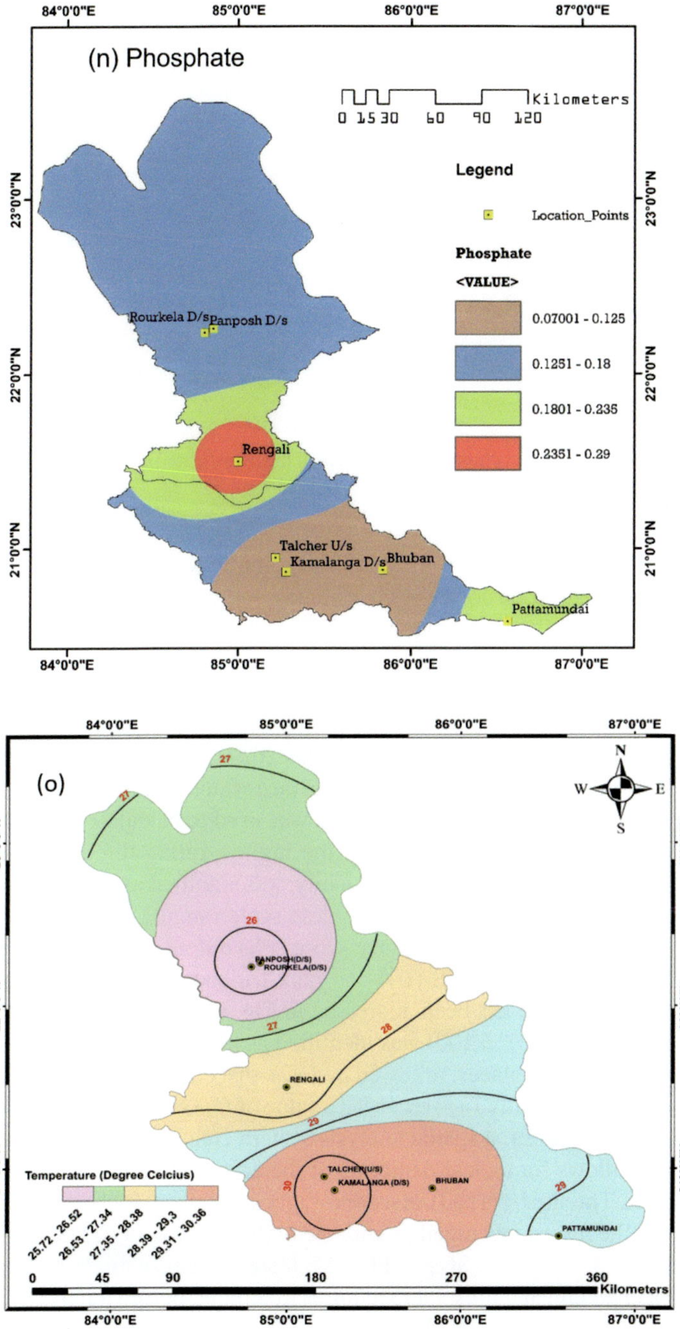

Fig. 5 (continued)

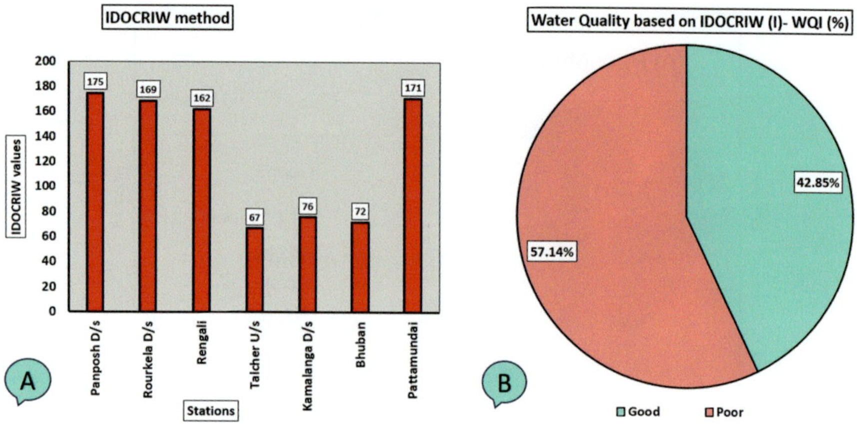

Fig. 6 Variation of the physicochemical parameters of the investigated surface water points **a** Score of IDOCRIW and **b** % of distribution

sites. Further, the spatial data utilized in this region was interpolated using ArcGIS 10.5's IDW tool based on location coordinates and quality settings. The water at J- (1), (2), (3) and (7) were categorized as being of poor quality, on the basis of $(I)_{IDOCRIW}$ coefficients. Thus, it may be said that at these locations, as highlighted in Fig. 7, the discharge of wastewater from fish farms, construction sites, agriculture, altered rainfall patterns, and construction projects has resulted in a reduction in the quality of the water (Wagh et al. 2019; Das 2024). This emphasizes the adverse effects of human activity on the quality of surface water (Yadav et al. 2018).

In this result's inquiry, COPRAS method was used to calculate, assess, or create overall rankings so that the station with the highest rating throughout each time would be the most contaminated. Additionally, it is signified as a helpful instrument for making decisions. A single dimensionless scale ranging from 0 to 1 has been created by normalizing the performance score (PS) output, as described by Hwang et al. (1993). Finally, overall PS (U_i) and rank for each site have been calculated using the technique, that has been addressed in Fig. 8. As you look at in Fig. 8, the sampling location J-1 (U_i = 100) was identified as the spot that was most impacted when compared to other places, followed by J-(2), that consists a score $U_i = 83.67$, J-(7) as $U_i = 61.34$ and finally J-(5) recorded a value of $U_i = 52.45$. Using the PS values of seven station locations, a spatial analysis system created an appropriate diagram in Fig. 9, that allows for a clear depiction of the areas classified under several WQ characteristics. The site J-(7) had the second-highest weight, which was obvious. This demonstrates the lowest DO together with the highest values of "TDS, EC, Na^+, K^+, Ca^{2+}, Mg^{2+}, F^-, Cl^-, NO_3^-, SO_4^{2-}, PO_4^{3-}". Here, the cause might be water with a greater critical weight and rank value requires treatment, before it may be consumed or used for other purposes. (Hedge et al. 1995). Meanwhile, Certain sites exhibit runt behaviour and are individually accountable for contributing to the pollution within the basin. However, the quality of surface water at these previously mentioned

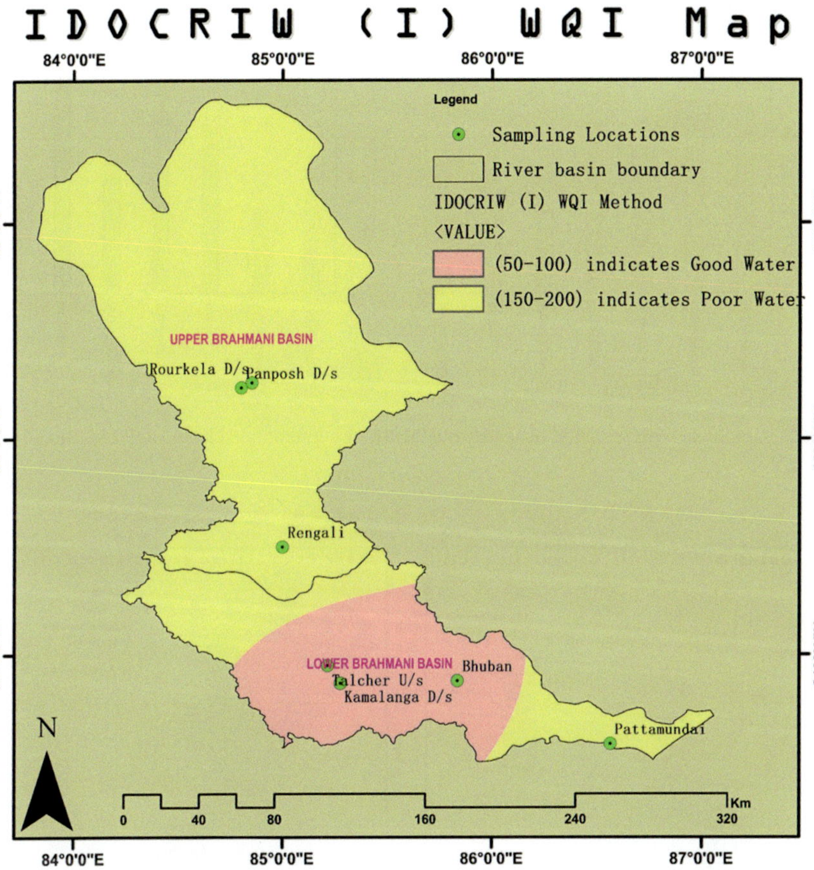

Fig. 7 Spatial distribution map based on IDOCRIW approach

locations has been negatively impacted by large agricultural activity, improper waste management, and urbanization (Egbueri 2020; Das 2023).

Therefore, it is confirmed that both strategies are crucial for acting as platforms for policy decisions and river restoration that can be given top priority (Alam et al. 2019; Das 2024). In this current topic, because of the availability of only 15 physicochemical indicators, both $(I)_{IDOCRIW}$ and COPRAS methods have been effectively carried out. The research determined that the work being done now is only for ten-year (2013–2023) timeframe. Therefore, in order to obtain a comparative result, seasonal fluctuation needs to be included in the next monitoring program.

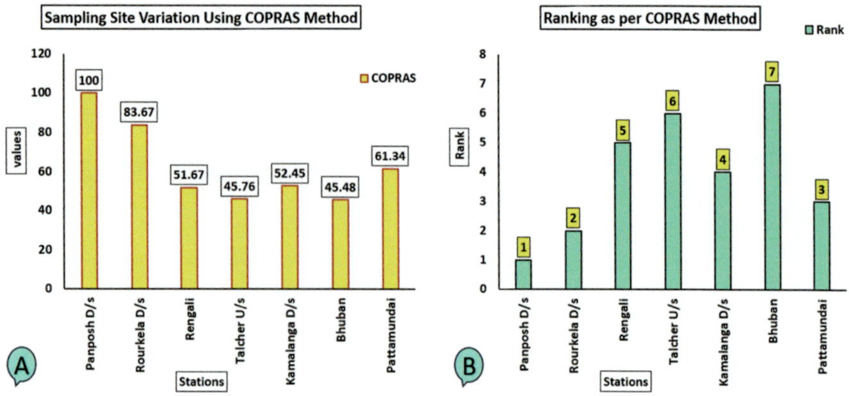

Fig. 8 Graphical representation of scores and their ranks commutated from COPRAS model

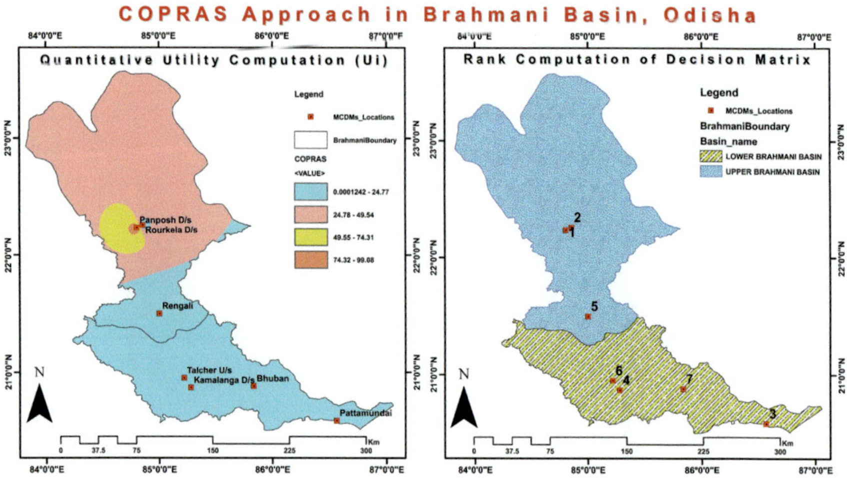

Fig. 9 Spatial distribution of COPRAS map in selected gauging points

6 Conclusions

A comprehensive surface water monitoring approach is necessary for the long-term sustainability of regional economies and livelihoods. The main source of ecosystem contamination has always been water. It is now necessary to evaluate the river's water quality due to the recent water crisis in the area. The primary hazards to the viability of surface water in this watershed are the salt accumulation of surface water, intensive agricultural methods, and the use of associated fertilizers. In the studied region, water samples from seven locations were considered to test 15 physicochemical parameters in the selected sampling sites for a duration of 10-year interval, i.e. 2013–2023. Therefore, samples reflecting the overall pollution load were gathered annually at discharge locations. This paper has brought attention to a surface water quality assessment for drinking purposes through the coordinated use of IDOCRIW-based WQI with reliability- based MCDMs such as COPRAS have been employed. It has been shown that the IDW interpolation technique, along with GIS-based spatial analysis, is a highly effective tool for representing the distribution of significant ions in the research area. In ArcGIS 10.5, the IDW technique was implemented. Using the same methodology, an I-WQI and COPRAS map was also produced to enhance comprehension of the water potability throughout space. From the resulting findings, DO was discovered to be a factor that improved the quality of the water. In addition, the river turned out to be a little alkaline. Here, complex data were evaluated with I-based WQ and COPRAS (U_i) methodology. The study reveals that I-WQI for the region varies from 67–175, indicating good to poor category. Based on this, it exhibited 42.85% (n = 3) water sources were of good quality and 57.15% (n = 4) showed poor water quality. It is found that J-(1), (2), (3) and (7) are the main observed poor water sites, that holds more impact on water quality than other stations. Higher TDS, EC, Na^+, K^+, Ca^{2+}, Mg^{2+}, F^-, Cl^-, NO_3^-, SO_4^{2-} and PO_4^{3-} values were also recorded. It was experiential that the degree of pollution at these stations was more closely linked to a wide range of expanding human activities, such as excessive water use, fertilizer effects, agricultural runoff, and industrial activity in and around the river corridor. Based on the results, the composition of surface water is mainly influenced and controlled by hydro geochemical processes such as sea water mixing, mineral phase dissolution, water rock interaction, and ion exchange. COPRAS methodology served as a useful technique for sorting sampling sites and indicated that J-(1) was the most polluted followed by (2), (7) and (5) in comparison to other sites. To reduce pollution levels in water bodies and stop them from changing the physical and chemical structure of drinkable water. It is crucial to monitor the treatment and disposal of home, industrial, and sewage waste. Thus, the adopted methods present an overview of the various aspects of water quality frameworks and their potential to solve the water management issue for drinking use. It also provides a practical guide for planning a water resource project. The limitation of this study: Details research may be required to analyse the impact of other land use/land cover, geological, and topographical existing in and around the catchment on a variation of surface water level in the river basin. For future research, in order to provide a more complete picture of the health of water resources, the WQIs can be utilized in conjunction with other metrics like biological integrity, trace metals, and artificial intelligence to improve WQ prediction. Conservation measures are necessary to protect this valuable resource. It is also needed to carry out a climate resilience- sustainability-vulnerability assessment on the behavioural and cultural aspects of the local population to develop a comprehensive study on the effects of climate and water quality change on household and small holder farmer's practices. Proper management and monitoring measures ought to have been implemented in the study area to ensure the quality of surface water resources and prevent salinization, thereby promoting their safe use.

6.1 Credit Authorship Contribution Statement

Abhijeet Das: Conceptualization, Data Curation, Methodology, Software, Validation, Formal analysis, Investigation, Resources, Project Administration, Funding Acquisition, Writing—review and editing, Visualization.

Acknowledgements The author extends their appreciation to State Pollution Control Board, Odisha. The author would like to acknowledge anonymous reviewers for their contribution to this manuscript. Author would also like to thank Rani Das, Ruchika Square, Bhubaneswar, for providing financial support for carrying out this innovative work.

References

Abbas DA, Ali KK (2020) Water quality of groundwater and Diyala River in Jisr Diyala area within Baghdad City—Iraq. Iraqi J Sci 584–590. https://doi.org/10.24996/ijs.2020.61.3.14

Akter T, Jhohura FT, Akter F, Chowdhury TR, Mistry SK, Dey D, Barua MK, Islam MA, Rahman M (2016) Water Quality Index for measuring drinking water quality in rural Bangladesh: a cross sectional study. J Health Popul Nutr 35(4):1–12

Alam MS, Mondal M (2019) Assessment of sanitation service quality in urban slums of Khulna city based on SERVQUAL and AHP model: a case study of railway slum, Khulna, Bangladesh. J Urban Managem 8(1):20–27

Alemu T, Mulugeta E, Tadese M (2017) Determination of physicochemical parameters of "Hora" natural mineral water and soil in Senkele Kebele, Oromia Region, Ethiopia. Cogent chem 3(1):1354800. https://doi.org/10.1080/23312009.2017.1354800

Ali SK (2018) Assessment of the Tigris River water quality in selected Iraqi governments. Int J Sci Res 7:500–504

Anand AK, Pradhan SP (2023) Evaluation of bivariate statistical and hybrid models for the preparation of flood hazard susceptibility maps in the Brahmani River Basin, India. Environ Earth Sci 82(16):389

APHA (2017) Standard Methods for examination of water and wastewater, 22nd edn. American Public Health Association, Washington, USA

Azhdarpoor A, Radfard M, Pakdel M, Abbasnia A, Badeenezhad A, Mohammadi AA, Yousefi M (2019) Assessing fluoride and nitrate contaminants in drinking water resources and their health risk assessment in a semiarid region of southwest Iran. Desalin Water Treat 149:43–51

Bhadra AK, Sahu B, Rout SP (2014) A study of water quality index (WQI) of the river Brahmani, Odisha (India) to assess its potability. Int J Curr EngTechnol 4(6):4270–4279

Brauers WK, Zavadskas EK (2006) The MOORA method and its application to privatization in a transition economy. Control Cybern 35(2):445–469

Bruce D, Hardeep S (2016) Electrical conductivity and pH guide for hydroponics

Bruce M, Limin M (2021) Recent advances on water pollution research in Africa: a critical review. Int J Scient Adv 2(3). https://doi.org/10.51542/ijscia.v2i3.23

Burrough PA, McDonnell RA (1998) Principles of geographical information systems. Oxford University Press, Oxford

Chakraborty B, Roy S, Bera A, Adhikary PP, Bera B, Sengupta D, Bhunia GS, Shit PK (2020) Cleaning the river Damodar (India): impact ofCOVID-19 lockdown on water quality and future rejuve-nation strategies. Environ Dev Sustain. https://doi.org/10.1007/s10668-020-01152-8

CPCB (2008) Guidelines for water quality management. Central Pol- lution Control Board, Parivesh Bhawan, East Arjun Nagar, DelhiDas, A. (2022). Multivariate statistical approach for the assessment of water quality of Mahanadi basin, Odisha. Mater Today: Proc 65:A1–A11

Das A (2022a) Multivariate statistical approach for the assessment of water quality of Mahanadi basin, Odisha. Mater Today: Proc 65:A1–A11

Das A (2023b) Assessment of Surface Water Quality by Means of MCDM Approach in Mahanadi River Basin (MRB), Odisha. Resilience 360:51

Das A (2023c) Identification of surface water contamination zones and its sources on Mahanadi River, Odisha using entropy-based WQI and MCDM techniques. Eng Res Transcripts 4:67–92

Das A (2023d) Anthropogenic effects on surface water quality assessment in Baitarani River Basin, Odisha using GIS and MCDM techniques. Eng Res Transcripts 5:37–64

Das A (2022) Water criteria evaluation for drinking purposes in Mahanadi River Basin, Odisha. In: International conference on trends and recent advances in civil engineering, August, Singapore, Springer Nature Singapore, pp 237–260

Das A (2023) Assessment of potability of surface water and its health implication in Mahanadi Basin, Odisha. Mater Today: Proc

Das A (2023) Modelling of surface water quality and spatial mapping in Mahanadi Basin, Odisha. J Mines, Metals Fuels 71(9)

Das A (2024) Water quality assessment using water quality index (WQI) under GIS framework in Brahmani Basin, Odisha. In: International conference innovation in smart and sustainable infrastructure. Springer, Singapore, pp 131–150

Das A (2024) Surface water quality modelling using water quality index (WQI) and geographic information system (GIS) on the Mahanadi Basin, Odisha. In: Saxena S, Shukla S, Mural PKS (eds) Emerging materials and technologies in water remediation and sensing. ICWT 2022. Lecture notes in civil engineering, vol 439. Springer, Singapore. https://doi.org/10.1007/978-981-99-6762-9_2

Das A, Goya A, Soni A (2023) Use of water quality indices and its evaluation to verify the impact of Mahanadi River basin, Odisha. In: AIP conference proceedings, July, vol 2721(1). AIP Publishing

Das A, Goyal A, Soni A (2023) Deciphering surface water quality for irrigation and domestic purposes: a case study in Baitarani Basin, Odisha. In: AIP conference proceedings, July, vol 2721(1). AIP Publishing

Dash S, Kalamdhad AS (2021) Discussion on the existing methodology of entropy-weights in water quality indexing and proposal for a modification of the expected conflicts. Environ Sci Pollut Res 28(38):53983–54001

Devi U, Bhattacharyya KG (2013) Trace elements in the rainwater runoff of the urban catchment of Guwahati, India. Adv Environ Res 2(2):99–118

Diakoulaki D, Mavrotas G, Lefteris P (1995) Determining objective weights in multiple criteria problems: the CRITIC method. Comp Operate Res 22(7):763–770

Egbueri JC (2020) Groundwater quality assessment using pollution index of groundwater (PIG), ecological risk index (ERI) and hierarchical cluster analysis (HCA):a case study. Groundwater for Sustain Developm 10:100292

El Baba M, Kayastha P, Huysmans M, De Smedt F (2020) Evaluation of the groundwater quality using the water quality index and geostatistical analysis in the Dier al-Balah Governorate, Gaza Strip, Palestine. Water 12(1):262

Fatoba JO, Sanuade OA, Hammed OS, Igboama WW (2017) The use of multivariate statistical analysis in the assessment of ground-water hydrochemistry in some parts of southwestern Nigeria. Arab J Geosci 10(15):1–11

Gao X, Zuo D, Xu Z, Cai S, Xianming H (2018) Evaluation of blue and green water resources in the upper Yellow River basin of China. Proc IAHS 379:159–167. https://doi.org/10.5194/piahs-379-159-2018

Goel P, Saxena A, Singh DS, Verma D (2018) Impact of rapid urbani- zation on water quality index in groundwater fed Gomati River, Lucknow, India. Curr Sci 114:650–654. https://doi.org/10.18520/cs/v114/i03/650-654

Gorgij AD, Wu J, Moghadam AA (2019) Groundwater quality ranking using the improved entropy TOPSIS method: a case study in Azar-shahr plain aquifer, east Azerbaijan. Iran Hum Ecol Risk Assess 25:176–190. https://doi.org/10.1080/10807039.2018.1564235

Górski J, Dragon K, Kaczmarek PMJ (2019) Nitrate pollution in the Warta River (Poland) between 1958 and 2016: trends and causes. Environ Sci Pollut Res 26(3):2038–2046

Gupta A, Rai DK, Pandey RS, Sharma B (2009) Analysis of some heavy metals in the riverine water, sediments and fish from river Ganges at Allahabad. Environ Monit Assess 157(1):449–458

Hasan MK, Shahriar A, Jim KU (2019) Water pollution in Bangladesh and its impact on public health. Heliyon 5(8):e02145

Hegde GR, Kale YS (1995) Quality of lentic waters of Dharwad district in North Karnataka. Ind J Environ Health 37(1):52–56

Hwang CL, Lai YJ, Liu TY (1993) A new approach for multiple objective decision making. Comput Oper Res 20(8):889–899

Islam MS, Mostafa MG (2021) Groundwater quality and risk assessment of heavy metal pollution in Middle-West part of Bangladesh. J Earth and Environ Sci Res 3(2):1–15. https://doi.org/10.47363/JEESR/2021(3)143

Jain CK (2004) Metal fractionation study on bed sediments of river Yamuna. India. Wat Res 38(3):569–578

Judeh T, Bian H, Shahrour I (2021) GIS-based spatiotemporal map- ping of groundwater potability and palatability indices in arid and semi-arid areas. Water 13(9):1323

Karunanidhi D, Aravinthasamy P, Subramani T, Wu J, Srinivasamoorthy K (2019) Potential health risk assessment for fluoride and nitrate contamination in hard rock aquifers ofShanmuganadhi River basin, South India. Hum Ecol Risk Assess Int J 25(1–2):250–270

Karuppannan S, Panneerselvam B, Haji M, Murugesan B, Shube H, Serre Kawo N (2022) Evaluation of groundwater quality and its suitability for drinking and cultivation practices in and around the deltaic regions of South India using DWQI, IWQI, and GIS. Climate Change Impact on Groundwater Resources. Springer, Cham, pp 181–200. https://doi.org/10.1007/978-3-031-04707-7_10

Khan MYA, Gani KM, Chakrapani GJ (2016) Assessment of surface water quality and its spatial variation. A case study of Ram-ganga River, Ganga Basin, India. Arab J Geosci 9:28

Kosal-Sahin S, Sivri N, Alver-Sahin U, Sonmez VZ, Seker DZ (2019) Quality assessment and chemical composition of surface water in the Akcadag Basin (Malatya), Turkey: a case study for irrigation and drinking purposes. Fresenius Environ. Bull. 28, 895–901. No. 19th International MESAEP Symposium on Environmental Pollution and Its Impact on Life in the Mediterranean Region

Laltu AD (2023) Drinking suitability and source apportionment of physical parameters in surface water in Mahanadi Basin, Odisha. Odisha (April 14, 2023)

Le TTH, Amiri V, Rezaei M, Sohrabi N (2014) Groundwater quality assessment using entropy weighted water quality index (EWQI) in Lenjanat. Iran Environ Earth Sci 72:3479–3490. https://doi.org/10.1007/s12665-014-3255-0

Le T-H, Truong T, Hoang TL, Pham TTV, Pham TPT, Tran LT (2022) Influences of anthropogenic activities on water quality in the Saigon River, Ho Chi Minh city. J Water and Health (in press)

Li B, Yang G, Wan R, Hörmann G, Huang J, Fohrer N, Zhang L (2017) Combining multivariate statistical techniques and random forests model to assess and diagnose the trophic status of Poyang Lake in China. Ecol Indic 83:74–83

Li P, Wu J, Qian H (2016) Hydrochemical appraisal of groundwater quality for drinking and irrigation purposes and the major influencing factors: a case study in and around Hua County, China. Arab J Geosci 9(1):15. https://doi.org/10.1007/s12517-015-2059-1

Li J, Gao Z, Li C (2021) Study on distribution characteristics of Sr element of karst water and its indication in Jinan. Groundwater 43(04), 5–8. (in Chinese) https://doi.org/10.19807/j.cnki.DXS.2021-04-002

Long-Ling O, Yun-Rong S, Jie-Qing Y, Shu-Jie M, Qi Y, Yun-Long W (2021) Water quality assessment and pollution source analysis of Yaojiang River Basin: a case study of inland rivers in Yuyao City, China. Water Supply 13:1–16

Lumb A, Sharma TC, Bibeault J-F (2011) A review of genesis and evolution of water quality index (WQI) and some future directions. Water Qual Expo Heal. https://doi.org/10.1007/s12403-011-0040-0

Machiwal D, Cloutier V, Güler C, Kazakis N (2018) A review of GIS-integrated statistical techniques for groundwater quality evaluation and protection. Environ Earth Sci 77:681

Mahadevan H, Krishnan KA, Pillai RR, Sudhakaran S (2020) Assessment of urban river water quality and developing strategies for phosphate removal from water and wastewaters: integrated monitoring and mitigation studies. SN Appl Sci 2:772. https://doi.org/10.1007/s42452-020-2571-0

Marghade D (2020) Detailed geochemical assessment and indexing of shallow groundwater resources in metropolitan city of Nagpur (western Maharashtra, India) with potential health risk assessment of nitrate enriched groundwater for sustainable development. Geochemistry. https://doi.org/10.1016/j.chemer.2020.125627

Mishra TK, Moharana JK, Nanda PM, Garnaik BK (2014) Study of water quality index of brahmani river water in Talcher-angul industrial vicinity, Odisha. Int J Curr Res 6(06):704 9–70 52

Mukate S, Wagh V, Panaskar D, Jacobs JA, Sawant A (2019) Development of new integrated water quality index (IWQI) model to evaluate the drinking suitability of water. Ecol Indic 101:348–354

Muniraj K, Panneerselvam B, Devaraj S, Jesudhas CJ, Sudalaimuthu K (2021) Evaluating the effectiveness of emissions reduction measures and ambient air quality variability through ground-based and Sentinel-5P obser- vations under the auspices of COVID pandemic lockdown in Tamil Nadu, India. Int J Environ

Muthusamy B, Sithu GDS, Ramamoorthy A, Shankar K, Gnanachandrasamy G, Monica C, Xiaozhong H (2022) Isotopic signatures, hydrochemical and multivariate statistical analysis of seawater intrusion in the coastal aquifers of Chennai and Tiruvallur District, Tamil Nadu, India. Mar Pollut Bull. https://doi.org/10.1016/j.marpolbul.2021.113232

Narsimha A (2020a) Controlling factors and mechanism of ground- water quality variation in semiarid region of South India: an approach of water quality index (WQI) and health risk assessment (HRA). Environ Geochem Health 42:1725–1752. https://doi.org/10.1007/s10653-019-00374-8

Narsimha A (2020b) Spatial distribution, exposure, and potential health risk assessment from nitrate in drinking water from semi-arid region of South India. Hum Ecol Risk Assess 26:310–314. https://doi.org/10.1080/10807039.2018.1508329

Narsimha A, Sudarshan V (2017b) Assessment of luoride contamination in groundwater from Basara, Adilabad District, Telangana State, India. Appl Water Sci 7:2717–2725. https://doi.org/10.1007/s13201-016-0489-x

Niyoyitungiye L, Giri A, Ndayisenga M (2020) Assessment of coliforms bacteria contamination in Lake Tanganyika as bioindicators of recreational and drinking water quality. South Asian J Res Microbiol 9–16. https://doi.org/10.9734/sajrm/2020/v6i330150

Nwankwo CB, Hoque MA, Islam MA, Dewan A (2020) Groundwater constituents and trace elements in the basement aquifers of Africa and sedimentary aquifers of Asia: medical hydrogeology of drinking water minerals and toxicants. Earth Syst Environ 4:369–384. https://doi.org/10.1007/s41748-020-00151-z

Parvin F, Tareq SM (2021) Impact of landfill leachate contamination on surface and groundwater of Bangladesh: a systematic review and possible public health risks assessment. Appl Water Sci 11:1–17

Parvin F, Haque MM, Tareq SM (2022) Recent status of water quality in Bangladesh: a systematic review, meta-analysis and health risk assessment. Environ Challenges 6(100416):1–13. https://doi.org/10.1016/j.envc.2021.100416

Peng F, Zhang Y, Li Q (2020) Multi-indicator evaluation and evolution characteristics analysis of water quality at the entry section of Guanting Reservoir. Environ Monit China 36:65–74

Pius A, Jerome C, Sharma N (2012) Evaluation of groundwater quality in and around Peenya industrial area of Bangalore, South India using GIS techniques. Environ Monit Assess 184(7):4067–4077

Ramesh S, Sukumaran N, Murugesan AG, Rajan MP (2010) An innovative approach of drinking water quality index: a case study from Southern Tamil Nadu India. Ecol Indic 10(4):857–868

Ravi R, Aravindan S, Shankar K, Balamurugan P (2020) Suitability of groundwater quality for irrigation in and around the main Gadilam river basin on the east coast of southern India. Arch Agricult Environ Sci 5(4):554–562. https://doi.org/10.26832/24566632.2020.0504019

S, Lorenz M, Meon G (2017) Multivariate statisti- cal assessment of a polluted river under nitrification inhibition in the tropics. Environ Sci Pollut Res 24(15):13845–13862

Sadat-Noori SM, Ebrahimi K, Liaghat AM (2014) Groundwater quality assessment using the water quality index and GIS in Saveh- Nobaran aquifer, Iran. Environ Earth Sci 71(9):3827–3843

Sarkar AM, Lutfor Rahman AKM, Samad A, Bhowmick AC, Islam JB (2019) Surface and ground water pollution in Bangladesh: a review. Asian Rev Environ Earth Sci 6(1):47–69. https://doi.org/10.20448/journal.506.2019.61.47.69

Schiavon M, Moore KK (2021) How to properly read your irrigation water analysis for turf and landscape. EDIS 6:1–3. https://doi.org/10.32473/edis-ep616-2021

Shikazono N, Tatewaki K, Mohiuddin KM, Nakano T, Zakir HM (2012) Sources, spatial variation and speciation of heavy metals in sediments of the Tamagawa river in central Japan. Environ Geochem Hlth 34(1):13–26

Singh AN, Shrivastava R, Mohan D, Kumar P (2018) Assessment of spatial and temporal variations in water quality dynamics of river Ganga in Varanasi. Pollution 4:239–250

Singh KR, Dutta R, Kalamdhad AS, Kumar B (2020) Review of exist- ing heavy metal contamination indices and development of an entropy-based improved indexing approach. Environ Dev Sustain 22(8):7847–7864

Solangi GS, Siyal AA, Babar MM, Siyal P (2019) Groundwater quality evaluation using the water quality index (WQI), the synthetic pollution index (SPI), and geospatial tools: a case study of Sujawal district, Pakistan. Hum Ecol Risk Assess Int J

Song K, Yang G, Wang F, Liu J, Liu D (2020) Application of geophysical and hydrogeochemical methods to the protection of drinking groundwater in Karst Regions. Int J Environ Res Public Health 17(10):3627. https://doi.org/10.3390/ijerph17103627

Soujanya Kamble B, Saxena PR, Kurakalva RM, Shankar K (2020) Evaluation of seasonal and temporal variations of groundwater quality around Jawaharnagar municipal solid waste dumpsite of Hyderabad city, India. SN Appl Sci 2:498. https://doi.org/10.1007/s42452-020-2199-0

Srivastava PK, Gupta M, Mukherjee S (2012) Mapping spatial distribution of pollutants in groundwater of a tropical area of India using remote sensing and GIS. Appl. Geomat. 4(1):21–32. https://doi.org/10.1007/s12518-011-0072-y

Su Z, Wu J, He X, Elumalai V (2020) Temporal changes of groundwa- ter quality within the groundwater depression cone and prediction of confined groundwater salinity using Grey Markov model in Yinchuan area of northwest China. Expo Health 12(3):447–468. https://doi.org/10.1007/s12403-020-00355-8

Swain SS, Kumar SB, Mishra A, Chatterjee C (2023) Sensitive or resilient catchment?: A Budyko-based modeling approach for climate change and anthropogenic stress under historical to CMIP6 future scenarios. J Hydrol 622:129651

Ukah BU, Egbueri JC, Unigwe CO, Ubido OE (2019) Extent of heavy metals pollution and health risk assessment of groundwater in a densely populated industrial area, Lagos, Nigeria. Int J Energy Water Resourc 3(4):291–303

Vashist A (2023) Appointment of arbitrator under arbitration and conciliation act, 1996 in Light of Brahmani River Pellets Ltd. vs Kamachi Industries Ltd. Civil Appeal No. 5850 of 2019. Supremo Amicus, vol 32. pp 93
Wagh VM, Panaskar DB, Mukate SV, Aamalawar ML, Laxman Sahu U (2019) Nitrate associated health risks from groundwater of Kadava river basin, Nashik, Maharashtra, India. Hum Ecol Risk Assess Int J. https://doi.org/10.1080/10807039.2018.1528861
Wang Z, Walker GW, Muir DCG, Nagatani-Yoshida K (2020) Toward a global understanding of chemical pollution: a first comprehensive analysis of national and regional chemical inventories. Environ Sci Technol 54(5):2575–2584
WHO Geneva (2017) Guidelines for drinking-water quality. World Health Organization 216:303–304
Yadav KK, Gupta N, Kumar V, Khan SA, Kumar A (2018a) Environ Int 111:80–108
Yadav KK, Gupta N, Kumar V, Choudhary P, Khan SA (2018b) GIS- based evaluation of groundwater geochemistry and statistical determination of the fate of contaminants in shallow aquifers from different functional areas of Agra city, India: levels and spatial distributions. RSC Adv 8(29):15876–15889
Yang B, Lei Y, Yan B (2015) Distributed multi-human location algorithm using naive bayes classifier for a binary pyroelectric infrared sensor tracking system. IEEE Sens J 16(1):216–223
Zakir HM, Islam MM, Hossain MS (2017) Heavy metal contents in sediments of an urban industrialized area-a case study of Tongi canal, Bangladesh. Asian J Wat Environ Pollut 14(1):59–68
Zango MS, Sunkari ED, Abu M, Lermi A (2019) Hydrogeochemical controls and human health risk assessment of groundwater fluoride and boron in the semi-arid North East region ofGhana. J Geochem Explor 207:106363. https://doi.org/10.1016/j.gexplo.2019.106363
Zhao Y, Guo L, Liang J, Zhang M (2016) Seasonal artificial neural network model for water quality prediction via a clustering analysis method in a wastewater treatment plant of China. Desalin Water Treat 57:3452–3465

Unraveling the Nexus Between Socioeconomic Facets and Land Use and Cover Change for Analyzing Agricultural Productivity Within Riverine Ecosystems

Gaurav Talukdar, Arup K. Sarma, and Rajib K. Bhattacharjya

Abstract Understanding the dynamics of agricultural systems and their impact on society is critical for making informed decisions and achieving long-term growth. This study emphasizes the importance of using integrated approaches in agricultural productivity for socioeconomic development, such as field investigations, data collection, questionnaire surveys, and personal interviews. Remote sensing allow efficient monitoring of large agricultural areas thereby assisting in land management decisions. The study utilized satellite imageries to categorize the Brahmaputra River Basin into four classes: waterbody, dry sandbar, cropland, and vegetation. The analysis showed a decrease in dry sandbar areas from ~ 36% to ~ 27% and an increase in vegetation cover from ~ 39% to ~ 50% over the years. Field investigations provide essential insights into the challenges and opportunities faced by farmers by closely analyzing agricultural methods and their on-ground repercussions. Through surveys and interviews, factors affecting production, such as access to resources, market conditions, transportation and climate change adaption techniques, were identified and these factors were vital in understanding the challenges faced by farming communities and developing interventions tailored to their needs. Therefore, for a comprehensive understanding of agricultural productivity and its impact on socioeconomic development, an integrated approach is critical. These strategies work in tandem to bridge the gap between ground-level reality and large-scale data analysis. Based on such detailed studies, informed decision-making is essential for developing policies and strategies that support sustainable agricultural development, food security, and overall socioeconomic progress.

G. Talukdar (✉)
Department of Civil Engineering, Indian Institute of Technology Delhi, Hauz Khas 110016, New Delhi, India
e-mail: gtalukdar@civil.iitd.ac.in

A. K. Sarma · R. K. Bhattacharjya
Department of Civil Engineering, Indian Institute of Technology, Guwahati, Assam 781039, India
e-mail: aks@iitg.ac.in

R. K. Bhattacharjya
e-mail: rkbc@iitg.ac.in

S. C. Pal and U. Chatterjee (eds.), *Surface, Sub-Surface Hydrology and Management*, Springer Geography, https://doi.org/10.1007/978-3-031-62376-9_10

Keywords Agriculture · Adaptation · Socioeconomic · Surveys · Farming

1 Introduction

Agricultural productivity is dependent on multifunctional agro-ecosystems that provide a variety of ecosystem services, including provisioning, regulation, support, and culture. Over time, various farming systems with distinct benefits and drawbacks have emerged, disappeared, and reappeared. These systems, integral to food systems, are primarily oriented towards food production, although often complicated (Shipley et al. 2020). Amidst global challenges, farming systems face various environmental, social, economic, and traditional obstacles. Today's agricultural systems face numerous environmental, economic, social, and institutional issues (Maye et al. 2018). Economic and social challenges include sudden changes in market access (Daugbjerg and Feindt 2017), shift toward less stable and less protective policy environments (Myers et al. 2016) and increasing controversy about agricultural practices (Guyton et al. 2015). Operating in this dynamic context requires stakeholders to anticipate and prepare for future issues by enhancing the resilience of farming systems with integrated approaches. Various literature (Stone and Rahimifard 2018; Meuwissen et al. 2019, Muhar et al. 2018) suggest the systematic and integrated analysis of the farming systems thereby encompassing a variety of challenges (Reeves et al. 2012) and could provide useful insights into the importance of diversity in farm activities. However, it remains unclear how these and other traits will be measured at the level of farming systems.

The current problem is to strike a balance between agricultural expansion and environmental protection while conserving agricultural output and natural resources and maintaining socioeconomic integrity. Recognizing tradeoffs and synergies between these competing functions has become increasingly crucial in the face of agricultural intensification and greater demands on natural habitats (Sayer et al. 2013). In the context of sustaining agriculture, particularly in floodplains and sandbar farming, the riverine ecosystem plays a pivotal role. Riverine ecosystems, characterized by the proximity of rivers and their associated floodplains, offer unique advantages for agriculture. Floodplains are fertile areas enriched by sediment deposits carried by rivers, fostering nutrient-rich soils that support robust crop growth. Sandbar farming, a distinctive practice along riverbanks, capitalizes on the nutrient-laden sediments deposited during river floods. The importance of riverine ecosystems in sustaining agriculture lies in their ability to replenish soil fertility through natural processes. This not only enhances agricultural productivity but also promotes environmental sustainability by minimizing the need for synthetic fertilizers (Talukdar et al. 2021). The riverine ecosystem contributes to the resilience of farming systems, although not much studies have been found on utilizing these sandbars for agricultural productivity and food systems. However, farmers in these regions have developed traditional knowledge and practices that align with the cyclical patterns of the river, allowing them to navigate the challenges posed by seasonal variations in water levels. But, as we look toward the future, factors such as increased urbanization and population growth may impact the behavior of riverine ecosystems. Zhang et al. 2021 recognizes that growing urbanization will impact the natural ecossyem and will affect the ecosystem services (ESs) supply and demand. It becomes imperative to assess and understand these changes to formulate sustainable agricultural practices that account for evolving environmental and socio-economic conditions. Preserving the delicate balance of riverine ecosystems is not only crucial for the resilience of farming systems but also for ensuring the long-term sustainability of agriculture in floodplains and sandbar farming.

Various researchers have discussed approaches that contribute to sustainable agriculture, addressing environmental, social, and economic dimensions (Muhie 2022; Romero-Lankao 2018). Recently, progress has leaned towards more participatory approaches, reflecting the demands of the more complex agricultural systems of the modern era brought about by climate change and other related global challenges (Carberry et al. 2002). Participatory approaches involve engaging various stakeholders, including farmers, researchers, policymakers, and local communities, in the decision-making process. These approaches recognize the importance of local knowledge and perspectives. In the context of agricultural planning, participatory approaches provide a valuable means of considering several possibilities (Delmotte et al. 2013; Walker et al. 2002). By involving diverse stakeholders, these approaches facilitate a more inclusive and comprehensive exploration of potential scenarios of farming systems. Since stakeholders are typically involved in follow-up procedures, their agreement with the problem definition and suggested action plan is crucial, making stakeholder participation crucial (Quist and Vergragt 2006; Talukdar et al. 2022). Because stakeholders are crucial elements of socio-ecological systems, their perspectives are especially valuable since they may explain or influence system dynamics (Nuno et al. 2014). As a result, using participatory approaches can offer a first investigation into the composition, workings, and conduct of farming systems in the future. Field based approaches provide an opportunity for direct interaction with farmers and other stakeholders. Engaging in conversations with those directly involved in agriculture helps researchers gain valuable insights into local knowledge, traditions, and community dynamics that influence agricultural systems. Field investigations also help identify challenges faced by farmers, such as soil degradation, water scarcity, or pest infestations, simultaneously highlighting opportunities for improvement, such as the adoption of sustainable farming practices or the introduction of new crop varieties. For farmers, approaches of Precision Agriculture (PA) has grown in importance, especially in resource-poor and risk-prone emerging nations. PA involves the use of geospatial techniques and remote sensing applications to observe, measure and respond to inter- and intra-field variability in crops (Burkart et al. 2018; Yost et al. 2017). According to Kpienbaareh et al. (2019), the SNAP Toolbox's vegetation index algorithms are capable of precisely determining the biophysical and growth characteristics of crops, such as their chlorophyll content, nitrogen status, infestation levels of pests and diseases, and water needs. By helping farmers see and respond to patterns that may be caused by illness, nutrient deficiency, or water stress, PA helps farmers respond to variability in crop growth in the face of resource shortage. This enables farmers to more precisely identify differences in their farmlands and implement varied treatments (Adhikari et al. 2018). Sustainable development in a number of areas, such as resource management, social progress, economic growth, and environmental preservation, depends on informed decision-making approaches. A number of crucial criteria motivate the requirement for well-informed decision-making in the context of sustainable development such as complex interconnected systems, resource scarcity and environmental challenges, community and stakeholder engagement, economic viability and social equity, technological advancements and innovation, climate change mitigation, LULC mapping, adaptation and ethical considerations and corporate social responsibility.

Land use and land cover (LULC) change is considered one of the important measurement to asses changes across a wide range of spatiotemporal scales (Lambin 1997). The information of LULC is vital for the planning and execution of various schemes to meet the increasing demands in different sectors resource allocation and planning, crop suitability and rotation, monitoring land degradation and conservation, policy development and land zoning etc. Various researchers have performed studies pertaining to LULC changes using satellite imagerics (Pandey et al. 2012; Rawat and Kumar 2015; Singh et al. 2014). LULC classification techniques are developed to classify, map, quantify, and monitor the extent of environmental changes. To classify and quantify identified land use changes, most research works on LULC have relied on either supervised or unsupervised classification algorithms (Chamling and Bera 2020). The information of LULC is vital for planning and execution of various schemes in order to meet the increasing demands considering various sectors. Environmentalists are concerned about the bio diversity and water ecology with the changing effects of LULC (Butt et al. 2015). Remote sensing studies are capable of classifying different landscapes on a greater scale and map the changes with different technologies (Ozesmi and Bauer 2002). There is a need for a new decision framework for agricultural land use planning that allows for the testing of various concurrent and interacting aspects via alternative scenarios.

Agricultural productivity plays a pivotal role in the socio-economic development of nations, particularly in agrarian economies. The study aims to investigate and understand the intricate relationship between socioeconomic facets and land use and cover change within riverine ecosystems, with a specific focus on its implications for agricultural productivity. The rationale for conducting this study is rooted in several key considerations. Understanding the dynamics of agricultural systems is critical for making educated decisions and achieving long-term socioeconomic growth. This chapter delves into the vital role of integrated methodologies in understanding the intricacies of agricultural productivity, such as field investigations, data gathering, questionnaire surveys, and personal interviews. Furthermore, the involvement of researchers in data collection is critical for developing sustainable practices, supporting economic development, and tackling the agricultural sector's complicated difficulties. By closely examining on-ground realities, these methodologies aim to inform targeted interventions and policies that enhance agricultural sustainability and contribute to broader socioeconomic progress. Furthermore, the study highlights the role of technology and innovation in improving agricultural output in riverine ecosystems. It aims to analyze the adoption of modern agricultural practices, such as precision farming and agroecological approaches, as well as their ability to reduce environmental degradation while improving yields. In addition, the study seeks to emphasize the significance of stakeholder participation and participatory approaches in agricultural development initiatives. By including local communities, policymakers, and agricultural practitioners in the research process, the goal is to encourage ownership and collaboration in implementing solutions that address the agricultural sector's requirements and issues and explores the social dimesnsions that enables the adaptive ecosystem based management.

2 Data Sources and Processing

2.1 Study Area

The dynamically braided Brahmaputra River rises from the Manasarovar Lake in the Himalayan Kailash range. The river is transboundary, traversing Tibet, India, Bangladesh, and culminating in the Bay of Bengal. The river basin spans 5,80,000 square kilometers, with 1,94,413 square kilometers of territory covered in India, Tibet, Bangladesh, and Bhutan. In India, the Brahmaputra River flows through the state of Assam, between 95°23′5.94" and 89°49′14.29" East longitudes and 27°49′15.82" to 25°43′40.38" North latitudes (Fig. 1). Over almost 2900 km, it drains through a wide variety of environments and topography. The soil types found in floodplains are mostly red loamy and alluvial, with some clayey, laterite, and sandy soils as well. According to Talukdar et al. (2004), upland soil has less than 28% sand at the top and less at deeper depths, which makes it sutabke for agrciltural and farming practices (Talukdar et al. 2023). In the basin, the silt percentage varies from 30 to 75 percent. The clay content increases from a depth of 28 to 49 cm in the soil of the middle, eastern, and western parts of the basin with increased rainfall. According to Chakravarty et al. (1978), the amount of organic matter in the soils increases as rainfall increases and decreases as temperature increases. The area has a humid climate with moderate to heavy rainfall that occurs from May to September. In the winter, the average temperature is between 10–15 °C, and in the summer, it is between 30–40 °C. Since agriculture sustains a large proportion of the population in Assam, it plays a critical part in the socioeconomic development of the marginizalized communities dependant on the river system. According to the Economic Survey Report of Assam, 2014–15, the industry employs about 53% of the workforce and provides employment for over 70% of the state's population. These characteristics of large dynamic rivers emphasize the necessity of understanding and preserving them, as well as utilizing them optimally for sustainable development.

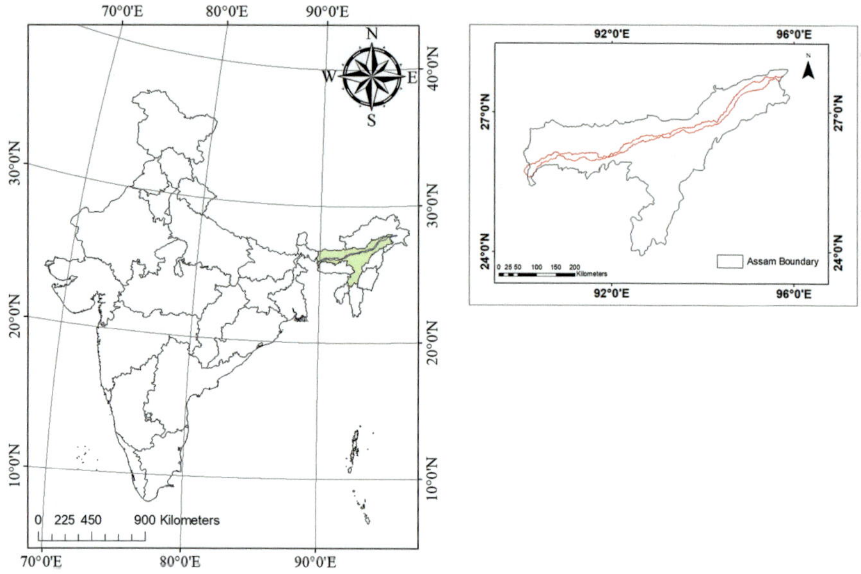

Fig. 1 The Brahmaputra River flowing through Assam

2.2 *Remote Sensing Data*

The study utilized georeferenced satellite imagery of the Brahmaputra River for the following years: 1993–95, 2003–04, 2008–11, and during the 2016–17 monsoon season. The National Remote Sensing Centre (NRSC) in Hyderabad provided the imagery for the years 1993–1995 (IRS 1B, LISS I) and 2003–2004 (IRS P6, LISS III). Satellite imegeries (Resourcesat-1, LISS III) were obtained from United States Geological Survey (USGS) for the years 2008–2011. To examine the alterations throughout the June–July 2016 flood season, RISAT-1 pictures were acquired. These datasets were mosaicked, and layer stacked using the ERDAS tool. The specifications of various satellite sensors are listed in Table 1.

Table 1 Satellite data sources

Sl. no	Data		Year of acquisition
	Satellite	Sensor	
1	IRS-1B	LISS-I	1993–95
2	IRS-P6	LISS-III	2003–04
3	Resourcesat 1	LISS-III	2008–11
4	RISAT-1	SAR	Monsoon 2016

2.3 Primary and Secondary Data

In the realm of research, data serves as the foundation for analysis, interpretation, and deriving significant findings. The study was carried out by gathering both primary and secondary data. An unstructured in-depth interview with farmers from various classes was undertaken to better understand the traditional practices prevalent in riverine sandbars as it is crucial to comprehend the shifts in the cropping pattern as well as the socioeconomic status of those living and working in sandbar farming. A satellite-based study was conducted to begin this work, which was then followed by field research and questionnaire surveys that collected information about the types of crops grown, their cropping patterns, the age of the farmers, the duration they were planted, their profits, their productivity, etc. The primary data collection (Table 2) included surveys and questionnaires, field observations, interviews, focus group discussions, farm diaries and logbooks while the secondary data included government agricultural reports, research papers and academic journals, market reports and price data which gives a sense of the features of agricultural production and marketing. Therefore, integrating primary and secondary data makes agricultural research more thorough. While secondary data offer larger perspectives and historical background, primary data offer in-depth, context-specific insights. Researchers, decision-makers, and agricultural practitioners can create focused interventions, make well-informed decisions, and support the sustainable growth of agriculture by integrating the two forms of data.

Table 2 Information collected during surveys

Particulars	
Vegetables* and potatoes	Dairy farming
Fixed cost	*Fixed cost*
Leased land	Leased Land
Operational cost	*Operational cost*
Seeds	Human force
Fertilizers	Cattle feed
Medicine	Medicines
Human force	
Animal force	
Machine power	
Transport services	

3 Methodology

Figure 2 shows the overall methodology developed for understanding the integrated approach of agricultural productivity and its impact on socioeconomic development, for developing policies and strategies to support sustainable agricultural development, food security, and overall socioeconomic progress.

3.1 Unveiling On-Ground Realities Through Field Investigations

Field investigations serve as the cornerstone of integrated approach, providing essential insights into the challenges and opportunities faced by farmers. This section explores how these investigations, through direct observation and analysis of agricultural methods, contribute to a nuanced understanding of on-ground effects. To get firsthand data, field studies entail researchers going into agricultural sites to observe practices and analyze conditions. This approach allows for a thorough comprehension of the context-specific difficulties that farmers face on a daily basis. Through engagement in the agricultural setting, researchers are able to acquire actual information that is essential for creating workable solutions, going beyond theoretical models. For example, the impact of traditional farming techniques on soil health or

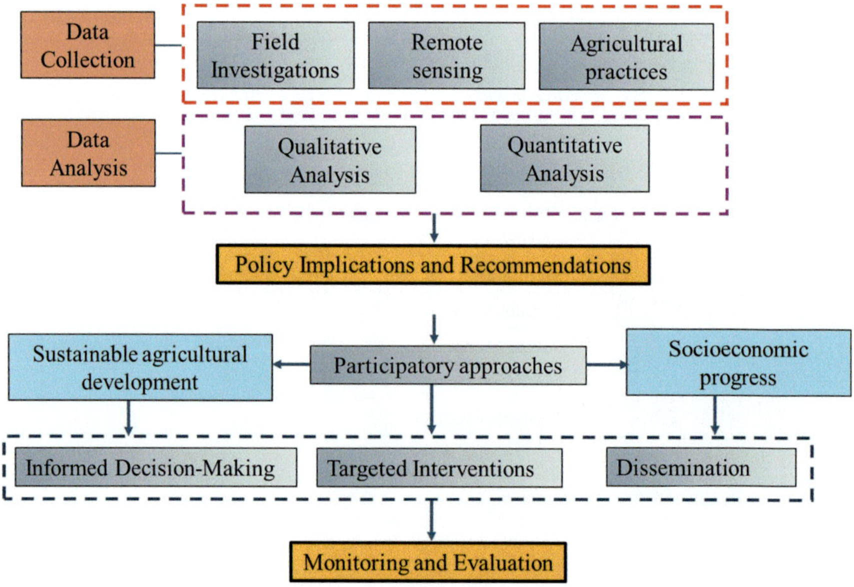

Fig. 2 Methodological framework developed for the study

the effectiveness of pest management practices becomes vividly clear through direct observation. While field investigations offer invaluable insights, they are not without challenges. The researchers often face logistical issues, including personnel shortages, seasonal variations in data collection, and access to remote locations. Despite these challenges, the benefits derived from on-ground observations far outweigh the difficulties, as they provide a detailed depth of understanding of in situ situations. Therefore, as a case study, field investigation and surveys were conducted in the sandbars and riverine areas within the Brahmaputra River. The focus of the investigation was to understand the types of farming activities, income, challenges faced by farmers and types of cropping practices. The field investigations involved direct observation, interviews with local farmers, and collaborative discussions with agricultural experts familiar with the regions. The team visited multiple sandbars and riverine areas, engaging with farmers to understand their experiences and challenges. Focussed group discussions and individual interviews were conducted to gather diverse perspectives on crop issues, seed quality, irrigation methods, transport facilities etc. Figure 3 shows the types of cropping practices and Table 3 shows the details of Information derived from survey on crops grown/activities practices in the riverine sandbars and floodplains.

Fig. 3 Personal interview with farmers regarding current practices and problems faced **a** sandbar cropping **b** eroded sandbars **c** damaged crops **d** field preparation

Table 3 Information derived from survey on crops grown/activities performed in the sandbars within the Brahmaputra River, Assam (Talukdar et al. 2022)

Sl. No	Crops/activities	Farmer class I (>60 years)		Farmer class II (30–60 years)	
		Cultivation performed in sandbars	Duration (months)	Cultivation performed in sandbars	Duration (months)
1	Paddy	×	3–4	✓	3
2	Maize	×	1.5–3	×	–
3	Pea	×	1.5–2	✓	1.5–2
4	Beans	×	1.5–2	×	–
5	Turnip	✓	2	×	–
6	Brinjal	✓	1.5–2	×	–
7	Bitter Gourd	✓	1.5–2	×	–
8	Pumpkin	✓	3	✓	3–4
9	Carrot	✓	3	✓	4
10	Chilli	✓	2–3	×	--
11	Potato	×	2–3	✓	2–3
12	Ground Nut	×	–	×	–
13	Mustard	×	1.5–3	✓	2
14	Capsicum	×	1–2	×	–
15	Cabbage	✓	2–3	✓	3
16	Cauliflower	×	1.5–2.5	×	–
17	Jute	×	–	×	–
18	Potato	×	–	×	–
19	Tomato	×	–	✓	6
20	Black gram	✓	2–3	✓	2–3
21	Dairy faming	×	–	✓	–

3.2 *Efficient Monitoring and Analysis with Remote Sensing*

Remote sensing, with its capacity to capture and interpret data from a distance, has revolutionized the way we understand and manage agricultural landscapes. It facilitates the analysis of changes in land usage and land cover (LULC), aiding in informed land management decisions. This information is crucial for understanding trends in agricultural expansion, deforestation, or urbanization. There are two main approaches that are widely explored: (i) Supervised classification and (ii) Unsupervised classification. Both methods, however, are established on different assumptions about the scene that needs to be classified. To determine the signature of distinct land cover classes in supervised classification, prior knowledge on the types of land cover is required. On the other hand, since it separates the scene into spectral clusters, such information is not necessary for unsupervised categorization. In this study, the land cover map of the study area was created using the unsupervised classification approach to show how the area changed in relation to the various classifications. Using the iso cluster unsupervised classification method, satellite imageries were categorized. Using a set of input raster bands, this tool combines the capabilities of maximum likelihood classification with iso cluster classification to conduct classification. The imageries of the Brahmaputra River that were taken between 1993–95, 2003–04, 2008–11 and July 2016 and were categorized into four classes: waterbody, dry sandbar, cropland, and vegetation.

The accuracy assessment of the observed changes in land use/cover was done in order to assess the performance of the classifiers. Calculating the error/confusion matrix is a standard method for assessing accuracy. Overall accuracy, which assesses the proportion of correctly classified classes, is seen to be a crucial metric for calculating the error. The kappa statistic, which was created by Cohen in 1960, is another significant statistic that is frequently used to measure the agreement between observed and projected values (Bishop et al. 1977).

$$\text{Estimated Kappa(K)} = \frac{N\sum_{i=1}^{r} -X_{ii} - \sum_{i=1}^{r}(x_{i+})(x_{+i})}{N^2 - \sum_{i=1}^{r}(x_{i+})(x_{+i})} \tag{10.1}$$

where, r represents the no. of rows in the matrix, N represents no. of observations, X_{ii} represents the no. of observations in i^{th} row and i^{th} column (diagonal elements) and x_{i+} and x_{+i} represents marginal totals of row r and column i, respectively.

3.3 Socioeconomic Dimensions Through Surveys and Interviews

Addressing the complex issues that farming communities face requires an understanding of the socioeconomic aspects of agriculture. Numerous elements fall under the category of socioeconomic considerations, such as the general well-being of agricultural households, market conditions, education, cultural and traditional habits, and resource accessibility. Surveys and interviews are effective instruments for developing focused interventions that cater to the unique requirements of farmers and advance the more general objective of sustainable agricultural development by revealing these features. It concerns how social ethics, dignity, and philosophy interact with economic science. The inhabitants who engage in farming and cultivation have a tendency to carry on their ancestors' practices. They consider themselves fortunate to be able to engage in these activities, and they typically don't drift from their traditional or typical ways of doing things. Such conventional methods, however, do not expand the possibilities for climate change adaptation. The primary goal is to identify and examine the socioeconomic factors that the farmers of agricultural region are dealing with. Moreover, noting or marking differences in market circumstances, educational opportunities, and resource accessibility are other prime factors that concern the socio-economic status of the marginalized communities within the riverine ecosystem. Other goals that need research or comprehension include determining how well farmers can adjust to shifting weather patterns, recognize how cultural factors affect farming methods, examine how social networks and community dynamics affect decision-making.

The selection criteria included representation of various farming methods, diversity in agroecological zones, and fluctuating socioeconomic circumstances to ensure representation across various agroclimatic and demographic groups. Quantitative data on market participation, land ownership, income, and demography were developed and gathered using structured questionnaires. As part of community engagement, teams of workers with cultural sensitivity, language skills, and respect for the interviewee collaborated with local leaders and community members to foster trust and streamline the survey process. The interview protocol included experienced farmers, community leaders, and agricultural extension workers. Semi-structured interviews were conducted to explore qualitative aspects of socioeconomic dimensions. The subject matter discussed the impact of culture, difficulties encountered, coping mechanisms used, and viewpoints on community dynamics. In addition, audio-recorded interviews were transcribed for detailed qualitative analysis. Through organized interviews, the procedure included market representatives, policymakers, and agricultural specialists to gather diverse perspectives. Most importantly, the survey ensured privacy of the participants through data anonymization and sensitive information protection thereby incorporating cultural norms and practices into survey. The questionnaire survey has helped in knowing the opinions and emotions of the people on the basis of their experience and nature of work.

1. A standard questionnaire provides quantifiable answers and is a practical way to gather data for the research topic.
2. The answers are relatively easy to compare between participants and concluding remarks can be made.
3. Specific onsite queries can be dealt related to real time present situations.
4. Large amounts of information can be collected in a short span of time in a cost-effective way.

3.4 Integrated Approaches for Comprehensive Understanding

The synergy of field research, remote sensing, and socioeconomic surveys aims to bridge the gap between on-ground realities and large-scale data analysis, providing nuanced insights into the dynamics of agricultural productivity and its socioeconomic dimensions. Through thorough field research, first efforts concentrate on identifying possibilities and constraints inherent in agricultural activity. A more sophisticated knowledge of the various elements influencing agricultural productivity is made possible by this field analysis. In order to track changes in land usage and land

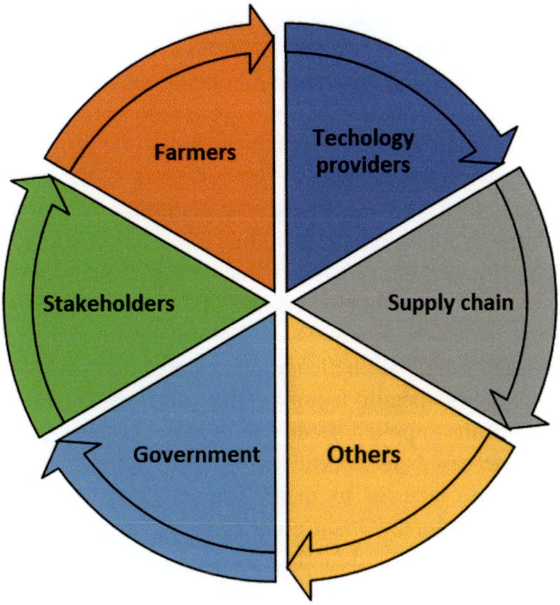

Fig. 4 Integrated approach towards sustainable agriculture

cover at the same time, the study uses remote sensing techniques. This allows for the collection of important data on how the spatial dynamics of agricultural landscapes are changing over time. Evidence-based policies, strategies, and interventions aimed at promoting sustainable development in the agricultural sector are heavily influenced by the synthesis of data. This methodology enables a comprehensive and informed approach to resolving the obstacles and maximizing the opportunities for sustainable agriculture practices by combining various data sources and analytical techniques (Fig. 4). This integrated approach leverages the benefits of field research,not limiting to remote sensing, and socioeconomic surveys to provide a nuanced and comprehensive understanding of agricultural systems. It intends to inform evidence-based decision-making, policy creation, and the development of interventions that support sustainable farming practices and contribute to socioeconomic success by combining data from these varied techniques.

4 Results

4.1 Unraveling Through Understanding

The case study highlighted the effectiveness of field investigations in identifying key issues, significant challenges and shaping interventions to boost agricultural productivity. The field surveys revealed a varied range of agricultural strategies used by farmers on the Brahmaputra River's sandbars and riverine zones. The variety of crops grown during different seasons as depicted in Fig. 5, demonstrates the durability and adaptability of agricultural operations in this ecosystem. Paddy cultivation, jute farming, and other vegetables were among the major cropping systems witnessed. The region's agroecological complexity and farmers' capacity to modify their operations to local conditions are reflected in the region's crop diversity. Concerns among farmers were widespread regarding the damage being shown to agricultural roots, especially in commodities like vegetables and paddy. Diverse opinions were expressed on the possibility that changing water levels and shifting river dynamics were causes of soil erosion, which destroys crop root systems and undermines crop stability. Farmers expressed that determining the precise source of root damage is a considerable challenge due to changing sediment patterns making it difficult to pinpoint specific factors. Some farmers expressed concerns about the effectiveness of the supplied seeds due to reports of low germination rates, variable yields, and increased susceptibility to pests and illnesses and demanded compensation for the losses incurred by them. The surveys revealed a wide range of income levels among farmers, with differences determined by factors such as land size, age group, crop selections, and market access. Furthermore, Farmers provided perceptions on their crop choices, including factors for soil quality, market demand, and traditional knowledge.

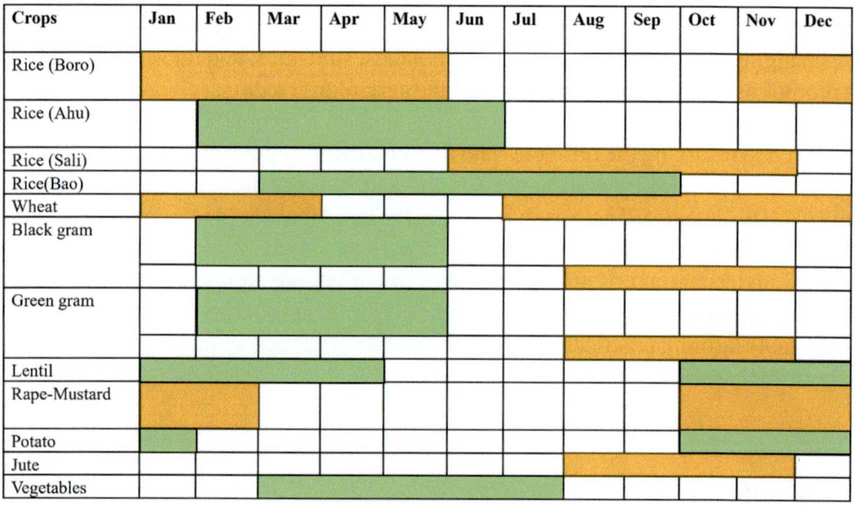

Fig. 5 Information on cropping pattern and crops grown in sandbars of large braided Brahaputra River in Assam

Table 4 Change in LULC area for various classes

Year	*Water* Body (%)	Dry sandbars (%)	Vegetated sandbar (%)	*Forest Area* (%)
1993–95	21.45	36.65	39.03	2.87
2003–04	18.67	28.65	50.21	2.47
2008–11	23.72	27.11	46.02	3.14
Monsoon	67.78	32.22%		

4.2 Exploring the Dynamics of Land Use Change

In this study, a comprehensive land cover map of the Brahmaputra River Basin was generated using an unsupervised classification approach. The iso cluster unsupervised classification method was employed, integrating satellite imageries from Landsat, IRS, and Resourcesat for the years 1993–95, 2003–04, 2008–11, and July 2016 (monsoon). The land cover map delineated four key classes: waterbody, dry sandbar, cropland, and vegetation. The accuracy assessment of the classified imageries was crucial in evaluating the performance of the classification method. The study assessed the reliability by comparing observed changes in land use/cover with ground truth data collected during the field investigations. The emphasis was on finding changes in the extent of waterbodies, dry sandbars, croplands, and vegetation cover. The classification work conducted on a decadal scale found that dry sandbars area has been decreasing and vegetation cover has been increasing. The dry sandbar areas can be seen decreasing from ~ 36% to ~ 27%, and vegetation cover increasing from ~ 39% to ~ 50% over the years. However, the increase in vegetation cover may suggest a positive trend in ecological resilience, increased production, livelihood generation, but conservation efforts should take into account preserving and developing regions that demonstrate such positive ecological indicators. Table 4 and Fig. 6 provides a quantitative overview of the ecosystem dynamics, showing the percentage changes in dry sandbar regions during the given time period.

5 Discussion

In agriculture, informed decision-making is a strategic process that requires a thorough grasp of on-the-ground realities, environmental dynamics, and socioeconomic considerations. It entails making wise use of data gathered through field research, remote sensing, and socioeconomic surveys. Informed decision-making enables farmers, policymakers, and stakeholders to make decisions that maximize resource use, increase production, and assure long-term sustainability. The integrated approach adopted in this study combines various data sources and analytical techniques to provide a comprehensive and informed understanding as well as addressing the challenges faced in the agricultural systems. By integrating field research, remote sensing, and socioeconomic surveys, the study aims to capture the complexity of

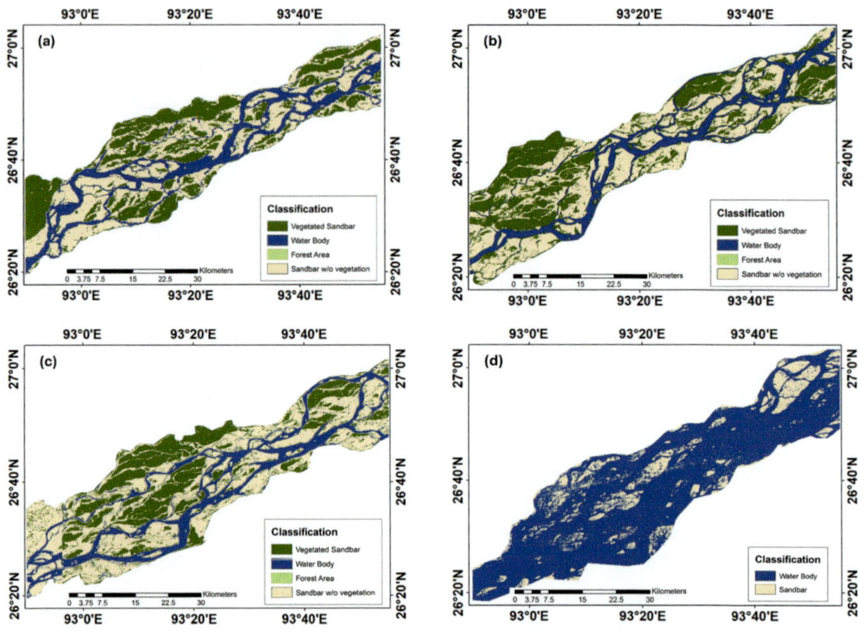

Fig. 6 Spatial and temporal distribution of LULc changes for years **a** 1993–95 **b** 2003–04 **c** 2008–11 **d** Monsoon period 2016

interactions between environmental, agronomic, and socio-economic factors influencing farming practices. Integrated farming systems (IFS) provides a holistic approach to farming that aims to address multiple objectives such as provide farm resilience, food security, farmer livelihoods, ecosystem services, optimal farming etc. thereby catering the requirements of small and marginal farmers and make farms adaptive and resilient. Farmers highlighted the difficulties in establishing the source of root damage, fluctuating water levels as well as changing sediment patterns which concerns the impact of environmental changes on crop health. The diversified socioeconomic environment among farmers highlights the importance of specific interventions and policies that consider the various issues that different groups of the farming community confront. Furthermore, the manpower requirement for diverse crops and livestock systems gives an opportunity for increased employment generation and keeps farm families involved in agriculture activities. The LULC classification highlighted the dynamic nature of the landscape and the importance of monitoring land cover over extended periods. The observed increase in vegetation cover in the study area may indicate a good trend in ecological resilience. Focusing on improving ecological resilience, on the other hand, helps to the overall sustainability of the agricultural environment. Policies for agricultural systems are critical for developing sustainable practices, guaranteeing food security, and sustaining farming communities' livelihoods. Environmental sustainability, economic viability, and social equality are all challenges that should be addressed by effective agricultural policies. Farmers are suggested to practice crop rotation with integrated farming, diversify crops to enhance soil fertility, pest reduction and minimize the risk of monoculture. Furthermore, they are encouraged to opt for agroecological methods that emphasize resilience, biodiversity, and the conservation of natural resources by incorporating ecological concepts into agricultural systems. To be aware of the future climate change adaptation policy studies should investigate the potential of agricultural integration for climate resilience by producing climate-resilient crop varieties that can endure extreme weather events and changing climatic conditions. Practices contributing to carbon sequestration, such as cover cropping and reduced tillage should be implemented. Policies that ensure farmers, particularly smallholders, to have access to credit, inputs, crop insurance and resources to boost productivity and livelihoods should be implemented including skill development programs that increase farmers' capacity to embrace modern agricultural methods and technologies. Invest in infrastructure such as transportation, storage, and market facilities to improve farmers' market access and reduce post-harvest losses are other key adaptations and sustainable practices for agricultural systems. The findings of the study are useful for policy formation in the areas of land use planning, water resource management, and environmental conservation. Each region have its unique environmental context and agricultural practices, although the impact of changing climatic variability could affect the food system globally. While comparing the study's findings with similar studies, it was found that crop productivity has been affected and farmers have migrated from mono cropping to different crop combinations within growing seasons. This compensates the losses from a single crop farming and highlights the need to implement adaptive strategies and ecison making to mitigate these challenges. The aligns with the significance of the our study contributing to evidence-based decision-making processes that support long-term agricultural development, integrated farming system and socioeconomic resilience in riverine ecosystems. Its comprehensive approach to understanding the intricate dynamics of agricultural productivity and land use change aims to give significant insights for policymakers, researchers, and practitioners working to ensure the long-term viability of agrarian economies.

6 Recommendations

1. According to direct observations and farmer interviews, environmental factors such as irregular weather patterns and floods events pose substantial dangers to crop health and production. Farmers are concerned about the increased frequency of extreme weather occurrences affecting their yields. Collaboration with agricultural specialists focused light on the need of climate-resilient farming techniques and providing farmers with adaptive strategies to lessen the impact of environmental issues.

2. The study emphasizes the importance of focused initiatives to improve the economic condition of smallholder farmers. In addition, the farmers expressed needs for improved infrastructure and support systems in order to increase their resilience to environmental and economic difficulties.
3. The rise in vegetation cover may indicate a positive trend in ecological resilience. Understanding the drivers behind this increase is crucial for assessing the health and adaptability of the riparian ecosystem. Conservation efforts should consider preserving, balancing human needs and enhancing areas exhibiting such positive ecological indicators.
4. Farmers had varying perspectives on the reasons of root damage, which included changing water levels, river morphology and sediment dynamics. The subjectivity of farmer views limits the ability to make definitive results and hence in-depth analysis to scientifically identify and assess the elements that contribute to root injury is recommended.
5. Optimization models and system dynamics models can take into account multiple possible futures in so-called explorations by using scenarios that capture uncertainty on socio-economic developments and climate change

7 Conclusion

Recognizing the difficulties connected with implementing integrated approaches and developing effective policies, the study investigates new trends and future directions desirable in agricultural research, technology uptake, and policy frameworks that are critical for guaranteeing long-term agricultural development. Field investigations serve as a lens through which researchers can identify both challenges and opportunities within agricultural systems. Challenges may range from soil degradation and water scarcity to the effects of changing weather patterns. Opportunities, on the other hand, could include the adoption of sustainable farming practices, the introduction of innovative technologies, or the identification of crop varieties better suited to local conditions. The analysis revealed a notable decline in dry sandbar areas, decreasing from approximately 36% to 27%, alongside a marked increase in vegetation cover, rising from about 39% to 50% over the observed timeframe. The increased vegetation indicates the fertility of the sandbars for crop cultivation and also suggests that there is a significant portion of fertile land available within the river basin that has not yet been fully exploited for agricultural purposes. Looking into these parameters can help with expansion and agricultural intensification potentially leading to food security and increased production. In addition, being formed within the river, these sandbars provide access to water for irrigation, which is critical for agricultural output, particularly during unpredictable rainfall patterns. Complementing this analysis, surveys and interviews were conducted to delve into the socioeconomic dimensions of agriculture, specifically addressing factors such as market conditions, education levels, and accessibility to resources. This nuanced understanding is essential for developing targeted interventions that address specific issues faced by farmers. In conclusion, field investigations play an indispensable role in unveiling on-ground realities within agricultural systems. The direct observation and analysis provide researchers with a holistic understanding of the challenges and opportunities faced by farmers. The insights derived from field investigations are instrumental in shaping targeted interventions and policies that contribute to the sustainable development of agriculture. As we navigate the complexities of modern agriculture, the significance of on-ground investigations remains persistent in ensuring informed and effective strategies for the future.

As we navigate the complexities of modern agriculture, the integration of remote sensing technologies remains fundamental in shaping informed and effective strategies for the future. Remote sensing stands as a transformative tool in the realm of agriculture, offering unparalleled insights into the dynamics of agricultural systems. The efficiency, scale, and versatility of remote sensing technologies make them indispensable for monitoring and analyzing diverse aspects of agricultural landscapes. While challenges exist, ongoing advancements and a commitment to capacity building can help unlock the full potential of remote sensing for sustainable agricultural development. The integrated approach of unsupervised classification using iso cluster methodology has proven effective in mapping and analyzing land cover changes in the Brahmaputra River Basin. The observed trends in dry sandbar reduction and vegetation increase provide valuable insights into the environmental dynamics of this critical region, emphasizing the importance of ongoing monitoring and proactive conservation measures.

There has been little research into the various types of production and their environmental consequences. In the IFS, assessing specific farm sizes, types of operations, and recycling technologies would allow for clearer identification of scale-specific correlations between farm size and environmental impacts. The lack of long-term data and/or changes in agricultural management, technology, or infrastructure that can obfuscate the identification of the impact of climate on agriculture make researching the impacts (or averted repercussions) of climate change on economic activity difficult. The integrated approaches combining field research, remote sensing, and socioeconomic surveys gives a solid platform for understanding and resolving the complexities of agricultural systems. The synthesis of data from these various methodologies not only informs evidence-based decision-making and policy development, but also helps to establish sustainable agricultural practices that promote both environmental resilience and socioeconomic success. This strategy highlights the necessity of multidisciplinary research in developing effective answers to the agriculture sector's difficulties. Informed decision-making, learnt by the insights gained from these integrated approaches, emerges as the hub for developing policies and strategies that support sustainable agricultural development, food security, and overall socioeconomic progress. This structured exploration of integrated approaches to agricultural productivity provides a comprehensive understanding of the agriculture system within the riverine environment for researchers, policymakers, and practitioners seeking to enhance their understanding of the intricate dynamics shaping sustainable agricultural development and its broader impact towards the society.

References

Adhikari P, Araya H, Aruna G, Balamatti A, Banerjee S, Baskaran P, Barah BC, Behera D, Berhe T, Boruah P, Dhar S (2018) System of crop intensification for more productive, resource-conserving, climate-resilient, and sustainable agriculture: experience with diverse crops in varying agroecologies. Int J Agric Sustain 16(1):1–28

Bishop YMM, Fienberg SE, Holland PW, Light RJ, Mosteller F (1977) Book review : discrete multivariate analysis: theory and practice. Appl Psychol Meas 1(2):297–306

Burkart A, Hecht VL, Kraska T, Rascher U (2018) Phenological analysis of unmanned aerial vehicle based time series of barley imagery with high temporal resolution. Precision Agric 19:134–146

Butt A, Shabbir R, Ahmad SS, Aziz N (2015) Land use change mapping and analysis using remote sensing and GIS: a case study of Simly watershed, Islamabad, Pakistan. The Egyptian J Remote Sens and Space Sci 18(2):251–259

Carberry PS, Hochman Z, McCown RL, Dalgliesh NP, Foale MA, Poulton PL et al (2002) The FARMSCAPE approach to decision support: Farmers', advisers', researchers' monitoring, simulation, communication and performance evaluation. Agric Syst 74(1):141–177

Chakravarty DN, Sehgal JL, Dev G (1978) Influence of climate and topography on the pedogenesis of alluvium-derived soils of Assam. Indian J Agric Chem 11:77–97

Chamling M, Bera B (2020) Spatio-temporal patterns of land use/land cover change in the Bhutan-Bengal foothill region between 1987 and 2019: study towards geospatial applications and policy making. Earth Syst and Environ 4(1):117–130

Daugbjerg C, Feindt PH (2017) Transforming public policies: post-exceptionalism in food and agricultural policy, introduction to the special issue. J Eur Public Policy 24(11):1–20

Delmotte S, Lopez-Ridaura S, Barbier JM, Wery J (2013) Prospective and participatory integrated assessment of agricultural systems from farm to regional scales: comparison of three modeling approaches. J Environ Manage 129:493–502

Guyton KZ, Loomis D, Grosse Y, El Ghissassi F, Benbrahim-Tallaa L, Guha N, Straif K (2015) Carcinogenicity of tetrachlorvinphos, parathion, malathion, diazinon, and glyphosate. Lancet Oncol 16(5):490–491

Harrison MT, Cullen BR, Rawnsley RP (2016) Modelling the sensitivity of agricultural systems to climate change and extreme climatic events. Agric Syst 148:135–148

Kpienbaareh D, Kansanga M, Luginaah I (2019) Examining the potential of open source remote sensing for building effective decision support systems for precision agriculture in resource-poor settings. GeoJournal 84:1481–1497

Lambin EF (1997) Modelling and monitoring land-cover change processes in tropical regions. Prog Phys Geogr 21(3):375–393

Maye D, Chiswell H, Vigani M, Kirwan J (2020) 'Present realities' and the need for a 'lived experience'perspective in Brexit agri-food governance. In: Brexit geographies. Routledge, pp 174–190

Meuwissen MP, Feindt PH, Spiegel A, Termeer CJ, Mathijs E, De Mey Y, Reidsma P (2019) A framework to assess the resilience of farming systems. Agric Syst 176:102656

Muhar A, Raymond CM, Van Den Born RJ, Bauer N, Böck K, Braito M, Van Riper CJ (2018) A model integrating social-cultural concepts of nature into frameworks of interaction between social and natural systems. J Environ Planning Manage 61(5–6):756–777

Muhie SH (2022) Novel approaches and practices to sustainable agriculture. J Agricul Food Res 100446

Myers JP, Antoniou MN, Blumberg B, Carroll L, Colborn T, Everett LG, Benbrook CM (2016) Concerns over use of glyphosate-based herbicides and risks associated with exposures: a consensus statement. Environ Health 15(1):19

Nuno A, Bunnefeld N, Milner-Gulland EJ (2014) Managing social–ecological systems under uncertainty: implementation in the real world. Ecol Soc 19(2)

Ozesmi SL, Bauer ME (2002) Satellite remote sensing of wetlands. Wetlands Ecol Manage 10(5):381–402

Pandey PC, Sharma LK, Nathawat MS (2012) Geospatial strategy for sustainable management of municipal solid waste for growing urban environment. Environ Monit Assess 184:2419–2431
Quist J, Vergragt P (2006) Past and future of backcasting: The shift to stakeholder participation and a proposal for a methodological framework. Futures 38(9):1027–1045
Rawat JS, Kumar M (2015) Monitoring land use/cover change using remote sensing and GIS techniques: a case study of Hawalbagh block, district Almora, Uttarakhand, India. The Egyptian J Remote Sens Space Sci 18(1):77–84
Reeves M, Haanaes K, Love C, Levin S (2012) Sustainability as adaptability. J Appl Corp Financ 24(2):14–22
Reidsma P, Bakker MM, Kanellopoulos A, Alam SJ, Paas W, Kros J, de Vries W (2015) Sustainable agricultural development in a rural area in the Netherlands? Assessing impacts of climate and socio-economic change at farm and landscape level. Agric Syst 141:160–173
Romero-Lankao P, Bulkeley H, Pelling M, Burch S, Gordon DJ, Gupta J, Munshi D (2018) Urban transformative potential in a changing climate. Nat Clim Chang 8(9):754–756
Sayer J, Sunderland T, Ghazoul J, Pfund JL, Sheil D, Meijaard E, Buck LE (2013) Ten principles for a landscape approach to reconciling agriculture, conservation, and other competing land uses. Proc Natl Acad Sci 110(21):8349–8356
Shipley NJ, Johnson DN, van Riper CJ, Stewart WP, Chu ML, Suski CD, Shew JJ (2020) A deliberative research approach to valuing agro-ecosystem services in a worked landscape. Ecosyst Serv 42:101083
Singh P, Gupta A, Singh M (2014) Hydrological inferences from watershed analysis for water resource management using remote sensing and GIS techniques. The Egyptian J Remote Sens Space Sci 17(2):111–121
Stone J, Rahimifard S (2018) Resilience in agri-food supply chains: a critical analysis of the literature and synthesis of a novel framework. Supply Chain Managem: an Int J 23(3):207–238
Talukdar G, Swain JB, Patra KC (2021) Flood inundation mapping and hazard assessment of Baitarani River basin using hydrologic and hydraulic model. Nat Hazards 109(1):389–403
Talukdar G, Sarma AK, Bhattacharjya RK (2022) Assessment of land use change in riverine ecosystem and utilizing it for socioeconomic benefit. Environ Monit Assess 194(11):841
Talukdar G, Bhattacharjya RK, Sarma AK (2023) Optimal cropping pattern based on short-term streamflow forecasts to improve agricultural economic benefits and crop productivity under uncertainty conditions. Hydrol Sci J 68(2):246–260
Talukdar NC, Bhattacharyya D, Hazarika S (2004) Soils and agriculture. In: The Brahmaputra basin water resources. Springer, Dordrecht, pp 35–71
Walker B, Carpenter S, Anderies J, Abel N, Cumming G, Janssen M, Pritchard R (2002) Resilience management in social-ecological systems: a working hypothesis for a participatory approach. Conservation Ecol 6(1)
Yost MA, Kitchen NR, Sudduth KA, Sadler EJ, Drummond ST, Volkmann MR (2017) Long-term impact of a precision agriculture system on grain crop production. Precision Agric 18:823–842
Zhang Z, Peng J, Xu Z, Wang X, Meersmans J (2021) Ecosystem services supply and demand response to urbanization: a case study of the Pearl River Delta, China. Ecosyst Services 49:101274

Geospatial Analysis of Flood and Drought Vulnerability in Rajshahi District: A Formal Regionalization and GIS-Based Approach for Upazila-Level Assessment

Md. Tahatul Azid and Jakia Alam

Abstract Bangladesh, a densely populated South Asian country, is grappling with recurring floods and droughts due to its geographical features, monsoon climate, and proximity to major rivers. In Rajshahi District, floods and droughts are frequent natural occurrences that have a substantial impact on the area's environment, economy, and residents' quality of life. There is still a lack of extensive research on this subject at the regional level. Addressing this gap is crucial for enhancing disaster-mitigating risks, and building resilience, as it helps identify unique vulnerabilities within each upazila. This study evaluates flood and drought vulnerability in Rajshahi District using regionalization techniques and GIS-based tools, providing local insights for better disaster mitigation and resource allocation. The study assesses the vulnerability of flood in Rajshahi district by focusing on the factors of flood frequency, riverine area, and low land area. And it also identifies the vulnerable areas of drought by considering the existing situation of groundwater table, irrigation, and soil type. It identifies high-risk areas for disaster preparedness and response planning. The study also examines temporal vulnerability, considering historical and projected climate trends. The resultant data analysis according to the method of formal regionalization and GIS-based analysis including primary and secondary data, indicates that Rajshahi is more prone to drought than flood. The findings of this study offer crucial insights for planning sustainable development, concentrating on the three most vulnerable upazilas, Bagha, Charhgat, and Godagari, ensuring water delivery, reducing disaster risk, and aiding in decision-making to enhance resilience and mitigate drought's negative effects.

Keywords Flood · Drought · Intra region · GIS · Risk reduction

Md. Tahatul Azid (✉) · J. Alam
Department of Urban and Regional Planning, Rajshahi University of Engineering and Technology, Rajshahi, Bangladesh
e-mail: litunurp18@gmail.com

J. Alam
e-mail: 1807001@student.ac.bd

S. C. Pal and U. Chatterjee (eds.), *Surface, Sub-Surface Hydrology and Management*, Springer Geography, https://doi.org/10.1007/978-3-031-62376-9_11

1 Introduction

Bangladesh, which is located in the Ganges-Brahmaputra-Meghna (GBM) basin delta, frequently deals with issues brought on by water-related calamities. Because of its unique topography and climate (Dilley et al. 2005). The northwest district of Rajshahi is especially vulnerable to floods and droughts. The district's varied socioeconomic and natural environments necessitate a thorough vulnerability assessment at the Upazila level. Extreme weather events brought on by climate change, such as droughts and floods, present serious problems for areas that are already susceptible. These occurrences affect infrastructure, agriculture, and communities. Understanding the spatial distribution of susceptibility is essential for efficient disaster management and sustainable development in Bangladesh, a nation vulnerable to these kinds of environmental disasters.

The Upazila-level study will offer more detailed information about the actual scenario which will help to mitigate the tragedy of root-level people as well as incorporate their needs and aspirations, acknowledging that vulnerability is frequently varied, even within a single district. The study intends to discover unique susceptibility patterns within Upazilas by using a formal regionalization technique that takes into account several elements including land use, geography, hydrological parameters, and socioeconomic indicators. With the use of Geographic Information System (GIS) technology, geospatial analysis provides a potent instrument for evaluating and illustrating the geographical patterns of vulnerability.

This research will offer a comprehensive knowledge of the elements leading to flood and drought risk at the Upazila level by integrating a variety of data sources, including climatic data, and socioeconomic indicators.

There are several steps in the research approach. Initially, the Rajshahi district will be divided into discrete sections according to pertinent environmental and socioeconomic factors using the formal regionalization method. Then, different datasets—such as those pertaining to land cover, population density, river basin features, and precipitation patterns—will be integrated and analyzed using GIS-based techniques. The project plans to create maps of vulnerability that identify Upazilas that are more vulnerable to droughts and floods. Local government officials, legislators, and community stakeholders will find great use for these maps as they work to establish plans focused on preparedness, mitigation, and adaptation for disasters.

Recent studies underscore the escalating threats posed by extreme weather events induced by climate change, which disproportionately impact regions like Rajshahi. The research emphasizes the escalating frequency and intensity of droughts and floods in the region, attributing them to climate change-induced shifts in precipitation patterns and hydrological cycles. Moreover, Ahmed et al. highlight the socioeconomic ramifications of these events, elucidating their ripple effects on livelihoods and food security in vulnerable communities.

To address these challenges, a meticulous Upazila-level study is imperative to discern distinct vulnerability patterns and tailor interventions accordingly. Leveraging formal regionalization techniques, encompassing factors such as land use, geography, hydrology, and socioeconomic indicators, is crucial in this endeavor. Integrating Geographic Information System (GIS) technology enables a spatially explicit analysis, facilitating a comprehensive understanding of vulnerability hotspots and their underlying drivers.

The primary objective of the study is to conduct a detailed examination of the hazardous conditions present in Rajshahi City, with a special focus on floods and droughts. These natural events are of significant concern due to their potential to disrupt lives, damage property, and negatively impact the local economy and environment. By assessing these conditions from multiple perspectives, the research aims to provide a holistic understanding of the challenges faced by the city. The study highlighted an ambitious endeavor aimed at addressing some of the most pressing environmental challenges faced by Rajshahi City. By evaluating the hazardous conditions related to floods and droughts, assessing their impact on the community, identifying areas of heightened vulnerability, and developing strategies for building resilience, this research has the potential to significantly improve any city's capacity to manage and mitigate the risks associated with these natural disasters. This, in turn, will contribute to the overall well-being and safety of its residents.

1.1 Formal Regionalization

Consistent characteristics such as language, landscape, and climate define an area with great uniformity and help it to develop a unified identity. The same language and culture, the same landscape, and the same weather patterns all exhibit this consistency. Within the area, there is a strong sense of solidarity due to the shared language and scenery (Arafat 2023). Before the Industrial Revolution, regionalization was the process of identifying an area by placing points on the surface of the Earth, observing uniformities, and developing theories, models, systems, and structures. A geographical area that meets certain characteristics, such as topography, climate, industrial type, or agricultural type, is referred to as a formal region (Chowdhury 2009). A formal region, according to J. Glasson, is a geographical area that meets certain requirements, such as terrain, climate, industrial type, or agricultural type, uniformly or homogeneously.

The ultimate goal of assessing flood vulnerability is to reduce the risk and impact of flooding events. This involves not only identifying areas and populations at risk but also implementing practical measures to reduce vulnerability and enhance resilience. Strategies may include infrastructure improvements, such as constructing flood barriers or improving drainage systems; implementing land-use planning strategies that avoid development in high-risk areas; and developing early warning systems and emergency evacuation plans. This translation is fundamental to effective flood risk management, aiming to minimize the adverse impacts of floods on communities and economies. The integration of concept in this research work is not only a goal but also a necessity for building resilience and safeguarding against future flood events.

1.2 Flood Vulnerability Measures

One of the most damaging natural disasters that Bangladesh is susceptible to is flooding. Anthropogenic climate change is expected to increase the frequency of intense rainfall events and the ensuing floods in the future (Schwarz and Kuleshov 2022). MJM Yusof in his study mentioned that the primary concept in flood risk management is vulnerability. Establishing a direct link between the theoretical ideas of flood risk and routine administrative procedures is one of the most important goals of flood vulnerability assessment (Nasiri et al. 2016). Every year, there are more and more people who are at risk, and most of them live in developing nations with high rates of poverty, which makes them more susceptible to natural catastrophes (UN/ISDR 2004). Because of the problem of climate change, the destructive effects of floods are progressively spreading around the world, especially to Bangladesh's riverine islands. Because of their location, riverine islands and coastal settlements are the main casualties (Sarker et al. 2022). Bangladesh's geographical location means that floods are a constant concern. Bangladesh, being a country with a riverine floodplain, is regarded as one of the most flood-prone countries in the world. Based on flood information that this tool presently has access to, the urban flood danger in the Rajshahi region is rated as low (Matallo n.d.). This indicates that there is a more than 1% risk that in the next ten years, possibly fatal and destructive river floods may occur (Think Hazard-Rajshahi-Urban Flood n.d.). Overall, while the Rajshahi region's current urban flood danger rating is low, the broader context of Bangladesh's flood risk underscores the importance of continued vigilance, adaptation, and innovation in disaster risk management. Through a combination of government initiatives, community action, and international support, Bangladesh is working to mitigate the impacts of its inherent geographical challenges.

1.3 Drought Vulnerability Measures

The term "drought" refers to the naturally occurring phenomena that result from precipitation that is much less than normally reported amounts, seriously disrupting hydrological balances and negatively affecting land resource production systems (UNCCD). Risk is the result of combining the likelihood of an event with its unfavorable effects (Das et al. 2023). Drought (Disaster) Risk is the possibility of death, deteriorated health, or loss of livelihoods, assets, and ecosystem services related to drought that might affect a specific group or society over a predetermined amount of time in the future (UN-ISDR 2009). The evaluation of drought vulnerability takes into account several variables beyond the geophysical effects of climate change and uses a variety of data sources, such as gridded estimations, modeling outcomes, and statistical data from different institutions (Wang and Sun 2023).

Figure 1 shows the procedure for determining, evaluating, and prioritizing the vulnerabilities in a situation including drought. Evaluating the risks that a possible drought poses to the environment, infrastructure, people, and so on, Vulnerability assessment (institutional and socioeconomic study), Calculating the exposure period (weather forecast), and Outlining the capabilities and actions that must be done (Fig. 1). One of the long-lasting and significant effects of global warming is drought, which has a dramatic influence on Bangladesh, especially on Barind tract. Climate change increases the likelihood of hotter days, heat waves, and longer dry spells than in the previous 50 years (Sultana 2016). One of the long-lasting and significant effects of global warming is drought, which has a dramatic influence on Bangladesh. Climate change increases the likelihood of hotter days, heat waves, and longer dry spells than in the previous 50 years. Bangladesh, a nation sensitive to climate change, is expected to have a rise in average temperature between 2030 and 2050, according to the Global Circulation Model (GCM) (Habiba et al. 2013). Drought ultimately spreads to other upazilas in the Rajshahi area due to insufficient rainfall Mamun et al. (2021).

The Disaster Management Plan of Rajshahi which was Published in 2014, showed 450,000–549,000 people were affected by sunray which can also be identified as drought as it leads to scarcity of potable water. The number of wretched people due to river erosion was 240,750–290,250 and 18,000–20,250 families were affected because of the layer of the water. Because of these reasons, Rajshahi district is a study area.

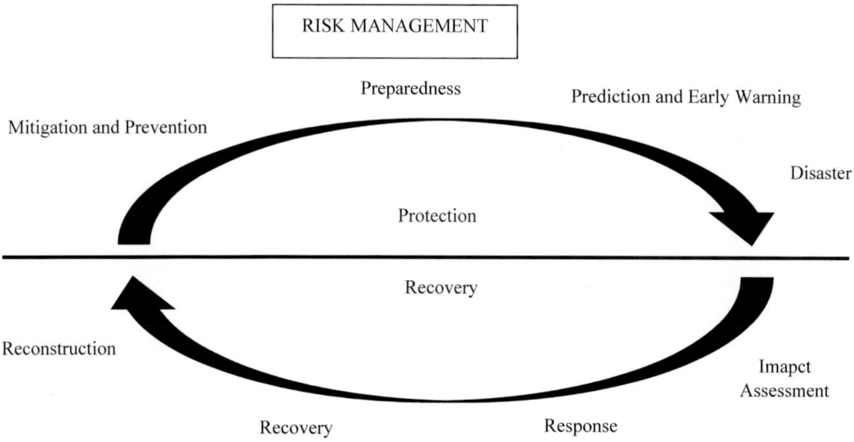

Fig. 1 Risk management in terms of drought vulnerability. *Source* Drought vulnerability and Risk Assessment by UNCCD

2 Formal Regionalization (Factor-Based Analysis)

The two approaches for formal delineation are the "Weighted index method," also known as the "Composite index method," and the "Factor analysis method." The "Composite Index" approach is being utilized in this study to improve comprehension and statistical analysis (Jewel et al. 2018). The issue was chosen based on the urbanization of every Upazila in Rajshahi District.

Initially, a basic literature analysis was done to understand the current scenario of flood and drought vulnerability in Rajshahi district under certain factors. Data was gathered from BBS in the second stage using the three formal regionalization variables and these are the flood frequencies, low land areas, and riverine areas for flood vulnerability measures, on the other hand, Ground Water Table (mm), Irrigation, and Soil Type for drought vulnerability (Table 1).

The data has been collected from the Bangladesh Bureau of Statistics (BBS) website for all the 10 upazilas of Rajshahi. The data shows the observation of 2011. The data has been collected and organized in an Excel worksheet to calculate the composite score of all three factors.

Formulas for calculation are shown below:

$$W_n = \frac{\text{mean of } \log_{10}(x_n)}{\text{standard deviation of } \log_{10}(x_n)}$$

where $n = 1, 2, 3$.

Source Glasson (1975)

The composite weighted index technique, which will be utilized to give each municipality a relative value, is represented by the following equation:

$$W = \frac{\log_{10}(x_1) \times W_1 + \log_{10}(x_2) \times W_2 + \log_{10}(x_3) \times W_3}{(W_1 + W_2 + W_3)}$$

Source Glasson (1975)

Using these formulas, the composite score for each upazila has been calculated. To determine class interval three methods named equal interval method, mean standard deviation Method, and arithmetic mean method are used.

The W score for a site displays the significance of that area about the aspects that impact it. Next, using the factor-based weighted data collected for the entire district,

Table 1 Composite weight calculation table for flood and drought vulnerability factors

Upazila	Vulnerability	Factor 1	Factor 2	Factor 3	Weight (W)
	Flood	Flood frequencies	Low land areas	Riverine areas	
	Drought	Ground water table (mm)	Irrigation	Soil type	

Source Author's preparation (2023)

the class spacing and frequency can be computed. To establish a baseline and evaluate the importance of each region, we employ the class interval as a statistic.

We can also derive the frequency—or roughly how many districts fall into each class—from the class interval. There are now three methods available for calculating the class interval. We will choose the solution from the three that offers the most precise and reliable class interval tactics. The three possibilities are equal class interval technique, mean standard deviation method, and arithmetic approach. The mean standard deviation strategy is the only one that, when used properly, even approaches perfect skewness. Compared to the earlier approaches, the class interval from the mean standard deviation methodology produces subgroups that are more cohesive with one another. To ascertain the best course of action, the importance of kurtosis will be investigated.

2.1 Selection of Factors and Justifications

We have classified it into 6 factors. We have selected these factors because they are highly connected with flood and drought vulnerability of different upazila in Rajshahi District. The factors are:

- **Flood Vulnerability Factors**

  Flood Frequencies: Vulnerability is determined by flood frequency, the area that is repeatedly affected by flood is more vulnerable than the one which is not. Recognizing past trends aids in evaluating and reducing possible effects on infrastructure and communities.

  Low Land Areas: Low-lying regions are inherently more vulnerable to flooding due to their proximity to water bodies and their susceptibility to inundation. These areas often face challenges in drainage and runoff management, making them prone to severe consequences during heavy rainfall or storm surges.

  Riverine Areas: Flooding danger is increased by rivers, particularly in communities that are close to rivers. Riverine flooding involves overflow, inundating lands and communities. Comprehensive strategies consider upstream and downstream conditions.

  Comprehending the elements that contribute to flood vulnerability is essential for disaster preparedness, response, and resilience planning, as it minimizes the effects of floods and protects people and property.

- **Drought Vulnerability Factors**

  Ground Water Table (mm): A region's water availability is indicated by the depth of its groundwater table; high levels show resilience in the face of low precipitation, while low levels suggest restricted supplies for activities requiring water.

  Irrigation: The effectiveness of irrigation systems has a major influence on drought sensitivity; established areas are more robust to droughts, while those that primarily rely on rain-fed agriculture are more vulnerable.

Soil Type: Poor soil types increase susceptibility, whereas soil type and quality have a major influence on water retention capacity, affecting vegetation's ability to support plants and lessen the impact of drought.

When variables are evaluated as a whole, drought susceptibility may be predicted. This information denotes the distribution of resources and preemptive actions to reduce the possible effects on ecosystems, agriculture, and communities.

2.2 *Selection of Class Interval*

Three approaches are used by class interval to measure upazila weights and frequency. For comparison, the most dependable class period is chosen.

- **Equal Class Interval Method**

  The class interval has been established using the equal class interval approach by following the formula

$$\text{Class Interval} = \frac{\text{Highest Value} - \text{Lowest Value}}{\text{Number of Classes}}$$

  Source Glasson (1975)

  From the equation, the highest value is – 0.01125, the lowest value is – 3.5477, the preferred class number is 5, and the class interval is 0.7073.

- **Mean Standard Deviation Method**

  In the mean standard deviation method, the mean and standard deviation of the composite score have been calculated and then the class is found by calculating a mean of 3.8 and standard deviation of 4.

  In this method, the class interval has been calculated using the equation. The composite score's mean and standard deviation have been computed, and the class is then determined by calculating the Mean ± 1st Standard Deviation, mean ± 2nd Standard Deviation, and Mean ± 3rd Standard Deviation:

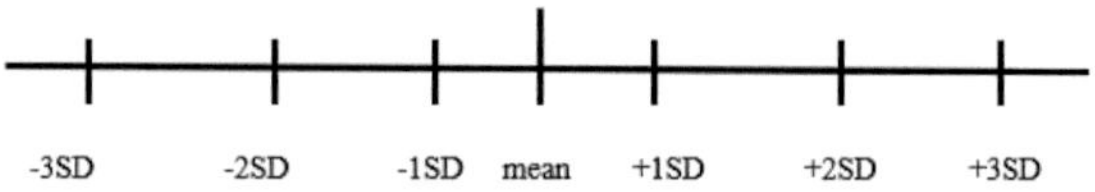

  Then the six classes have been determined. After calculating the frequency, the data has been plotted graphically to compare the histograms for getting a class distribution with a normal distribution curve.

- **Arithmetic Method**

  The following equation is utilized while applying this approach.

$$A + X + 2X + 3X = B$$

  where

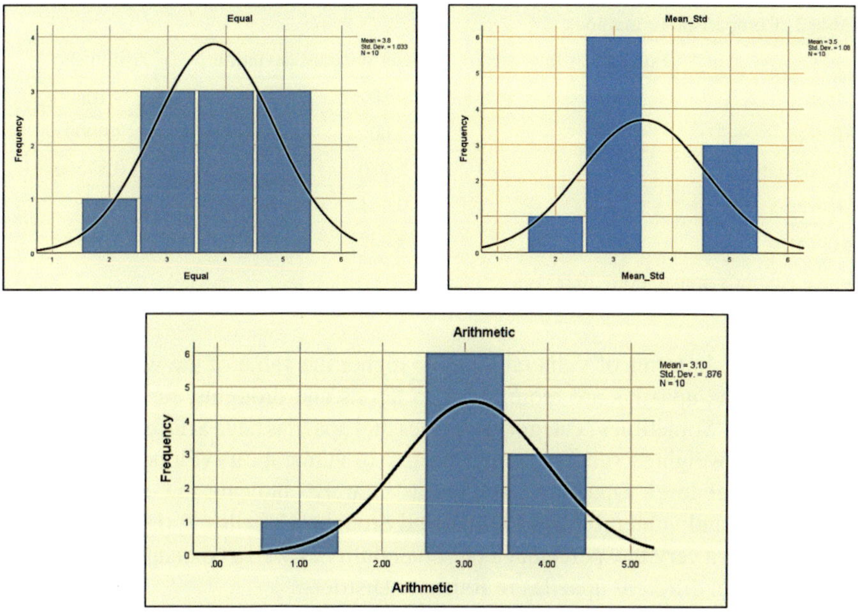

Fig. 2 Histogram of equal class interval, standard deviation method, arithmetic method. *Source* Prepared by Author (2023)

A Minimum Value
B Maximum Value
X Class Interval.

Then the four classes have been determined. After calculating the frequency, the data has been plotted graphically to compare the histograms for getting a class distribution with a normal distribution curve.

Finding each class's lowest cut point is the basic idea. After that, the cut points are added to the SPSS program, which produces the frequency, mean, median, mode, skewness, and kurtosis.

Histogram of the three methods is shown in Fig. 2. Among the three, the equal class interval method is the most suitable one for the data set as it is bell-shaped and the skewness and kurtosis values both are closer to zero than the other two arithmetic and mean standard deviation methods (Table 2).

2.3 GIS-Based Categorization (Upazila-Wise Integration)

Based on the degree of vulnerability, the 10 Upazila that make up Rajshahi's region are split into 4 groups. They are "Never," "Rarely," "Sometimes," and "Often." Upazilas falling under the "Often" category have greater levels of Flood and Drought

Table 2 Comparison of methods

	Equal class interval	Mean standard deviation	Arithmetic
Mean	3.80	3.50	3.1000
Median	4.00	3.00	3.0000
Std. deviation	1.033	1.080	0.87560
Skewness	– 0.272	0.661	– 1.465
Kurtosis	– 0.896	– 1.032	3.613

Source Author's preparation (2023)

Vulnerability. In terms of vulnerability, the higher the value of the weighted index, means that the upazilas are more prone to floods and droughts, resulting in more damage. The "Sometimes" category indicates that upazilas have a different number of range for the weighted value in the perspective of vulnerability (Flood and Drought) than the upper level. Upazilas classified as "Rarely" indicate low average wages for damages and vulnerabilities to flood and drought. Upazilas in the last category, "Never" have a very low percentage of vulnerability as well as damages due to flood and drought in different upazilas in Rajshahi Districts.

The composite values of W, which indicate the frequency of a certain quality, show a wide range of variation throughout the region's Upazilas. The characteristic in Godagari is defined as a never occurrence, which is one single incident with a 10% frequency. In a little higher range that includes Rajshahi City Corporation, Tanore, and Mohanpur, the feature is considered unusual because it appears three times and makes up 30% of the data. Five Upazilas (Paba, Baghmara, Durgapur, Puthia, and Bagha) account for 50% of the manifestations of the attribute. Lastly, Charghat records a higher frequency of observation for the characteristic, marking it as frequently as 10% of all occurrences. This subtle distribution of composite values throughout the Upazilas offers a thorough comprehension of the attribute's fluctuating frequencies, which helps develop a more detailed analysis of the region's characteristics (Table 3).

Table 3 Formal regionalization based on flood vulnerability for Rajshahi district

Range of composite value W	Attribute	Frequency	Percentage	Upazilas
– 3.54 to – 2.67	Never	1	10.0	Godagari
– 2.66 to – 1.78	Rarely	3	30.0	RCC, Tanore, Mohanpur
– 1.77 to – 0.89	Sometimes	5	50.0	Paba, Baghmara, Durgapur, Puthia, Bagha
– 0.88+	Often	1	10.0	Charghat

Source Author's preparation (2023)

Table 4 Formal regionalization based on drought vulnerability for Rajshahi district

Range of composite value *W*	Attribute	Frequency	Percentage	Upazilas
– 3.54 to – 2.67	Never	1	10.0	RCC
– 2.66 to – 1.78	Rarely	3	30.0	Durgapur, Puthia, Tanore
– 1.77 to – 0.89	Sometimes	4	40.0	Charghat, Paba, Mohanpur, Baghmara
– 0.88+	Often	2	20.0	Bagha, Godagari

Source Author's preparation (2023)

Based on drought risk, the Rajshahi District has been formally regionalized, with different Upazilas showing diverse degrees of susceptibility. "Never," the first category, is connected to RCC, while "Rarely" is connected to Tanore, Puthia, and Durgapur. The term "sometimes" is found in Charghat, Paba, Mohanpur, and Baghmara, making up 40.0% of all instances. "Often" is most common in Godagari and Bagha (20.0%) (Table 4).

3 Flood Vulnerability Assessment (Formal Regionalization)

The area of each district is represented by the Upazila (measured in square kilometers) for Rajshahi Districts. Additionally, these places have parameters as mentioned in the previous for measuring flood and drought vulnerability (Population and Housing Census 2011). A detailed analysis showing the reason behind different areas being in various ranges has been shown in Sects. 3 and 4.

According to the map in Fig. 3, Charghat Upazila is the most affected by the flood. But the actual scenario is quite different in various portions of Charghat Upazilla such as Yusufpur Union is in a vulnerable position in terms of flood. Whereas the people living in Charghat Union, mostly in Chandansahar village, people are terrified of river erosion. When asked about their problems almost all of them complained about river erosion, and wanted a permanent solution to the matter. There are some government projects going on to strengthen afforestation to adapt to the situation but still, it's inadequate in accordance with what people have to face losing their lands, and valuable assets because of river erosion. Along with river erosion, there are also other notable hazards like Kal-boishakhi, and storm surges that are creating problems in the day-to-day lives of people.

To validate the results gathered from the GIS output, it was compared with existing disaster management plans of Bangladesh showing vulnerable areas of flood in Fig. 4, representing various portions of Rajshahi which also shows the same results as the map prepared by the authors in Fig. 3.

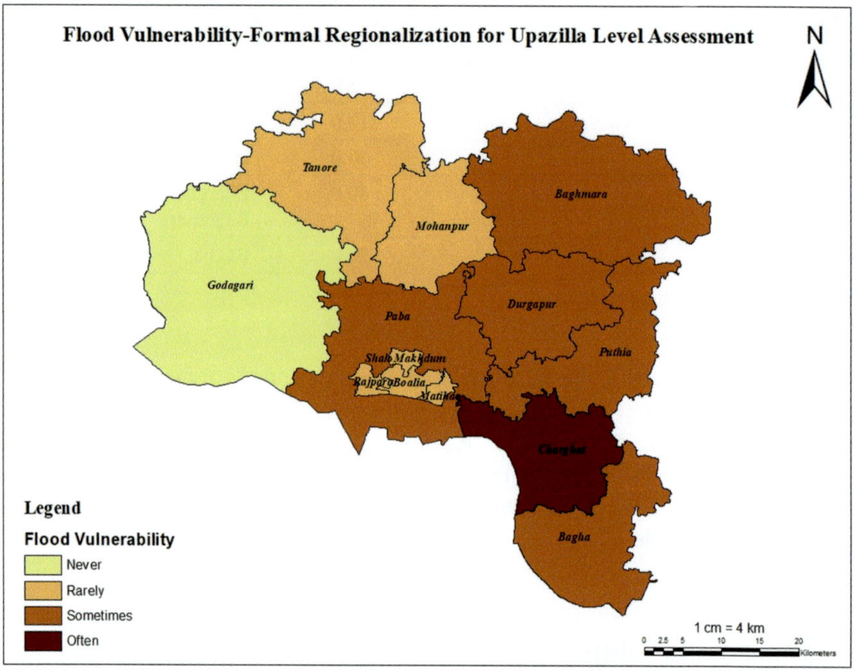

Fig. 3 Flood vulnerability (formal regionalization for Upazila-Level Assessment). *Source* Author's preparation (2023)

Among the other three criteria, which include the areas that are never affected by flood, rarely affected by flood and sometimes affected by flood; Godagari indicates the areas that are never affected by flood but are the most prone in terms of drought.

Tanore, Mohonpur, Rajpara, Boalia, Matihar, and Shah Makkdum are the unions that are rarely affected by the flood. The reason may be defined as most of the areas are built-up areas having many administration areas and are far from the river banks.

Baghmara, Puthia, Durgapur, Bagha, and Paba are the areas that are sometimes affected by floods. Actually, Rajshahi, the Barind region is known for its decreasing water level and its scarcity of water, so, flood is quite rare in Rajshahi (Fig. 4). The only areas that are affected by the flood are the areas that are near the river, which is in the Charghat upazila that also can be described as quite lowly and only in Yusufpur (Fig. 5).

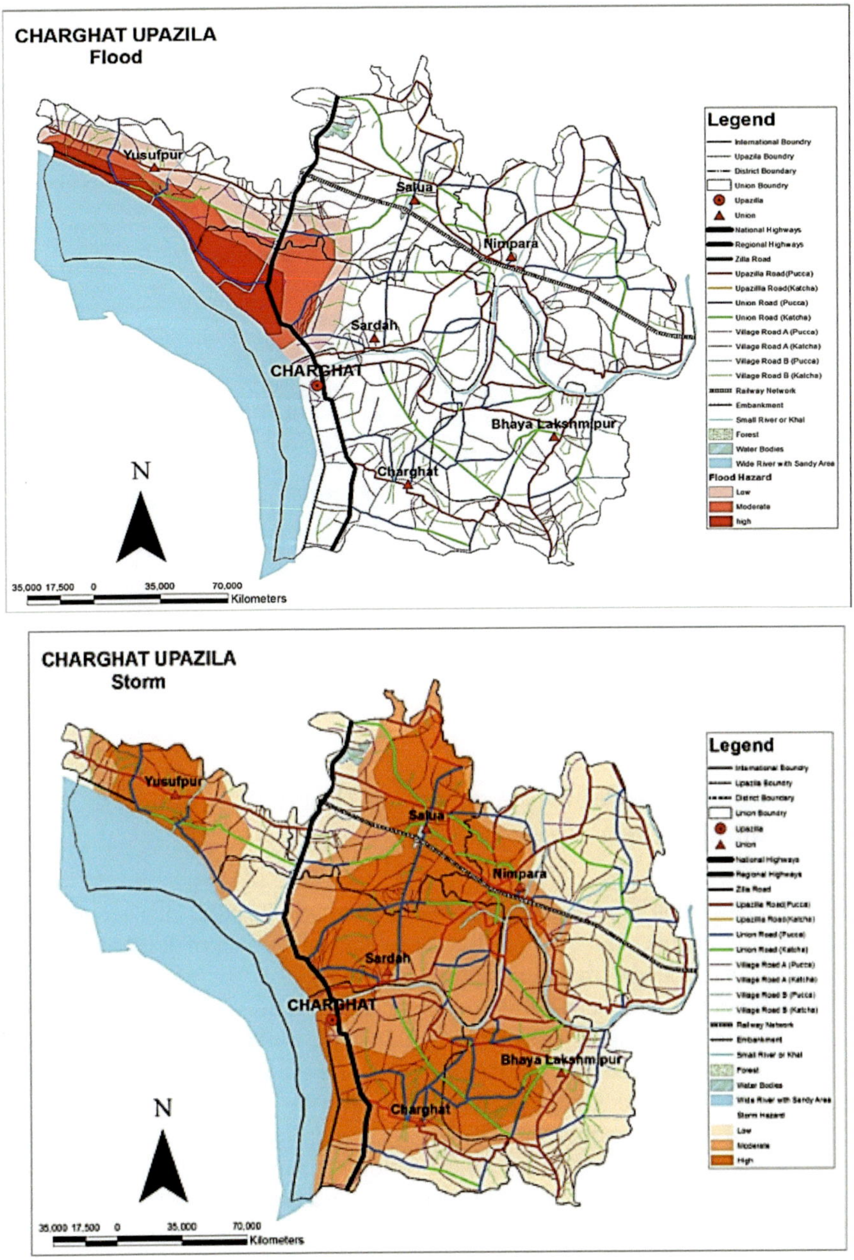

Fig. 4 Flood vulnerability at Charghat Upazilla in Rajshahi district. *Source* Upazilla Management Plan for Rajshahi District-2014

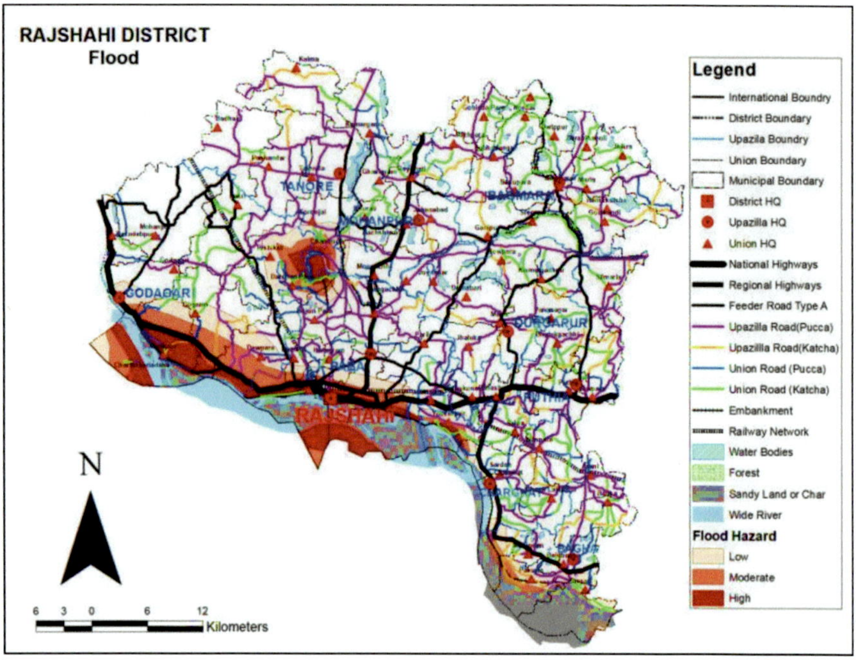

Fig. 5 Overall flood condition in Rajshahi district. *Source* Upazilla Management Plan for Rajshahi District-2014

4 Drought Vulnerability Assessment (Formal Regionalization)

Figure 6 shows that Godagari and Bagha are the two Upazilas that are the most affected by droughts. The existing disaster management plan of Rajshahi (Fig. 7) also supports the output from the map.

Paba, Charghat, Mohonpur, and Baghmara are the areas where people are sometimes affected by drought. Then Tanore, Durgapur, Puthia, Rajpara, Boalia, and Matihar are rarely affected by drought. As most of these areas are somewhat developed and the Rajshahi WASA is responsible for the distribution of water; the occurrence of drought is quite rare.

So, it can be concluded that not all the factors are uniformly important, the first two factors are directly interrelated like the groundwater table. The second factor of irrigated land is also important as in the city corporation most areas are urbanized and not many people are engaged in agriculture. As a result less irrigated land would indicate less impact of drought. The third factor sandy soil was selected because of its low water and low nutrient-holding capacity. Sandy soils have low water and nutrient-holding capacity and struggle to retain sufficient amounts of crops. Shallow-rooted crops are more susceptible to drought stress in sandy soils, as they may experience water deficits that hinder their growth and yield.

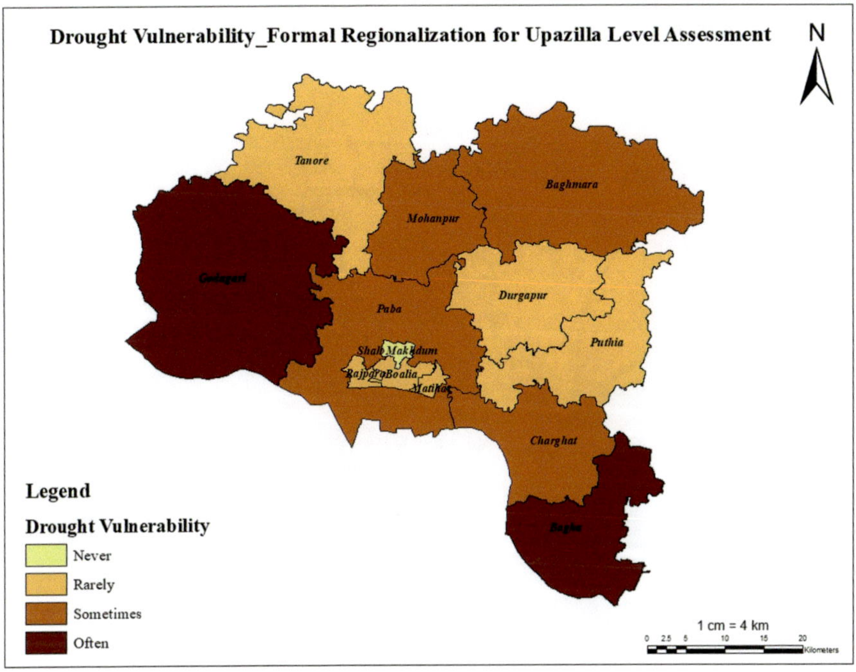

Fig. 6 Drought vulnerability (formal regionalization for Upazila-Level Assessment)

The people of Bagha urged us to have their problems solved. In a focus group discussion, they said, that although they have been facing drought for quite some time, but last year the problem reached its peak. People didn't have water in their tube wells for up to 3 months during summer which was also the month of Ramadan. The duration of the hazard according to the month has been shown in Sect. 5. People are living in fear as no organizations reached out to them to solve their problems till now (Fig. 7).

5 Upazila-Based Flood and Drought Vulnerability

5.1 Bagha Upazilla

Data on the drought conditions in Bagha Upazilla are available in the 2014 Upazila Disaster Management Plan, with two scenarios of damage which are: "Very Much" and "Middle." 1976, 1979, 1989, 1992, 1996, 1999, 2004, 2005, 2007, 2010, 2011, and 2012 are the years that are impacted. While fish and cattle were impacted in the "Middle" scenario, agricultural resources and health services were also impacted by the damage scenarios (Table 5).

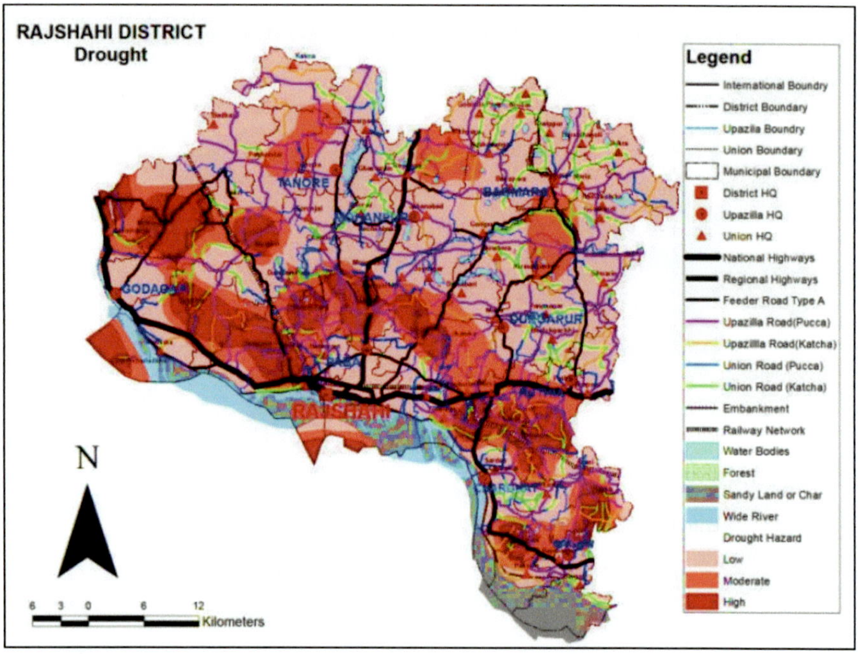

Fig. 7 Overall drought condition in Rajshahi district. *Source* Upazilla Management Plan for Rajshahi District-2014

Table 5 Year-wise drought condition (scenario) in Bagha Upazilla

Name of disaster	Year	Scenario of damage	Name of damaged goods
Drought/Sunray	1976, 1979, 1989, 1992, 1996, 1999, 2004, 2005, 2007, 2010, 2011, 2012	Very much	Agricultural resources. Health services
		Middle	Fishes. Cattle

Source Upazila Disaster Management Plan, Bagha, Rajshahi (2014)

The Upazila Disaster Management Plan for Bagha in Rajshahi (2014) emphasizes how susceptible the area is to solar radiation and drought, which can lead to agricultural damage, a risk to fish populations, and a scarcity of water. Bagha Upazila contains resources like 45 Tube-Wells, one health complex, and seven health sub-centers despite these difficulties (Table 6).

A drought similar to the one that struck in 2001 may destroy the crops on about 16,454 hectares and seriously impair the financial standing of 4392 households. The different causes of drought are shown in Table 7.

Every year, the people who live in Bagha Upazila face a wide range of difficulties that pose unique barriers to their day-to-day existence. The most prominent of these challenges is the drought, which poses a serious threat to the community's health.

Table 6 Drought affliction and capabilities in Bagha Upazilla

Catastrophes	Affliction	Capability
Drought/Sunray	Unsettling solar radiation destroys farmed crops Infants, disabled, and expectant Women and the elderly must deal with the issue Fish stocks are in jeopardy There will soon be a shortage of drinking water An obstacle enters the communication network	Bagha Upazila has forty-five tube-wells The Upazila is home to one health complex and seven sub-centers of health

Source Upazila Disaster Management Plan, Bagha, Rajshahi (2014)

Table 7 Causes of drought in different unions in Bagha Upazilla

Affliction	Most affliction place	Causes of affliction	Qnty. of wretched people
Drought	Monigram, Arani, Bowsha, Baju Bagha and Gorgari	Scarcity of potable water May have to face various diseases Obstruction of Cultivation crops appear Lack of water production of crops becomes decrease	10,000–12,200 families

Source Upazila Disaster Management Plan, Bagha, Rajshahi (2014)

Drought has a significant influence on fundamental areas of everyday life as well as agricultural activities. The people's already difficult circumstances are made worse by the sun's unrelenting beams and the prolonged dry period. Once-thriving fields are now left dry and bare by the nourishing touch of rain. The situation is made worse by the dry season's water shortage, which forces the people to fight constantly for access to this essential resource. The account provided by the locals highlights how serious the situation is. Their stories powerfully depict communities coming together to overcome the problems presented by drought, demonstrating perseverance in the face of adversity (Table 8). The repercussions of the drought are felt in every aspect of life in Bagha Upazila, from agricultural losses to the demand for water sources. This emphasizes the urgent need for lasting solutions and assistance for people who are struggling with nature's harsh grasp.

5.2 Charghat Upazilla

The Charghat Upazila flood conditions are described in the 2014 Upazila Disaster Management Plan. The plan highlights maximum flood scenarios in 1981, 1988, and

Table 8 Monthly drought conditions in Bagha Upazilla

Catastrophe	Season											
	April	May	June	July	August	September	October	November	December	January	February	March
	Boi	Joi	Ash	Srab	Vad	Ash	Kart	Aghr	Poush	Magh	Fal	Choi
Drought												

Source Upazila Disaster Management Plan, Bagha, Rajshahi (2014)

1998, as well as moderate impacts in 1986, 1991, and 1995, which predominantly affect cattle, agriculture, and fisheries (Table 9).

The devastating effects of floods in Charghat Upazilla are highlighted in the 2014 Upazila Disaster Management Plan. These effects include destruction of agriculture, submersion of graveyards, damage to infrastructure, loss of fisheries, water shortage, livestock destruction, and increased vulnerability for vulnerable groups. Notwithstanding these obstacles, the area has a number of tools at its disposal to lessen the effects, including 96 high graveyards, 28 drinking water tanks, 10 embankments, 4 sluice gates, and 980 high-based tubewells (Tables 10 and 11).

Table 12 shows different intensity of flood during various months in Bangladesh.

Table 9 Year-wise flood condition (scenario) in Charghat Upazilla

Name of disaster	Year	Scenario of damage	Name of damaged goods
Flood	1981, 1988, 1998	Maximum	Agriculture, fisheries, livestock, human resource, infrastructure, plantation
	1986, 1991, 1995	Moderate	Agriculture, fisheries, livestock

Source Upazila Disaster Management Plan, Charghat, Rajshahi (2014)

Table 10 Flood affliction and capabilities in Charghat Upazilla

Catastrophes	Affliction	Capability
Flood	It ruins agricultural products Graveyards go underwater It harms human resources It damages infrastructure It washes away fisheries It causes a lack of safe drinking water It destroys livestock It endangers children, the pregnant, the disabled, and the old	There are 96 high graveyards There are 28 tanks of pure drinking water There are 10 (158 ha) effective embankments and cross dams There are 4 sluice gates There are 980 high-based tubewells

Source Upazila Disaster Management Plan, Charghat, Rajshahi (2014)

Table 11 Causes of flood in different unions in Charghat Upazilla

Affliction	Most affliction place	Causes of affliction	Qnty. of wretched people
Flood	Yusufpur and areas adjacent to rivers	Many a piece of agricultural land are being swallowed by flood in these areas	10,000–12,200 families

Source Upazila Disaster Management Plan, Charghat, Rajshahi (2014)

Table 12 Monthly flood conditions in Charghat Upazilla

Catastrophe	Season											
	April	May	June	July	August	September	October	November	December	January	February	March
	Boi	Joi	Ash	Srab	Vad	Ash	Kart	Aghr	Poush	Magh	Fal	Choi
Flood												

Source Upazila Disaster Management Plan, Charghat, Rajshahi (2014)

5.3 *Godagari Upazilla*

The lack of precipitation during the dry season and the effects of climate change produce sunrays in this area. The usual circumstance in this place is a sunray invasion. The two times a year, when the people suffer from the unbearable heat of the sun, are during Chaitra to Jaistta and Vadra to Kartic, which leads to several issues that have an impact on land farming, fisheries, forests, wildlife, and birds. Every kind of channel, a little river dried up someplace. The groundwater layer descends farther. This location would become perilous if this confusing situation persisted (Table 13).

Table 14 shows different intensity of drought during various months in Bangladesh.

Table 13 Drought affliction and capabilities in Godagari Upazilla

Catastrophes	Affliction	Capability
Drought	Decreasing ground water table (mm) hampers the crop production Problems must be faced by elderly people, pregnant women, disabled people, and newborns Fish stocks are in jeopardy There will be a shortage of drinking water soon A barrier enters the communication system	• 3490 ponds • 5 bil (Chorai Bil) • 1 livestock hospital • 52 NGOs running project work

Source Upazila Disaster Management Plan, Godagari, Rajshahi (2014)

Table 14 Monthly drought conditions in Godagari Upazilla

Catastrophe	Season											
	April	May	June	July	August	September	October	November	December	January	February	March
	Boi	Joi	Ash	Srab	Vad	Ash	Kart	Aghr	Poush	Magh	Fal	Choi
Drought												

Source Upazila Disaster Management Plan, Godagari, Rajshahi (2014)

6 Comparison Between the Upazilas

Capabilities of the three upazila Godagari, Charghat, and Bagha were analyzed in both terms of flood and drought. Only these three areas were selected as they were proven more vulnerable than the other upazillas in the previous section. Here the term capability identifies the power that helps in preventing the disaster. Well, natural disasters cannot be totally prevented but factors that are shown here can help to mitigate the aftermath of the disaster to a great extent. All the information shown in Figs. 5 and 7 was gathered from different secondary resources. The number of sluice gates, ponds, forest areas, tube wells, embankments, and flood shelters were gathered from the Upazilla Disaster Management Plan of Rajshahi and BBS. And the number of NGOs was collected from the National Web Portal of NGOs.

Sluice gate helps stop the flow of water; NGOs help strengthen the community and their miscellaneous projects also help to improve the condition of the area. From Fig. 8, Godagari has the highest number of sluice gates and flood shelters and a moderate number of NGOs making it more resilient in terms of flood than the other two. Although Bagha is showing poor results in this manner but it is not prone to flood according to the map prepared by authors in Fig. 3 and the already existing map of vulnerability in the Upazilla level Disaster Management Plan of Rajshahi.

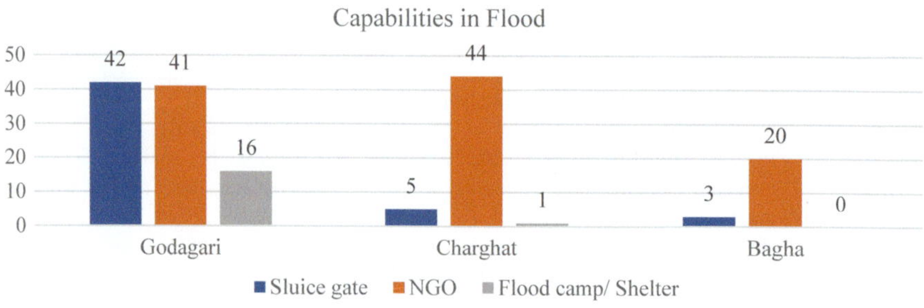

Fig. 8 Resilience against flood. *Source* Author's preparation

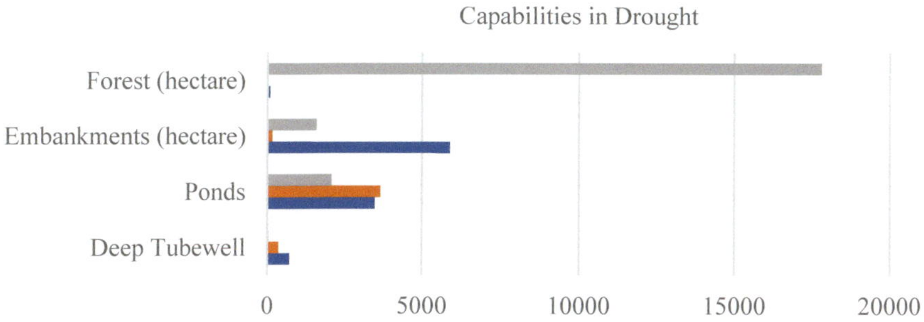

	Deep Tubewell	Ponds	Embankments (hectare)	Forest (hectare)
Bagha	45	2096	1600	17800
Charghat	358	3668	158	
Godagari	716	3490	5900	90

Bagha Charghat Godagari

Fig. 9 Capabilities in drought among the Upazilas. *Source* Author's preparation (2023)

Flood vulnerability varies differently in the areas of Charghat. Usufpur is prone to flood whereas in the village of Chandansohor, the authors found people are mostly affected by river erosion. The condition of flood shelters is bad in both Charghat and Bagha, so these two have a high scope of improvement.

Resilience in Drought

Ponds and Deep tubwells directly help in drought/help in building resilience against drought. But the embankment and forest areas are not directly related to drought but they help to build resilience against river erosion and affect the layer of the water. Figure 9 shows Bagha as the most vulnerable in terms of Drought susceptibility having the lowest number of both deep tubewell and ponds. But high number of forest areas and embankments builds its adaptive capacity in flood and river erosion as shown in Fig. 6.

Godagari has the highest number of embankments and deep tube wells. But still is prone to drought. The number of forested areas in Charghat could not be found in the secondary resources but when surveying the areas authors found the afforestation initiatives taken by the government. Which can also be seen in Fig. 10.

7 Limitations of the Study

The main limitation of the study is the timeframe in which it was conducted. The study shows the vulnerability in terms of flood and drought. The time when drought is at its peak was pinpointed from the Focus Group Discussion but due to the time

Fig. 10 Government afforestation plan in Charghat Upazila. *Source* Author's preparation (2023)

limit the authors did not get a chance to figure out the extent of their sufferings first hand or survey the exact time of the disaster.

8 Conclusions

With regard to Rajshahi City's differing susceptibilities to drought and flood, the extensive research offers insightful information about the existing condition in the Upazillas focusing on the root-level people. The results establish the fact that drought is a more danger to Rajshahi than floods among these natural disasters highlighting the most vulnerable areas.

Areas close to rivers are often susceptible to flood as shown from the output which was in Charghat. But the primary data collection pointed out that along with flood some areas of Charghat are more prone to River erosion. Among the other upazillas of Rajshahi City Godagari and Bagha is vulnerable to drought. A comparative analysis among the three most vulnerable areas was also shown indicating the strength and their resilience against the disaster. The study highlights how crucial it is to take proactive steps in order to manage and prepare for the upcoming drought scenario. Drought can have disastrous effects if deliberate actions are not taken.

The factors that are used to delineate the regions and the reason for selecting those factors were also analyzed in the chapter. The study's research identifies particular causes that make Rajshahi City more susceptible to these natural occurrences. Significantly, the river's position turns out to be a critical element affecting flood risk, and the water table and land irrigation status are major factors aggravating the city's drought vulnerability.

In summary, the study offers policymakers, urban planners, and community stakeholders a comprehensive knowledge of the unique issues presented by drought and flood in Rajshahi City, making it an essential tool for their work. The utilization of a GIS-based strategy and a focus on geographic analysis improves the accuracy of the results, providing a useful foundation for focused interventions and adaptable tactics to lessen the effects of these natural disasters.

Acknowledgements We sincerely thank Allah Ta'ala for everything. We are incredibly grateful to my parents, whose unwavering support and selfless acts have served as my cornerstones. We would like to thank the Rajshahi University of Engineering and Technology's eminent professors and lecturers in Urban and Regional Planning for inspiring our common interest in and understanding of the field. Our development and success have been greatly impacted by their priceless efforts.

References

Arafat RA (2023) Regionalization | definition, examples & principles-video & lesson…, regionalization. Available at: https://study.com/academy/lesson/the-regionalization-process-definition-examples.html. Accessed 30 Dec 2023

Chowdhury M (2009) The population challenge facing Bangladesh. ResearchGate. https://www.researchgate.net/publication/266251152_The_Population_Challenge_Facing_Bangladesh

Das A, Bhattacharjee S, Ahmad B (2023) Drought assessment based on monthly TRMM rainfall data using standardized precipitation index for Rajshahi. ResearchGate. https://www.researchgate.net/publication/370418908_Drought_Assessment_Based_on_Monthly_TRMM_Rainfall_data_using_Standardized_Precipitation_Index_for_Rajshahi_Division_Bangladesh_Keywords?enrichId=rgreq-5bbd3df7dac588f54d4c222532b03219-XXX&enrichSource=Y292ZXJQYWdlOzM3MDQxODkwODtBUzoxMTQzMTI4MTE1NDYwNjk0NkAxNjgyOTI4NjI0MTI1&el=1_x_3&_esc=publicationCoverPdf

Dilley M, Chen RS, Deichmann U, Lerner-Lam AL, Arnold M, Agwe J et al (2005) Natural disasters hotspots: a global risk analysis. Synthesis report. The World Bank, Hazard Management Unit, Washington, DC

DM Plan Paba Upazila Rajshahi District_English Version-2014 (n.d.). Scribd. https://www.scribd.com/document/261610465/DM-Plan-Paba-Upazila-Rajshahi-District-English-Version-2014

Glasson J (1975) Introduction to regional planning, 2nd edn

Habiba U, Hassan AWR, Shaw R (2013) Livelihood adaptation in the drought prone areas of Bangladesh. In: Disaster risk reduction, pp 227–252. https://doi.org/10.1007/978-4-431-54249-0_13

Jewel MDaA, Emon SR, Das A, Zubayer S (2018) Formal region delineation of Bangladesh: comparatively urbanized districts of Bangladesh. ResearchGate. https://www.researchgate.net/publication/329541824_Formal_Region_Delineation_of_Bangladesh_Comparatively_Urbanized_Districts_of_Bangladesh?enrichId=rgreq-9434972f7986c2777a2f96c09f223cd8-XXX&enrichSource=Y292ZXJQYWdlOzMyOTU0MTgyNDtBUzo5MTY1NDc0NTEyMjQwNjRAMTU5NTUzMzM0OTUwMQ%3D%3D&el=1_x_3&_esc=publicationCoverPdf

Mamun MaA, Rahman SM, Uddin MN, Sultan-Ul-Islam M, Arifin MNK, Rahman SA (2021) Rainfall and drought tendencies in Rajshahi division, Bangladesh. Geogr Environ Sustain 14(1):209–218. https://doi.org/10.24057/2071-9388-2020-124

Matallo H (n.d.) Capacity development to support national drought management policies, 1st edn. United Nations Convention to Combat Desertification, Saudi Arabia. https://www.ais.unwater.org/ais/pluginfile.php/571/mod_page/content/88/UNCCD.pdf

Ministry of Disaster Management and Relief-Government of the People's Republic of Bangladesh (n.d.). http://203.76.123.197/egls/documents/270/search-detail

Ministry of Disaster Management and Relief-Government of the People's Republic of Bangladesh (n.d.). http://203.76.123.197/egls/documents/313/search-detail

Ministry of Disaster Management and Relief-Government of the People's Republic of Bangladesh (n.d.). http://203.76.123.197/egls/documents/249/search-detail

Nasiri H, Yusof MJM, Ali TaM (2016) An overview to flood vulnerability assessment methods. Sustain Water Resour Manag 2(3):331–336. https://doi.org/10.1007/s40899-016-0051-x

Population and Housing Census (2011) In: Bangladesh Bureau of Statistics

Sarker NI, Alam GMM, Firdaus RBR, Biswas JC, Islam ARMT, Raihan L, Hattori T, Alam K, Joshi NP, Shaw R (2022) Assessment of flood vulnerability of riverine island community using a composite flood vulnerability index. Int J Disaster Risk Reduction 82:103306. https://doi.org/10.1016/j.ijdrr.2022.103306

Schwarz I, Kuleshov Y (2022) Flood vulnerability assessment and mapping: a case study for Australia's Hawkesbury-Nepean catchment. Remote Sens 14(19):4894. https://doi.org/10.3390/rs14194894

Sultana N (2016) Vulnerability assessment of drought prone areas in Bangladesh through extreme temperature modeling. J Sci Technol Environ Inform 3(1):151–160. https://doi.org/10.18801/jstei.030116.17

Think Hazard-Rajshahi-Urban flood (n.d.). https://thinkhazard.org/en/report/61362-bangladesh-rajshahi/UF

UN/ISDR (2004) Living with risk: a global review of disaster reduction initiatives. United Nations, Geneva

Wang T, Sun F (2023) Integrated drought vulnerability and risk assessment for future scenarios: an indicator based analysis. Sci Total Environ 900:165591. https://doi.org/10.1016/j.scitotenv.2023.165591

Water Quality Assessment and Pollution Source Identification of a Semi-urban Water Body Near to Waste Dumping Zone, Katwa, India, Using Pollution Indices and Geostatistical Methods

Krishnendu Kumar Pobi, Sayantani Chakraborty, Ritwik Chakraborty, Gourisankar Panda, and Sumanta Nayek

Abstract In our present investigation, the water quality and trophic status of semi-urban water body near solid waste dumping site at Katwa, West Bengal, India were estimated using various pollution indices and represented via geospatial techniques. In relation, the potential sources of pollution were evaluated by using multivariate statistical applications. The appraisal of the examined water body reveals that the pH is neutral to slightly alkaline in nature (pH 7.2–8) and the other physicochemical parameters are found well under the recommended value of standard guidelines for drinking water (BIS (Bureau of Indian Standards) IS: 10500 in Indian standard for drinking water specification, second revision. Bureau of Indian Standards, Government of India, New Delhi, 2012) except Fe. The estimated water quality index (WQI) value of water body varies from 32.43 to 42.13 and is classified as "Good quality" which is intended for drinking/domestic purposes. The calculated organic pollution index (OPI) value of the water samples ranged from 1.90 to 2.03 and lies under "Lightly Polluted" to "Begin to contaminated" category. But the geospatial representation of eutrophication index (EI) values ranged from 7.08 to 8.60, suggesting the condition of water body is "eutrophic." The multivariate statistical investigations such as Pearson's correlation matrix (PCM), principal component analysis (PCA), and cluster analysis (CA) clearly distinct the sources of pollution in monitored water

K. K. Pobi · G. Panda
Department of Basic Science and Humanities (Chemistry), Asansol Engineering College, Asansol, West Bengal 713305, India

S. Chakraborty
Department of Bachelor of Computer Application, Asansol Engineering College, Asansol, West Bengal 713305, India

R. Chakraborty
Department of Bachelor of Civil Engineering, Asansol Engineering College, Asansol, West Bengal 713305, India

S. Nayek (✉)
Department of Environmental Science, AIES, Amity University, Kolkata, West Bengal, India
e-mail: sumanta.nayek@gmail.com

S. C. Pal and U. Chatterjee (eds.), *Surface, Sub-Surface Hydrology and Management*, Springer Geography, https://doi.org/10.1007/978-3-031-62376-9_12

body which are mainly derived from several anthropogenic and natural origins like leachate from solid waste dumpsite, atmospheric disposition, soil erosion, surface runoff and municipal waste disposal, etc. Thus, the current investigation also serves requisite suggestion to take necessary actions for restoration and rejuvenation of water body and implement the proper solid management policies to safeguard the environment.

Keywords Semi-urban water body · Water quality index (WQI) · Organic pollution index (OPI) · Multivariate analysis · PCA

1 Introduction

Surface water bodies play a pivotal role as vital natural resources for both humanity and the biotic communities inhabiting Earth's landscape. Lakes and ponds enhance the quality of life in urban and semi-urban areas by offering numerous amenities, recreational facilities, and educational opportunities. Furthermore, they contribute to climate regulation in these urban and semi-urban environments (Arroyo and Jáuregui 2000; Persson 2012). Beyond these aspects, in urban and semi-urban settings, lakes and ponds play a crucial role in preserving regional biodiversity, providing socioeconomic advantages, managing stormwater drainage, and delivering various other ecosystem services (Gledhill and James 2012; Oertli and Parris 2019). The swift urbanization, industrialization, and growing population impose a scarcity of inland water resources. Furthermore, the deposition of solid waste and leachate from areas where solid waste is dumped contributes to the degradation of the natural water quality in inland water bodies. However, generation of huge amount of Municipal solid waste (MSW) due to rapid population growth and industrialization has increased significantly, owing to the people's changing lifestyles and social standing. India generates about 100,000 metric tons of solid garbage every day, of which 90% is disposed of in open spaces, according to estimates from the Ministry of Urban Affairs, Government of India (MoUD 2000). Lack of improper solid waste management practices is the key route for surface and groundwater contamination. The leachate derived from open solid waste dumping sites contains various organic constituents, heavy metals, and xenobiotic organic substances which ultimately alter the physicochemical characteristic of surface as well as groundwater resources (Naveen et al. 2017). Along with poor solid waste management practices coupled with warm climatic environments adversely affect the environment both locally and globally (Sumathi et al. 2008). Kanmani and Gandhimathi (2013) described the leachate obtained from open dumping sites is potential for surface soil and ground water contamination. Further, Nartey et al. (2012) revealed the surface water resources in the Accra Metropolitan Area, Ghana were severely affected solid waste dumping sites due to various pollutants such as organic waste, toxic metals, total and fecal coliforms. The seasonal and temporal variations in groundwater quality around MSW dumping sites in Hyderabad City, India are dramatically affected by leachates containing both organic and inorganic constituents as mentioned by Kamble et al. (2020). Additionally, various human activities, such as wastewater discharge, sewerage dumping, construction debris disposal, and the runoff of nutrients from nearby crop fields, along with recreational activities, collectively degrade the physicochemical characteristics of semi-urban water bodies (Sharma et al. 2010). Water quality indices are crucial analytical tools that utilize mathematical frameworks to assess shifts in water quality, as demonstrated by several earlier researchers in the field of surface and groundwater quality assessment (Panda et al. 2021; Howladar et al. 2021). Such data proves invaluable for water management authorities, enabling them to effectively safeguard the health of surface water resources and implement proper water management practices. Moreover, they facilitate the monitoring of water quality changes over time, and subsequent management steps allow for the maintenance of ecological and climatological balance across various scales, ranging from local to global (Şimşek and Mutlu 2023).

The eutrophic state in semi-urban water is notably impacted by soil erosion, as the influx of nutrients and sediments from the watershed poses a serious threat to aquatic life (Auddy et al. 2023). Therefore, safeguarding and preserving the integrity of semi-urban water bodies are deemed top priorities due to the limited availability of freshwater resources. It is also crucial to minimize human interference, which can have detrimental effects on the quality of surface inland water (Parween and Ramanathan 2019).

To evaluate the overall quality of the water, a number of pollution indices are commonly used, including the Water Quality Index (WQI), Organic Pollution Index (OPI), and Eutrophication Index (EI). These indices are useful for establishing the acceptability of water for certain reasons such as residential or drinking use, as well as for providing insight into the potential impact on aquatic life survival. Multivariate statistical analysis is utilized on extensive datasets to assess correlations among variables. Various algorithms and techniques are then employed to uncover interdependencies within the interpreted dataset. Common statistical analysis methods, such as the Pearson correlation matrix, Principal Component Analysis (PCA), and Hierarchical Cluster Analysis (HCA), serve to identify influential sources of contamination (Singh et al. 2017). Multivariate analysis also differentiates between point and non-point sources of pollution in the study area. The Geographic information system (GIS) combined with IDW interpolation technique is used severally to assess and monitor the surface and groundwater quality as described by earlier researchers in their studies (Panda et al. 2021; Kamble et al. 2020).

The Ministry of Housing and Urban Affairs, Government of India has launched "Mission Amrit Sarovar-Jal Dharohar Sanrakshan" project envisioned protecting traditionally important semi-urban water bodies for ensuring water security of cities by involving youth and the community. There the main aim includes the preservation, restoration, and rejuvenation of the semi-urban water bodies in various cities as selected by state authorities of India. Thus this chapter describes the evaluation of degree of contamination, interrelationships, and interdependencies among the measured physicochemical water quality parameters of selected water body near MSW dumping site in Katwa Municipal area, West Bengal, India using various pollution indices and multivariate statistical techniques. Additionally, it assesses the

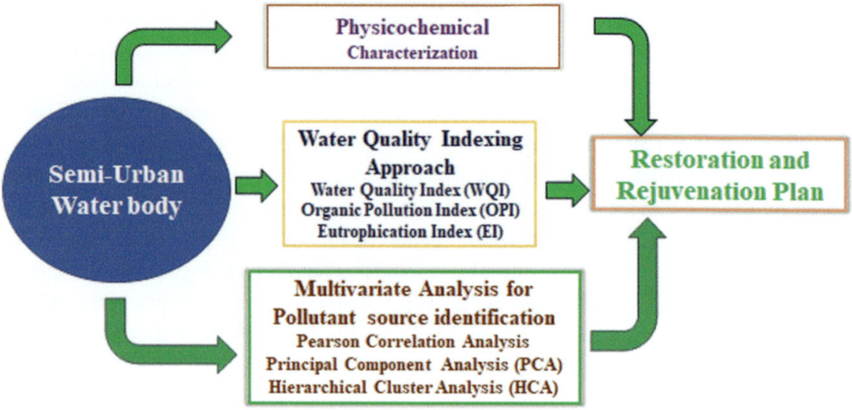

Fig. 1 Schematic flow chart of the study

levels of organic load, trophic conditions, and water quality intended for domestic or drinking purposes of the selected water body near Katwa MSW dumping site. The chapter also provides the GIS mapping based on contamination level in order to visualize the spatial pattern of pollution and connect the origin of various pollutants into the monitored water body. The schematic flow of the study is illustrated in Fig. 1.

2 Material and Methods

2.1 Description of Study Area

The semi-urban water body is situated within Katwa Stadium in Katwa Municipality near MSW duping site, Purba Bardhaman district, West Bengal, India, positioned at 23.636308° N 88.129998° E (Fig. 2). Katwa is located 158 km away from Kolkata and is only 56 km from Bardhaman city in the Purba Bardhaman district. This water body is relatively new among the freshwater bodies in Katwa Municipality and appears to have a rectangular shape. It extends approximately 50 m from east to west and 25 m from north to south, with an average depth of 1.5 m. The Katwa Municipality is bounded on the east, west, and north by the Ajay and Bhagirathi rivers, and the territory has tropical climates. The charm of the studied water body is enhanced by the presence of numerous evergreen trees in its vicinity.

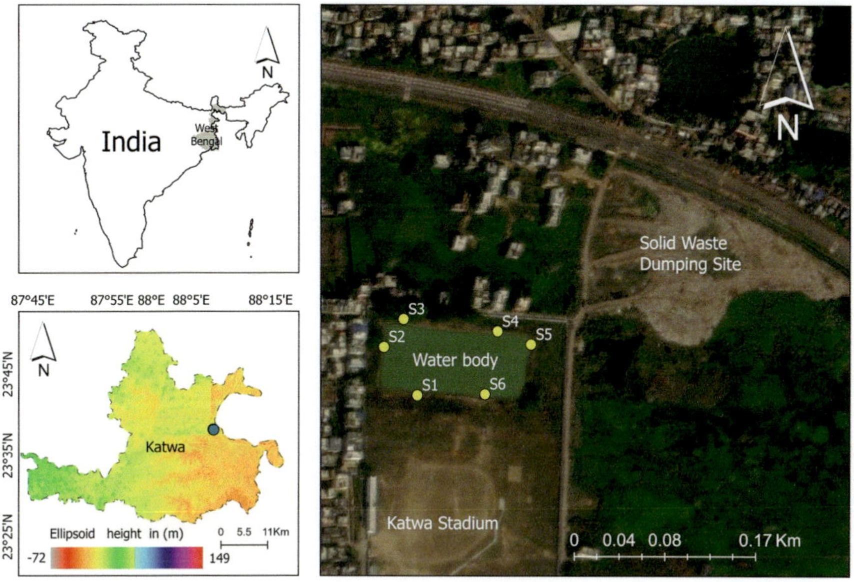

Fig. 2 Geographical location of water body with sampling locations

2.2 *Sampling Strategies and Physicochemical Analysis of the Sample Water*

To assess water quality, six specific sampling sites were chosen to collect water samples from the selected water body in August 2022. These sampling locations were deliberately selected to obtain water samples from the peripheral areas of the inland water body. Temperature, pH, electrical conductivity (EC), total dissolved solids (TDS), and dissolved oxygen were measured onfield immediately after collection using Eutech Instruments portable analyzers. To avoid any contamination, the collected samples were carefully refrigerated under 4 °C. Other physicochemical parameters were analyzed within 48 h of sample collection using standard methods (APHA 2005). To assess overall water quality, the analyzed results were compared to drinking water guidelines recommended by the World Health Organisation (WHO 2006) and the Bureau of Indian Standards (BIS 2012).

2.3 *Quantification of Water Quality of Semi-urban Water Body Using Various Indices*

Water Quality Index (WQI)

The WQI is computed through a weighted arithmetic technique to assess the suitability of water for human consumption purposes. This index captures the combined impact of measured physicochemical variables and assesses the influence of both anthropogenic and geogenic factors on several key parameters, ultimately reflecting the water quality for drinking purposes (BIS 2012). To evaluate WQI value, 10 significant physicochemical parameters, including pH, EC, TDS, TH, TA, DO, Cl^-, SO_4^{2-}, NO_3^-, and PO_4^{3-}, along with their recommended limits by the WHO (2006) and the BIS (2012), were taken into account for calculating the WQI. Each parameter is assigned a specific weightage (W_a) ranging from 1 to 5 based on its relative impact on the overall water quality (Table 1). Initially, the relative weight (W_r) is determined using the following equation.

$$W_r = W_{a_i} / \sum_{i=1}^{n} W_{a_i}$$

where n signifies the number of variables, and W_r and W_a stand for the relative weightage and assigned weightage of each parameter, respectively. After that, each parameter's quality rating scale (Q_i) is determined using the given formula:

$$Q_i = [(C_i - V_i)/(S_i - V_i)] \times 100$$

where C_i represents the concentration of individual chemical parameters; Q_i signifies the quality rating; V_i denotes the ideal value for each parameter, set at 0 for all parameters except pH (7) and DO (14.6). Here also, S_i represents the standard permissible limit established by the WHO (2006) and BIS (2012) for the specific parameter.

Certainly, in the last step, the sub-indices (SI) are first established for each parameter to calculate the WQI value. These sub-indices are subsequently employed in the computation of the overall WQI, as outlined in the following equations:

Table 1 Weightage for calculating WQI

Parameters	WHO Standards (2006)	BIS Standards (2012)		Weight (W_i)	Relative weight (W_r)
		Desirable	Permissible		
pH	6.5–8.5	6.5–8.5	NX	4	0.118
EC (μS/cm)	2000	2000	NX	3	0.088
TDS	1000	500	2000	4	0.118
TH	–	200	600	3	0.088
TA	200	200	600	2	0.059
DO	5	≥ 5	NX	2	0.059
Cl^-	250	250	1000	4	0.118
SO_4^{2-}	200	200	400	4	0.118
NO_3^-	45	45	NX	5	0.147
PO_4^{3-}	0.3	0.3	NX	3	0.088
				$\sum W_i = 34$	$\sum W_r = 1.000$

$$\mathrm{SI}_i = W_\mathrm{r} * Q_i$$

$$\mathrm{WQI} = \sum \mathrm{SI}_i$$

Now, the classification of WQI followed by Auddy et al. (2023) is given as: excellent (WQI < 25), good (WQI: 25–50), moderate (WQI: 51–75), poor (WQI: 76–100), and very poor (WQI > 100).

Organic Pollution Index (OPI)

The Organic Pollution Index is formulated to encompass the collective effects of chemical oxygen demand (COD), dissolved inorganic nitrogen (DIN), dissolved inorganic phosphorus (DIP), and dissolved oxygen (DO) on the quality of inland water. The formula for calculating the OPI is provided below:

$$\mathrm{OPI} = \frac{\mathrm{COD}}{\mathrm{COD_S}} + \frac{\mathrm{DIN}}{\mathrm{DIN_S}} + \frac{\mathrm{DIP}}{\mathrm{DIP_S}} + \frac{\mathrm{DO}}{\mathrm{DO_S}}$$

where $\mathrm{COD_S}$, $\mathrm{DIN_S}$, $\mathrm{DIP_S}$, and $\mathrm{DO_S}$ denote the standard values for the mentioned parameters, as mentioned in IS:2296 (1982). The classification of the Organic Pollution Index (OPI) is as follows: less than 1 indicates uncontaminated (Class I), 1–2 signifies the beginning of contamination (Class II), 2–3 represents lightly polluted (Class III), 3–4 indicates moderate pollution (Class IV), and greater than 4 denotes heavy pollution (Class V).

Eutrophication Index (EI)

The computed Eutrophication Index (EI) serves as an indicator of the trophic level in surface water quality, determined by three key parameters: Chemical Oxygen Demand (COD), Dissolved Inorganic Nitrogen (DIN), and Dissolved Inorganic Phosphorus (DIP). The EI calculation method follows the approach outlined by Liu et al. (2011), and the formulation is detailed below:

$$\mathrm{EI} = \frac{\mathrm{COD} \times \mathrm{DIN} \times \mathrm{DIP}}{4500} \times 10^6$$

Following Son et al. (2020), the classification for EI values as: (i) EI < 0 indicates no eutrophication, and (ii) EI > 0 signifies eutrophic conditions. Higher EI values correspond to a pronounced level of eutrophication in surface inland water.

3 Results and Discussion

3.1 Assessment of Physiochemical Properties of Semi-urban Water Body

The analysis of limnological parameters in examined water body has provided valuable insights into the water quality and its purposeful uses. The descriptive statistics of analyzed parameters of examined water body and drinking water quality guideline as per BIS (2012) and WHO (2006) standard are mentioned in Table 2. The recorded surface water temperature falls within the range of 31.1–31.4 °C. The pH values of the collected water samples indicate a slightly alkaline nature, ranging between 7.2 and 8 because of mixing of leachate water from MSW dumping site and similar findings stated by Kamble et al. (2020). The electrical conductivity (EC) and total dissolved solids (TDS) of the sample water vary between 697–976 μS/cm and 339–477 mg/L, respectively. Importantly, the values of total hardness (TH), total alkalinity (TA), dissolved oxygen (DO), chloride (Cl^-), sulfate (SO_4^{2-}), nitrate (NO_3^-), phosphate (PO_4^{3-}), and iron (Fe) concentrations are well within the recommended standards for drinking water as per the guidelines set by BIS (2012).

Table 2 Statistical summary of physicochemical parameters of semi-urban water body

Water parameters	Semi-urban water body				WHO (2006)	BIS (2012), drinking water standards	
	Min	Max	SD	Mean		Desirable	Permissible
Temp. (°C)	31.1	31.4	0.14	31.23	–	–	
pH	7.2	8	0.35	7.70	6.5–8.5	6.5–8.5	NX
EC (μS/cm at 25 °C)	697	976	101.19	781.50	–	2000	NX
TDS (mg/L)	339	477	49.63	386.50	1000	200	600
Turbidity (NTU)	1.6	2.2	0.21	1.87	5	1	5
TH (mg/L)	192	198	2.35	194.50	600.00		–
TA (mg/L)	200	295	36.42	247.00	200	200	600
DO (mg/L)	6.1	6.7	0.22	6.30	–	–	–
BOD (mg/L)	9.6	13.25	1.43	11.47	–	–	–
COD (mg/L)	31.45	38.25	2.23	35.24	–	–	–
Cl^- (mg/L)	113.47	129.15	6.65	122.86	250	250	1000
SO_4^{2-} (mg/L)	14.72	16.04	0.50	15.44	200.00	200	400.00
NO_3^- (mg/L)	0.32	0.39	0.03	0.36	45	45	NX
PO_4^{3-} (mg/L)	0.06	0.07	0.01	0.07	0.3	–	–
Fe (mg/L)	21.32	27.02	2.28	23.91	0.3	0.3	NX

NX No relaxation

3.2 *Assessing Water Quality Using Water Quality Index (WQI)*

The WQI is a consolidated metric that condenses various water parameters into a single value, providing a comprehensive assessment of the studied water body (Kumar et al. 2019). For the computation of the WQI of examined water body, ten physiochemical parameters (as listed in Table 1) were used. WQI scores for water bodies range from 32.43 to 42.13. Despite falling within the "Good" quality category on the scale (26–50), as depicted in Fig. 3, it is noteworthy that all sample water from studied water body is categorized as "Good." However, it is recommended to consider the water for safe drinking purposes only after undergoing appropriate microbial treatment. The GIS map based on WQI value of the studied water body indicated the path of origin of pollutants and allowed to provide needed information to take proper management for its restoration and conservation focusing on the surrounding public health and environment. For this purpose, watershed management techniques must be expanded to stop sediments and leachate from entering the catchment upstream. Sediment dams can be used to regulate areas that are prone to erosion (Issaka and Ashraf 2017).

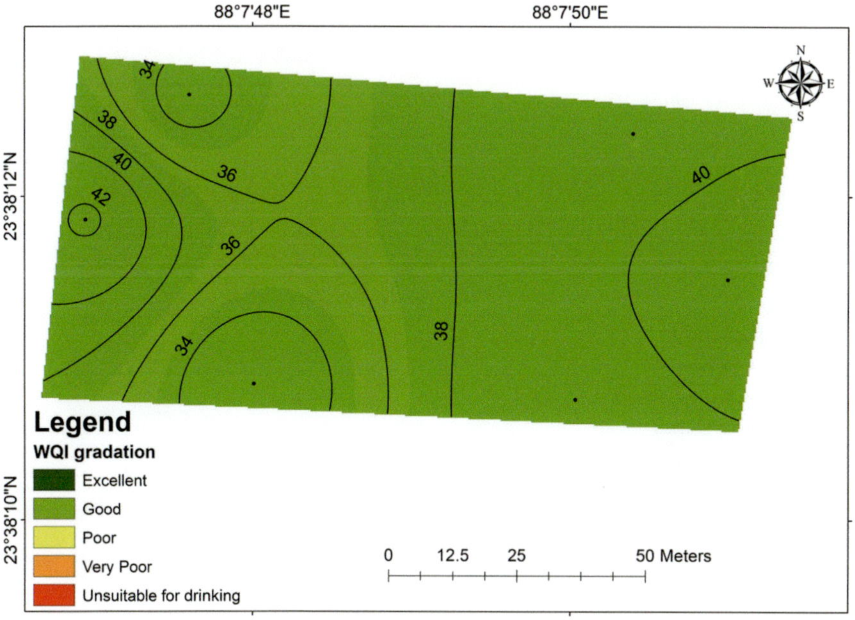

Fig. 3 Spatial distribution of WQI

3.3 Appraisal of Organic Pollution Index (OPI)

The OPI is a comprehensive index used to evaluate the combined influence of COD, DIN, DIP, and DO on quality of water. In this chapter, the OPI values, as depicted in Fig. 4, range between 1.90 and 2.03, with a mean value of 1.94. The majority of the collected sample water, based on the OPI value, falls under the "Begin to contaminated" category, except for Sampling site 6, where the OPI value is 2.03, classifying it as "Lightly Polluted." The spatial distribution maps showed patterns of organic pollutants derived from various anthropogenic and geogenic sources, connected with leachate from a solid waste dumping site and surface runoff from the watershed area. Furthermore, the geointerpolation map indicated the dumping of wastewater from residential areas is also considered a potential source of organic pollution (Schaider et al. 2017). This outcome suggests the initiation of aquatic weed growth, emphasizing the need for immediate preventive measures to control and manage the situation effectively.

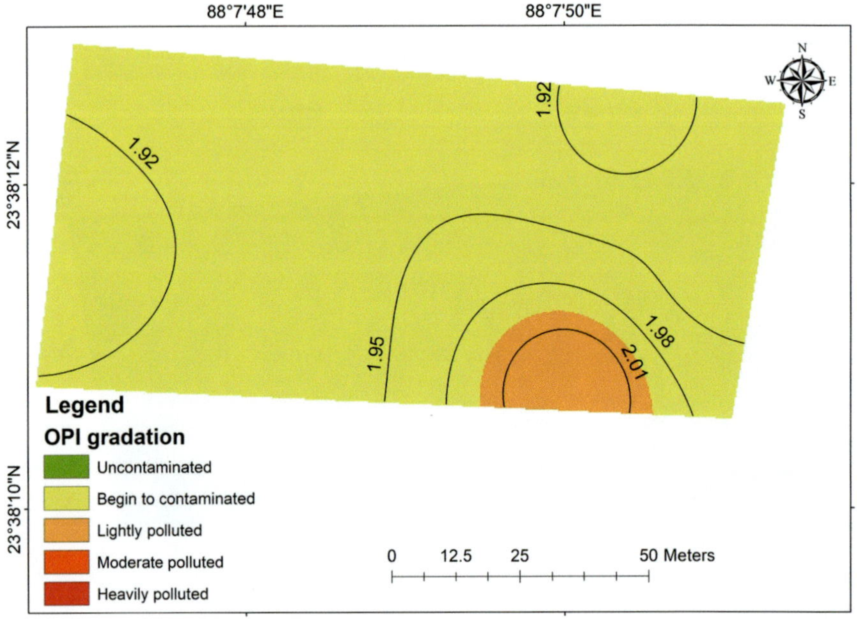

Fig. 4 Distribution of OPI value of monitored inland water body

3.4 Evaluation of Trophic State Using Eutrophication Index (EI)

The Eutrophication index (EI) determines the trophic condition of water quality relying on three key parameters (COD, DIN, and DIP) as established by Liu et al. (2011). For examined water body, the EI values range from 7.08 to 8.60, with a mean reading of 7.93 (Fig. 5). Based on the EI classification, the water body is categorized as "eutrophic." This situation is attributed to a significant accumulation of organic pollutants and nutrient concentrations carried by surface runoff water into the water body. This, in turn, enhances algae productivity, contributing to an elevated level of eutrophication status. Furthermore, the GIS map based on EI values allows for the provision of baseline data for monitoring the changes in water quality and the long-term trophic state of the studied water body, enabling the adoption of ongoing preventative measures to stop pollutants from entering the water body from upstream contaminant sources.

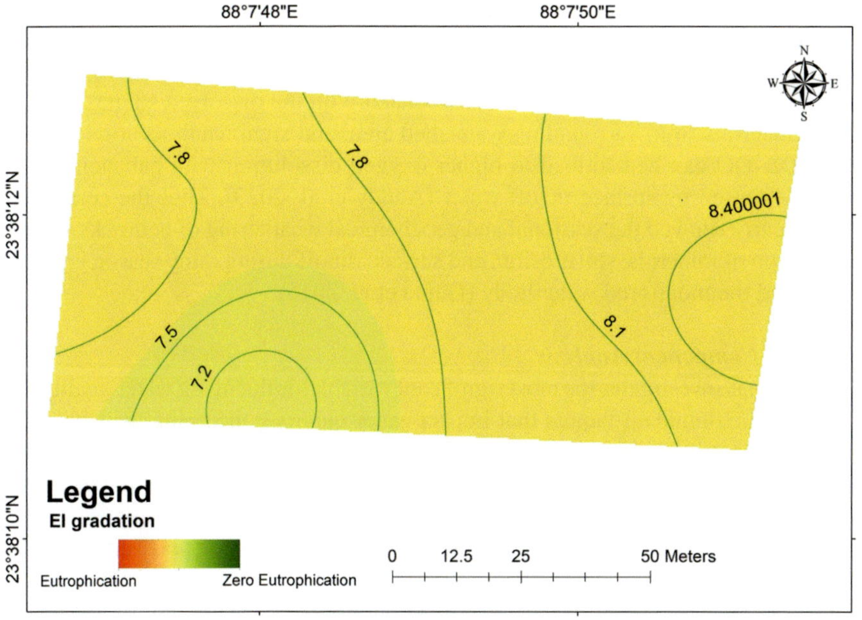

Fig. 5 Spatial distribution of EI of the inland water body

3.5 *Multivariate Statistical Approaches for Pollutants Source Identification*

Multivariate analyses, including correlation analysis, Principal Component Analysis (PCA), and Cluster Analysis (CA), were conducted on the dataset to discern potential sources of contamination and distinguish influencing factors such as natural as well as anthropogenic activities on the semi-urban water ecosystem (Panda et al. 2021).

Pearson's Correlation Analysis

The Pearson's correlation coefficient, with values close to + 1 or − 1, indicates the strength and direction of the correlation among variables, signifying either a strong or weak relationship and highlighting both interrelationship and interdependency among them (Panda et al. 2021). The correlation analysis among the variables of examined water body is presented in Table 3. From the results, strong positive correlation is noted among pH versus TA, TH and COD, EC versus TDS, TA versus TH, Temp. versus DO, COD versus TH, TA, and BOD, and strong negative correlation is observed among SO_4^{2-} versus EC and TDS, Cl^- versus PO_4^{3-} which indicates that the monitored water body get influenced by human activities such as dumping of wastewater, leachate from MSW dumping site many more. The strong positive correlation among EC and TDS may associated with the dissolved solutes through erode materials from surrounding watershed area; and significant negation correlation TDS–DO can be ascribed to higher oxygen dissolution with enhancing TDS contents caused by surface runoff water (Auddy et al. 2023). Thus the correlation study clearly showed the sessional and geochemical weathering of bedrocks such as dissolution of minerals, soil erosion, and surface runoff during rainy season is highly influenced the monitored water body (Gupta et al. 2016).

Principal Component Analysis

PCA analysis investigates the most significant variables influencing water quality and identifies the dominant factors that impact water quality concerning their potential sources of contamination (Gupta et al. 2016). The results of PCA analysis, following varimax rotation, along with eigenvalues and total cumulative variance, are presented in Table 4. The PCA analysis identified four distinct principal components (PCs) with eigenvalues exceeding 1. PC1 exhibits higher loadings for pH, Turbidity, TH, TA, BOD, and COD, suggesting a link to human influence and external contributions. The PC2 is associated with EC, TDS, and SO_4^{3-}. PC3 demonstrates higher loadings for Temp., DO, Cl^-, and PO_4^{3-}, indicating influences from temporal and geogenic factors, as well as leachate from municipal dumping areas and sewerage disposal into the water body. PC4, characterized by higher loadings of NO_3^- and Fe, implies contributions from soil erosion and mineral dissolution in the watershed of the examined water body.

Table 3 Pearson correlation matrix among the analyzed water quality variables

Variables	Temp.	pH	EC	TDS	Turbidity	TH	TA	DO	BOD	COD	Cl^-	SO_4^{2-}	NO_3^-	PO_4^{3-}	Fe
Temp.	1	– 0.390	– 0.531	– 0.475	– 0.520	– 0.562	– 0.269	0.802	– 0.060	– 0.281	0.402	0.224	0.038	– 0.535	0.042
pH		1	0.147	0.050	0.697	0.892*	0.923**	0.016	0.647	0.840*	0.332	0.341	0.564	– 0.036	0.461
EC			1	0.990**	0.626	0.096	– 0.152	– 0.450	– 0.499	– 0.095	0.081	– 0.842*	– 0.369	– 0.265	0.584
TDS				1	0.550	– 0.021	– 0.264	– 0.484	– 0.523	– 0.154	0.035	– 0.899*	– 0.423	– 0.254	0.541
Turbidity					1	0.743	0.580	– 0.177	0.163	0.644	0.547	– 0.304	– 0.175	– 0.354	0.442
TH						1	0.934**	– 0.078	0.574	0.829*	0.303	0.362	0.330	0.078	0.160
TA							1	0.193	0.721	0.861*	0.398	0.591	0.544	– 0.010	0.209
DO								1	– 0.010	– 0.059	0.664	0.432	0.247	– 0.667	0.287
BOD									1	0.856*	0.062	0.670	0.531	0.315	– 0.207
COD										1	0.376	0.376	0.282	0.029	0.035
Cl^-											1	– 0.022	– 0.176	– 0.891*	0.431
SO_4^{2-}												1	0.716	0.308	– 0.280
NO_3^-													1	0.283	0.301

(continued)

Table 3 (continued)

Variables	Temp.	pH	EC	TDS	Turbidity	TH	TA	DO	BOD	COD	Cl^-	SO_4^{2-}	NO_3^-	PO_4^{3-}	Fe
PO_4^{3-}														1	−0.593
Fe															1

*Correlation is significant at the 0.05 level (2-tailed)
**Correlation is significant at the 0.01 level (2-tailed)

Table 4 Varimax rotated using Kaiser normalization principal component loadings matrix for water quality parameters

Rotated component matrix[a]				
Variables	Component			
	PC 1	PC 2	PC 3	PC 4
Temperature	– 0.445	– 0.478	0.711	0.052
pH	0.908	0.107	0.042	0.400
EC	0.045	0.998	– 0.050	0.002
TDS	– 0.046	0.990	– 0.066	– 0.042
Turbidity	0.760	0.602	0.180	– 0.156
TH	0.955	0.052	– 0.069	0.097
TA	0.929	– 0.189	0.110	0.259
DO	– 0.075	– 0.404	0.856	0.254
BOD	0.722	– 0.543	– 0.145	0.082
COD	0.955	– 0.139	0.023	– 0.068
Cl^-	0.384	0.110	0.902	– 0.161
$SO_4{}^{2-}$	0.340	– 0.859	– 0.010	0.338
$NO_3{}^-$	0.288	– 0.391	– 0.114	0.865
$PO_4{}^{3-}$	0.018	– 0.314	– 0.947	0.056
Fe	0.132	0.600	0.464	0.630
Eigen values	**5.493**	**4.474**	**3.268**	**1.289**
% of variance	**36.619**	**29.829**	**21.789**	**8.591**
Cumulative %	**36.619**	**66.448**	**88.237**	**96.827**

Extraction method: Principal component analysis
[a]Rotation method: Varimax with Kaiser normalization

3.6 *Cluster Analysis (CA)*

Cluster Analysis (CA) facilitates the grouping of water variables or sample locations by identifying similarities in their chemical compositions and interdependencies (Routroy et al. 2013). The dendrogram derived from the CA analysis reveals that the water variables can be categorized into three statistically significant distinct clusters, as depicted in Fig. 6. The first cluster (C_1) is strongly associated with pH, Temp., turbidity, DO, BOD, COD, $SO_4{}^{2-}$, $NO_3{}^-$, $PO_4{}^{3-}$, and Fe, and defines their occurrence because of several anthropogenic activities such as direct discharge of wastewater and sewerage disposal, accompanied by temporal fluctuations. The second cluster C_2 is strongly accompanied by TA, TDS, TH, and Cl^-, which is introduced by the surface runoff from the surrounding area and leachate derived from Katwa MSW site. Cluster three (C_3) has shown higher loading for EC, which can be attributed to weathering and erosion.

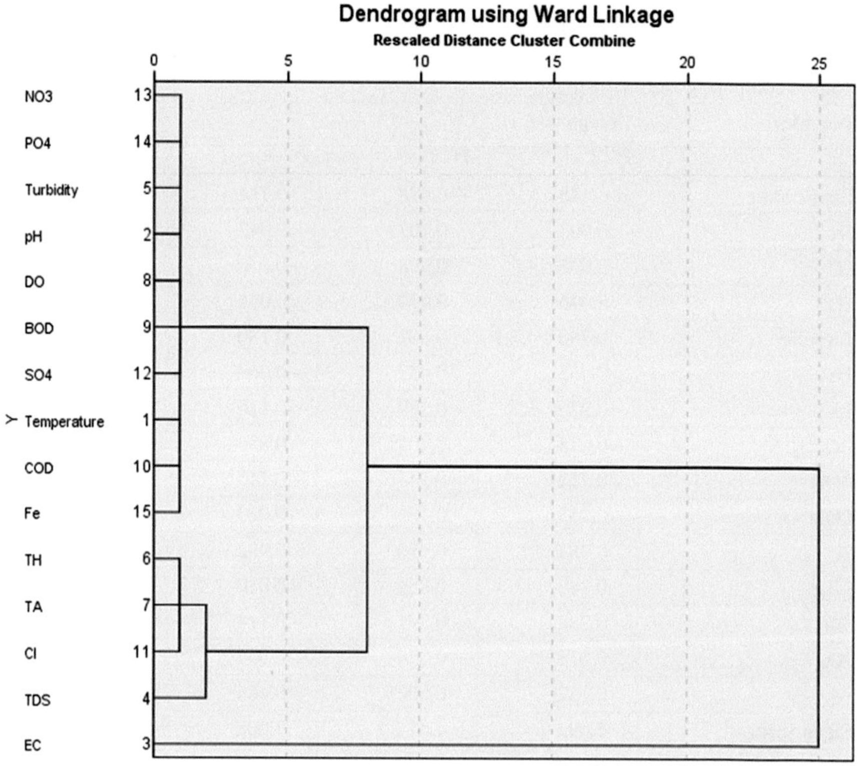

Fig. 6 The dendrogram of the analyzed water quality parameters and the sampling locations

4 Conclusion

To comprehend global water dynamics and quality, it is crucial to investigate urban and semi-urban inland water bodies, which serve as vital freshwater resources for both humans and ecosystems. The majority of the physicochemical parameters in the examined surface water body adhere well to the guidelines set by WHO (2006) and BIS (2012). Elevated concentrations of TA, TH, and Fe can be attributed to sewage contributions from nearby human settlements and deposits of leachate from municipal solid waste dumping zones. The computed Water Quality Index (WQI) categorizes the studied water body as "Good," although it is recommended for use after suitable treatment. The GIS contour map of OPI and EI distinctly classifies it as "Lightly Polluted" to "Begin to Contaminated" and "Eutrophic," respectively. Correlation studies, PCA, and HCA have played a crucial role in delineating the comparative interdependency of water variables and identifying potential sources. The results from correlation analysis, HCA, and PCA offer clear insights into the potential sources of pollution for the water body, encompassing both anthropogenic and geogenic factors. These factors include surface runoff, sewerage disposal, leachate from MSW dumping site, seasonal weathering, and soil erosion. Based on these observations, it is imperative for the government to implement relevant management strategies and restoration plans to enhance the ecological condition of the studied inland water body. According to this investigation, in order to lessen contamination, the leachates produced by the dumpsite should be appropriately handled before being disposed of on land. The public should be made aware of the solid waste management hierarchy by regulatory bodies, as this will help control the level of surface water pollution and mitigate public health risks.

Acknowledgements We sincerely acknowledge the Asansol Engineering College, West Bengal for providing infrastructural facilities throughout the Mission Amrit Sarovar Project. We also greatly acknowledge All India Council for Technical Education (AICTE), Government of India, Ministry of Housing and Urban Affairs and National Education Alliance for Technology (NEAT) for financial support. All members of the Team also are grateful to Mr. Kingshuk Banerjee, Asst. Engineer, and others Authority of Katwa Municipality Corporation.

References

APHA (2005) Standard methods for the examination of water and wastewater, 21st edn. American Public Health Association/American Water Works Association/Water Environment Federation, Washington, DC

Auddy N, Sinha R, Bhattacharya SD, Pobi KK, Dutta S, Nayek S (2023) Application of water quality indices and geostatistical methods for analyzing mountain lakes in relation to anthropogenic influences and catchment features: a case study in East Sikkim, India. Water Pract Technol 18(11):2638–2652

BIS (Bureau of Indian Standards) IS: 10500 (2012) Indian Standard for drinking water specification, 2nd revision. Bureau of Indian Standards, Govt. of India, New Delhi

Gledhill DG, James P (2012) Socio-economic variables as indicators of pond conservation value in an urban landscape. Urban Ecosyst 15:849–861

Gupta S, Nayek S, Chakraborty D (2016) Hydrochemical evaluation of Rangit river, Sikkim, India: using water quality index and multivariate statistics. Environ Earth Sci 75(7):567. https://doi.org/10.1007/s12665-015-5223-8

Howladar MF, Chakma E, Jahan Koley N, Islam S, Numanbakth MA, Ahmed Z, Chowdhury TR, Akter S (2021) The water quality and pollution sources assessment of Surma river, Bangladesh using, hydrochemical, multivariate statistical and water quality index methods. Groundwater Sustain Dev 12:100523

IS 2296 (1982) Tolerance limits for inland surface water subject to pollution. Bureau of Indian Standard, New Delhi

Issaka S, Ashraf MA (2017) Impact of soil erosion and degradation on water quality: a review. Geol Ecol Landscape 1:1–11

Kamble BS, Saxena PR, Kurakalva RM, Shankar K (2020) Evaluation of seasonal and temporal variations of groundwater quality around Jawaharnagar municipal solid waste dumpsite of Hyderabad city, India. SN Appl Sci 2:498. https://doi.org/10.1007/s42452-020-2199-0

Kanmani S, Gandhimathi R (2013) Assessment of heavy metal contamination in soil due to leachate migration from an open dumping site. Appl Water Sci 3:193–205

Kumar P, Meena NK, Mahajan AK (2019) Major ion chemistry, catchment weathering and water quality of Renuka Lake, north-west Himalaya, India. Environ Earth Sci 78(10):1–16

Liu S, Lou S, Kuang C, Huang W, Chen W, Zhang J, Zhong G (2011) Water quality assessment by pollution-index method in the coastal waters of Hebei Province in western Bohai Sea, China. Mar Pollut Bull 62(10):2220–2229

Martínez-Arroyo A, Jáuregui E (2000) On the environmental role of urban lakes in Mexico City. Urban Ecosyst 4:145–166

MoUD (2000) Manual on municipal solid waste management, the expert committee constituted by Ministry of Urban Development. Government of India

Nartey VK, Hayford EK, Ametsi SK (2012) Assessment of the impact of solid waste dumpsites on some surface water systems in the Accra metropolitan area, Ghana. J Water Resour Prot 4(8):605–615

Naveen BP, Mahapatra DM, Sitharam TG, Sivapullaiah PV, Ramachandra TV (2017) Physico-chemical and biological characterization of urban municipal landfill leachate. Environ Pollut 220:1–12

Oertli B, Parris KM (2019) Review: toward management of urban ponds for freshwater biodiversity. Ecosphere 10:02810. https://doi.org/10.1002/ecs2.2810

Panda GS, Pobi KK, Gangopadhyay S, Gope M, Rai AK, Nayek S (2021) Contamination level, source identification and health risk evaluation of potentially toxic elements (PTEs) in groundwater of an industrial city in eastern India. Environ Geochem Health 44:2685–2709

Parween M, Ramanathan AL (2019) Wastewater management to environmental materials management. In: Hussain C (ed) Handbook of environmental materials management. Springer, Berlin. https://doi.org/10.1007/978-3-319-73645-7_72

Persson J (2012) Urban lakes and ponds. In: Bengtsson L, Herschy RW, Fairbridge RW (eds) Encyclopedia of lakes and reservoirs. Springer, Dordrecht

Routroy S, Harichandran R, Mohanty JK, Panda CR (2013) A statistical appraisal to hydrogeochemistry of fluoride contaminated ground water in Nayagarh District, Odisha. J Geol Soc India 81:350–360

Schaider LA, Rodgers KM, Rudel RA (2017) Review of organic wastewater compound concentrations and removal in onsite wastewater treatment systems. Environ Sci Technol 51(13):7304–7317

Sharma MP, Kumar A, Rajvanshi S (2010) Assessment of trophic state of lakes: a case of Mansi Ganga lake in India. Hydro Nepal J Water Energy Environ 6:65–72

Şimşek A, Mutlu E (2023) Assessment of the water quality of Bartın Kışla (Kozcağız) Dam by using geographical information system (GIS) and water quality indices (WQI). Environ Sci Pollut Res 30:58796–58812. https://doi.org/10.1007/s11356-023-26568-3

Singh R, Venkatesh AS, Syed TH, Reddy AGS, Kumar M, Kurakalva RM (2017) Assessment of potentially toxic trace elements contamination in groundwater resources of the coal mining area of the Korba Coalfield, Central India. Environ Earth Sci 76(16):1–17

Son CT, Giang NTH, Thao TP, Nui NH, Lam NT, Cong VH (2020) Assessment of Cau river water quality assessment using a combination of water quality and pollution indices. J Water Supply Res Technol AQUA 69(2):160–172

Sumathi VR, Natesan U, Sarkar C (2008) GIS-based approach for optimized siting of municipal solid waste landfill. Waste Manag 28:2146–2160

WHO (2006) Drinking water guidelines. World Health Organization Standard, Geneva, p 6

Estimation of Morphometric and Morphotectonic Indices of the Kanhan River Basin, Central India, Implication for Susceptibility of Soil Erosion and Groundwater Potential

Suresh Chandra Bhatt, M. M. Singh, Pallvi Rana, Adesh Patel, Narendra Kumar Rana, Sudhir Kumar Singh, and Kaushal Kishor

Abstract The Kanhan River, the Wainganga's longest tributary, runs through the central Indian districts of Chhindwara and Nagpur. The SRTM, remote sensing, and GIS data, were used to investigate morphometric and morphotectonic indices of the Kanhan River Basin and these parameters (linear, areal, and relief) were further employed to assess the vulnerability of the basin to soil erosion, flood hazards, groundwater potential, and tectonic activity. Several morphotectonic parameters including lineament, lineament density, sinuosity, hypsometric integral, and drainage

S. C. Bhatt (✉) · M. M. Singh · P. Rana · A. Patel · K. Kishor
Institute of Earth Sciences, Bundelkhand University, Jhansi, Uttar Pradesh, India
e-mail: scbhatt@bujhansi.ac.in

M. M. Singh
e-mail: mmsgeol@bujhansi.ac.in

P. Rana
e-mail: pallvi4444@gmail.com

A. Patel
e-mail: adeshpatel777@gmail.com

K. Kishor
e-mail: kaushalk@rgipt.ac.in

N. K. Rana
Department of Geography, Institute of Science, Banaras Hindu University, Varanasi, Uttar Pradesh, India
e-mail: nkrana.in@gmail.com

S. K. Singh
K. Banerjee Centre of Atmospheric and Ocean Studies, IIDS, Nehru Science Centre, University of Allahabad, Prayagraj, Uttar Pradesh, India
e-mail: sudhirinjnu@gmail.com

K. Kishor
Rajiv Gandhi Institute of Petroleum Technology, Jais, Amethi, India

S. C. Pal and U. Chatterjee (eds.), *Surface, Sub-Surface Hydrology and Management*, Springer Geography, https://doi.org/10.1007/978-3-031-62376-9_13

basin asymmetry were estimated for the Kanhan River Basin (KRB). The lineament density was validated across five classes ranging from 0 to 0.91 km/km^2. We observed that a high lineament density indicates a high ground water potentiality. The hypsometric integral value (HI = 0.315) suggests that the Kanhan River is in its old or monadnock stage. The Standard Sinuosity Index (SSI) infers that the river follows a sinuous course. The Hydraulic Sinuosity Index (HSI = 61%) suggests that the river has developed flood plains during excessive flooding. The transverse topographic symmetry (TTS) ranges from 0.07 to 0.57, which shows that the basin has an asymmetric nature. Based on these findings, we can interpret that the elongated basin dominated by the old stage is represented by low stream frequency, high permeability, mild slope, and low surface runoff.

Keywords Kanhan River · River morphometry · Morphotectonic · Remote sensing

1 Introduction

The configuration, shape, and size of the Earth's surface along with its landforms were assessed mathematically using morphometry (Horton 1945; Clarke 1996; Patel et al. 2022; Barman et al. 2021a, b). The geomorphological and geological processes play vital roles in the study of the landforms and physical as well as erosional properties of soil (Barman et al. 2021a, b; Kaur et al. 2014; Bhatt et al. 2020, 2021a, c; Bhatt 2020). This resource offers details on drainage characteristics, topographic patterns, landform dimensions, and lithological and structural influences on drainage patterns. It also informs about erosion-dominant areas and the potential for groundwater and surface resources (Pareta and Pareta 2011; Bhatt et al. 2017; Bhatt et al. 2021b, d). The Digital Elevation Model (DEM) is an important database for understanding morphometric, morphotectonic, and hydrological studies (Patel et al. 2022; Prakash et al. 2017; Chauhan et al. 2022; Gaurav Singh and Singh 2022). Conventional methods have been utilized to study the drainage and basin characteristics of numerous rivers across various regions of the world (Krishnamurthy et al. 1996). By analyzing the geomorphic features linked to morphometric parameters, we can estimate the flow intensity and surface runoff of the drainage system (Ozdemir and Bird 2009). Recently the morphometric indices were used by Memon et al. (2020) to explore the flood vulnerability in the Rel River catchment of Gujrat in India. These models were applied by Kumar and Singh (2021) and Pandey et al. (2023) to estimate the quantitative variations of erosion causing the topographic changes in different river basins. Kumar Pradhan et al. (2020) employed morphometric parameters to find out the soil and water conservation in Koshi river basin in India. Bhatt (2020), Bhatt et al. (2021a, b, c, d) and Bhatt et al. (2024) evaluated morphometric and morphotectonic attributes to explore the erosion susceptibility, flood conditions, basin characteristics, and vulnerability to tectonic activity in Pahuj, Jamini, Sind, Rohini and Ken basins of Central India.

To evaluate morphometric parameters, several drainage parameters need to be analyzed (Singh et al. 2023). Additionally, analyzing the drainage system quantitatively is crucial in understanding the watershed's characteristics (Strahler 1957, 1964). When conducting a hydrological investigation (assessing groundwater potential, managing groundwater, basin management, or environmental assessment), it's crucial to consider the characteristics of drainage basins worldwide (Jodhani et al. 2023; Pandey et al. 2023a, b). These characteristics were studied using traditional methods (Horton 1945).

Morphotectonic refers to the form and shape that have developed over time due to tectonic activity, whether recent or in the past (Yadav and Singh 2022). The most prominent evidence of past tectonic activity can be found in drainage lines, which follow physical structures and provide insight into the area's history (Bhatt et al. 2024). Geomorphological mapping is an essential component of terrain assessment and is particularly crucial in slope-stability evaluations (Yadav and Singh 2022). When mapping landslides, it's best to use standard field-based techniques for morphometry and morphology (Bisht et al. 2023). These methods involve analyzing terrain features such as the shape of slopes, drainage, and vegetation to determine areas of instability (Bromhead 1992; Dikau et al. 1997).

For analyzing the morphometry of a drainage basin, and hydrology remote sensing using satellite images is a convenient technique (Mahanta et al. 2023). These images give a view over a large area, making them useful for this purpose. In this study, the DEM of the Kanhan basin was generated in a GIS environment to determine morphometric and morphotectonic parameters. By combining remote sensing data for hydrological along with spatial analysis, it becomes easier to identify as well as differentiate the drainage area (Pirasteh et al. 2010). The study examines how structural activities and lithology impact the formation of landforms, particularly its relation to the stream network characteristics of the Kanhan basin. Understanding the drainage system is crucial in comprehending factors such as the initial gradient, differences in rock durability, structural influence, vulnerability to soil erosion, and geological as well as geomorphological background of the drainage basin or watershed. The main purpose of our research is to quantify the significance of linear, areal, and relief parameters in the evolution of drainage patterns and landscapes and to explore the susceptibility of soil erosion, flood conditions, and influence of tectonics in the KRB. The output of this research may be useful in the study of other river basins of India and the world.

2 Study Area

The Kanhan River is a right-bank tributary of the Wainganga River. Its basin lies between 21° 30′–25° 78′ N latitude and 78° 25′–79° 30′ E longitudes in the south of the Satpura range in Central India (Fig. 1). It has a 275 km run, passing through the states of Maharashtra and Madhya Pradesh. Pench the largest tributary of the Kanhan River is flowing in the border areas of Chhindwara and Seoni districts and is supplying water to Nagpur city. Jam and Kulbehra, other tributaries of the Kanhan River flow through the Sausar region and Chhindwara and Mohkhed areas, respectively. The study area experienced a hot and dry climate during all seasons except the southwest monsoon period. The average annual precipitation varies from 850.20 to 1064.10 mm.

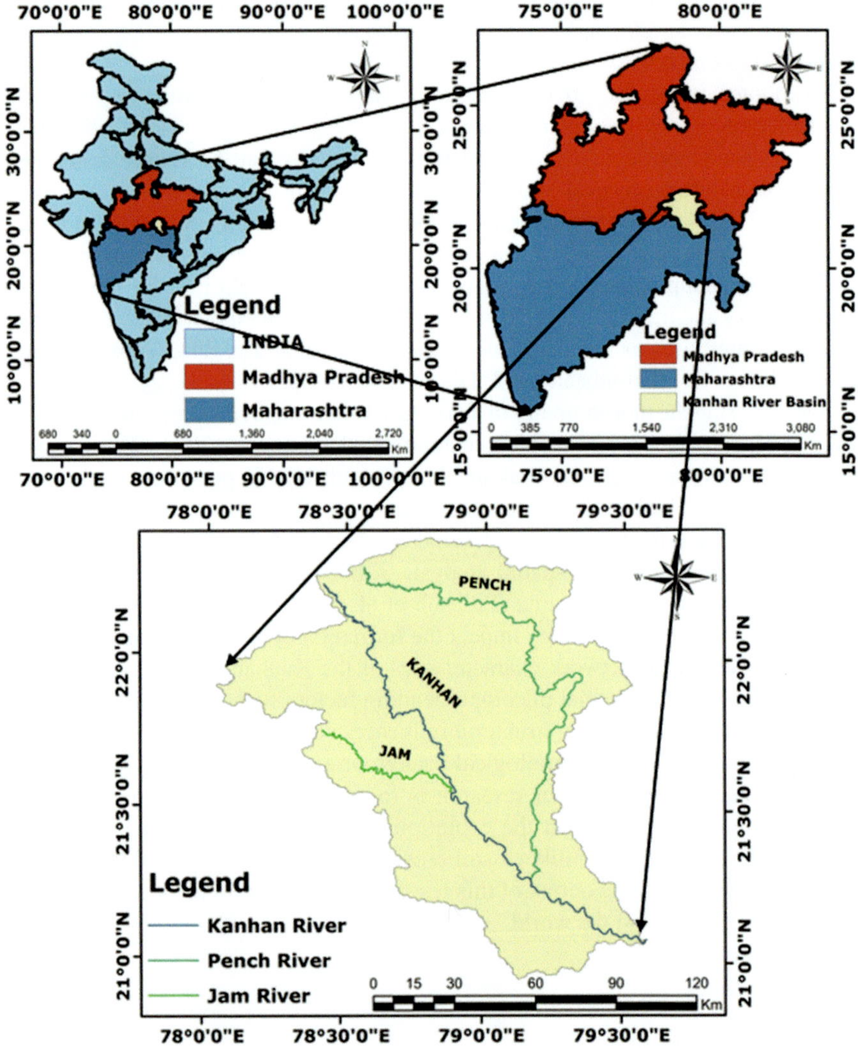

Fig. 1 Location map of the Kanhan basin

2.1 Geological Setup

The Kanhan River Valley, which runs from north to south, features a distinctive cross-section of Precambrian terrain in Central India. It passes through the East-West trending Proterozoic Betul Belt, located in the north direction and it has the Sausar Belt which is located in the south, where gneisses are visible between the two belts. The major Gavilgarh-Tan Shear Zone marks the southern boundary of the Betul belt (Golani et al. 2001). The Gondwana Basin is in Central India and covers an area of approx. 14,700 km^2. It has a faulty relationship with the Betul-Chhindwara crystallines to the south and the Son-Narmada Lineament (SNL) to the north (Bhatt and Hussain 2008; Bhatt and Gupta 2009; Bhatt and Gupta 2014). The Bundelkhand craton separated from the southern Indian block by Central Tectonic Zone (CITZ) is traversed by several shear zones (Bhatt and Mahmood 2012; Bhatt and Hussain 2012; Bhatt 2014; Bhatt and Singh 2021). The sub-latitudinal Satpura having ENE-WSW trend along with the Godavari NW-SE drain are the two tectonic trends that meet to generate this fracture basin (Peters and Singh 2001). The northern sub-basin is the Denwaand sub-basin Pench-Kanhan in the south is separated by the Malni-Tamia high. The Betul-Chhindwara crystalline complex, a significant Central Indian Tectonic Zone (CITZ) litho-tectonic unit, serves as the foundation for Gondwana deposits. Several ductile shear zones trending ENE-WSW pass through these complexes (Roy and Prasad 2001).

Geologically, the KRB is divided into eight groups of rocks (Fig. 2). The youngest basaltic lava flow associated with inter-trappean sediments belongs to the Deccan Traps of the Upper Cretaceous age to the Eocene age. Two types of basaltic lava flow; massive unit (confined to the lower part of the flow) and vesicular unit (confined to the upper part of the flow) widely occurred in the study area (Fig. 2). These rocks are exposed in the northeastern and the southwestern fringes of the area. The Upper Gondwana Group (Mesozoic) consists of calcareous sandstone and limestone, coarse-grained sandstone and conglomerates, sandstone-mudstone alternations, and the Lower Gondwana Group (Permian to Carboniferous) comprises sandstones and carbonaceous shale, and red clay, carbonaceous shale, and coalbeds, polymictic boulder conglomerate of Talcher Formation, and olive-greenshale are occupying the area in upper reaches of the basin (Fig. 2).

The Lower Gondwana series, which was deposited over the Paleoproterozoic Betul-Chhindwara crystalline complex, is found in the study region in some areas of the Chhindwara district of Madhya Pradesh. The Talcher, Bijori, Barakar, and Motur Formations, which are primarily exposed in the Pench-Kanhan Sub-basin, are part of this sequence, which is capped by Deccan Traps. Contrarily, the Pachmarhi, Denwa, along with Bagra Formations of the Upper Gondwana sequence are found in the Denwa Sub-basin (Crookshank 1936; Chakraborty et al. 2003; Pascoe 1973; Ghosh et al. 2004; Dutta 2002; Chakraborty and Ghosh 2005).

Numerous significant geological units, such as the Pelitic to Psammitic gneisses, calc-silicate gneiss, and amphibolites, make up the Sausar Group. This group, which dates from the Paleo to the Meso-Proterozoic, is found in the southeast region of

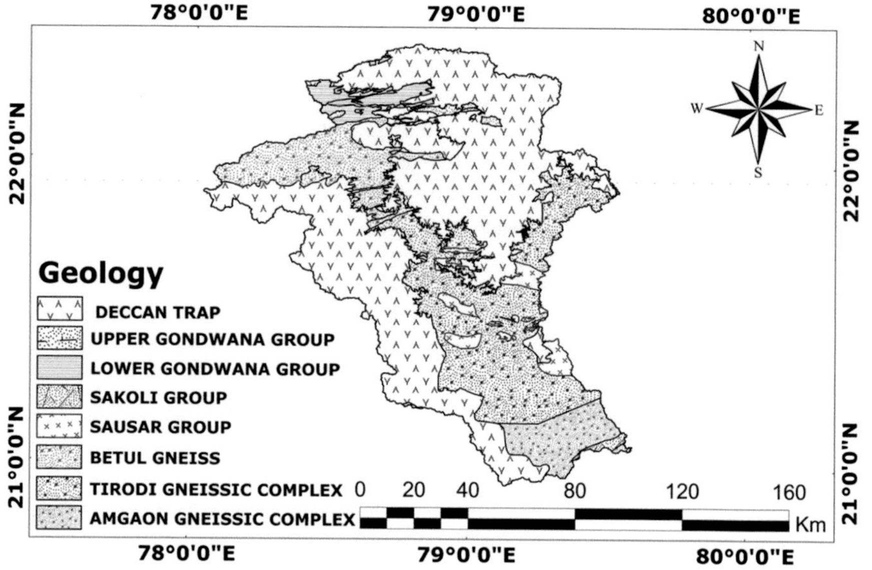

Fig. 2 Geological map of the Kanhan basin

the research area, close to the southern edge of the CITZ. From south to north, the Sausar Belt is made up of basically four main geologic components: the southern granulite belt, with Sausar Group's supracrustal metasedimentary rocks, and the Sakoli Group, which is composed of Archean-aged, medium-to-coarse-grained, well-foliated, greyish to greenish-gray mica schist. The eastern and western edges of the shear zones are produced by the heavily deformed Sakoli and Sausar belts, respectively. The Sakoli Group, which is Meso-Proterozoic in age and consists of felsic volcanic with accompanying tuff, meta-basalt with related tuff, meta-rhyolite, and mica schist, is in the southern portion. The Betul belt, composed of quartzite, metapelite, and bimodal volcanic rocks (also known as basalt-rhyolite), with calc-silicate rocks, as well as banded iron formation, is situated in the study areas far west. The Amgaon Gneissic Complex is located mainly in the southeastern region of the Kanhan basin and is comprised of pinkish-green to grey, medium to coarse-grained, hard, and compact granitic gneisses. The central as well as eastern margins of the basin contain granite, aplite, and pegmatite veins, as well as biotite gneiss, biotite-muscovite schist, hornblende schist and gneiss, and hornblende schist and gneiss. Between the Bundelkhand craton in the Northern Indian region and the Singhbhum, Baster, and Dharwar cratons in Southern India, the Proterozoic Sakoli and Sausar mobile belts make up the bulk of the CITZ. Between the Sausar Belt, located in the south (1.5–1.0 Ga), and the Mahakoshal Belt, located in the north (2.4–1.8 Ga), the Meso-Proterozoic Betul inlier creates a small litho-tectonic unit in the CITZ. The southeastern portion is home to the Archaean to Paleoproterozoic Amgaon Gneissic Complex, which is made up of quartzite, ultramafic, migmatite gneisses, calc-silicates, granitic gneisses, and amphibolite. Two generations of granite are found mainly in the Tirodi Gneissic complex along with the Sausar Group.

2.2 Geomorphology

Using a digital elevation model in ArcGIS, a geomorphological map of the basin was created (see Fig. 3). The study area is made up of pediplain complexes that are clearly defined by lateral erosion (see Fig. 3).

The main geomorphic features that can be seen in the Kanhan basin are Satpura Plateau and Nagpur Plateau. The northern region of the Kanhan basin comprises hill ranges, belonging to the Satpura Mountains stretching nearly East-West. The occurrence of active comprises hill ranges, belonging to the Satpura Mountains stretching nearly east along the Major-river courses, such as Kanhan River, and its two major tributaries Jam and Pench. The structurally dissected valleys, plateaus, and hills covered by Gondwanas, can be seen easily in the middle reaches as well as the northern part of the basin (Fig. 3).The denudational hills of sandstone belonging to buried plains commonly occur in the central Kanhan basin and the surroundings of Sausar. The dissected Deccan plateau is widely spreading in the eastern and north-eastern parts of the basin (Fig. 3). The ponds, mine dumps active quarries, abandoned quarries, active and older flood plains, older and younger alluvial plains, dams and reservoirs, and water bodies are peculiar geomorphic features present in the Kanhan basin.

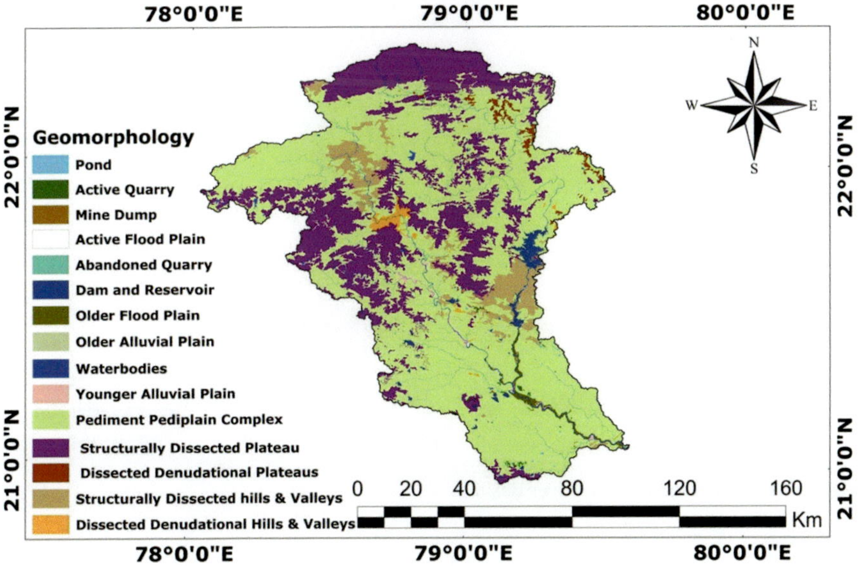

Fig. 3 Geomorphological map of the Kanhan River Basin

2.3 Soil in the Kanhan Basin

Three types of soils (lithosols, vertic cambisols, and chromic vertisols) are identified in KRB (Fig. 4). The thin soil consisting of imperfectly weathered rock fragments is called Lithosol. Lithosols (Azonal soils) lacking well-developed horizons due to immaturity are occupying the larger area in the central Kanhan basin. These soils are widely occurring in the surrounding Sausar, Pandhurna, and Pench National Park areas. Lithosols are present in Bichhua and Pandhurana.

The vertisols containing high clay content and commonly showing shrinking and swelling with changes in moisture content are covering larger areas in the southern as well as northeastern and northwestern regions of the basin. The expansion and shrinking of vertisols developed deep and wide cracks during dry and wet periods may cause serious problems. In the Chhindwara district, the chromic vertisols can be found in Mohkhed, Chourai, Amarwara, Parasia, and Umreth tehsils, as well as in Saoner, Parshivni, Ramtek, Kalameshwar, and Mauda tehsils of the Nagpur district. Cambisols, which lack deposited clay, humus, soluble salts, iron, and aluminum oxides, cover the northeastern and northwestern margins of the basin. These soils' aggregate structure, color, clay content, carbonate content, and other characteristics set them apart from their unweathered parent material. Vertic Cambisols are found in the Tamia and Damua regions.

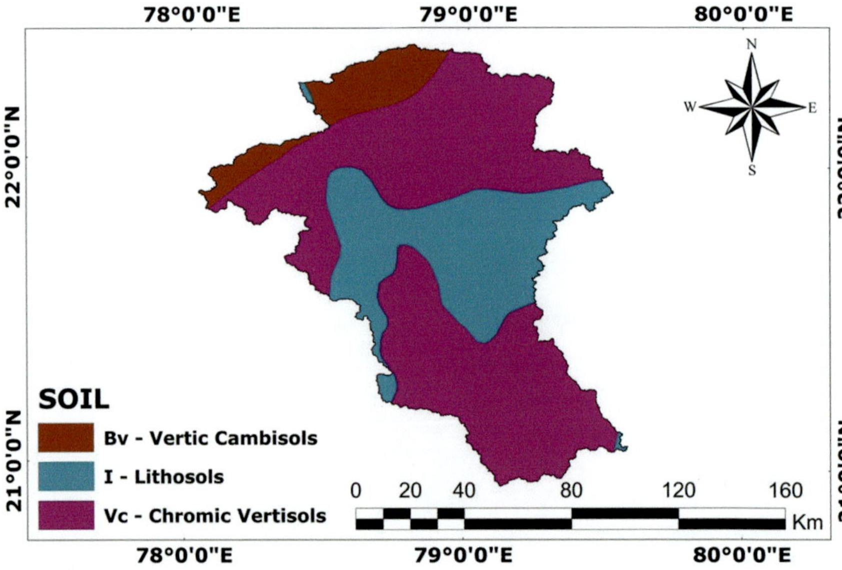

Fig. 4 Soil map of the Kanhan River Basin

2.4 *Landuse and Landcover (LULC)*

LULC is highly important for managing and planning landscapes in both urban and rural areas (Singh et al. 2015; Kumar et al. 2023). The Kanhan basin's LULC map was created using GIS (Fig. 5).

In the Kanhan basin, most of the land is occupied by cropland, covering about 34% of the area, which spans over 266.18 km^2. Trees make up about 12% of the land, with a total area of 91.85 km^2, while scrub/shrubs cover 8% of the land, occupying 57.83 km^2 (Table 1). This suggests that the soil in the basin is highly fertile. Water bodies make up about 33% of the area, spanning over 257.90 km^2. Settlements are found in both rural and urban areas, covering about 12% of the total area, including cities such as Nagpur, Mauda, Saoner, and Chhindwara.

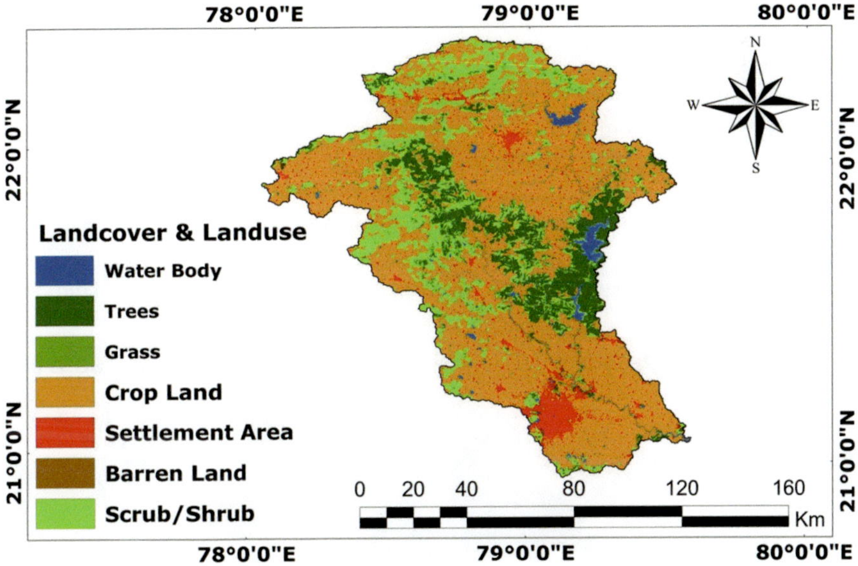

Fig. 5 Landuse and landcover map of the Kanhan River Basin

Table 1 Landuse and landcover data of the Kanhan basin

S. No.	Name	Area (in km^2)	Percentage (in %)
1	Water body	257.90	33
2	Trees	91.85	12
3	Grass	2.23	< 1
4	Cropland	266.18	34
5	Settlement area	96.70	12
6	Barren land	10.61	1
7	Scrub/Shrub	57.83	8

3 Methods

Morphometry and morphotectonics of the Kanhan basin with the help of GIS and SRTM DEM data were carried out (Fig. 6). The DEM data of the study are downloaded from open topography. The DEM that was processed has been effectively utilized in creating the network of streams and many other complementary layers (Mesa 2006; Magesh et al. 2011). To create longitudinal profiles downstream of the river, the STRM DEM was utilized. In addition, Strahler (1964) identified stream orders, and Horton's law (Horton 1945) was used to analyze the drainage network the linear, areal, and relief characteristics of the Kanhan basin were calculated using conventional techniques (Horton 1932, 1945; Strahler 1952, 1956; Melton 1957; Miller 1953; Sreedevi et al. 2005; Kale and Gupta 2001; Reddy et al. 2004). In this study, the Horton (1945) methodology was used to analyze the network of watersheds with the drainage patterns, and the Strahler (1964) approach for stream ordering was used.

We used ArcGIS software to measure the length of the stream along with the KRB basin area and generated the stream order using Strahler's (1952) system and the ArcHydro tool. From ArcGIS, we extracted basic standard parameters like the individual stream length, and basin area along with its perimeter, which were used to calculate morphometric parameters (linear, areal, relief, as well as tectonic facets) with the support of various equations. To calculate morphotectonic parameters such as hypsometric integral (Strahler 1952), drainage basin asymmetry, and sinuosity analysis, we used standard methods as given by Mueller (1968). In this study, the

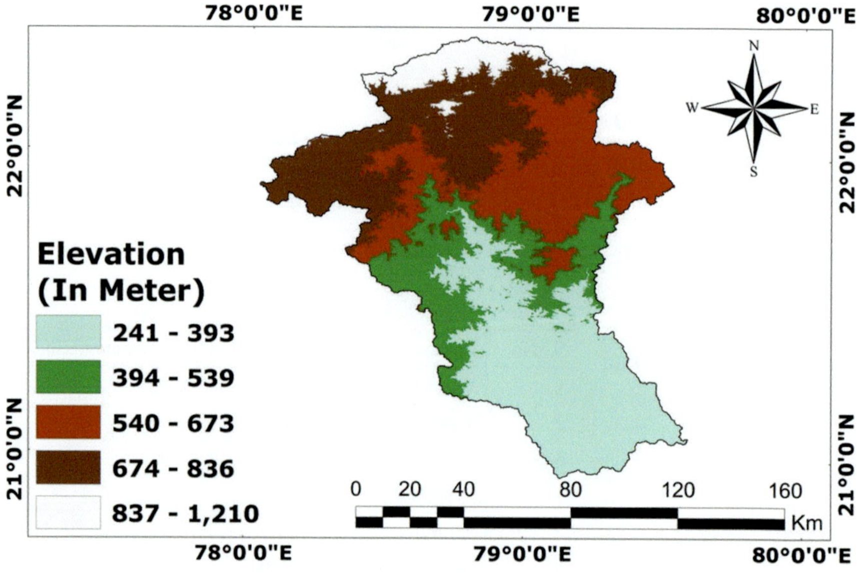

Fig. 6 Digital elevation model (DEM) of the Kanhan basin

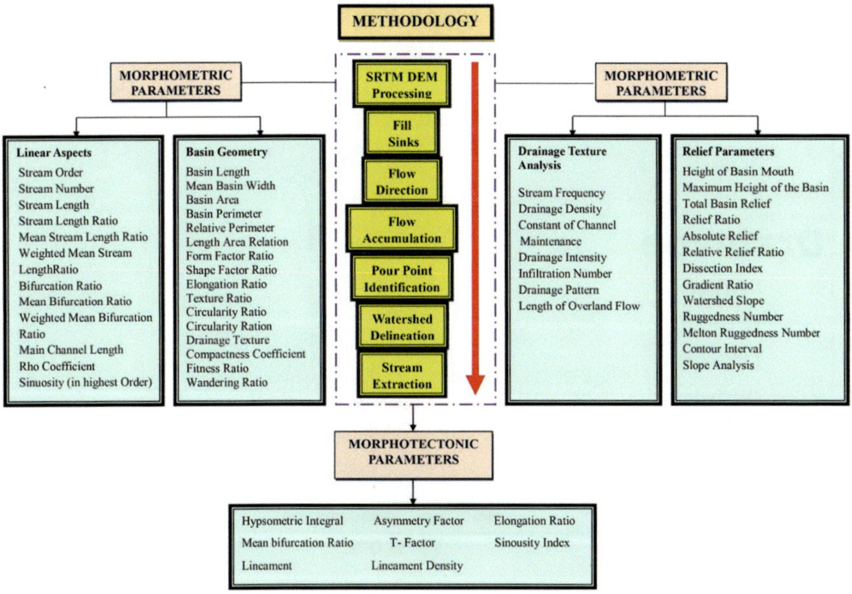

Fig. 7 Flowchart showing the flow of study in the Kanhan River Basin

properties of DEM in ArcGIS software were used. Each step of the methods used in this study is shown in Fig. 7.

4 Results and Discussion

4.1 Morphometric Parameters

4.1.1 Drainage Pattern

The way natural streams are arranged is called the drainage pattern (Barman et al. 2021a, b; Rathore et al. 2023). This pattern shows how slope, lithology, and structure affect the streams and helps us understand how erosion works. In the basin, most streams have a dendritic to sub-dendritic pattern, as shown in Fig. 8. The basin also has seven different orders of rivers. The dominant drainage orientation in the basin is a polymodal distribution.

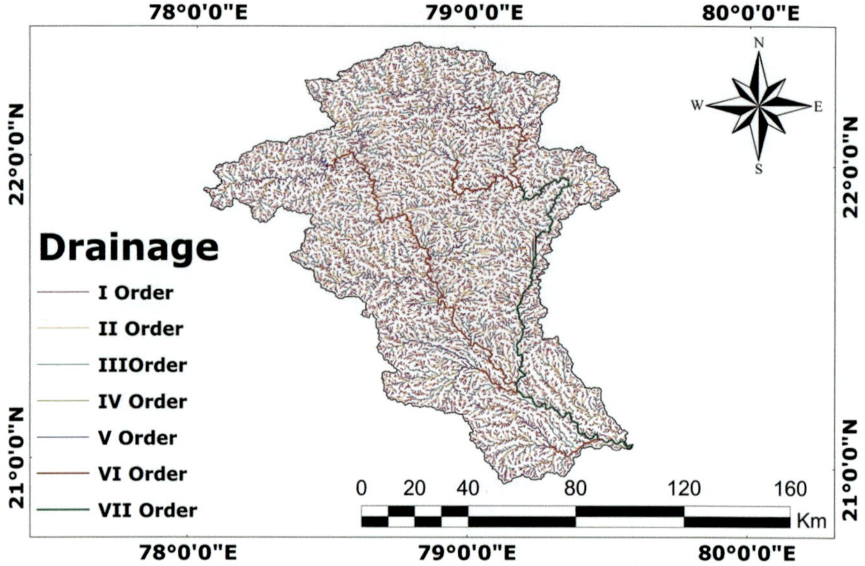

Fig. 8 Drainage map of the Kanhan basin

4.1.2 Linear Parameters

For this analysis, the Strahler method from 1957 was used to determine stream order. The Kanhan basin has a total stream length of 14,346.8 km (Table 2). First-order streams make up approximately 50% of the total length, while second to seventh orders contribute 24.7%, 13%, 6%, 2.6%, 2.4%, and 1.3%, respectively. The study identified a total of 7192 and 3544 first-order and second-order streams respectively. While third to sixth-order streams have 1799, 902, 379, and 194 streams, respectively (Fig. 8).

Increasing stream order results in decreases in stream length, with maximum length observed in first-order streams (Horton 1945). Lower-order streams generally have shorter lengths but a higher number of streams, whereas higher-order streams have longer lengths but fewer streams. R_b is the proportion of streams of a certain order to the number of stream segments of the next higher order (Bhatt et al. 2021d). The R_b values greater than 5.0 indicate enough structural control along with low infiltration as well as moderate to steep slope of the basin whereas a lower R_b value (< 3.0) indicates plain surface, low runoff as well as high permeability of bedrocks (Yadav et al. 2014). The values of the R_b estimated for II, III, IV, V, VI, and VII order are 4.72, 4.62, 4.43, 4.76, 3.4, and 5, respectively (Table 2). R_b for II, III, IV, V, and VII order, suggests that the Kanhan drainage basin is structurally controlled, and exhibits low infiltration and moderate steep slope. The value of R_b for the VI order is suggestive of a plain surface, low runoff, and high permeability. The value of R_{bm} = 4.4 specifically describes that the drainage of the Kanhan basin was coordinated

Table 2 Morphometric parameters calculated for the Kanhan basin

S. No.	Morphometric parameter	Formula	Result
A	*Linear aspects*		
1	Stream order (S_u)	Hierarchical rank	1–7
2	1st order stream (S_{uf})	$S_{uf} = N_1$	7846
3	Stream number (N_u)	$N_u = N_1 + N_2 + \cdots N_n$	9969
4	Stream length (L_u) km	$L_u = L_1 + L_2 \ldots L_n$	14,346.8
5	Bifurcation ratio (R_b)	$R_b = N_{un}/N_{un} + 1$	3.4–5
6	Mean bifurcation ratio (R_{bm})	R_{bm}	4.48
7	Weighted mean bifurcation ratio (R_{bwm})	R_{bwm}	4.69
8	Main channel length (C_l) km	GIS software	273.04
9	Sinuosity index	$S = \dfrac{\text{Length of stream channel of highest order}}{\text{Length of straight line distance of same stream}}$	1.88
B	*Basin geometry*		
1	Main channel length	GIS software analysis	273.04
2	Basin length (L_b) km	GIS software analysis	180.64
3	Mean basin width (W_b)	$W_b = A/L_b$	71.11
4	Basin area (A) km^2	GIS software analysis	12,846.45 km^2
5	Basin perimeter (P) km	GIS software analysis	903.11
6	Form factor ratio (R_f)	$F_f = A/L_b^2$	0.39
7	Shape factor ratio (R_s)	$S_f = L_b^2/A$	2.54
8	Elongation ratio (R_e)	$R_e = \sqrt{4A/\pi L_b^2}$ A = basin area (km^2), $\pi = 3.14$, L_b = maximum basin length (km)	0.07
9	Texture ratio (R_t)	$R_t = N_1/P$	8.68
10	Circularity ratio (R_c)	$R_c = 12.57 * \left(A/P^2\right)$	0.19

(continued)

Table 2 (continued)

S. No.	Morphometric parameter	Formula	Result
11	Circularity ration (R_{cn})	$R_{cn} = A/P$	14.22
12	Drainage texture (D_t)	$D_t = N_u/P$ N_u = total number of streams, P = basin perimeter (km)	11.04
13	Compactness coefficient (C_c)	$C_c = 0.2841 * P/A^{0.5}$	2.26
14	Fitness ratio (R_f)	$R_f = C_l/P$	0.30
15	Wandering ratio (R_w)	$R_w = C_l/L_b$	1.51
C	*Drainage texture analysis*		
1	Stream frequency (F_s)	$F_s = N_u/A$	0.77
2	Drainage density (D_d) km/km^2	$D_d = L_u/A$	1.11
3	Constant of channel maintenance (km^2/km)	$C = 1/D_d$	0.90
4	Drainage intensity (D_i)	$D_i = F_s/D_d$	0.69
5	Length of overland flow (L_g) km	$L_g = A/2 * L_u$	0.44
D	*Relief parameters*		
1	Height of basin mouth (z) m	GIS analysis/DEM	241.40 m
2	Maximum height of the basin (Z) m	GIS analysis/DEM	1212 m
3	Total basin relief (H) m	$H = Z - z$	971
4	Relief ratio (R_{hl})	$R_{hl} = H/L_b$	0.005
5	Relative relief ratio (R_{hp})	$R_{hp} = H * 100/P$	0.10
6	Gradient ratio (R_g)	$R_g = (Z - z)/L_b$	0.005
7	Watershed slope (S_w)	$S_w = H/L_b$	0.005
8	Ruggedness number (R_n)	$R_n = D_d * (H/1000)$	0.001
9	Melton ruggedness number (MR_n)	$MR_n = H/A^{0.5}$	0.008
10	Hypsometric integral (HI)	$HI = (H_{mean} - H_{min})/(H_{max} - H_{min})$ H_{mean} = mean elevation, H_{min} = minimum elevation (z), H_{max} = maximum elevation (Z)	0.315

by homogenous geological and lithology structures. The R_{bwm} value (4.69) of the Kanhan basin infers a runoff-induced soil erosion-sensitive river basin.

The sinuosity is a numerical measure that compares the length of a channel to the distance it travels downstream. Typically, it ranges from 1 to 4 or higher. This index is important for understanding the role of streams in shaping landscapes. In 1968, Mueller proposed important calculations for measuring the sinuosity index. He also distinguished between two types of sinuosity: topographic sinuosity, which relates to the flow of natural stream courses, and hydraulic sinuosity, which concerns the development of floodplains. According to Leopold et al. (2020), if the sinuosity index (S) is 1.5, the stream is sinuous, and if S is greater than 1.5, the stream is meandering. If S falls between 1.25 and 1.5, the stream is considered twisty, while a value of less than 1.05 indicates a straight stream. In the case of the Kanhan River watershed, the sinuosity value was calculated to be 1.88 (Table 2), indicating that the river has a meandering nature.

4.1.3 Areal Parameters

The basin perimeter is used in order to determine the shape and size of a watershed. The measurement of watershed size and shape is determined by the distance between the watersheds. The Kanhan River Basin (KRB) has a perimeter of 903.11 km and a total area of 12,846.45 km^2 (Table 2). Its form factor is calculated by dividing the basin area by the square of the basin length (Horton 1932). It was observed that a perfect circular watershed will always have the F_f be less than 0.754. A watershed with a greater F_f will have high peak flows of shorter duration, while an elongated watershed with a low form factor will have flows for a longer duration and a range from 0.33, indicating an elongated shape. The Kanhan River Basin has a form factor (F_f) value of 0.39 (Table 2), revealing an elongated shaped basin. The circularity ratio is the proportional computation of the watershed area to the area of a circle which has as same perimeter as the watershed (Strahler 1964). It ranges from 0 to 1, indicating the presence of an elongated basin with a highly permeable, homogeneous geological terrain. If the value of the R_c is close to 1, the selected basin is more circular in shape. In (Table 2), the value of the R_c (0.19) shows that the Kanhan basin has an elongated shape.

The drainage texture refers to the spacing between drainage lines, which is influenced by the underlying lithology as well as the infiltration capacity and relief (Kumar et al. 2018). It is calculated by dividing the total stream segments which are present in all orders by the calculated perimeter of the drainage basin (Horton 1945). Smith (1950) was a pioneer in order to classifying drainage texture into four classes; coarse (< 4), medium (4–10), fine (10–15), and very fine (> 15). The fine texture (11.03) of the Kanhan basin (Table 2) refers to high erodible ground. Based on the research of Horton (1932) and Strahler (1964), D_d measures the total computed stream length in a particular area. The density is categorized as extremely low (from 0 to 1), low (from 1 to 2), moderate (from 2 to 4), moderately high (from 4 to 6), and high (> 6). In the case of the Kanhan basin, the low density (1.11) suggests that the basin was formed in permeable geological conditions with a thick layer of vegetation, resulting in less erosion and low relief. This indicates that the higher density is mainly found in the middle as well as submarginal reaches of the Kanhan basin, represented by high-order streams (Fig. 9). The drainage frequency (or channel/stream frequency) is the calculated number of stream segments, found per unit area (Horton 1932). Following Horton (1945), F_s can be categorized into five groups based on their size. These categories are low (0–5 km^2), moderate (5–19 km^2), moderately high (10–15 km^2), high (15–20 km^2), and very high (20–25 km^2). The Kanhan River Basin has low F_s of 0.7 km^2, which suggests the presence of more vegetation, high permeability, and relatively gentle slopes. Faniran's (1968) expression of D_i measures the ratio of F_s to D_d. The Kanhan River Basin has a D_i value of 0.69, indicating that D_d and F_s have a minimal surface impact, i.e., lowered by agents of denudation. Despite low values for F_s, D_d along with D_i, the basin is highly prone to flooding as well as gully erosion due to slow surface runoff removal. The hill shade map shows the hill areas of the study area (Fig. 10).

The compactness coefficient (C_c) is influenced by the steepness of the slope rather than the size of the watershed. A high C_c value ($C_c > 1$) indicates that the drainage basin is not circular while a lower value suggests that the basin is elongated. If a basin is circular, it has a quick concentration of discharge, but if it's elongated, it has a slow concentration of discharge. Based on Table 2, the Kanhan basin has a C_c value of 2.26, which means that the drainage basin is noncircular shaped.

The texture ratio, on the other hand, represents the proportions of the first-order streams (i.e., N_1) to the perimeter (P) of the selected basin. Its estimated value (R_t = 8.68) in this study area shows that it has a high runoff condition. Schumm (1956)

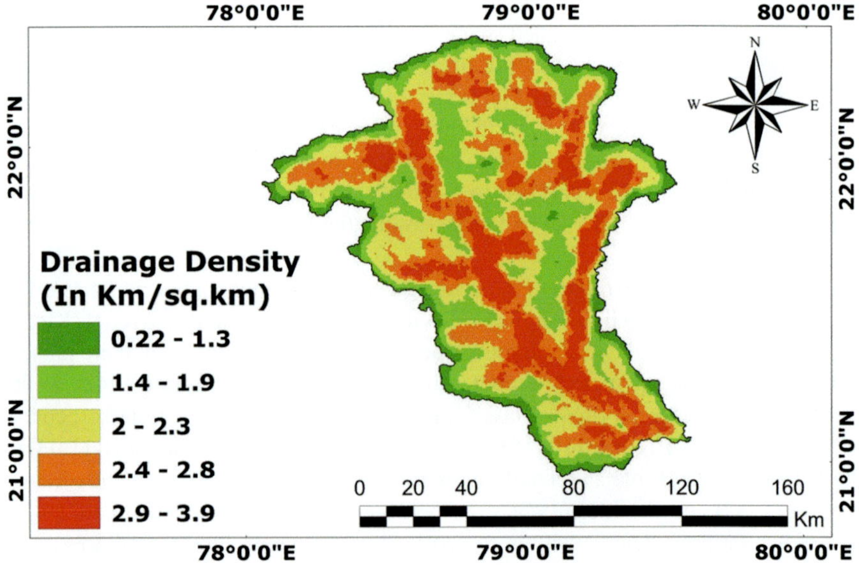

Fig. 9 Drainage density map of the Kanhan basin

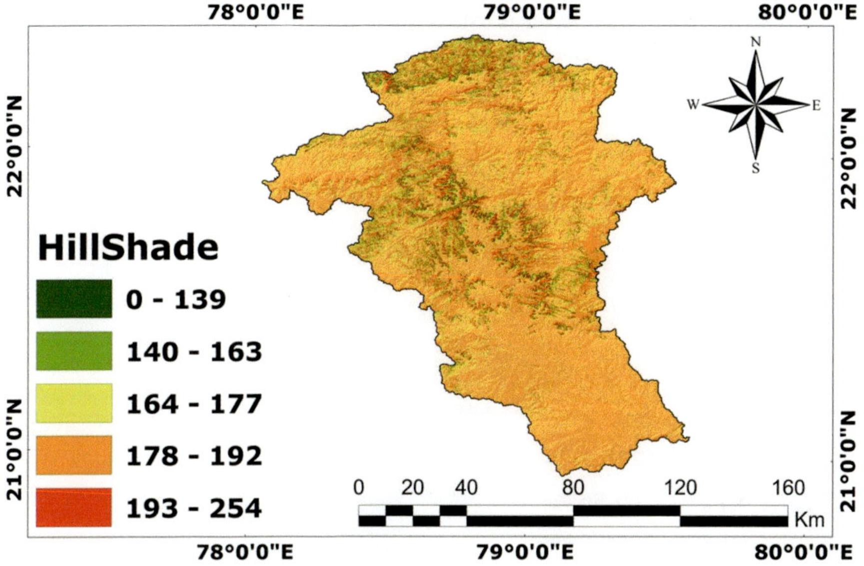

Fig. 10 Hillshade map of the Kanhan basin

introduced a concept called the constant of channel maintenance (C) to describe landforms. This standard parameter represents the basin surface area needed to maintain a channel that is one kilometer long. Kanhan basin has a C value of 0.90, indicating that it experiences minimal structural disturbances and has low levels of runoff (Table 2).

4.1.4 Relief Parameters

Schumm (1956) defines, the ratio between a drainage basin's total relief and its longest dimension parallel to the main drainage line as called relief ratio (R_{hl}). Schumm discovered that the amount of sediment lost per unit area is closely linked to the R_{hl}. A low R_{hl} indicates a gentle slope, while a high R_{hl} indicates a steep slope. The R_{hl} is also linked to a soil's vulnerability to erosion, and thus, a high R_{hl} reveals the susceptibility of the basin towards soil erosion. In the Kanhan basin, the low R_{hl} (0.005) suggests less susceptibility to soil erosion (Table 2). In the Kanhan River Basin, the Relative Relief value is 0.10 (Table 2). In moderately high terrain, the R_{hl} is representative of the presence of a steep slope. It depicts the degree of erosional processes present in the drainage basin as well as the overall steepness of the river basin (Schumm 1956). The stream gradient affects flood patterns along with total sediment volume that could be transported (Hadley and Schumn 1961).

A slope's steepness is expressed as a ratio between its horizontal equivalent and its vertical intervals that have been reduced to unity. The result obtained is utilized to calculate the tangent of the basin's slope angle (Fig. 11). The R_g is a channel slope characteristic that makes it easier to calculate the runoff volume (Sreedevi et al.

2005). The R_g for Kanhan River is estimated at 0.005 (Table 2). The contour map for the study area was generated in ArcGIS (Fig. 12).

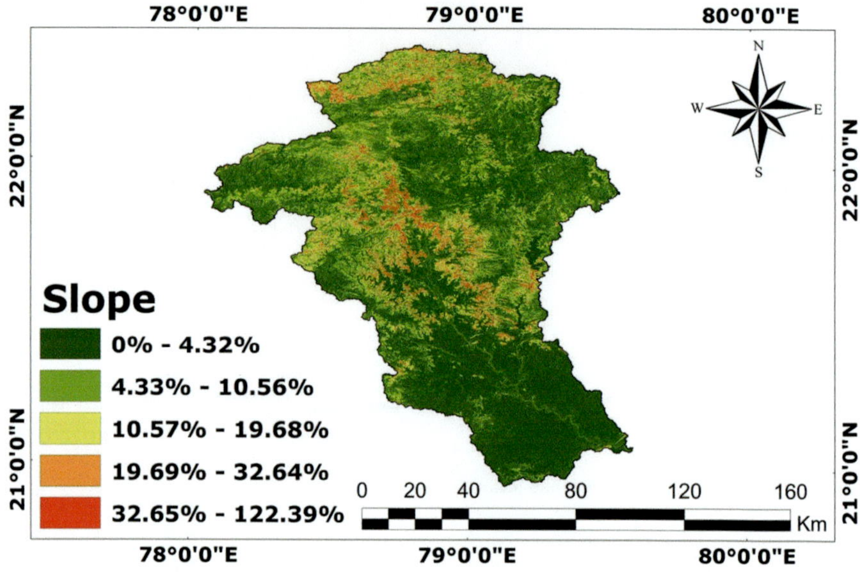

Fig. 11 Slope map of the Kanhan River Basin

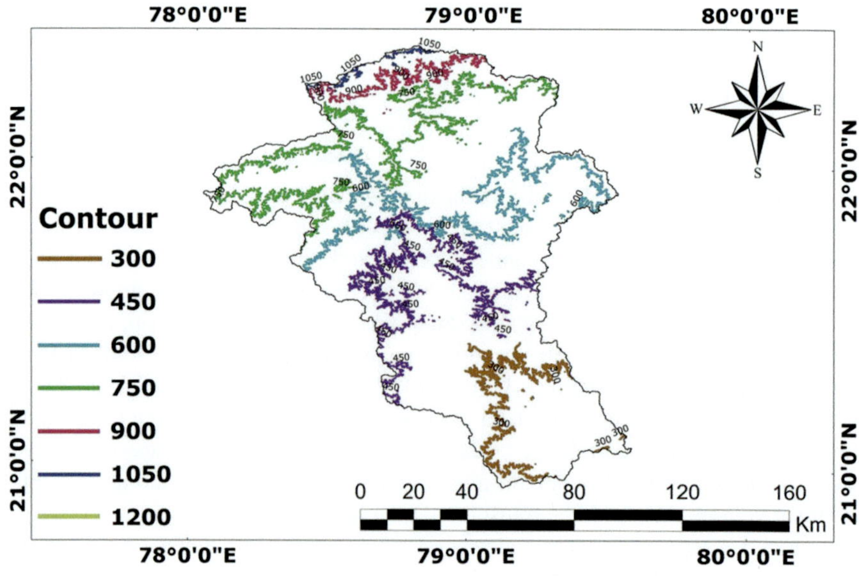

Fig.12 Contour map of the Kanhan River Basin

The Kanhan basin's contours range from 300 m (lowest) and 1050 m (highest), respectively. ArcGIS was used to create the drainage basin's slope map (Fig. 11). An essential factor that enables the evaluation of runoff as well as soil erosion from a specific drainage basin is the basin slope. The R_n is the dimensionless parameter. The R_n describes the degree to which the terrain is smooth and rough and susceptible to soil erosion (Selvan et al. 2011). There are four categories of (R_n), subdued morphology (< 0.1), slight morphology (from 0.1 to 0.4), moderate morphology (from 0.4 to 0.7), and sharp morphology (from 0.7 to 1.0). The R_n ranges from zero to one. The basin terrain is smoother if the R_n value is 0. The terrain is more rugged, however, if the toughness rating is more than zero or equal to zero. The value of the roughness number increases if the D_d, as well as relief values, are very high. It suggests that the ground is mild in a long trough, not steep (Strahlar 1956). The R_n value of the Kanhan River Basin (0.001) shows subdued morphology and moderately rugged terrain and is less prone to erosion (Table 2). A slope index called the Melton Ruggedness Number (MR_n) offers a specific illustration of relief ruggedness (Melton 1965). The MR_n (0.008) estimated for the Kanhan basin has less susceptibility of the basin toward erosion. The S_w offers information about the watershed topography. It is considered as an independent variable. The values of the rainfall-induced runoff along with its period of concentration are significantly impacted by the average slope of a watershed (Figs. 11 and 13). The S_w provides information about the topography. For the Kanhan basin, the S_w value is 0.005 (Table 2).

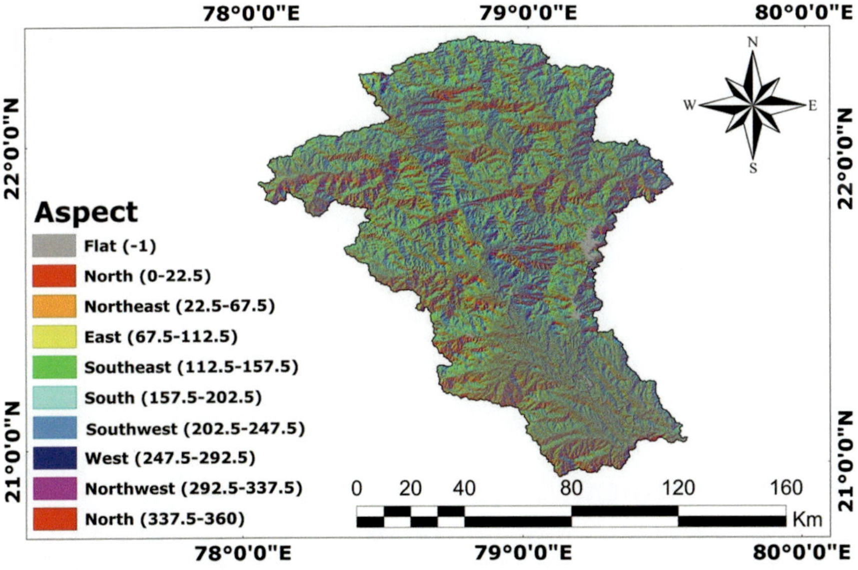

Fig. 13 Aspect map of the Kanhan River Basin

4.2 Morphotectonic Parameters

4.2.1 Lineament and Lineament Density

Linear structures on the Earth's surface known as lineaments indicate a generic expression of subsurface fractures (Pradhan et al. 2006; Pradhan and Youssef 2010; Singh et al. 2010; Murmu et al. 2019). Nur (1982) has defined lineaments as structurally controlled nonlinear features, also called curvilinear features, usually identified with the help of satellite imagery by their relatively linear arrangements. ArcGIS was used to create the lineament map of the Kanhan basin (Fig. 14). The lineaments are trending in the Kanhan basin in the following directions: NE-SW, ENE-WSW, NW-SE, and WNW-ESE (Fig. 14).

The lineament density is calculated by dividing the total recorded lineament length by the basin area under consideration. The Kanhan basin's lineament density map was created and shows five classes: 0–0.079, 0.08–0.21, 0.22–0.35, 0.36–0.53, and 0.54–0.91 km/km^2 (Fig. 15). A lineament density can be used to inadvertently explore a region's potential for groundwater. A permeable zone may indicate the presence of lineaments, and locations with a high lineament density are thought to have strong groundwater potential zones. The northeastern, northwestern, and lower reaches of KRB shows higher density of lineaments (Fig. 15).

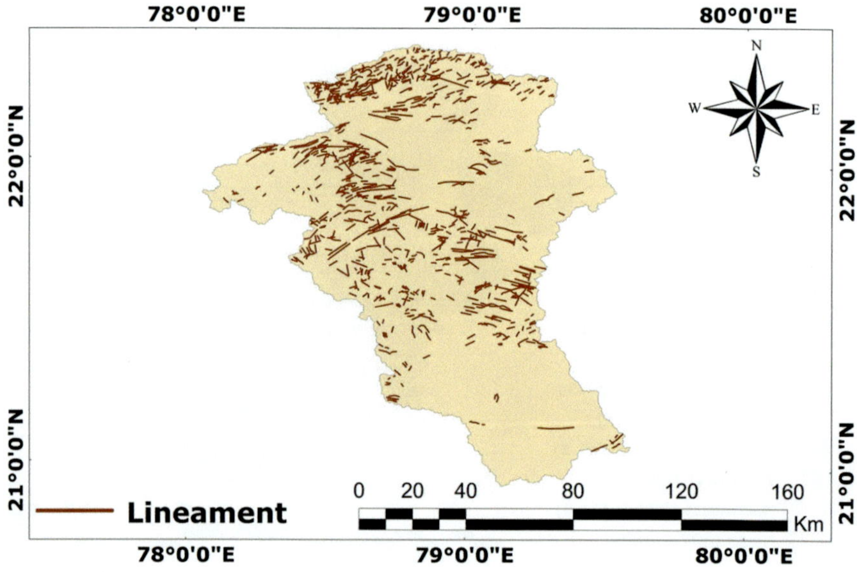

Fig. 14 Lineament map of the Kanhan River Basin

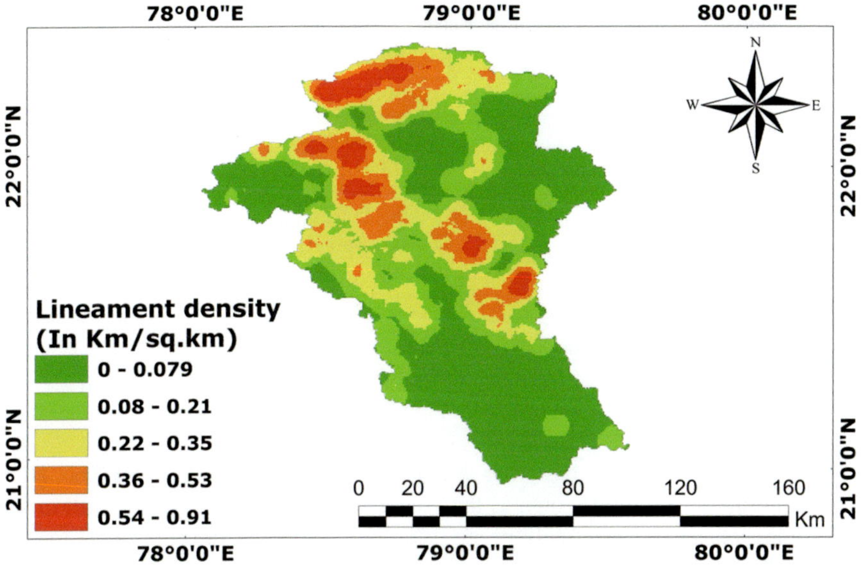

Fig. 15 Lineament density of the Kanhan River Basin

4.2.2 Hypsometric Integral (HI)

Measurement of the interactions between basin area and basin altitude is referred to as hypsometric analysis (Singh and Singh 2018). It has been used to explain how tectonic, geologic, and climatic forces affect topography changes.

$$\mathrm{HI} = \frac{E_{\mathrm{mean}} - E_{\mathrm{min}}}{E_{\mathrm{max}} - E_{\mathrm{min}}}$$

where

E_{mean} Mean elevation
E_{min} Minimum elevation (z)
E_{max} Maximum elevation (Z).

The hypsometric curve (HC) and hypsometric integral (HI) can be used to determine the hypsometry. In the examination of landforms, it is crucial to consider the HC shapes and HI values. These can be described in terms of relative landform age along with the degree of landscape dissection. The HI also serves as an indication of the erosion cycle (Strahler 1952; Kumar and Singh 2021). The length of time needed to bring a land topological unit to reach its base level or lowest level is known popularly as the cycle of erosion. Three categories, each of which corresponds to one of the three major stages of the geomorphic cycle, can be used to group the entire period or cycle of erosion: The monadnock stage, also known as the old stage, is where the basin reaches a height of 0.35 HI. The equilibrium also known as the mature stage,

Table 3 Morphotectonic parameters of the Kanhan River Basin

		Formulas	Results
D	*Tectonic parameters*		
1	Hypsometric integral (HI)	$HI = (h_{mean} - h_{min})/(h_{max} - h_{min})$	0.315
2	Drainage basin asymmetry (A_F)	$A_F = (A_r/A_t) * 100$	41.48%
E	*Sinuosity parameters*		
3	Channel index (CI)	$CI = C_L/A_L$	1.54
4	Valley index (VI)	$VI = V_L/A_L$	1.21
5	Standard sinuosity index (SSI)	$SSI = C_L/V_L$	1.27
6	Topographic sinuosity index (TSI)	$TSI = (VI - 1)/(CI - 1)$	0.38
7	Hydraulic sinuosity index (HSI)	$HSI = (CI - VI)/(CI - 1) * 100$	61%

in which the growth of the basin has attained its steady state, is 0.35 HI 0.60. HI > 0.60; The in-equilibrium (known as the young stage), when the basin is still in the developing phase and extremely prone to erosion (Strahler 1952). The HI for the Kanhan River Basin was computed at 0.315 (Tables 2 and 3), which indicates that the river belongs to the old stage or monadnock stage.

4.2.3 Drainage Basin Asymmetry

The position of the originating stream in relation to the left and right water divides is related to the asymmetry of a drainage basin. Structural control imposes asymmetry on the stream network in the form of tilting or dips (Bloom 2003). The asymmetry of the drainage basin is calculated with the help of the formula by Goudie (2004): $A_F = (A_r/A_t) * 100$ where, A_F = Asymmetry Factor, A_r = Basin area of the right side (facing downstream of the trunk stream), A_t = Total area of the basin. The A_F value 50 shows basin is symmetric. The divergence from this value reveals a greater degree of tilt or dip. The Kanhan River Basin is 41.48% asymmetry (Table 3).

4.2.4 Sinuosity Analysis

The investigation of the effects of the river sequence over the terrain and the topography over the river sequence is helped by the sinuosity study (Panda and Bora 1992). The ratio of the channel length to the distance down the valley is known as the sinuosity, and it describes the channel pattern of a drainage basin. Sinuous rivers are those with a sinuosity of 1.5 or less, and meandering rivers are those with a sinuosity of more than 1.5 (Leopold et al. 2020). For understanding the significance of streams in the formation of landscapes, it is a significant quantitative index. Mueller (1968) divided sinuosity into three categories and provided a few key computations for the measurement of the sinuosity index (Tables 2 and 3). The Hydraulic Sinuosity Index is concerned with the formation of floodplains, and topographic sinuosity, which is related to the flow of natural stream channels. The SSI identifies the kind of river course, which might be straight (SSI = 1.00), sinuous (SSI = 1.00–1.50), or meandering (SSI > 1.50). The Kanhan River Basin's SSI rating of 1.27 indicates that the river's channel is sinuous. As it identifies the stage of basin growth and the governing aspect of sinuosity, the TSI is a useful tool for drainage morphometry (Mueller 1968). In the Kanhan basin, the VI is 1.21 and the CI is 1.54 (Table 1). The Kanhan River's watershed has a TSI of 0.39 (or 39%) and an HSI of 0.61 (or 61%), both of which show that the river has begun to form floodplains.

4.2.5 Transverse Topographic Symmetry (*T* or TTS)

According to Cox (1994), a drainage basin's TTS is expressed by the ratio TTS = D_a/D_d, where D_a is the distance between the drainage basin's midline and the active meander belt's midline and D_d is the distance between the basin midline and the basin divide (Fig. 15).

It is a vector with magnitude and direction that is also used to ascertain the potential tilt direction. For a perfectly symmetrical basin, *T* is equal to 0, and as asymmetry increases, *T* rises and gets closer to the value of 1.0. The Kanhan basin is divided into 20 segments, each with a number from 1 to 20, and these segments are used to calculate the *T* values for the basin. The value of *T* was subsequently determined for each section, as illustrated in Fig. 16. According to Table 4, the predicted *T* values of Kanhan basin range from 0.07 to 0.57, demonstrating the basin's asymmetrical structure.

4.2.6 Elongation Ratio

According to Strahler (1964), the R_e value fluctuates from 0.5 to 1.0 across a broad range of climatic as well as geologic types. The index of R_e is used to categorize different watershed shapes: circular (from 0.9 to 1.0), oval (from 0.8 to 0.9), less elongated (from 0.7 to 0.8), elongated (from 0.5 to 0.7), and more elongated (up to 0.5) (Choudhari et al. 2018). According to the calculated R_e for the Kanhan basin ($R_e = 0.07$), the basin is not significantly impacted by tectonic activity.

5 Discussion

The Kanha River basin (KRB) forming a part of Wainganga River basin in Central India is represented by homogenous geological and diverse characteristics of morphometric and morphotectonic indices. The slightly semicircular to more elongated KRB is validated by seven orders of streams. The basin has inherent control over a variety of lithological and structural aspects and is vulnerable to runoff and

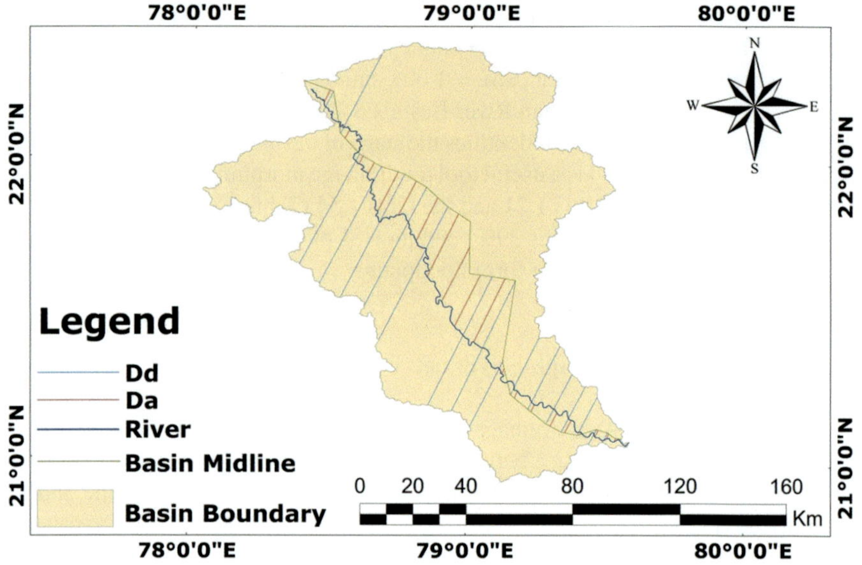

Fig. 16 Showing TTS of the Kanhan River Basin

soil erosion, according to the bifurcation and weighted mean bifurcation ratio. The lithosols, vertic cambisols, and chromic vertisols are common soils found in the KRB and are cover pediments, plains, and valleys. These soils are products of older metamorphic, sedimentary of Gondwana rocks and basaltic flows of Deccan Traps. The form factor (F_f), circulatory and elongation ratio, and sinuosity indicate that the basin is elongated and meandering in nature. The low relief (R_{hl}) value of KRB suggests that the basin is less susceptible to soil erosion and dominant by gentle slopes. The fine texture, low density, and low frequency indicate that the permeable soil with dense vegetation is dominant in the flat plains and gentler slopes of KRB. The slow surface runoff revealed by low constant of channel maintenance is significantly increasing the intensive flood conditions in KRB. The low susceptibility of soil erosion in moderately rugged terrain of KRB is reflected by low ruggedness (R_n) and Melton ruggedness (MR_n) values.

The dominance of NE-SW trending lineaments resembling the directions of low-order streams exhibits the tectonic control on drainage. The high-order streams following the NW-SE directions are guided by ENE-WSW and NW-SE trending lineaments. The high density of lineaments in the NE, NW, and lower reaches of the KRB is marked by the rich potential zones for groundwater. The KRB showing predominance of old stage is not tilted much and does not show influence of active tectonics. Its sinuous channel is continuously performing geological work in the evolution of flood plains and other erosional and depositional features of the river.

Table 4 Transverse topographic symmetry (T) of Kanhan River Basin

Segment number	D_d (km)	D_a (km)	Transverse topographic symmetry (TTS)
1	3.98	1.61	0.40
2	4.28	2.04	0.47
3	6.45	2.74	0.42
4	15.94	0.88	0.05
5	18.84	5.13	0.27
6	20.63	5.43	0.26
7	25.99	2.79	0.10
8	23.45	2.77	0.11
9	26.90	2.11	0.07
10	53.59	26.22	0.48
11	53.09	21.22	0.40
12	53.53	22.49	0.42
13	60.82	25.89	0.42
14	55.27	17.75	0.32
15	49.39	14.14	0.28
16	51.45	16.39	0.31
17	48.83	5.18	0.10
18	46.19	3.25	0.07
19	32.87	3.78	0.11
20	8.29	4.98	0.57

6 Conclusions

Based on the findings of the morphometric and morphotectonic analysis of KRB the following conclusion can be made. Kanhan an elongated basin constitutes of older Proterozoic metamorphic, Gondwana rocks and flows of Deccan Traps and represents the dendritic to subdendritic drainage pattern and less eroded flood plains. It has predominance of pediment pediplains, dissected hills, and alluvium flood plains and is largely covered by cropland and trees. Its sinuous course representing old stage is marked by low runoff and low susceptibility to erosion. The Kanha is an asymmetric basin and is not influenced by active tectonics. High groundwater potentiality and structural discontinuities are indicators of the high lineament density found in the northeastern as well as lower parts of the basin. The planner can make strategies for preventing extreme flood conditions and soil loss in the study area. The retaining walls, check dams, drop structures and the diversion of channels, contour bunds, and porous barriers can be constructed to mitigate flood and soil loss. These research findings may be beneficial to pursue such research in the similar river basins of India and worldwide.

Acknowledgements We are extremely thankful to the Institute of Earth Sciences, Bundelkhand University Jhansi, and IoE, Banaras Hindu University, Varanasi, India for providing the necessary laboratory facilities.

References

Barman BK, Rao CUB, Rao KS, Patel A, Kushwaha K, Singh SK (2021a) Geomorphic analysis, morphometric-based prioritization and tectonic implications in Chite Lui river, Northeast India. J Geol Soc India 97:385–395

Barman BK, Bawri GR, Rao KS, Singh SK, Patel D (2021b) Drainage network analysis to understand the morphotectonic significance in upper Tuirial watershed, Aizawl, Mizoram. In: Agricultural water management. Academic Press, USA, pp 349–373

Bhatt SC (2014) Geological and tectonic aspects of Bundelkhand craton, Central India. Angel Publication, Bhagwati Publishers and Distributors C-8/77-B, Keshav Puram, New Delhi, p 190

Bhatt SC (2020) Natural resources, water harvesting and drought in Central India. Shree Publishers & Distributers, New Delhi, pp 1–130

Bhatt SC, Husain A (2008) Structural history and fold analysis of basement rocks around Kuraicha and adjoining areas, Bundelkhand Massif Central India. J. Geol. Soc India 72:331–347

Bhatt SC, Gupta MK (2009) Tectonic significance of shear indicator in the evolution of Dinara-Garhmau shear zone, Bundelkhand Massif, Central India, MTM. Macmillian Publisher India Ltd., New Delhi, pp 122–132

Bhatt SC, Hussain A (2012) Shear indicators and strain pattern in quartz mylonites of Chituad–Deori shear zone, Bundelkhand Massif, Central India. Earth Sci India 5(II):60–78

Bhatt SC, Mahmood K (2012) Deformation pattern and microstructural analysis of sheared gneissic complex and mylonitic metavolcanics of Babina-Prithvipursector Bundelkhand Massif, Central India. Indian J Geosci 66:79–90

Bhatt SC, Gupta MK (2014) Microstructural analysis and strain pattern in mylonites and implications of shear sense indicators in evolution of Dinara-Garhmau shear zone, Bundelkhand Massif, Central India. J Mineral Soc India 48(2):186–206

Bhatt SC, Singh VK (2021) Tectonic evolution of Babina–Prithvipur crustal shear zones, Bundelkhand craton, India: implications of shear indicators. In: Geological and geo-environmental processes on earth. Springer, Singapore, pp 165–174

Bhatt SC, Hussain A, Suresh M, Balasooriya NWB (2017) Geological structural control on Sukhnai Basin and land use/land cover pattern in Mauranipur and adjoining areas, Bundelkhand craton, Central India. J Geol Soc Sri Lanka 18(2):53–61

Bhatt SC, Singh R, Ansari MA, Bhatt S (2020) Quantitative morphometric and morphotectonic analysis of Pahuj catchment basin, Central India. J Geol Soc India 96:513–520

Bhatt SC, Singh R, Singh R, Saif M, Singh MM (2021a) A GIS-based approach for morphometric analysis of Jamini Basin and its Subwatersheds: implication for conservation of soil and water resources. In: Geological and geo-environmental processes on earth. Springer, Singapore, pp 243–257

Bhatt SC, Singh R, Singh R, Singh VK, Patel A (2021b) Morphometric and morphotectonic studies of Sindh Basin, Central India, using advance techniques of remote sensing and GIS. In: Geological and geo-environmental processes on earth. Springer, Singapore, pp 259–275

Bhatt SC, Mishra S, Singh R, Bhatt S (2021c) Morphometric analysis of Rohni Watershed, Upper Betwa Basin, Bundelkhand Region, Central India. In: Geological and geo-environmental processes on earth. Springer, Singapore, pp 277–291

Bhatt SC, Singh R, Taiyab M, Bhatt S (2021d) Quantitative morphometric analysis of watershed of upper Pahuj Basin, Bundelkhand Region, Central India. In: Geological and geo-environmental processes on earth. Springer, Singapore, pp 293–304

Bhatt SC, Patel A, Pradhan SR, Singh SK, Singh VK, Tripathi GP, Kishor K (2024) Morphometric and morphotectonic attributes of Ken basin, Central India: depicting status of soil erosion, and tectonic activities. Total Environ Adv 9:1–14

Bisht S, Rawat KS, Singh SK (2023) Earth observation data and GIS based landslide susceptibility analysis through frequency ratio model in lesser Himalayan region, India. Quat Sci Adv 100141

Bloom AL (2003) Geomorphology: a systematic analysis of late Cenozoic landforms. Pearson Education Inc., Indian print, New Delhi, 482p

Bromhead E (1992) The stability of slopes. CRC Press, USA

Chakraborty T, Ghosh AK (2005) Need and prospect of R&R in Jharia coalfield. In: Conference on technological advancement and environmental challenges in mining and allied industries in the 21st century, Rourkela, pp 5–6

Chakraborty C, Mandal N, Ghosh SK (2003) Kinematics of the Gondwana basins of peninsular India. Tectonophysics 377(3–4):299–324

Chauhan P, Malviya D, Sain K, Ray RL, Singh SK, Singh D (2022) Assessing the vulnerability of watersheds to environmental degradation in the lesser Himalayan region using satellite data and a series of models. Geocarto Int 37(27):18372–18399

Choudhari PP, Nigam GK, Singh SK, Thakur S (2018) Morphometric based prioritization of watershed for groundwater potential of Mula river basin, Maharashtra, India. Geol Ecol Landscapes 2(4):256–267

Clarke JI (1996) Morphometry from maps. Essay in geomorphology. Elsevier Publication Co., New York, pp 235–274. Geol Survey India 117(2):61–124

Cox RT (1994) Analysis of drainage-basin symmetry as a rapid technique to identify areas of possible quaternary tilt-block tectonics: an example from the Mississippi Embayment. Geol Soc Am Bull 106(5):571–581

Crookshank H (1936) Geology of the northern slopes of the Satpur as between the Mor and the Sher Rivers. Central Book Departmentôt in Komm

Dikau R, Hennrich K, Schmidt J, Schröder M (1997) Regionalization of geomorphometric parameters in hydrologic modelling using GIS. In: International symposium regionalization in hydrology, Braunschweig

Dutta P (2002) Gondwana lithostratigraphy of peninsular India. Gondwana Res 5(2):540–553

Faniran A (1968) The index of drainage intensity—a provisional new drainage factor. Aust J Sci 31:328–330

Gaurav Singh V, Singh SK (2022) Analysis of geo-morphometric and topo-hydrological indices using COP-DEM: a case study of Betwa River Basin, Central India. Geol Ecol Landscapes 1–28

Ghosh SK, Chakraborty C, Chakraborty T (2004) Influence of fluvial-tidal interactions on the nature of cross-stratified packages in a deltaic setting: examples from the Barakar Coal Measure (Permian), Satpura Gondwana Basin, Central India. Geol J 39:1–17

Golani PR, Dora ML, Bandyopadhyay BK (2001) Base metal mineralization associated with hydrothermal alteration in felsic volcanic rocks in Proterozoic belt at Bhuyari, Chhindwara district, Madhya Pradesh. J Geol Soc India 68(5):797–808

Goudie SA (2004) Encyclopedia of geomorphology, 1st edn. Routledge, London, pp 748–749

Hadley R, Schumm S (1961) Sediment sources and drainage basin characteristics in upper Cheyenne River basin. USGS Water Supply Paper, 1531-B, W, 198. Washington, DC

Horton RE (1932) Drainage basin characteristics. Trans Am Geophys Union 13:350–361

Horton RE (1945) Erosional development of streams and their drainage basins: a hydrophysical approach to quantitative morphology. Geol Soc Am Bull 56(3):275–370

Jodhani KH, Patel D, Madhavan N, Singh SK (2023) Soil erosion assessment by RUSLE, Google earth engine, and geospatial techniques over Rel River watershed, Gujarat, India. Water Conserv Sci Eng 8(1):1–17

Kale VS, Gupta A (2001) Introduction to geomorphology. Orient Longman Ltd., UK

Kaur M, Singh S, Verma VK, Pateriy B (2014) Quantitative geomorphological analysis and land use/land cover change detection of two subwatershed in northeast region of Punjab. Int Arch Photogramm Remote Sens Spatial Inf Sci V:XI–XI8

Krishnamurthy J, Srinivas G, Jayaram V, Chandrasekhar MG (1996) Influence of rock types and structures in the development of drainage networks in typical hardrock terrain. ITC J 3–4:252–259

Kumar N, Singh SK (2021) Soil erosion assessment using earth observation data in a trans-boundary river basin. Nat Hazards 107:1–34

Kumar N, Singh SK, Pandey HK (2018) Drainage morphometric analysis using open access earth observation datasets in a drought-affected part of Bundelkhand, India. Appl Geomatics 10:173–189

Kumar N, Singh VG, Singh SK, Behera DK, Gašparović M (2023) Modeling of land use change under the recent climate projections of CMIP6: a case study of Indian river basin. Environ Sci Pollut Res 1–17

Kumar Pradhan R, Srivastava PK, Maurya S, Kumar Singh S, Patel DP (2020) Integrated framework for soil and water conservation in Kosi River Basin. Geocarto Int 35(4):391–410

Leopold LB, Wolman MG, Miller JP, Wohl EE (2020) Fluvial processes in geomorphology. Courier Dover Publications

Magesh NS, Chandrasekar N, Soundranayagam JP (2011) Morphometric evaluation of Papanasam and Manimuthar watersheds, parts of Western Ghats, Tirunelveli district, Tamil Nadu, India: a GIS approach. Environ Earth Sci 64(2):373–381

Mahanta AR, Rawat KS, Kumar N, Szabo S, Srivastava PK, Singh SK (2023) Assessment of multi-source satellite products using hydrological modelling approach. Phys Chem Earth Parts A/B/C 103507

Melton MA (1957) An analysis of the relations among elements of climate, surface properties and geomorphology

Melton MA (1965) The geomorphic and paleoclimatic significance of alluvial deposits in southern Arizona. J Geol 73:1–38

Memon N, Patel DP, Bhatt N, Patel SB (2020) Integrated framework for flood relief package 637 (FRP) allocation in semiarid region: a case of Rel River flood, Gujarat India. Nat Hazards 638(100):279–311

Mesa LM (2006) Morphometric analysis of a subtropical Andean basin (Tucuman, Argentina). Environ Geol 50(8):1235–1242

Miller VC (1953) A quantitative geomorphic study of drainage basin characteristics on the Clinch mountain area. Virginia and Tennessee. Proj., NR 389-402, Tech Rep 3, Columbia University, Department of Geology, ONR, New York

Mueller JE (1968) An introduction to the hydraulic and topographic sinuosity indexes. Ann Assoc Am Geogr 58(2):371–385

Murmu P, Kumar M, Lal D, Sonker I, Singh SK (2019) Delineation of groundwater potential zones using geospatial techniques and analytical hierarchy process in Dumka district, Jharkhand India. Groundwater Sustain Dev 9:100239

Nur A (1982) The origin of tensile fracture lineaments. J Struct Geol 4:31–40

Ozdemir H, Bird D (2009) Evaluation of morphometric parameters of drainage networks derived from topographic maps and DEM in point of floods. Environ Geol 56:1405–1415

Panda PC, Bora HN (1992) A study of sinuosity index of Siang River and its major tributaries: Arunachal Pradesh. Environ Manag 1:97–101

Pandey HK, Singh VK, Singh SK, Sharma SK (2023a) Mapping and validation of groundwater dependent ecosystems (GDEs) in drought-affected part of Bundelkhand region, India. Groundwater Sustain Dev 100979

Pandey HK, Singh VK, Singh RP, Singh SK (2023b) Soil loss estimation using RUSLE in hard rock terrain: a case study of Bundelkhand, India. Water Conserv Sci Eng 8:55. https://doi.org/10.1007/s41101-023-00229-5

Pareta K, Pareta U (2011) Quantitative morphometric analysis of a watershed of Yamuna basin, India using ASTER (DEM) data and GIS. Int J Geomatics Geosci 2(1):248–269

Pascoe EH (1973) A manual of the geology of India and Burma, vol 1. Controller of Publications

Patel A, Singh MM, Singh SK, Kushwaha K, Singh R (2022) AHP and TOPSIS based sub-watershed prioritization and tectonic analysis of Ami River Basin, Uttar Pradesh. J Geol Soc India 98(3):423–430

Peters J, Singh SK (2001) Satpura Basin-an example of pre-rift, syn-rift and post-rift Gondwana sedimentation in India. J Geol Soc India 309–320

Pirasteh S, Rizvi SMA, Ayazi MH, Safari H, Ramli FM, Pradhan B, Shattri M, Mahmoodzadeh A, Pirateh S, Rizvi SMA (2010) Using ERS-1 synthetic aperture radar for flood delineation, Bhuj taluk, Kuchch district-Gujarat India. Int Geoinform Res Dev J 1:13–22

Pradhan B, Youssef AM (2010) Manifestation of remote sensing data and GIS on landslide hazard analysis using spatial-based statistical models. Arab J Geosci 3:319–326

Pradhan B, Singh RP, Buchroithner MF (2006) Estimation of stress and its use in evaluation of landslide prone regions using remote sensing data. Adv Space Res 37(4):698–709

Prakash K, Mohanty T, Pati JK, Singh S, Chaubey K (2017) Morphotectonics of Jamini River basin, Bundelkhand craton, Central India; using remote sensing and GIS technique. Appl Water Sci 7:3767–3782

Rathore VKS, Verma PK, Singh AK, Patel A, Singh MM, Bhatt SC, Singh SK (2023) Channel dynamics of a middle reach of Rapti River, Eastern Indo-Gangetic plain, India. J Geol Soc India 99(7):995–1005

Reddy GPO, Maji AK, Gajbhiye KS (2004) Drainage morphometry and its influence on landform characteristics in a basaltic terrain, Central India—a remote sensing and GIS approach. Int J Appl Earth Obs Geoinf 6(1):1–16

Roy A, Prasad MH (2001) Precambrian of central India: a possible tectonic model. Geol Surv India Spec Publ 64:177–197

Schumm SA (1956) Evolution of drainage systems and slopes in badlands at Perth Amboy, New Jersey. Geol Soc Am Bull 67(5):597–646

Selvan MT, Ahmad S, Rashid SM (2011) Analysis of the geomorphometric parameters in high altitude glacierised terrain using SRTM DEM data in Central Himalaya, India. ARPN J Sci Technol 1(1):22–27

Singh V, Singh SK (2018) Hypsometric analysis using microwave satellite data and GIS of Naina–Gorma river basin (Rewa district, Madhya Pradesh, India). Water Conserv Sci Eng 3:221–234

Singh S, Singh C, Mukherjee S (2010) Impact of land-use and land-cover change on groundwater quality in the Lower Shiwalik hills: a remote sensing and GIS based approach. Open Geosci 2(2):124–131

Singh SK, Mustak S, Srivastava PK, Szabó S, Islam T (2015) Predicting spatial and decadal LULC changes through cellular automata Markov chain models using earth observation datasets and geo-information. Environ Process 2:61–78

Singh D, Karan K, Singh SK, Chauhan P, Berndtsson R (2023) Fusing morphometric characteristics with extreme precipitation indices for identifying the most vulnerable sub-basin at risk of flooding. J Hydro Environ Res 50:44–56

Smith KG (1950) Standards for grading texture of erosional topography. Am J Sci 248(9):655–668

Sreedevi PD, Subrahmanyam K, Ahmed S (2005) Integrated approach for delineating potential zones to explore for groundwater in the Pageru River basin, Cuddapah District, Andhra Pradesh, India. Hydrgeol J 13:534–543

Strahler AN (1952) Hypsometric analysis of erosional topography. Bull Geol Soc Am 63:1117–1142

Strahler AN (1956) Quantitative slope analysis. Bull Geol Soc Am 67:571–596

Strahler AN (1957) Quantitative analysis of watershed geomorphology. Eos, Trans Am Geophys Union 38(6):913–920

Strahler AN (1964) Quantitative geomorphology of drainage basin and channel network. In: Handbook of applied hydrology, pp 39–76

Yadav SK, Singh SK (2022) Morpho-tectonic assessment of Central Northern escarpment of Peninsular India, based on tectonically sensitive geomorphic indices. Phys Geogr 43(6):753–783

Yadav SK, Singh SK, Gupta M, Srivastava PK (2014) Morphometric analysis of Upper Tons basin from Northern Foreland of Peninsular India using CARTOSAT satellite and GIS. Geocarto Int 29(8):895–914

Waterlogging Susceptibility Zone Assessment in Kolkata by the Flow Accumulation Method

Moumita Mondal, Parama Banerjee, and Nayan Dey

Abstract Urban waterlogging refers to the phenomenon when there is a rainstorm or short time heavy rain, which surpasses the capacity of the urban drainage system. Due to the absence of a properly planned drainage system, urban are facing waterlogging conditions for heavy rainfall. Rainfall irregularities inundate more waterlogging conditions. Waterlogged roads not only lead to traffic snarls but can also affect health and safety, as it can increase the risk of fatal accidents, skidding, damage to vehicle engines, damage to properties, etc. Kolkata city has been chosen for this study. Here, flow accumulation is used to determine the waterlogging point. The digital elevation model (DEM) can be used to gradient for any spot on the ground surface. Various studies have used the computational hydrodynamics (CHD) base technique to determine the direction and gradient of various spots in a hilly area. The gradient determination of Kolkata is very complicated due to its deltaic phenomena. Determination of the runoff, subsequent flow accumulation, and waterlogging areas of Kolkata is the prime objective of this chapter. On the digital elevation model (DEM) map of Kolkata, 300 m $\times$ 300 m grid sizes are used to calculate the threshold value of surface runoff patterns using a D8 algorithm. After computing the gradient, the surface runoff direction or pattern is traced which helps to determine or frame the waterlogged susceptible zone in Kolkata Municipal Corporation (KMC). Based on the threshold value, five types of pour points are detected, i.e. very small pour points (6–14), small pour points (15–24), Medium Pour Points (25–35), large pour points (36–51) and very large pour points (Above 51). Large to very large pour points assist near the eastern and southern segments of Kolkata Municipal Corporation (KMC). Propensity based on waterlogging conditions is also pronounced by the distance from the canal and the distance from the pumping station. Roads near Cossipur, College

M. Mondal
Department of Geography, Rabindra Bharati University, Emerald Bower Campus, 56A, BT Road, Kolkata 700050, India

P. Banerjee
Department of Geography, Naba Barrackpore Prafulla Chandra Mahavidyalaya, Kolkata, India

N. Dey (✉)
Department of Geography, Pt. Ravishankar Shukla University, Raipur, India
e-mail: dey.nayanrbu@gmail.com

S. C. Pal and U. Chatterjee (eds.), *Surface, Sub-Surface Hydrology and Management*, Springer Geography, https://doi.org/10.1007/978-3-031-62376-9_14

Street, and Garden reach MG Road are experiencing moderate to high vulnerable conditions.

Keywords Urban waterlogging · Drainage congestion · Computational hydrodynamics · Flow accumulation · Digital elevation model · Gradient · Deltaic plain · Threshold value

1 Introduction

Waterlogging is an ungovernable leading crux for urban dwellers day by day. The infrastructure of the urban centres (i.e. drainage, road, swear facilities) is not developing with the rapid growth of population. The huge population pressure breaks the sequential sewage system. From the previous studies, it has been revealed consensually that waterlogging in urban areas is caused by the high intensive rainfall. The waterlogging scenarios are significantly different in different urban. Therefore, various types of urban waterlogging have been distinguished. It is obvious significance that precipitation is prime for waterlogging. It is considered a hydrological urban disaster such as a landslide, flood, etc. The chapter is for the comprehensive assessment of the impact of rainfall intensity and accumulation for waterlogging in urban regions (Ma et al. 2020). Rapid unplanned urbanisation is transforming a city into a concrete world that chocks the natural rainwater infiltration and spontaneously increases the surface runoff (Singh and Singh 2011; Radford and James 2013; Hassan and Nazem 2016; Shao et al. 2019; Shukla and Jain 2019; Nowak and Greenfield 2020). The depressed zones of urban areas accumulate rainwater and streets act as streams due to the increase in the rate of surface runoff and it became a most hydrodynamic predisposing controller for this situation (Sajikumar and Remya 2015; Tam and Nga 2018; Subrina and Chowdhury 2018; Rahmati et al. 2020). From the previous study, it has been depicted that high intense rainfall is not only the prime factor but also other factors, i.e. depressed landscape profile, unscientifically landuse and landcover change, unplanned urban growth with inadequate drainage system as well as mismanagement occurs the waterlogging situations (Quan et al. 2010; Ning et al. 2017; Aktar et al. 2017; Dhiman et al. 2019; Shao et al. 2019). Kolkata is an ancient metropolitan city that was established along the left bank of the lower Ganga. With the vertical growth, the land has been disrupted and sentiment has become suppressed. In the initial period, the Kolkata sewerage drainage network built by the British with the adjacent canals was for the then sparse population. But this system gradually falling down due to an increase in unscientific land use for the name of urban expansion. The 'waterlogging' is an alarming problem for the city of Joy. The old sewage system of the British era is failing to equalize drainage requirements. Therefore several segments of Kolkata Municipal Corporation (KMC) are facing the problems of waterlogging which disrupts the natural movement of these regions (Roy and Dhali 2016). Waterlogging has become a serious problem in numerous metropolitan cities. Most of the metropolitan cities in developing and underdeveloped countries are experiencing waterlogging problems due to the special obedience of poor drainage systems. Consequently, civilians in those urban centres face various environmental problems, i.e. water pollution, waterborne diseases, and traffic congestion. The previous study also recommends clearing the drainage system of any obstruction or blockage and establishing organised management institutions for effective mitigation (Alom and Khan 2014). Research on urban floods in India from 1988 to 2007 is carried out in four metropolitan cities Kolkata, Chennai, Delhi and Mumbai. Singh and Pandey (2013) identified the factors behind it, observed how urban civilians are affected by these conditions and provided some suggestions for mitigation. Mycosphaerlapinodes infection could spread with waterlogging that depletes the plant growth (McDonald and Dean 1996). New Delhi has faced a miserable traffic jam situation due to heavy rain during monsoon. The thirty-five causes are depicted behind it (The India Express, August 10, 2015, 10.50 a.m.). Mumbai's Dadar area has experienced widespread waterlogging due to massive rainfall that brings difficulties normal life of daily commuters. This incident has exposed the government's failure and mismanagement (Mumbai, ANI, August 21, 2014, 11.57 a.m.). Stormwater pumping stations at Love Grove and Cleveland Bunda failed to deliver as promised during heavy rains in Mumbai. (The India Express, Tanushree Venkartraman June 20, 2015, 10.50 a.m.)

Several research papers have been published on the proposed topic, waterlogging. They depict the climatic factors and various socio-economic factors viz. low rate of rainwater infiltration, high population density, lack of drainage capacity and length of road. In addition, mitigation planning for urban waterlogging has been suggested. Rahman et al. (2009) researched the analysis of the waterlogging problem and its environmental effects in Khulna city using GIS approaches. This research work identified that waterlogging is caused by high-intensity storms, rainfall and runoff in the city area that is inundated due to lack of proper drainage system and inefficient management. In addition, the said research work examined the causes of waterlogging, the direct and indirect damages of waterlogging and the environmental condition of the waterlogged area of the city based on the above-mentioned factors. Mitra and Banerji (2016) researched the pluvial flooding in some selected parts of Kolkata. Through this research work natural slope patterns including clogging of drainage lines, situation of the existing canals and shrinking of the wetlands of the city. In addition, the research work examined the rate of flooding and its consequences based on the above-mentioned factors. Roy and Dhali (2016) researched the seasonal water logging problems in Kolkata, India. They plotted the inundated waterlogging area of Kolkata and revealed the cause behind it. The said research paper has tried to assess how canal profiles control waterlogging and how they impact traffic. In addition, they also suggest some plans regarding to new pumping station and sewage facilities. Joarder (2016) and Dutta and Sarkar (2020) researched the causes of waterlogging in the southwestern suburban area of Kolkata. Through this research work factors of waterlogging like ground configuration and drainage system, changing pattern of land use, unplanned built-up area and shortage of road spaces, disappearing wetlands, ponds and canals and silting of canals or outfall channels are analysed. In addition, the said research work examined that during the rainy season, the daily activities of the economic and social life of the local residents are disrupted by the above-mentioned factors. Adhikary et al. (2017) researched to identify waterlogging problems in Ghatal Municipality through mapping. This chapter identified that waterlogging is caused by excessive rainfall in the rainy season, lowland characteristics of the region, sedimentation in riverbeds, and poor drainage system. In addition, the said research work examined the causes and duration of waterlogging, local people faced with different problems and waterlogging hazard vulnerability zones based on the above-mentioned factors. Mukhopadhyay (2004), Paul (2009) and Banerjee (2018) researched the urban flooding scenario of Kolkata city. Through this assessment, it was found that the causes of flooding in the city area are climate change, excessive rainfall and some parts of geographical texture. In addition, the said research work examined the scenario of different parts of the city during flood or waterlogging. Panchal et al. (2019) researched on problems of waterlogging in Nadiad City and its probable analytical solutions. Through this research, the railway underpass is measured and the waterlogging problems are identified. In addition, the said research work examined the current scenario of the underpass of the city and provided possible solutions to remove this problem based on the above-mentioned factor. However, water logging assessment by employing ArcGIS on CHD-based flow accumulation models (FAM) and vulnerable zone and road assessment due to waterlogging in Kolkata, has still not been advocated. In this chapter, the slope of each grid point of the digital elevation model (DEM) point, runoff and waterlogging patterns are calculated. In addition, the said chapter examined a real view of Kolkata's waterlogging based on the above-mentioned factors.

2 Objectives

I. To identify the most of water logging zones in Kolkata.
II. To assess the most vulnerable area and road for water logging pockets in Kolkata.

3 Study Area

The delta city, Kolkata is the capital of West Bengal, located on the left bank of the Ganga-Bhagirathi River from 22° 28′ to 22° 37′ N latitude and 88° 17′ 30″ E to 88° 25′ E longitude covering an area of 185 km^2 (Fig. 1). Compression by urbanisation has led to allowing the compaction of the sediments development of the city at an average elevation of around 9 m from sea-level (NASA 1999; Bhaumik and Das 2021). The Kolkata Municipal Corporation (KMC) is formed with 144 wards and twelve boroughs. Three more wards have been added recently. Kolkata belongs to the tropical monsoonal climatic region and experiences a maximum annual temperature of 40 °C in Summer (April–mid-June) and a minimum temperature is about 10 °C in the winter season (December–January). Kolkata receives an average annual rainfall of 165 cm (70–80% in Mid June–September) (John and Das 2020). Most of the area is sheltered by the Hooghly River in the north-west the Bagjola, Beliaghata, and Central Circular Canals in the north and numerous canals such as the Adi-Ganga and ToliNalla in the south, but they have been silted up (Bhaumik and Das 2021). According to the Census of India (2011), the total population of Kolkata is 4,496,694 people and the metropolitan population is 14,112,536 people.

4 Methodology

4.1 Data and Materials

This study is based on the scheduled survey as a source of primary data, by the interaction with local people, such as other dignitaries have been interacted with directly or indirectly. About 101 samples have been taken as a scheduled survey from the study area with the simple random sampling method. Secondary data has been collected for this chapter by accessing archival records, government documents, census reports, etc. In addition, political maps and satellite imagery have been downloaded from the following website (Table 1).

4.2 Waterlogging Point Assessment

Step-I (Flow Accumulation)

Flow accumulation is especially important to understand topographic controls on water, carbon, nutrient and sediment flows within and over full watersheds (Do et al. 2011). For the hydrological and topographic analysis, (viz. rill, gully, stream extraction from digital elevation modelling (DEM), Stream hierarchy, stream catchment area delineation) the flow accumulation (FA) is an essential tool (Zhou et al. 2019). The flow accumulation could be derived from the digital elevation modelling (DEM) directly (Arge et al. 2003; Bai et al. 2015). The direction of the stream flow matrix is the GIS product of flow accumulation. The single-flow D8 Method has been employed to determine the water streaming direction from the flow accumulation (Barnes et al. 2014; Garbrecht and Martz 1997; Nardi et al. 2008; O'Callaghan and Mark 1984). After creating the DEM from Google Earth Pro, Flow Direction and Flow Accumulation tools of the ArcGIS are used to determine the runoff pattern of Kolkata. A 300 m * 300 m grid has been used to make the map and based on this the water logging points of Kolkata have been identified.

Step-II (Threshold Value Estimation)

The threshold zones are recognised as waterlogging zones depending on water invasion from the number of surrounding cells. If the rainwater streams towards the

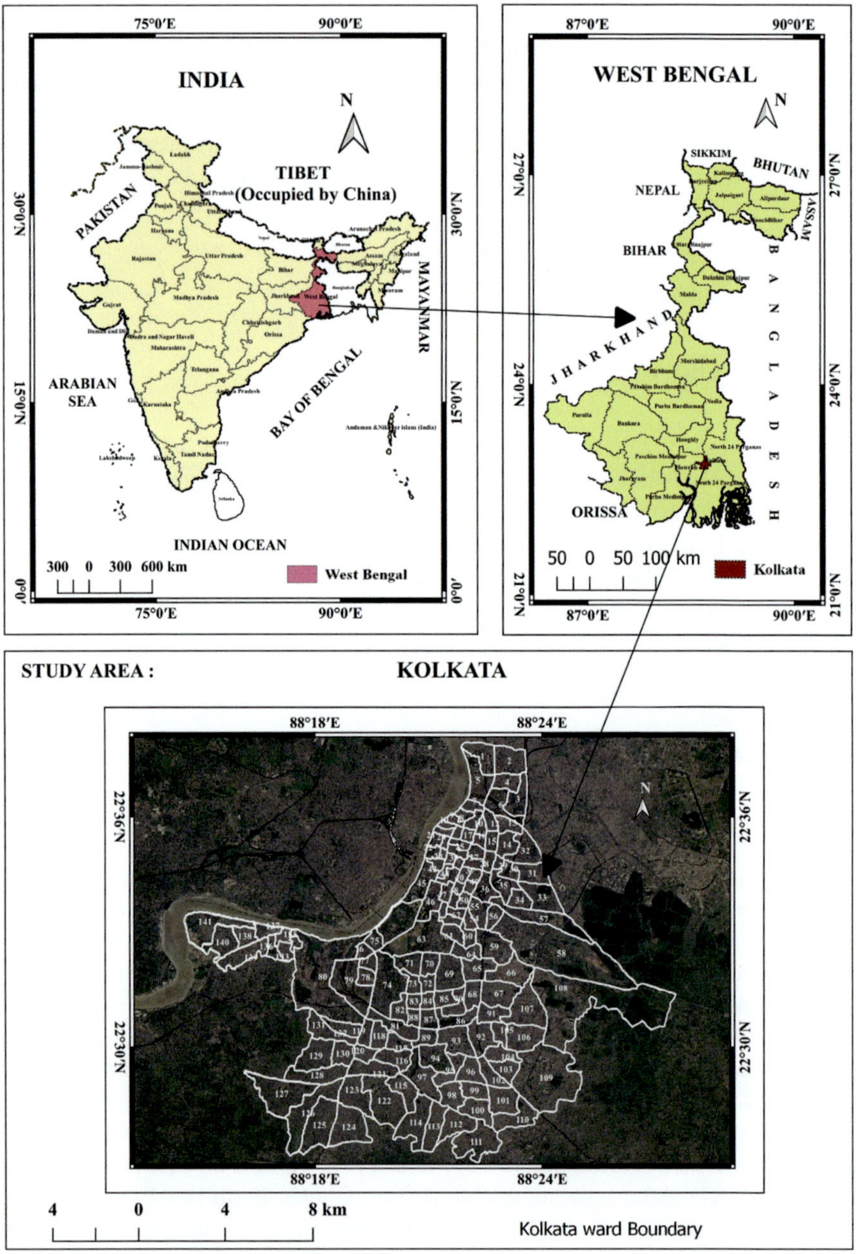

Fig. 1 Location map

Table 1 Data

Sl. No.	Secondary data	Source
1	Satellite imagery	LANDSAT 4–5, LANDSAT 8–9
2	Maps of Kolkata	NATMO, Kolkata
3	Waterlogging data	Research paper, K.M.C.
4	Demographic data	Census of India, 2011
5	Climate data	Indian Meteorological Department (IMD)

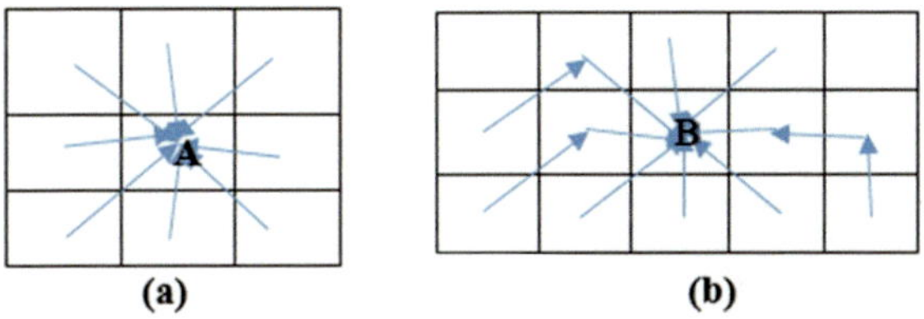

Fig. 2 Flow accumulation potential zone **a** location "A" has flow accumulation probability of 8 and **b** location "B" is 11

middle cell from adjacent cells, it becomes waterlogged. The computation of flow direction from the nearest adjoining eight (8) cells towards a particular cell is called threshold value of flow accumulation and is determined by D8 algorithm (Fig. 2a). After flow accumulation from the adjacent eight cells to the next cell, if the water is streamed towards the other cell then it is also added to the inclusive tally to calculate the waterlogging potential zone. Therefore 'B' scored 12 in defining the relative flow accumulation potential zone (Fig. 2b). From the flow catchment potential assessment, the waterlogging zone of Kolkata is detected by considering runoff from more cells, and more flow storage is identified (Figs. 2 and 3).

4.3 Weighted Sum Method

Weighted sum method is a multi-criterion decision-making method in which there will be multiple alternatives and we have to determine the best alternative based on multiple criteria.

Criteria are defined, *step-1* Ranks are based on the value of each criterion (The one whose value is less being rank 1), *step-2* After ranking, the ranks of all criteria have been summed up according to water logging points, *step-3* After ranking, they are divided into five classes, *step-4* It is shown on the map.

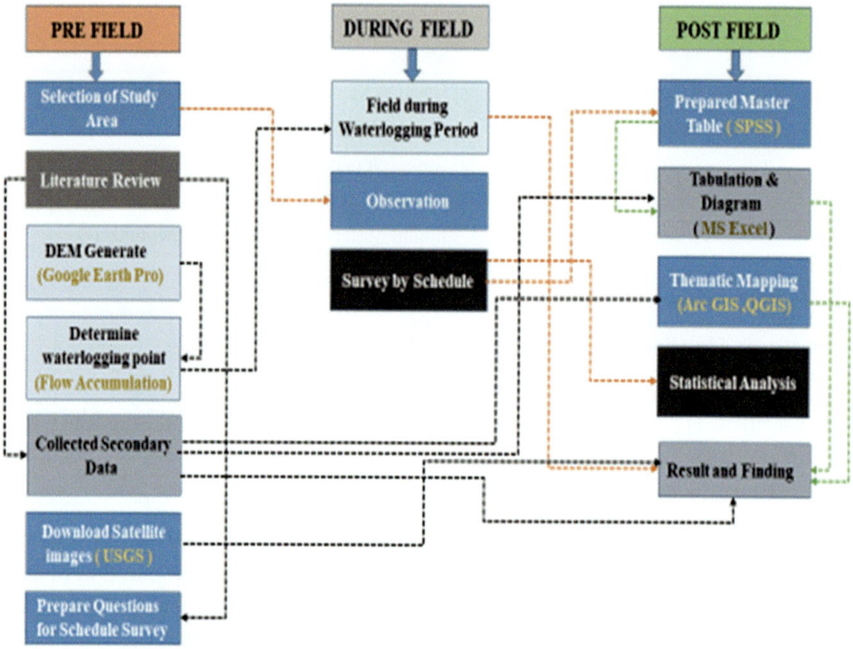

Fig. 3 Flow chart of methodology

4.4 Kappa Coefficient

The Kappa coefficient is a statistical assessment employed to measure the level of agreement status between the user raster and producer raster in this chapter. The feasibility of the level of agreement is assessed by the correlations between inter-raster relationships. The accuracy level of maps is assessed by the Kappa coefficient. The said chapter also uses the Kappa coefficient for the study of agreement based on the disparities of the classified raster with the original.

Calculation of the Kappa coefficient (κ)

(I) Calculation for overall accuracy (OA)

$$OA = \frac{\sum CCP}{\sum RP} \times 100$$

where

CCP Total No. of Correctly Classified Pixel
RP Total No. of Reference Pixel

(II) User Accuracy Calculation (UAC)

$$\mathrm{UAC} = \frac{\mathrm{CCP_E}}{\sum \mathrm{CP_E}} \times 100$$

where

$\mathrm{CCP_E}$ No. of Correctly Classified Pixel in each Category
$\mathrm{CP_E}$ Total No. of Classified Pixel in that category

(III) Producer Accuracy Calculation (PAC)

$$\mathrm{PCA} = \frac{\mathrm{CCP_E}}{\sum \mathrm{RP_E}} \times 100$$

where

$\mathrm{CCP_E}$ No. of Correctly Classified Pixel in each Category
$\mathrm{RP_E}$ Total No. of Reference Pixel in that Category

(IV) Kappa Coefficient (K)

$$K = \frac{(\mathrm{TS} \times \mathrm{TCS}) - \sum (\mathrm{Colum\ total} \times \mathrm{Row\ total})}{(\mathrm{TS})^2 - \sum (\mathrm{Colum\ total} \times \mathrm{Row\ total})}$$

where

TS Total Sample
TCS Total Corrected Sample

Indexation of the Kappa coefficient (κ)

The Kappa coefficient ranges from − 1 to 1, where − 1 represents the discard of the agreement, 0 represents the agreement with no better than chance and + 1 represents the perfect agreement. The coefficient value is about 0.8 which indicates good agreement.

5 Result and Discussion

5.1 Waterlogging Zone Assessment by Flow Accumulation Method

Step I Waterlogging area assessment using DEM

Digital elevation models (DEM) are widely used for automatically delineating the watershed, stream flow direction assessment, slope analysis, aspect, flow accumulation, and flow capacity assessment via geographical information system (GIS) (Do et al. 2011). DEM of Kolkata is generated using Google Earth Pro and ArcGIS. The DEM of Kolkata split up into 300 m × 300 m grid (Fig. 4).In this diagram, the darkest Pixels have low elevation and white is visible in the case of high elevation. Thus, the pixel's colour becomes whiter or darker as the elevation increases or decreases. The DEM of Kolkata shows that the highest elevation here is about 30.39 m and the lowest elevation is about 0.96 m. From the geomorphological overview, natural level of a river gain more elevation than the adjoining flood plain in the lower course. Kolkata city is belongs to the lower course of the Ganges River. Therefore, buffer along the river bank experiences relatively higher than the other portion of Kolkata Municipal Corporation (KMC). In addition, the middle segment also gains higher elevation in dissected manner due to its deltaic configuration.

From the digital elevation modelling (DEM), it has been evacuated that Kolkata is characterised as pan-shaped. The runoff pattern map is a sketch of the direction of surface runoff based on elevation profile (Fig. 5). The flow direction (FR) tool of

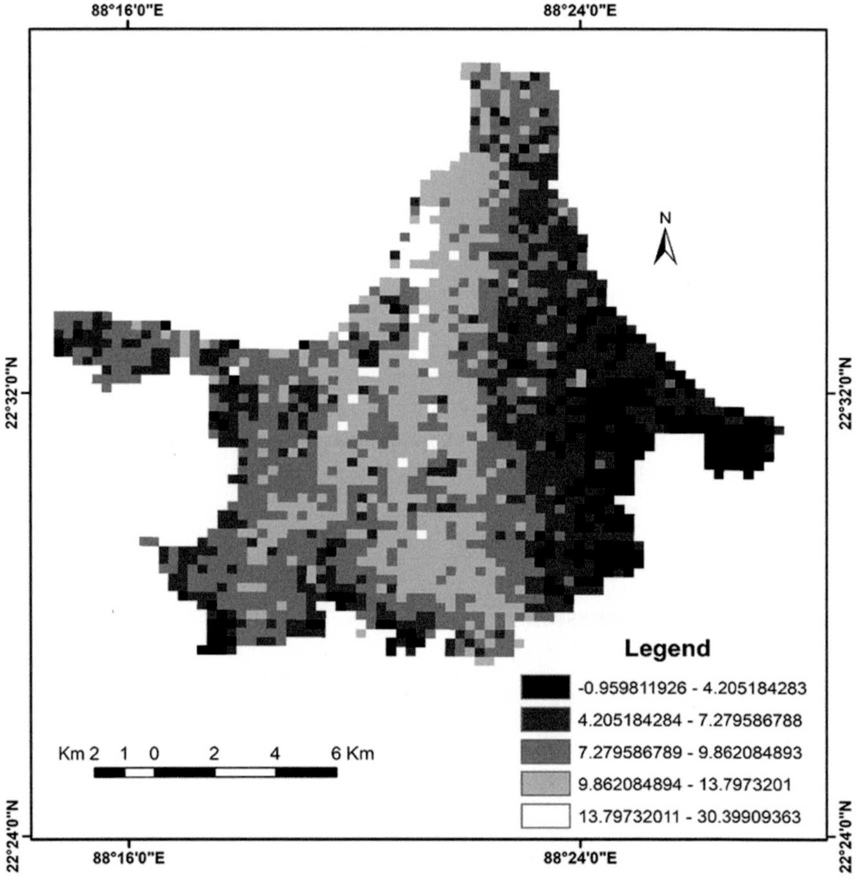

Fig. 4 DEM of Kolkata using 300 m * 300 m grid in ArcGIS

Hydrology is used to calculate the streaming direction. Surface elevation is the key input for the flow direction raster, which assesses the steepest way to stream direction or surface runoff to the downslope of each pixel. The flow direction or surface runoff direction could be determined by the three statistical models, viz. D8 Method (that traces streaming direction from adjacent eight sides), multi-flow direction (MFD) method and D-Infinity (DINF) method. The flow accumulation raster presents the sum of surface runoff from upstream (adjoining cells) to each pixel. The magnitude of waterlogging is determined by the flow accumulated value. After creating the DEM, the ArcGIS tools Flow Direction and Flow Accumulation are used to determine the runoff pattern of Kolkata (Fig. 5). From this study, it has been concluded that water streams inward directions concerning its geomorphological setup. The inward movement of surface runoff occurs in several waterlogging pockets. The volume of waterlogging depends on the catchment area, from where water is drained.

A 300 m * 300 m grid has been used to make the map and based on this the water logging points of Kolkata have been identified. Figure 6 shows that there are multiple water logging areas in Kolkata. The water logging zones are defined as pour point. In this chapter, it has been assessed that maximum pour point occurs at the north and middle sections of Kolkata Municipal Corporation (KMC).

Step II Waterlogging potential postulation zone assessment by the threshold value

Based on the threshold value waterlogged zones are classed into three, viz., mild, moderate and massive (Table 2). It depends on how many cells contribute water to a particular cell. If the number of neighbouring cells increases, then the maximum amount of water streams towards the cell centre that concluded the centre cell region becomes more waterlogged (Fig. 7).

The diagram (Fig. 7) shows that the flow accumulation potential of some areas water logging areas of Kolkata is high to very high, there are some areas where the flow accumulation potential is medium and some areas where flow accumulation potential is low. Consequently, a large portion of Kolkata Municipal Corporation (KMC) is flooded for an effective period which greatly disrupts the life of the city. Water logging potential zone assessment has been pronounced by the threshold value or the water streaming from the number of grids (Fig. 7). Flow accumulation of Hydrology in Arc GIS is used to plot the waterlogging potential zone. From this assessment, it has been revealed that the mild waterlogging potential zone is found in North Kolkata, viz., Block-B, Cossipore, Tala, Belgachia, Girish Park, and Maniktala (Bagmari). The moderate pour points occur near the middle segment of Kolkata Municipal Corporation (KMC), viz., Fort William, Gorer Math, College Street, Rajabazar, Maidan, and Hastings. The massive waterlogging point seasonally developed near the east and south Kolkata, Viz., Alipore, Behala, Patuli, and Rabindra Sarabar.

Step III Vulnerability Assessment of waterlogging zone due to water logging by weighted sum method

Waterlogging is not only depends on the geomorphologic configuration but also pronounced by the distance from the canal, and distance from the pumping station.

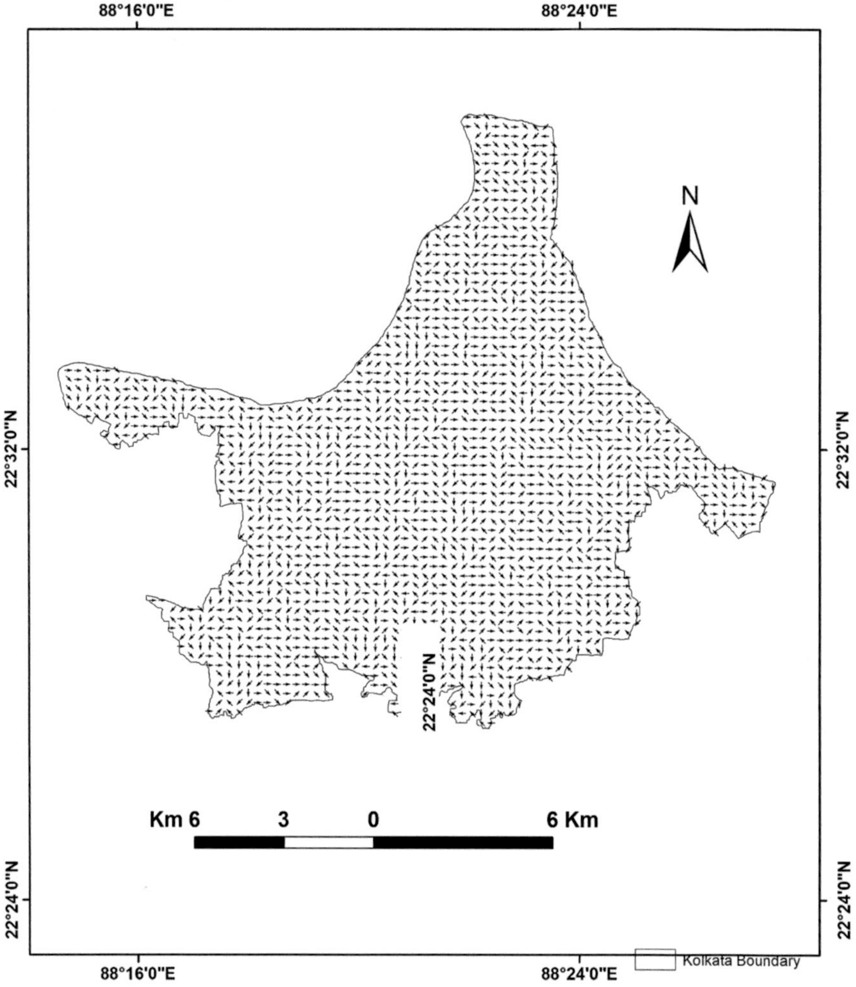

Fig. 5 Runoff pattern using flow accumulation tools in ArcGIS

The weighted sum method has been employed to determine the waterlogged propensity zone in Kolkata (Fig. 8). In this chapter, it has been traced several roads that are seasonally submerged by waterlogging and resultant traffic congestion. Five types of vulnerable zones (i.e. very low, low, medium, high and very high) have been assessed based on the amount of water accumulation and waterlogging period.

Unplanned urban development has destroyed the physical configuration of the city of Kolkata. The CBD of the city (mostly North and Central Kolkata) has survived with massive population pressure and high road density which has already become an influencer of waterlogging. The map (Fig. 8) shows that Middle and South Kolkata are more affected by water logging. The very vulnerable high zones are the Girish Park area of North Kolkata and the Gobra area of Middle Kolkata. Because the

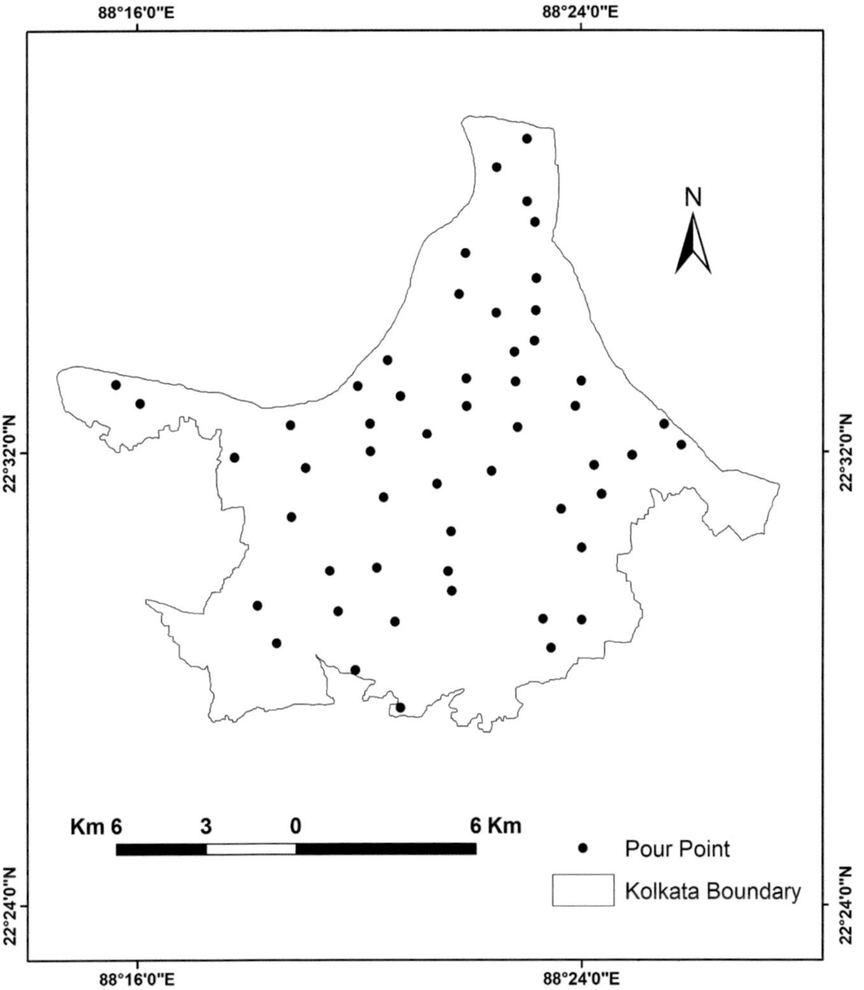

Fig. 6 Waterlogging point of Kolkata

Table 2 Types of waterlogging

Classification of waterlogging zones	Postulation
Mild	If water streams from adjacent cell numbers 6–14 to a middle cell
Moderate	If water flows from adjacent cell numbers 15–35 to a middle cell
Massive	If water flows from adjacent cell numbers 25–75 to a middle cell

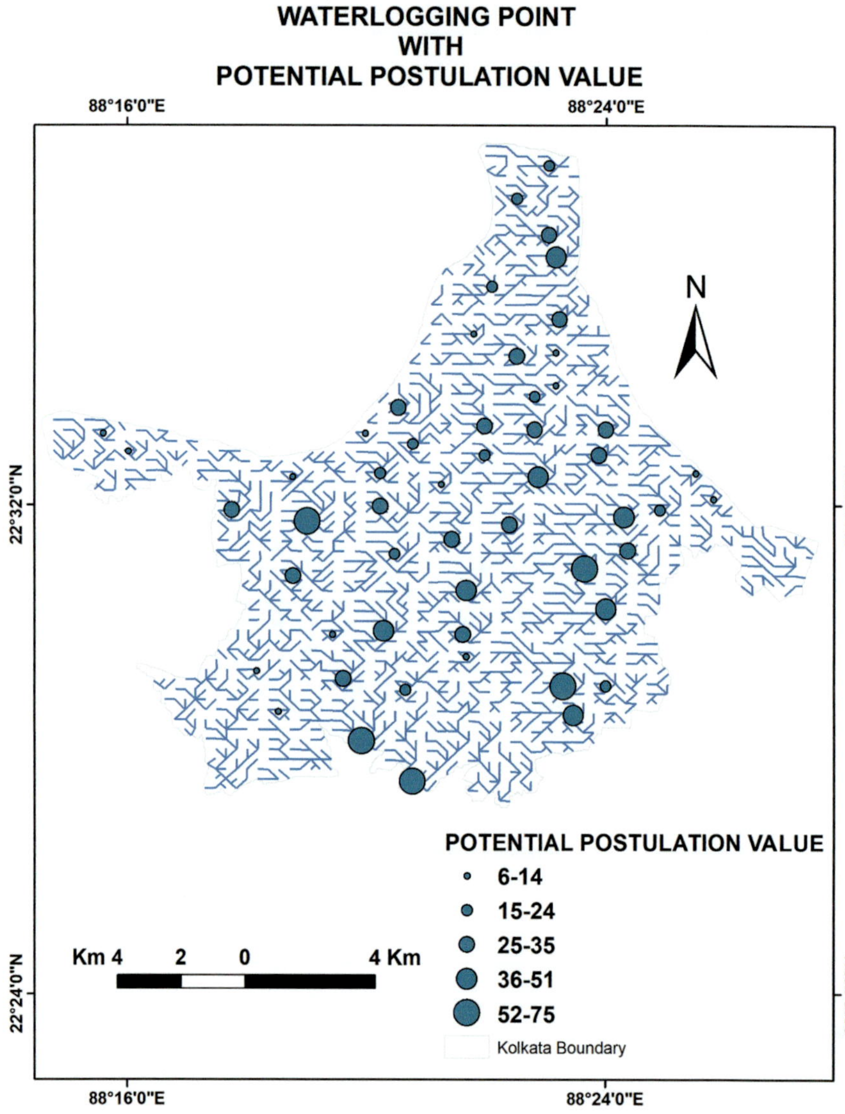

Fig. 7 Type of waterlogging area

household density in North Kolkata is very high and in Gobra which is a slum area, the density is high as well as the drainage system is not good. It is also seen that most of the area of Central Kolkata is in a High Water Logging Zone and West Kolkata is in a Medium Water Logging Zone (Table 3).

Below one-hour waterlogged roads are Basanti Highway, and Beleghata Main Road. One to three hours water seasonally waterlogged roads are Jossore Road in North Kolkata and Hazra Road. Tartala Road, Maniktala Main Road, and Garden

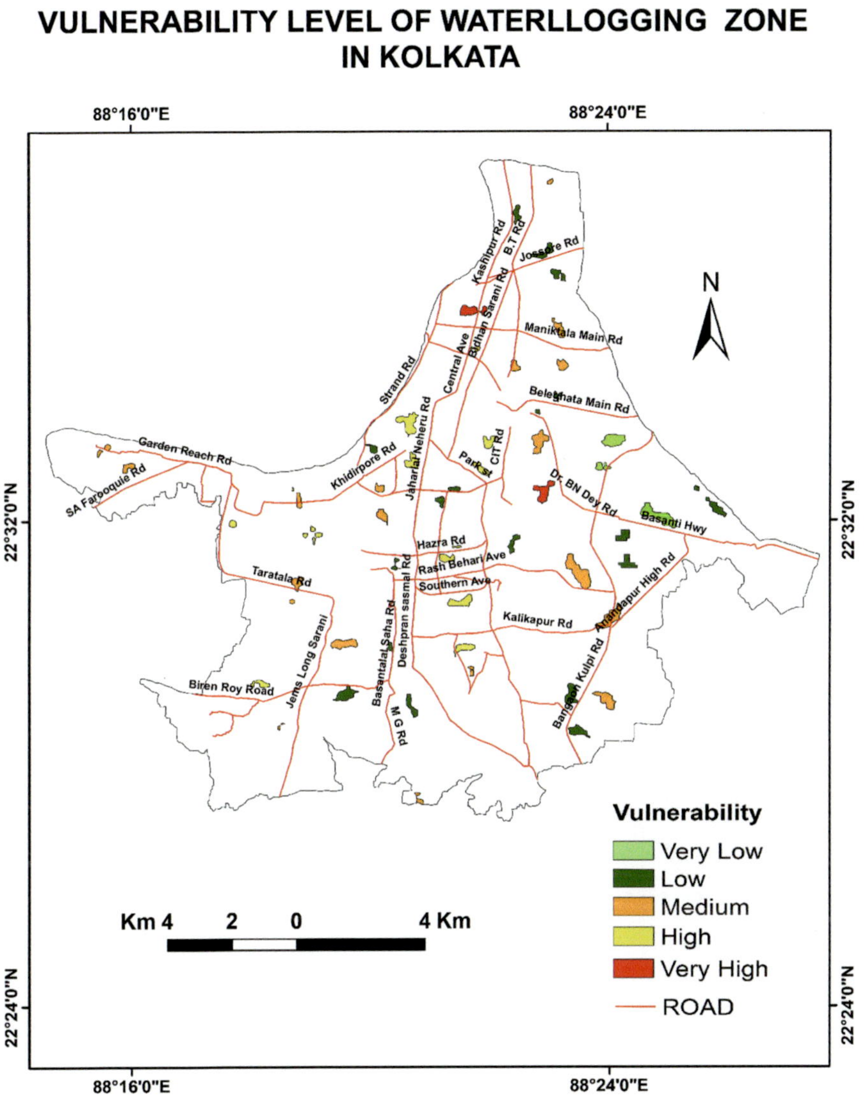

Fig. 8 Vulnerable zone of waterlogging in Kolkata

Reach Road come under the moderately vulnerable zone which means three hours to six hours waterlogged road during the rainy season. Jawaharlal Nehru Rd, Park Street, CIT Rd, and Central Ave are experiencing 6 h to one-day water logging conditions occasionally. Dr BN Dey Rd, Cossipur Rd facing waterlogged conditions for more than one day due to poor drainage conditions. After calculation, it is found that the overall accuracy of waterlogging validation is 71.70% and the Kappa coefficient is 0.616 (substantial agreement) (Table 4). That means the vulnerability level of the

Table 3 Waterlogging susceptibility zone assessment

Name of pour point	Nearest pumping station	Elevation in m (pour point to pumping station)	Rank	Distance in m (pour point to pumping station)	Rank	Distance in m (pour point to canal)	Rank	Distance in m (pumping station to canal)	Rank	Total rank
Block-B	Beer Para	1	3	1526.78	2	1850.02	4	43.51	1	10
Cossipore	Ps 2	– 2	2	769.41	1	661.48	3	93.23	2	8
Tala	Begachia	– 6	2	822.19	1	884.85	3	62.59	2	8
Belgachia	Begachia	– 2	2	936.19	1	272.91	2	62.59	2	7
Girish park	Hririkesh Park	2	4	1281.06	2	1311.24	4	821.04	4	14
Bagmari	Maniktala	– 7	1	1005.47	2	716.73	3	265.99	3	9
College Street	Hririkesh Park	– 2	2	834.9	1	1583.97	4	821.04	4	11
Rajabazar	Hririkesh Park	– 1	3	923.59	1	257.66	2	821.04	4	10
Kankurgachi	Maniktala	– 3	2	1482.28	2	924.62	3	265.99	3	10
Tangra	Palmer Bridge	– 1	3	231.57	1	18.56	1	145.07	2	7
Taltala	Palmer Bridge	0	3	1571.65	2	1707.16	4	145.07	2	11
Fort Willam Cantten	Palmer Bridge	– 2	2	3695.94	4	609.61	3	145.07	2	11
Garer Math	Mominpore	– 5	2	2941.43	3	1678.21	4	200.77	2	11
Beniapukur	Ballygunge	2	4	1623.27	2	1632.97	4	37.64	1	11
Seal Lane	Palmer Bridge	– 4	2	1091.74	2	831.12	3	145.07	2	9
Tangra	Pagla Danga	– 4	2	339.75	1	71.69	1	16.42	1	5
Tangra	Pagla Danga	– 2	2	838.89	1	41.58	1	16.42	1	5
Dhapa	Chowbaga Old	– 1	3	1610.23	2	277.66	2	23.86	1	8

(continued)

Table 3 (continued)

Name of pour point	Nearest pumping station	Elevation in m (pour point to pumping station)	Rank	Distance in m (pour point to pumping station)	Rank	Distance in m (pour point to canal)	Rank	Distance in m (pumping station to canal)	Rank	Total rank
West Chowbaga	Chowbhaga Old	– 2	2	286.27	1	234.31	2	23.86	1	6
Purbachal Kalitala	Chowbhaga 1	– 1	3	3491.47	4	176.49	2	34.3	1	10
Shantipally	Chowbhaga 1	0	3	3014.74	4	324.27	2	34.3	1	10
Gulshan Colony	Chowbhaga 1	– 2	2	504	1	14.48	1	34.3	1	5
Ballygunge	Ballygunge	– 1	3	891.61	1	985.27	3	37.64	1	8
Gobra	Topsia	– 1	3	634.66	1	423.45	2	157.31	2	8
Sreepally	Kalighat	4	4	2227.07	3	1564.63	4	111.54	2	13
Alipore	Mominpore	– 2	2	1019.5	2	35.04	1	200.77	2	“7
Maidan, Hastings	Mominpore	– 4	2	2021.14	3	406.8	2	200.77	2	9
Fort William	Mominpore	– 1	3	2704.42	3	427.32	2	200.77	2	10
Mominpore	Mominpore	0	3	467.78	1	336.36	2	200.77	2	8
Garden Reach	Nimakmahal	– 5	2	1381.54	2	1093.5	3	837.66	4	11
Metiabruz	Dhankheti Ls	0	3	2222.71	3	810.04	3	12.94	1	10
Cpt Colony	Mominpore	1	3	1988.2	2	354.91	2	200.77	2	9
Behala	Ls 2	– 2	2	1286.53	2	2106.15	4	672.48	4	12
Indrajit Pally	Ls 3	– 2	2	1905.51	2	1784.91	4	12.38	1	9

(continued)

Table 3 (continued)

Name of pour point	Nearest pumping station	Elevation in m (pour point to pumping station)	Rank	Distance in m (pour point to pumping station)	Rank	Distance in m (pour point to canal)	Rank	Distance in m (pumping station to canal)	Rank	Total rank
Sukanta Pally, Barisha	Ls 3	− 7	1	2373.35	3	931.23	3	12.38	1	8
Silpara	Ls 2	− 2	2	953.5	1	994.61	3	672.48	4	10
Banerjee Para	Ls 5	0	3	2147.84	2	16.5	1	0.12	1	7
Golf Green	Ls-2b	− 1	3	1804.38	3	1181.84	3	12.44	1	10
Bikramgarh	Jodhpur Park	− 5	2	1242.5	2	1521.15	4	1310.11	4	12
New Alipore	Ls 3	− 2	2	1162.1	2	309.25	2	12.38	1	7
Kalighat	Kalighat	− 2	2	537.39	1	185.08	2	111.54	2	7
Hazra Road	Chetla Lock	− 3	2	1800.45	2	1733.33	4	1256.16	4	12
Santoshpur	Ls-1a	− 9	1	2464.29	3	245.18	2	49.22	1	7
Patuli	Kamdahari	− 7	1	1922.52	2	439.89	2	61.86	2	7
Adarsha Nagar	Chowbhaga 1	− 2	2	1133.12	2	437.29	2	34.3	1	7
Rabindra Sarabar	Jodhpur Park	− 4	2	642.72	1	1939.15	4	1310.11	4	11
Badartala	Dhankheti Ls	− 1	2	5029.16	4	383.29	2	12.94	1	9
Khanaberia	Chowbhaga Old	− 3	2	1616.28	2	131.46	2	23.86	1	7
Paschim Putiary	KestopukurMps	− 5	2	434.41	1	523.85	2	765.26	4	9
Purba Putiary	KestopukurMps	− 4	2	1424.84	2	384.17	2	765.26	4	10
Beleghata	Palmer Bridge	− 1	2	883.29	1	373.87	2	145.07	2	7

(continued)

Table 3 (continued)

Name of pour point	Nearest pumping station	Elevation in m (pour point to pumping station)	Rank	Distance in m (pour point to pumping station)	Rank	Distance in m (pour point to canal)	Rank	Distance in m (pumping station to canal)	Rank	Total rank
Garden Reach	Nimakmahal	− 4	2	648.86	1	644.38	3	837.66	4	10
ChakGaria	Kamdahari	− 8	1	3260.93	4	200	2	61.86	2	9
	Max	4		5029.16		2106.15		1310.11		14
	Min	− 9		231.57		14.48		0.12		5

Source Researcher's computation

Table 4 Validation of waterlogging using Kappa coefficient

Accuracy level measurement						
Classes	1 (very low)	2 (low)	3 (medium)	4 (high)	5 (very high)	Total (user)
Bellow 1 h	3	0	1	1	0	5
1–3 h	0	13	2	0	0	15
3–6 h	0	3	12	1	0	16
6 h–1 day	0	2	3	8	0	13
Above 1 days	0	0	0	2	2	4
Total (producer)	3	18	18	12	2	53

Source Researcher's computation

water logging zone in Kolkata on the map and the perception of respondents on waterlogging do not match in some particular areas.

6 Conclusion

Water logging is a prime labyrinth in Kolkata and has a bad effect on the locals. The evolution of Calcutta is very interesting both physically and culturally. The physical setup here like active clay soil, climate, poor drainage and vegetation on the other hand social aspects like urbanisation, population density, road density, and condition of unorganised waste dumping are accountable for this situation. The said chapter has identified the water logging zone, of that area, the waterlogged road and the people (mainly in the slum people) who are affected in various ways. Kolkata Municipal Corporation (KMC) should pay attention to cleaning the sewage in the water logging area at least once a week and maintaining the drainages and sewerage system. The pumping capacity of the pump station should be increased during monsoon. In addition, pumping stations should be placed in low-elevation areas where water is logged so that it takes less time and energy to pump out the stormwater. However due to the lack of interest of KMC, proper maintenance is not done and this terrible situation is seen in some places (Sealdah, Khidirpur, Gobra). Again, there are some places (Bhawanipur, some parts of Tangra, Mayo Road) where new high drains have been constructed or are being constructed, regular drain cleaning results in less or no water logging conditions.

References

Adhikary B, Rath KC, Das N (2017) Mapping of water logging problems in urban area: a case study at Ghatal Municipality of West Bengal. J Indian Cartography Assoc (INCA) 37:378–386

Aktar A, Mohit SA, Chowdhury MAH (2017) Predicting urban strom water-logging for Chittagong city in Bangladesh. Int J Sustain Built Environ 6(1):238–249. https://doi.org/10.1016/j.ijsbe.2017.01.005

Alom MM, Khan MZH (2014) Environmental and social impact due to urban drainage problems in Dhaka City, Bangladesh. Int J Eng Adv Technol 3(6):128–132

Arge L, Chase JS, Halpin P, Toma L, VitterJS, Urban D, Wickremesinghe R (2003) Efficient flow computation on massive grid terrain datasets. Geoinformatica 7(4):283–313. http://dx.doi.org/10.1023/A:1025526421410

Bai R, Li T, Huyang Y, Li J, Wang G (2015) An efficient and comprehensive method for drainage network extraction from DEM with billions of pixels using a size-balanced binary search tree. Geomorphol 238:5667

Banerjee S (2018) An analysis of urban flooding scenario in the city of Kolkata, West Bengal. Int J Innov Knowl Concept 6(5):1–9. 11.25835/IJIK-24

Barnes R, Lehman C, Mulla D (2014) An efficient assignment of drainage direction over flat surfaces in raster digital elevation models. Comput Geosci 62:128135

Bhaumik K, Das S (2021) An approach to identify urban waterlogging on a deltaic plain using ArcGIS on CHD based flow accumulation models. Research Square 8–20. https://doi.org/10.21203/rs.3.rs-230413/v1

Census of India (2011) Primary census abstract, census of India. The Government of India, Registrar General and Census Commissioner of India, Ministry of Home Affairs, New Delhi, India

Dhiman R, Vishnuradhan R, Eidho TI, Inamdar A (2019) Flood risk and adaptation in Indian coastal cities: recent scenarios. Appl Water Sci 9(1). https://doi.org/10.1007/s13201-018-0881-9

Do HT, Limet S, Melin E (2011) Parallel computing flow accumulation in large digital elevation models. Procedia Comput Sci 4:2277–2286. https://doi.org/10.1016/j.procs.2011.04.248

Dutta S, Sarkar S (2020) Canal-oriented development: integrating an urban canal front with the city 13. https://doi.org/10.6092/2281-4574/6954

Garbrecht J, Martz LW (1997) The assignment of drainage direction over flat surfaces in raster digital elevation models. J Hydrol 193:204–213. http://dx.doi.org/10.1016/S0022-1694(96)03138-1

Hassan MM, Nazem MNI (2016) Examination of land use/land cover changes, urban growth dynamics, and environmental sustainability in Chittagong city, Bangladesh. Environ Dev Sustain 18(3):697–716. https://doi.org/10.1007/s10668-015-9672-8

Joarder R (2016) A study on causes of waterlogging in Southwestern suburban area of Kolkata, India. Knowl Multi e-J 1(1)

John B, Das S (2020) Identification of risk zone area of declining piezometric level in the urbanized regions around the city of Kolkata based on ground investigation and GIS techniques. Groundwater Sustain Dev 11:100354. https://doi.org/10.1016/j.gsd.2020.100354

Kolkata district—2011–2023 (n.d.) Sensus2011.co.in. https://www.census2011.co.in/census/district/16-kolkata.html#:~:text=The%20initial%20provisional%20data%20released,km

Kolkata population 2022/2023 (n.d.) Population.com https://www.populationu.com/cities/kolkata-population#history

Kolkata population 2023 (n.d.) World population review.com. https://worldpopulationreview.com/world-cities/kolkata-population

Ma B, Wu Z, Wang H, Guo Y (2020) Study on the classification of urban waterlogging rainstorm and rainfall thresholds in cities lacking actual data. Water 12(12):3328. https://doi.org/10.3390/w12123328

McDonald GK, Dean G (1996) Effect of waterlogging on the severity of disease caused by Mycosphaerella pinodes in peas (*Pisum Sativum* L.). Aust J Exp Agric 36(2):219–222. https://doi.org/10.1071/EA9960219

Mitra D, Banerji S (2016) Pluvial flooding in Kolkata: a case study of the major streets in Northern, Southern and Central Kolkata. Int J Curr Res Aca Rev 4(5):51–65. https://doi.org/10.20546/ijcrar.2016.405.007

Mukhopadhyay A (2004) Geomorphology of Kolkata municipal corporation KMC its impact 7 on urban functions. PhD thesis, Department of Geography, University of Calcutta, 762 India. http://hdl.handle.net/10603/159507

Nardi F, Grimaldi S, Santini M, Petroselli A, Ubertini L (2008) Hydrogeomorphic properties of simulated drainage patterns using digital elevation models. The flat area issue. Hydrol Sci J 53(6):11761193

NASA (1999) PIA01844: space radar image of Calcutta, West Bengal, India. Photojournal. Retrieved on: 15 Jan 2012. Accessed from: http://photojournal.jpl.nasa.gov/catalog/PIA01844

Ning YF, Dong WY, Lin LS, Zhang Q (2017) Analyzing the causes of urban waterlogging and sponge city technology in China. IOP Conf Ser: Earth Environ Sci 59:12047

Nowak DJ, Greenfield EJ (2020) The increase of impervious cover and decrease of tree cover within urban areas globally (2012–2017). Urban For Urban Greening 49:126638. https://doi.org/10.1016/j.ufug.2020.126638

Nuissl H, Haase D, Lanzendorf M, Wittmer H (2009) Environmental impact assessment of urban land use transitions—a context-sensitive approach. Land Use Policy 26(2):414–424

O'callaghan JF, Mark DM (1984) The extraction of drainage networks from digital elevation data. Comput Vis, Graph Image Proc 28:328–344. http://dx.doi.org/10.1016/S0734-189X(84)80011-0

Panchal R, Bhavsar SN, Parmar AS, Raval KC, Patel DKD (2019) An overview of water logging problems in an urban area of Nadiad city and probable analytical solutions. Int J Pure Appl Res Eng Technol 7(6):80–88

Paul A (2009) Assessment of waterlogging in C M C and its social impact. PhD thesis, Department of Geography, University of Calcutta, India. http://hdl.handle.net/10603/154829

Quan R, Liu M, Lu M et al (2010) Waterlogging risk assessment based on land use/cover change: a case study in Pudong New Area, Shanghai. Environ Earth Sci 61(6):1113–1121. https://doi.org/10.1007/s12665-009-0431-8

Rahmati O, Dharabi H, Panahi M, Kalantari Z et al (2020) Development of novel hybridized models for urban flood susceptibility mapping. Sci Rep 10(1):1–19. https://doi.org/10.1038/s41598-020-69703-7

Radford KG, James P (2013) Changes in the value of ecosystem services along a rural—urban gradient: a case study of Greater Manchester, UK. Landscape Urban Plann 109(1):117–127. https://doi.org/10.1016/j.landurbplan.20

Rahman MM, Akteruzzaman AKM, Khan MMH, Jobber A, Rahman MM (2009) Analysis of water logging problem and its environmental effects using GIS approaches in Khulna city of Bangladesh. J Soc Res Dev 6(2):572–577

Roy R, Dhali KMd (2016) Seasonal water logging problem in a mega city: a study of Kolkata, India. J Res Humanit Soc Sci 4(4):1–9. www.questjournals.org

Sajikumar N, Remya RS (2015) Impact of land cover and land use change on runoff characteristics. Environ Manage 161:460-468. https://doi.org/10.1016/j.jenvman.2014.12.041

Sajjad H (2014) Living standards and health problems of lesser fortunate slum dwellers: evidence from an Indian city. Int J Environ Protect Policy 2(2):54–63. https://doi.org/10.11648/j.ijepp.20140202.13

Shao Z, Fu H, Li D, Altan O, Cheng T (2019) Remote sensing monitoring of multi-scale watersheds impermeability for urban hydrological evaluation. Rem Sens Environ 232:111338. https://doi.org/10.1016/j.rse.2019.111338

Shukla A, Jain K (2019) Critical analysis of rural-urban transitions and transformations in Lucknow city, India. Remote Sens Appl Soc Environ 13:445–456. https://doi.org/10.1016/j.rsase.2019.01.001

Singh RB, Singh S (2011) Rapid urbanization and induced flood risk in Noida, India. Asian Geogr 28(2):147–169. https://doi.org/10.1080/10225706.2011.629417

Singh SK, Pandey AC (2013) Geomorphology and the controls of geohydrology on waterlogging in Gangetic plains, North Bihar, India. Environ Earth Sci 71:1561–1579. https://doi.org/10.1007/s12665-013-2562-1

Subrina D, Chowdhury FK (2018) Urban dynamics: an undervalued issue for water logging disaster risk management in case of Dhaka city. Procedia Eng 212:801–808. https://doi.org/10.1016/j.proeng.2018.01.103

Tam VT, Nga TTV (2018) Assessment of urbanization impact on groundwater resources in Hanoi, Vietnam. Environ Manage 227:107–116. https://doi.org/10.1016/j.jenvman.2018.08.087

What is climate change? (n.d.) NASA Climate Kids. https://climatekids.nasa.gov

Zhou G, Dong W, Wei H (2019) A fast and simple algorithm for calculating flow accumulation matrices from raster digital elevation models. Int Cartographic Assoc (13):317–326. https://doi.org/10.5194/ica-abs-1-434-2019

Vulnerability of Reservoir to Sequestrated Solutes from Feeder Streams: A Case Study of Opa in Southwestern Nigeria

Alexandra Simi Ogbole and Adebayo Oluwole Eludoyin

Abstract Reservoirs are water sources in many communities and therefore deserve to be studied. This chapter is therefore focused on a reservoir as a hydrologic unit of study for an evaluation of spatiotemporal characteristics of solute at a typical reservoir in southwestern Nigeria. The reservoir feeder streams and spillway were instrumented with staff gauges and monitored for variations in discharge via velocity—cross-section area method for the 2017–2018 hydrological year. Water samples were collected biweekly, and daytime storm events sampled 54 times were obtained to determine total dissolved solids, pH, and electrical conductivity. Results showed that the catchment is characterized by 70–440 μS/cm of specific conductivity, 6.5–8.3 pH units, and 9.1–1344.5 tonnes of TDS. Total inputs of TDS from the tributaries into the reservoir exceeded the exported by 760.75 tonnes, and 47.2% of the total dissolved solids were sequestrated in the reservoir. The study concluded that reservoirs are vulnerable to the effects of land use/cover changes in the parts of the region, and therefore recommends the need to prioritize ecological restoration of contaminated reservoirs/streams as an important part of water quality control.

Keywords Solute yield · Sequestration · Opa reservoir · Feeder streams · Land use/cover · Reservoir

1 Introduction

Recent studies have raised concerns for water quality in sub-Saharan Africa, which with fluctuating water supply has been implicated in increasing cases of waterborne and other water-related diseases in the sub-region (Eludoyin and Olanrewaju 2020;

A. S. Ogbole
Department of Geography, Plateau State University, Bokkos, Nigeria

A. O. Eludoyin (✉)
Department of Geography, Obafemi Awolowo University, Ile-Ife, Nigeria
e-mail: oaeludoyin@yahoo.com

S. C. Pal and U. Chatterjee (eds.), *Surface, Sub-Surface Hydrology and Management*, Springer Geography, https://doi.org/10.1007/978-3-031-62376-9_15

Gebrechorkos et al. 2022). Although there are many sources of domestic water, waterworks—which often involve treating dammed surface water in reservoirs for quality control and supplying sufficiently treated water for community consumption—is an important method of providing sustainable and quality water at community levels (Onwusu et al. 2022; Nshemereirwe et al. 2022). Many reservoirs of varying sizes, usefulness, and importance exist in the West African region and are all vulnerable to the influence of land use/cover change, human activities, natural disasters, and climate extreme or change (Atulley et al. 2022; Gebrechorkos et al. 2022). Waterworks are known to supply potable water to communities, and this is an important community water system in Nigeria, probably until the 1990s when the water industry in the country began to be severely affected by political problems and poor governance, causing poor functioning and abandonment of pipe-borne water supply in many communities (Akoni 2019; Adeniran and Daniell 2021). Consequently, residents of many urban areas depend on supplies from rainwater harvesting and groundwater resources, whose quality and supply may vary with the socio-economic and demographic ratings of the residents (Oyeku and Eludoyin 2010; Eludoyin et al. 2021; Eludoyin and Olanrewaju 2020), probably because there is no standard directive and recommendations on private water sourcing and use for Nigerian residents.

Waterworks are generally known to comprise a reservoir system that is typically built across streams and valleys where they alter flow velocities, and consequently promote dissolved solids sequestration that may be adsorbed to sediment particles or sequestered on the reservoir bed (Newbold 1992; Troitsky et al. 2019; Pagnozzi et al. 2020; Yin et al. 2022). Dissolved solids or solutes are defined as dissolved ionic and non-ionic matter that include organic matter that has become an integral part of the water moving through the stream channel (van Niekerk et al. 2014; Akaahan et al. 2017). The dissolved solids are an important water quality parameter and a measure of chemical denudation, wet and dry deposition, and anthropogenic leachates in catchment hydrology (Ogunkoya and Jeje 1987; Jia et al. 2021; Burns et al. 2021; Siddiqui and Pandey 2022; Bufe et al. 2022). Reservoir contamination from dissolved solids is of concern because it impairs water quality (Carpenter et al. 1998) and increases the cost of water treatment (Dodds and Whiles 2004; Walling 2005). Excess dissolved solids in reservoirs may also accelerate eutrophication and result in algal blooms (May et al. 2020; Bakker 2021; Vantarakis 2021; Abell et al. 2022), causing death and decomposition of algae population, thus reducing freshwater eco-biodiversity (Imevbore et al. 1988; Nikolai and Dzialowski 2014). Chapman (1998) documents many physical, chemical, and biological parameters for water quality assessment, which studies have revealed to be expensive (Ahmed et al. 2019) requiring large budgets which may be a problem for many researchers, especially those from developing countries, including Nigeria. Consequently, many studies have adopted the use of composite parameters, especially pH, conductivity and total dissolved solids (TDS) have provided low-cost, rapid assessment and reliable advantage for scientists from many developing countries, including those in sub-Saharan Africa (Bonvin et al. 2013; Eludoyin 2013; Olsen and Cormier 2019; Shivayogimath et al. 2012; Patil et al. 2022; Kumar et al. 2022; Aleksander-Kwaterczak and Plenzler 2019; Saalidong et al. 2022).

The Opa reservoir in Ile-Ife, Nigeria is a typical medium-sized reservoir with a waterworks scheme that supplies water to the Obafemi Awolowo University communities that include more than 35,000 resident students and staff who live in the University hostels and staff quarters (Babatimehin et al. 2020; Esan et al. 2020). Eludoyin et al (2004) reported that the reservoir was receiving stream-borne market effluents through major tributaries and recommended the investigation of other tributaries of Opa River, which drain effluents from other parts of the University town. The present chapter is however focused on the investigation of the solute content of the reservoir from a system approach, such that all the prominent tributaries of the river are investigated. Existing knowledge in the study area and many other reservoirs in sub-Saharan Africa has been limited to the effects rather than the processes (Ogunkoya 2013). Also, existing studies on the study area (e.g., Eludoyin et al. 2004; Akindele and Adeniyi 2013; Olabanji and Oluyemi 2014; Adesakin et al. 2017) can be at best described to exist in piecemeal, lacking catchment-scale input–output approach of a system-focused analysis that has been advanced for catchment studies (Sivapalan et al. 2003; De la Fuente et al. 2021). Specific objectives are therefore to assess temporal trends and changes in selected composite chemical parameters (pH, conductivity, and TDS) at the input and output (at the spillway, just before water treatment) sources.

2 Materials and Methods

2.1 Study Area

The study area, Opa reservoir is within the Opa catchment, which covers an area of 110 km^2 in Ile-Ife, southwest Nigeria (Fig. 1). The reservoir was built on the Opa River, with tributaries that include Rivers Amafa, Ibu, Esimirin, Opa, Ose, and Olojumeta, for water supply needs in 1978. The tributaries have varying flow characteristics, which are generally undocumented; the reservoir is nonetheless documented as occupying a 66-ha land area, with 0.80 km at the widest point and 2.5 km long when it was first built (Ogunkoya 2013; Ogunkoya and Ogbole 2024).

The catchment's archetype vegetation was the lowland tropical rainforest, with several canopies and lianas (Adediji 2005). However, arable and cash crops like yam, cocoyam, plantain, cassava, maize, cocoa, and kolanut have encroached on the watershed, severely disrupting the archetypal vegetation. Rapid urbanization has occurred around the reservoir, resulting in a poorly developed urban environment (Ogunkoya and Aboyeji 2014). The catchment area is characterized by rainfall seasonality (wet and dry seasons occur in April–November and December–March, respectively), bimodal rainfall distribution with peaks in July–August–September, and a little dry season in August (Odekunle and Adejuwon 2007). Mean annual rainfall, temperature, and relative humidity vary as 1250–1500 mm, 20–33.6 °C, and 75.8–86%, respectively (Eludoyin and Adewole 2020). The catchment (Opa) is underlain by rocks

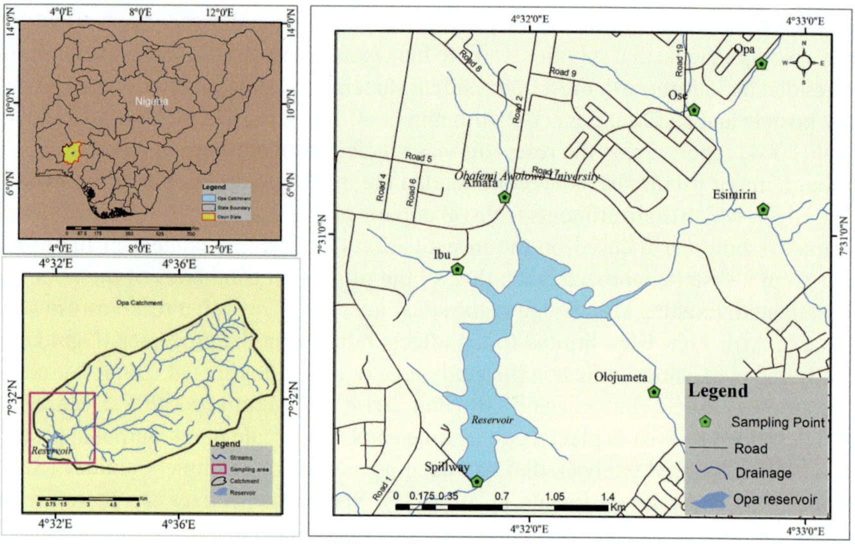

Fig. 1 The study area, Opa catchment in Osun State, Nigeria

from the Precambrian Basement complex, comprising of schists, granite-gneiss, and amphibolites, with undifferentiated schists and pegmatites covering about half of the catchment and granite-gneiss forming the prominent inselbergs (Adejuwon and Jeje 1975).

The six tributaries of the Opa reservoir (Esimirin, Opa, Olojumeta, Ibu, Amafa, and Ose) vary in terms of catchment area and characteristics. Esimirin stream, which covers 33.55 km^2 drains the main urban market (Oja-titun), and major commercial and residential areas in Ile-Ife; including Moore, Idita, Igboya, Sabo and Obafemi Awolowo University Teaching Hospital. The areas are known to generate substantial household and food waste materials that are generally assumed to be flushed into the reservoir, especially in the early wet season (see Eludoyin et al. 2004; Akindele and Adeniyi 2013; Olabanji et al. 2015; Ogunkoya and Ogbole 2024). River Opa flows into the reservoir from Osu—a nearby medium-sized community that is characterized by household, commercial, farming, and small-scale artisanal mining activities. Its area is 35.54 km^2. The catchment areas of Rivers Amafa, Ose, and Ibu are covered by the low-density senior staff housing of the University and are 2.17, 2.19, and 0.82 km^2, respectively. Olojumeta catchment is characterized by settlements with urban activities and the part closer to the reservoir, is essentially covered by peasant farms planted with a variety of tropical crops. The area covered by river Olojumeta is 1.69 km^2.

2.2 Data

The six tributaries of Opa reservoir and the outlet (spillway) of the reservoir were equipped with well-labeled staff gauges, each, to monitor the water level twice daily between May 1, 2017, and April 30, 2018, which makes a complete hydrological year in the southwestern Nigeria, based on rainfall regime (Odekunle and Adejuwon 2007). A Current meter (OTT Kempten) was used to determine the river velocity at the gauge station as described by Camnasio and Orsi (2011). Following this procedure, specific equations for the specified Propeller types 5 and 6 (based on the manufacturer's advice for streams of headwater streams) were adopted for periodic calculation of discharge for the period for which measurements were taken in Eqs. 1 and 2.

$$V = 0.0545n + 0.047 \tag{1}$$

$$V = 0.1035n + 0.018 \tag{2}$$

where

V velocity or displacement per unit time;
n number of revolutions completed by the propeller.

Consequently, the discharge-stage relationship was established using a logarithmic linear equation, and this was used to predict the period for which the current meter was not used and the period of floods when it was not safe to measure (Buchanan and Somers 1969) for further information on discharge measurements using the current meter. Furthermore, water samples were obtained in 2L polyethylene bottles following a depth-surface integrated approach (Gitto et al. 2017) for on-site determination of water pH, Conductivity, and TDS Hanna HI 9813-6 Portable pH/EC/TDS/°C meter was first calibrated using distilled water as described in the instruments' manual before it was used to determine the values of respective water parameters in the sampled stream water. Non-event water samples were collected twice in a week and 54 event water samples were obtained within the study period.

Discharge-TDS relationship was determined using a log-transformed linear regression relationship while the three water quality parameters (pH, conductivity, and TDS) were compared using the stepwise regression method. TDS/Solute yields were determined as the product of total flow and TDS per unit area of a specific catchment as described (Alexandrov et al. 2009; Aulenbach et al. 2016). The values of TDS, conductivity, and pH were compared in the study for a composite explanation of their variability using the simple bivariate (when only two of the variables were compared), stepwise regression (when the three variables were compared) as well as wavelet and modeled using the kriging function as described by Cressie (1988).

3 Results and Discussion

3.1 Characteristics of the Streams Feeding the Opa Reservoir

3.1.1 Discharge and Selected Solute Parameters

Table 1 shows the discharge values of the tributaries of the Opa reservoir. Discharge in the rainy season was more than it was in the dry season in the catchment. Discharge increased in the order of Esinmirin > Opa > Amafa > Olojumeta > Ose > Ibu, and variability was high in all the seasons, as expected due to seasonality. Peak discharge occurred in the pattern of the average discharge except that the Ibu stream experienced more discharge at peak than Ose. Analysis of the relationship between discharge and basin area (Eq. 3) suggests that the discharge is well explained by the catchment area ($R^2 = 0.7$).

$$y = 3.2949x + 1.6369\left(R^2 = 0.7\right) \tag{3}$$

where

y discharge,
x basin area (km^2).

In terms of the solute characteristics, although the mean pH across the selected stream water does not suggest extreme acidity or alkalinity, River Esinmirin exhibited the highest concentration of total dissolved solids and measures of conductivity. The concentration of dissolved solids in Opa River, which was next to Esinmirin was 88.7% lesser. Conductivity at Esinmirin also almost tripled the value at the other streams and pH values generally ranged between 6.5 and 8.3 units; conductivity ranged between 70 and 440 μS/cm in the rivers.

Table 1 Discharge and solute concentrations at the streams feeding the Opa reservoir

Sub-catchment	Discharge			TDS		Conductivity (μS/cm)		pH (no unit)	
	Mean	Peak	CV	Mean	CV	Mean	CV	Mean	CV
Amafa	5.8	120.2	104.9	1.5	106.2	104.9	25.7	7.3	4.08
Opa	63.0	1422.4	102.1	13.6	50.0	99.0	7.8	7.2	3.34
Esinmirin	171.5	1625.9	96.7	112.0	101.7	305.4	24.1	7.4	4.31
Olojumeta	12.0	114.0	120.9	5.4	80.9	186.9	37.4	7.3	3.99
Ose	4.1	66.3	100.4	1.2	100.2	162.8	25.2	7.3	4.73
Ibu	3.7	93.0	98.7	0.8	113.9	104.4	21.4	7.3	4.41

3.1.2 Relationships Among Selected Solute Characteristics and Solute-Discharge Relationship

Results of the analysis of wavelet patterns in the characteristics of the selected parameters showed different patterns of variability; the pH showed more uniform periodicity at the mid-scale than either the TDS or conductivity (Fig. 2). The red band, which indicates high disturbance in signals was conspicuously differently patterned in TDS and conductivity, suggesting an impression of contrasts. But the blue band which indicates low signals appears similarly partitioned in the lower band of the wavelet. The linear relationship between TDS and conductivity, and with pH in the streams revealed a fairly strong link with or without the consideration of pH (compares Eqs. 4a and 4b).

$$y = 138.2 + 0.997x_1\left(R^2 = 0.492;\ F = 67.7, p = 0.001\right) \tag{4a}$$

$$y = -237.61 + 0.65x_1 + 0.178x_2\left(R^2 = 0.52;\ F = 37.4, p = 0.001\right) \tag{4b}$$

y conductivity (μS/cm),
x_1 TDS (tonnes),
x_2 pH (no unit).

In general terms, the relationship between TDS and conductivity appears consistent and varies with changes in pH, especially within the small pH difference observed in the study (Fig. 3). When compared with discharge, the results revealed a negative and weak relationship with TDS, suggesting a relatively low but significant impact of discharge on TDS distribution in the study area (Table 2).

3.1.3 Solute and Discharge Characteristics of the Opa Reservoir at Spillway

Results of the analysis of the selected solute variables at the Opa reservoir showed pH, TDS, and conductivity values at the reservoir as 7.3 ± 0.3 no unit, 71.1 ± 73.7 tonnes, and 199.6 ± 56.1 μS/cm, respectively. When compared with the feeder streams, the solute characteristics of the reservoir were closer to that of Esinmirin River than others (Fig. 4), probably because they both (Esinmirin River and the reservoir) exhibited higher concentrations of TDS and conductivity than the other sites (Fig. 5). When compared, the monthly distribution of solute yields (in tonnes) across the study period showed expectedly more solute yields in the dry season period (November–March) and early wet season (April–May) in all the stations (Fig. 6). Esinmirin River was the greatest of all the sources of solute yields into the reservoir when compared with other feeder streams. The spillways recorded a total of 852.8 tonnes; 276.7 and 576.1 tonnes in the wet and dry seasons, respectively. The monthly distribution showed that dissolved solids yield was lowest in July and gradually increased in

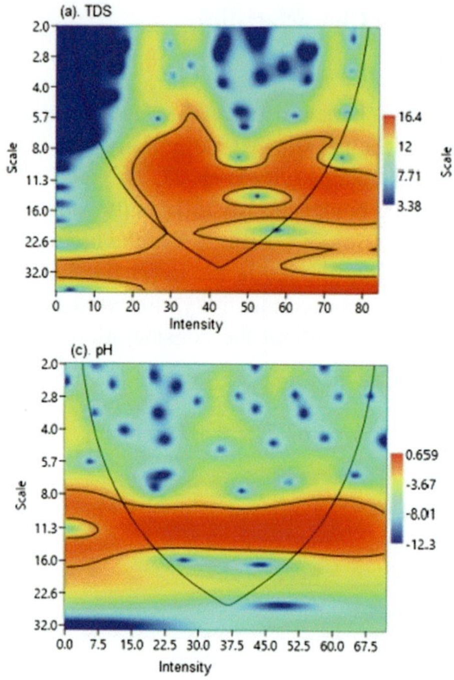

Fig. 2 Wavelet transform of conductivity, total dissolved solids, and pH in streams feeding Opa reservoir, Ile-Ife, Nigeria

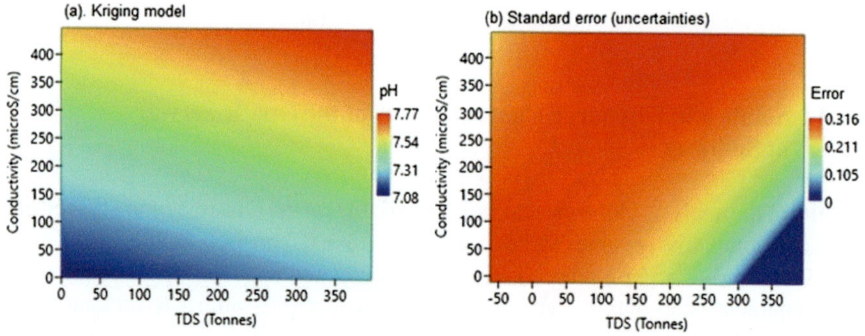

Fig. 3 Kriging-based modeled relationship between conductivity, total dissolved solids, and pH in streams feeding Opa reservoir, Ile-Ife, Nigeria

August before it declined in October for the wet period (Fig. 6). The yield was higher throughout the dry season and peaked in February (218.1 tonnes). The solute yields in Opa were more than those of other feeder streams (Ibu, Ose, and Olojumeta) but these were much lower than what was obtained at Esinmirin and the spillway and

Table 2 Trend characteristics of the relationship between discharge (independent variable) and dissolved solid loads (as dependent variable) across the sub-catchments

Sub-catchment	Linear trend ($a + bx$)	Coefficient of determination (R^2)	Pearson's correlation (r)
Amafa	$1.971 - 0.127x$	0.240	0.490
Opa	$1.974 - 0.149x$	0.269	0.518
Esinmirin	$1.718 - 0.167x$	0.344	0.586
Olojumeta	$2.153 - 0.256x$	0.389	0.624
Ose	$1.783 - 0.122x$	0.433	0.658
Ibu	$1.203 - 0.125x$	0.325	0.570

therefore could be interpreted that only Esinmirin River contributed significantly to the quality of the reservoir.

The low contribution of other tributaries to the dissolved yield of the reservoir can be attributed to the low contribution of subsurface flow to streamflow in the catchments. Also, comparatively lower solute yields in wet season months can be linked to

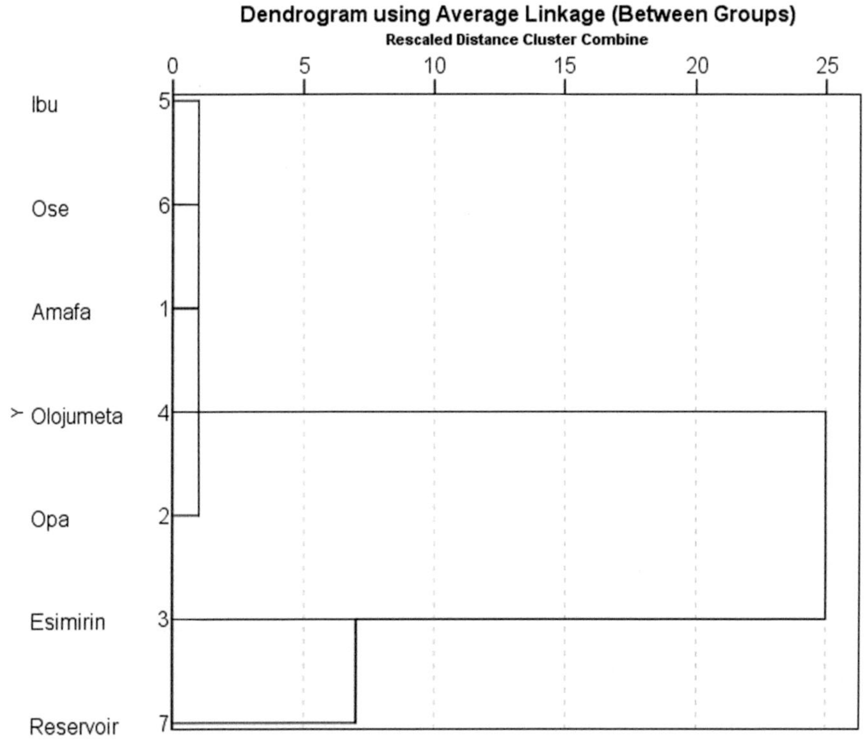

Fig. 4 Hierarchical classification of the Opa river sub-catchments and reservoir based on the concentrations of dissolved solids

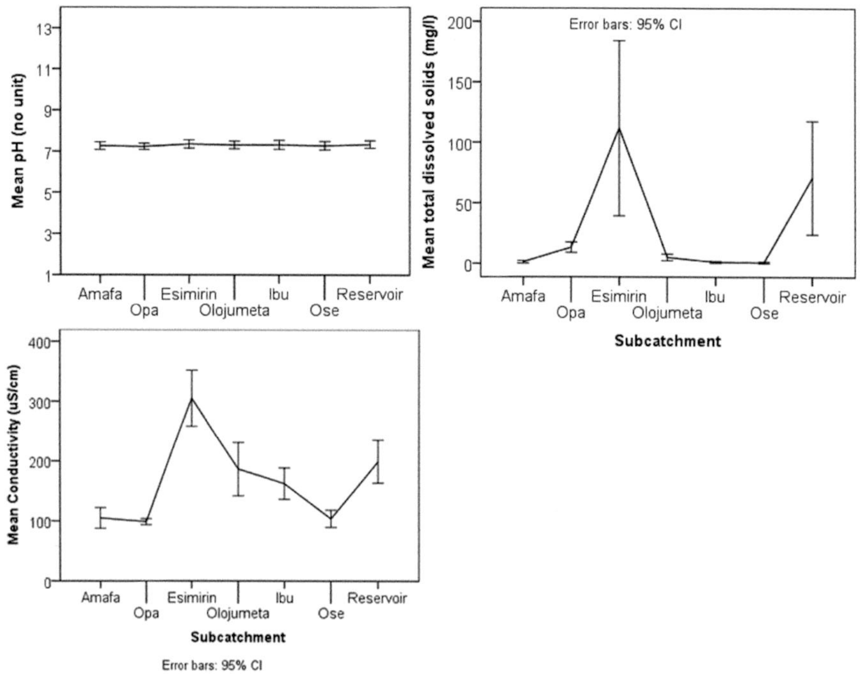

Fig. 5 Mean variations in pH, conductivity, and TDS at selected streams and the Opa reservoir

rapid flow pathways—overland and shallow subsurface, which have short residence time in the catchment and thus lower ability to entrain dissolved matter. Whereas Olojumeta receives wastes from unregulated deposition of domestic effluents and waste into the river channel, disposal of wastes into Ose stream is well regulated, being located in the Senior Staff Quarters of the Obafemi Awolowo University, and both catchments receive limited enrichments from farmlands and commercial activities, unlike Esinmirin and Opa rivers.

3.1.4 Import (Feeder Streams)—Export (Spillways/Reservoir) Comparison

Both monthly and annual solute yields at the feeder streams (Esinmirin, Opa, Amafa, Olojumeta, Ose, and Ibu) were summed to derive the input solute yields and these were compared to what was exported out of the Opa reservoir through the spillway, and compared for excess or deficit assessments. The total solute yield was 1613.6 tonnes (mean $\pm$ standard deviation $=$ 134.5 $\pm$ 126.4 tonnes) at the feeder streams compared to 852.8 tonnes (mean $\pm$ standard deviation $=$ 71.1 $\pm$ 73.6 tonnes), indicating an excess of an annual total 760.8 tonnes of the solute loads were sequestered in the reservoir. The potential for excess solute yield also prevailed

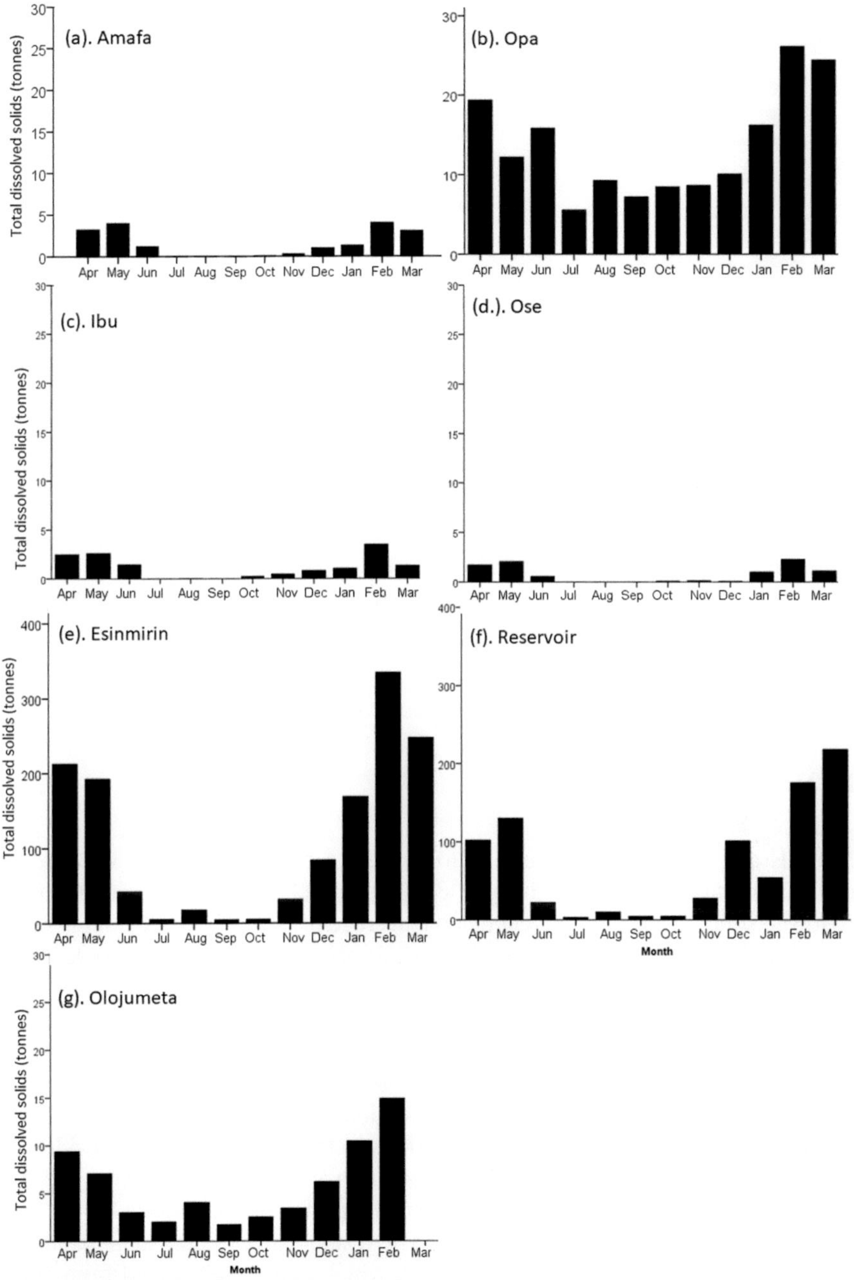

Fig. 6 Monthly variations in the concentrations of dissolved solids in the sub-catchments and reservoir in the study area

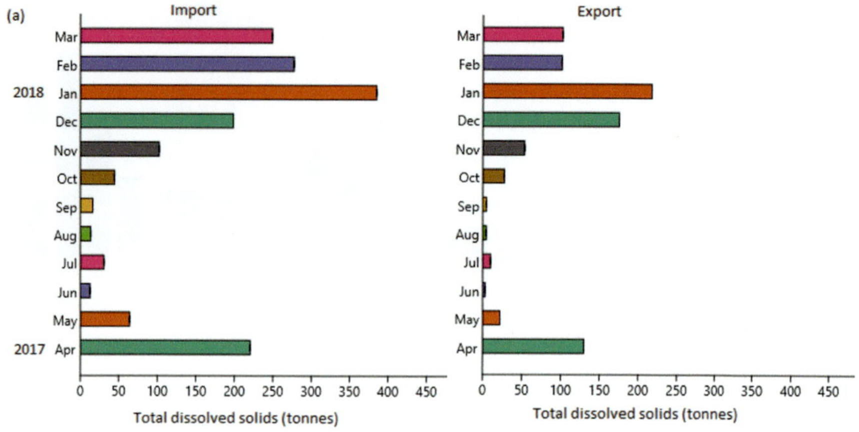

Fig. 7 Monthly dissolved solids import from the feeder streams and export at the Opa reservoir spillway

across different months in the period of study; albeit at varying amounts (Fig. 7). Sequestrated solutes have been linked with severe implications for the eutrophication of reservoirs in different parts of the world (see, e.g., Torres et al. 2007; Ismail and Najib 2011).

4 Discussion

Both pH and conductivity, according to studies (e.g., Dadi et al. 2018; Bearcock et al. 2019; Imran et al. 2020; Zhang et al. 2020) are related to the sum of all ionized dissolved solids. The results reveal that the average pH range of 6.5–8.3 units in the sub-catchments is within the reported (6.5–8.5) pH range for most freshwater lakes and streams in the world (Asante et al. 2008; Keremah et al. 2014; Prusty et al. 2020; Imran et al. 2020) as well as Opa reservoir in Nigeria (Akinbuwa 1988; Akindele and Adeniyi 2013) and Prado Reservoir in Colombia (Guevera 2009). Generally, the pH result shows that Opa catchment water is safe for domestic use when compared with the standard limit guideline (6.5–8.5) for domestic water quality (World Health Organisation 2016). The values of conductivity in the sub-catchments (70–440 μS/cm) however varied; being less than the observation of Ogunkoya and Aboyeji (2014) (661–1201 μS/cm) but slightly higher than that reported by Akinbuwa (1988) (150–309 μS/cm) in the same Opa catchment. The results depict a non-static pattern in the values of conductivity in the catchment, which may have resulted from either seasonal patterns or environmental changes. For example, studies have shown that changes in water conductivity may reflect dry–wet seasonal variations and attendant exhaustion, dilution, and flush effect as well as landuse/cover change (Yu et al. 2016;

Qin et al. 2020; McGrane et al. 2017). The Opa catchment has undergone land use/cover change since the study by Akinbuwa (1988) till now.

Generally, dissolved solids yield in the catchment is low, this may be due to the low subsurface contribution to the streamflow which is manifested by the rapid recession of the streamflow. Effluents from the Oja-titun market and the Obafemi Awolowo University Teaching Hospital Complex (OAUTHC) hospital effluents discharged into a tributary of the Esinmirin River. Also, there is evidence of effluents heavy with dissolved solids from the gold mining sites in the Kosere area of Ife Township, thus accounting for the high dissolved solid load observed in R. Esinmirin. However, the low yield from Ose River is because solute sources are only limited to the rocks and saprolite of the catchment, coupled with atmospheric deposition.

Seasonal variability in the dissolved solid yield varied with discharge in the sub-catchments. All of the investigated sub-catchments recorded relatively lower discharge of dissolved solid load in the rainy seasons than dry season. Dissolved solids are sourced through washing out of accumulated loads during the dry season, especially in the early rains, as reported in Osun and Owena drainage basins (Jeje et al. 1999). The low dissolved solid load in the rainy season, particularly during peak discharges as experienced in all the sub-catchments may be linked with short residence time occasioned by increased saturation in the season (Gwenzi et al. 2017; Clark et al. 2017; Benettin et al. 2020; Botter et al. 2018; Knapp et al. 2020). Also, dilution due to increased inflows into the river channel could result in a decrease in concentration as the rainy season progresses, thus the low dissolved solids concentration (Gwenzi et al. 2017; Clark et al. 2017; Benettin et al. 2020; Botter et al. 2018). A decrease in the concentration with an increase in discharge can also be an indication of a reduction in the supply of dissolved solids or its removal from solution by chemical reaction or solution. This occurs under anaerobic conditions where dissolved solids adsorbed to sediments become reduced thereby enabling their removal from solution. However, under aerobic conditions dissolved solids adsorption to sediment is re-established as the solubility of dissolved solids decreases. The dominant land use activities in the Opa catchment were more related to urban development (Eludoyin and Adewole 2020; Ogunkoya and Ogbole 2024) than agriculture where farm inputs, such as fertilizers or other crop additives may be the main contributor of nutrients in catchments within agricultural (arable) areas, thus accounting for the generally low dissolved solids in the catchment.

The positive budget suggests sequestration of dissolved solids in Opa reservoir; of the total amount of dissolved solid imported into the reservoir, 53% was transported through and exported out of the reservoir. This information may indicate that more dissolved solids are readily discharged through the spillway during flood events, thus showing the relative dominance of exported dissolved solids than are sequestered in the reservoir. The value of the dissolved solids budget of Opa Reservoir was higher than that reported in Bukit Merah Reservoir, Malaysia (Ismail and Najib 2011) and Pampulha Reservoir, Brazil (Torres et al. 2007). These studies attributed the positive dissolved solids budget to the possibility of metabolic activities occurring in the reservoirs suggesting problems of pollution reported in the reservoir in these studies. Also, high dissolved solid yield in water bodies has been associated with nutrient enrichment, eutrophication in reservoirs, and other oxygen-depleting activities that have been linked with deterioration of water quality, negative impact on aquatic life, and consequent increased cost of water treatment (Ogunkoya and Aboyeji 2014) which all affects the sustainable use of the reservoir. The problem of nutrient enrichment may be sorted if inflows from the sub-catchments are controlled or better managed (Ogunkoya 2013).

5 Conclusions

No doubt increased human activities, population growth, urbanization, and natural influences negatively impact the water quality in reservoirs, especially in sub-Saharan Africa. As typical of many reservoirs in developing countries, the Opa reservoir in southwestern Nigeria is not protected from receiving effluents from its tributaries in the face of poor environmental management and enhanced anthropogenic activities as the urban population increases. This may indicate a growing tendency for accelerated eutrophication which impacts water contamination in the reservoirs. In all, the study concluded on the need to prioritize ecological restoration of contaminated tributaries and headwater streams with a reservoir catchment in the part of the region as an important part of water quality control in the sub-region.

References

Abell JM, Ozkundakci D, Hamilton DP, Reeves P (2022) Restoring shallow lakes impaired by eutrophication: approaches, outcomes and challenges. Crit Rev Environ Sci Technol 52(7):1199–1246

Adediji A (2005) Reservoir sedimentation: the case of the Opa reservoir catchment Southwestern Nigeria. S Afr Geogr J 87(2):123–128

Adejuwon JO, Jeje LK (1975) Land element of the environmental system of Ife area. Environmental Resource Base Project Publication No. 2, Department of Geography, Obafemi Awolowo University, Ile Ife, Nigeria

Adeniran AB, Daniell KA (2021) Transqua: power, political change and the transnational politics of a water megaproject. Int J Water Res Dev 37(2):234–255

Adesakin TA, Adedĳi AA, Aduwo AI, Taiwo FT (2017) Effects of discharges from re-channeled rivers and municipal runoff on water quality of Opa reservoir, Ile-Ife, southwestern, Nigeria. Afr J Environ Sci Technol 11(1):56–70

Ahmed U, Mumtaz R, Anwar H, Shah AA, Irfan R, Garcia-Nieto J (2019) Efficient water quality prediction using supervised machine learning. Water 11(1):2210

Akaahan TJA, Olabanji FM, Azua ET (2017) Studies on regression analysis between electrical conductivity and total dissolved solids as environmental variables in lower River Benue, Makurdi, Nigeria. Afr J Water Conserv Sustain 5(5):221–229

Akinbuwa O (1988) The study of the physico-chemical factors and Rotifera fauna of Opa reservoir, Ile-Ife, Nigeria. Unpublished MSc thesis, Department of Zoology, University of Ife, Ile-Ife

Akindele EO, Adeniyi IF (2013) A study of the physico-chemical water quality, hydrology and zooplankton fauna of Opa reservoir catchment area, Ile-Ife, Nigeria. Afr J Environ Sci Technol 7(5):192–203

Akoni MM (2019) Citizenship in Ghana: understanding its cultural and political construction. Doctoral dissertation, Universidade do Minho, Portugal

Aleksander-Kwaterczak U, Plenzler D (2019) Contamination of small urban watercourses on the example of a stream in Krakow (Poland). Environ Earth Sci 78(16):530

Alexandrov YC, Laronne JB, Reid I (2009) Suspended sediment load, bedload and dissolved load yields from a semi-arid drainage basin: a 15-year study. Water Resour Res 45(8)

Asante KA, Quarcoopome T, Amevenku FYK (2008) Water quality of the Weija reservoir after 28 years of impoundment. West Afr J Appl Ecol 13(1):125–131

Atulley JK, Onwusu-Ansah ED, Ampofo S, Jacob A, Nii OS, (2022) Modelling the impact of past and future land cover changes on a reservoir catchment hydrology in semi-arid, Africa

Aulenbach BT, Burns DA, Shanley JB, Yanai RD, Bea K, Wild AD, Yang Y, Yi D (2016) Approaches to stream solute load estimation for solutes with varying dynamics from five diverse small watersheds. Ecosphere 7(6):e01298

Babatimehin OI, Eludoyin AO, Ekundayo OO, Ekwere NJ (2020) Water quality and perception on use in Obafemi Awolowo University, Ile-Ife, Nigeria. Interdisc Environ Rev 20(2):100–117

Bakker C (2021) Nutrients and biota in a lake system before and after restoration; a data analysis of the Swedish eutrophication case studies Vaxjosjon

Bearcock JM, Smedley PL, Fordyce FM, Everett PA, Ander EL (2019) Controls on surface water quality in the River Clyde catchment, Scotland, UK, with particular reference to chromium and lead. Earth Environ Sci Trans R Soc Edinb 108(2–3):249–267

Benettin P, Fovet O, Li L (2020) Nitrate removal and young stream water fractions at the catchment scale. Hydrol Process 34(12):2725–2738

Bonvin F, Razmi AM, Barry DA, Kohn T (2013) Micropollutant dynamics in Vidy Bay: a coupled hydrodynamic-photolysis model to assess the spatial extent of ecotoxicological risk. Environ Sci Technol 47(16):9207–9216

Botter M, Burlando P, Fatichi S (2018) Anthropogenic and catchment characteristic signatures in the water quality of Swiss rivers: a quantitative assessment. Hydrol Earth Syst Sci

Buchanan TJ, Somers WP (1969) Discharge measurement at gaging stations. US Government Printing Office, Washington, DC

Bufe A, Cook KL, Galy A, Wittmann H, Hovius N (2022) The effect of lithology on the relationship between denudation rate and chemical weathering pathways-evidence from the eastern Tibetan Plateau. Earth Surf Dyn 3:513–530

Burns DA, Bhatt G, Linker LC, Bash JO, Capel PD, Shenk GW (2021) Atmospheric nitrogen deposition in the Chesapeake Bay watershed: a history of change. Atmos Environ 15(251):118277

Camnasio E, Orsi E (2011) Calibration methods of current meters. J Hydraul Eng 137(3):386–392

Carpenter SR, Bolgrien D, Lathrop RC, Stow CA, Reed T, Wilson MA (1998) Ecological and economic analysis of lake eutrophication by non-point pollution. Aust J Ecol 23(1):68–79

Chapman MG (1998) Improving sampling designs for measuring restoration in aquatic habitats. J Aquat Ecosyst Stress Recover 6:235–251

Clark KE, Shanley JB, Scholl MA, Perdrial N, Perdrial JN, Plante AF, McDowell WH (2017) Tropical river suspended sediment and solute dynamics in storms during an extreme drought. Water Resour Res 53(5):3695–3712

Cressie N (1988) Spatial prediction and ordinary kriging. Math Geol 20:405–421

Dadi D, Mengistie E, Terefe G, Getahun T, Haddis A, Birke W, Beyene A, Luis P, Van der Bruggen B (2018) Assessment of the effluent quality of wet coffee processing wastewater and its influence on downstream water quality. Ecohydrol Hydrobiol 18(2):201–211

De la Fuente L, Gupta HV, Condon LE (2021) Towards a representational approach to prediction, understanding and discovery in hydrology

Dodds WK, Whiles MR (2004) Quality and quantity of suspended sediment particles in rivers: continent-scale patterns in the United States. Environ Manag 33:355–367

Eludoyin AO (2013) Modelling the hysteretic patterns of solute concentration-discharge relationships and their significance for hydrological pathways at the farm-scale. University of Exeter, United Kingdom
Eludoyin OA, Adewole OA (2020) A remotely sensing-based evaluation of an ungauged drainage basin in southwestern Nigeria. Int J River Basin Manag 18(3):307–319
Eludoyin AO, Olarewaju OE (2020) Water supply and quality in the sub-Saharan Africa. In: Clean water and sanitation, pp 1–7
Eludoyin AO, Ofoezie IE, Ogunkoya OO (2004) The effect of Oja-titun market effluent on the chemical quality of receiving Opa reservoir in Ile-Ife, Nigeria. J Environ Manag 72:249–259
Eludoyin AO, Eludoyin O, Martins T, Oyinloye M, Eslamian S (2021) Water security using rainwater harvesting. In: Handbook of water harvesting and conservation: basic concepts and fundamentals, pp 57–67
Esan OT, Akinyemi AP, Bakare TA, Balogun YL, Ogunwusi AO (2020) Determinant of uptake of periodic medical examination among students of college of health sciences, Obafemi Awolowo University, Ile-Ife, southwestern Nigeria. Niger J Med 29(4):575–581
Gebrechorkos SH, Pan M, Anghileri D, Forstythe N, Pritchard DM, Fowler HJ, Obuobie E, Darko D, Sheffield J (2022) Variability and changes in hydrological drought in the Volta basin, West Africa. J Hydrol Reg Stud 42:101143
Gitto AB, Venditti JG, Kostaschuk R, Church M (2017) Representative point-integrated suspended sediment sampling in rivers. Water Resour Res 53(4):2956–2971
Guevera G (2009) Horizontal and seasonal pattern of tropical zooplankton from the eutrophic Prado in Colombia. Limnol Ecol Manag Inland Water 39:128–139
Gwenzi W, Chinyama SR, Togarepi S (2017) Concentration-discharge patterns in a small urban headwater stream in a seasonally dry water-limited tropical environment. J Hydrol 550:12–25
Imevbore AMA, Ogunkoya OO, Sagua VO (1988) Effects of soil erosion on the aquatic ecosystem in Nigeria. In: Sagua VO, Enabor EE, Ofomata GEK, Ologe KO, Oyebande L (eds) Ecological disasters in Nigeria: soil erosion. Federal Ministry of Science and Technology, Lagos, pp 247–261
Imran U, Weidhaas J, Ullah A, Shaikh K (2020) Risk associated with spatio-temporal variations in trace metals and a metalloid in a major freshwater reservoir of Pakistan. Hum Ecol Risk Assess Int J 1:1–20
Ismail WR, Najib SAM (2011) Sediment and nutrients balance of Bukit Merah Reservoir, Perak, Malaysia. Lakes Reserv Res Manag 16:179–184
Jeje LK, Ogunkoya OO, Oluwatimelehim JM (1999) Solute load concentration in some streams in the upper Osun and Owena drainage basins, central western Nigeria. J Afr Earth Sci 29(4):799–807
Jia H, Qu W, Ren W, Qian H (2021) Impact of chemical weathering and human perturbations on dissolved loads of the Wei River, the Yellow River catchment. J Hydrol 603:126950
Keremah RI, Davies OA, Abezi ID (2014) Physico-chemical analysis of fish pond water in freshwater areas of Bayelsa State Nigeria. Greener J Biol Sci 4(2):033–038
Knapp JL, von Freyberg J, Studer B, Kiewiet L, Kirchner JW (2020) Concentration–discharge relationships vary among hydrological events, reflecting differences in event characteristics. Hydrol Earth Syst Sci 24(5):2561–2576
Kumar A, Tripathi VK, Sanchan P, Rakshit A, Singh RM, Shukla SK, Pandey R, Vishwakarma A, Panda KC (2022) Sources of ions in the river ecosystem. In: Ecological significance river ecosystem, pp 187–202
May L, Olszewska J, Gunn IDM, Meis S, Spear BM (2020) Eutrophication and restoration in temperate lakes. IOP Conf Ser Earth Environ Sci 535(1):012001
McGrane SJ, Hutchins MG, Miller JD, Bussi G, Kjeldsen TR, Loewenthal M (2017) During a winter of storms in a small UK catchment, hydrology and water quality responses follow a clear rural-urban gradient. J Hydrol 545:463–477
Newbold JD (1992) Cycles and spirals of nutrients. In: The handbook, vol 1, pp 379–408

Nikolai SJ, Dzialowski AR (2014) Effects of internal phosphorus loading on nutrient limitation in a eutrophic reservoir. Limnologica 49:33–41
Nshemereirwe A, Zewge F, Malambala E (2022) Evaluation of formation and health risks of disinfection by-products in drinking water supply of Ggaba waterworks, Kampala Uganda. J Water Health 20(2):560–574
Odekunle TO, Adejuwon SA (2007) Assessing changes in the rainfall regime in Nigeria between 1961 and 2004. Geo J 70(2):145–159
Ogunkoya OO (2013) All rivers run into the sea; yet the sea is not full. Inaugural lecture series 256th, Obafemi Awolowo University, Ile-Ife Nigeria
Ogunkoya OO, Aboyeji OS (2014) Wetland sustainability and urban growth in Southwestern Nigeria: an example from Ile-Ife. Niger Field 79:3–24
Ogunkoya OO, Jeje LK (1987) Sediment and solutes yield of some third order basins on the basement complex rocks in southwestern Nigeria. Catena 14:383–397
Ogunkoya OO, Ogbole AS (2024) Estimation of suspended sediment balance of a small catchment in Southwestern Nigeria. Environ Monit Assess 196(3):292
Olson JR, Cormier SM (2019) Modeling spatial and temporal variation in natural background specific conductivity. Environ Sci Technol 53(8):4316–4325
Olabanji IO, Oluyemi EA (2014) Preliminary assessment of heavy metal pollution of Opa reservoir, Ile-Ife, southwestern Nigeria using *Mormyrus rume* and *Tilapia zillii*. IFE J Sci 16(1):35–43
Olabanji IO, Oluyemi EA, Fakoya TO, Eludoyin AO, Makinde WO (2015) Effects of wastes disposal system on metal composition of hospital dumpsite soil in Ilesa, Southwestern Nigeria. J Int Environ Appl Sci 10(3):319–330
Onwusu A, Mul M, Strauch M, van der Zaag P, Volk M, Slinger J (2022) The clam and the dam: a Bayesian belief network approach to environmental flow assessment in a data scare region. Sci Total Environ 810:151315
Oyeku OT, Eludoyin AO (2010) Heavy metal contamination of groundwater resources in a Nigerian urban settlement. Afr J Environ Sci Technol 4(4)
Pagnozzi G, Carroll S, Reible DD, Millerick K (2020) Biological natural attenuation and contaminant oxidation in sediment caps: recent advances and future opportunities. Curr Pollut Rep 6(3):281–294
Patil C, Bandekar SS, Hosamana SS, Adavimath A (2022) Assessment of surface water quality parameters of Panchganga River. In: Sustainability trends and challenges in civil engineering, pp 781–797
Prusty P, Farooq SH, Swain D, Chandrasekharam D (2020) Association of geomorphic features with groundwater quality and freshwater availability in coastal regions. Int J Environ Sci Technol 3:1–16
Qin C, Li SL, Waldron S, Yue FJ, Wang ZJ, Zhong J, Ding H, Liu CQ (2020) High-frequency monitoring reveals how hydrochemistry and dissolved carbon respond to rainstorms at a karstic critical zone, Southwestern China. Sci Total Environ 714:136833
Saalidong BM, Aram SA, Otu S, Lartey PO (2022) Examining the dynamics of the relationship between water pH and other water quality parameters in ground and surface water systems. PLoS One 17(1):e0262117
Shivayogimath CB, Kalburgi PB, Deshannavar UB, Virupakshaiah DBM (2012) Water quality evaluation of river Ghataprabha, India. Int Res J Environ Sci 1(1):12–18
Siddiqui EP, Pandey JA (2022) Atmospheric deposition: an important determinant of nutrients and heavy metal level in urban surface runoff reaching to the Ganges River. Arch Environ Contam Toxicol 82(2):191–205
Sivapalan M, Bloschl G, Zhang L, Vertessy R (2003) Downward approach to hydrological prediction. Hydrol Process 17(11):2101–2111
Torres C, Resck RP, Pinto-Coelho RM (2007) Mass balance estimation of nitrogen, carbon, phosphorus and total suspended solids in the urban eutrophic, Pampulh reservoir Brazil. Acta Limnol Bras 19(1):79–91

Troitsky B, Zhu DZ, van Duin B, Mahmood K (2019) Nutrient processes and modeling in urban stormwater ponds and constructed wetlands. Can Water Res J/Rev Can Res Hydriques 44(3):230–247
Van Niekerk H, Silberbauer MJ, Maluleke M (2014) Geographical differences in the relationship between total dissolved solids and electrical conductivity in South African rivers. Water 40(1):133–138
Vantarakis A (2021) Eutrophication and public health. In: Chemical lake restoration, pp 23–47
Walling DE (2005) Tracing suspended sediment sources in catchments and river systems. Sci Total Environ 344:159–184
World Health Organization (2016) Guidelines for drinking water quality, 3rd edn. Vol 1: Recommendations. WHO, Geneva
Yin Y, Zhang W, Tang J, Chen X, Zhang Y, Cao X, Li Q (2022) Impact of river dams on phosphorus migration; a case of the Pubugou reservoir on the Dadu River in China. Sci Total Environ 25(809):151092
Yu S, Xu Z, Wu W, Zuo D (2016) Effect of land use types on stream water quality under seasonal variation and topographic characteristics in the Wei River basin, China. Ecol Ind 60:202–212
Zhang Z, Jia W, Zhu G, Ma X, Xu X, Yuan R, Shi Y, Yang L, Xiong H (2020) Hydrochemical characteristics and ion sources of precipitation in the upper reaches of the Shiyang River China. Water 12(5):1442

Soil Erosion Risk and Vulnerable Zone Identification Using Remote Sensing and GIS Techniques

Nilkamal Konra

Abstract The natural process of soil erosion occurs when the Earth's surface soil is gradually worn down and carried by a variety of forces, including gravity, wind, water, and ice. Even while erosion is a natural occurrence, human activity and land use practices can cause it to occur more quickly, which poses a serious environmental risk. Soil erosion, in particular, is a serious environmental problem that can have far-reaching effects on the economy, ecology, and society. Therefore, identifying erosion-prone sites is an essential first step in adopting proper erosion mitigation and soil conservation strategies. This study set out to (i) analyze soil erosion vulnerability and model soil erosion vulnerability using the Analytic Hierarchy Process (AHP) approach. Ascertain the suitable remedial actions and long-term therapies to halt soil erosion in the study area. Rainfall, slope, ruggedness, NDVI, erodibility, vegetation cover, and land use/cover ratios are among the geo-environmental elements that contribute to soil erosion. These parameters were mapped out using the Analytic Hierarchy Process (AHP), along with their percentage weights. The soil erosion vulnerability model that was developed can be a helpful planning tool for ranking the areas that should be prioritized for soil conservation and erosion control techniques including sustainable agriculture and bioengineering interventions. The study highlights three key factors—slope steepness, ruggedness, and stream power index—that collectively contribute to 57.1% of soil erosion vulnerability. Steep slopes increase runoff, rugged terrain enhances sediment transport, and higher stream power indicates greater erosive potential. Understanding and managing these factors are crucial for mitigating erosion risks and preserving landscape health.

Keywords Remote sensing · GIS · Stream Power Index · Terrain Ruggedness Index · AHP · Geomorphology

N. Konra (✉)
Agriculture and Soils Department, Indian Institute of Remote Sensing-Indian Space Research Organisation, Dehradun, India
e-mail: nilkamalkonra99@gmail.com

S. C. Pal and U. Chatterjee (eds.), *Surface, Sub-Surface Hydrology and Management*, Springer Geography, https://doi.org/10.1007/978-3-031-62376-9_16

1 Introduction

The Himalayan watershed region is facing a significant challenge in the form of soil erosion, given its distinctive terrain, abundant biodiversity, and ecological importance. This issue not only threatens the fragile ecosystems thriving in these mountains but also poses a risk to agricultural output. Recognizing the urgency of managing soil erosion in this critical area, our project focuses on developing a Soil Erosion Risk Vulnerability Map tailored specifically to the Himalayan watershed, often referred to as the "Third Pole." The Himalayan environment is characterized by its steep peaks, diverse weather patterns, and complex hydrological and geological systems. These inherent qualities, coupled with mounting human pressures, make the region highly susceptible to soil erosion. Left unchecked, soil erosion can lead to far-reaching consequences beyond just the loss of fertile topsoil. To address this pressing issue, our project leverages cutting-edge technologies such as machine learning, remote sensing, and Geographic Information System (GIS) capabilities. By integrating a plethora of environmental data, including topography, precipitation, land cover, and soil characteristics, we aim to create a comprehensive and accurate Soil Erosion Risk Vulnerability Map for the Himalayan watershed. Our primary objective is to pinpoint areas at high risk of soil erosion, thereby closing the knowledge gap that exists between scientific understanding and practical conservation efforts in this region. While numerous studies have explored soil erosion globally, our project stands out by focusing specifically on the unique qualities of the Himalayan area. Through our research, we aspire to lay the groundwork for sustainable land use planning and environmental stewardship in the Himalayan watershed. By shedding light on the nuances of soil erosion vulnerability, we aim to protect this vital area for the benefit of present and future generations. Soil erosion presents a critical environmental challenge, significantly reducing soil productivity by depleting clay content, organic matter, pedon thickness, and the volume of soil essential for providing nutrients and water to plants (Srinivasan et al. 2021; Quinton et al. 2010). Moreover, it disrupts sediment flows and geomorphic processes, exacerbating the issue of rising sedimentation and siltation in rivers and reservoirs (Swarnkar et al. 2018; Pandey et al. 2007). Research has shown that soil erosion is further intensified and accelerated by the combined effects of land use practices and climate change (Lal 2003; Paustian et al. 2016). This degradation of soil features, alongside biodiversity loss and disruptions in biogeochemical cycles, poses a significant risk to ecological processes and soil health (Wall et al. 2015).

In mountainous regions like the Himalayas, steep topography, inadequate land use management, and the impacts of climate change contribute to the heightened severity of soil erosion (Tarolli and Straffelini 2020). Therefore, understanding the influence of climate change on erosion is crucial for comprehending its broader implications for ecosystem health and food security (Lal 2005; Pimentel 2006). An effective and systematic method of decision-making, the Analytic Hierarchy Process (AHP) emphasizes a hierarchical arrangement of objectives, selection criteria, and options at several levels (Saaty 1990). These days, AHP is widely used in many different areas of the planet, especially when it comes to solving issues with several criteria. Among other uses, it is noteworthy that it has been widely used in the evaluation of soil erosion risk (Wu and Wang 2007; Ni et al. 2008; Rahman et al. 2009; Pradeep et al. 2015). Studying a catchment's physical behavior aids in understanding hydrologic and geomorphic issues like flooding, erosion, and mass movement. Morphometric analysis quantitatively describes the drainage system, crucial for watershed characterization. By measuring linear, aerial, and relief aspects, alongside slope contribution, this analysis helps predict watershed behavior when coupled with geomorphology and geology. The hydrological response is closely linked to physiographic characteristics such as size, shape, slope, drainage density, and stream length of the drainage basin, influencing erosion patterns (Bassey Eze and Efiong 2010; Esper Angillieri 2008; Nag and Chakraborty 2003).

2 Study Area

The study area is the intermediate catchment of Tehri dam and the area is 1320 km^2, latitude is 78° 9′ 15″ to 79° 24′ 55″ E, and longitude is 29° 33′ 40″ to 30° 51′ 0″ N. Tehri Gharwal, Uttarkashi this two district fall in this study area. Elevation of this area is 570–3242 m. Rainfall is mainly occurring during Jun to Sep. average annual rainfall is 1028.5 mm (Fig. 1).

The Tehri catchment has been divided into four sub-catchments (Fig. 2).

(1) Bhagirathi up to Maneri Bhali II (~4693 Km^2)
(2) Bhilangana up to Ghansali (~792 Km^2),
(3) Balganga up to Sarasgaon (~488 Km^2),
(4) Intermediate catchment (~1320 Km^2).

The intermediate catchment consists primarily of forest, agricultural land, water body, barren land or scrubland, and settlement. According to the tree canopy, the forest cover was divided into two categories: dense forest and open forest, including open forest pine and oak forests. According to the Survey, Wheat is cereal crop and Mustard is oil seed crop and different types of millet and fruit trees like Malta, banana, etc., are major cultivating crops. Resettlement, forestation programme, the scrubland, and arid land on the upper slopes gradually changed into agricultural terrain.

3 Materials and Data Used

Sentinel-2A/2B

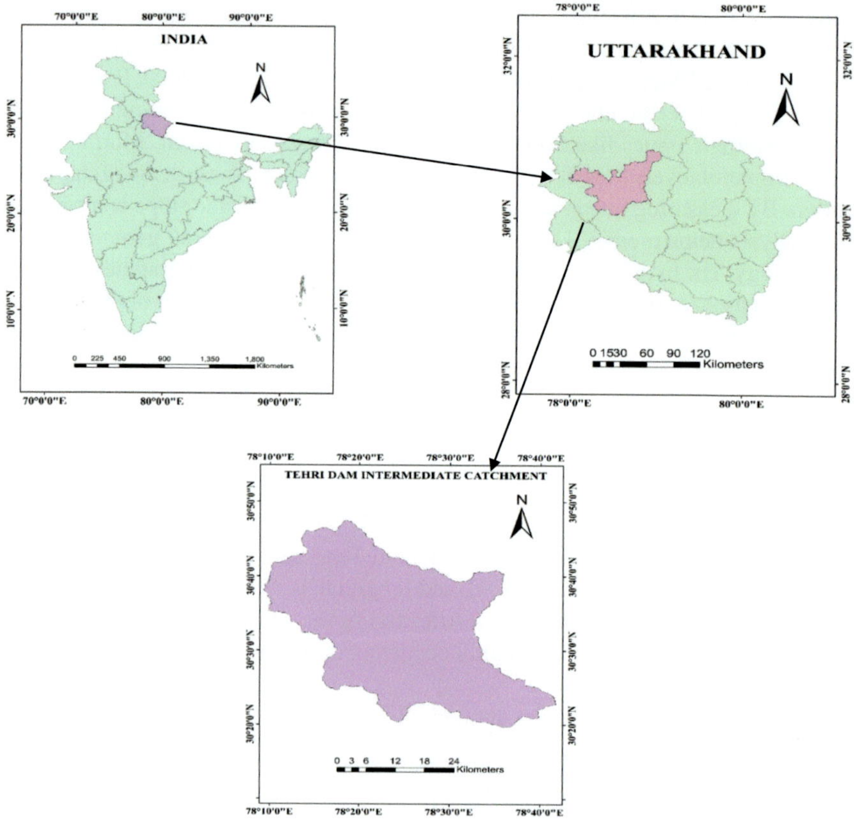

Fig. 1 Study area

Sentinel 2A/2B configurations	
Spectral resolution	13 bands
Spatial resolution	10 m (Four visible and NIR bands)
	20 m (Six red edges and shortwave IR bands)
	60 m (Three atmospheric bands)
Revisit cycle	5 days
Swath	290 km

SRTM DEM

Projection	Geographic
Horizontal datum	WGS 84
Vertical datum	EGM96

(continued)

(continued)

Projection	Geographic
Vertical units	Meters
Spatial resolution	1 Arc second for global coverage (30 m) 3 Arc second for global coverage (90 m)
Raster size	1 degree tiles
C band wavelength	5.6 cm

Source www.usgs.gov
Rainfall data: ESDAC

3.1 Software Used

3.1.1 Google Earth Engine (GEE)

Google Earth Engine was used for satellite data downloading and processing of data. It provided all satellite data in a single platform and helps in processing also (https://code.earthengine.google.com/).

3.1.2 ERDAS Imagine Version 2018

ERDAS imagine software with a raster graphics editor designed by ERDAS geospatial applications. It was used for image analysis and this software provides the opportunity for processing satellite images.

3.1.3 ArcGIS Version 10.3

ArcGIS consists of a group of geographical information system software products produced by ESRI. ArcGIS is a system for working with maps and geographic information. It was used in the creation of mapping, compiling geospatial data, extraction of geospatial data, and managing data in geographical information.

3.1.4 QGIS 3.34

QGIS consists of group geographical information system software.

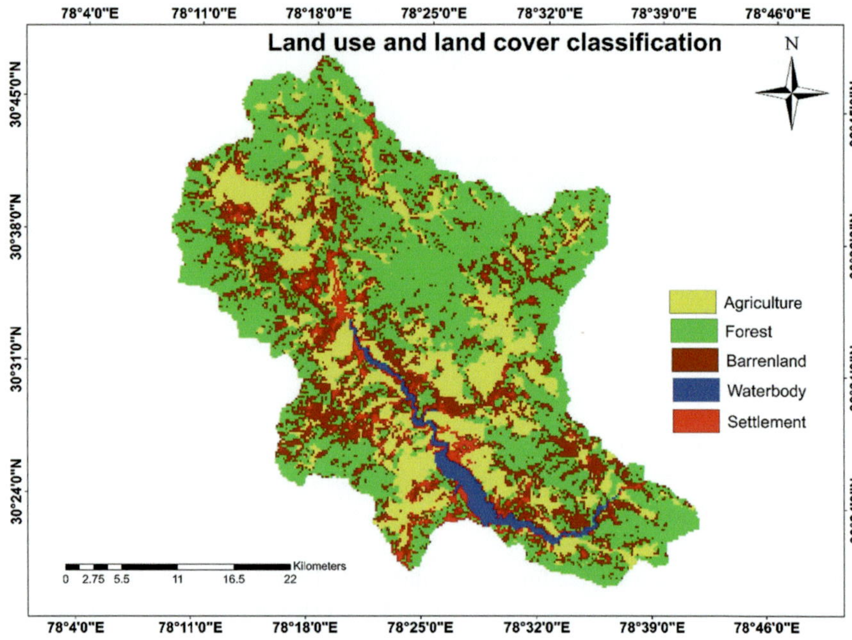

Fig. 2 Land use and land cover classification

3.2 *Geo-environmental Parameters*

3.2.1 *R* Factor

In the field of soil conservation and erosion control, the *R* factor is often associated with the Rainfall Factor in the Universal Soil Loss Equation (USLE). The USLE is used to estimate soil erosion, and the *R* factor represents the erosivity of rainfall. It takes into account the amount and intensity of rainfall in a particular location. The *R* factor is a crucial component in predicting soil loss due to water erosion (Fig. 3).

3.2.2 *K* Factor

The erodibility factor, or "*K* factor" in soil erosion, is a key element in the Universal Soil Loss Equation (USLE) and its more sophisticated version, the Revised Universal Soil Loss Equation (RUSLE). The erodibility factor, or "*K*," represents a soil's vulnerability to erosion processes. It measures the ways in which the characteristics of the soil affect the detachment and transportation by water, taking into account the soil's natural erodibility as well as its susceptibility to erosion under certain environmental circumstances. Knowing the *K* factor is essential because it shows how susceptible a certain type of soil is to erosion. Land managers can successfully prevent soil loss

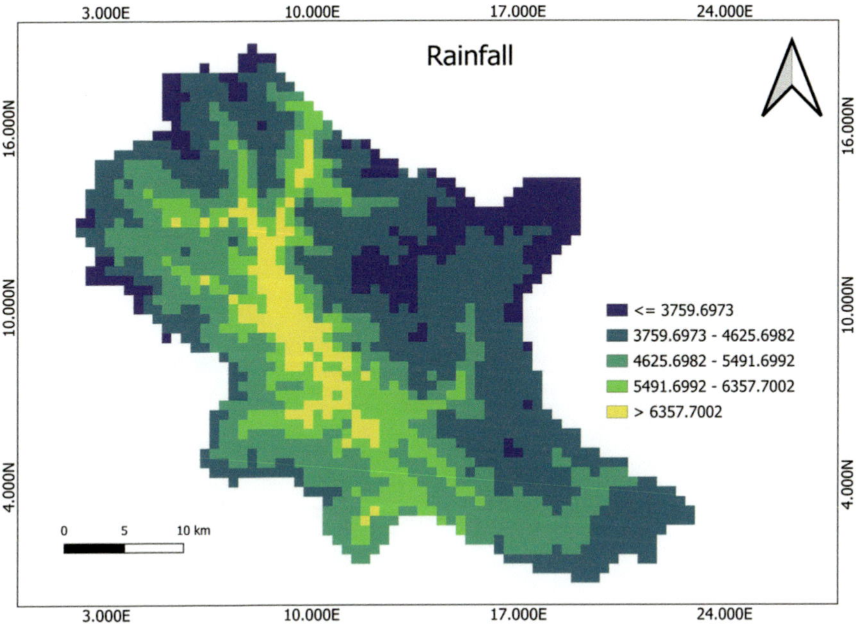

Fig. 3 Rainfall factor

by anticipating erosion hazards and customizing solutions based on this assessment. As a result, the *K* factor is essential to sustainable land management techniques and soil conservation initiatives (Fig. 4).

3.2.3 *C* Factor

This factor considers a range of agricultural practices such as tillage methods, crop residue management, cover cropping, and contour farming, among others. Each practice affects soil erosion differently based on factors like soil structure, moisture retention, and surface roughness. It takes into account the impact of factors such as crop cover, residue management, and tillage practices on the soil's susceptibility to erosion. The *C* factor is a dimensionless value ranging from 0 to 1, where 0 indicates minimal erosion potential (high cover or conservation practices), and 1 indicates high erosion potential (little or no cover, intensive tillage). Here I have used *C* factor using NDVI (Fig. 5).

3.2.4 Slope Map

A slope map is a thematic map that visually represents the steepness or gradient of the terrain across a geographic area. It provides a clear and spatially explicit depiction

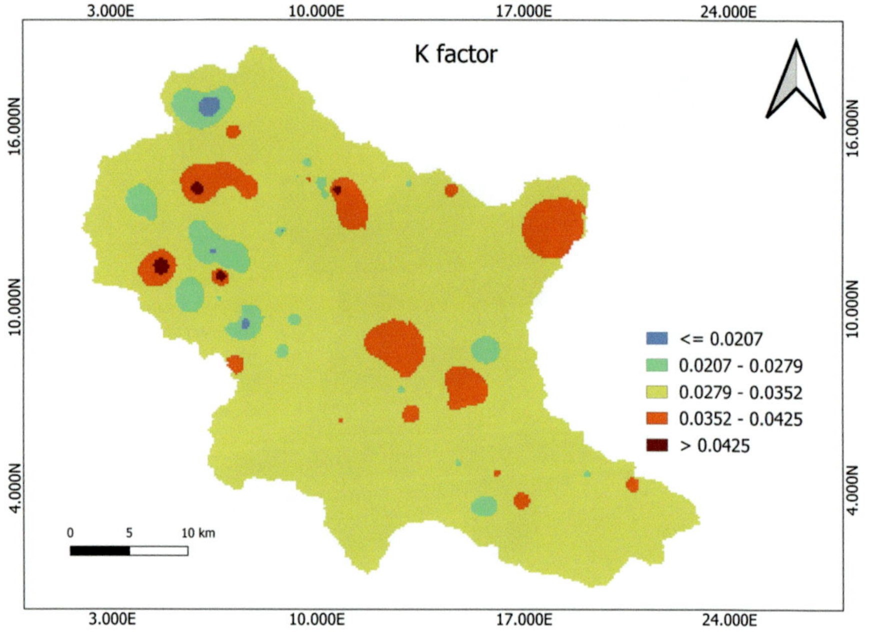

Fig. 4 K factor

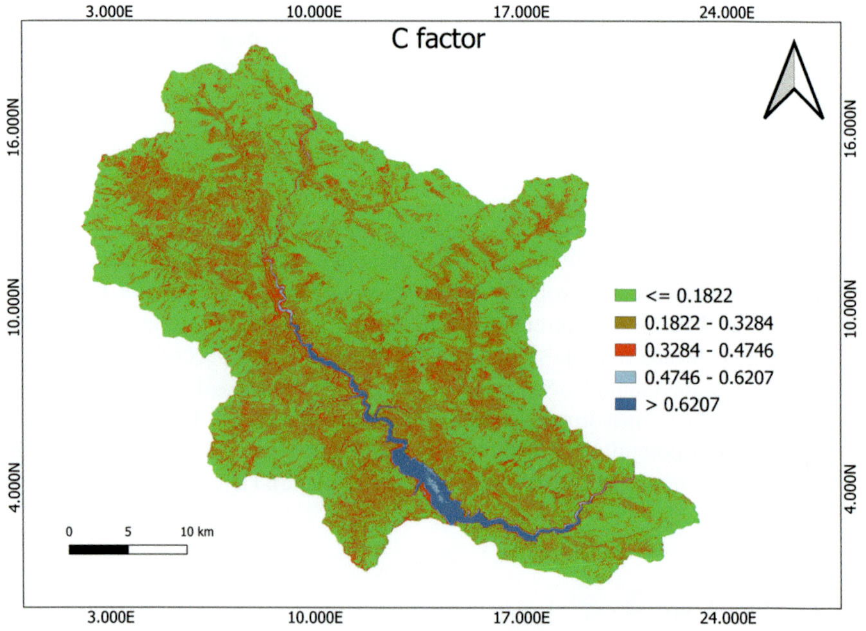

Fig. 5 *C* factor

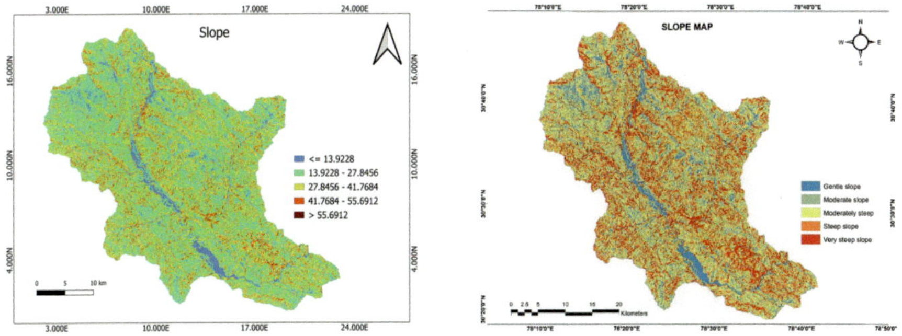

Fig. 6 Slope

of the slope characteristics, helping to understand the topography of the landscape. Slope maps are commonly used in various fields, including geography, geology, environmental science, forestry, and urban planning (Fig. 6).

3.2.5 NDVI

NDVI is a crucial metric for assessing the density and health of plant cover in environmental research and remote sensing applications. It is obtained by comparing the reflectance of red and near-infrared (NIR) light wavelengths through the study of satellite or aerial photography.

Higher NDVI values, which range from -1 to $+1$, suggest denser and healthier vegetation, whereas values that are getting close to zero or going negative indicate sparse or non-vegetated regions.

$$\mathrm{NDVI} = (\mathrm{NIR} - \mathrm{Red})/(\mathrm{NIR} + \mathrm{Red})$$

where:

NIR is the reflectance in the near-infrared part of the electromagnetic spectrum.

RED is the reflectance in the red part of the electromagnetic spectrum (Fig. 7).

3.2.6 Ruggedness Index

The Terrain Ruggedness Index (TRI) is a numerical measure used to quantify the ruggedness or roughness of a terrain. It provides an assessment of the variation in elevation within a specified area. The Terrain Ruggedness Index is particularly useful in terrain analysis, geomorphology, and environmental studies to characterize landscapes based on their topographic complexity. The Terrain Ruggedness Index is a

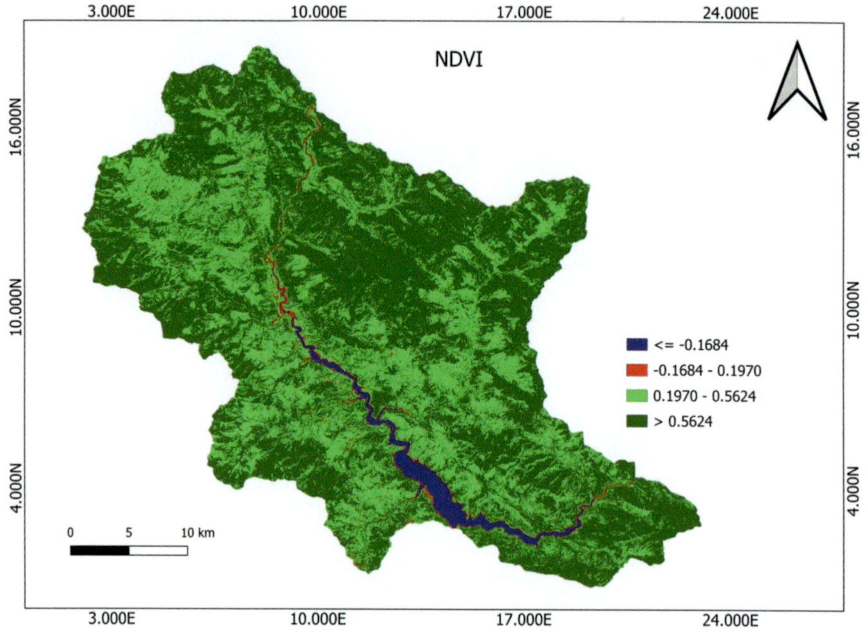

Fig. 7 NDVI

versatile tool for analyzing and characterizing terrain complexity, providing valuable insights for various scientific and applied purposes (Fig. 8).

3.2.7 Stream Power Index

The Stream Power Index (SPI) is a geomorphic index used in the field of fluvial geomorphology to assess the erosive power of streams and rivers. It quantifies the potential of flowing water to transport sediment and shape the landscape based on variables such as slope, drainage area, and channel gradient. The Stream Power Index is a valuable tool for geomorphologists, hydrologists, and environmental scientists studying river systems. It aids in the assessment of erosive power, sediment transport dynamics, and the overall geomorphic behavior of rivers (Fig. 9).

3.2.8 Drainage

It is a measure of how well or poorly a landscape is drained by rivers and streams. Drainage density is often used in hydrology, geomorphology, and environmental science to characterize the drainage patterns of a region and understand its hydrological behavior. The calculation of drainage density involves the ratio of total stream length to the total area of the watershed or drainage basin (Fig. 10).

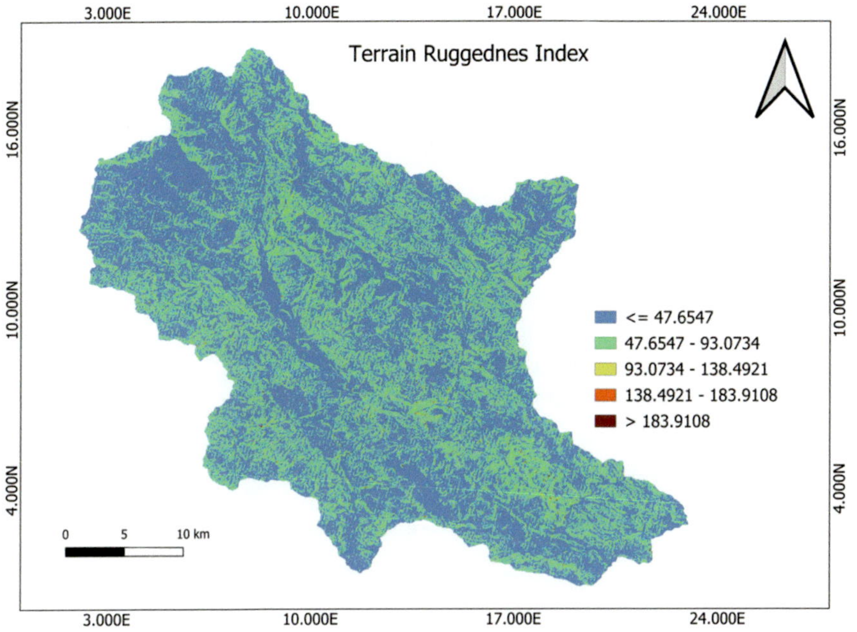

Fig. 8 Terrain Ruggedness Index

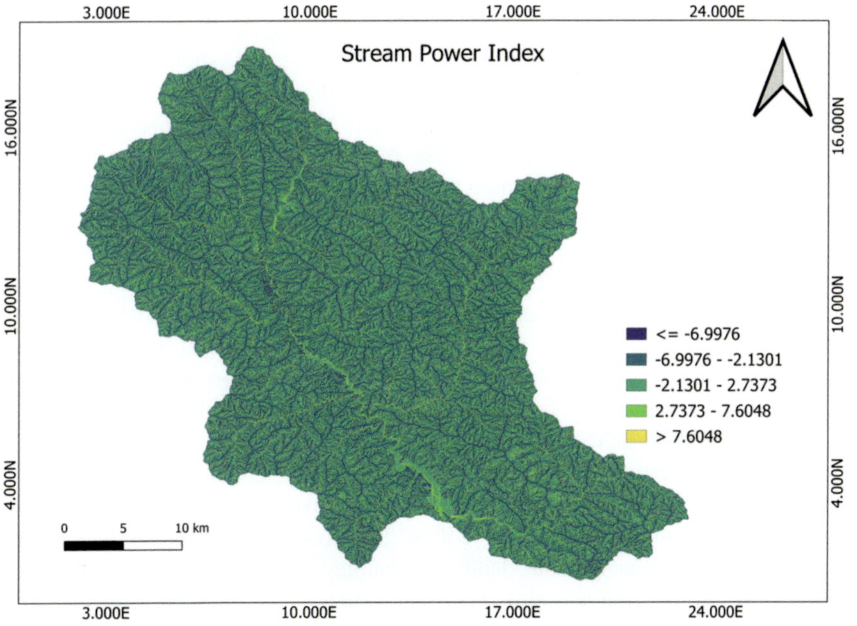

Fig. 9 Stream Power Index

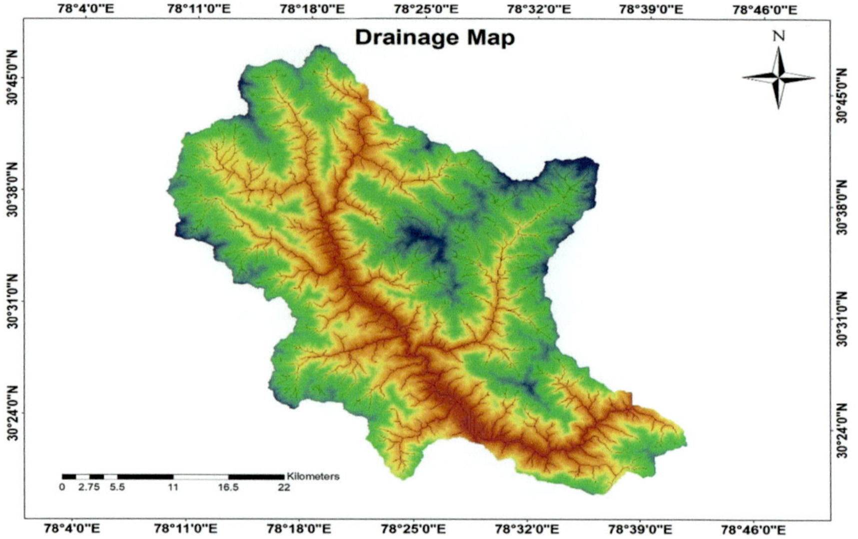

Fig. 10 Drainage density

4 Methodology

The methodology for assessing soil erosion risk using the Analytical Hierarchy Process (AHP) and geo-environmental factors involves a systematic approach to integrate spatial information, expert knowledge, and multi-criteria decision-making. SRTM DEM stands for Shuttle Radar Topography Mission Digital Elevation Model. The SRTM mission was a joint project of NASA (National Aeronautics and Space Administration) (www.jpl.nasa.gov) and the National Geospatial-Intelligence Agency (NGA). The mission aimed to collect high-resolution topographic data for most of the Earth's land surfaces. We have downloaded the SRTM data with resolution of 30 m. Then, generate the different indices for terrain analysis. We have generated land use land cover using sentinel-2 image. With the help of Google Earth Engine, we classified the area into five different classes like agriculture, water body, settlement, scrub, and forest. We have chosen different geo-environmental parameters; geo-environmental parameters refer to the physical, chemical, and biological factors in the Earth's environment that influence or are influenced by geological processes and conditions. These parameters play a crucial role in understanding the Earth's surface, subsurface, and the interactions between geological, hydrological, and environmental systems. Understanding and analyzing these geo-environmental factors are essential for environmental management, land use planning, and sustainable development. Geographic Information Systems (GIS) and other spatial analysis tools are often employed to integrate and analyze these factors for comprehensive environmental assessments. We have generated the slope, aspect, drainage, etc. We have taken seven parameters like slope, NDVI, Terrain Ruggedness Index, Stream power Index, Rainfall, Erodibility factor, crop management for assessing erosion risk of this area using AHP. The Analytical Hierarchy Process (AHP) is a decision-making methodology developed by Thomas L. Saaty in the late 1970s. It is designed to handle complex decisions involving multiple criteria and alternatives. Key features of AHP include the creation of a hierarchical structure, pair-wise comparisons to assess preferences, consistency checks to ensure reliability, and the use of mathematical calculations to derive weights for criteria and alternatives. The method synthesizes these weights to generate a comprehensive ranking or scoring of alternatives, aiding decision-makers in choosing the most favorable option.

In summary, AHP provides a structured and systematic approach to decision analysis, allowing decision-makers to prioritize and choose the most suitable alternatives based on a set of defined criteria. While it has proven valuable in various contexts, users should be mindful of its limitations and the need for careful consideration in the decision-making process (Fig. 11).

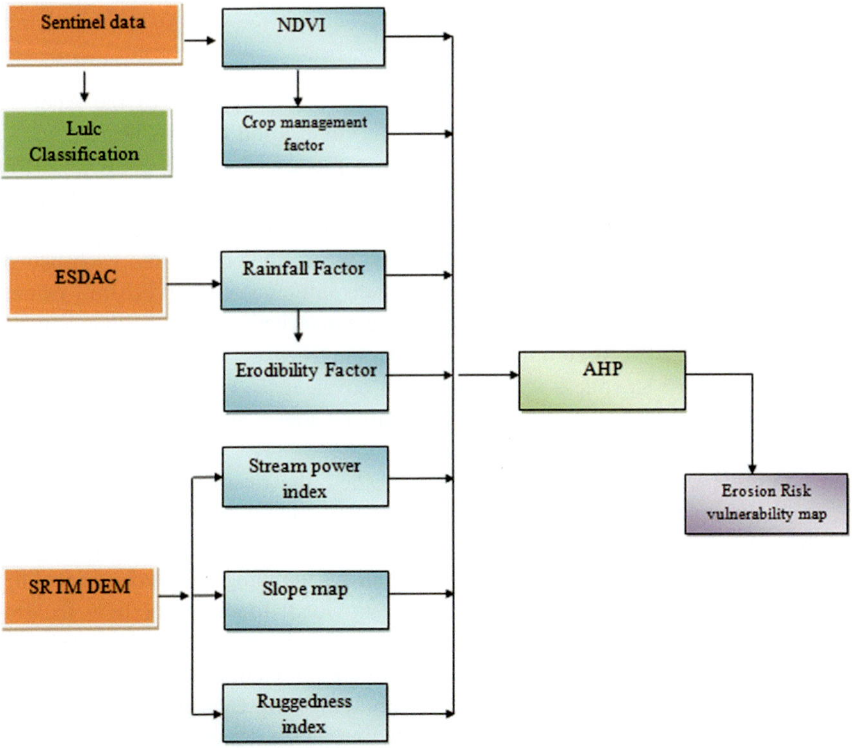

Fig. 11 Methodology flowchart

4.1 Analytical Model

The study employed the Analytic Hierarchy Process (AHP) model, there were multiple steps involved in developing the model. Initially, the AHP methodology was used to create a Pair-Wise Comparison Matrix for each of the seven geo-environmental criteria. With the use of this matrix, researchers were able to methodically assess how much each characteristic contributed to the vulnerability of soil erosion. On the basis of the pair-wise comparisons, the relative weights of the parameters were then determined. These weights revealed how much of an impact each criterion had on the vulnerability assessment as a whole.

In order to evaluate the consistency of the decisions made throughout the pair-wise comparisons, a consistency ratio (CR) was calculated to guarantee the model's resilience. Enhancing the accuracy and dependability of the model results was the goal of this step.

4.2 Pair-Wise Comparison Matrix

The Pair-Wise Comparison Matrix is a fundamental component of the Analytic Hierarchy Process (AHP) methodology used to assess the relative importance of different criteria or factors in decision-making processes. In the context of this study on soil erosion vulnerability, the Pair-Wise Comparison Matrix was utilized to systematically compare the significance of each of the seven parameters.

Researchers engaged in a structured process where they compared each pair of parameters against one another, considering their influence on soil erosion vulnerability. These comparisons involved assigning numerical values to represent the relative importance or preference of one parameter over another. For instance, if comparing slope steepness with vegetation cover, researchers would determine whether slope steepness has a higher, equal, or lower impact on soil erosion vulnerability compared to vegetation cover.

Upon completion of all pair-wise comparisons, the resulting matrix offered a thorough summary of each parameter's relative significance with respect to the others. The basis for determining the respective weights of the parameters was this matrix.

In further investigations, including vulnerability mapping and decision-making procedures, these weights were vital. They provided direction for how to allocate resources and prioritize activities to reduce the danger of soil erosion in the research area. The AHP model enabled a detailed knowledge of the intricate relationships causing soil erosion risk by precisely weighting each parameter, guiding more focused and efficient management strategies. Make it precise.

4.3 Consistency Ratio (CR) Calculation

The Consistency Ratio (CR) in the Analytic Hierarchy Process (AHP) is calculated using the formula:

$$\mathrm{CR} = \mathrm{CI}/\mathrm{RI}$$

CI is the Consistency Index, calculated as

$$CI = \frac{\lambda_{\max} - n}{n - 1}$$

With $\lambda_{\max}$ being the maximum eigenvalue of the Pair-Wise Comparison Matrix, and n being the number of criteria.

RI is the Random Index, a reference value based on the size of the Pair-Wise Comparison Matrix.

The CR is compared against a predefined threshold (usually 0.1) to assess the consistency of pair-wise comparisons. If CR < 0.1, comparisons are considered consistent; otherwise, they may need to be reviewed.

	R	*K*	Slope	*C* Factor	SPI	NDVI	Ruggedness
R	1	1	1/5	3	1/3	1	1/3
K	1	1	1/3	1	1	1/2	1
Slope	5	3	1	3	3	1	1
C factor	1/3	1	1/3	1	1	1/3	1
SPI	3	1	1/3	1	1	2	1
NDVI	1	2	1	3	1/2	1	1/2
Ruggedness	3	1	1	1	1	2	1
Total	14.33	10	4.20	13	7.83	7.83	5.83

Pair-wise comparison matrix (excel template *Source* www.scbuk.com)

R	*K*	Slope	*C*-Factor	SPI	NDVI	Ruggedness
0.069767442	0.1	0.047619048	0.230769231	0.042553191	0.127659574	0.057142857
0.069767442	0.1	0.079365079	0.076923077	0.127659574	0.063829787	0.171428571
0.348837209	0.3	0.238095238	0.230769231	0.382978723	0.127659574	0.171428571
0.023255814	0.1	0.079365079	0.076923077	0.127659574	0.042553191	0.171428571
0.209302326	0.1	0.079365079	0.076923077	0.127659574	0.255319149	0.171428571
0.069767442	0.2	0.238095238	0.230769231	0.063829787	0.127659574	0.085714286
0.209302326	0.1	0.238095238	0.076923077	0.127659574	0.255319149	0.171428571

Normalized (excel template *Source* www.scbuk.com)

AHP	
0.097	9.7%
0.098	9.8%
0.257	25.7%
0.089	8.9%
0.146	14.6%
0.145	14.5%
0.168	16.8%

Consistency check = 11%

Fundamental scale	
Extremely less important	1/9
	1/8
Very strongly less important	1/7
	1/6
Strongly less important	1/5
	1/4
Moderately less important	1/3
	1/2
Equal importance	1
	2
Moderately more important	3
	4
Strongly more important	5
	6
Very strongly more important	7
	8
Extremely more important	9

(Template *Source* www.scbuk.com)

AHP-1	CA
0.097	1.38319
0.098	0.98425
0.257	1.07986
0.089	1.15363
0.146	1.14143
0.145	1.13677
0.168	0.98227

Randomness index	
3	0.58
4	0.9
5	1.12
6	1.24
7	1.32
8	1.41
9	1.45
10	1.51
11	1.52
12	1.54
13	1.56
14	1.58

Lambda—7.861396507
CI—0.1435661
RI—1.32
CI/RI—0.108762185

5 Result

The study delved into the complexities of soil erosion vulnerability within the intermediate catchment, examining seven key geo-environmental parameters. These parameters were scrutinized to understand their individual contributions to the overall vulnerability of the area. The parameters contribution is in terms of percentage like rainfall contribute (9.7%), soil erodibilty contribute (9.8%), slope steepness (25.7%), crop management factor (8.9%), Stream Power Index (14.6%), NDVI (14.5%), and ruggedness contribute (16. 8%).Among them, slope steepness, ruggedness, and Stream Power Index (total 57.1%) emerged as the most influential factors, each exerting a significant impact on soil erosion vulnerability. One important factor that has been shown to influence erosion susceptibility is slope steepness, which is a measure of the terrain's inclination. Since steeper slopes are more likely to experience erosion because of greater surface runoff and gravitational forces, erosion prevention techniques must take these slopes into account. The study's spatial distribution map clarified the uneven distribution of soil erosion risk throughout the terrain. Soil erosion vulnerability ranges from very high to moderate in the northern and central sections of intermediate catchment due to the combined influence of steep slopes, plentiful rainfall, and scant vegetation. These regions make up about half of the research area, and in order to prevent future environmental deterioration, specific erosion control techniques are needed in these locations. On the other hand, areas that are less vulnerable to erosion, which are primarily found in the southern part of catchment and the eastern and western edges, have different environmental characteristics. These places show promise as conservation zones where risks of soil erosion can be effectively reduced by managing and preserving natural features (Figs. 12, 13 and 14).

Fig. 12 Factors' contribution in terms of percentage in soil erosion risk

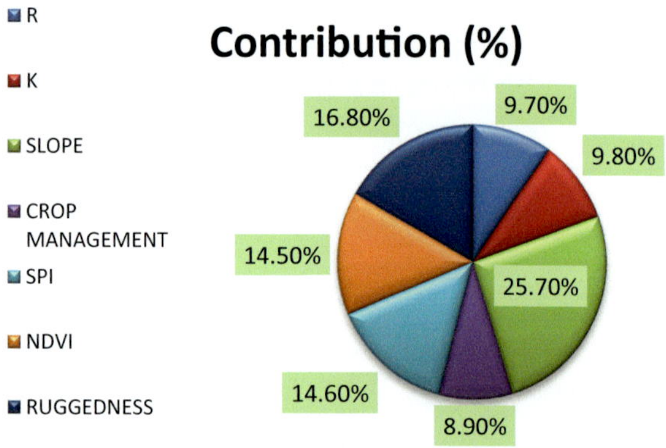

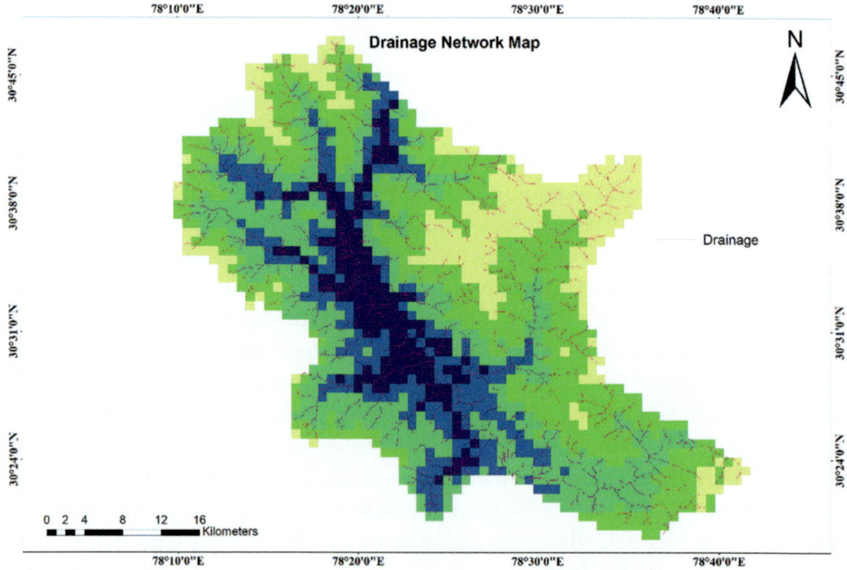

Fig. 13 Drainage network

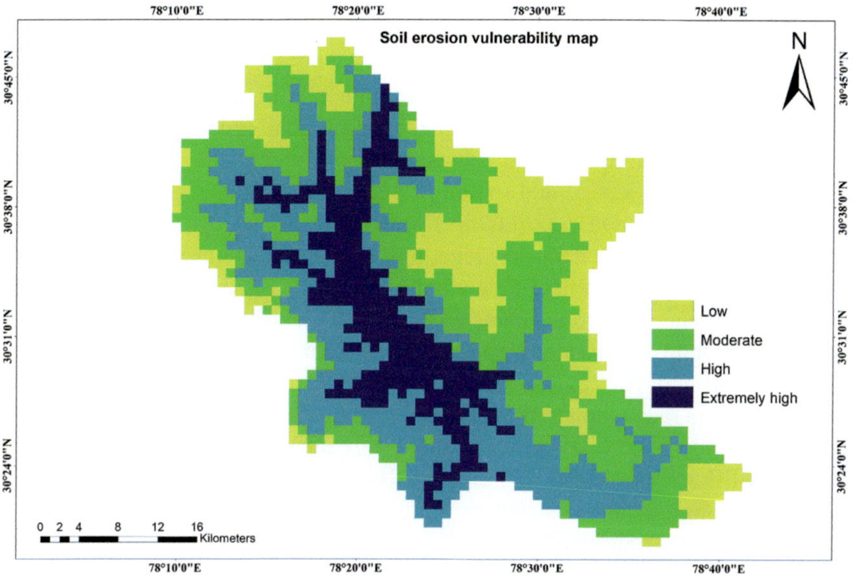

Fig. 14 Soil erosion risk vulnerability map

6 Discussion

The yellow color shows very low risk and the dark blue color shows very high risk of erosion. The results of soil erosion risk vulnerability mapping provide valuable insights into the areas prone to soil erosion within a given region. The mapping process involves integrating various factors that contribute to soil erosion vulnerability, such as topography, land cover, and climate. The results may include the classification of areas into different risk classes or zones. These classes could be defined based on the severity of erosion risk, such as low, moderate, high, or very high risk zones. This zonation provides a clear understanding of the varying degrees of vulnerability. The results may provide insights into the specific factors contributing to soil erosion vulnerability in different areas. For example, the mapping might reveal that steep slopes, bare soil, or intensive land use practices are major contributors to erosion risk. The results should be effectively communicated to stakeholders, including local communities, agricultural practitioners, and environmental agencies. Visualizations, such as maps and reports, can aid in conveying the information and garnering support for sustainable land management practices. The mapping results can be integrated into land use planning processes to guide development activities in a manner that minimizes soil erosion risk. This integration supports sustainable land use practices and natural resource management.

7 Conclusions

In conclusion, the study's comprehensive examination of soil erosion risk and its mapping underscores its critical significance for sustainable land management and environmental well-being. By analyzing key geo-environmental parameters such as slope steepness, ruggedness, and stream power index, we gain crucial insights into vulnerability patterns and contributing factors. This knowledge empowers stakeholders to make informed decisions regarding land use and conservation strategies, prioritizing interventions where they are most needed. The research highlights the significant influence of terrain characteristics, rainfall, and vegetation cover on soil erosion vulnerability, with steep slopes, rugged terrain, and high stream power index emerging as primary contributors. Spatial distribution maps reveal the uneven distribution of erosion risk, emphasizing the need for targeted interventions in areas with high vulnerability, particularly in the northern and central sections of the intermediate catchment. Efforts to mitigate soil erosion should focus on implementing tailored erosion prevention techniques such as afforestation, contour plowing, and erosion control structures. Additionally, identifying conservation zones in less vulnerable areas offers opportunities to preserve natural features and reduce erosion risks effectively. Overall, the findings underscore the importance of proactive management strategies guided by scientific understanding. By taking decisive action, stakeholders can work towards sustainable land management practices that safeguard soil health and promote environmental sustainability, ensuring the resilience of landscapes and the well-being of both ecosystems and human communities.

References

Bassey Eze E, Efiong J (2010) Morphometric Parameters of the Calabar River Basin: implication for Hydrologic Processes. J Geogr Geol 2(1):18–26. https://doi.org/10.5539/jgg.v2n1p18
Esper Angillieri MY (2008) Morphometric analysis of Colangüil river basin and flash flood hazard, San Juan, Argentina. Environ Geol 55(1):107–111. https://doi.org/10.1007/s00254-007-0969-2
Lal R (2003) Soil erosion and the global carbon budget. Environ Int 29(4):437–450. https://doi.org/10.1016/S0160-4120(02)00192-7
Lal R (2005) Soil erosion and carbon dynamics. Soil Tillage Res 81(2):137–142. https://doi.org/10.1016/j.still.2004.09.002
Nag SK, Chakraborty S (2003) Influence of rock types and structures in the development of drainage network in hard rock area. J Indian Soc Remote Sens 31(1):25–35. https://doi.org/10.1007/bf03030749
Ni J-R, Li X-X, Borthwick AGL (2008) Soil erosion assessment based on minimum polygons in the Yellow River basin, China. Geomorphology 93(3):233–252. https://doi.org/10.1016/j.geomorph.2007.02.015
Pandey A, Chowdary VM, Mal BC (2007) Identification of critical erosion prone areas in the small agricultural watershed using USLE, GIS and remote sensing. Water Resour Manage 21(4):729–746. https://doi.org/10.1007/s11269-006-9061-z
Paustian K, Lehmann J, Ogle S, Reay D, Robertson GP, Smith P (2016) Climate-smart soils. Nature 532(7597):49–57. https://doi.org/10.1038/nature17174

Pimentel D (2006) Soil erosion: a food and environmental threat. Environ Dev Sustain 8(1):119–137. https://doi.org/10.1007/s10668-005-1262-8

Pradeep GS, Krishnan MVN, Vijith H (2015) Identification of critical soil erosion prone areas and annual average soil loss in an upland agricultural watershed of Western Ghats, using analytical hierarchy process (AHP) and RUSLE techniques. Arab J Geosci 8(6):3697–3711. https://doi.org/10.1007/s12517-014-1460-5

Quinton JN, Govers G, Van Oost K, Bardgett RD (2010) The impact of agricultural soil erosion on biogeochemical cycling. Nat Geosci 3(5):311–314. https://doi.org/10.1038/ngeo838

Rahman MR, Shi ZH, Chongfa C (2009) Soil erosion hazard evaluation—an integrated use of remote sensing, GIS and statistical approaches with biophysical parameters towards management strategies. Ecol Modell 220(13):1724–1734. https://doi.org/10.1016/j.ecolmodel.2009.04.004

Saaty TL (1990) How to make a decision: the analytic hierarchy process. Eur J Oper Res 48(1):9–26. https://doi.org/10.1016/0377-2217(90)90057-I

Srinivasan R, Karthika KS, Suputhra SA, Chandrakala M, Hegde R (2021) Mapping of soil erosion and probability zones using remote sensing and GIS in arid part of South Deccan Plateau, India. J Indian Soc Remote Sens 49(10):2407–2423. https://doi.org/10.1007/s12524-021-01396-5

Swarnkar S, Malini A, Tripathi S, Sinha R (2018) Assessment of uncertainties in soil erosion and sediment yield estimates at ungauged basins: an application to the Garra River basin, India. Hydrol Earth Syst Sci 22(4):2471–2485. https://doi.org/10.5194/hess-22-2471-2018

Tarolli P, Straffelini E (2020) Agriculture in hilly and mountainous landscapes: threats, monitoring and sustainable management. Geogr Sustain 1(1):70–76. https://doi.org/10.1016/j.geosus.2020.03.003

Wall DH, Nielsen UN, Six J (2015) Soil biodiversity and human health. Nature 528(7580):69–76. https://doi.org/10.1038/nature15744

Wu Q, Wang M (2007) A framework for risk assessment on soil erosion by water using an integrated and systematic approach. J Hydrol 337(1):11–21. https://doi.org/10.1016/j.jhydrol.2007.01.022

Performance Evaluation of a Reservoir During Flooding

Monjeet Kakati, Wajahat Annayat, and Iftekhar Gull

Abstract This chapter presents a combination of goal programming and mixed-integer programming models formulation for a reservoir system. Only gauged flow parameter is taken to determine optimum reservoir storage at any given time and the least downstream flow to minimize the flood water flow in the downstream section of the reservoir. To determine the optimal reservoir release at a particular time flow, the downstream and storage of the reservoir are considered as objective functions in accordance with their priority where downstream flow is given the highest priority or weight. In this chapter, we have executed various hypothetical hydrographs considering two models determining the stability and efficiency of a certain reservoir to minimize flood damage.

Keywords Goal programming · Mixed-integer programming · Weights · Hypothetical hydrographs · Flood damage

1 Introduction

From earlier times, rivers have played a significant role in establishing human settlements (Annayat and Sil 2020a, b, 2021; Annayat et al. 2018, 2021, 2022a, b; Debnath et al. 2023). Area around rivers used to have highest population density as compared to other locations and those residential still exist in the same way (Annayat et al. 2022a, b; Sil et al., 2023; Wajahat et al. 2018). Whenever any hydraulic structure such as dam is constructed on the river, water gets collected on the upstream side known

M. Kakati
Department of Civil Engineering, National Institute of Technology Silchar, Silchar, Assam, India

W. Annayat (✉)
Department of Civil Engineering, National Institute of Technology Srinagar, Srinagar, Jammu and Kashmir, India
e-mail: Wajahatannayat123@gmail.com

I. Gull
Department of Civil Engineering, Institute of Technology, University of Kashmir, Srinagar, Jammu and Kashmir, India

S. C. Pal and U. Chatterjee (eds.), *Surface, Sub-Surface Hydrology and Management*, Springer Geography, https://doi.org/10.1007/978-3-031-62376-9_17

as reservoir (Chaube et al. 2023). A reservoir system comprises of a lot of physical components such as reservoir, outlet, spillway and tunnels and each reservoir system has to deal with an enormous amount of water over which the lives of humans and animals depend (Morris 2020). Human life and their ecology and environment are badly affected by the floods. Flood water is retained by the reservoirs. Whenever there is a peak flow it is retained and flattened by reservoirs. In such a case dealing with uncertain weather and flood water inflow, irrigation requirements and other requirements becomes a complicated task. If the incoming flow from the catchment to the reservoir is uncertain then there will be uncertainty regarding the storage in the reservoir, because if the water level in the storage reservoir increases beyond its permissible value then there may be a case of failure of the dam (Wu et al. 2023). On the other hand, if the water released through the spillways and outlets are very irregular then due to frequent changes of flow discharge scouring may take place and it might cause flooding in the downstream reach (Sun et al. 2023). So we have to restrict the storage of the reservoir up to a certain level as well as restrict the flow discharge through the outlet or the spillway. Sometimes there is a shortage of water at that time there is a need to operate multipurpose reservoir in order to maximize the possible water (Jain et al. 2023).

Optimization of reservoirs for flood control has been a popular area of research for more than 40 years (Wan et al. 2023). In the field of optimization, dynamic programming is extensively used. It is being used because it can take a non-linear and stochastic feature which describes most of our reservoir system. Operation of a flood reservoir is a usual multi-objective problem, which is balanced by both, i.e. benefit and damage (Wan et al 2016; Glavan et al 2020). Whenever there is a multi-reservoir system, it is really difficult to highlight the flood risk for tributaries and main-river so as to assign the flood storage capacity among reservoirs (Lei et al. 2018). These individualities pose substantial modelling and optimization encounters in the flood procedure of multi-reservoir systems. From the literature review, various simulation and optimization techniques developed by Ibrahim et al. (1975) are being used for the modelling of reservoir operations. Various techniques of optimization such as linear programming developed by Divi et al. (1980), Marino and Mohammadi (1981) and Yazicigil et al (1983), dynamic programming by Fults et al. (1972), Tauxe et al. (1980), Turgeon (1980) and Yeh et al. (1980). The objectives of all the models were to maximize the benefits based on the reservoir storage function. However, other operation models are provided in the literature but a few were developed to optimize reservoir operation during flooding. USACE (1975) developed HEC-5 and HEC-5C for reservoir operation for flood control. Needham et al. (2000), developed a mixed-integer LP (Linear Programming) model for flood control in Iowa and Des-Moines River in the US. Needham's model was based on Windsor (1973) that includes release and storage constraints along with one additional parameter, penalty (Unver and Mays 1990). Pre-emptive goal programming model incorporating multiple water inflow forecasting was also presented by Choudhury (2010).

From the literature survey authors found that no study has been conducted based on the stability of reservoir for heavy inflows, Therefore our prime aim of this chapter was to determine the performance, and stability of the reservoir at any time during the heavy inflow from the catchment for two different models taking several different possible inflow hydrographs. In each inflow hydrograph, the rate of flood was different and every hydrograph was of different shape for the same time period. The model was applied to Rani Avanti Bai Sagar reservoir on the Narmada River in India.

2 Study Area

The Narmada River Basin in India (Fig. 1), with a drainage area of roughly 99,000 km^2, lies between 72°32′ E to 81°45′ E longitude and 21°20′ and 23°45′ N latitude. There are three important reservoirs on it, of which Rani Avanti Bai Sagar (RABS) is on the main stretch of the Narmada River. The main purpose of RABS is for irrigation, hydroelectric generation and water supply. A record flood in the year 1999, in the city of Hoshangabad downstream of RABS that caused enormous loss in the city changed the perspective of how these reservoirs are operated. The characteristics of the reservoir used are shown in Table 1.

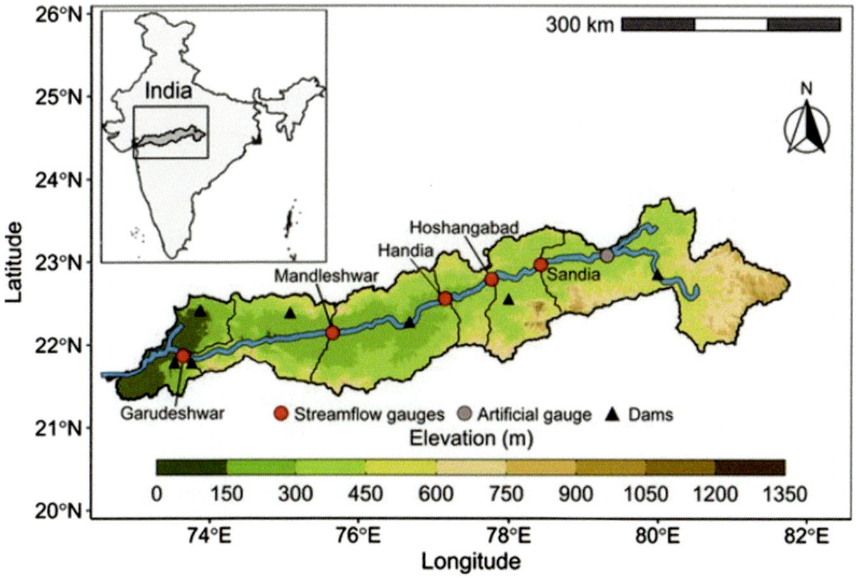

Fig. 1 Map showing the study area

Table 1 Reservoir characteristics

Parameters	Rani Avanti Bai Sagar reservoir
Maximum water level (MWL)	425.70 m
Full reservoir level (FRL)	422.76 m
Dead storage level (DSL)	403.55 m
Maximum capacity (MC)	4806.00 mcm
Full reservoir capacity (FRC)	3920.00 mcm
Dead storage capacity (DSC)	740.00 mcm
Drainage area (DA)	14,556.00 sq.m

3 Methodology (Model Formulation)

We have considered two different models based on the works of Needham et al. (2000) and Choudhury (2010). Both the models were run for different hypothetical hydrographs. The model, Needham et al. (2000) prepared was used in flood control in the Iowa and Des-Moines River in the USA, while the model developed by Choudhury was used in flood control operation of Narmada River. The model comprises of the following constraints and objective functions:

(a) Continuity equation.
(b) Channel routing constraint.
(c) Release restriction constraint.
(d) Variable restriction constraint.
(e) Objective function.

A methodological flowchart is shown in Fig. 2.

3.1 *Model 1*

3.1.1 Continuity Equation

In a reservoir, the continuity equation at time period t is formulated as:

$$I - R = \frac{\mathrm{d}S}{\mathrm{d}t}$$
$$I\mathrm{d}t - R\mathrm{d}t = S_{(t-1)} + S_t$$
$$\sum_{l=1}^{Z} S_{t,l} - R_t = \sum_{l=1}^{Z} S(t-1), l + I_t \tag{1}$$

Here, $R_t = \left(\frac{r_{(t-1)} + r_t}{2}\right)\Delta t$

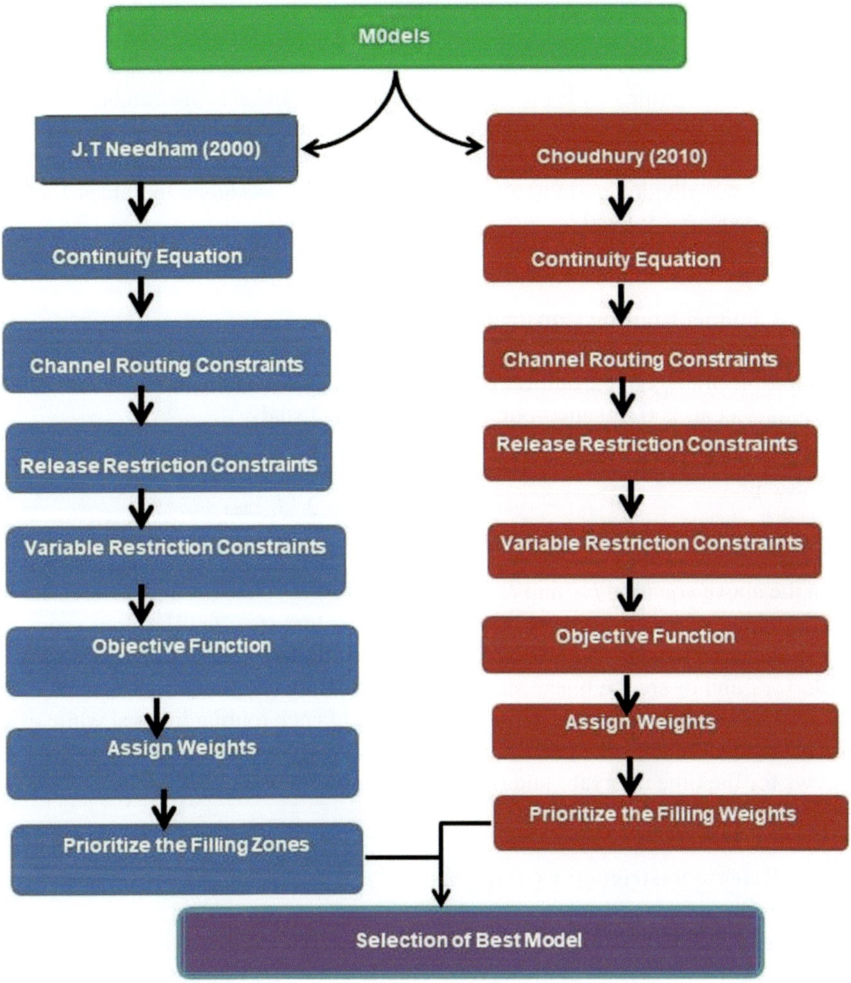

Fig. 2 Flowchart of methodology

$$It = \left(\frac{i(t-1) + it}{2}\right)\Delta t$$

$$\sum_{l=1}^{Z} S_{t,l} - \sum_{l=1}^{Z} S(t-1), l + R_t = I_t \tag{2}$$

where $t = 2, 3, 4, 5, 6, 7, 8, 9, 10, 11, 12$.

I = Inflow to the reservoir.

R = Release from the reservoir.

$\frac{dS}{dt}$ = Rate of change of storage.

where t is time steps, r_t is the release made by the reservoir at time t, i_t is the inflow to the reservoir at time t, S_t is the storage at time step t and Z is the number of zones in the storage. Here the flood hydrograph was taken for a total time interval of 12 h hence the model was run for a total time steps, t equal to 12. The storage reservoir of the dam was divided into three zones denoted by 'Z' as shown in Tables 1 and 2 in Supplementary Material.

3.1.2 Channel Routing Constraint

The multiple inflows single outflow extended Muskingum routing model (Khan 1993; Choudhury et al. 2002) is incorporated in the present study.

$$\sum_{l=1}^{N} f_{t,l} - c_2 r_t = c_1 r_{(t-1)} + c_3 \sum_{l=1}^{N} f_{(t-1),l} \tag{3}$$

In the above equation $f_{t,l,}$ and $f_{(t-1),l}$ are the flows at zones $l = 1, 2, \ldots, N$, where l is the index of the flow zones at time intervals t and $(t - 1)$. The flow zones or the downstream section was also divided into two zones denoted by 'N' as shown in Table 1. c_1 and c_2 are the linear multiple inflow routing coefficients associated with the reservoir release at the beginning and the end of each routing interval while c_3 is the routing coefficient associated with outflow. i_t and $i_{(t-1)}$ are respective unregulated inflows for the time interval t and $(t - 1)$ at gauged uncontrolled channels.

3.1.3 Release Restriction Constraints

Needham et al (2000), expressed the maximum possible release as a piecewise linear function of the storage in the reservoir. This is considered in this model with the storage expressed in terms of zones.

$$r_t - \sum_{l=1}^{Z} \frac{\beta_l}{2}\left(S_{(t-1),l} + S_{t,l}\right) \leq 0 \tag{4}$$

where $S_{t,l}$ is the storage at time t, β_l is the slope of the storage–discharge curve for respective zones. It is well known that storage–discharge curve is non-linear, non-convex function for a given method of gate opening. For the present study 10.70, 10.90 and 11.45 are the values assigned to β_1, β_2 and β_3, respectively (Choudhury 2000).

3.1.4 Variable Restriction Constraints

In the present study in order to handle the decision variables properly mixed-integer programming method is incorporated. In mixed-integer programming, certain variables are restricted to be integers. Needham represented the non-convex storage–discharge relation using logical and binary constraints. The following constraints are added to ensure that water gets filled in zone 1 first then in zone 2 and 3. Similarly, on the downstream section water flows through zone 1 first then it flows through zone 2.

$$S_{t,1} \geq y1 * (\text{Max}S_1) \tag{5}$$

$$S_{t,2} \leq y1 * (\text{Max}S_2) \tag{6}$$

$$\sum_{l=1}^{2} S_{t,l} \geq y2 * \sum_{l=1}^{2} \text{Max}S1 \tag{7}$$

$$S_{t,3} \leq y2 * (\text{Max}S_3) \tag{8}$$

$$f_{t,1 \geq} y3 * (\text{Max}f_1) \tag{9}$$

$$f_{t,2} \leq y3 * (\text{Max}f_2) \tag{10}$$

$$y1,\ y2,\ y3 = \{0,\ 1\}$$

$\text{Max}S_1$, $\text{Max}S_2$, $\text{Max}S_3$ $\text{Max}f_{1,}$ $\text{Max}f_2$ are the maximum storage and maximum flows at storage zones 1, 2, 3 and flow zones 1, 2.

Additionally, the storage zones designated by the index l are constrained as:

$$S_{t,l} \leq \text{Max}S_l \tag{11}$$

While, for the downstream flow at controlled point, each zone is constrained as:

$$f_{t,l} \leq \text{Max}f_l \tag{12}$$

3.1.5 Maximum Permissible Change of Release Constraint

If the release from the reservoir is too rapidly changed then it causes a negative impact such as bank sloughing, erosion, scouring of channel bed and inadequate response time to changing conditions downstream. Changes in release rates may also be limited by the equipment available to change gate or outlet settings. To impose this, the LP model includes a set of constraints that limit the maximum release rate for a particular period based on the last period's release. The max permissible change was allowed at 3000 m^3/s.

$$\left| r_{(t-1)} - r_t \right| \leq 3000 \tag{13}$$

3.1.6 Objective Function

The objective function can be formulated by minimizing the storage and flow downstream attaching some weights on storage and flow value.

$$\text{Minimize Objective} = \sum_{t=1}^{T} \sum_{l=1}^{Z} (W_l S_{t,l}) + \sum_{t=1}^{T} \sum_{l=1}^{N} (X_l f_{t,l}) \tag{14}$$

where, W_l is the penalty associated with each zone of reservoir storage defined by the zone index $l = 1, 2, \ldots, Z$ for the time period t. For the downstream flow zones, X_l is the penalty associated with it for the respective zones specified by the zone index $l = 1, 2, \ldots, N$. Where Z and N are the respective total number of zones in the reservoir storage and the downstream flow point or the downstream section.

Weights are assigned in order to prioritize the filling up of zones. Maximum weight is assigned to 2nd zone and 3rd zones as program would minimize the storage in 2nd and 3rd zones the most. Similarly, the function will also minimize the 2nd zone in the downstream flow and the weight assigned will be the maximum there. These weights are taken on the basis of the importance of a particular zone, for example, the 3rd zone, i.e. the zone above the maximum capacity of the reservoir, it the very undesirable if the water fills up to that level because a sudden change in inflow to the reservoir at this stage would mean failure of the reservoir.

3.2 Model 2

3.2.1 Continuity Constraint

The continuity equation for a reservoir at time period t is given by:

$$S_{t+1} - S_t + 0.5(r_t + r_{t+1}) = 0.5(i_t + i_{t+1}) \quad t = 2, 3, 4, \ldots, t \tag{15}$$

where *St* and *St* + 1 are water storage at time *t* and *t* + 1, *rt* and *rt* + 1 are the release from the reservoir at time *t* and *t* + 1; similarly *it* and *it* + 1 are the inflow at time *t* and *t* + 1.

3.2.2 Channel Routing Constraint

The Muskingum routing model

$$f_{t+1} - \sum_{i=1}^{n} (d_1 r_t + d_2 r_{t+1}) - d_3 f_t = 0 \tag{16}$$

d_1, d_2 and d_3 are the Muskinghum Routing Coefficients (Choudhury 2010). f_t is the flow at downstream whereas r_t is the reservoir release.

3.2.3 Goal Constraints

The goal constraints for reservoir water storage and downstream control points are calculated as:

$$St - \sum_{l=1}^{\text{NZS1}} \overset{+}{\text{DS}} + \sum_{l=1}^{\text{NZS2}} \overset{-}{\text{DS}} = \text{TS} \tag{17}$$

And

$$ft - \sum_{l=1}^{\text{NZF1}} \overset{+}{\text{Df}} + \sum_{l=1}^{\text{NZF2}} \overset{-}{\text{Df}} = \text{TF} \tag{18}$$

"Where NZS1 and NZF1 = number of storage zones and flow zones, respectively, above the target. NZS2 and NZF2 = number of storage and flow zones, respectively, below the target." $\overset{+}{\text{Ds}}$ and $\overset{+}{\text{Df}}$ are the positive deviation of storage and flow above the target while $\overset{-}{\text{Ds}}$ and $\overset{-}{\text{Df}}$ are the negative deviation of storage and flow below the target, respectively. The positive and negative deviation of storage are constrained as:

$$\underset{l}{\overset{+}{\text{DS}}} \leq S \max, l \tag{19}$$

$$\underset{l}{\overset{-}{DS}} \leq S \max, l \tag{20}$$

Similarly, the control point flow deviations are constrained as:

$$\overset{+}{\text{Df}} l \leq f \max, l \tag{21}$$

$$\overset{+}{\text{Df}} l \leq f \max, l \tag{22}$$

3.2.4 Release Restriction Constraints

For a piecewise linear convex storage–discharge function, the maximum release rate for a reservoir can be written as:

$$\begin{aligned} r_t \le {} & \frac{\beta_1}{2}\left(S_{\max,1} - \mathrm{DS}_{3,t} + S_{\max,1} - \mathrm{DS}_{3,(t-1)}\right) \\ & + \frac{\beta_2}{2}\left(S_{\max,2} - \mathrm{DS}_{2,t} + S_{\max,2} - \mathrm{DS}_{2,(t-1)}\right) \\ & + \frac{\beta_3}{2}\left(\mathrm{DS}_{1,t} + \mathrm{DS}_{2,(t-1)}\right) \end{aligned} \tag{23}$$

Which can be simplified into

$$\begin{aligned} r_t & + \frac{\beta_1}{2}\left(\mathrm{DS}_{3,t} + \mathrm{DS}_{3,(t-1)}\right) + \frac{\beta_2}{2}\left(\mathrm{DS}_{2,t} + \mathrm{DS}_{2,(t-1)}\right) \\ & + \frac{\beta_3}{2}\left(\mathrm{DS}_{1,t} + \mathrm{DS}_{2,(t-1)}\right) \le \frac{\beta 1}{2}\left(2S_{\max,1}\right) + \frac{\beta 2}{2}\left(2S_{\max,2}\right) \end{aligned} \tag{24}$$

where $t = 2, 3, 4, \ldots, T$.

Here $\mathrm{Ds}_{1,t}$, $\mathrm{Ds}_{2,t}$ and $\mathrm{Ds}_{3,t}$ are the deviation in storage in zones 1, 2 and 3, respectively. Slope of the storage–discharge relationship is represented by β_1, β_2, β_3, for three zones of the storage and its values are taken as 10.70, 10.90 and 11.45, respectively (Choudhury 2010). r_t is the maximum allowable release and $S_{\max,\,l}$ is the maximum storage in zones 1, 2 and 3.

3.2.5 Maximum Permissible Change of Release Constraint

The max permissible change was allowed at 3000 m^3/s as explained in Model 1.

$$\left|r_{(t-1)} - r_t\right| \le 3000 \tag{25}$$

3.2.6 Objective Function

"An optimal resolution of the conflict requires the use of objective function, with combination of weighted storage and flow variables". The objective function with the weighted goals in the first priority level can be written as:

$$\sum_{t=1}^{T}\sum_{l=1}^{Z} W_l \cdot \mathrm{DS} + \sum_{t=1}^{T}\sum_{l=1}^{N} X_l \cdot \mathrm{Dq} \tag{26}$$

For $t = 2, 3, 4, \ldots, T$,

where W_l and X_l are the weights given to the variation variables.

4 Results and Discussions

For the present study, three hypothetical hydrographs are considered and the model was configured to run for the flood period with an operating interval of 12 h. The proposed model was run using the software LINGO 16.0. The damage site taken in Barmanghat which is a downstream point in the river Narmada. Flood flow at Barmanghat is due to controlled releases from the RABS reservoir. The storage in the reservoir RABS (Bargi) is divided into three zones. First zone represents the storage from the dead water level to the FRL, second zone represents the storage from FRL to MWL. The third zone represents the water level from MWL to the water level at 4900 MCM. The downstream control flow point has a maximum flow of 11,000 m^3/s in the first zone and the second zone has a maximum flow of 14,000 m^3/s. The flow in the first zone represents complete safe flow at the downstream points while flow in the second zone represents flood flow at damage site. The critical flow was considered to be around 25,000 m^3/s. "During flood control operation of a multi-purpose reservoir, the objectives are best served if the reservoirs are operated with full water conservation storage, producing non-damaging flow at the downstream control points". Reservoir operation performance may be regarded as being worst when a reservoir operating below FRL results in damaging flows at the downstream points. "To optimize different goals in the same priority level using PGP technique, the specification of relative weights by the modeller is essential, since the weights decide the relative priority in achieving the different goals stated by the objective". Thus in case of storage maximum penalty is given in zone three, i.e. 30 on the other hand in the case of downstream flow the maximum penalty was given in zone two, i.e. 60. The Muskingum routing coefficients C_1, C_2 and C_3 were taken as 0.014, 0.025 and 0.961, respectively (Choudhury 2010). In the problem, the model was run for both the models, for the same inflow hydrograph taking initial storage level at the FRL, i.e. 3922 MCM, initial release as 10,000 m^3/s and initial downstream flow as 10,000 m^3/s also.

4.1 *Triangular-Shaped Hydrograph (Case 1)*

For triangular-shaped hydrograph shown in Figs. 3, 4 and 5, hydrograph 3, in the case of Model 1, the reservoir approaches its full capacity and then recedes at $t =$ 12. Clearly, this situation is undesirable as it would risk the failure of the reservoir completely if more flood water comes abruptly. For triangular hydrograph 3 in the case of Model 2, the system increases the release to around 37,000 m^3/s as compared to Model 1 which kept its release at around 10,000 m^3/s. The downstream flow resulted in case of Model 2 was taken roughly.

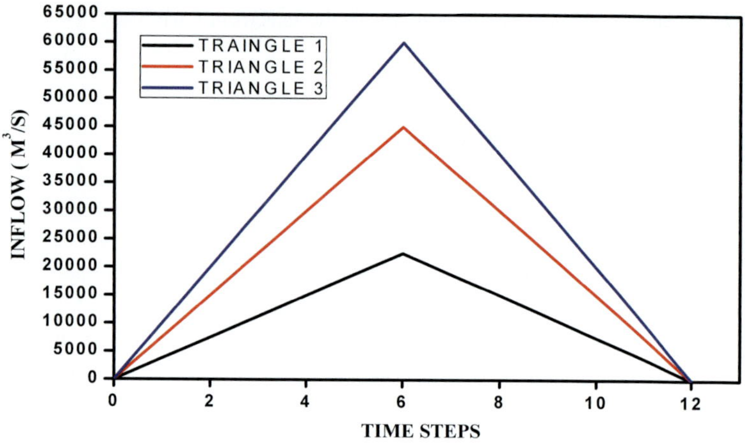

Fig. 3 Triangular inflow hydrograph

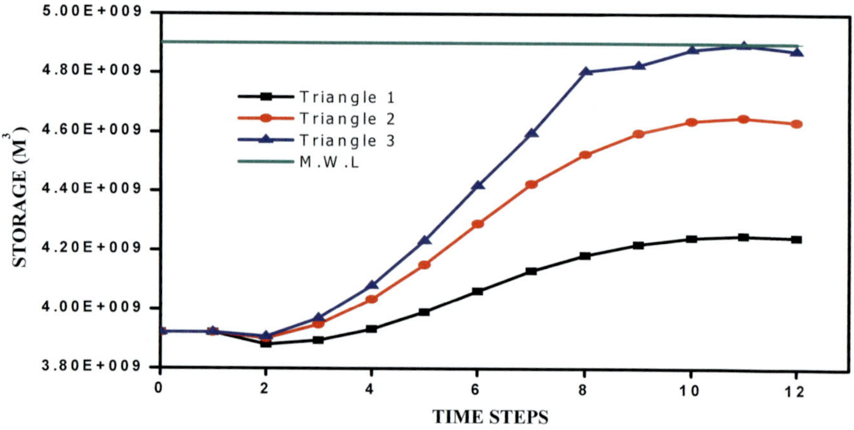

Fig. 4 Storage versus time graph for triangular hydrograph (Model 1)

We considered the target at 3922 MCM for storage and 11,000 m^3/s for flow and the difference of the target and water level at each time period is added to get the total deviation. From the storage deviation, it is seen that for Model 1 deviation is more than the deviation for Model 2 for each of the three inflow hydrographs (Fig. 6).

4.2 *Trapezoidal-Shaped Hydrograph (Case 2)*

For trapezoidal-shaped hydrograph shown in (Figs. 7, 8 and 9) the storage and flow for Model 2 was well within the limits, however in the case of Model 1 for the same

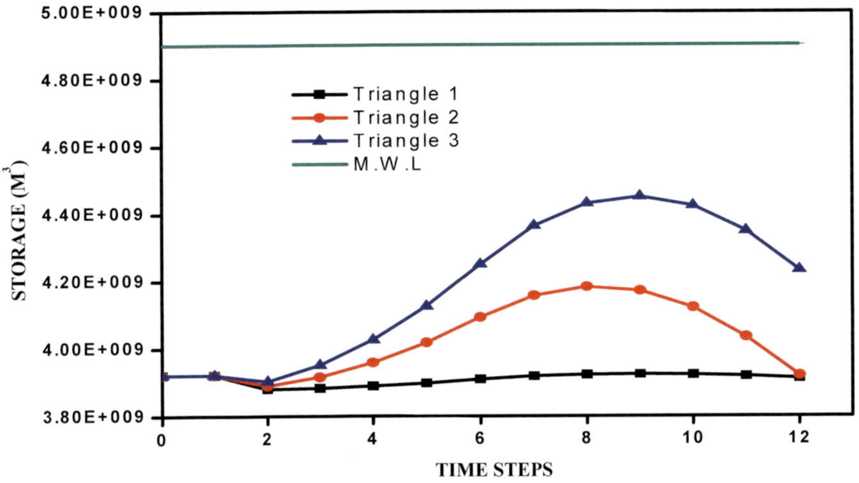

Fig. 5 Storage versus time graph for triangular hydrograph (Model 2)

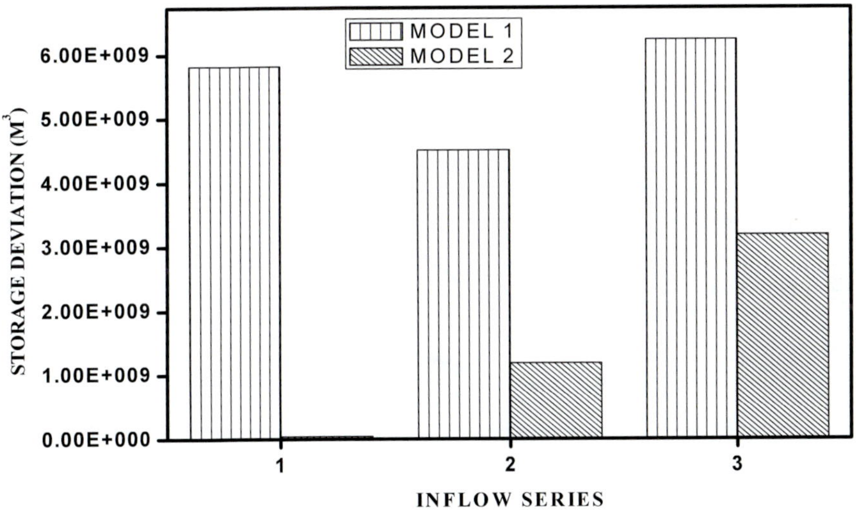

Fig. 6 Total storage deviation for triangular hydrograph

hydrograph the storage dangerously reaches near the maximum level which is quite undesirable. In the third series of the trapezoidal shaped hydrograph the maximum downstream flow for Model 1 is 8916.4 m^3/s and for Model 2 is 15,081.24 m^3/s.

The maximum amount of deviation difference can be seen in series 1 of the storage deviation graph (Fig. 10).

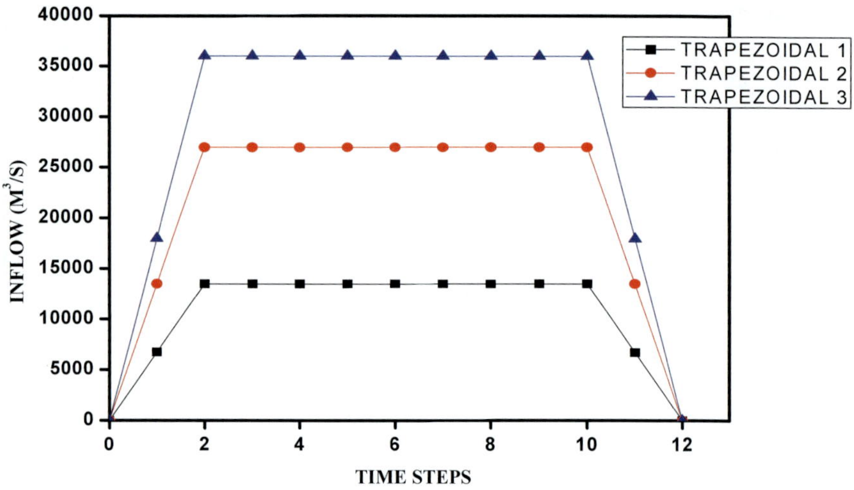

Fig. 7 Trapezoidal shaped hydrograph

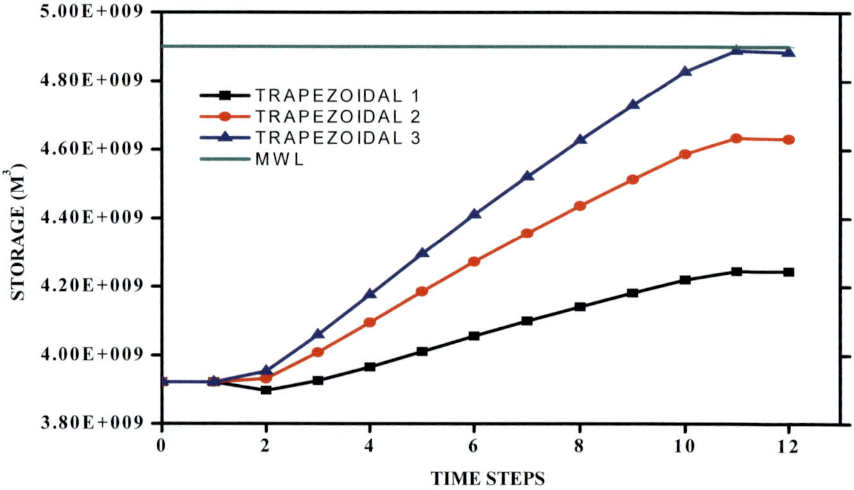

Fig. 8 Storage versus time for trapezoidal hydrograph (Model 1)

4.3 Irregular-Shaped Hydrograph (Case 3)

For irregular-shaped hydrograph shown in Figs. 11, 12 and 13. In the third hydrograph, the maximum storage for Model 1 peaked at 4856.3 MCM while for Model 2 it peaked at 4394.2 MCM. Model 1 gave results more near to the MWL than Model 2 thus Model 2 is performing quite well. The downstream flow for the same above hydrograph, the maximum downstream flow for Model 1 peaked at 9984 m^3/s after

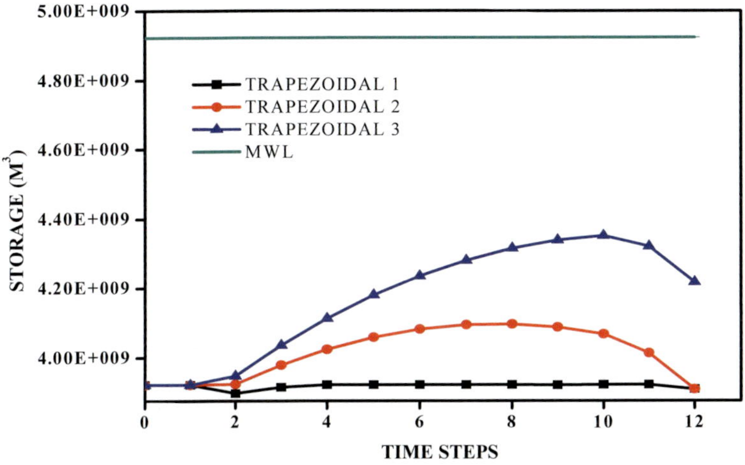

Fig. 9 Storage versus time for trapezoidal hydrograph (Model 2)

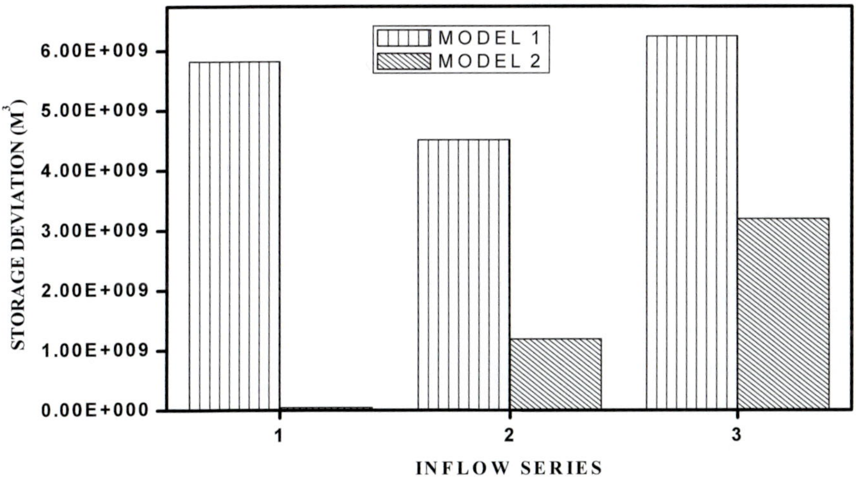

Fig. 10 Comparison of deviation of storage for both models

simulation not considering the initial flow of 10,000 m^3/s. while for Model 2 the same yielded a result of 15,083 m^3/s. In Model 2 the maximum downstream flow is more than there is in Model 1 (Fig. 14).

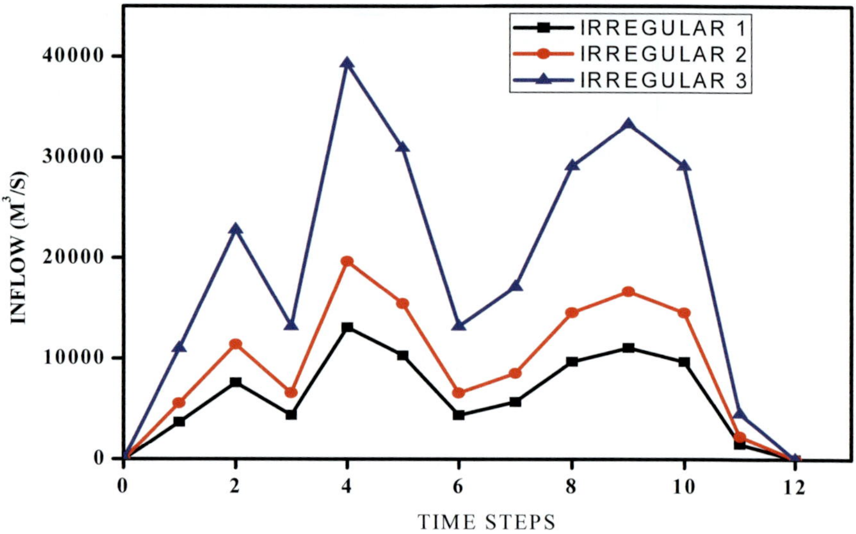

Fig. 11 Irregular-shaped hydrograph

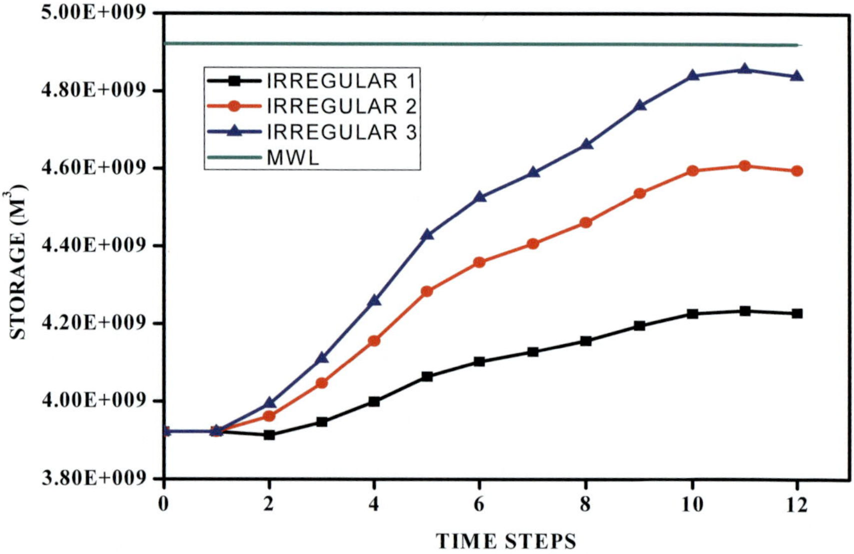

Fig. 12 Storage versus time for irregular hydrograph (Model 1)

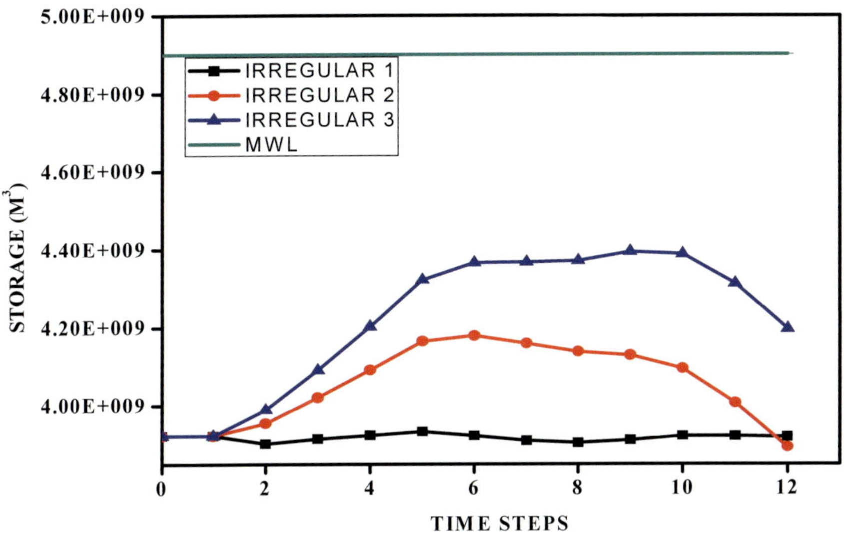

Fig. 13 Storage versus time for irregular hydrograph (Model 2)

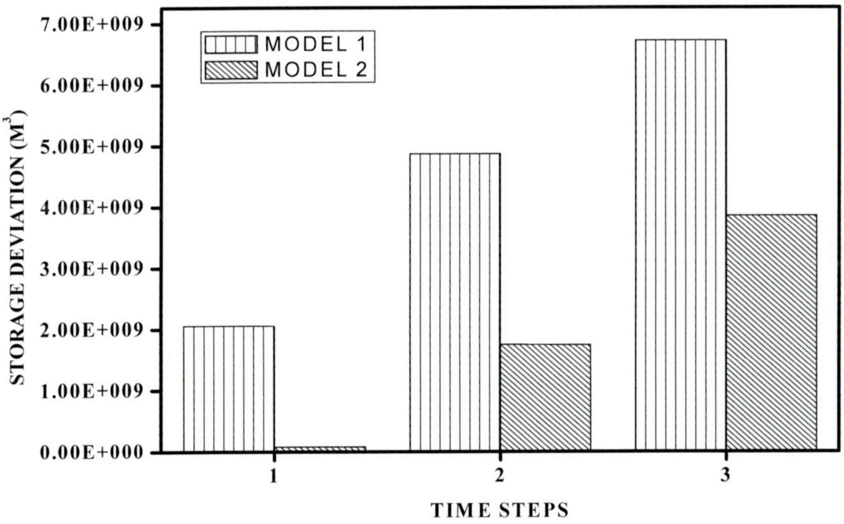

Fig. 14 Storage deviation for irregular hydrograph

5 Challenges and Limitations of the Study

As far as the stability of the dam is considered, even after simulating the reservoir for various different hydrographs we found out that the reservoir won't fail for the given hydrographs. However, if the rate of inflow is increased significantly the model won't be able to optimize. "Consistent with all reservoir operation model applications, the results obtained from this studied network are limited by the approximations made to model the system". All relationships in the model, including maximum possible water releases, the routing equation and the penalty functions, were approximated with linear or piecewise linear functions. Local water flows to the damage site from the intervening catchments and inflows from ungauged tributaries were not considered. Thus, the use of non-linear functions and detailed modelling of local flow are essential for accurate model results.

6 Conclusions

A characteristic comparison of two different types of weighted models for coordinated reservoir operation was presented in this study. The models combined with various reservoir regulation policies can be used for reservoir operation. Multiple objectives and flow variables are used to obtain optimal water release decisions for a reservoir system. In all the model runs it is seen that for Model 1 the water approaches dangerously to the MWL of the reservoir and the flow downstream is also below 10,000 m^3/s which is quite safe. On the other hand, for Model 2, the storage water barely could approach its MWL instead recedes quickly, this is due to the fact that the downstream flow increases within the safe limit of 25,000 m^3/s. For Model 1 priority was given more to reduce the downstream flow variables rather than storage, due to which the storage approaches the MWL while the flow remains under 10,000 m^3/s. whereas for Model 2, priority was based on decreasing the deviation of flow and storage from their respective targets. So it is concluded that if equal priority is given to the flow and storage variable of Model 1, we might be able to obtain a model which can minimize more efficiently.

Supplementary Material

Model Application

See Tables 1 and 2.

Table 1 Parameters of Rani Avanti Bai Sagar reservoir

Parameter	MWL (m)	FRL (m)	DSL (m)	MC (MCM)	FRC (MCM)	DSC (MCM)	DA (sq km)
Rani Avanti Bai Sagar	425.70	422.76	403.55	4806	3922	740	14,556

The division of zones and weights are given in the following manner

Table 2 Reservoir storage zones and control point flow zones and associated weights

Zones, *Z*	Reservoir storage/control flow points and associated weights			
	Storage (mcm)	*W* (Storage weights)	Flow (cumec)	*X* (Flow weight)
1	0–3922	7	0–11,000	10
2	3922–4806	25	11,000–25,000	60
3	4806–4900	30	–	–

References

Abraham CE (1975) Computer use for river regulation. J Hydraul Div ASCE 101(HY2):291–297

Annayat W, Sil BS (2020) Assessing channel morphology and prediction of centerline channel migration of the Barak River using geospatial techniques. Bull Eng Geol Environ 79(10):5161–5183

Annayat W, Sil BS (2020b) Changes in morphometric meander parameters and prediction of meander channel migration for the alluvial part of the Barak River. J Geol Soc India 96(3):279–291

Annayat W, Sil BS, Ajay G (2018) Development of rainfall-runoff model using FFBPNN and LRNN for Silchar city—a case study. Dis Adv 11:19–23

Annayat W, Ashwini K, Sil BS (2022a) Channel migration detection and landuse/landcover changes for Barak River in Northeast India. J Geol Soc India 98:981–990

Annayat W, Sil BS (2021) Predicting meander migration of the Barak River by empirical and time sequence methods. In: Water Security and Sustainability: Proceedings of Down To Earth 2019. Springer Singapore, pp 113–129

Annayat W, Gupta A, Prakash KR, Sil BS (2021) Application of artificial neural networks and multiple linear regression for rainfall–runoff modeling. In: Communication Software and Networks: Proceedings of INDIA 2019. Springer, Singapore, pp 719–727

Annayat W, Ashwini K, Sil BS (2022) Monitoring land use and land cover analysis of the Barak Basin using geospatial techniques. In: Anthropogeomorphology. Springer, Cham, pp 427–441

Can EK, Houck MH (1984) Real-time reservoir operation by goal programming. J Water Resour Plan Manage 110(3):297–309

Chaube UC, Pandey A, Singh VP (2023) Operation of dams and barrages. In: Canal irrigation systems in India: operation, maintenance, and management. Springer Nature, Cham, pp 149–169

Che D, Mays LW (2015) Development of an optimization/simulation model for real-time flood-control operation of river-reservoirs systems. Water Resour Manage 29(11):3987–4005

Choudhury P (2010) Reservoir flood control operation model incorporating multiple uncontrolled water flows. Lake Reserv Res Manage 15(2):153–163

Choudhury P, Rakesh Kumar Shrivastava, Sandeep M. Narulkar (2002) Flood routing in river networks using equivalent Muskingum inflow. J Hydrol Eng 7(6):413–419

Choudhury P et al (2002) Flood routing in river networks using equivalent Muskinghum inflow. J Hydrol Eng 7(6):413–419

Debnath J, Sahariah D, Saikia A, Meraj G, Nath N, Lahon D, Annayat W, Kumar P, Chand K, Singh SK, Kanga S (2023) Shifting sands: assessing bankline shift using an automated approach in the Jia Bharali River, India. Land 12(3):703

Divi R et al (1980) A short term optimization model for operation of a large multi-reservoir hydroelectric system. In: Presented at the TIMS/ORSA Joint National Meeting, Washington, D.C., May

Fults DM, Hancock LF (1972) Optimum operations model for Shasta-Trinity system. J Hydr Div 98(9):1497–1514

Glavan M, Cvejić R, Zupanc V, Knapič M, Pintar M (2020) Agricultural production and flood control dry detention reservoirs: Example from Lower Savinja Valley, Slovenia. Environ Sci Policy 114:394–402

Hsu NS, Wei CC (2007) A multipurpose reservoir real-time operation model for flood control during typhoon invasion. J Hydrol 336(3–4):282–293

Ibrahim BB (1975) Closed loop control of hydrostatic drive farm tractors. Kansas State University

Jain SK, Shilpa LS, Rani D, Sudheer KP (2023) State-of-the-art review: operation of multi-purpose reservoirs during flood season. J Hydrol 618:129165

Johnson WK (1975) Analysis of structural and nonstructural flood control measures using computer program HEC5C. No. 7. US Army, Corps of Engineers

Khan MH (1993) Muskingum flood routing model for multiple tributaries. Water Resour Res 29(4):1057–1062

Kim J, Wei-Chung H, Pen-Chung Y |(1986) COBRA: An adaptive runtime binary optimization framework for multithreaded applications. In: 2007 International conference on parallel processing (ICPP 2007). IEEE, 2007

Lei X, Zhang J, Wang H, Wang M, Khu ST, Li Z, Tan Q (2018) Deriving mixed reservoir operating rules for flood control based on weighted non-dominated sorting genetic algorithm II. J Hydrol 564:967–983

Loganathan GV, Bhattacharya D (1990) Goal programming techniques for optimal reservoir operations. J Water Resour Plann Manage 116(6):820–838

Marino MA, Mohammadi B (1981) Multiobjective optimization in real-time operation of a single reservoir. In: Proceedings, International Symposium on Real-Time Operation of Hydrpsystems, University of Waterloo, Canada, June, pp 534–549

Morris GL (2020) Classification of management alternatives to combat reservoir sedimentation. Water 12(3):861

Needham JT, Watkins DW, Lund JR (2000) Linear programming for flood control in Iowa and Des Moines River. J Water Resour Plann Manage 126(3):118–127

Sil BS, Ashwini K, Annayat W, Debnath J, Farooq M, Meraj G (2023) Observing spatiotemporal inconsistency of erosion and accretion in the Barak river using remote sensing and GIS techniques. Conservation 3(1):14–31

Sun Z, An S, Zhou X, Li Z, Zou L (2023) A framework approach to address the trend and causes of flood stage change in a river reach downstream of a dam influenced by tributaries. Int J Sedim Res 38(5):662–672

Tauxe GW, Mades DM, Inman RR (1980) Multiple objectives in reservoir operation. J Water Resour Plann Manage Div ASCE 106(WR1):225–238

Turgeon A (1980) Optimal operation of multireservoir power system with stochastic inflows. Water Resour Res 16(2):275–282

Unver OI, Mays LW (1990) Model for Real-Time flood control operation Reservoir operation. Water Resour Manage 4(1):21–46

Wajahat A, Sundar SB (2018) Estimation and analysis of possible flood for the Silchar city—a case study. Disaster Adv 11(1)

Wan W, Zhao J, Lund JR, Zhao T, Lei X, Wang H (2016) Optimal hedging rule for reservoir refill. J Water Resour Plan Manag 142(11):04016051

Wan W, Liu Y, Zheng H, Zhao J, Zhao F, Lu Y (2023) Optimization of multi-reservoir flood control operating rules: a case study for the Chaobai river basin in China. Water 15(15):2817

Wasimi SA, Kitanidis PK (1983) Real-time forecasting and daily operation of multi-reservoir system during floods by Linear Quadratic Gaussian Control. Water Resour Res 19(6):1511–1522
Windsor JS (1973) Optimization models for the operation of flood control system. Water Resour Res 9(5):1219–1226
Wu W, Eamen L, Dandy G, Razavi S, Kuczera G, Maier HR (2023) Beyond engineering: a review of reservoir management through the lens of wickedness, competing objectives and uncertainty. Environ Modell Softw 167:105777
Yazicigil H, Houck MH, Toebes GH (1983) Daily operation of a multipurpose reservoir system. Water Resour Res 19(1):1–13
Yeh WW-G (1985) Reservoir management and operation model: a state-of-the-art review. Water Resour Res 21(12):1797–1818
Yeh WWG (1986) Review of parameter identification procedures in groundwater hydrology: the inverse problem. Water Resour Res 22(2): 95–108
Yeh WW-G, Becker L, Chu WS (1979) Real-time hourly reservoir operation. J Water Resour Plann Manage Div 105(WR2):187–202

Applications of Water Quality Index (WQI) and Geostatistical Methods for Long-Term Assessment of Mountain Lake: A Source for Drinking Water in Darjeeling Hill Town

Krishnendu Kumar Pobi, Sayantika Mukherjee, Dipanwita Das, Amrita Saha, Subhankar Dutta, and Sumanta Nayek

Abstract The present study aims to assess the water quality of Senchal Lake, the main source of drinking water for Darjeeling Hill town, covering the time period of 2011–2020. A total of 14 water quality parameters were considered to evaluate lake water quality. The annual mean values of measured water parameters were within the desirable limits for drinking water (BIS, in 2012 Indian standard for drinking water specification 2012 second revision. Bureau of Indian Standards, New Delhi, 2012; WHO, in Drinking water guidelines World Health Organization standard 2006. World Health Organization, Geneva, 2006). Water Quality Index (WQI) was employed to assess the lake's health and to classify the lake water quality. The calculated WQI values ranged between 13.21 and 105.58 during the period of 2011–2020. Year-wise WQI values revealed mostly 'Excellent to Good' types of water quality, except in 2013, when water quality was recorded to be 'very poor'. Various statistical tools such as Pearson's correlation study, principal component analysis (PCA), and cluster analysis (CA) were employed in order to identify the factors which influence the lake water quality. The results of statistical analyses (PCA and CA) have demonstrated

K. K. Pobi
Department of Chemistry, Asansol Engineering College, Asansol, West Bengal, India

S. Mukherjee · D. Das · A. Saha · S. Nayek (✉)
Department of Environmental Science, Amity Institute of Environmental Sciences, Amity University, Kolkata, West Bengal, India
e-mail: sumanta.nayek@gmail.com

S. Mukherjee
e-mail: sgmukherjee@kol.amity.edu

D. Das
e-mail: ddas@kol.amity.edu

A. Saha
e-mail: asaha@kol.amity.edu

S. Dutta
Amity Research Center, Amity University, Kolkata, West Bengal, India

S. C. Pal and U. Chatterjee (eds.), *Surface, Sub-Surface Hydrology and Management*, Springer Geography, https://doi.org/10.1007/978-3-031-62376-9_18

that geogenic sources, catchment features and temporal variations are the major factors which influence the lake water characteristics, with limited contributions from anthropogenic sources. The observation has shown an increasing trend of nutrient enrichment (NO_3^- 0.27 ± 0.15; PO_4^{3-} 0.17 ± 0.38) which may lead to the onset of eutrophication in the near future. Therefore, it is important to identify the sources of nutrients in lake water and address this on an urgent basis. Additionally, regular monitoring of the water quality and the proper execution of effective management plans are essential to protect this lake and its water resources.

Keywords Water Quality Index (WQI) · Environmental factors · Geostatistical applications · Drinking water resource · Sustainable management

1 Introduction

The Himalayan Mountain range is one of the world's largest suppliers of fresh water and habitat for a number of natural freshwater lakes. Different rivers, that serve as a great water resource for Asian countries, including India, originate from Himalayan lakes and glaciers. Darjeeling, popularly recognized as the Queen of Hill stations in the Eastern Himalayas, is located in the northern part of the State of West Bengal, India. Darjeeling Municipality is one of India's oldest municipalities which was established in 1850. In 2011, the municipality featured a population of about 120,000, inhabiting an area of approximately 13.81 sq.km (Rai et al. 2016). To provide drinking water supply to its population, Darjeeling Municipality established a centralized water management infrastructure through Senchal Lake. It is 15 km away from the main town with an elevation of approximately 8160 feet above sea level. It is a beautiful water body situated within the territory of Senchal Wildlife Sanctuary. The water reservoir is actually constituted of two lakes, namely, North Senchal Lake (1910) and South Senchal Lake (1932). The 26 springs of the Senchal Forest and catchment areas of wildlife sanctuary are tapped into a masonary conduit channel as part of Darjeeling town's water supply system. Accordingly, the discharge water flows into Senchal Lake. The total capacity of twin lakes is about 48 million gallons, which provides portable water to Darjeeling city (Ghatani and Khawas 2020; Rai et al. 2016).

From these lakes, water flows to the filter house situated at Jorebunglow, where filtration is done through pressure filters, and then water is conveyed through large water mains to reservoirs at St. Paul's School estate and Rock Ville. However, this water supply was established to support a population of about 15,000 during the years 1910–1915, afterward a number of water supplies were installed, like the Khong Khola pumping station, the Rambi water line, Bokshi Jhora, and BangiaKhola, but these could not cope up with the rapid increase in population, as a result of which acute scarcity of potable water persists, especially the period between December to May (Shah and Badiger 2020). Apart from population explosion, rapid escalation of urbanization, increased rate of evaporation losses from landscapes, rate of runoff losses, lack of infiltration potential, lowering down of ground water table, and increase organic pollution load as a consequence of both tourism pressure and climatological changes triggering water scarcity as well as diminution of water quality. As per the reports of the Waterworks Department (2014), there is a 30–35% loss of water in transmission from Senchel Lake to the 11 settlement areas around (Drew and Rai 2016).

Earlier, researchers documented different assessments of the water quality of several mountain lakes throughout the entire Himalayan region in India. Kumari et al. (2023) worked on freshwater lakes in the Western Himalayan region which is currently facing various challenges due to anthropogenic activities, climate change, and natural processes. Moreover, Kumar et al. (2022) assessed the water quality of Dal Lake, Kashmir. In addition, Auddy et al. (2023) portrayed the qualitative assessment of the glacial sacred lake in Sikkim Himalaya. Potential indicator species for water quality of high altitude wetlands of Garhwal Himalayan zones have been assessed by Singh and Sharma (2020) as well.

Auddy et al. (2022, 2023) investigated the major ion chemistry of post-glacial lake, Khechiopalri Lake, in Sikkim Himalaya; and also evaluated the water quality and pollution load of popular tourist destination Mirik Lake in Darjeeling Himalayan region. They also concluded that different anthropogenic, climatological, and weathering factors contributed to the current eutrophic status of the lake. A similar study was conducted by Kumar and Sharma (2019), who assessed the water quality of the Glacier-fed Neel Tal Lake of Garhwal Himalaya.

Water resource management largely relies on the analysis, evaluation, and monitoring of water quality since these methods provide a comprehensive depiction of the state of the water. Siqueira et al. (2023) examined the correlation between physicochemical characteristics and water quality. Further, they also analyzed the impact of several spatial scales on water quality. In many other scientific applications along with water quality evaluation, techniques like cluster analysis (CA), and principal component analysis (PCA) are frequently used to investigate structure and correlations in multivariate data (Singh and Kataria 2008; Fraga et al. 2020).

Those approaches help in the evaluation of geographical and temporal variations in water quality and distinguish between monitoring stations with similar and dissimilar water quality parameters in order to accomplish efficient management of water resources (Li et al. 2020). This study aims to study long-term water quality data for Senchal Lake using the water quality index (WQI) and geostatistical methods to track pollution sources and create a more clearly comprehendible estimate of water quality, as well as to develop effective management strategies to protect this water resource.

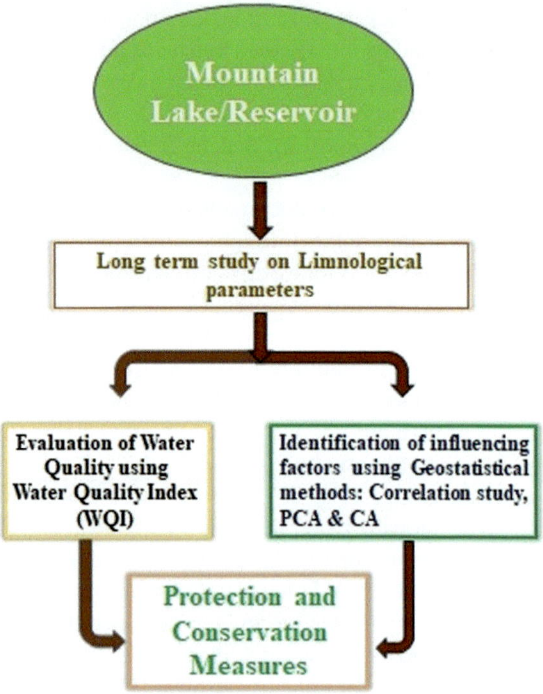

Fig. 1 Flowchart of methodology

2 Methodology

2.1 Sampling Strategies

The monitoring and investigation of Senchal Lake are based on secondary data on limnological parameters of lake water obtained from the West Bengal Pollution Control Board: (http://emis.wbpcb.gov.in/waterquality/viewsampledatacitizen.do) for the time period of 2011–2020. To avoid unforeseen errors, the monthly recorded data was collated and statistically standardized. The annual mean values were used for further analysis, graphical representations, and result construction using the indexing approach, traditional statistical method, geostatistical analysis, and interpretation. The schematic representation of the methodology is shown in Fig. 1.

2.2 Evaluation of Water Quality Index (WQI)

The water quality index has been evaluated by using the drinking water quality standard suggested by the BIS (2012) and WHO (2006). The calculation of WQI was initially developed by Brown et al. (1972) and severally used in earlier many literature (Singh et al. 2018; Kumar et al. 2019) following the weighted arithmetic method to quantify the feasibility of the tested water for human consumption or not. For this purpose, 11 water quality parameters were considered in this study and look forward to observing their combined effects through a single numerical WQI value. Based on their relative influence on overall water quality, the certain weightage (W_a) for each measured variable is assigned between 1 and 5 (Table 1).

At first, the relative weight (W_r) of each parameter is formulated by using the below equation.

$$W_r = W_{ai} \Big/ \sum_{i=1}^{n} W_{ai}$$

where, W_a is represented the assigned weightage of each parameter and n defines the number of considered parameters.

After that, the quality rating scale (Q_i) of each water quality parameter is estimated by the following equation:

$$Q_i = \left[(C_i - V_i) \big/ (S_i - V_i) \right] \times 100$$

where, C_i and V_i represent the measured concentration and the ideal value of each parameter. The V_i for all parameters is taken as zero, except pH (7) and DO (14.6).

Table 1 WQI weightage calculation using WHO (2006) and BIS (2012) standards for drinking water

Parameters	WHO (2006)	BIS standards (2012)		Weight (w_i)	Relative Weight (W_r)
		Desirable	Permissible		
pH	6.5–8.5	6.5–8.5	NX	4	0.103
EC (μS/cm at 25 °C)	2000	250	NX	5	0.128
TDS (mg L^{-1})	1000	500	2000	4	0.103
Turbidity	–	1	5	3	0.077
TA (mg L^{-1})	200	200	600	2	0.051
TH (mg L^{-1})	-	200	600	2	0.051
DO (mg L^{-1})	5	≥ 5	NX	5	0.128
Cl^- (mg L^{-1})	250	250	1000	3	0.077
SO_4^{2-} (mg L^{-1})	200	200	400	4	0.103
NO_3^- (mg L^{-1})	50	45	NX	5	0.128
PO_4^{3-} (mg L^{-1})	0.3	–	–	2	0.051
				$\sum W_a = 39$	$\sum W_r = 1.000$

NX No relaxation

The S_i is the standard permissible limit of each parameter as recommended by WHO (2006) and BIS (2012).

Finally, the WQI is calculated by sum of the sub-indices (SI) value for each physicochemical parameter following the below equation:

$$\mathrm{SI}_i = W_\mathrm{r} * Q_i$$

$$\mathrm{WQI} = \sum \mathrm{SI}_i$$

The classification of WQI value is owing by Brown et al. (1972) and Chatterjee and Raziuddin (2002) and mentioned as excellent (WQI < 25), good (WQI: 25–50), moderate (WQI: 51–75), poor (WQI: 76–100), and very poor (WQI > 100).

2.3 Multivariate Analysis

Multivariate methods provide reliable information on water quality and ecological status by interpreting many parameters and their variability (Markad et al. 2021). Examining multiple variables to identify any potential relationships between them is the primary focus of multivariate analysis. The multivariate statistical analyses comprising of correlation study, and cluster analysis have been carried out with the help of IBM SPSS Statistics software (Version 21).

3 Result and Discussion

3.1 General Characterization of Lake Water

From the study, it is noted that the recorded values of water parameters are within the permissible range of WHO (2006) and BIS (2012) drinking water standards, which represent good quality of lake water, while some parameters are exhibiting some notable changes in the time period of 2011–20 which may indicate some alarming changes in near future. The temperature ranges between 6.82 and 14.37 °C. pH varies between 4.95 and 7.63 which slightly differs from the standard value (6.5–8.5). EC values are very much lower than the standard values of WHO. The value of electrical conductivity (EC) of water affects its taste. Lower the conductivity, greater is the taste. Total dissolved solids (TDS) values range between 15.55 and 66.08 which is also far lower than the WHO limit (although abrupt increase was noticed in 2012) considering the lake water quality is good. EC and TDS are closely related to each other. EC and TDS concentrations are directly proportional to each other. Total hardness is also an important parameter to detect water quality. Its value also ranges between 10.49 and 21.34, between the permissible range of WHO (2006) standard. At such higher altitudes, there is no collision of water with the rocks and other minerals so very few cations are dissolved in water which causes less hardness. Alkalinity, a total measure of substances present in the water that have 'acid-neutralizing' ability is not considered as a pollutant for assessing water quality. The values range between 14.46 and 36.28, which is within the permissible limit of WHO standards. Dissolve oxygen (DO) is an indicator of good health of water bodies and the values also signify good water condition. DO level exhibits a decreasing trend from 2011 to 2020 which can be an alarming signal. From the dataset, there are no such noteworthy changes recorded in BOD and COD levels in the lake water. Low BOD and COD are indicators of less organic pollution load. Nitrate (NO_3^-), sulphate (SO_4^{2-}) and phosphate (PO_4^{3-}) are the parameters which indicate the eutrophic condition of any water bodies. Cl^-, SO_4^{2-} and NO_3^- concentrations in Senchal Lake water are showing a significant escalation from 2011 to 2020, which supports eutrophic nature in the near future, although all values are within the permissible range of recommended WHO (2006) standards (Table 2).

Table 2 Physicochemical characterization of lake water 2011–2020

Parameter	Min	Max	Mean ± SD	Mean standard error	Skewness	Kurtosis
Temperature (°C)	6.82	14.37	10.45 ± 2.63	0.88	0.008	– 1.69
pH	4.95	7.63	7.04 ± 0.72	0.24	– 2.75	8.19
EC (μS/Cm)	18.22	36.8	26.49 ± 5.9	1.96	0.35	– 1.06
TDS (mg/L)	15.55	66.08	35.31 ± 13.4	4.47	0.86	1.81
Turbidity (NTU)	0.95	60.02	8.92 ± 17.2	5.73	3.06	9.53
TH (mg/L)	10.49	21.23	17.26 ± 3.72	1.24	– 0.390	– 1.41
TA (mg/L)	14.46	36.28	23.15 ± 6.53	2.18	0.830	– 0.075
DO (mg/L)	5.22	11.43	8.02 ± 1.41	0.47	0.741	4.27
BOD (mg/L)	0.97	3.46	1.55 ± 0.68	0.23	2.35	6.27
COD (mg/L)	4.78	12.75	8.36 ± 2.7	0.90	0.04	– 1.50
Cl^- (mg/L)	2.53	15.72	7.94 ± 4.46	1.49	0.20	– 1.30
SO_4^{2-} (mg/L)	0.92	15.81	2.95 ± 4.32	1.44	3.07	9.54
NO_3^- (mg/L)	0.12	0.6	0.27 ± 0.15	0.05	1.08	0.07
PO_4^{3-} (mg/L)	0.01	1.32	0.17 ± 0.38	0.13	3.12	9.79

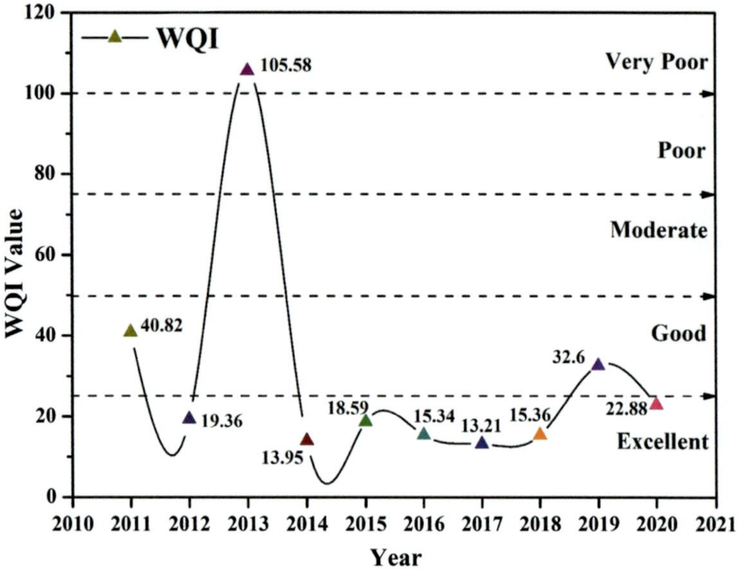

Fig. 2 Year wise WQI values and classification of Lake water

3.2 Evaluation of WQI

The water quality index (WQI) is an effective tool for determining the quality of water for drinking and human use (Singh et al. 2018). WQI is a quality rating indexing approach that uses a single numerical number to replicate the cumulative influence of possible water parameters. The criteria selected for WQI calculation are based on their suitability and relevance for drinking and human uses (Aminiyan et al. 2018).

The year-wise estimated WQI values reveal that the water quality of the monitored lake for the observation period is mostly recognized as 'Excellent' and Good' type, and therefore considered as very much suitable for human uses and consumption for drinking purposes. However, a drastic decline in water quality was recorded for 2013, which indicated 'very poor' type of water with respect to drinking water standards (Fig. 2). Taking consideration into situation, it is therefore required to take proper preventative measures, and follow the effective treatment to ensure the quality of water as per recommended guidelines (WHO 2006; BIS 2012) before supplying to the municipality, residential and settlements areas for human uses.

3.3 Multivariate Analysis

The objective of the multivariate statistical approach is to understand the relations between a large number of independent variables and the dependent variables that are correlated with each other. In this study, multivariate analysis is carried out between 14 water quality parameters of Senchal Lake. In the multivariate analysis, correlation study, Principal component analysis (PCA) and cluster analysis (CA) were performed separately.

Correlation Study

In order to find out whether a relationship exists between the 14 water quality parameters, the correlation analysis has been carried out and simultaneously, observe the magnitude and action of those relationships. From the correlation matrix, the relationships between 14 variables were observed (Table 3). The results have shown that temperature is positively correlated with TDS, and negatively correlated with COD, and NO_3^-, respectively. Similarly, pH and TDS are negatively correlated with NO_3^-, respectively. TA is positively correlated with TH and Cl^-, respectively. DO is positively correlated with BOD and both DO and BOD are positively correlated with $PO^4{}_{3-}$, respectively. COD is positively correlated with NO_3^-. Strong positive correlations between the measured variables infer an identical trend or a common origin, whereas negative correlations may be attributed to their unparalleled nature.

Principal Component Analysis (PCA)

In order to delineate the pollution sources, and identify the environmental factors that influence lake water, Principal Component Analysis (PCA) was carried out. A total of 14 water quality parameters were selected for the analysis. The PCA test was performed based on selecting the rotation method, which was Varimax with Kaiser normalization. The rotation was converged in nine iterations. The rotated component matrix was obtained after executing the PCA test. The vast majority of water parameters in the matrix were separated into five main components based on weightage or loading (Table 4). The first component is comprised of temperature, pH, EC, TDS, COD, and NO_3^-, which can be attributed to the combined influence of temporal factors and wastewater discharge. The second component is denoted by DO, BOD, PO_4^{3-}, and the third component is denoted by TH, TA, and Cl^-. From this observation, the second component direct towards organic waste loading from adjacent forest-dominated areas, while the third component can be linked to anthropogenic contributions in lake catchment areas. However, the fourth and fifth components are distinguished by the higher loadings of SO_4^{2-} and turbidity, respectively, which might be due to weathering and erosion in the lake catchment area. The results of PCA reveal that geogenic sources, watershed features, and temporal fluctuations are the primary determinants of lake water characteristics, while anthropogenic inputs have little impact on lake water quality.

Cluster Analysis (CA)

The cluster analysis has been carried out based on 14 water quality parameters studied for Senchal Lake, namely, NO_3^-, PO_4^{3-}, BOD, SO_4^{2-}, pH, DO, COD, temperature, Cl^-, turbidity, TH, TA, EC, and TDS. After obtaining the result of cluster analysis on the basis of Dendrogram, it has been observed that 14 variables were grouped under three clusters, i.e. Cluster I, Cluster II, and Cluster III.

Table 3 Correlation analysis of studied parameters of Lake water

		Temperature	pH	EC	TDS	Turbidity	TH	TA	DO	BOD	COD	Cl^-	SO_4^{2-}	NO_3^-	PO_4^{3-}
Temperature	Pearson Correlation	1	.484	.545	.706*	-.130	.246	-.262	.288	.361	-.663*	-.448	-.311	-.859**	.110
	Sig. (2-tailed)		.157	.104	.022	.719	.494	.465	.420	.305	.037	.194	.382	.001	.762
pH	Pearson Correlation	.484	1	.423	.538	.173	.586	.081	.575	.184	-.355	-.194	.109	-.748*	.037
	Sig. (2-tailed)	.157		.223	.109	.634	.075	.824	.082	.610	.315	.591	.765	.013	.918
EC	Pearson Correlation	.545	.423	1	.443	.193	.569	.088	.333	.383	-.437	-.053	-.274	-.514	.224
	Sig. (2-tailed)	.104	.223		.199	.592	.086	.809	.348	.274	.206	.884	.444	.129	.534
TDS	Pearson Correlation	.706*	.538	.443	1	-.015	.054	-.365	.396	.397	-.514	-.513	.075	-.657*	.139
	Sig. (2-tailed)	.022	.109	.199		.968	.882	.300	.258	.256	.128	.130	.836	.039	.701
Turbidity	Pearson Correlation	-.130	.173	.193	-.015	1	-.299	-.487	.063	-.075	-.357	-.453	-.053	-.189	-.029
	Sig. (2-tailed)	.719	.634	.592	.968		.401	.154	.862	.836	.311	.188	.884	.600	.938
TH	Pearson Correlation	.246	.586	.569	.054	-.299	1	.717*	.485	.300	.088	.451	-.139	-.312	.217
	Sig. (2-tailed)	.494	.075	.086	.882	.401		.020	.155	.400	.809	.190	.701	.380	.546
TA	Pearson Correlation	-.262	.081	.088	-.365	-.487	.717*	1	-.013	-.113	.404	.907**	-.130	.313	-.158
	Sig. (2-tailed)	.465	.824	.809	.300	.154	.020		.973	.757	.247	.000	.720	.379	.664

(continued)

Table 3 (continued)

DO	Pearson Correlation	.288	.575	.333	.396	.063	.485	-.013	1	.860**	-.021	-.390	-.134	-.560	.812**
	Sig. (2-tailed)	.420	.082	.348	.258	.862	.155	.973		.001	.954	.265	.711	.092	.004
BOD	Pearson Correlation	.361	.184	.383	.397	-.075	.300	-.113	.860**	1	.002	-.420	-.315	-.427	.930**
	Sig. (2-tailed)	.305	.610	.274	.256	.836	.400	.757	.001		.997	.227	.375	.218	.000
COD	Pearson Correlation	-.663*	-.355	-.437	-.514	-.357	.088	.404	-.021	.002	1	.502	.466	.708*	.207
	Sig. (2-tailed)	.037	.315	.206	.128	.311	.809	.247	.954	.997		.139	.175	.022	.566
Cl^-	Pearson Correlation	-.448	-.194	-.053	-.513	-.453	.451	.907**	-.390	-.420	.502	1	.060	.602	-.404
	Sig. (2-tailed)	.194	.591	.884	.130	.188	.190	.000	.265	.227	.139		.868	.066	.247
SO_4^{2-}	Pearson Correlation	-.311	.109	-.274	.075	-.053	-.139	-.130	-.134	-.315	.466	.060	1	.322	-.174
	Sig. (2-tailed)	.382	.765	.444	.836	.884	.701	.720	.711	.375	.175	.868		.365	.631
NO_3^-	Pearson Correlation	-.859**	-.748*	-.514	-.657*	-.189	-.312	.313	-.560	-.427	.708*	.602	.322	1	-.233
	Sig. (2-tailed)	.001	.013	.129	.039	.600	.380	.379	.092	.218	.022	.066	.365		.516
PO_4^{3-}	Pearson Correlation	.110	.037	.224	.139	-.029	.217	-.158	.812**	.930**	.207	-.404	-.174	-.233	1
	Sig. (2-tailed)	.762	.918	.534	.701	.938	.546	.664	.004	.000	.566	.247	.631	.516	

*. Correlation is significant at the 0.05 level (2-tailed).

**. Correlation is significant at the 0.01 level (2-tailed).

Table 4 Outcome of PCA test on composite dataset of Lake water parameters

Rotated Component Matrix					
	Component				
	1	2	3	4	5
Temperature	0.902	0.110	– 0.078	– 0.241	– 0.211
pH	0.732	0.157	0.378	0.345	0.327
EC	0.568	0.215	0.379	– 0.284	0.252
TDS	0.833	0.190	– 0.213	0.190	– 0.133
Turbidity	0.027	– 0.024	– 0.295	– 0.057	0.940
TH	0.287	0.289	0.902	– 0.003	– 0.022
TA	– 0.244	– 0.087	0.918	– 0.068	– 0.234
DO	0.323	0.885	0.168	0.105	0.155
BOD	0.219	0.943	– 0.024	– 0.176	– 0.103
COD	– 0.737	0.229	0.250	0.448	– 0.272
Cl^-	– 0.436	– 0.371	0.767	0.003	– 0.256
SO_4^{2-}	– 0.117	– 0.145	– 0.087	0.942	– 0.033
NO_3^-	– 0.891	– 0.272	0.041	0.146	– 0.186
PO_4^{3-}	– 0.042	0.983	– 0.079	– 0.078	– 0.026

Rotation Method: Varimax with Kaiser Normalization
Rotation converged in 9 iterations

In the Cluster I, four variables, namely, NO_3^-, PO_4^{3-}, BOD, and SO_4^{2-} were grouped together. Consequently, based on four grouped water quality parameters, the Cluster I highlight the deposition of organic materials from forest-dominated catchment areas. Similarly, in the Cluster II, five water quality parameters, namely, pH, DO, COD, temperature, and Cl^- were grouped together. The Cluster II reflects the climatic or temporal influence and anthropogenic inputs. Finally, the Cluster III comprises of the remaining five water quality parameters, including turbidity, TH, TA, EC, and TDS. The Cluster III reveals the soil erosion and geological weathering in the Senchal Lake watershed (Fig. 3).

3.4 Conservation Strategies for Sustainable Management of Lake Water Resources

According to the National Water Policy (2012), it is necessary to maintain sustainable environmental practices in the Himalayan region, including lake water conservation, which is the need of the hour. In this context, the water quality of Senchal Lake would require preservation approaches such as developing an integrated management system from both centralized and decentralized methods. Considering the current

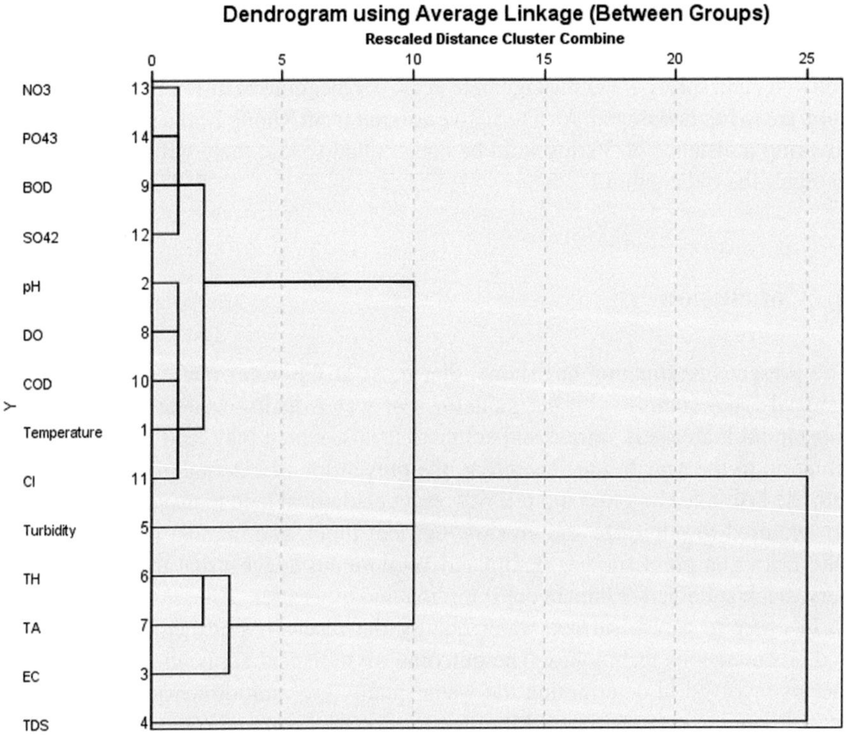

Fig. 3 Cluster analysis (CA) of lake water parameters

status of the lake's condition, some sustainable management approaches should be taken to protect its surface water resources in the future. As the total dissolved solids are showing an escalation, a thorough and consistent programme for forest management, including reforestation, should be introduced so that soil erosion can be prevented. Moreover, recently developed mountain tourism activities in and around Senchal Lake catchment areas should have some restrictions; otherwise, it may affect the lake's health in the near future. As the nitrate level is significantly increasing in the lake water, a constant monitoring programme needs to be established to detect changes in water quality, as it is an early sign of eutrophication. To address this issue, local residents in the nearby villages in the catchment areas should be encouraged to follow organic farming practices and use organic fertilizers instead of chemical fertilizers to restrict the inputs of inorganic nutrients (nitrogen and phosphate) in the lake water.

In order to address the rapid development of human settlements and climate change crises in the Himalayan Mountain region, including the Senchal Lake watershed area, long-term strategies would be beneficial for combating such challenges and simultaneously ensuring the improvement of Senchal Lake water quality in the future (Rai et al. 2016). Additionally, effective and proper treatment of water and an adequate supply of drinking water to meet the demand during peak seasons or during crisis periods have to be ensured. Considering the water crisis during the summer months, rainwater and spring water management at proper locations in different types of reservoirs are to be encouraged. An alternative conduit from Khong Khola to Senchal Lake covering a distance of 5 km should be constructed so that there will be no problem to repair the old conduit.

4 Conclusion

The present investigation has shown that most of the water quality parameters of Senchal Lake comply with the guidelines of water quality standards except for a substantial increase in nitrate and sulphate levels, which may lead to an eutrophic situation in the near future. Therefore, the prevention of external nutrient intrusion into the lake must be given top priority. Strict ecological land management practices are required to minimize soil erosion, nutrient input, and organic loading into the lake from non-point sources. Estimated WQI values suggest that the lake water is very much suitable for human consumption and other uses.

In order to assess surface water quality the research study employed reliable and advantageous techniques. The outcome of statistical applications confers that factors involved in determining the water quality are multidimensional, as several aspects need to be considered. Moreover, the presence of anthropic, geological and meteorological external elements influence the spatial and temporal variation in water quality as well as composition of lake water. Results of the study can be helpful, and regions of India and other countries with comparable features may benefit from similar approaches for the management and conservation of water resources.

A long-term preventive strategy to control the causes of lake degradation is needed to address the issue. A strict law controlling human activity in the lake catchment would be beneficial in this context. To conserve the natural lake condition, sustainable management in terms of watershed management, streamlining management, and continuous water quality monitoring are required to stabilize the existing water demand in the Darjeeling area. These sustainable measures not only maintain the lake condition of Senchal Wildlife Sanctuary but also secure water supplies for the district's growing population.

Acknowledgements The authors would like to thank Amity University Kolkata for providing the resources needed to carry out this research, as well as the West Bengal Pollution Control Board for sharing the data at no cost.

References

Aminiyan MM, Aitkenhead-Peterson J, Aminiyan FM (2018) Evaluation of multiple water quality indices for drinking and irrigation purposes for the Karoon river, Iran. Environ Geochem Health 40:2707–2728

Auddy N, Pobi KR, Chakraborty S, Nayek S (2022) Ion chemistry and quality assessment of a post-glacial sacred lake in Sikkim Himalaya: source identification and conservation measures. Int J Energy Water Resour 6:303–314

Auddy N, Kumar Rai A, Chatterjee S, Pobi K, Dutta S, Nayek S (2023) Trophic classification and assessment of lake health using indexing approach and geostatistical methods for sustainable management of water resources. Water Pract Technol 18(4):967–980

BIS (2012) Indian Standard for drinking water specification 2012 Second Revision. Bureau of Indian Standards, New Delhi

Brown RM, Mccleiland NJ, Deiniger RA, O'Connor MF (1972) Water quality index-crossing the physical barrier. Res Jerusalem 6:787–797

Chatterjee C, Raziuddin M (2002) Determination of Water Quality Index (WQI) of a degraded river in Asansol industrial area (West Bengal). Nat Environ PollutTechnol 1:181–189

Drew G, Rai RP (2016) Water management in post-colonial Darjeeling: the promise and limits of decentralised resource provision. Asian Stud Rev 40(3):321–339

Fraga MDS, Reis GB, da Silva DD, Guedes HAS, Elesbon AAA (2020) Use of multivariate statistical methods to analyze the monitoring of surface water quality in the Doce River basin, Minas Gerais, Brazil. Environ Sci Pollut Res 27:35303–35318

Ghatani S, Khawas V (2020) Some aspects of urban water management in Darjeeling, Himalaya—a study of Darjeeling town. In: Singh MA, Singha K (eds) Understanding urbanisation in Northeast India. Routledge, p 14

Kumar R, Sharma R (2019) Assessment of the water quality of Glacier-fed lake Neel Tal of Garhwal Himalaya, India. Water Sci 33(1):22–28. https://doi.org/10.1080/11104929.2019.1631554

Kumar P, Meena NK, Mahajan AK (2019) Major ion chemistry, catchment weathering and water quality of Renuka Lake, north-west Himalaya, India. Environ Earth Sci 78(10):1–16

Kumar R, Parvaze S, Huda MB, Allaie SP (2022) The changing water quality of lakes—a case study of Dal Lake, Kashmir Valley. Environ Monit Assess 194 (3)

Kumari N, Thakur K, Kumar R, Kumar S, Mahajan D, Brar B, Sharma D, Sharma AK (2023) Freshwater lakes in the Western Himalayan Region: an analysis of the present situation. Water Energy Nexus 6:18–31

Li B, Wan R, Yang G, Wang S, Wagner PD (2020) Exploring the spa tiotemporal water quality variations and their influencing factors in a large floodplain lake in China. Ecol Indic 115:106454

Markad AT, Landge AT, Nayak BB, Inamdar AB, Mishra AK (2021) A multivariate statistical approach for the evaluation of spatial and temporal dynamics of surface water quality from the small reservoir located in the drought-prone area of Southwest India: a case study of Tiru reservoir (India). Environ Sci Pollut Res 28:31013–31031

National Water Policy (2012) as recommended by National Water Board. Ministry of Water Resources: Retrieved 15 August 2012, Government of India

Rai RP, Shrestha P, Sunar D, Rao P, Tamang L (2016) Water security in the Darjeeling Himalaya, unravelling the seen and unseen forces. In: Sustainable Mountain Development Summit V, 20th to 22nd September 2016, pp 1–25

Shah R, Badiger S (2020) Conundrum or paradox: deconstructing the spurious case of water scarcity in the Himalayan Region through an institutional economics narrative. Water Policy 22(S1):146–161

Singh Y, Kataria HC (2008) Studies on water quality of Kaliyasot Dam, Madhya Pradesh, India. Curr World Environ 3:147–152. https://doi.org/10.12944/CWE.3.1.22

Singh S, Sharma RC (2020) Zooplankton diversity and potential indicator species for assessment water quality of high altitude wetland, Dodi Tal of Garhwal Himalaya, India. Acad Arena 12(5):1–16

Singh KR, Dutta R, Kalamdhad AS, Kumar B (2018) Risk characterization and surface water quality assessment of Manas River, Assam (India) with an emphasis on the TOPSIS method of multi-objective decision making. Environ Earth Sci 77(23):1–10

Siqueira TDS, Pessoa LA, Vieira L, Cionek VDM, Singh SK, Benedito E, do Couto EV (2023) Evaluating land use impacts on water quality: perspectives for watershed management. Sustain Water Res Manag 9(6):192. https://doi.org/10.1007/s40899-023-00968-2

WHO (2006) Drinking water guidelines World Health Organization standard 2006. World Health Organization, Geneva

Assessment of Soil Erosion Dynamics and Flood Zonation Using Integrated RS and GIS-Based RUSLE Model for Prioritization of Sub-watersheds Morphometry: A Case Study of Shilabati River Basin, West Bengal, India

Sudip Dey and Suprakash Pan

Abstract This research, conducted in the Shilabati River Basin (SRB) in West Bengal, India, addresses the pressing challenges of soil erosion and floods, which have profound implications for both the environment and the economy. Utilizing integrated remote sensing (RS) and geographic information system (GIS) techniques, the study employs the Revised Universal Soil Loss Equation (RUSLE) to comprehensively assess soil erosion dynamics, sediment yield rates. Flood impact exacerbates soil erosion, with predicted soil loss zones ranging from 0 to 346.9 T/ha^{-1}/year^{-1} and an average annual soil loss of 20 T/ha^{-1}/year^{-1}. Approximately, 79.74% of the basin experiences high to very high erosion severity, highlighting the urgent need for erosion control measures to sustain the ecosystem. Particularly, high soil loss and flood-prone areas are identified in sub-basins like Ghatal and Chandrakona. The findings underscore that agricultural development, land use changes, climate change, population dynamics, and rill/gully erosion significantly contribute to land degradation in the SRB. The study used Sub-Watershed Prioritization Tool (SWPT) for analyzing morphometric and topo-hydrological factors which automatically identify critical and priority of the sub-watersheds. Statistical analyses reveal the positive influence of runoff on soil loss, but other contributing factors are also evident. This research enhances understanding of soil erosion, flood susceptibility, and their correlation with drainage basin morphometry, offering crucial insights for policymakers and stakeholders in developing effective erosion control and management strategies for sustainable land use planning and environmental conservation amidst future climate change scenarios.

S. Dey
ICAR-National Bureau of Soil Survey and Land Use Planning, Regional Center, Kolkata, West Bengal 700091, India

S. Pan (✉)
Department of Geography, Haringhata Mahavidyalaya, University of Kalyani, Kalyani, West Bengal, India
e-mail: pan.suprakash@gmail.com

S. C. Pal and U. Chatterjee (eds.), *Surface, Sub-Surface Hydrology and Management*, Springer Geography, https://doi.org/10.1007/978-3-031-62376-9_19

Keywords Land degradation · Soil loss erosion · RUSLE · Shilabati River · SWPT · Watershed prioritization

1 Introduction

Soil erosion is a pervasive and intricate environmental challenge that significantly impacts land resources, morphometry, hydrology alteration, agricultural productivity, and ecological sustainability worldwide. This phenomenon involves complex processes (Weslati and Serbaji 2024). On the other hand, the analysis of the quantitative characteristics of a drainage morphodynamic network plays a crucial role in assessing soil erosion susceptibility (Strahler 1964).

The Shilabati River Basin (SRB) in West Bengal, India, stands as a poignant example of a region grappling with the formidable threats posed by soil erosion (Mahala 2020). Characterized by high rainfall and vulnerable topography, the SRB faces substantial challenges to its land resources and agricultural productivity. The degradation of land, arising from soil erosion and accompanying floods, gives rise to numerous detrimental consequences (Das and Bandyopadhyay 2015). These include diminished soil fertility, the loss of valuable topsoil, reduced water-holding capacity, and increased sedimentation in water bodies. These cascading effects not only impede agricultural productivity but also jeopardize ecosystem services, posing a profound risk to the overall sustainability of the region environment (Wu et al. 2018). In response to these challenges, this research endeavors to conduct a comprehensive assessment of soil erosion dynamics and flood zonation in the SRB, employing an assimilated remote sensing (RS) and geographic information system (GIS)-based approach (Al Shoumik et al. 2023). The integration of the Revised Universal Soil Loss Equation (RUSLE) with Automated Sub-Watershed Prioritization Tool (SWPT) is instrumental in unraveling the complexities of soil erosion processes and flood susceptibility within a basin. Several factors influence soil erosion susceptibility, including rainfall intensity, land use practices, soil type, slope gradient, and drainage characteristics. This research aims to bridge the knowledge gap by meticulously examining the interplay between soil erosion dynamics, flood zonation, and drainage basin morphometry from high-resolution imageries in the SRB and also elaborating potential avenues for future research. The integration of RS and GIS technologies, coupled with advanced modeling tools, empowers this study to provide valuable insights into the spatial patterns and controlling factors of erosion vulnerability (Pandey et al. 2021) within the SRB. By understanding these dynamics, stakeholders can make informed decisions for sustainable land use zoning, ecosystem restoration, and the conservation of vital water resources. The objectives of this present research article are—(i) to assess soil erosion susceptibility in the SRB by identifying most vulnerable sub-watersheds for soil water conservation and (ii) to examine drainage basin morphometry and explore the relationship with soil erosion rate and recent flood scenarios. In essence, this research contributes not only to the scientific understanding of soil erosion and flood processes but also provides practical implications for the development of evidence-based strategies to address these challenges. The case study of the SRB serves as a microcosm, offering transferable knowledge applicable to regions facing analogous environmental stressors, and paves the way for resilient, sustainable, and ecologically sound land management practices.

2 Study Area

The study area encompasses the SRB, an integral component of the Ganga River system, situated in the Indian state of West Bengal. The Shilabati River (Shilai) originates in the Chotanagpur plateau near Chak Gopalpur village in the Purulia district and flows south-eastward through the districts of Bankura and Paschim Medinipur with tributaries including Buriganga, Kubai, Tamal, and Donai (Roy et al. 2021). The river, spanning 300 km, converges with River Kankai, forming the Rupnarayan River near Ghatal, West Medinipur, ultimately joining the Hooghly River and discharging into the Bay of Bengal. The river basin is characterized by dendritic to sub-dendritic type of drainage pattern is which controlled by relief, rock structure, and regional slope of the area. The SRB is also home to diverse ecosystems, including forests, wetlands, etc. The SRB extended between 22°36′ to 23°12′ North latitude and 86°38′ to 87°43′ East longitudes, covering approximately 3500 km^2 (Fig. 1). Geographically, the upper basin exhibits undulating terrain, representing the proper plateau region, while the lower basin transitions to the peripheral belt of the Chotanagpur plateau at an altitudinal range of − 4 to 230 m. Geological compositions vary, with the upper basin predominantly composed of gneiss and schist, while the lower basin surrounded by younger alluvium. The region experiences a sub-humid tropical climate, characterized by average annual rainfall ranging from 100 to 150 cm and average temperatures between 32 °C and 37 °C (Ghosh and Mondal 2019). The floodplain areas of the Shilabati basin, particularly in the middle and lower sections, are extensively utilized for agriculture, supporting a high-population density of over 10 lakh residents across 23 C.D. blocks. This detailed characterization forms the foundation for the comprehensive assessment of soil erosion dynamics and flood zonation conducted in this research.

3 Materials and Methods

3.1 Data Acquisition

This article utilized freely accessible geospatial datasets from reputable sources for research purposes. Data sources for estimating these factors are outlined in Table 1. Government reports, both published and unpublished, were employed. Digital elevation model (DEM) tiles of SRTM were downloaded from USGS with a resolution

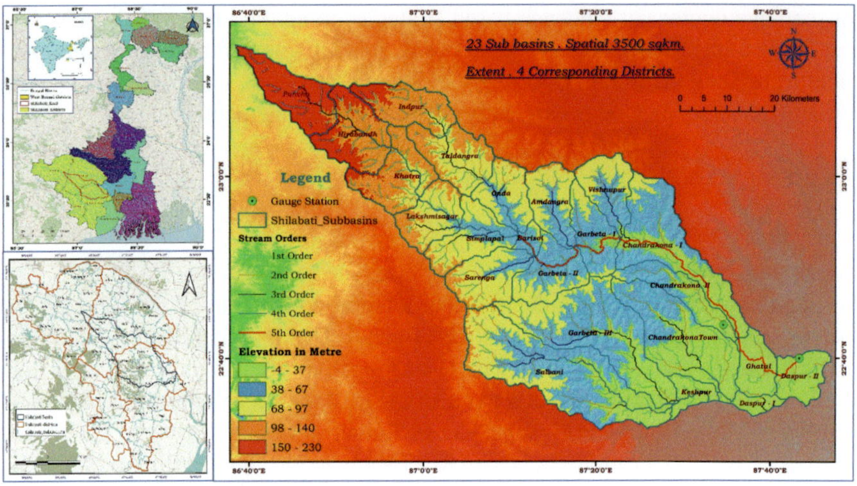

Fig. 1 Location of SRB

of 30 m. The slope map was prepared with the help of same SRTM DEM data extracted from the USGS Earth Explorer. For this study, Landsat 9 images of 30-m resolution were utilized for LULC pattern recognition. Geomorphological Shapefiles were extracted through Bhuvan. The latest 2018 flood events were collected from Dartmouth Flood Observatory database (Tellman et al. 2021). Additionally, field investigations were conducted to validate and verify ground conditions in the SRB.

3.2 Sub-watershed Delineation and Stream Network Generation

The initial step involved mosaicking and projecting DEM tiles covering the SRB onto the WGS-1984 coordinate system, zone 45N. ArcGIS 10.7.1, specifically the Arc SWAT tool, facilitated the generation of stream networks and sub-basins (Luo et al. 2011). The "D8" flow model, using filled sinks as inputs, established flow direction, which the "flow accumulation" function then employed to create a stream network. Subsequently, the flow accumulation grid was then converted into a stream grid via the "Stream Definition" function. This involved applying a threshold value to identify areas with sufficient flow to be considered streams. Stream segmentation connected stream grids to form stream networks, producing vectorized representations. The "Catchment Polygon Processing" function delineated sub-watershed boundaries in vector form. Generated figures depict the resulting sub-watersheds and stream networks, illustrating the methodology employed for comprehensive assessment.

Table 1 Data source for soil erosion susceptibility

Sl. no	Factors	Data type	Source	Unit	Description	References
1	Rainfall erosivity (R)	Annual rainfall data	NASA power access	MJ mm hm^{-2} h^{-1} a^{-1}	Rainfall data for 2011–2021 with 20 rain gauge stations	Singh et al. (1992)
2	Soil erodibility (K)	Soil data	FAO soil data digital shapefile	t h MJ^{-1} mm^{-1}	Soil map based on soil textural categories of 2023	Williams (1995)
3	Slope length and steepness (LS)	Digital elevation model	USGS Earth Explorer	Dimensionless	SRTM DEM (30 m resolution)	Griffin et al. (1988)
4	Crop management (C)	Satellite image	USGS Earth Explorer	Dimensionless	Landsat 9 image (30 m resolution)	Multiple authors
5	Support practice (P)	Agricultural resistivity	USGS Earth Explorer	Dimensionless	SRTM DEM (30 m resolution)	Wischmeier and Smith (1978)

3.3 Theoretical Underpinnings of the Morphometric Indices

Analyzing the drainage morphometry of a river basin helps to understand its hydrological characteristics, predict soil erosion potential, and inform effective water resource management strategies (Faniran 1968). The Sub-Watershed Prioritization Tool (SWPT) was developed to systematically prioritize sub-watersheds within the study area, crucial for assessing runoff/peak discharge and erosion risk. Prioritization was driven by morphometric and topo-hydrological factors, reflecting well-established methods (Abdulkareem et al. 2018). The complex interplay of morphometric features within the drainage area necessitated the precise measurement of channel network gradients and ground slopes (Thakkar and Dhiman 2007). The study incorporated these parameters for sub-watershed prioritization, including (1) areal aspects [drainage density (D), drainage texture (R_t), stream frequency (F_s), constant of channel maintenance (C)], (2) shape aspects [compactness coefficient (C_c), form factor (R_f), circularity ratio (R_c), elongation ratio (R_e)]; and (3) linear aspects [bifurcation ratio (R_b)]; and (4) topo-hydrological attributes [topographic wetness index (TWI), sediment transport index (STI), and stream power index (SPI)] highlighted in Table 2. By leveraging a high-resolution DEM with 30-m pixel size, the SWPT extension tool automatically calculated crucial soil and water conservation factors.

The study leverages weighted sum analysis (WSA), combining powerful geospatial and statistical tools, to precisely rank hydrological units. Using weights based on statistical correlations, WSA pinpoints relative significance of morphometric and topo-hydraulic factors. The prioritization formula (Aher et al. 2014) is defined as

Table 2 Computation of morphometric and topological parameters

Aspects	Parameters	Computation formula	Description	References
Areal	Stream frequency (F_s)	$F_s = N_u/A$ where N_u is total number of stream segments of order "u" and A is area of the drainage basin	F_s refers to the total number of stream segments per unit area of the basin. A high F_s value indicates a dense network of streams, while a low F_s value indicates a sparse network	Horton (1932)
	Drainage texture ratio (R_t)	$R_t = L \cdot \text{Total}/P$ where, L·Total = total length of crenulations in the basin contour, P = the length of the perimeter of the drainage basin	R_t tells about the relative spacing of drainage lines within a basin. It's essentially measuring how efficiently a basin collects and drains water through its stream network	Horton (1945)
	Constant of channel maintenance (C)	$C = 1/D$ where 1 = Schumm constant, D = drainage density	It indicates the average area of land required to maintain one unit of channel length	Schumm (1956)
	Drainage density (D)	$D = \Sigma L_u/A$ where L_u is total length of channels of order "u"	It is used to quantify the degree of dissection by streams in a drainage basin	Horton (1932)
Shape	Elongation ratio (R_e)	$\text{Re} = 4A/\pi Lb^2$, where, A is the drainage basin area, Lb is the basin length	It indicates how closely the basin approximates a circle	Schumm (1956)
	Circularity ratio (R_c)	$R_c = (4 * \pi * A)/P^2$, where: R_c is the circularity ratio, A is the drainage basin area, P is the basin perimeter	R_c tells how closely the shape of a basin resembles a perfect circle, providing insights into its internal drainage efficiency and hydrological behavior	Miller (1953)
	Form factor (R_f)	$R_f = A/L_2$, where, A = area of the basin (km^2), L = basin length (km)	R_f has been calculated as the ratio between the basin area to the square of basin length	Horton (1945)

(continued)

Table 2 (continued)

Aspects	Parameters	Computation formula	Description	References
	Compactness constant (C_c)	$C_c = 0.2821 \times P/$ $A0.5$ where $P =$ basin perimeter,	It is calculated as one-fourth of the ratio between the perimeter of a shape (P) and the square root of twice the product of π (pi) and the area of the shape (A)	Horton (1945)
Linear	Bifurcation ratio (R_b)	$R_b = (N_u/N_u + 1)$ where $N_u + 1 =$ Number of streams of next higher order '$u + 1$'	R_b quantifies the average number of streams of one order that combine to form a stream of the next higher order. It tells about how frequently streams branch within the basin	Schumm (1956)
Topo-hydraulic	Topographic wetness index (TWI)	$\text{TWI} = \ln (A_s/\tan(\beta))$ where $A_s =$ specific catchment area per unit stream length (m^2/m), $\tan(\beta) =$ Local slope angle (°)	It essentially indicates the likelihood of a particular point on the landscape or soil becoming saturated with water	Miller (1953)
	Stream power index (SPI)	$\text{SPI} = A_s * \tan(\beta)$ where $A_s =$ specific catchment area per unit stream length (m^2/m), $\tan(\beta) =$ local slope angle (°)	SPI refers to capability of streams to modify topographical surface through gully erosion and transportation	Whipple and Tucker (1999)
	Stream transport index (STI)	$\text{STI} = (m + 1) * A_s/22.13\text{^}m * \sin(\beta)/0.0896\text{^}n$, where $m =$ the contributing area exponent, $\sin(\beta) =$ local slope angle (°), and $n =$ the slope exponent	It complements other indices like TWI and SPI for a comprehensive assessment of hydrological and geomorphic processes	Moore and Burch (1986)

$$\text{Prioritization} = \sum_{i=1}^{n} Wi * Xi,$$

where *Wi* represents the weight of each morphometric parameter, determined using the WSA approach, while *Xi* denotes the corresponding value.

Though effective, traditional methods involve laborious data analysis and require specialized software. Addressing these constraints, SWPT overcomes traditional hurdles, streamlining analysis for better efficiency and accessibility (Marowka 2018; Rahmati et al. 2016). The conceptual methodology of the SWPT is presented in Fig. 2.

3.4 Estimation of Soil Erosion Susceptibility and Preparation of Thematic Maps

SRB's annual soil loss was estimated through the Revised Universal Soil Loss Equation (RUSLE). To address certain limitations in the Universal Soil Loss Equation (Wischmeier and Smith 1978) presented in the United States Department of Agriculture (USDA) handbook, RUSLE employed tailored adjustments, resulting in a more accurate and effective tool for soil erosion prediction. The RUSLE model estimates mean annual soil degradation over time (A) by considering parameters such as runoff, cropping diversity, topography, and soil conservation. This can be mathematically expressed as (Renard et al. 1997):

$$\mathbf{A} = R * K * L * S * C * P,$$

where A = the average potential soil loss per unit area (tons/ha^{-1}); R = the rainfall erosivity factor (MJ mm hm^{-2} h^{-1} a^{-1}); K = the soil erodibility factor (t h MJ^{-1} mm^{-1}); L = the slope length factor; S = the slope steepness factor; C = the cover and management factor; P = the support practices factor. Table 1 serves to outline the data sources utilized for the estimation of RUSLE conditioning factors. Thematic maps of the SRB were generated for illustrating the spatial distribution of these conditioning factors. The study also focused on assessing drainage basin morphometry and soil erosion in the SRB, contributing to a comprehensive understanding of erosion dynamics in the region.

3.4.1 Rainfall Erosivity Factor (R)

Rainfall erosivity captures the potential of rain to detach and transport soil particles. It is calculated by multiplying the kinetic energy of raindrops, determined by their size, intensity, and fall velocity (Pandey et al. 2007). An empirical equation (Singh et al. 1992) has been used for estimating annual and seasonal R factors in Indian context, obtained as follows:

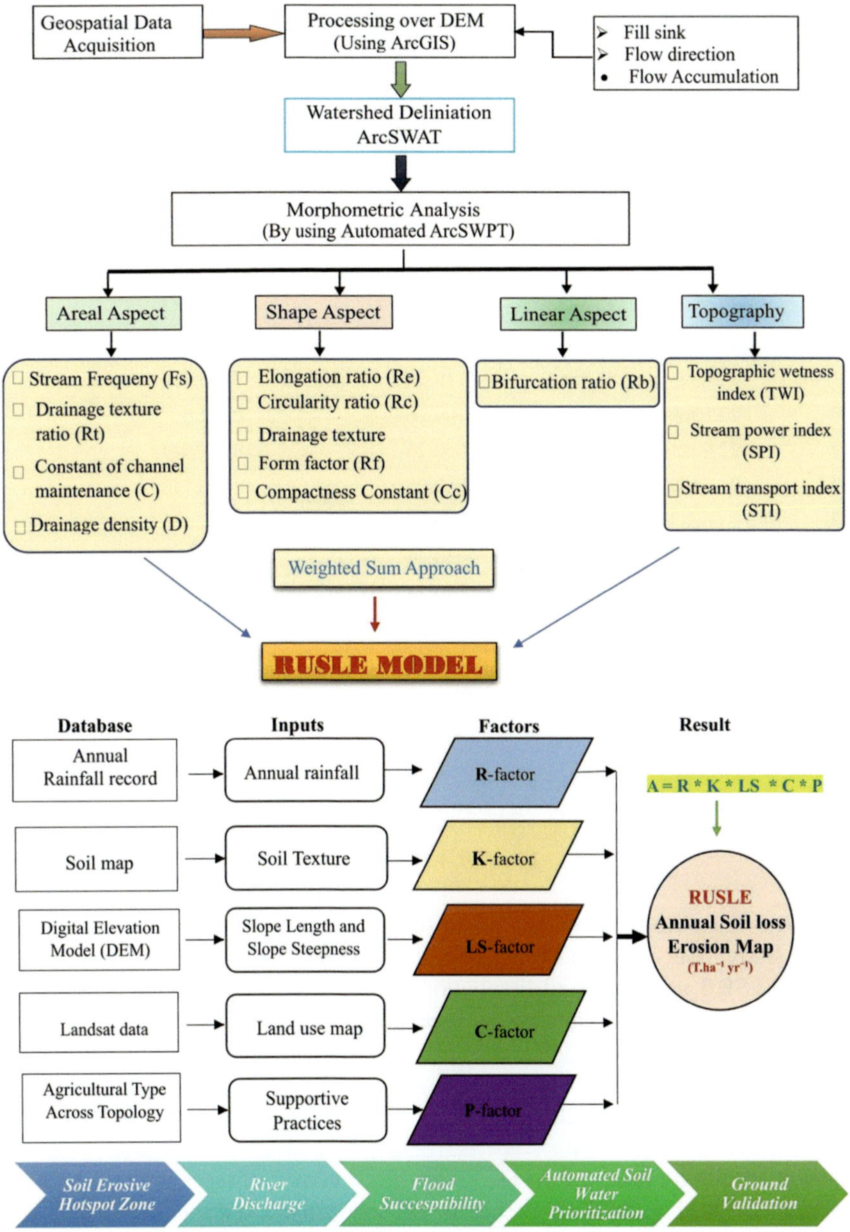

Fig. 2 Flowchart of methodology

Table 3 Average annual rainfall (mm) and calculated *R* value for the selected stations

Station	Latitude	Longitude	Annual rainfall 2021	Rainfall erosivity factor
1	22.6144	87.7001	2214.61	882.9034
2	22.5297	87.6349	2210.61	881.4514
3	22.6359	87.3251	2214.12	882.7256
4	22.6488	87.5966	2211.61	881.8144
5	22.6637	87.7372	2214.61	882.9034
6	22.6707	87.7995	2213.24	883.4061
7	22.7582	87.161	2050.26	823.2444
8	22.7663	87.553	2119.26	848.2914
9	22.832	87.4864	2119.26	848.2914
10	22.8569	87.4095	2109.26	844.6614
11	22.9221	87.1095	2050.26	823.2444
12	23.0169	87.174	2050.06	823.1718
13	23.0283	86.8609	2050.26	823.2444
14	23.1471	86.7332	2030.26	815.9844
15	23.1508	86.9309	2040.26	819.6144
16	23.2127	86.6892	2050.26	823.2444
17	22.7247	86.9546	2121.13	848.9702
18	23.1064	86.3912	2080.41	834.1888
19	23.3378	86.3904	1949.13	786.5342
20	23.3523	86.7879	1896.71	767.5057

$$\mathbf{R} = 79 + 0.363 * P,$$

where $R =$ the average annual rainfall erosivity factor; and $P =$ the recorded rainfall. In this study, R is computed using data from 20 rain gauge stations from in and around the SRB, and the spatial distribution of R factors estimated through the inverse distance weighting (IDW) interpolation method, as presented in Table 3.

3.4.2 Soil Erodibility Factor (K)

The soil erodibility index factor (K) values for diverse soil types in the SRB were determined based on the FAO soil map. Subsequently, a soil erodibility map was generated by plotting the K values of each map unit. The K factor, ranging from 0 to 1, signifies the susceptibility to soil erosion, with values closer to 0 indicating lower vulnerability and those closer to 1 indicating higher susceptibility. A detailed table was compiled, considering distinct features of various soil types, and the K values of surface soil were computed. The K factor was estimated in SI units using a specific equation (Williams 1995).

$$K = f_{\mathrm{csand}} \times f_{\mathrm{cl}} - si \times f_{\mathrm{orgc}} \times f_{\mathrm{hisand}}$$

$$f_{\mathrm{csand}} = \left[0.2 + 0.3 \times \exp(-0.256 \times m_{\mathrm{s}} \times (1 - m_{\mathrm{silt}}/100))\right]$$

$$f_{\mathrm{cl\text{-}si}} = \frac{(m_{\mathrm{silt}})3}{(m_{\mathrm{c}} + m_{\mathrm{silt}})}$$

$$f_{\mathrm{orgc}} = \left(1 - 0.25 \times \mathrm{orgC}/\left(\mathrm{orgC} + \exp\left[3.75 - 2.95 \times \mathrm{orgC}\right]\right)\right)$$

$$f_{\mathrm{hisand}} = \left(1 - 0.7 \times (1 - m_{\mathrm{s}}/100)/\left((1 - m_{\mathrm{s}}/100) + \exp\left[\begin{array}{c} -5.51 + 22.9 \\ \times(1 - m_{\mathrm{s}}/100) \end{array}\right]\right)\right),$$

where f_{csand} = the sand factor, $f_{\mathrm{cl\text{-}si}}$ = the silt and clay factor, f_{orgc} = the organic carbon factor, f_{hisand} = the high sand factor, m_{s} is the percentage of sand in topsoil, m_{silt} is the percentage of silt in topsoil, m_{c} is the percentage of clay content in topsoil, orgC = topsoil percentage of the organic content.

3.4.3 Topographic Erosivity Factor (LS)

The topographic erosivity factor (LS) stands as a key parameter in RUSLE analysis, exerting significant influence on soil erosion by water. It captures the intricate interplay of slope characteristics and their impact on erosivity. As slope length increases, surface runoff accumulates, amplifying soil erosion (Cerdà and Doerr 2008). The LS factor computed from slope gradient and length, derived from processed DEM data, including fill and flow direction. Utilizing the ArcGIS spatial analysis raster calculator function, the LS map derived with an empirical formula (Griffin et al. 1988).

$$\begin{aligned} \boldsymbol{LS} &= \mathrm{Power}(\text{“Flow Accumulation”} * \text{Cell size}/22.13, 0.4) \\ &\quad * \mathrm{Power}(\mathrm{Sin}(\text{“SLOPE”} * 0.01745/0.0896), 1.4) * 1.4. \end{aligned}$$

3.4.4 Crop Management Factor (C)

The *C* factor, as per RUSLE, plays a pivotal role in determining erosion rates based on cropping patterns (Kumar et al. 2015). To compute the *C* factor, the land use-land cover (LULC) map of the SRB was utilized. Seven major LULC classes were classified from Landsat 9 Satellite imagery using supervised classification and the maximum likelihood method in ARCGIS 10.7.1. Subsequently, the LULC map was reclassified, assigning C factor values to each class to generate the crop management factor map. Landsat 9 data, processed through supervised classification, facilitated

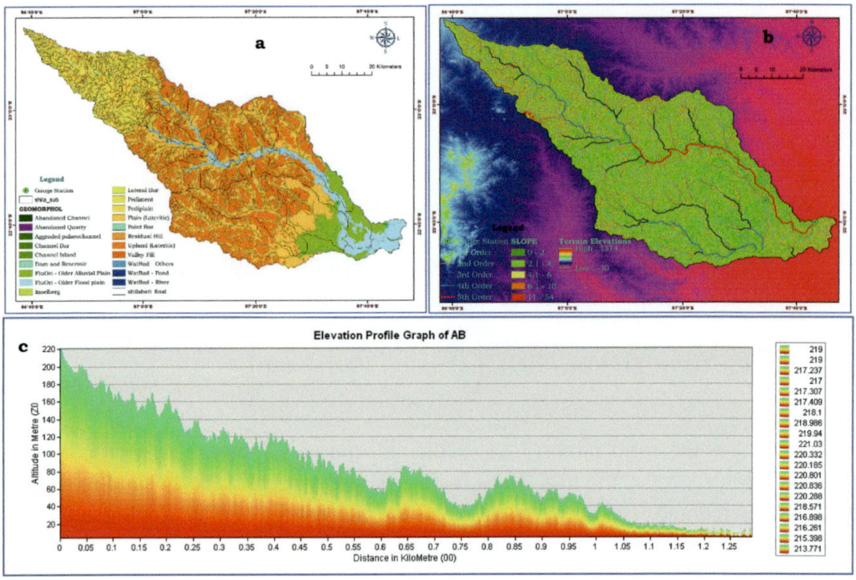

Fig. 3 Topographical attributes of SRB **a** Basin geomorphology, **b** Slope, **c** Elevation profile

the conversion of the LULC raster map to a vector format. The empirical scheme proposed by various scientists determined the *C*-factor values for each LULC class.

3.4.5 Conservation Supporting Practice Factor (P)

The conservation practice factor (*P*) reflects the ratio of soil loss under a support practice compared to that in conventional straight-row farming up and down the slope. This factor acknowledges the positive impact of supportive practices (Renard et al. 1997). In this study, the attribute value of the *P* factor was determined based on observed contouring practices in the agricultural land of the SRB (Wischmeier and Smith 1978). The *P* factor associates control practices that mitigate runoff erosion through their impact of drainage morphometry, runoff concentration, velocity on soil type. With values ranging from 0 to 1, closer to 0 indicates effective conservation practices, while nearing 1 signifies inadequate conservation practices (Fig. 3; Table 4).

3.5 Prioritization of Sub-watersheds

Prioritization of sub-watersheds is a pivotal step in integrated watershed management. The SWPT-generated correlation matrix provides a comprehensive overview of factor

Table 4 *P* factor values as per agricultural practice after Wischmeier and Smith (1978)

Slope (%)	Conservation support practice factor (*P*)		
	Contouring practice	Strip cropping practice	Terracing practice
0.0–7.0	0.55	0.27	0.10
7.0–11.3	0.60	0.30	0.12
11.3–17.6	0.80	0.40	0.16
17.6–26.8	0.90	0.45	0.18
> 26.8	1.00	0.50	0.20

interactions, facilitating prioritization decisions based on their relative importance. The WSA index, calculated by SWPT, acts as a decision-making compass, steering sub-watershed prioritization based on their unique characteristics (Rahmati et al. 2016). The tool pinpoints the most vulnerable sub-watersheds by ranking them in descending order of susceptibility to runoff and erosion, with the most critical areas at the top of the list.

4 Results and Discussions

4.1 Morphometric and Topo-hydrological Parameters

The prioritization of sub-watersheds was accomplished through SWPT. The results, summarized in Table 5, reveal significant geomorphic parameters. It has been observed that rugged landscapes, fast-flowing streams, and flood potentiality are closely intertwined. Morphometric diversity of the watersheds and its influence on SRB's landform characteristics are shown in Fig. 5. The frequency of streams (F_s) exhibits variation, ranging from 0.000000171 in Ghatal, Salbani, to 0.0000139 in Daspur-I. Bifurcation ratio (R_b) values ranged from the highest in Garbeta-III (3.8333) to the lowest in Onda (1.75). Elongation ratio (R_e) and form factor (R_f) show similar trends, with Daspur-I having the highest Re (0.970569) and Daspur-I and Garbeta-III having the highest and lowest R_f values, respectively (0.739827 and 0.221011). Circularity ratio (R_c) values ranged from the highest in Ghatal (0.440847) to the lowest in Chandrakona-I (0.188079). Drainage density (D) ranks Daspur-I and Garbeta-III at the first and last positions, while compactness coefficient (C_c) values range from the highest in Chandrakona-I (2.3053) to the lowest in Ghatal (1.505754). In the present study, TWI varied between 11.490 and 11.930 where lower reaches have presented higher value of TWI. SPI ranged between 4.35 and 4.59, showing that most parts of the basin have no or less surface erosion. All of these parameters were very important in determining and delineating soil erosion and flood-prone regions.

Table 5 Morphometric and topo-hydrological attributes of the sub-basins

Sub-basin	Sub-basin name	F_s	R_b	R_f	R_e	R_c	D	R_t	C_c	C	TWI	SPI	STI
1	Daspur-II	3.68E−06	2.232143	0.610922	0.881971	0.417539	0.00217	0.006231	1.54721	460.7723	11.93531	4.562643	13.73767
2	Daspur-I	1.39E−05	2.561111	0.739827	0.970569	0.318593	0.004305	0.011934	1.77125	232.312	11.84934	4.450677	12.46199
3	Ghatal	1.71E−06	2.66523	0.638374	0.901569	0.440847	0.001375	0.002969	1.505754	727.5175	11.91259	4.420123	13.70853
4	Keshpur	2.97E−06	2.673851	0.519609	0.813392	0.26651	0.002029	0.003052	1.936606	492.8574	11.93043	4.350023	13.20433
5	Chandrakona Town	2.52E−07	1.910256	0.293987	0.611823	0.322437	0.00064	0.000687	1.760661	1563.493	11.8993	4.450686	15.38757
6	Chandrakona -II	1.32E−06	3.319444	0.449195	0.756274	0.188693	0.001238	0.001702	2.30155	807.4464	11.78048	4.403362	13.81628
7	Chandrakona-I	2.77E−07	2.259615	0.290698	0.60839	0.188079	0.000649	0.000558	2.3053	1541.572	11.76992	4.468636	14.8695
8	Salbani	1.71E−07	2.133333	0.248927	0.562986	0.242103	0.000517	0.0005	2.031878	1934.034	11.73976	4.489719	15.56013
9	Garbeta-III	1.7E−07	3.833333	0.221011	0.530479	0.23555	0.000516	0.000441	2.059948	1938.889	11.73562	4.479978	15.39702
10	Garbeta-I	6.81E−07	2.340278	0.517508	0.811746	0.350595	0.00102	0.001217	1.688476	980.1949	11.67927	4.505678	14.64422
11	Vishnupur	5.99E−07	2.102273	0.401886	0.715341	0.377034	0.000923	0.001105	1.628199	1083.2	11.60571	4.539292	14.5993
12	Amdangra	6.11E−07	1.8375	0.378096	0.693845	0.393087	0.000879	0.001189	1.594607	1137.593	11.57706	4.570606	14.52463
13	Garbeta-II	5E−07	1.848485	0.497183	0.795646	0.378718	0.000837	0.001005	1.624576	1194.245	11.68558	4.514108	14.4457
14	Sarenga	5.49E−07	2.261364	0.408269	0.720999	0.434279	0.000844	0.001193	1.517098	1185.187	11.66674	4.500932	14.52096
15	Barisol	3.93E−06	2.270833	0.449359	0.756412	0.39617	0.002136	0.003451	1.58839	468.2249	11.63339	4.433715	13.1324
16	Simplapal	5.04E−07	2.147727	0.379018	0.694691	0.304521	0.000849	0.000924	1.811712	1178.526	11.67475	4.502468	14.69757
17	Onda	6.21E−07	1.75	0.438343	0.747082	0.396154	0.000979	0.001154	1.588422	1021.799	11.60826	4.555799	14.53343
18	Lakshmisagar	6.15E−07	1.967949	0.686198	0.93473	0.434453	0.000913	0.001271	1.516794	1094.858	11.62167	4.56146	14.38504
19	Taldangra	5.44E−07	1.954167	0.400215	0.713852	0.338257	0.000805	0.001089	1.718994	1241.512	11.59452	4.553265	14.44133

(continued)

Table 5 (continued)

Sub-basin	Sub-basin name	F_s	R_b	R_f	R_e	R_c	D	R_t	C_c	C	TWI	SPI	STI
20	Khatra	6.26E−07	3.375	0.628543	0.8946	0.319561	0.000994	0.001262	1.768564	1006.501	11.49136	4.657791	14.50111
21	Indpur	5.37E−07	2.033333	0.303933	0.622086	0.320886	0.000898	0.000994	1.764909	1114.063	11.56609	4.595087	14.67648
22	Hirabandh	4.02E−07	1.801136	0.579037	0.858647	0.333709	0.000793	0.000833	1.730668	1260.334	11.51018	4.672784	14.56979
23	Puncha	5.13E−07	2.096591	0.262229	0.577832	0.288681	0.000839	0.000959	1.860754	1191.945	11.49044	4.689069	14.73904

4.2 *Rainfall Erosivity Factor* (R)

In the study period from 2011 to 2021, the annual average rainfall ranged between 1896.71 and 2214.61 mm. The calculated R factor values ranged varied from 767.50 to 882.90 MJ mm hm^{-2} h^{-1} a^{-1}/year. Utilizing the natural breaks method, the area was categorized into five natural classes. The average R factor for the entire region was computed as 849.23 MJ mm hm^{-2} h^{-1} a^{-1}/year. Notably, the R factor values exhibited an increasing trend from the upper to lower domains of the study area. Almost every year Shilabati River foam up with rain, particularly near Banka, Khipai, and Ghatal areas in respective Hooghly catchments. Addressing rainfall erosivity through real-time data is crucial for protecting soil erosion and flood susceptibility in the SRB.

4.3 *Soil Erodibility Factor* (K)

The soil erodibility factor (*K*) is controlled by pedological attributes, i.e., soil organic matter, pH, and texture of sand, silt, clay (Mallick and Sinha 2013). *K* factor values were assigned based on these characteristics in the soil map, generating the soil erodibility map for the SRB (Table 6, Fig. 4). The soil cover of the SRB characterized by sub-tropical humid climate. The *K* factor exhibited a range of values, spanning from 0 to 0.02, with lower values associated with soils exhibiting low permeability and antecedent moisture content. The spatial distribution of surface soil *K* values revealed higher values (0.01–0.02) in low-relief areas like Ghatal, Chandrakona covered by alluvial plains, floodplains, and northern ridges. Floodplain soils, characterized by sandy loam texture and low organic matter content, showed higher susceptibility to erosion. The middle domain of SRB exhibited high infiltration rates and low surface runoff, indicating lower susceptibility to soil erosion.

Table 6 Top-soil erodibility status and *K* factors value

Soil unit symbol	Sand % topsoil	Silt % topsoil	Clay % topsoil	pH water topsoil	*K* factor values
Ferric Luvisols (LF)	74.6	9.6	15.9	6.4	0.02
Lithosols (I)	58.9	16.2	24.9	7.1	0.017
Eutric Cambisols (BE)	36.4	37.2	26.4	6.9	0.02
Orthic Luvisols (Lo)	76	9.9	14.1	6.4	0.017

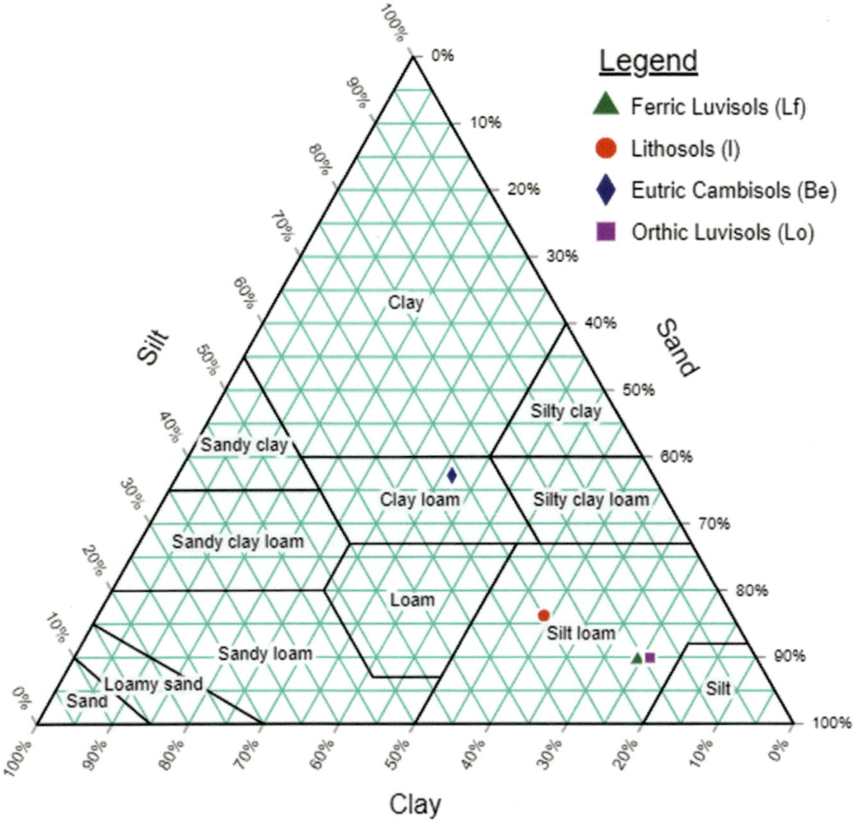

Fig. 4 Soil Taxonomy of SRB

4.4 *Topographic Factor (LS)*

The LS factor, representing the overall influence of slope length and steepness on the erosional process, computed from flow accumulation and slope data and extracted through the DEM. The topographic erosivity factor in the SRB ranged from 0 to 5.61, showing an increase with higher flow accumulation and slope values (Fig. 5). The slope length factor (L) and its steepness (S) presented the influence of topography on soil erosion, as it incorporates factors influencing channelization and surface runoff direction. The higher LS factor in SRB leads to increased overland flow velocities and consequently, greater erosivity. While the basin generally exhibits lower LS factor values, the upper domain shows scattered and isolated high LS factor values, reflecting the rugged topography through which a significant portion of the basin flows.

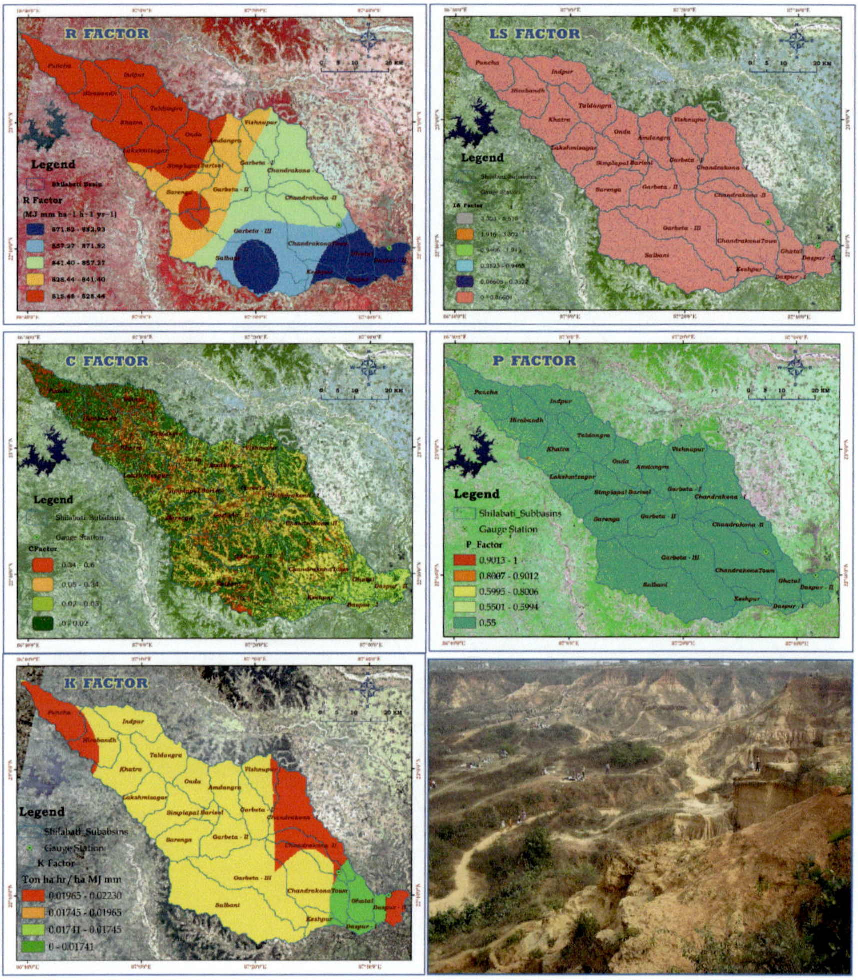

Fig. 5 Component factors of RUSLE model in SRB and gully Erosion in Gangani Badland

4.5 *Crop Management Factor (C)*

The analysis of land use yields valuable insights into utilization patterns, encompassing cropping, forestry, wasteland, and surface water bodies. The study area was categorized into six land use classes, and different land use patterns were assigned crop management factors as per Table 7. The cover management factor (C) assumes a crucial role against soil erosion, can easily adjust to minimize environmental degradation. C factor value in the SRB ranges from 0 to 0.6, with lower values (0.008–0.02) predominant in the eastern part where dense forests and plantations cover a significant portion (33.2% area). Agricultural areas in the central basin exhibit moderate

Table 7 Computed *C* values for Shilabati River Basin area

Sl. no	LULC class	Area (sq.km)	Area (%)	*C*-factor	References
1	Mixed forest	496.69	14.19	0.05	Tirkey et al. (2013)
2	Water	324.75	9.28	0	Erdogan et al. (2006)
3	Dense forest	670.99	19.17	0.008	Kumar and Kushwaha (2013)
4	Plantation	719.55	20.56	0.02	Kumar and Kushwaha (2013)
5	Barren land	477.92	13.65	0.6	Biswal (2015)
6	Built up area	460.89	13.17	0.2	Hurni (1985)
7	Agriculture	349.20	9.98	0.34	Kumar and Kushwaha (2013)

C factor values, while the upper SRB areas, classified as wasteland and barren land, show higher *C* factor values (0.34–0.6), as depicted in Fig. 5.

4.6 Conservation Supporting Practice Factor (P)

Effective land management reduces erosivity, as evidenced by lower *P* factor values in well-managed areas (Bera 2017). The calculation of the conservation practice factor (*P*) involves evaluating slope conditions and assigning corresponding *P* factor values based on slope classes. The *P* factor is crucial in the RUSLE method, representing the ratio of soil loss under a specific support practice to that under traditional up and downslope cultivation. Ranging from 0 to 1, a *P* value of 0 indicates effective erosion-resistant practices, while a value of 1 signifies the absence of such measures. SRB region has hilly terrain, making certain areas prone to soil erosion. Rapid urbanization along the riverbanks and intensive agriculture without proper soil conservation measures can exacerbate erosion. Importantly, the effectiveness of conservation practices is particularly notable in the lower domain of the basin.

4.7 Soil Erosion Potential Hotspot Zones

Identification and mapping of soil erosion vulnerability and prioritizing hotspot zones with the highest erosion potential in the SRB were conducted by integrating various thematic maps in GIS. The potential soil loss in SRB ranges from 0 to 346.9 tons/ha^{-1}/year^{-1}. Estimated soil loss in this region has been categorized into four categories viz. low (0–30.01 tons/ha^{-1}/year^{-1}), medium (30.01–110.2 tons/ha^{-1}/year^{-1}), high (110.3–208.1 tons/ha^{-1}/year^{-1}), very high (208.2–346.9 tons/ha^{-1}/year^{-1}), as illustrated in Table 8 and Fig. 6. Very high erosivity is attributed to steep gradients, significant rainfall, and coarse soil texture region. The lower central watershed falls into the extremely very high potential soil loss zone due to these factors, necessitating immediate sustainable land management. Surrounding this zone are areas with

Table 8 Areal distribution of different potential soil erosion classes

Soil Erosion Zones	Potential Soil Erosion (T/ha^{-1}/Year^{-1})	Pixels Area Count	Area in sq.km	Area in Percentage	Non-structural Measures	Structural Measures
Low	0-30.01	26998	24.30	0.69	Plantation of soil erosion controlling grasses in fallow and waste lands	Diversion channels, interception ditches, Green belts are need to be installed
Medium	30.01-110.2	760842	684.76	19.56	Mulching in the agricultural field	Construction of permanent embankments
High	110.3-208.2	2613340	2352.01	67.20	Crop rotation, forestation, restoration of forest	Site specific structures, Channel modulation, Channel routing
Very High	208.2-346.9	487696	438.93	12.54	Drainage linkage system (DLS), Flood proofing and forecasting	Check dam, flood design, Storage reservoirs

high to moderate soil loss, emphasizing the urgency of intervention. Low predicted soil loss zone (0–30 tons/ha^{-1}/year^{-1}) is identified in the upper SRB, owing to fine loamy soil, lower rainfall intensity. Another essential observation that can be deciphered from Fig. 6 is that the annual soil loss in the SRB demonstrates a decreasing trend from the lower to the upper domain of the region. By understanding the factors contributing to soil erosion, identifying hotspot zones, and implementing appropriate conservation strategies, the SRB can be protected from the detrimental effects of soil loss.

4.8 Flood Zonation and River Discharge

The Shilabati River, typically a tranquil meander through the lush plains of West Bengal, undergoes a dramatic transformation during the monsoon season. Intense rainfall turns this once serene river into a formidable force, causing it to breach its banks and initiate floods that wreak havoc on villages and agricultural fields. The region is particularly susceptible to flooding due to the heavy monsoon and post-monsoonal precipitation (Annual Flood Report 2022). Monsoonal floods, occurring from June to September or October, result from persistent heavy rainfall leading to the overflow of the river. Hydrographs depicting river discharge during monsoonal floods, as illustrated in Fig. 7, highlight the heightened danger levels, particularly in November. Prolonged periods of excessive rainfall play a pivotal role, and the absence of such conditions limits the duration of each flood event. Hydromet river gauge data from two key stations, Banka and Gadghat, were sourced from the Irrigation and Waterways Department, Government of West Bengal (2023). Analysis of the latest Dartmouth Flood database identifies the lower part of the basin as more vulnerable to flood inundation (Fig. 8). Specifically, areas such as Ghatal, Keshpur,

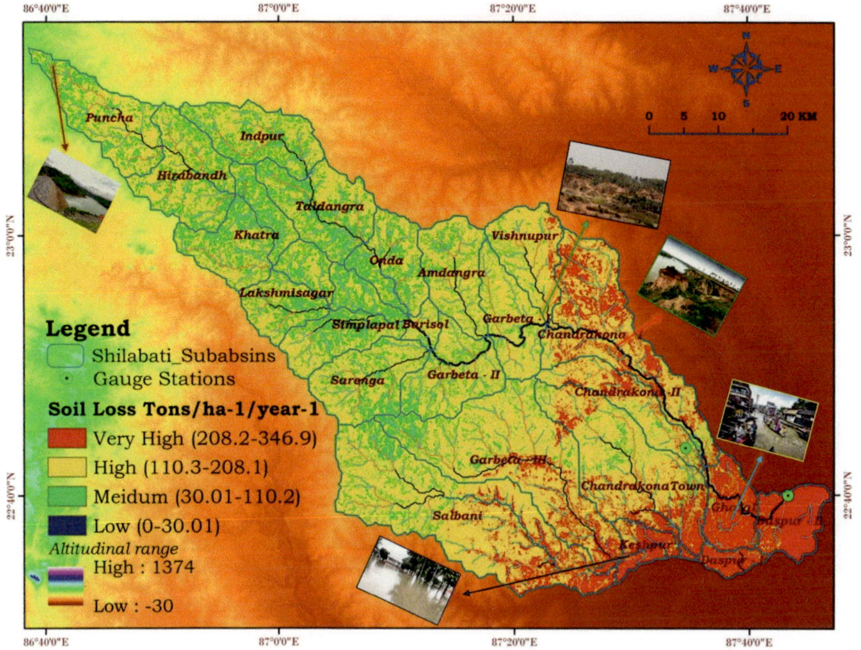

Fig. 6 Soil erosion potential zones of SRB

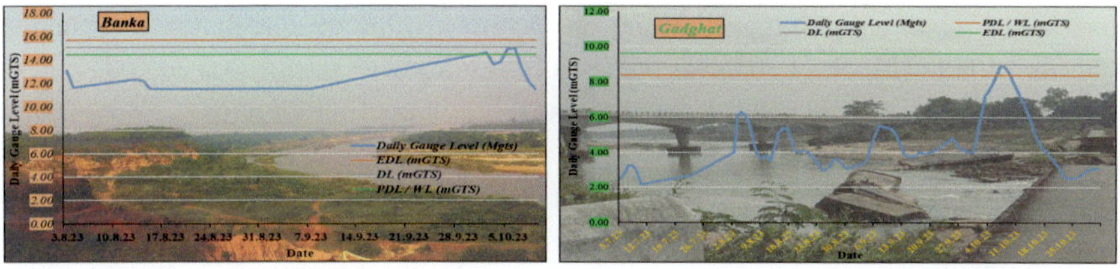

Fig. 7 Discharge hydrographs for the year 2023 of the flood-prone gauge stations of SRB

and Chandrakona Town face significant threats. The impact of uncontrolled development, marked by the encroachment of concrete structures onto the river floodplain, exacerbates the existing challenges of land degradation.

4.9 *Interlinking Morphometric Parameters with Soil Erosion Susceptibility of Sub-watersheds*

In the current research, hydro-geomorphological aspects are used to analyze flood susceptibility in Shilabati River Basin. Table 9 displays the correlation matrix derived

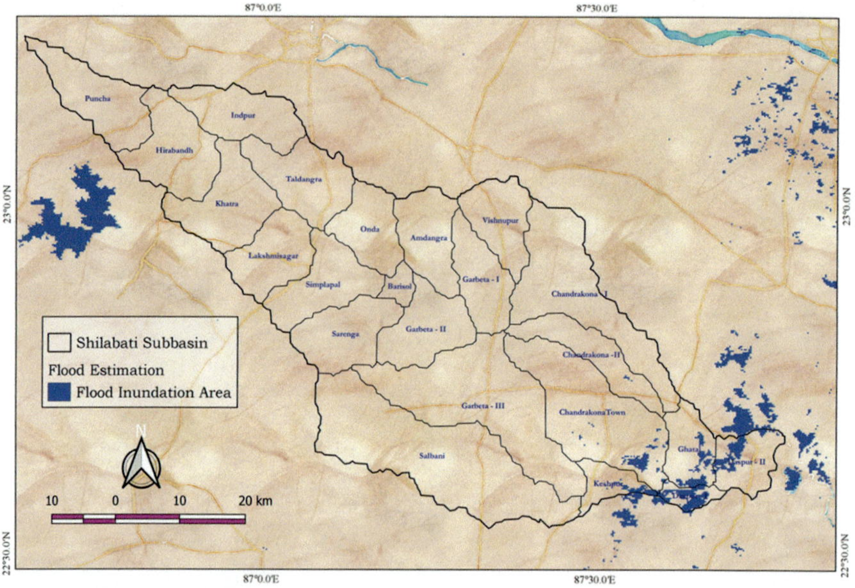

Fig. 8 Flood inundation map of SRB

from the WSA approach, showcasing the morphometric properties of the sub-watersheds. The highlighted results for the correlation coefficient (r) were more than 0.6. F_s has a significant correlation, positive value of correlation coefficient, with R_b ($r = 0.13$), R_f ($r = 0.54$), R_e ($r = 0.51$), R_c ($r = 0.05$), D ($r = 0.97$), R_t ($r = 0.87$), TWI ($r = 0.38$), and a negative correlation coefficient with C_c (r = -0.08), C ($r = -0.67$), SPI ($r = -0.3$) and STI ($r = -0.78$). R_b only correlates with F_s among the morphometric parameters. While R_f has a high and positive correlation ($r = 1.0, 0.61$) with R_e, D, and R_t, it only shows high negative correlation with C ($r = -0.72$). Strong positive correlations were observed between D, R_t with TWI, while significant negative correlations displayed in C, SPI, STI. TWI has a high and negative correlation with SPI ($r = -0.78$), and STI ($r = -0.33$). Despite SPI's high negative correlations with F_s, D, R_t, and TWI, a moderate positive relationship with STI ($r = 0.37$) is evident. The flow hydraulics reveals that channel gradient reduces considerably toward downstream. Channel shift and bank erosion are fairly common in the lower reach of SRB. As per flood inundation map (Fig. 9), the lower basin area is more susceptible to flood. In this area, the river has lost stream erosive power due to low channel gradient and huge sedimentation. Whenever precipitation is externally added as a form of water during southwest monsoon season, the river foams up and develops flood and paedogenic hazard. Therefore, south easterly oriented meander loops are far more susceptible to erosion than those with other orientations.

Table 9 Correlation matrix of morphometric properties for the sub-watersheds

	Fs	Rb	Rf	Re	Rc	D	Rt	Cc	C	TWI	SPI	STI
Fs	1.00	0.13	0.54	0.51	0.05	0.97	0.97	-0.08	-0.67	0.38	-0.30	-0.78
Rb	0.13	1.00	0.05	0.02	-0.45	0.14	0.12	0.47	-0.03	0.23	-0.31	-0.13
Rf	0.54	0.05	1.00	1.00	0.48	0.61	0.61	-0.46	-0.72	0.17	-0.01	-0.70
Re	0.51	0.02	1.00	1.00	0.50	0.59	0.58	-0.48	-0.74	0.16	-0.02	-0.70
Rc	0.05	-0.45	0.48	0.50	1.00	0.12	0.15	-0.98	-0.37	-0.11	0.21	-0.23
D	0.97	0.14	0.61	0.59	0.12	1.00	0.97	-0.14	-0.81	0.42	-0.33	-0.89
Rt	0.97	0.12	0.61	0.58	0.15	0.97	1.00	-0.16	-0.73	0.46	-0.26	-0.80
Cc	-0.08	0.47	-0.46	-0.48	-0.98	-0.14	-0.16	1.00	0.36	0.17	-0.28	0.21
C	-0.67	-0.03	-0.72	-0.74	-0.37	-0.81	-0.73	0.36	1.00	-0.26	0.22	0.94
TWI	0.38	0.23	0.17	0.16	-0.11	0.42	0.46	0.17	-0.26	1.00	-0.78	-0.33
SPI	-0.30	-0.31	-0.01	-0.02	0.21	-0.33	-0.26	-0.28	0.22	-0.78	1.00	0.37
STI	-0.78	-0.13	-0.70	-0.70	-0.23	-0.89	-0.80	0.21	0.94	-0.33	0.37	1.00

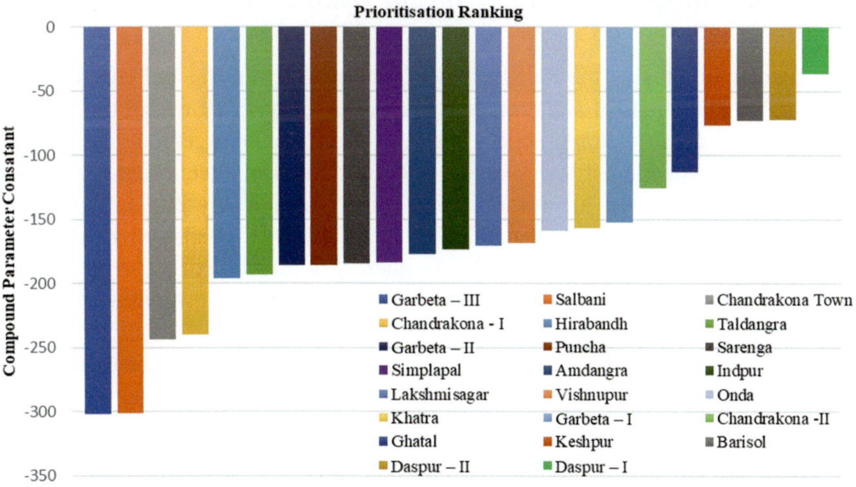

Fig. 9 Ranking of sub-watersheds for soil water prioritization

4.10 Automated Prioritization and Performance Assessment of Sub-watersheds

SRB faces various challenges like soil erosion, groundwater depletion, and monsoonal flood. To address these issues effectively, prioritizing the sub-watersheds within the basin for intervention is crucial. The prioritization and ranking of 23 sub-watersheds rely on the Compound Parameter Values (CPV) as shown in Fig. 9. Those with the lowest CPV are ranked highest susceptible, and the others are ranked lower based on their CPV (Aher et al. 2014). Notably, sub-watershed Garbeta-III attains the top priority with a CPV of − 301.75, while sub-watershed Daspur-I holds the lowest priority with a CPV of 36.28. To validate the practical relevance of sub-watershed prioritization contemporary geohazard, i.e., latest flood inventories from the West Bengal Irrigation Department were visualized. Flow information from two-gauge stations in the Shilabati River Basin was examined for the period spanning 2022 to 2023. Historical flood event records (Applied Economics and Statistics 2015) indicate that Jharul, Bagpata, Palpukur, Banka, Manik Kundu, Kaskuli, Khirpai, Gadghat, Barada, Jadupur, Kharar, Dalpatipur, Baropol, Khasbar, Mansuko, Kushpata and Bandar areas of Garbeta-III, Chandrakona Town, Keshpur, Ghatal, and Salbani were identified as most critical sub-basins. This aligns with the prioritization results obtained from the SWPT, confirming an accuracy exceeding 80% based on available records.

4.11 Overall Assessment of Erosion Dynamicity

This chapter, employing the RUSLE model with remote sensing and GIS data, has identified critical areas of soil erosion dynamics and flood risk within the SRB by informing soil conservation practices and long-term land management in the basin. The high erosion zones of lower basins require immediate attention, as confirmed by most of the reviewed literature findings (Bhandari and Roy 2021). The identified flood-affected blocks like Ghatal, Keshpur, Daspur-II, Daspur-I located at the lower part of the basin have contributed frequently flood problems. The findings on identified flood-prone areas can be coupled with sedimentation estimates to identify infrastructure at risk from flood events and sedimentation. This information is crucial for prioritizing maintenance and potential upgrades to bridges, dams, and other critical infrastructure. The analysis likely focuses on the current state. However, climate change projections may suggest altered rainfall patterns and increased frequency of extreme weather events. Future research should incorporate climate change scenarios into the RUSLE model to predict how soil erosion patterns might evolve, allowing for proactive adaptation strategies. This knowledge can be used to promote development in less vulnerable areas and encourage land use practices that minimize soil erosion and flood frequencies. The research findings can also be compared with similar hydro-climatic conditions and land use patterns in other regions that have

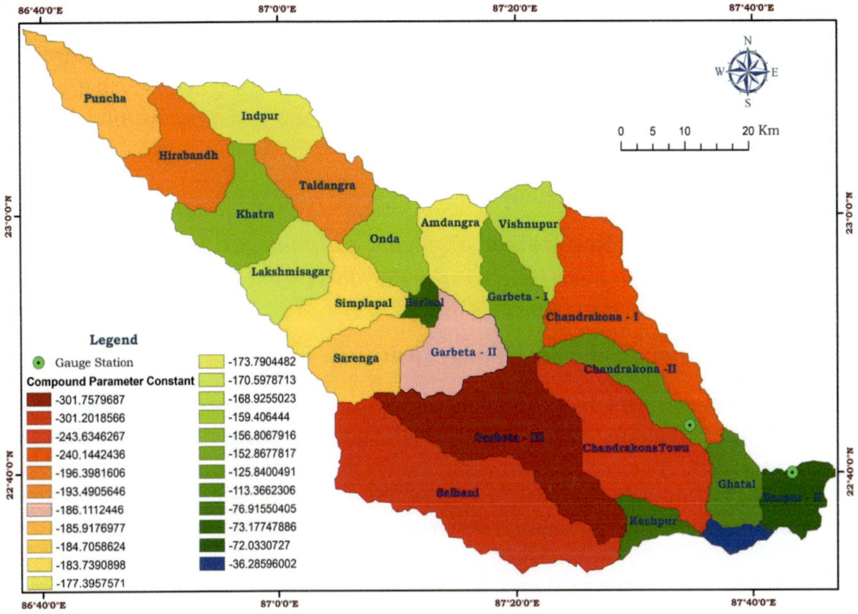

Fig. 10 Prioritization of sub-watersheds based on ArcSWPT

applied the RUSLE model. Factors like basin size, topography, local knowledge, and historical flood events may influence this comparison. This comparison can help validate the model's effectiveness in different geographical settings and identify potential areas for further calibration specific to the SRB's unique characteristics. Remote sensing-based mapping approach of the study area for flood risk analysis is a cost-effective way and applicable to other severe flood-prone areas (Fig. 10).

5 Recommendations

For sustainable watershed development in the SRB, prioritizing soil management practices like control grazing, afforestation, conservation tillage practices, terrace farming, and contour bunding in the lower reaches is essential. The recommendations detailed in Table 8 delineate specific measures tailored to each erosion class (Navarro-Hevia et al. 2016). For areas with excessive soil erosion susceptibility (Class-III and Class-IV), effective and scientific conservation practice is an immediate requirement. Conversely, regions with low and medium soil erosion classes (Class-I and Class-II) necessitate minimal conservation efforts. Here are some strategies for sustainable land, soil, and water conservation (Boardman et al. 2003):

1. **Sustainable Soil Conservation Practices**:

- **Contour Plowing, Cover Crops, and Agroforestry**: Implementing these practices mitigates soil erosion, excessive runoff, enhances soil structure, and promotes biodiversity. This could involve land use regulations that restrict development in high-risk zones, promoting the construction of levees or embankments for flood protection.

2. **Water Conservation Techniques**:
   - **Riparian Zone Restoration**: Dredging of sediments from river beds and restoring riparian zones enhances groundwater potentiality in SRB, supporting overall water conservation practices.

3. **Land Use Monitoring and Management**:
   - **Integrated Land Use Planning**: Considering soil erosion susceptibility in land use planning helps identify and manage vulnerable area of the SRB catchment effectively.
   - **Sustainable Agricultural Practices**: Promoting sustainable green technologies reduces environmental impact and maintains soil health.

4. **Community Engagement and Awareness**:
   - **Community Participation in Water Management**: Encouraging community participation (river festivals, awareness drives, etc.), in water resource management ensures holistic and inclusive conservation.

6 Conclusion

This study has delved into the intricate dynamics of soil erosion and flood zonation within the SRB in West Bengal utilizing an integrated RS and GIS-based approach. The research has highlighted the substantial threats posed by soil erosion to land resources and agricultural productivity in SRB characterized by high rainfall and vulnerable topography. This research has specifically focused on elucidating the roles of various factors, including rainfall intensity, land use practices, soil type, slope gradient, and drainage characteristics, in shaping soil erosion susceptibility. Notably, the study has emphasized the crucial regulatory role of drainage basin morphometry, covering basin area, slope length, drainage density, and topographic ruggedness, in influencing landscape erosive forces through water runoff pathways. By addressing the existing knowledge gap on the specific relationship between drainage basin morphometry and soil erosion susceptibility in the SRB, this research has provided valuable insights into the spatial patterns and controlling factors of erosion vulnerability to the growing body of research on soil erosion dynamics and flood zonation. The outcomes contribute to effective land management, erosion control, and sustainable land use planning in the SRB, with transferable knowledge applicable to similar river basins facing analogous challenges. Construction of a dam from Western part of Ghatal to Chandrakona road on the right bank of Shilabati River and reformation of the abandoned channel Ghatal circuit may become a holistic strategy for effective soil water conservation. In conclusion, this study unlocks new insights into soil erosion dynamics and flood zones, paving the way for more effective erosion control strategies. The broader implications extend to supporting sustainable development in the SRB and comparable regions, highlighting the pivotal role of integrated RS and GIS-based models in guiding erosion management strategies and fostering environmental resilience.

References

Abdulkareem JH, Pradhan B, Sulaiman WNA, Jamil NR (2018) Quantification of runoff as influenced by morphometric characteristics in a rural complex catchment. Earth Syst Environ 2(1):145–162. https://doi.org/10.1007/s41748-018-0043-0

Aher PD, Adinarayana J, Gorantiwar SD (2014) Quantification of morphometric characterization and prioritization for management planning in semi-arid tropics of India: a remote sensing and GIS approach. J Hydrol 511:850–860

Al Shoumik BA, Khan MZ, Islam MS (2023) Soil erosion estimation by RUSLE model using GIS and remote sensing techniques: a case study of the tertiary hilly regions in Bangladesh from 2017 to 2021. Environ Monit Assess 195(9):1096. https://doi.org/10.1007/s10661-023-11699-4

Applied Economics and Statistics, Government of West Bengal (2015) District statistical handbook. Paschim Medinipur

Bera A (2017) Estimation of soil loss by USLE model using GIS and remote sensing techniques: a case study of Muhuri River Basin, Tripura, India. Eurasian J Soil Sci 6(3):206–215

Bhandari U, Roy M (2021) Assessment of GIS and remote sensing-based flood vulnerability of Shilabati River basin using AHP method. Int J Curr Res Technol (IJCRT) 11(35):364–368

Biswal SK (2015) Mathematical model for flow and sediment yield estimation on Tel river basin of India. A dissertation submitted in partial fulfilment of the requirements for the award of the degree of master of technology in civil engineering with specialization in water resources engineering. Department of Civil Engineering National Institute of Technology Rourkela, India

Boardman J, Poesen J, Evans R, Bierkens M (2003) Soil erosion in Europe. Wiley

Cerdà A, Doerr SH (2008) The effect of ash and needle cover on surface runoff and erosion in the immediate post-fire period. CATENA 74(3):256–263

Das B, Bandyopadhyay A (2015) Flood risk reduction of Rupnarayana River, towards disaster management—a case study at Bandar of Ghatal Block in Gangetic Delta. J Geogr Nat Disasters 5:1–6

Faniran A (1968) The index of drainage intensity: a provisional new drainage factor. Aust J Sci 31(9):326–330

Ghosh A, Mondal S (2019) Application of multivariate statistics towards the geochemical evaluation of fluoride enrichment in groundwater at Shilabati river bank, West Bengal, India. Environ Eng Res 24(2):279–288

Government of West Bengal, Irrigation and Waterways Department (2022) Annual Flood Report. https://www.wbiwd.gov.in/index.php/applications/anual_flood_report

Government of West Bengal, Irrigation and Waterways Department (2023) Hydromet river gauge data. https://www.wbiwd.gov.in/index.php/applications/hydrometData

Griffin ML, Beasley DB, Fletcher JJ, Foster GR (1988) Estimating soil loss on topographically non uniform field and farm units. J Soil Water Conserv 43(4):326–331

Horton RE (1932) Drainage-basin characteristics. Trans Am Geophys Union 13(1):350–361

Horton RE (1945) Erosional development of streams and their drainage basins; hydrophysical approach to quantitative morphology. Geol Soc Am Bull 56(3):275–370

Hurni H (1985) Erosion-productivity conservation systems in Ethiopia. In: Proceedings of 4th International Conference on Soil Conservation, Maracay, Venezuela, 3–9 November 1985, pp 654–674

Kumar S, Kushwaha SPS (2013) Modelling soil erosion risk based on RUSLE-3D using GIS in a Shivalik sub-watershed. J Earth Syst Sci 122:389–398

Kumar V, Roy DK, Dharminder KR, Kumar S, Hans H (2015) Influence of crop establishment methods and different weed management practices on growth, yield and quality of direct seeded rice (*Oryza Sativa L.*). The Ecoscan 10(1&2):249–252

Luo Y, Su B, Yuan J, Li H, Zhang Q (2011) GIS techniques for watershed delineation of SWAT model in plain polders. Proc Environ Sci 10:2050–2057

Mahala A (2020) The significance of morphometric analysis to understand the hydrological and morphological characteristics in two different morpho-climatic settings. Appl Water Sci 10(1):1–16

Mallick S, Sinha R (2013) Soil erosion modelling using RUSLE and GIS in upper catchment of Damodar Valley basin, Jharkhand, India. Environ Earth Sci 68(2):519–532

Marowka A (2018) On parallel software engineering education using python. Educ Inf Technol 23(1):357–372

Miller VC (1953) A quantitative geomorphic study of drainage basin characteristics in the Clinch Mountain Area, Virginia and Tennessee. Department of Geology, Columbia University, New York, pp 389–402

Moore I, Burch G (1986) Modelling erosion and deposition: topographic effects. Trans ASAE 29(6):1624–1630

Navarro-Hevia J, Lima-Farias TR, de Araújo JC, Osorio-Peláez C, Pando V (2016) Soil erosion in steep road cut slopes in Palencia (Spain). Land Degrad Dev 27(2):190–199. https://doi.org/10.1002/ldr.2459

Pandey A, Chowdary VM, Mal BC (2007) Identification of critical erosion prone areas in the small agricultural watershed using USLE, GIS and remote sensing. Water Resour Manage 21:729–746

Pandey S, Kumar P, Zlatic M, Nautiyal R, Panwar VP (2021) Recent advances in assessment of soil erosion vulnerability in a watershed. Int Soil Water Conserv Res 9(3):305–318

Rahmati O, Haghizadeh A, Stefanidis S (2016) Assessing the accuracy of GIS-based analytical hierarchy process for watershed prioritization; Gorganrood River Basin, Iran. Water Resour Manage 30:1131–1150

Renard KG, Foster GR, Weesies GA, McCool DK, Yoder DC (1997) Predicting soil erosion by water: a guide to conservation planning with the revised universal soil loss equation (RUSLE). Agriculture Handbook No. 703, USDA, Washington DC

Roy M, Shamim F, Chatterjee S (2021) Evaluation of physicochemical and Biological parameters on the water quality of Shilabati River, West Bengal, India. Water Sci 35(1):71–81

Schumm SA (1956) Evolution of drainage systems and slopes in Badlands at Perth Amboy, New Jersey. Geol Soc Am Bull 67(5):597–646

Singh G, Babu R, Narain P, Bhushan LS, Abrol IP (1992) Soil erosion rates in India. J Soil Water Conserv 47(1):97–99

Strahler AN (1964) Part II. Quantitative geomorphology of drainage basins and channel networks. In: Handbook of Applied Hydrology, vol 39(4). McGraw-Hill, New York, pp 4–76

Tellman B, Sullivan JA, Kuhn C, Kettner AJ, Doyle CS, Brackenridge GR, Erickson T, Slayback DA (2018) Satellites observe increasing proportion of population exposed to floods. Nature. https://doi.org/10.1038/s41586-021-03695-w

Thakkar AK, Dhiman SD (2007) Morphometric analysis and prioritization of mini watersheds in Mohr watershed, Gujarat using remote sensing and GIS techniques. J Indian Soc Remote Sens 35:313–321
Tirkey AS, Pandey AC, Nathawat MS (2013) Use of satellite data, GIS and RUSLE for estimation of average annual soil loss in Daltonganj watershed of Jharkhand (India). J Remote Sens Technol 1(1):20–30
Weslati O, Serbaji MM (2024) Spatial assessment of soil erosion by water using RUSLE model, remote sensing and GIS: a case study of Mellegue Watershed, Algeria-Tunisia. Environ Monit Assess 196:14. https://doi.org/10.1007/s10661-023-12163-z
Whipple KX, Tucker GE (1999) Dynamics of the stream-power river incision model: implications for height limits of mountain ranges, landscape response timescales, and research needs. J Geophys Res Solid Earth 104(B8):17661–17674
Williams JR (1995) Chapter 25: The EPIC model. In: Singh VP (ed) Computer models of watershed hydrology, vol 1130. Water Resources Publications, Highlands Ranch, CO
Wischmeier WH, Smith D (1978) Predicting rainfall erosion losses: a guide to conservation planning. In: Agriculture Handbook No. 537. USDA, Washington, DC
Wu J, Chen L, Shi Z, Huang D, Zhang G, Wei W (2018) Soil erosion and conservation in global agriculture. Sci Total Environ 624:117–128

Geo-spatial Modeling of Potential Soil Erosion Estimation for Better Conservation Planning

Fatemeh Mohammadyari, Khodayar Abdollahi, Mohsen Tavakoli, and Jurate Suziedelyte Visockiene

Abstract Geo-spatial modeling aids in better soil erosion estimation by considering factors like topography and land use. It assists in identifying erosion hotspots for targeted conservation efforts, integrating diverse datasets for comprehensive analysis, and predicting future erosion trends to plan sustainable conservation strategies efficiently. This study harnesses the applicability of geo-spatial technologies to quantify soil erosion within the Ilam dam watershed, a critical step toward sustainable ecosystem management. By integrating the InVEST-SDR model with Geographic Information System (GIS), we offer a novel approach to mapping SE potential, crucial for the strategic implementation of conservation measures. Our findings reveal a stark spatial variability in soil erosion across the watershed, governed by land use patterns, topography, and human activities. The annual erosion potential is estimated at 202.7 t/ha, underscoring the urgency for targeted conservation efforts. This study not only identifies areas with heightened erosion risk but also classifies sub-watersheds into severity classes, providing a blueprint for prioritizing interventions. The geo-spatial modeling conducted herein is pivotal for decision-makers and land

F. Mohammadyari (✉) · K. Abdollahi
Department of Environmental Engineering, Faculty of Natural Resources and Earth Sciences, Shahrekord University, Shahrekord, Iran
e-mail: mohammadyari.f@sku.ac.ir

K. Abdollahi
e-mail: kabdolla@sku.ac.ir

K. Abdollahi
Department of Nature Engineering, Faculty of Natural Resources and Earth Sciences, Shahrekord University, Shahrekord, Iran

M. Tavakoli
Department of Rangeland and Watershed Management, Faculty of Agriculture, Ilam University, Ilam, Iran
e-mail: m.tavakoli@ilam.ac.ir

J. S. Visockiene
Department of Geodesy and Cadaster, Vilnius Gediminas Technical University, Sauletekio Av. 11, 10223 Vilnius, Lithuania
e-mail: jurate.visockiene@vgtu.lt

S. C. Pal and U. Chatterjee (eds.), *Surface, Sub-Surface Hydrology and Management*, Springer Geography, https://doi.org/10.1007/978-3-031-62376-9_20

managers, equipping them with the insights needed to design appropriate soil conservation measures. Despite data limitations, the flexibility of the InVEST-SDR model allows for a comprehensive assessment of soil erosion, highlighting the importance of geo-spatial analysis in conservation planning. This research contributes significantly to the global understanding of soil erosion dynamics, offering a foundation for informed decision-making and fostering sustainable land management practices in watersheds worldwide. The experimental results can help to evaluate the areas sensitive to erosion in the Ilam dam watershed. On the other hand, the decision-makers and managers can better design appropriate soil conservation measures.

Keywords Soil erosion · Sediment export · Land management strategies · Ecosystem services and disservices · InVEST-SDR model · Watershed

1 Introduction

Soil, a vital component of the biosphere, underpins human survival and development through its roles in food production, material recycling, and organic matter storage (Saha et al. 2022; Qiao et al. 2023). Soil erosion (SE), a pressing environmental and socioeconomic challenge, is recognized as one of the ten major soil threats, contributing to degradation, biodiversity loss, organic carbon depletion, food security issues, land degradation, and mass human migration (Kumarasiri et al. 2023; Lefèvre et al. 2020; Getu et al. 2022). SE also exacerbates off-site problems such as sedimentation in downstream areas and eutrophication processes (Gashaw et al. 2021).

SE is the ecological process where soil materials are separated, transported, and deposited by erosive agents, primarily gravity, wind, and rainfall (Tamire et al. 2022). Human activities, interacting with natural parameters, are significant geophysical processes shaping earth's terrain, with land resource disturbance being a dominant factor in SE (Gomiero 2016; Li et al. 2022). Approximately, 2 billion hectares of soil worldwide are affected by human-induced degradation (Ougougdal et al. 2020), with SE rates expected to rise due to agricultural activities, land use/cover changes (LULC), and global warming (Naipal et al. 2018; Guerra et al. 2020).

LULC transitions, reflecting natural and ecological processes, influence SE dynamics through changes in vegetation cover, microtopography, soil properties, and runoff rates (Borrelli et al. 2017). These transitions alter hydrological flow paths, affecting rainfall infiltration, surface runoff interception, and groundwater replenishment, thereby accelerating soil erosion and altering sediment delivery paths (Bhattacharya et al. 2024).

The impact of LULC on SE has garnered continuous attention due to its implications for sustainable development and the intensification of human-land conflicts (Aneseyee et al. 2020). SE diminishes soil productivity and the natural environment, adversely affecting economic development, ecosystem services (ESs), and functions (Berihun et al. 2019; Fenta et al. 2020, 2021; Yohannes et al. 2021), thus becoming a global concern. While ESs contribute to well-being, regulation, support, and cultural services (Tavakoli and Mohammadyari 2023; Mohammadyari et al. 2023a), ecosystem disservices (EDS), such as SE, lead to undesirable environmental, ecological, and human impacts (Paudel and States 2023; Maliqi and Singh 2019). Despite the encouragement to integrate both ES and EDS into research frameworks, EDS remains less discussed than ES (Carucci et al. 2022).

Traditional soil sampling and erosion accumulation methods provide valuable information but are time-consuming, expensive, and challenging to generalize at the basin scale (Berihun et al. 2020). Decision-makers in watershed management face the challenge of accurately modeling the complexity of watershed processes with available information (Lamane et al. 2022). Specialized tools have been developed to assess the intensity of SE and soil erosion patterns (De Vente et al. 2013). Among the most commonly used models for spatio-temporal prediction of SE at the watershed scale are EUROSEM, MUSLE, WaTEM/SEDEM, WEPP, LISEM, LEAM, SWAT, USLE, RUSLE, EPM, IntErO, and Integrated Valuation of Ecosystem Services and Tradeoffs (InVEST).

The USLE/RUSLE model is an experimental SE prediction model whose applicability and reliability have been validated in previous studies (Andriyani et al. 2021). The InVEST Sediment Delivery Ratio (SDR) model, calibrated and verified in diverse topographies, land use types, soil types, and climates, offers advantages over the RUSLE model, including the ability to calculate SE rates, sediment retention, and sediment export (Gashaw et al. 2021; Tamire et al. 2022). The InVEST model, developed by Stanford University, the Natural Capital Project, TNC, and WWF, aims to harmonize environmental conservation objectives with socioeconomic forces for informed natural resource management (Zarandian et al. 2023). It utilizes data from GIS and RS for modeling (Hu and Gao 2020), playing an essential role in soil and water conservation (Kumar and Singh 2021).

The Ilam dam watershed, the primary drinking water source for Ilam city, faces rapid environmental crises and LULC changes. Unreasonable land use is a principal cause of ecological degradation and SE. Understanding the relationship between SE and land use dynamics is crucial (Qiao et al. 2023). Identifying SE processes and sediment export is necessary for evaluating management scenarios, designing superior techniques, and prioritizing conservation planning (Roba et al. 2021).

This study's novelty lies in integrating GIS models, satellite data, and InVEST-SDR to assess the annual SE rate in Ilam province's critical watershed. The objectives are to: (1) demonstrate the applicability of the InVEST-SDR model for SE and sediment export mapping across the Ilam dam watershed; and (2) evaluate the effects of LULC on SE at the watershed scale. The InVEST-SDR model was selected for its simulation capabilities in various environments and the availability of necessary data for its application in the Ilam dam watershed. The results aim to inform ecological interventions before irreversible damage and provide managers with insights into soil and water management in the watershed.

2 Materials and Methods

2.1 Description of the Study Area

The Ilam dam watershed (Fig. 1) is located (33° 26′ N to 33° 38′ N; 46° 38′ E to 46° 32′ E) in the Ilam province and has an area of 47,538.69 ha. Its elevation ranges from 968 to 2604 m. The average annual temperature and precipitation are16 °C and 787 mm, respectively. Based on the Köppen climate classification, the climate of the watershed is semi-arid Mediterranean. Loamy-clay is the dominant soil texture of this area (Tavakoli and Mohammadyari 2023) and the main LULC types are forest land and pasture. The geology of the watershed is characterized by shale, limestone, and marn, and it comprises 13 sub-basins and 22 urban and rural areas.

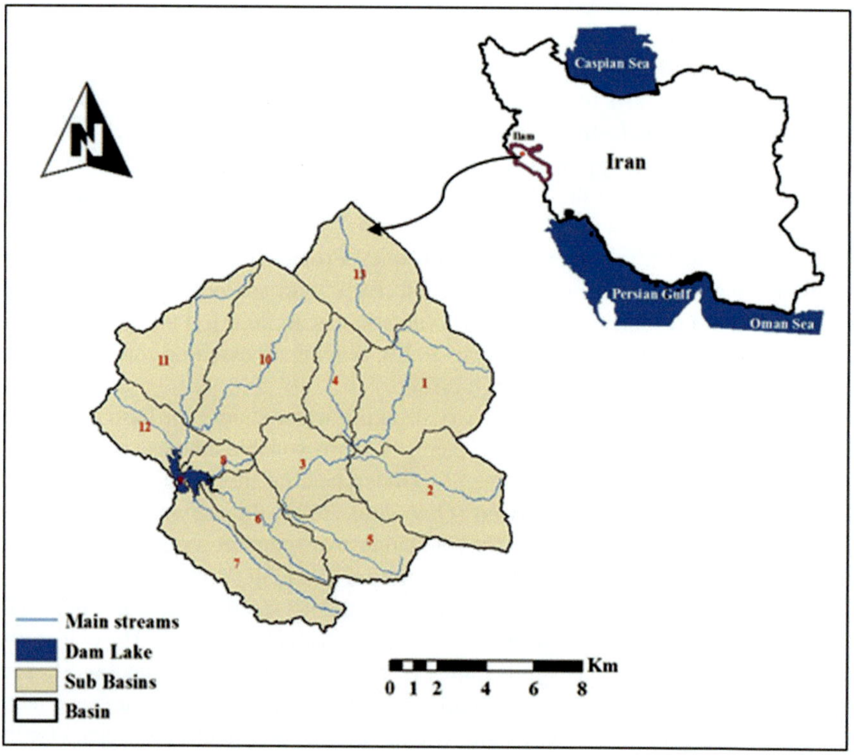

Fig. 1 Map of the Ilam dam watershed

2.2 Modeling Potential SE and Sediment Export in the InVEST-SDR Model

The InVEST-SDR model is used to estimate the potential for soil loss by sediment deposition and water that reaches the stream in the basin. With global change in mind, this information can be useful to study the service of soil retention in a watershed (Sharp et al. 2020; Tamire et al. 2022). This is important for instream water quality and reservoir management, both of which may have economic value (Sharp et al. 2020). This model is less data-intensive and more flexible than other hydrological models. The InVEST-SDR model simulates erosion from each land use type using the RUSLE equation (Tamire et al. 2022).

Erosion and sediment retention on the surface of the earth are natural processes that affect the concentration of sediment in streams. Land erosion (soil particles are separated and transported by overland waterflow and rain), landslides (mass erosion), coastal erosion, and gullies (channels that concentrate runoff) are the main sources of sediment. Sinks include instream and floodplain deposition, reservoir retention, and on-slope deposition.

2.2.1 SE Estimation

The estimation of SE was simulated using the RUSLE model (Eq. 1). The design of this model was according the loss of soil caused by runoff from certain fields in specific cultivation and management systems (Panditharathne et al. 2019). The RUSLE model for estimating annual SE rates and carrying out risk assessments by integrating R (precipitation erosivity) parameters (MJ mm ha^{-1} year^{-1}), K (soil erodibility) (Mg h MJ^{-1} mm^{-1}), LS (slope length gradient), C (cover management), and P (conservation practice) was implemented by Bagwan and SopanGavali (2021).

$$\text{usle}_i = R_i \times K_i \times \text{LS}_i \times C \times P. \tag{1}$$

The LS_i parameter was calculated using Eq. (2) and expanded using the method in Desmet and Govers (1996).

$$L = \frac{(A_{i\text{-in}} + D^2)^{m+1} - A_{i\text{-in}}^{m+1}}{D^{m+2} \cdot x_i^m \cdot (22.13)^m}, \tag{2}$$

where

- S_i is the slope parameter for the grid pixel, which is estimated as a function of the slope radians.
- $\theta < 9\%$, $S = 10.8 \cdot \text{Sine}(\theta) + 0.03$.
- $\theta \geq 9\%$, $S = 16.8 \cdot \text{Sine}(\theta) - 0.50$.

- $A_{i\text{-in}}$ is the area which has the contribution (m^2) in the input of a pixel, calculated by the flow direction model.
- D refers to the linear dimension of the grid cell (m).
- $x_i = ||sin\ ai| + |\cos \alpha i|$; αi represents the direction of pixel i.
- m refers to the RUSLE length exponent parameter.

In order not to overestimate the LS parameter in heterogeneous landscapes, the length of long slopes is limited to 333 m (Sharp et al. 2020).

2.2.2 Sediment Delivery Ratio (SDR)

SDR indicates the proportion of sediment produced in a given region that goes downstream, as a function of the downslope flow path and upslope region for each pixel (Ahmadi Mirghaed et al. 2018). In this method, the connection index model (IC) is calculated, according to the study of Borselli et al. (2008) (Eq. 3).

$$\mathrm{IC} = \mathrm{Log}\ 10(D_{\mathrm{up}}/D_{\mathrm{dn}}). \tag{3}$$

The SDR for each pixel depends on the downslope flow path and the upslope area (Eqs. 4, 5, and 6) (Sharp et al. 2020), where D_{up} and D_{dn} are upslope and downslope components, respectively, and are calculated based on Eqs. (4) and (5):

$$D_{\mathrm{up}} = SC\sqrt{A} \tag{4}$$

where S is average slope gradient of the upslope participating region (mm). Furthermore, C is average of the C parameter (mm). A refers to the upslope participating region (m^2).

$$D_{\mathrm{dn}} = \sum_i \frac{d_i}{C_i S_i}, \tag{5}$$

whereas

- D_i from pixel i to flow, it shows the average length of the flow path in the downslope direction (m).
- C_i is the C parameter of the pixel i.
- S_i is the slope gradient of the pixel i.

The next step is to calculate the SDR ratio for a pixel i from the conductivity index IC according to the work of Vigiak et al. (2012) (Eq. 6).

$$\mathrm{SDR}_i = \frac{\mathrm{SDR}_{\max}}{1 + \exp\left(\frac{\mathrm{IC}_0 - \mathrm{IC}_i}{k}\right)}, \tag{6}$$

whereas SDR_{max} is the maximum theoretical SDR, with a default value of 0.08 (Vigiak et al. 2012). K and IC_0 are calibration parameters (Hamel et al. 2015).

The sediment export from a certain pixel i, which is represented by E_i in terms of tons per hectare per year (ton ha^{-1} year^{-1}), was calculated according to Eq. (7), and the total sediment export, represented by E, was estimated based on Eq. (8)

$$E_i = \text{usle}_i \times \text{SDR}_i \tag{7}$$

$$E = \sum_i E_i. \tag{8}$$

2.3 *Input Data for InVEST-SDR Model*

The basic maps required for the implementation of the model include the DEM, land use, K parameter, R parameter, P parameter, and C parameter (in raster format), as well as the maps of the watershed and sub-watershed (in vector format). Furthermore, a biophysical table, including model information related to each land use class, should be designed. The methodology and data used for modeling in this chapter are presented in Fig. 2 and Table 1, respectively.

The details of the preparation of the required data are presented.

2.3.1 Land Use Map

The land use map for the year 2021 was extracted from the Landsat satellite image and the OLI sensor, with a resolution of 30 m, from the website of the United States Geological Survey (USGS). After the necessary pre-processing, and geometric and atmospheric corrections, the image was classified by the support vector machine algorithm (SVM) in ENVI software (Mirsanjari et al. 2021), and the land use map was prepared with 13 classes: low-semi- and dense forest, farmland, dry farmland, garden, low-semi- and dense grassland, lake, settlement, barren, and road. To assess the accuracy of the classification, the kappa coefficient and overall accuracy of agreement were calculated. Values of 90 and 94% were obtained for Kappa coefficient and overall accuracy, respectively. As mentioned, the dominant coverage of the area is forest land, which is 62% of the area assigned. In contrast, garden and farmland classes make up < 1% of the area.

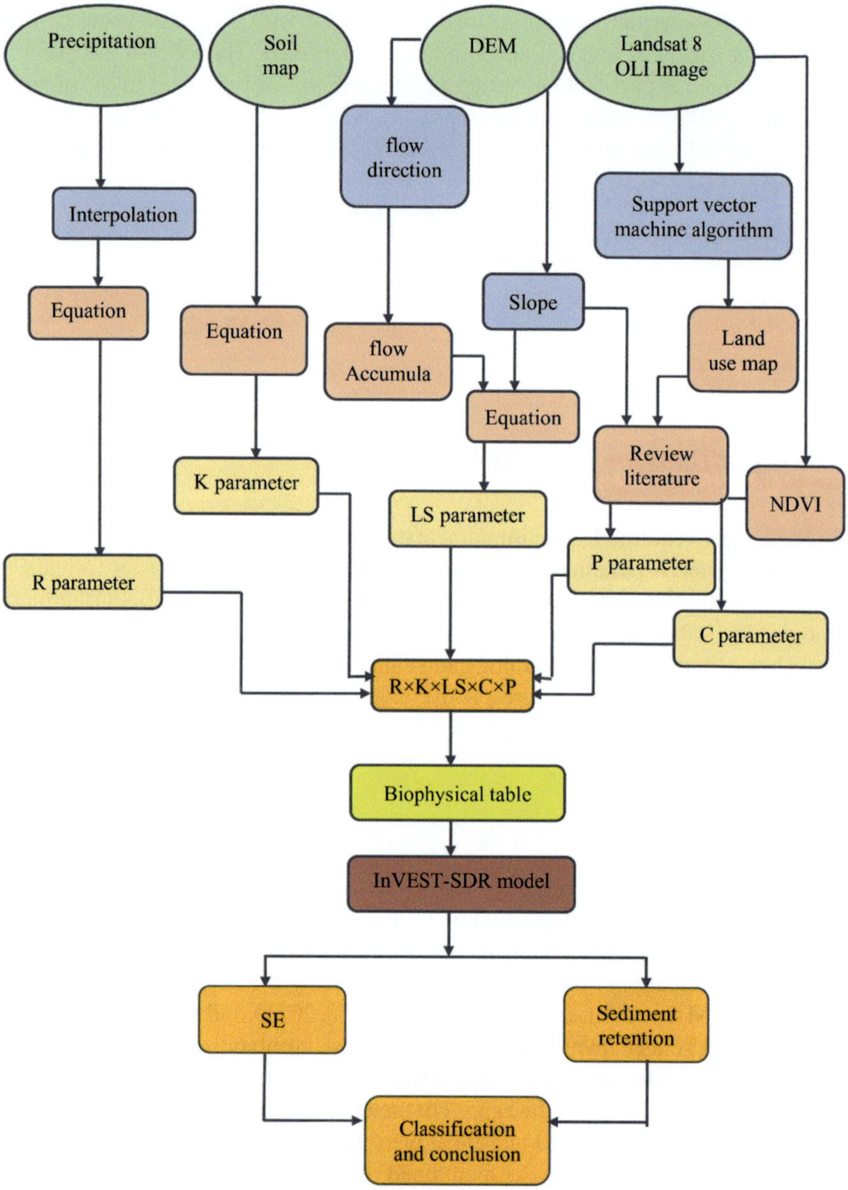

Fig. 2 Flowchart of estimated potential of SE and sediment retention

Table 1 Input data for InVEST-SDR model

Input	Type	Source and value	Unit
Land use	Raster (30 m)	Landsat 8 (OLI) image	–
DEM	Raster (30 m)	Aster image	m
R parameter	Raster (30 m)	Monthly precipitation data	MJ/mm $(ha/h)^{-1}$
K parameter	Raster (30 m)	Soil organic and soil texture material	t/ha/h $(MJ/ha/mm)^{-1}$
LS parameter	Raster (30 m)	DEM	–
C parameter	Raster (30 m)	NDVI	–
P parameter	Integer	Land uses	–
k	Decimal	2	–
IC_0	Decimal	0.5	–
SDR_{max}	Decimal	0.8	Micrometers

2.3.2 DEM

In this chapter, the Shuttle Radar Topography Mission (SRTM) DEM layer was used, with a spatial resolution of 30 m and provided by the USGS. The model's high-accuracy and extensive coverage make it the most popular choice for environmental and terrain analysis (Liu et al. 2021).

2.3.3 *R* Parameter

The potential for rainstorms to cause SE is described by the *R* parameter (Zhou et al. 2019). This parameter indicates the numerical strength of rainfall in the region. SE is determined by the amount, intensity, and distribution of rainfall (Tamire et al. 2022). The *R* parameter was estimated from the product of kinetic energy of rain (*E*) in the maximum precipitation intensity of 30 min (I30) (Endalamaw et al. 2021) (Eq. 9).

$$R = E \cdot 130 = \left(210 + \log_{10} 130\right) * 130. \tag{9}$$

Due to the lack of access to 130 in our study region, the *R* parameter was calculated using the Fournier index (Eq. 10), according to the monthly and annual precipitation at meteorological stations over a 30-year period (Tavakoli and Mohammadyari 2023).

$$F = \frac{\sum_{i=1}^{12} pi^2}{\sum_{i=1}^{12} pi} \tag{10}$$

whereas P_i is the average precipitation (mm) in month *i* and *p* is the average annual precipitation (mm). In the next step, using Eqs. (11) and (12), the *R* parameter value was estimated for index stations. Eventually, the *R* index was drawn using the kriging interpolation method (Tavakoli and Mohammadyari 2023; Zhou et al. 2019; Ahmadi

Mirghaed et al. 2018) using a GIS software.

$$R \text{ parameter} = \left(0.07397 * F^{1.847}\right) \quad F < 55\,\text{mm} \tag{11}$$

$$R \text{ parameter} = \left(95.77 - 6.081 * F + 0.447 * F^2\right) \quad F > 55\,\text{mm}. \tag{12}$$

2.3.4 *K* Parameter

One of the main parameters governing SE is the *K* parameter (Girma and Gebre 2020). This parameter shows the inherent sensitivity of the soil to erosion by precipitation and runoff (Mazigh et al. 2022). The *K* parameter is calculated based on Eq. (13) (Renard et al. 1997).

$$K = \frac{2.1 \cdot 10^{-4}(12 - a)M^{1.14} + 3.25(b - 2) + 2.5(c - 3)}{759}, \tag{13}$$

where

- *K* is the soil erodibility parameter (t ha hr $(\text{MJ mm ha})^{-1}$).
- *M* is the very fine sand (%) + silt (%) (100-clay (%)).
- *A* is the organic matter (%).
- *B* is the structure code: from very structured, relatively structured, to slightly and solid structured, and indicated by codes 1–4, respectively.
- *C* is the profile permeability code: fast (1), medium to fast (2), medium (3), medium to slow (4), slow (5), and very slow (6) (Sharp et al. 2020). In the case of a lack of access to profile structure and permeability, soil erodibility was calculated according to organic matter content and soil texture (Table 2).

2.3.5 LS Parameter

The LS coefficient was extracted from a DEM. The topographic parameter (LS) is obtained from the integration of the *L* (slope length) and *S* (slope steepness) parameters (Tamire et al. 2022). The rate and volume of SE are determined by the combination of these two parameters (Degife et al. 2021). A higher value of LS coefficient indicates a high SE potential (Wischmeier and Smith 1978).

2.3.6 *C* Parameter

Parameter *C* is vital for soil and water conservation, to which natural and human activities are both sensitive (Zhou et al. 2019). This parameter represents the impact

Table 2 Estimation of *K* parameter

	Average OMC	OMC < 2%	OMC > 2%
Clay	0.22	0.24	0.21
Clay loam	0.30	0.33	0.28
Coarse sandy loam	0.07	0.07	0.07
Fine sand	0.08	0.09	0.06
Fine sandy loam	0.18	0.22	0.17
Heavy clay	0.17	0.19	0.15
Loam	0.30	0.34	0.26
Loamy fine sand	0.11	0.15	0.09
Loamy sand	0.04	0.05	0.04
Loamy very fine sand	0.39	0.44	0.25
Sand	0.02	0.03	0.01
Sandy clay loam	0.20	0.20	0.20
Sandy loam	0.13	0.14	0.12
Silt loam	0.38	0.41	0.37
Silty clay	0.26	0.27	0.26
Silty clay loam	0.32	0.35	0.30
Very fine sand	0.43	0.46	0.37
Very fine sandy loam	0.35	0.41	0.33

Reference: Sharp et al. (2020)

of vegetation cover on reducing erosion and also the effect of management and cultivation activities on agricultural lands (Ahmadi Mirghaed et al. 2018). Here, this parameter was estimated using the Normalized Difference Vegetation Index (NDVI) (Eq. 14). NDVI index data was obtained from the Landsat 8 (OLI) images in 2021. NDVI is related to biomass, carbon decomposition, plant water stress, and biodiversity and is suitable for monitoring changes in vegetation cover and investigating the responses of vegetation cover to climate change (Mohammadyari et al. 2019).

$$\mathrm{C} = (1 - \mathrm{NDVI}/2) \tag{14}$$

2.3.7 *P* Parameter

The *P* parameter is the role of land management and conservation practices in minimizing SE (Tamire et al. 2022). Herein, the values of this parameter were determined by reclassifying land cover types according to the literature review. Overall, the range of this parameter is close to zero for soils without support operations and one when appropriate erosion control measures are implemented (Getu et al. 2022).

2.3.8 Biophysical Table

This is a table in CVS format, which contains three columns: land use code, coverage management coefficient, and management measures coefficient, for each of the land use map classes.

2.3.9 Watershed and Sub-watershed Boundary

The watershed and sub-watershed boundary are extracted from DEM using the GIS software.

3 Results

The spatial distribution of the model input parameters is presented in Fig. 3. The ecology of the north and east of the Ilam dam watershed is mainly affected by forest and grassland classes, while the middle part is occupied by settlements, farmland, and barren lands (Fig. 3a). The R parameter values (Fig. 3c) are in the numerical range 201–204.8 MJ mm ha^{-1} year^{-1}. Likewise, the value of the K parameter (Fig. 3d) in the watershed ranges from 0.00 to 0.28 Mg h MJ^{-1} mm^{-1}, and the most value of this parameter was observed in the central parts. The LS parameter (Fig. 3e) is between 0 and 77. The value of the C parameter (Fig. 3f) is 0.22–0.66, where low values indicate the accumulation of areas with vegetation and high values indicate areas without vegetation. Additionally, the P parameter (Fig. 3g) is equal to one due to non-implementation of protection and management measures in the area.

After the calibration and implementation of the InVEST-SDR model, the annual potential of SE and sedimentation was estimated (in 30 $\times$ 30-pixel dimensions) for the sub-watershed (Fig. 4). It can be seen that the erosion potential (Fig. 4a) varies from 0 to 147 tons per pixel per year. On the other hand, the lowest and highest amounts of sediment export (Fig. 4b) are 0 and 13 tons per pixel per year, respectively. In terms of spatial distribution, the northern, central, and southeastern parts have higher levels of SE potential and sediment export. The findings show that the annual erosion potential for the entire watershed was estimated to be 9,640,723.51 tons per pixel, or 202.7 tons per hectare, per year. Likewise, sediment export for the entire basin is 462,087.41 tons per pixel, equivalent to 9.72 tons per hectare per year. It is worth mentioning that sub-watersheds SW2 (1,875,020.82 ton/ha), SW1 (1,644,441.19 ton/ha), and SW10 (1,315,261.74 ton/ha) have the highest potential for erosion and sediment export, respectively. It is clear that these sub-watersheds need more urgent protection measures than other sub-watersheds. In contrast, the lowest erosion and sediment export potential were observed in sub-watersheds SW9 (19,920.31 ton/ha) and SW8 (80,233.73 ton/ha). The Ilam dam lake is the dominant use of this watershed. Overall, the obtained results indicate that SE is influenced by various parameters, especially the LULC pattern. That is to say, the amount of

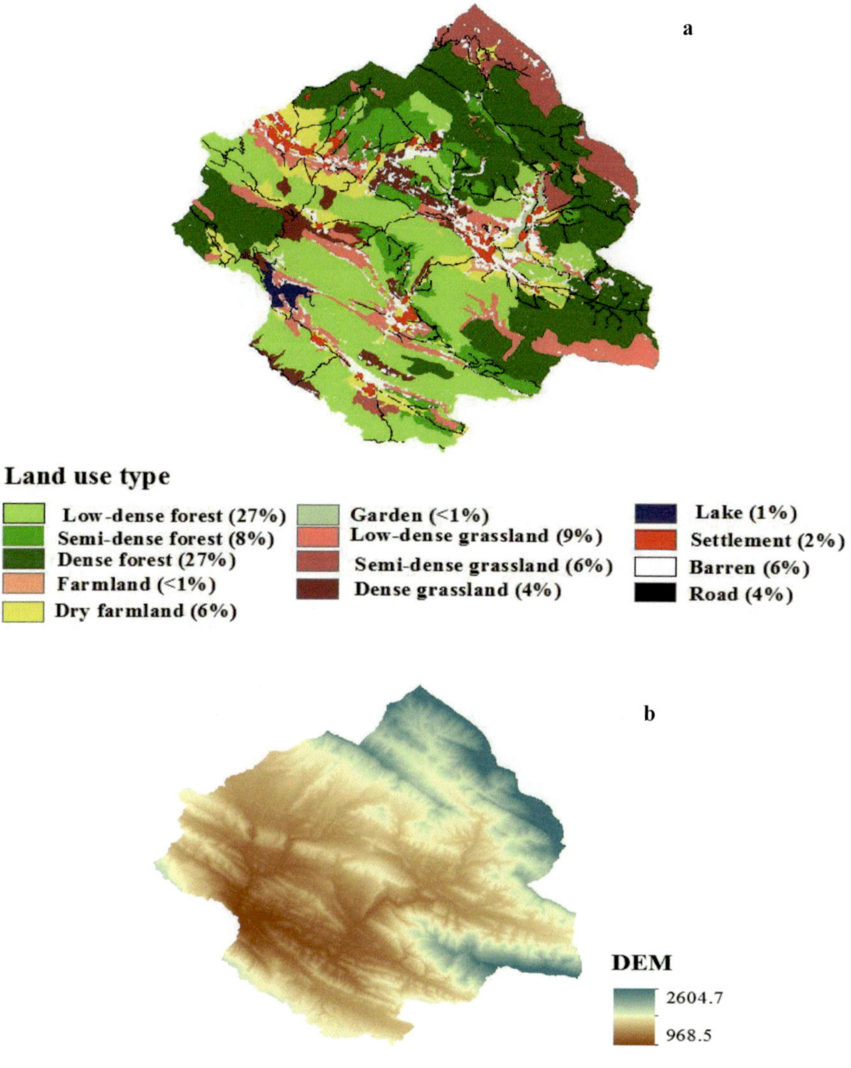

Fig. 3 Input layers of InVEST-SDR model

SE in areas with low vegetation and places with dominant human activity is more than in other places. In contrast, in areas with forest cover, especially thick cover, the highest soil retention, and the least erosion and sedimentation were observed. In line with other studies, it can be confirmed that SE is highly dependent on the spatial pattern of LULC type (e.g., Kusi et al. 2020; Yohannes et al. 2021; Sun et al. 2022; Mohammadyari et al. 2023b). Therefore, prevention of LULC changes is an important parameter in maintaining the ecosystem service of soil retention and preventing SE. As illustrated before, high and low rates of SE and sediment

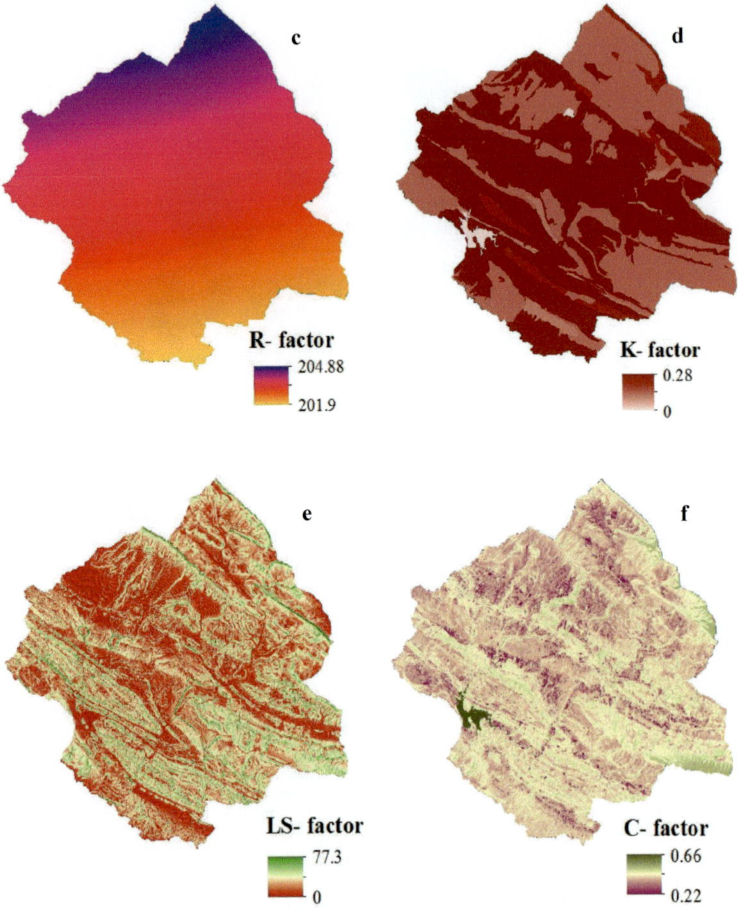

Fig. 3 (continued)

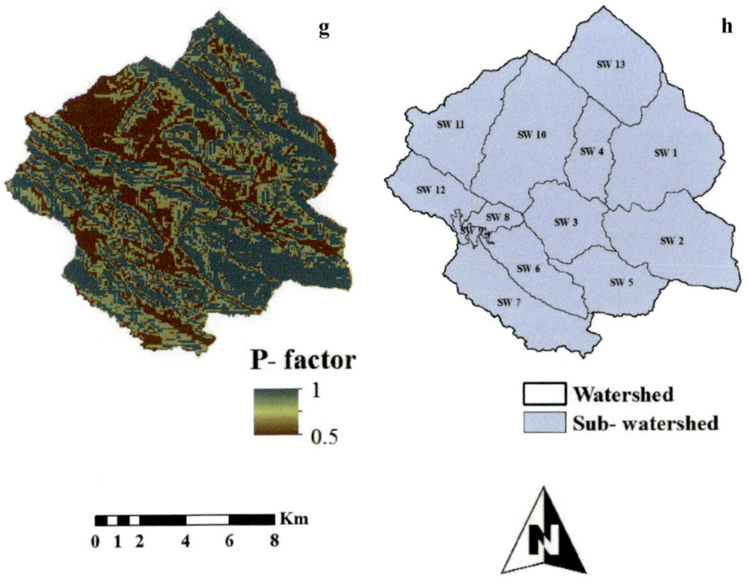

Fig. 3 (continued)

export were similar in the sub-watersheds. This is because sediment export is the multiplication of the rate of annual sediment delivery ratio and SE in our study (Tamire et al. 2022).

Due to the variety of input parameters such as topography land use, soil type, and different water and soil management activities, SE is not uniform throughout the basin (Bekele et al. 2021). Considering that it is not possible to implement protective measures in the entire area at once (Mohammed et al. 2020), it is crucial to prioritize high-risk areas when applying protective measures (Woldemariam et al. 2018). To that end, the SE hotspots were determined, based on comparing the annual soil loss quantified in each sub-watershed. Next, the sub-watersheds were classified into five erosion severity classes. In Fig. 5, the severity classes were very slight (medium lime in color), slight (medium olive), moderate (yellow), high (orange), and very high (red).

It can be seen that two sub-watersheds (SW8 and SW9) were classified as very slight class (V), one sub-watershed (SW12) was classified as slight (IV), seven sub-watersheds (SW3 to SW7–SW11, and SW13) were classified as moderate (III) class, two sub-watersheds (SW1 and SW10) were classified as high (II) class, and finally, one sub-watershed (SW2) was classified as very high (I) class. Consequently, the prioritization of protection measures in the sub-basins should be: SW2 > SW1, and then SW10 > SW7–SW13–SW11–SW3–SW5–SW4–SW6 > SW12 > SW8, and then SW9. The findings show that there was spatial variability of SE in the Ilam dam watershed. From this point of view, the watersheds that are categorized as being of high and very high erosion potential should be considered as a priority class for

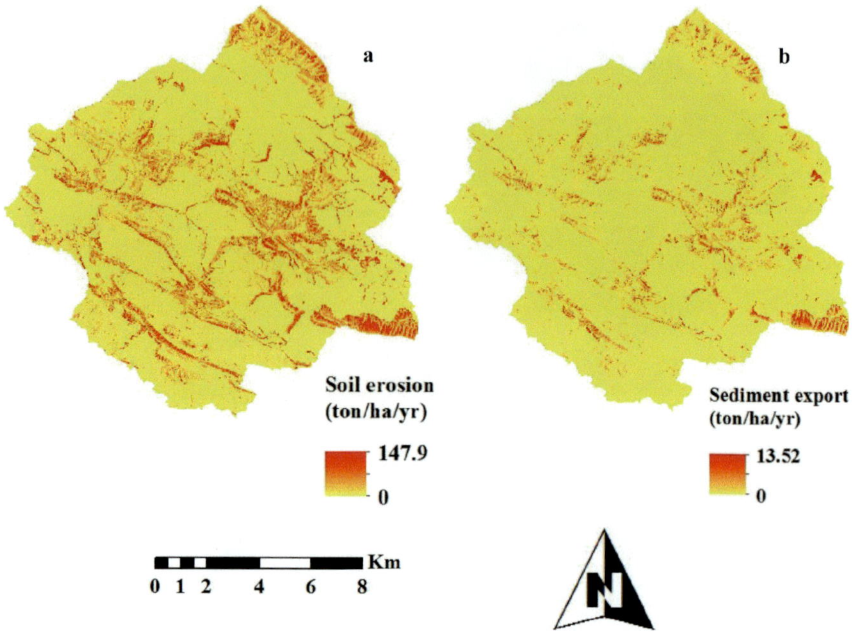

Fig. 4 Output maps of InVEST-SDR model

conservation planning. Although most of the watershed is in the moderate class, in terms of erosion, and about 29% of the basin is in the high and very high erosion class, these two sub-watersheds need more attention and protection operations. Clearly, this does not mean that other sub-basins do not need conservation programs at all, however.

4 Discussion

This study, through the lens of geo-spatial techniques, has estimated the soil erosion potential in the Ilam dam watershed. The integration of the Revised Universal Soil Loss Equation (RUSLE) with Geographic Information System (GIS) tools has been pivotal in providing actionable insights for conservation planning. The analysis has uncovered a significant spatial variability in soil erosion across the watershed, influenced by an interplay of land use/land cover patterns, topography, and human activities. The results have pointed to a clear dichotomy: areas characterized by sparse vegetation and intense human activity are hotspots for higher soil erosion potential, whereas regions with dense forest cover have showcased commendable soil retention capabilities. This dichotomy is particularly pronounced in sub-watersheds SW2, SW1, and SW10, which have emerged as areas with the highest potential for erosion

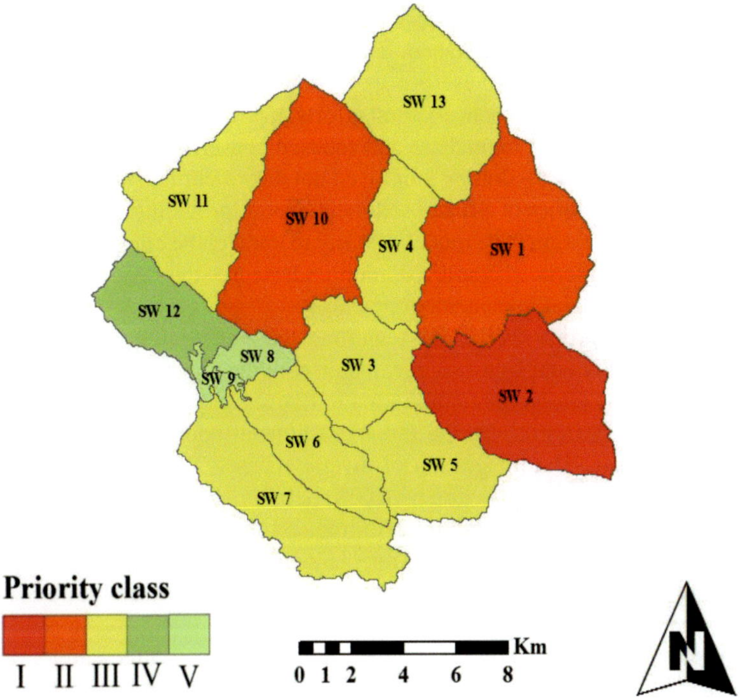

Fig. 5 SE severity class map

and sediment export, underscoring the urgency for protection measures. Conversely, sub-watersheds SW9 and SW8 have been identified as having the lowest erosion and sediment export potential. In an effort to prioritize conservation actions, the study has classified sub-watersheds into five erosion severity classes, ranging from very slight to very high. This classification system serves as a crucial tool for directing protection measures toward high-risk areas to effectively mitigate soil loss. The findings have reiterated the critical need to prevent land use/land cover changes to preserve ecosystem services such as soil retention and to minimize soil erosion.

In essence, the research has highlighted the non-uniform distribution of soil erosion throughout the basin, driven by a multitude of influencing factors. It has emphasized the importance of prioritizing conservation strategies in areas identified as high-risk to address soil erosion in a targeted manner. By delineating hotspots of soil erosion and categorizing sub-watersheds based on severity levels, the study has provided valuable insights for targeted conservation planning and management practices within the Ilam dam watershed. The core of this work lies in its commitment to utilizing geo-spatial technologies for the prediction and management of soil erosion (SE) and sediment export. The study has illuminated the spatial complexities of SE and sediment export, which are integral to developing sustainable land management strategies and effective conservation programs. By adeptly merging the InVEST-SDR model with GIS, our study has successfully mapped the varying degrees of SE and sediment export across the watershed. The heightened vulnerability of the middle and southwestern regions to these processes, due to factors such as steep terrain, unregulated grazing, limited conservation efforts, rapid LULC changes, and intense rainfall erosivity, calls for immediate and focused conservation interventions. The geo-spatial modeling methodology employed not only identifies high-risk zones but also quantifies the severity of SE and sediment export, providing clear guidance for conservation prioritization. The model's ability to unravel the complex relationships between land use patterns and soil retention underscores the need for policies that promote sustainable land management practices.

Moreover, the study's revelations regarding the adverse impacts of SE on the Ilam dam reservoir, a vital water source, underscore the urgent need for soil conservation measures. By adopting rehabilitation initiatives and sustainable soil management practices, we can mitigate the threats of human-induced erosion to our land and water resources. The integration of diverse datasets within the InVEST-SDR modeling framework, in conjunction with GIS spatial analysis, emerges as a powerful tool for identifying SE hotspots and prioritizing sub-watersheds for conservation actions. This capability is invaluable for planners and decision-makers, enabling them to implement targeted conservation strategies and anticipate future SE trends as a proactive measure.

Despite challenges posed by limited field-laboratory data and a scarcity of information, the proficiency of the InVEST-SDR model in pinpointing erosion focal points with minimal input data offers a solution to these constraints. The model has effectively provided a comprehensive SE assessment in the Ilam dam watershed, aiming to prevent irreversible damage and establish a robust dataset for regional policy formulation and decision-making support. In conclusion, this study underscores the pivotal role of geo-spatial modeling in understanding and addressing soil erosion for conservation planning. It advocates for a holistic approach that integrates scientific research, policy development, and community involvement to combat soil erosion and sediment export. As we move forward, refining methodologies, expanding datasets, and fostering international cooperation are essential to safeguarding watersheds for present and future generations. The insights from this study serve as a roadmap for action, guiding the responsible management of natural resources in a manner that harmonizes human progress with ecological preservation. This comprehensive analysis enhances our understanding of soil erosion dynamics in complex landscapes and provides a solid foundation for informed decision-making toward sustainable land management practices and conservation efforts in similar watersheds globally.

5 Conclusions

Geo-spatial modeling is a powerful technology to predict and manage soil erosion and sediment transport. This study, set within the Ilam dam watershed, has illuminated the spatial complexities of soil erosion and sediment export, which are pivotal for crafting sustainable land management strategies and implementing effective protection programs. The geo-spatial modeling approach adopted in this study not only identifies the areas at risk but also quantifies the severity of soil erosion and sediment export, providing a clear directive for prioritizing conservation actions. The model's ability to discern the intricate patterns of land use and its impact on soil retention underscores the need for policies that promote sustainable land management practices. The comprehensive study of soil erosion and sediment export within the Ilam dam watershed, as presented in this paper, underscores the critical importance of geo-spatial modeling for conservation planning and sustainable land management. The integration of the InVEST-SDR model with GIS has provided a nuanced understanding of the spatial distribution of these environmental degradation processes, revealing a heterogeneous pattern across the watershed. The middle and southwestern sections, characterized by steep slopes, rapid land use and land cover (LULC) changes, uncontrolled human activities, and high rainfall erosivity, emerged as areas with heightened potential for soil erosion and sediment export. The implications of these findings are manifold. Firstly, they highlight the vulnerability of the Ilam dam watershed to erosive processes, which, if left unchecked, could compromise the integrity of the Ilam dam reservoir—a critical source of drinking water for the city's residents. This threat necessitates the implementation of protective and corrective soil management measures aimed at mitigating the adverse effects of soil erosion. The adoption of soil conservation practices, rehabilitation programs, and sustainable soil management measures can significantly reduce the rate of soil erosion, particularly in areas most susceptible to human-induced erosion. The role of sustainable soil management measures is particularly salient. By adopting appropriate policies that encourage the preservation of soil integrity, the rate of soil erosion caused by anthropogenic activities can be curtailed. This study has demonstrated that a multi-source dataset within the InVEST-SDR modeling framework, when applied in GIS spatial analysis, can effectively identify soil erosion hotspots and prioritize sub-watersheds for conservation planning. Such a tool serves as a valuable guide for planners and decision-makers in introducing targeted conservation measures and can also be utilized to predict future soil erosion as a preventative measure. Despite the robustness of the InVEST-SDR model in identifying areas of erosion with minimal input data, our study faced limitations due to the lack of field-laboratory measurements, constrained by funding and time, as well as a dearth of data for comprehensive soil erosion assessments. Nevertheless, the model's capability to pinpoint focal areas of erosion offers a viable solution to overcome these limitations. The general situation of soil erosion assessment in the Ilam dam watershed has been clearly delineated, with the aim of addressing the problem before it becomes irreversible. This approach can create a robust dataset for regional policies and decision-makers, providing essential information for watershed managers and stakeholders. It can also serve as a reference for optimizing ecological construction and land use patterns in the watershed. In light of the new patterns of LULC that arise due to continuously expanding human activities and the destruction of vegetation, several measures are recommended for the Ilam dam watershed. These include the enforcement of strict forest land protection policies, the implementation of soil protection and ecological forest protection projects, the adoption of sustainable watershed management approaches, and the execution of effective engineering and biological measures to control erosion and sediment export. Additionally, regional government officials, NGOs, concerned organizations, and local communities should consider the further degradation of watersheds through an awareness of land use, advocating for the sustainable use of natural conservation methods without compromising its utility for future generations. The sharing of scientific knowledge between international partners and the availability of databases with high spatial resolution are pivotal for enhancing our understanding of soil erosion and for making fundamental improvements in its estimation. It is anticipated that such collaborative efforts will lead to the adoption of appropriate policies that align with the principles of environmental sustainability and resilience. As the global community grapples with the challenges posed by environmental degradation, the insights gleaned from this study offer a blueprint for action, guiding the stewardship of natural resources in a manner that harmonizes human development with ecological preservation. This study presented herein provides a compelling case for the integration of advanced modeling techniques with field data to inform conservation strategies. It emphasizes the need for a multi-faceted approach that combines policy intervention, community engagement, and scientific research to effectively combat soil erosion and sediment export. As we move forward, it is imperative that we continue to refine our methodologies, expand our datasets, and foster international collaboration to safeguard our watersheds for the benefit of current and future generations.

References

Ahmadi Mirghaed F, Souri B, Mohammadzadeh M, Salmanmahiny A, Mirkarimi SH (2018) Evaluation of the relationship between soil erosion and landscape metrics across Gorgan Watershed in northern Iran. Environ Monit Assess 190(11):1–14

Andriyani I, Indarto I, Soekarno S, Pradana MP (2021) Analysis of rainfall erosivity factor (R) on prediction of erosion yield using USLE and RUSLE Model's: a case study in Mayang Watershed, Jember Regency, Indonesia. SAINS TANAH-J. Soil Sci Agroclimatology 21(1):64-73

Aneseyee AB, Elias E, Soromessa T, Feyisa GL (2020) Land use/land cover change effect on soil erosion and sediment delivery in the Winike watershed, Omo Gibe Basin, Ethiopia. Sci Total Environ 728:138776. https://doi.org/10.1016/j.scitotenv.2020.138776

Bagwan WA, SopanGavali R (2021) Dam-triggered land use land cover change detection and comparison (transition matrix method) of urmodi river watershed of Maharashtra, India: a remote sensing and GIS approach. Geol Ecol Landsc 1–9

Bekele DA, Gella GW, Ejigu MA (2021) Erosion risk assessment: a contribution for conservation priority area identification in the sub-basin of Lake Tana, north-western Ethiopia. Int Soil Water Conserv Res. https://doi.org/10.1016/j.iswcr.2021.04.010

Berihun ML, Tsunekawa A, Haregeweyn N, Meshesha DT, Adgo E, Tsubo M, Masunaga T, Fenta AA, Sultan D, Yibeltal M, Ebabu K (2019) Hydrological responses to land use/land cover change and climate variability in contrasting agroecological environments of the Upper Blue Nile basin, Ethiopia. Sci Total Environ 689:347–365

Berihun ML, Tsunekawa A, Haregeweyn N, Dile YT, Tsubo M, Fenta AA, Meshesha DT, Ebabu K, Sultan D, Srinivasan R (2020) Evaluating runoff and sediment responses to soil and water conservation practices by employing alternative modelling approaches. Sci Total Environ 747:141118

Bhattacharya RK, Chatterjee ND, Das K (2024) Modelling of soil erosion susceptibility incorporating sediment connectivity and export at landscape scale using integrated machine learning, InVEST-SDR and Fragstats. J Environ Manage 353:120164

Borrelli P, Robinson DA, Fleischer LR, Lugato E, Ballabio C, Alewell C, Meusburger K, Modugno S, Schutt B, Ferro V, Bagarello V, Van Oost K, Montanarella L, Panagos P (2017) An assessment of the global impact of 21st century land use change on soil erosion. Nat Commun 8:2013. https://doi.org/10.1038/s41467-017-02142-7

Borselli L, Cassi P, Torri D (2008) Prolegomena to sediment and flow connectivity in the landscape: a GIS and field numerical assessment. Catena 75:268–277

Carucci T, Whitehouse-Tedd K, Yarnell RW, Collins A, Fitzpatrick F, Botha A, Santangeli A (2022) Ecosystem services and disservices associated with vultures: a systematic review and evidence assessment. Ecosyst Serv 56:101447

De Vente J, Poesen J, Verstraeten G, Govers G, Vanmaercke M, Van Rompaey A, Arabkhedri M, Boix-Fayos C (2013) Predicting soil erosion and sediment yield at regional scales: where do we stand? Earth Sci Rev 127:16–29

Degife A, Worku H, Gizaw S (2021) Environmental implications of soil erosion and sediment yield in Lake Hawassa watershed, south-central Ethiopia. Environ Syst Res 10(1)

Desmet PJJ, Govers G (1996) A GIs procedure for automatically calculating the USLE LS factor on topographically complex landscape units. J Soil Water Conserv 51:427–433

Endalamaw NT, Moges MA, Kebede YS, Alehegn BM, Sinshaw BG (2021) Potential soil loss estimation for conservation planning, upper Blue Nile Basin, Ethiopia. Environ Challenges 5:100224

Fenta AA, Tsunekawa A, Haregeweyn N, Tsubo M, Yasuda H, Shimizu K, Kawai T, Ebabu K, Berihun ML, Sultan D, Belay AS (2020) Cropland expansion outweighs the monetary effect of declining natural vegetation on ecosystem services in sub-Saharan Africa. Ecosyst Serv 45:101154

Fenta AA, Tsunekawa A, Haregeweyn N, Tsubo M, Yasuda H, Kawai T, Ebabu K, Berihun ML, Belay AS, Sultan D (2021) Agroecology-based soil erosion assessment for better conservation planning in Ethiopian river basins. Environ Res 195:110786

Gashaw T, Bantider A, Zeleke G, Alamirew T, Jemberu W, Worqlul AW, Dile YT, Bewket W, Meshesha DT, Adem AA, Addisu S (2021) Evaluating InVEST model for estimating soil loss and sediment export in data scarce regions of the Abbay (Upper Blue Nile) Basin: implications for land managers. Environ Challenge 5:10038. https://doi.org/10.1016/j.envc.2021.100381

Getu LA, Nagy A, Addis HK (2022) Soil loss estimation and severity mapping using the RUSLE model and GIS in Megech watershed, Ethiopia. Environ Challenges 8:100560

Girma R, Gebre E (2020) Spatial modeling of erosion hotspots using GIS-RUSLE interface in Omo-Gibe river basin, Southern Ethiopia: implication for soil and water conservation planning. Environ Syst Res 9(1):1–14

Gomiero T (2016) Soil degradation, land scarcity and food security: reviewing a complex challenge. Sustainability 8:281. https://doi.org/10.3390/su8030281

Guerra CA, Rosa IM, Valentini E, Wolf F, Filipponi F, Karger DN, Nguyen Xuan A, Mathieu J, Lavelle P, Eisenhauer N (2020) Global vulnerability of soil ecosystems to erosion. Landsc Ecol 35(4):823–842. https://doi.org/10.1007/s10980-020-00984-z

Hamel P, Chaplin-Kramer R, Sim S, Mueller C (2015) A new approach to modeling the sediment retention service (InVEST 3.0): case study of the Cape Fear catchment, North Carolina, USA. Sci Total Environ 524–525:166–177

Hu Y, Gao M (2020) Evaluations of water yield and soil erosion in the Shaanxi-Gansu Loess Plateau under different land use and climate change scenarios. Environ Dev 34:100488

Kumar N, Singh SK (2021) Soil erosion assessment using earth observation data in a trans-boundary river basin. Nat Hazards 107:1–34

Kumarasiri AD, Udayakumara EP, Jayawardana JM (2023) Modeling soil erosion: Samanalawewa watershed, Sri Lanka. In: Multi-Hazard Vulnerability and Resilience Building, pp 205–220. Elsevier

Kusi KK, Khattabi A, Mhammdi N, Lahssini S (2020) Prospective evaluation of the impact of land use change on ecosystem services in the Ourika watershed, Morocco. Land Use Policy 97:104796

Lamane H, Moussadek R, Baghdad B, Mouhir L, Briak H, Laghlimi M, Zouahri A (2022) Soil water erosion assessment in Morocco through modeling and fingerprinting applications: a review. Heliyon. 8:e10209

Lefèvre C, Cruse RM, dos Anjos LHC, Calzolari C, Haregeweyn N (2020) Guest editorial–soil erosion assessment, tools and data: a special issue from the Global Symposium on soil Erosion 2019. Int Soil Water Conserv Res 8(4):333–336

Li N, Zhang Y, Wang T, Li J, Yang J, Have LM (2022) Have anthropogenic factors mitigated or intensified soil erosion over the past three decades in South China? J Environ Manage 302:114093

Liu X, Xin L, Lu Y (2021) National scale assessment of the soil erosion and conservation function of terraces in China. Ecol Indic 129:107940

Maliqi E, Singh SK (2019) Quantitative estimation of soil erosion using open-access earth observation data sets and erosion potential model. Water Conserv Sci Eng 4(4):187–200

Mazigh N, Taleb A, El Bilali A, Ballah A (2022) The effect of erosion control practices on the vulnerability of soil degradation in Oued EL Malleh catchment using the USLE model integrated into GIS, Morocco. Trends Sci 19(2):2059–2059

Mirsanjari MM, Sužiedelytė Visockienė J, Mohammadyari F, Zarandian A (2021) Modelling of expansion changes of Vilnius city area and impacts on landscape patterns using an Artificial Neural Network. Ecol Chem Eng S28(3):429–447. https://doi.org/10.2478/eces-2021-0029

Mohammadyari F, Mirsanjari MM, Zarandian A (2019) Monitoring of vegetation changes in Karaj watershed using NDVI index and gradient analysis. J RS GIS Nat Resour 9(4):55–72

Mohammadyari F, Tavakoli M, Zarandian A, Abdollahi S (2023a) Optimization land use based on multi-scenario simulation of ecosystem service for sustainable landscape planning in a mixed urban-Forest watershed. Ecol Model 483:110440

Mohammadyari F, Zarandian A, Mirsanjari MM, SuziedelyteVisockiene J, Tumeliene E (2023) Modelling impact of urban expansion on ecosystem services: a scenario-based approach in a mixed natural/urbanised landscape. Land 12(2):291

Mohammed S, Alsafadi K, Talukdar S, Kiwan S, Hennawi S, Alshihabi O, Sharaf M, Harsanyie E (2020) Estimation of soil erosion risk in Southern part of Syria by using RUSLE integrating geo informatics approach. Remote Sens Appl Soc Environ 20:100375. https://doi.org/10.1016/j.rsase.2020.100375

Naipal V, Ciais P, Wang Y, Lauerwald R, Guenet B, Van Oost K (2018) Global soil organic carbon removal by water erosion under climate change and land use change during AD 1850–2005. Biogeosciences 15(14):4459–4480

Ougougdal HA, Khebiza MY, Messouli M, Bounoua L (2020) Delineation of vulnerable areas to water erosion in a mountain region using SDR-InVEST model: a case study of the Ourika watershed, Morocco. Sci African 10:e00646

Panditharathne DL, Abeysingha NS, Nirmanee KG, Mallawatantri A (2019) Application of revised universal soil loss equation (Rusle) model to assess soil erosion in "Kalu Ganga" River Basin in Sri Lanka. Appl Environ Soil Sci 2019:1–15

Paudel S, States S (2023) Urban green spaces and sustainability: exploring the ecosystem services and disservices of grassy lawns versus floral meadows. Urban For Urban Green 84:127932

Qiao X, Li Z, Lin J, Wang H, Zheng S, Yang S (2023) Assessing current and future soil erosion under changing land use based on InVEST and FLUS models in the Yihe River Basin, North China. Int Soil Water Conserv Res. https://doi.org/10.1016/j.iswcr.2023.07.001

Renard K, Foster G, Weesies G, McCool D, Yoder D (1997) Predicting soil erosion by water: a guide to conservation planning with the revised soil loss equation

Roba NT, Kassa AK, Geleta DY, Harka AE (2021) Streamflow and sediment yield estimation, and area prioritization for better conservation planning in the Dawe River watershed of the Wabi Shebelle River Basin, Ethiopia. Heliyon 7(12):e08509

Saha M, Sauda SS, Real HR, Mahmud M (2022) Estimation of annual rate and spatial distribution of soil erosion in the Jamuna basin using RUSLE model: a geospatial approach. Environ Chall 8:100524

Sharp ER, Douglass J, Wolny S, Authors C, Arkema K, Bernhardt J, Bierbower W, Chaumont N, Denu D, Fisher D, Glowinski K, Griffin R, Guannel G, Guerry A, Guerry-lawler C, Johnson J, Hamel P, Kennedy C, Kim C (2020) InVEST user's guide

Sun Y, Liu D, Wang P (2022) Urban simulation incorporating coordination relationships of multiple ecosystem services. Sustain Cities Soc 76:103432

Tamire C, Elias E, Argaw M (2022) Spatiotemporal dynamics of soil loss and sediment export in Upper Bilate River Catchment (UBRC), Central Rift Valley of Ethiopia. Heliyon 8(11):e11220

Tavakoli M, Mohammadyari F (2023) Modeling the spatial distribution of multiple ecosystem services in Ilam dam watershed, Western Iran: identification of areas for spatial planning. Urban Ecosyst 1–20

Vigiak O, Borselli L, Newham LT, McInnes J, Roberts AM (2012) Comparison of conceptual landscape metrics to define hillslope-scale sediment delivery ratio. Geomorphology 138:74–88

Wischmeier WH, Smith DD (1978) Predicting rainfall erosion losses—a guide to conservation planning. Sci Educ Admin 46:34–38

Woldemariam G, Iguala A, Tekalign S, Reddy R (2018) Spatial modeling of soil erosion risk and its implication for conservation planning: the case of the gobele watershed, East Hararghe zone, Ethiopia. Land 7(1):25. https://doi.org/10.3390/land7010025

Yohannes H, Soromessa T, Argaw M, Dewan A (2021) Impact of landscape pattern changes on hydrological ecosystem services in the Beressa watershed of the Blue Nile Basin in Ethiopia. Sci Total Environ 793:148559

Zarandian A, Mohammadyari F, Mirsanjari MM, Visockiene JS (2023) Scenario modeling to predict changes in land use/cover using Land Change Modeler and InVEST model: a case study of Karaj Metropolis, Iran. Environ Monitor Assess 195(2):273

Zhou M, Deng J, Lin Y, Belete M, Wang K, Comber A, Gan M, Haung L (2019) Identifying the effects of land use change on sediment export: integrating sediment source and sediment delivery in the Qiantang River Basin, China. Sci Total Environ 686:38-49

Digital Terrain Analysis for Characterization of Terrain Variables Governing Soil Erosion and Watershed Hydrology

Suresh Kumar, Anu David Raj, Kalambukattu Justin George, and Uday Chatterjee

Abstract The study investigates the use of satellite-derived digital elevation models (DEMs) and remote sensing methods to analyse the influence of terrain on various environmental factors crucial for comprehending and understanding land degradation phenomena. It assesses erosion risk and runoff intensity by integrating the examination of topographic characteristics with hydrologic indices derived from DEMs. The primary aim is to assess the effectiveness of geomorphometric factors and data obtained through remote sensing in gauging the vulnerability of ecosystems to water-induced erosion, thereby evaluating their fragility. An integrated modelling approach based on Geographic Information Systems (GIS), incorporating soil, terrain, and climatic parameters that influence soil erosion, was assessed for its practical effectiveness in evaluating soil erosion hazards. The study generated potential and actual soil erosion risk maps, providing valuable tools for implementing optimal conservation measures to mitigate soil erosion. This approach is easy to implement in hilly and mountainous regions, where soil erosion is primarily influenced by topography, and lack of sufficient data which hinders the soil erosion modelling. Furthermore, this could be utilized to identify potential soil erosion hotspots that areas require detailed study. Additionally, the study underscored the crucial role of land use and land cover in mitigating the impact of topography on soil erosion.

Keywords RS · GIS · Primary and secondary terrain attributes · Soil erosion risk potential

S. Kumar (✉)
Agriculture, Forestry and Ecology Group, Indian Institute of Remote Sensing, Indian Space Research Organisation (ISRO), Dehradun 248001, India
e-mail: sureshkumar100@gmail.com

A. David Raj · K. Justin George
Agriculture and Soils Department, Indian Institute of Remote Sensing, Indian Space Research Organisation (ISRO), Dehradun 248001, India

U. Chatterjee
Department of Geography, Bhatter College in Dantan, Paschim Medinipur, West Bengal 721426, India

S. C. Pal and U. Chatterjee (eds.), *Surface, Sub-Surface Hydrology and Management*, Springer Geography, https://doi.org/10.1007/978-3-031-62376-9_21

1 Introduction

Soil erosion stands as the foremost concern regarding land degradation, particularly in hilly and mountainous regions, posing a significant threat to both agricultural productivity and soil quality (Lal 2001). Particularly in the Himalayas, soil erosion is recognized as a critical issue, with detrimental effects on agricultural output and environmental sustainability (David Raj et al. 2023). Furthermore, the exacerbation of this issue may be witnessed due to the ramifications of climate change (Sooryamol et al. 2022; David Raj et al. 2022). The process of soil erosion is intricate, and influenced by a multitude of factors including vegetation, soil composition, topography, parent material, drainage patterns, rainfall intensity, and land usage (Morgan 2009). Among these factors, topography emerges as the predominant influencer, governing the magnitude and significance of soil erosion (Kumar and Gupta 2016). The topography of an area dictates the flow of water across the landscape surface, thereby shaping runoff patterns and erosion mechanisms, which are primary contributors to soil erosion. Moreover, topography also affects soil characteristics, impacting various facets of erosion, runoff dynamics, and moisture retention (Moore et al. 1993; Pickup and Chewings 1996; Wood 1996). Key topographic attributes such as slope gradient, curvature, and upslope contributing area assume crucial roles in estimating and predicting soil erosion. They serve as vital indicators for assessing the vulnerability of a landscape to erosion processes (Chaplot 2013; Huang et al. 2023).

Topographic attributes (TAs) are instrumental in providing quantitative descriptions of landforms and their structural features (Youfua et al. 2008). Digital terrain characterization is the quantitative assessment of land surfaces. DEM represent the topography of the area and is used often in geographic information systems to generate digital relief maps. Digital Elevation Models (DEMs) can be obtained using a variety of methods, including photogrammetry, LiDAR, InSAR, land surveying, etc. (Li et al. 2005). It represents spatially registered set of elevation points. The automation of terrain analysis, coupled with the utilization of DEMs, has facilitated the direct quantification of topographic attributes. This encompasses the assessment of terrain features like contributing area, slope, aspect, and plan curvature, which are instrumental in estimating the distribution patterns of soil water content and soil erosion across spatial landscapes (Beven and Kirkby 1979; Wilson and Gallant 1996).

Digital terrain analysis has emerged as a valuable tool for predicting erosion risk (Kumar and Gupta 2016). Utilizing digital elevation models (DEMs), researchers can investigate how erosional processes shape landscape morphology and evolution, often described through generalized quantitative relationships referred to as 'erosion laws' (Montgomery 2003). Additionally, various terrain-based indices, such as the Terrain Wetness Index (TWI) and Stream Power Index (SPI), serve as secondary terrain parameters used to assess soil moisture and terrain susceptibility (Pradhan and Kim 2014). These combined factors, known as hydrologic indices, can serve as indicators of drainage conditions, overland flow intensity, and erosion potential. Topography plays a significant role in hydrological behaviour, especially concerning soil moisture in gently sloped watersheds within humid, temperate regions (Beven and Wood 1983). The Terrain Wetness Index (TWI) is particularly valuable in delineating areas of water divergence and convergence based on slope gradient. In areas with low slope gradients, soil tends to retain moisture as water is not swiftly drained to lower elevations. Moore et al. (1988) noted a strong correlation between these indices and the allocation of surface soil water content. Similarly, Gessler et al. (1995) observed that TWI, combined with plan curvature, is a reliable predictor of soil attributes such as horizon depth and solum depth. These indices aid in characterizing water flow over the surface and, consequently, help predict soil erosion risk potential.

This study is an attempt to combine hydrologic indices with topographic analysis to assess erosion risk and runoff intensity; it also delivers the integration of satellite-derived digital elevation models and remote sensing techniques to analyse the influence of terrain on environmental factors crucial for understanding land degradation; it generates maps of soil erosion risk that are useful for conservation measures, especially in hilly and mountainous regions where data is scarce; and it highlights the important role that land use and cover play in mitigating the impact of topography on soil erosion. Hence, the objective of the current study is to characterize the terrain of the watershed based on indices derived from digital terrain models (DTMs). Thus, the study focuses on determining secondary topographic parameters such as Plan and Profile Curvature, slope, and aspect, along with relevant terrain indices like Terrain Wetness Index (TWI), Stream Power Index (SPI), and LS factor in predicting soil erosion conditions within the watershed.

2 Terrain Attributes

Terrain attributes encompass diverse aspects or qualities of the natural terrain or landscape. These qualities encompass a range of factors, including but not limited to elevation, slope, aspect (the direction in which a slope faces), curvature (whether the terrain is convex or concave in shape), roughness, and other derived metrics like the topographic wetness index or terrain ruggedness index.

2.1 Primary Attributes

Primary terrain attributes represent fundamental characteristics of the landscape, providing direct descriptions of its physical properties.

2.1.1 Elevation

The height of the land surface above a reference point, typically above the mean sea level.

2.1.2 Slope

The gradient of the terrain, typically quantified as the angle of incline from the horizontal plane, delineates how elevation alters across a specified horizontal span. Slope can be articulated in degrees, percentages, or represented as a ratio (rise over run). It constitutes a fundamental terrain attribute and holds pivotal significance in numerous natural phenomena, including water runoff patterns, soil erosion dynamics, and the management of natural resources (Fang 2021).

2.1.3 Aspect

Aspect denotes the orientation of a slope concerning the cardinal directions (north, south, east, west). It is conventionally gauged in degrees clockwise from north, spanning from 0° to 360°. For instance, an aspect of 90° indicates an east-facing slope, whereas 180° suggests a south-facing one. Aspect furnishes valuable insights into the alignment of terrain surfaces and can impact various factors such as exposure to solar radiation, distribution of temperature, and arrangement of vegetation patterns (Bennie et al. 2008; Yang et al. 2020).

2.1.4 Curvature

Terrain surface shape can manifest as either convex (outward bulging) or concave (inward scooping). Surface curvature delineates the degree of convexity or concavity and is typically categorized into profile and plan curvature. Profile curvature illustrates the surface along the direction of the steepest slope, while plan curvature describes a surface perpendicular to the steepest slope's direction. Also referred to as horizontal and vertical curvature, respectively, these curvatures are derived as topographic attributes based on the partial derivatives of elevation concerning the Cartesian plan coordinates x and y (Florinsky 1998).

2.1.5 Specific Catchment Area (Upslope Drainage Area)

Specific contributing area, also known as upslope contributing area per unit width of contour line, pertains to the amount of terrain surface that contributes runoff to a particular location. In a topographic context, upslope drainage area is typically delineated based on terrain slope and flow direction. Each point on the terrain surface channels water towards a downstream point, accumulating more drainage area as it descends. This attribute influences drainage patterns and the formation of gullies (Xu et al. 2017).

2.2 Secondary Attributes

Secondary terrain attributes are derived from primary terrain attributes or represent more specific characteristics of the terrain. They offer supplementary information beyond the fundamental terrain features and are frequently employed for more intricate analysis or modelling uses.

2.2.1 Terrain Wetness Index (TWI)

The quantification of water accumulation or flow potential based on terrain surface features is facilitated through a calculation method utilizing terrain slope and upslope contributing area. This method, known as the Topographic Wetness Index (TWI), has been extensively utilized to explore the effects of spatial scales on hydrological processes (Sörensen et al. 2006). The TWI serves as an indicator of soil properties and the spatial distribution of soil moisture across different landscape positions (Beven and Kirkby 1993).

It's calculation involves a combination of the standardized upslope contributing area (As) and the slope (β) as follows:

$$\mathrm{TWI} = \ln(\mathrm{As}/\tan\beta)$$

2.2.2 Stream Power Index (SPI)

The Stream Power Index (SPI) is a method used to quantify the erosive potential of water flow, based on the assumption that discharge is directly related to specific catchment area (Moore et al. 1993; Wilson and Gallant 2000). This index serves as a valuable tool for understanding both the morphology of rivers and erosion processes (Ghunowa et al. 2021). In the fields of geomorphology and hydrology, SPI is a quantitative measure employed to assess the erosive capacity of flowing water within river channels. It is derived from the hydraulic geometry of the stream and represents the potential energy available for sediment transport and channel formation. The computation of SPI involves the following equation:

$$\mathrm{SPI} = \mathrm{As} * \tan\beta$$

2.2.3 Sediment Transport Index

The Sediment Transport Index (STI) is extensively employed for characterizing water flows across landscapes and is particularly adept at identifying erosion phenomena. It encompasses the influence of topography on erosion dynamics and is determined by factors such as local upslope contributing area (slope length) and the steepness of the slope. It takes into account various factors such as channel slope, flow velocity, channel width, and sediment characteristics to estimate the capacity of flowing water to transport sediment particles (De Roo 1998). It shows how a particular slope and steepness is erosive in reference to 22.13 m length and 9% slope plot whose value is 1. It can be described as:

$$\text{STI} = (A_s/22.13)^{0.6}(\sin\beta/0.0896)^{1.3}$$

3 Major Sources for Acquiring Terrain Attributes

The recent availability of terrain attribute data through various online platforms offers researchers and professionals a vast resource for a wide array of applications. Websites like USGS Earth Explorer, NASA Earthdata, ISRO-Bhuvan/BhooNidhi/VISTA, and ESA Earth Observation Data grant access to high-quality Digital Elevation Models (DEMs) and associated terrain attributes. These platforms empower users to retrieve critical information such as elevation, slope, aspect, relief, and even secondary attributes like hillshade, slope position, and drainage density. By leveraging these readily accessible and comprehensive datasets, individuals can conduct in-depth analyses, modelling exercises, and decision-making processes across disciplines ranging from hydrology and geomorphology to land use planning and environmental management. Table 1 provides sources for acquiring terrain attribute data. Table 2 provides some common software that can be used for terrain analysis.

4 Terrain Characterization for Watershed Hydrology

A watershed is delineated as a land area where all the rainfall received flows towards a common outlet, such as a particular water body or low-lying topographic feature. This flow occurs due to the natural slope as well as contours of the land. Essentially, a watershed's limits are determined by its highest elevation points. Identifying watersheds is relatively simple in areas with hilly and mountainous regions since their boundaries are typically demarcated by ridges (Fig. 1). Hydrology is a scientific discipline focused on investigating the spatial and temporal dimensions of Earth's water resources, encompassing both their quantity and quality. It encompasses a broad

Table 1 Types and sources of terrain attribute data

Type of data	Description	Parameters generated	Source
Digital elevation models (DEMs)	Gridded representations of Earth's surface elevation, obtained from government agencies, research institutions, or commercial vendors	Elevation, slope, aspect, relief	USGS Earth Explorer, NASA Earthdata, ESA Earth Observation Data, BHUVAN-ISRO
LiDAR data	Remote sensing technology using laser pulses to measure surface elevation, collected by government agencies, research institutions, or private companies	High-resolution elevation, slope, aspect, landform features	USGS 3DEP
Aerial photography	Images captured from aircraft or drones, used for photogrammetric terrain analysis	Elevation, slope, aspect	N/A
Satellite imagery	Images from satellites, including high-resolution and stereo-capable ones, used for photogrammetric terrain analysis	Elevation, slope, aspect	USGS Earth Explorer, NASA Earthdata, ESA Earth Observation Data
Secondary attribute websites	Websites providing secondary attribute information derived from primary terrain attributes	Hillshade, slope position, drainage density, terrain stability, relief ratio	USGS Earth Explorer, NASA Earthdata, ESA Earth Observation Data

Table 2 List of commonly used terrain analysis software

S. No.	Software	Acronym
1	ArcGIS	Arc Geographic Information System
2	QGIS	Quantum Geographic Information System
3	GRASS GIS	Geographic Resources Analysis Support System Geographic Information System
4	SAGA GIS	System for Automated Geoscientific Analyses Geographic Information System
5	ILWIS	Integrated Land and Water Information System

spectrum of phenomena, including the occurrence, movement, distribution, circulation, storage, utilization, development, and conservation of water. These aspects are influenced by the complex relationship between water and the Earth's various components (Edwards et al. 2015). The process of transforming rainfall into runoff throughout a watershed is complex, exhibiting high nonlinearity and showcasing both temporal and spatial variability (ASCE 2000). Comprehending the utilization and circulation of water within a watershed lays the preliminaries for comprehending and elucidating the dynamics between landscapes and water.

Precise anticipation and understanding of the spatial distribution of wetness are crucial for effective water and land resource management planning. To address this need, the topographic wetness index (TWI) was devised to predict relative surface wetness and consequent hydrological behaviours across a watershed. It operates on the principle that shallow slope-parallel flow significantly influences the movement and dispersion of soil water. The hydrological characteristics of a watershed profoundly shape the physical landscape. For instance, valleys tend to accumulate water, leading to elevated moisture levels, while slopes promote water divergence. Leveraging the Digital Terrain Model (DTM), the Soil Wetness Index (SWI) provides a unique opportunity to assess the spatial variability of soil hydrological conditions, which dictate runoff generation (Kumar and Gupta 2016). There are

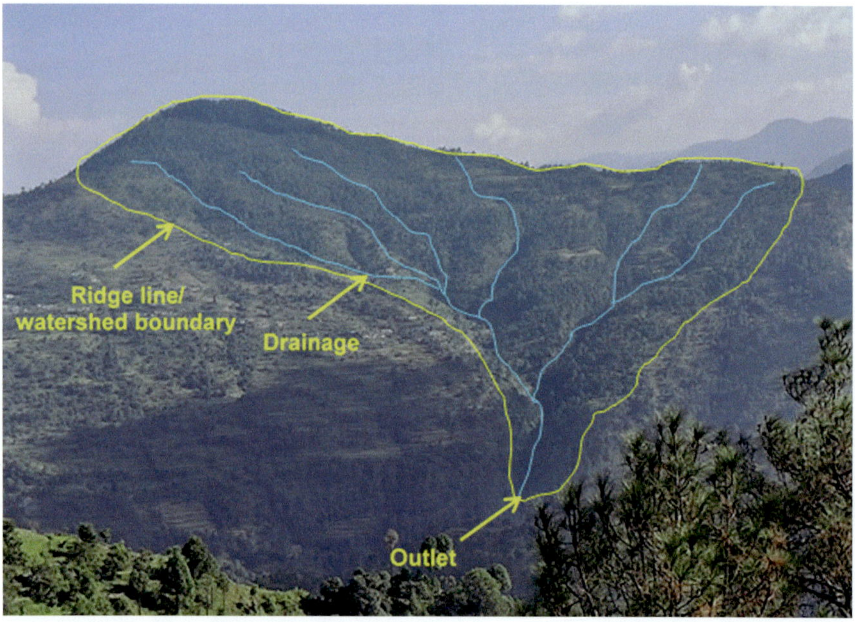

Fig. 1 A watershed located in the middle Himalayas, Uttarkashi, Uttarakhand. The boundary of the watershed is delineated by the highest points of elevation, such as ridges, as highlighted in yellow where drainage is highlighted in blue colour. Photo taken by A. David Raj

several secondary attributes which provide the explicit characterization of watershed hydrology in terms of terrain attributes.

The TWI has shown efficacy in regions where subsurface flow assumptions apply, such as mountainous terrain and areas characterized by shallow interflow, where surface runoff contributing areas are limited to small, saturated zones at the foot of hillslopes. Utilizing digital elevation models (DEMs), TWIs are computed to provide insights into soil moisture levels. The accuracy of TWI, akin to any hydrological model, is contingent upon surface topography. Initially introduced to establish a hydrological similarity index on a catchment scale (Kirkby and Weyman 1974; Kirkby 1978), TWI later became integrated with soil transitivity at saturation (Beven 1986), leading to the development of the widely utilized TOPMODEL. TWI has diverse applications, including global-scale land surface modelling, wetland area estimation, and spatial mapping of soil moisture. Higher TWI values typically manifest at ridge tops, where relative wetness is minimal. Soil moisture, which significantly regulates hydrological and biogeochemical processes, underscores the pivotal role of TWI (Buchanan et al. 2014). Since its introduction, the TWI concept has been assimilated into numerous hydrologic models, such as SWAT-VSA (Schneiderman et al. 2007; Kumar and Gupta 2016). Various researchers have refined TWI by incorporating indicators of soil storage capacity, such as saturated hydraulic conductivity and soil depth (Walter et al. 2002). Furthermore, integrating topographic factors and hillslope position as weighting factors has been explored to enhance the TWI model's portrayal of soil moisture content (Ma et al. 2010).

5 Digital Terrain Characterization in Predicting Soil Erosion Risk Using Satellite-Derived DEM and GIS—A Case Study

5.1 Study Area

The area is situated between 77°53′–77°55′ longitude and 30°26′–30°29′ latitude (Fig. 2). This study was conducted in Sitla Rao watershed of 8.05 km^2. Cartosat-derived DEM was utilized for terrain analysis, and the combination of Cartosat-1 imagery and Toposheet 53F/15 was employed to delineate the watershed boundary following a thorough understanding of the study area acquired during field visits. The climate is characterized by hot summers that transition to a humid to sub-tropical nature from the Doon Valley to the higher hilly regions, along with cold winters. The mean annual temperature ranges from 15 °C in winters to 35 °C in summers. Annual rainfall occurs from mid-June to mid-August, with amounts ranging from 1600 to 2500 mm depending on the elevation of the area. From a physiographic standpoint, the area is categorized into three principal units delineated by relief, slope, geology, and the configuration of landform features: hillside, upper piedmont, and middle piedmont. The soils mainly comprise Alfisols, Entisols, and Inceptisols,

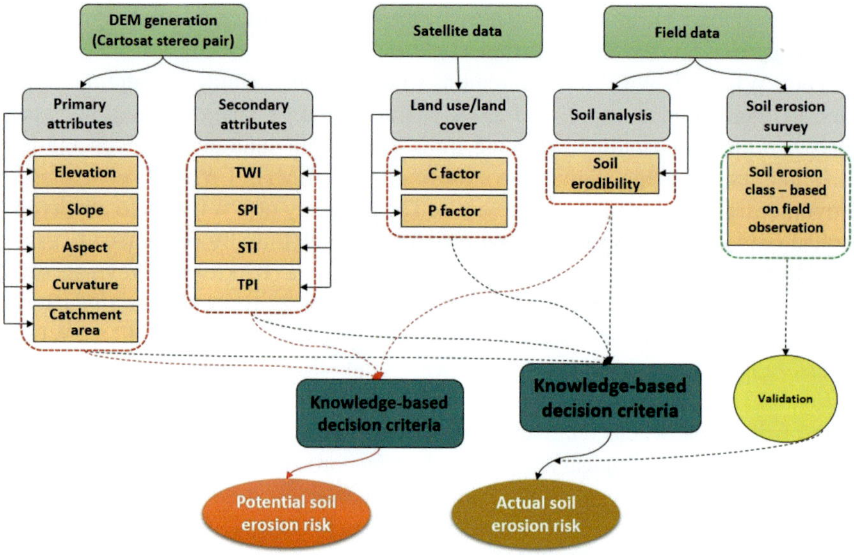

Fig. 2 Workflow of the study

with depths varying from shallow to deep. The predominant soil texture is sandy loam. The primary land use/land cover types in the study area are forests, agriculture, settlements, and scrublands. The overview of methodology illustrated in Fig. 2.

5.2 Materials and Method

5.2.1 DEM Generation

The Cartosat Stereo data (Fig. 3) was processed using the ERDAS [module/software] to produce a Digital Terrain Model (DTM). A total of 23 Ground Control Points (GCPs) were collected using DGPS. The resulting DTM achieved a vertical accuracy with a Root Mean Square Error (RMSE) of 4.48 m. Subsequently, the DTM was processed using GIS software to extract primary and secondary topographic attributes, as outlined by Güntner et al. (2004). These attributes were employed to characterize the terrain conditions for assessing erosion risk potential within the watershed. A qualitative approach, based on a knowledge base of criteria, was employed to predict erosion potential. The accuracy of these predictions was estimated through evaluation.

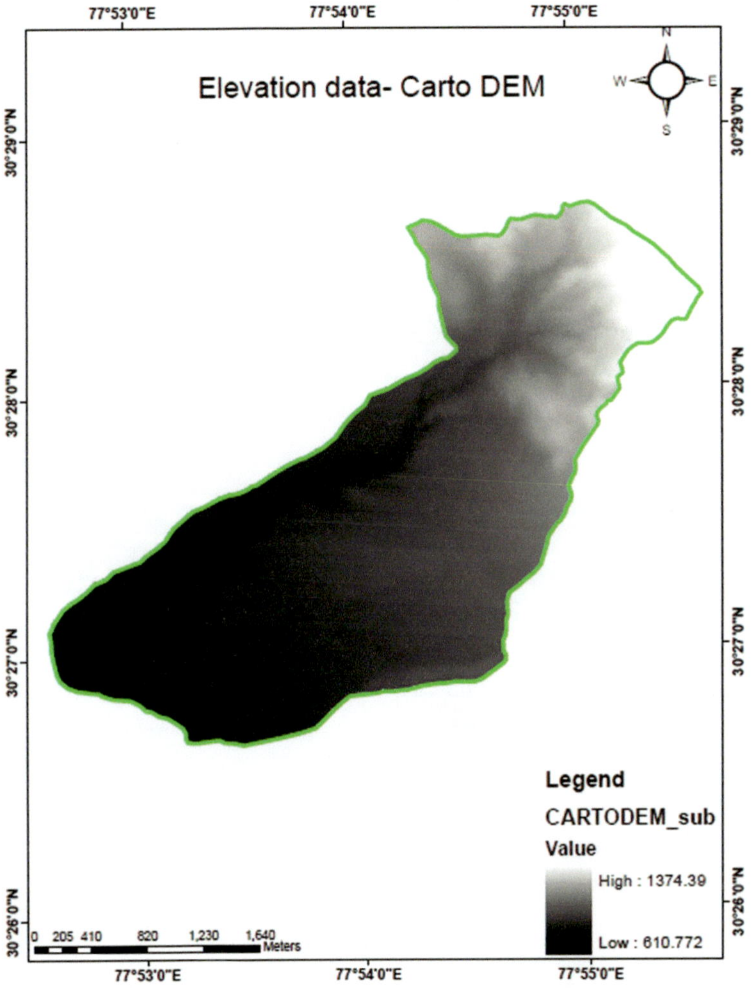

Fig. 3 Location of the study area

5.2.2 Predicting Soil Erosion Risk

Soil erosion is primarily influenced by climate (rainfall erosivity), soil characteristics (soil erodibility), topography (LS factor), and vegetation cover (NDVI/CP factor) as established by Wischmeier and Smith (1978). To assess soil erosion risk in the watershed, factor rating maps were created for terrain index (LS), soil susceptibility to erosion, and vegetation cover (CP), depicting spatial variations across the area. These factor maps were then integrated to derive a rating index value, which was categorized into potential soil erosion risk classes.

The size of the watershed is small (8.05 km^2) and with negligible rainfall variation, rainfall erosivity was assumed to be constant. Soil erodibility was assessed for soil map units based on dominant soil types, taking into account soil texture, structure, organic matter content, and soil permeability (Gupta et al. 2024). The secondary terrain attribute LS factor map, derived from the DTM, was utilized in the analysis (Figs. 4 and 5) The land use/land cover (LULC) map was made through supervised classification using IRS LISS IV (5.8 m spatial resolution) satellite data, and land management practices observed during field surveys were incorporated to categorize the LULC map into erosion risk classes.

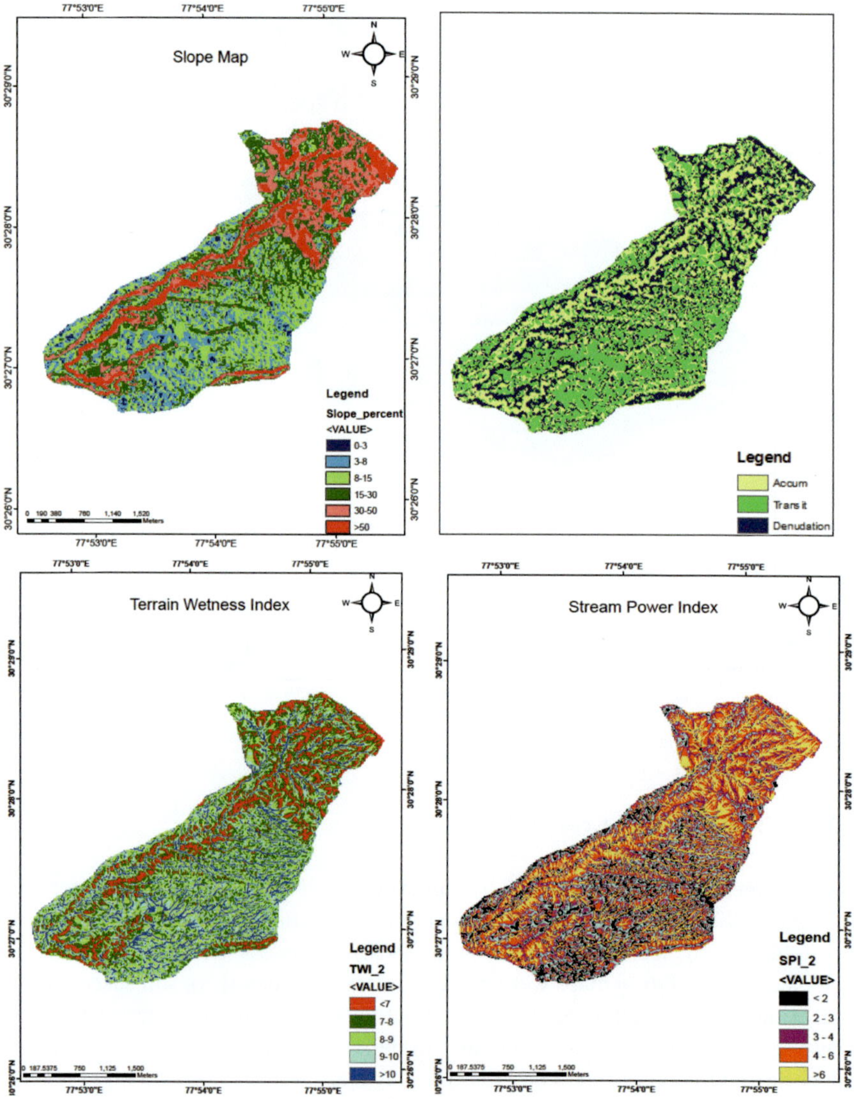

Fig. 4 Terrain attributes **a** slope, **b** curvature, **c** TWI, and **d** SPI

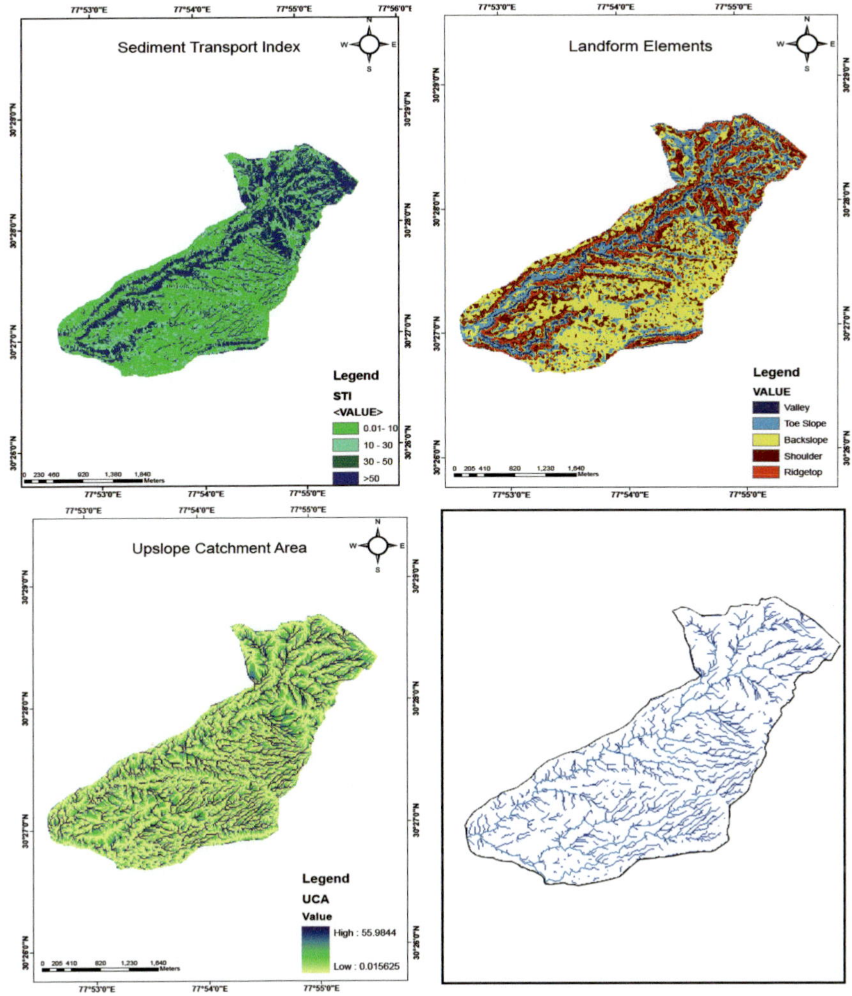

Fig. 5 Terrain attributes **a** STI, **b** land form elements, **c** upslope catchment area, **d** drainage

These factors were integrated based on a knowledge based criteria to predict areas of potential soil erosion within the watershed. Additionally, the erosion map was combined with the CP factor map to determine the actual erosion risk in different areas. Initially, soil erodibility, erosivity, and topography layers were overlaid to compute the potential risk of soil erosion. The formula for determining the actual soil erosion risk involves overlying these parameters along with the land use/land cover map. Subsequently, the actual soil erosion risk map was obtained and was classified into three distinct categories: low, moderate, and high, based on predetermined criteria.

5.2.3 Erosion Survey

A field survey was conducted to evaluate erosion conditions within the watershed. A total of 113 field sites were investigated, and their geographic coordinates were recorded using Garmin GPS device. These sites were assessed to determine soil erosion classes across different land use/land cover categories. Observations included assessments of soil colour, texture, depth, organic matter content, as well as erosion features such as rills and gullies. Additionally, vegetation cover and slope were considered during the characterization of each site. Table 3 field indicator and characteristics used to assess soil erosion classes at the sites.

Table 3 Field indicator used to assess soil erosion classes at the sites

Assessment factors	Class limit			
	Slight (< 10 t/ha/yr)	Moderate (10–20)	Severe (20–40)	Very severe (> 40 t/ha/yr)
Surface. status (% of gravels)	< 10	10–25	25–50	> 50
Soil depth (cm)	> 90	50–90	10–50	< 10
Erosion types	Sheet (soil color is lighter than surrounding soils)	Rills (conspicuous tiny rivulets or finger-shaped channels)	Gullies (well defined and permanent Incised land neither cultivable nor traversable)	Gullies (well defined and permanent incised land neither cultivable nor reversible. Network of very deep gullies)
		Moderate loss of topsoil typically occurs, often accompanied by some degree of dissection caused by runoff channels or gullies	Significant loss of topsoil is observed, usually accompanied by pronounced dissection caused by runoff channels or gullies	The sub-soil (B horizon) becomes exposed, and/or extensive and intricate dissection occurs due to runoff channels or gullies
	Some crusting of soil surface	Slight to moderate crusting, wash marks, splash pedestals, and exposed tree roots mark the level of former surface	Tree root exposure	Continuous network of channels with gullies every 10–15 m
Land cover	Crop land	Crop land/ fallows/land with/ without scrub/ degraded forests	Predominant in crop lands/fallows followed by land without scrub/ scrub land degraded forest	Mostly land with/ without scrub

5.3 *Validation*

The predicted erosion risk within the watershed was validated through field observations to evaluate the accuracy of the erosion risk map generated by the integrated analysis of thematic maps. Erosion risk classes were assigned to 3 × 3 pixel areas surrounding observed field sites. Various quantitative performance criteria were developed to assess the agreement between the simulated and observed spatial patterns. The efficacy of these criteria for comparing such patterns was analysed to determine their suitability for evaluating the accuracy of the erosion risk predictions.

5.4 *Results and Discussion*

5.4.1 Input Parameter Generation

GIS technology is widely employed in hydrological and water quality modelling. Traditionally, hillslope hydrologists have operated under the assumption that the movement of water downslope can be effectively captured by surface topography, given the predominant influence of gravitational potential. Consequently, with the increased obtainability of Digital Terrain Models (DTMs), surface topography plays a vital role in many popular hydrological models. Slope is a significant parameter governing numerous movements within landscapes. Slope steepness, in particular, is an underlying parameter in most soil erosion models and plays a crucial role in controlling redistribution and translocation processes within landscapes. In the study area, the majority of slopes fall within the 8–15% class (26.7%), followed by the 15–30% class (24.7%), and the 30–50% class (17.2%). Approximately 12.6% of the area exhibits slopes exceeding 50%, while 18.7% of the area features slopes less than 8% (Fig. 4). The land use/cover, soil erodibility and CP factor map are provided in Fig. 6.

Plan curvature characterizes the convergence and divergence of the surface, with negative values indicating divergence and positive values indicating convergence. This attribute delineates features such as ridges, gullies, and valley areas, which are instrumental in predicting erosion within the study area. The range of plan curvature in the study area varies from − 16.7 to 12.8. Curvatures, both profile and plan, are closely related to the concavity and convexity of the land surface. They impact the relative speeding up or slowing down of overland flow (vertical curvature) and its convergence or divergence (horizontal curvature). By overlaying profile and plan curvature, areas can be classified based on their potential for accumulation, deposition, or denudation. The range of profile curvature in the study area varies from − 13.7 to 20.4. Total curvature, which combines the effects of horizontal and vertical curvatures, indicates areas of accumulation, deposition, or denudation. When both curvatures are negative, accumulation and deposition occur, while positive values indicate denudation or removal. Areas where curvatures have opposite

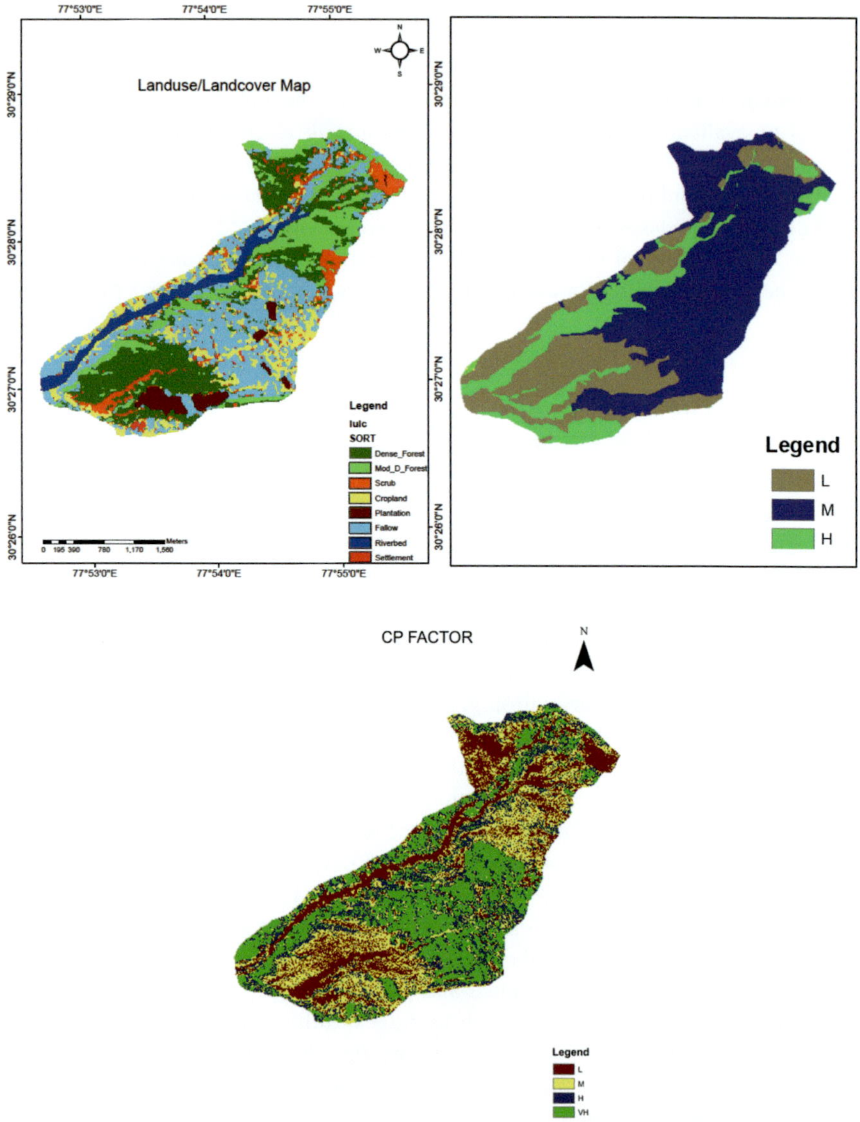

Fig. 6 Land use/land cover, soil erodibility and CP factor map of the study area

signs are considered transit zones. In the study area, the zones of erosion, deposition, and transition are distributed almost equally, with shares of 33.0%, 32.6%, and 34.4%, respectively. Profile/vertical curvature indicates the rate of change of potential gradient in the direction of maximum slope and defines the concavity or convexity of the terrain. Negative values signify concave surfaces, while positive values denote convex surfaces, influencing the acceleration or deceleration of water flows and, consequently, erosion.

The wetness index, particularly evident in valleys and gullies, indicates higher susceptibility to erosion in these areas. Conversely, backslopes, with their steeper gradients, experience greater erosion compared to shoulders and toeslopes. There is a linear correlation between the wetness index (TWI) and erosion, making it a well-studied indicator of soil properties and moisture distribution across different landscape positions. TWI quantifies the topographic control on hydrologic processes, with higher TWI values indicating water surplus and increased soil moisture. In the study area, TWI ranges from 4.61 to 17.19. Another significant index, the Stream Power Index (SPI), exhibits variation across landscape units, with lower values at ridgetops and higher values in valleys and backslopes. However, the LS factor demonstrates an increase in values from ridgetops to valleys. The SPI ranges from 0.19 to 11.90 in the study area. Landform analysis delineates various hillslope positions, including ridge, shoulder, backslope, toeslope, and valley. Backslope accounts for the majority of the area (30%), followed by valleys and ridgetops (20% each). Shoulders contribute 17% of the area, while toeslopes represent a smaller proportion at 13%. The primary land use/land cover types in the study area consist of forests, croplands, orchards, plantations, scrublands, settlements, and water bodies. Densely and moderately dense forest regions cover approximately 42% of the area, followed by croplands (including fallow land) at 40%. Scrublands cover 8%, riverbeds 6%, orchards 4%, and settlements a mere 1%. The C factor, calculated using NDVI values, and the P factor, based on field knowledge, contribute to the Combined CP factor, which is generated by assigning different weights accordingly.

5.4.2 Soil Erosion Risk Assessment

Employing knowledge-based decision criteria encompassing soil erodibility, and slope layers (Table 4; Figs. 7 and 8), generated a potential erosion risk map. Subsequently, merging this map with the land cover map created the actual soil erosion risk map. Analysis of the results, depicted in Fig. 6, revealed that approximately 15% of the watershed had a low potential erosion risk, while the rest exhibited moderate to high potential erosion risk levels. Specifically, the majority of the area (34%) fell under medium potential soil erosion risk, followed by high (27%) and very high (24%) levels. Regarding the actual soil erosion risk, the study area demonstrated 38% moderate, 33% high, 13% very high, and 16% low erosion risk levels.

Furthermore, the findings indicated that areas characterized by moderate and high erosion risk were mainly located in the north and northeast regions of the watershed, characterized by steep slopes with scrubland. In contrast, regions with low erosion risk were primarily situated in the southwest and east portions, featuring medium slopes with forest and paddy cultivation.

Highlighting the substantial impact of vegetation cover on soil erosion dynamics, the difference between regions identified as potential and actual erosion risk is emphasized. In the initial potential erosion risk map, areas designated as very high erosion

Table 4 Knowledge-based decision criteria encompassing soil erodibility (K), topographic attribute (LS), and cover and practice factor (CP)

LS\K	L	M	H
L	**1**	**1**	**2**
M	**2**	**2**	**3**
H	**3**	**4**	**4**

LS_K\CP	L	M	H	VH
E1	**2**	**2**	1	**1**
E2	**3**	**2**	1	**1**
E3	**4**	**3**	2	**2**
E4	**4**	**4**	3	**3**

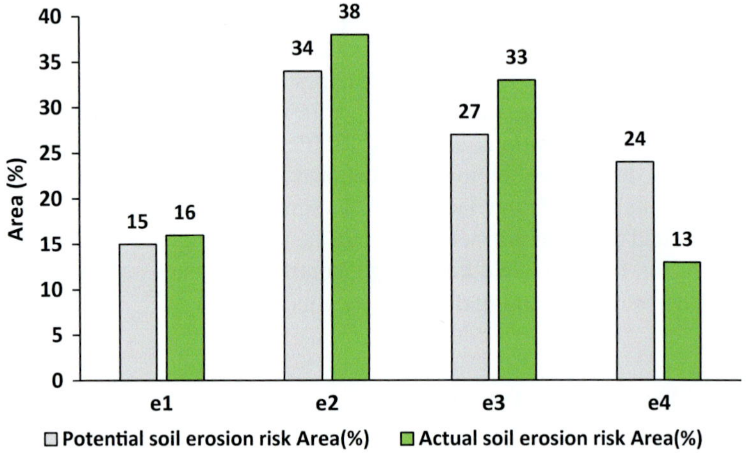

Fig. 7 Areal contribution of different soil erosion risk classes

risk decreased from 24 to 13% in the actual soil erosion risk map following the inclusion of the vegetation layer. This observation underscores that regions susceptible to high erosion risk are primarily covered by forest vegetation, as depicted in Fig. 4. Conversely, the combined areas classified as low and moderate erosion risk in the potential erosion risk map increased from 76 to 87% in the actual soil erosion risk map.

5.4.3 Accuracy Assessment with Reference to Field Data

The majority of the area is predominantly classified as moderately and marginally eroded lands, in contrast to highly eroded and negligibly eroded lands. An accuracy assessment was conducted with reference to 113 field locations to evaluate the

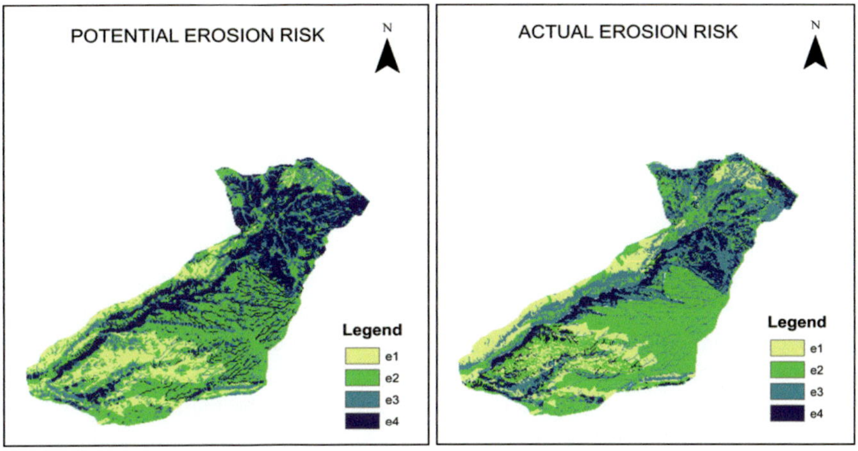

Fig. 8 Spatial distribution of potential and actual soil erosion risk

precision of various classifications. Producer accuracy ranges from 60 (E1) to 82.9% (E3), while user accuracy is found to be between 74 (E3) and 81% (E4). The overall accuracy stands at 76.99% (Table 5). Moreover, the validation process, conducted using soil survey data from 113 locations, revealed an approximate accuracy of 77% for the actual erosion risk map. This indicates a substantial correspondence between the actual soil erosion risk map and the soil survey data, validating its reliability.

Soil erosion is significantly influenced by topography (Kiani-Harchegani et al. 2022; Michalopoulou et al. 2022). Through the processes of erosion and deposition, it influences the movement of soil. Estimates of soil organic carbon (SOC) stocks in agricultural fields are aided by topography features like planform, profile curvature and percent slope (Dalzell et al. 2022). Different topographies and slope gradients have an impact on runoff, sediment output, soil and water loss (Chen et al. 2022). For the purpose of simulating soil erosion, topographic data, such as Digital Elevation Models (DEM), must be accurate. Our study also indicates that the soil erosion risk based on terrain attributes provides good consensus with the field validation and soil erosion survey. It indicates the suitability of this method in identifying the potential

Table 5 Accuracy assessment table

P\U	e1	e2	e3	e4	Total	Producers accuracy
e1	9	3	3	0	**15**	**60**
e2	3	31	5	0	**39**	**79.487**
e3	0	4	34	3	**41**	**82.927**
e4	0	1	4	13	**18**	**72.222**
Total	**12**	**39**	**46**	**16**	**113**	
Users accuracy	**75**	**79.487**	**73.913**	**81.25**		

soil erosion hotspots in rugged and inaccessible hilly and mountainous regions at low cost.

6 Conclusions

The effectiveness of utilizing Remote Sensing (RS) and Geographic Information System (GIS) technology for terrain parameterization to map erosion risk is highlighted in this chapter. It demonstrates that this approach enables a swift and cost-effective evaluation of soil erosion over larger watersheds. Despite the requirement for progressed RS and GIS technologies and intricate computer processing, individuals with basic computer skills can independently apply the methodology without specialized assistance. It is important to emphasize that while this method does not aim to quantify soil loss and sediment yield, it efficiently produces erosion risk maps for planning and environmental protection analyses. Furthermore, the model offers decision-makers valuable insights into areas prone to erosion, aiding in the development of comprehensive soil and water conservation strategies. Moreover, it facilitates detailed erosion studies in regions identified with high erosion risk, assisting in targeted conservation efforts.

References

ASCE (2000) Artificial neural networks in hydrology. I: Preliminary concepts. J Hydrol Eng 5(2):115–123

Bennie J, Huntley B, Wiltshire A, Hill MO, Baxter R (2008) Slope, aspect and climate: spatially explicit and implicit models of topographic microclimate in chalk grassland. Ecol Model 216(1):47–59

Beven KJ, Wood EF (1983) Catchment geomorphology and the dynamics of runoff contributing areas. J Hydrol 65:139–158

Beven KJ (1986) Runoff production and flood frequency in catchments of order n: an alternative approach. In: Gupta VK, Rodriguez-Iturbe I, Wood EF (eds) Scale problems in hydrology. Reidel, Dordrecht, pp 107–131

Beven K, Kirkby MJ (1993) Channel network hydrology. Wiley

Beven K, Kirkby MJ (1979) A physically based, variable contributing area model of basin hydrology. Hydrol Sci Bull 24:43–69

Buchanan BP, Fleming M, Schneider RL, Richards BK, Archibald J, Qiu Z, Walter MT (2014) Evaluating topographic wetness indices across central New York agricultural landscapes. Hydrol Earth Syst Sci 18(8):3279–3299

Chaplot V (2013) Impact of terrain attributes, parent material and soil types on gully erosion. Geomorphology 186:1–11

Chen T, Shu J, Han L, Tian G, Yang G, Lv J (2022) Modeling the effects of topography and slope gradient of an artificially formed slope on runoff, sediment yield, water and soil loss of sandy soil. CATENA 212:106060

Dalzell BJ, Fissore C, Nater EA (2022) Topography and land use impact erosion and soil organic carbon burial over decadal timescales. CATENA 218:106578

David Raj A, Kumar S, Sooryamol KR (2022) Modelling climate change impact on soil loss and erosion vulnerability in a watershed of Shiwalik Himalayas. CATENA 214:106279

David Raj A, Kumar S, Regina M, Sooryamol KR, Singh AK (2023) Calibrating APEX model for predicting surface runoff and sediment loss in a watershed—a case study in Shivalik region of India. Int J Hydrol Sci Technol 15(2):177–206

De Roo APJ (1998) Modelling runoff and sediment transport in catchments using GIS. Hydrol Process 12(6):905–922

Edwards PJ, Williard KW, Schoonover JE (2015) Fundamentals of watershed hydrology. J Contemp Water Res Educ 154(1):3–20

Fang H (2021) Responses of runoff and soil loss on slopes to land use management and rainfall characteristics in Northern China. Int J Environ Res Public Health 18(18):9583

Florinsky IV (1995) Flow denudation, transit and accumulation zones of the part of the Crimean peninsula and adjacent sea bottom. In: Mathematical methods of pattern recognition. Proceedings of the 7th conference, Pushchino, 25–30 Sept 1995. Computing Centre of the Russian Academy of Sciences, Moscow, pp 166–168

Florinsky IV (1998) Accuracy of local topographic variables derived from digital elevation models. Int J Geogr Inf Sci 12(1):47–61

Gessler PE, Moore ID, McKenzie NJ, Ryan PJ (1995) Soil-landscape modelling and spatial prediction of soil attributes. Int J Geogr Inf Syst 9(4):421–432

Ghunowa K, MacVicar BJ, Ashmore P (2021) Stream power index for networks (SPIN) toolbox for decision support in urbanizing watersheds. Environ Model Softw 144:105185

Güntner A, Seibert J, Uhlenbrook S (2004) Modeling spatial patterns of saturated areas: an evaluation of different terrain indices. Water Resour Res 40(5)

Gupta S, Borrelli P, Panagos P, Alewell C (2024) An advanced global soil erodibility (K) assessment including the effects of saturated hydraulic conductivity. Sci Total Environ 908:168249

Horton RE (1945) Erosional development of streams and their drainage basins: a hydrophysical approach to quantitative morphology. Bull Geol Soc Am 56:275–370

Huang D, Su L, Zhou L, Tian Y, Fan H (2023) Assessment of gully erosion susceptibility using different DEM-derived topographic factors in the black soil region of Northeast China. Int Soil Water Conserv Res 11(1):97–111

Kiani-Harchegani M, Talebi A, Asgari E, Rodrigo-Comino J (2022) Topographical features and soil erosion processes. In: Computers in earth and environmental sciences. Elsevier, pp 117–126

Kirkby MJ, Weyman DR (1974) Measurements of contributing area in very small drainage basins. Seminar Series B, No. 3. Department of Geography, University of Bristol, Bristol

Kirkby MJ (1978) Implications for sediment transport. In: Kirkby MJ (ed) Hillslope hydrology. Wiley, London, pp 325–363

Kumar S, Gupta S (2016) Geospatial approach in mapping soil erodibility using CartoDEM—a case study in hilly watershed of Lower Himalayan Range. J Earth Syst Sci 125:1463–1472

Lal R (2001) Soil degradation by erosion. Land Degrad Dev 12(6):519–539

Li Z, Zhu Q, Gold C (2005) Digital terrain modeling: principles and methodologyl. CRC Press, Boca Raton

Ma J, Lin G, Chen J, Yang L (2010) An improved topographic wetness index considering topographic position. In: Proceeding of the 18th international conference on geoinformatics: GIScience in change, geoinformatics. Peking University, Beijing, 18–20 June 2010

Marthews TR, Dadson SJ, Lehner B, Abele S, Gedney N (2014) A high-resolution global dataset of topographic index values for use in large-scale hydrological modelling. Hydrol Earth Syst Sci Discuss 11:6139–6166

Michalopoulou M, Depountis N, Nikolakopoulos K, Boumpoulis V (2022) The significance of digital elevation models in the calculation of LS factor and soil erosion. Land 11(9):1592

Montgomery DR (2003) Previsions des taux d'erosion a l'echelle du paysage par l'utilisation de modeles numeriques d'altitude. Comptes Rendus-Geoscience 16(335):1121–1130

Moore ID, Burch GJ, Mackenzie DH (1988) Topographic effects on the distribution of surface soil water and the location of ephemeral gullies. Trans ASAE 31(4):1098–1107

Moore ID, Wilson JP (1992) Length-slope factors for the revised universal soil loss equation: simplified method of estimation. J Soil Water Conserv 47:423–428

Moore ID, Gessler PE, Nielson GA, Peterson GA (1993) Soil attribute prediction using terrain analysis. Soil Sci Soc Am J 2:443–452

Morgan RPC (2009) Soil erosion and conservation. Wiley

Pickup G, Chewings VH (1996) Correlations between DEM-derived topographic indices and remotely sensed vegetation

Pradhan AMS, Kim YT (2014) Relative effect method of landslide susceptibility zonation in weathered granite soil: a case study in Deokjeok-ri Creek, South Korea. Nat Hazards 72:1189–1217

Quinn PFBJ, Beven K, Chevallier P, Planchon O (1991) The prediction of hillslope flow paths for distributed hydrological modelling using digital terrain models. Hydrol Process 5(1):59–79

Schneiderman EM, Steenhuis TS, Thongs DJ, Easton ZM, Zion MS, Neal AL, Mendoza GF, Todd Walter, M (2007) Incorporating variable source area hydrology into a curve-number-based watershed model. Hydrol Process: Int J 21(25):3420–3430

Sharma A (2010) Integrating terrain and vegetation indices for identifying potential soil erosion risk area. Geo-Spatial in Sci 13(3):201–209

Sooryamol KR, Kumar S, Regina M, David Raj A (2022) Modelling climate change impact on soil erosion in a watershed of north-western lesser Himalayan region. J Sediment Environ 7(2):125–146

Sörensen R, Zinko U, Seibert J (2006) On the calculation of the topographic wetness index: evaluation of different methods based on field observations. Hydrol Earth Syst Sci 10(1):101–112

Walter MT, Steenhuis TS, Mehta VK, Thongs D, Zion M, Schneiderman E (2002) A refined conceptualization of TOPMODEL for shallow-subsurface flows. Hydrol Process 16:2041–2046

Wilson JP, Gallant JC (1996) EROS: a grid-based program for estimating spatially-distributed erosion indices. Comput Geosci 22(7):707–712

Wilson JP, Gallant JC (2000) Digital terrain analysis. Terrain Anal.: Principles Appl 6(12):1–27

Wischmeier WH, Smith DD (1978) Predicting rainfall erosion losses: a guide to conservation planning (No. 537). Department of Agriculture, Science and Education Administration

Wood J (1996) The geomorphological characterisation of digital elevation models. Ph.D. Thesis, University of Leicester, Leicester (United Kingdom), 107 p

Xu X, Zheng F, Wilson GV, Wu M (2017) Upslope inflow, hillslope gradient and rainfall intensity impacts on ephemeral gully erosion. Land Degrad Dev 28(8):2623–2635

Yang J, El-Kassaby YA, Guan W (2020) The effect of slope aspect on vegetation attributes in a mountainous dry valley, Southwest China. Sci Rep 10(1):16465

Youfua DONG, Guoana TANG, Tinga ZHANG (2008) A systematic classification research of topographic descriptive attribute in digital terrain analysis. ISPRS Arch 37(2):357–362

Yuksel A, Gundogan R, Akay AE (2008) Using the remote sensing and GIS technology for erosion risk mapping of Kartalkaya dam watershed in Kahramanmaras, Turkey. Sensors 8(8):4851–4865

Climate Change Impact on Flood Frequencies Using Geospatial Modeling

Kanhu Charan Panda, Ram Mandir Singh, Pradosh Kumar Paramaguru, Uday Pratap Singh, Sudhir Kumar Mishra, and Gaurav Singh Vishen

Abstract It is anticipated that climate change will significantly affect flood frequencies, with more frequent and intense floods likely to occur in many parts of the world. Geospatial modeling can be used to assess possible effects of climate change on flooding frequencies, providing valuable information for flood risk management and adaptation planning. The chapter depicts the current state of knowledge on the use of geospatial modeling to assess the climate change impact on flood frequencies. It discusses the key mechanisms through which climate change can affect floods, the range of geospatial modeling approaches that can be used to assess the impact of climate change on flood frequencies, and the challenges and limitations of using geospatial modeling for this purpose. A case study was included in the chapter to demonstrate the use of advanced geospatial technique to access the impact of climate change on flood frequencies. The chapter concludes by discussing the future directions for research on the use of geospatial modeling to assess the climate change impact on flood frequencies. This includes the development of more sophisticated modeling approaches, the use of ensemble modeling to account for uncertainty, and the integration of geospatial modeling with other risk assessment tools.

Keywords Climate change · Flood frequencies · Geospatial modelling · Flood risk management · Flood adaptation planning

K. C. Panda (✉) · U. P. Singh · S. K. Mishra · G. S. Vishen
National PG College (Barhalganj), DDU Gorakhpur University, Gorakhpur, UP 273402, India
e-mail: kanhucharan.bm@gmail.com

K. C. Panda · R. M. Singh
Department of Agricultural Engineering, Institute of Agricultural Sciences, Banaras Hindu University, Varanasi, UP 273402, India
e-mail: ram.singh@bhu.ac.in

P. K. Paramaguru
ICAR—National Institute of Secondary Agriculture, Namkum, Ranchi, Jharkhand 834010, India

S. C. Pal and U. Chatterjee (eds.), *Surface, Sub-Surface Hydrology and Management*, Springer Geography, https://doi.org/10.1007/978-3-031-62376-9_22

1 Introduction

The twenty-first century has witnessed an unmistakable shift in our planet's climate. Driven by anthropogenic activities, primarily the relentless ascent of greenhouse gas emissions, Earth's atmosphere is warming at an alarming rate, disrupting delicate natural equilibriums and reshaping weather patterns globally (Panda et al. 2022). This transformation transcends mere shifts in average temperatures; it manifests in intensifying storms, altered precipitation patterns, and a worrying rise in sea levels. Among the most immediate and tangible consequences is the amplification of flooding events, posing a growing threat to communities, ecosystems, and economies worldwide (Panda et al. 2023). Rising global temperatures directly influence the hydrological cycle, fueling heavier precipitation events, and melting snowpacks at an accelerated pace. These factors translate into overflowing rivers, inundating coastal areas, and flash floods wreaking havoc on vulnerable populations. The Intergovernmental Panel on Climate Change (IPCC) underscores the urgency of this issue, projecting an increase in the frequency and intensity of extreme precipitation events in many regions across the globe (Singh et al. 2022a). The impacts of these amplified floods are multifaceted and profound. Lives are tragically lost, properties are destroyed, and infrastructure crumbles under the immense pressure of surging waters (Singh et al. 2022b). Agricultural lands are rendered unproductive, disrupting food security and livelihoods (Panda et al. 2021). Ecosystems are thrown into disarray, compromising delicate ecological balances and biodiversity. Perhaps most concerning is the displacement of populations, with the World Bank estimating that river floods alone could displace a staggering 200 million people globally by 2050. These escalating risks highlight the crucial need for a deeper understanding and proactive management of flood frequencies in a changing climate (Saroughi et al. 2023).

In the face of intensifying flood events, the need for robust prediction and proactive measures becomes paramount (Sahoo and Panda 2020). This is where understanding flood frequencies, or the likelihood of floods of different magnitudes occurring over a specific period, plays a vital role in risk management and mitigation strategies (Tulla et al. 2024). By analyzing historical data and leveraging climate models, researchers can estimate the probability of future floods under various climate scenarios. This information becomes invaluable for various applications:

- Planning and design: Accurate flood frequency estimates inform the design and construction of critical infrastructure, such as bridges, dikes, and floodplains, ensuring these structures are resilient to future events. For example, knowing the 100-year flood frequency allows engineers to design bridges high enough to withstand the water level associated with such a rare event.
- Early warning systems: Understanding flood probability patterns enables the development of more effective early warning systems. Knowing when and where floods are likely to occur, even with some lead time, allows communities to prepare evacuation plans, secure valuables, and move to higher ground, potentially saving lives and minimizing damage.

- Resource allocation: Informed flood frequency assessments help policymakers and resource managers allocate resources efficiently for flood risk reduction measures. Prioritizing vulnerable areas and populations ensures a targeted and impactful approach to mitigating flood risks.
- Investment decisions: Investors and insurance companies rely on flood frequency data to make informed decisions about infrastructure development and risk mitigation strategies in flood-prone regions. Having a deeper understanding of flood risks allows them to assess potential losses and allocate resources judiciously.

In essence, studying flood frequencies provides the foundation for informed decision-making, empowering communities, policymakers, and stakeholders to navigate the rising tide of flood risk in a changing climate.

Traditional methods of analyzing flood frequencies often relied on statistical techniques and historical data, offering limited insights into the complex interplay of factors influencing these events (Wright et al. 2020). However, the emergence of geospatial modeling has revolutionized this field, offering powerful tools for understanding and predicting flood risks. Geospatial modeling involves creating digital representations of the Earth's surface, encompassing landscapes, rivers, vegetation, and human settlements. By integrating diverse datasets, such as elevation data, rainfall records, soil properties, and land-use maps, these models can simulate the intricate hydrological processes leading to floods (Panda et al. 2024). Geospatial modeling offers a powerful approach to analyzing the impact of climate change on flood frequencies. Geospatial models combine geographic information systems (GIS) with computational techniques to integrate and analyze various spatial datasets (Bill et al. 2022). By incorporating climate data, topographic information, land cover data, and hydrological parameters, geospatial models provide a comprehensive framework for understanding the complex dynamics of floods. Geospatial modeling enables the examination of flood patterns and their relationship with climate variables across different spatial scales (Aceves et al. 2023). It allows researchers to simulate and predict floods by considering the interactions between environmental factors and climate change scenarios. By utilizing advanced modeling techniques, such as hydrological models and machine learning algorithms, geospatial modeling can capture the spatial variability of flood frequencies and provide valuable insights into future flood risk. Moreover, geospatial modeling facilitates effective flood risk assessment and management. It enables the identification of vulnerable areas, the evaluation of infrastructure resilience, and the optimization of flood mitigation measures. By integrating geospatial modeling with decision support systems, policymakers and stakeholders can make informed decisions to minimize the impacts of floods and enhance community resilience.

In this chapter, we delve into the topic of climate change impact on flood frequencies using geospatial modeling. We explore the underlying concepts, methodologies, and applications of geospatial modeling in flood frequency analysis. By examining the interplay between climate change and flood frequencies, we aim to enhance our understanding of the complex dynamics involved and provide insights into the future patterns of flooding. Through this research, we strive to contribute to the development of effective flood management strategies, policy decisions, and adaptation measures to mitigate the adverse effects of climate change on flood-prone regions. In the subsequent sections of this chapter, we will discuss the fundamentals of flood frequency analysis, the relationship between climate change and flood frequencies, and the role of geospatial modeling in this context. We will explore the data acquisition and preprocessing techniques required for geospatial modeling, as well as various modeling approaches and their applications. Additionally, we will examine case studies and highlight the challenges and future directions in this field. By the end of this chapter, readers will gain a comprehensive understanding of the potential impacts of climate change on flood frequencies and the valuable insights that geospatial modeling can provide in addressing this critical issue.

Most of the studies conducted for analysis of flood frequency analysis employed traditional approaches for its estimation. Limited studies used an advanced machine learning approach for the prediction of the frequency and magnitude of floods. Thus, the current study aimed to address two research questions: (1) Will machine learning (ML) models be able to predict flood magnitude for tropical river basins? and (2) Which ML model suits best for the tropical river basin? Based on the research questions, the study used four different ML approaches, viz., Multiple Linear Regression (MLR), Support Vector Machine (SVM), Random Forest (RF), and Multivariate Adaptive Regression Spline (MARS) to predict flood runoff in the Gola River basin.

2 Materials and Methods

2.1 Link Between Climate Change and Increased Flood Frequencies

The relationship between climate change and increased flood frequencies is well-established, with scientific evidence indicating a clear connection between the two. Due to anthropogenic activities like fossil fuel combustion and deforestation, the Earth's climate is continuing to rise. This causes substantial changes in the hydrological cycle, which in turn affects precipitation patterns and increases the danger of floods. The relationship between rising flood frequency and climate change is due to several important reasons. One of the primary mechanisms through which climate change affects flood frequencies is the intensification of rainfall events (Hirabayashi et al. 2021). Increased evaporation rates brought on by warmer temperatures raise the atmospheric moisture content. This increased moisture availability might fuel more intense rainfall events, resulting in larger volumes of water being discharged into river systems and increasing the likelihood of flooding. Studies have shown a clear trend of increasing extreme precipitation events in various regions around the world, which might be directly linked to climate change. For instance, the Intergovernmental Panel on Climate Change (IPCC) has demonstrated an increased frequency and intensity of heavy precipitation events over the last century (Wasko et al. 2021).

The increased occurrence of extreme precipitation events contributes to higher flood frequencies, as the excess water overwhelms drainage systems and river channels, leading to more frequent and severe flooding events.

Climate change also impacts the timing and magnitude of snowmelt, which is a significant contributor to river runoff in many regions. As temperatures rise, snowfall patterns may be altered, resulting in a shift from snow accumulation to more rain-dominated precipitation. This can lead to accelerated snowmelt, causing a rapid increase in river flows and an increased risk of flooding during the spring season (Qin et al. 2020). In regions where snowmelt is a crucial component of the hydrological regime, such as mountainous areas or high-latitude regions, changes in snow accumulation and melt patterns can significantly impact flood frequencies. Studies have observed changes in snowmelt dynamics, including earlier snowmelt and reduced snowpack, in many regions due to climate change. These changes can lead to increased spring runoff and contribute to higher flood frequencies (Fischer and Schumann 2021). Altered Storm Tracks and Weather Patterns: Climate Change can influence atmospheric circulation patterns, leading to changes in storm tracks and weather patterns. This can result in a redistribution of precipitation across regions, with some areas experiencing increased rainfall while others face prolonged dry spells. When areas already prone to flooding receive enhanced rainfall, the risk of flooding is further exacerbated. Changes in atmospheric circulation patterns can be attributed to various factors, including the warming of the atmosphere, changes in sea surface temperatures, and shifts in large-scale climatic oscillations such as El Niño and La Niña (Cai et al. 2021). These changes can lead to alterations in the frequency, intensity, and tracks of storms, affecting the spatial and temporal patterns of precipitation and increasing flood frequencies in certain regions.

Rising sea levels brought on by climate change put coastal areas in danger of storm surge flooding. Sea levels increase because of ice caps and glaciers melting and saltwater expanding due to rising global temperatures (Kumar et al. 2021). As sea levels increase, the baseline elevation for flood events also rises, making coastal areas more vulnerable to inundation during extreme weather events such as hurricanes or intense coastal storms. Coastal areas are home to a large population and vital infrastructure, making them highly susceptible to the impacts of sea-level rise and coastal flooding. Studies have shown that the combination of sea-level rise and increased storm intensity due to climate change has led to a higher frequency of coastal flooding events. The increased risk of coastal flooding further adds to the overall increase in flood frequencies associated with climate change. In conclusion, there is a clear link between climate change and increased flood frequencies. The warming of the Earth's climate leads to increased extreme precipitation events, changes in snowmelt dynamics, alterations in storm tracks and weather patterns, and a rise in sea levels, all of which contribute to higher flood frequencies (Walsh et al. 2020). These changes in flood frequencies have significant implications for human populations, ecosystems, and infrastructure, emphasizing the urgent need for effective adaptation and mitigation strategies to address the impacts of climate change on flood-prone regions.

2.2 *Impact of Temperature Rise on Flood Frequencies*

The rise in global temperatures due to climate change has a direct impact on flood frequencies through various mechanisms. Changes in temperature influence the timing and magnitude of precipitation, snowmelt, and evaporation rates, altering the hydrological balance and increasing the risk of flooding (Wasko and Sharma 2017). Understanding the impact of temperature rise on flood frequencies is crucial for effective flood risk management and adaptation strategies. Several key factors contribute to the relationship between temperature rise and flood frequencies. Warmer temperatures lead to increased evaporation rates from land surfaces and water bodies. This results in higher moisture content in the atmosphere, providing the necessary fuel for more intense rainfall events. The increased availability of atmospheric moisture amplifies the potential for heavy rainfall and subsequent flooding. As temperatures rise, more water evaporates from oceans, lakes, rivers, and moist surfaces, leading to an increased amount of water vapor in the atmosphere (Wang et al. 2018). This increased moisture availability sets the stage for more intense precipitation events. When extreme weather systems, such as storms or frontal systems, interact with this moisture-rich atmosphere, they can produce heavy rainfall over shorter durations, resulting in a higher likelihood of flooding.

Temperature plays a critical role in determining the form of precipitation. As temperatures rise, there is a shift from snowfall to rain in regions that experience winter snow accumulation. This can lead to accelerated snowmelt, as warmer temperatures promote faster melting rates. The rapid release of snowmelt water can overwhelm river systems, contributing to increased flood frequencies (Wyżga et al. 2016). In regions with a significant snowmelt contribution to river flows, such as mountainous areas or high-latitude regions, temperature rise can alter the timing and magnitude of snowmelt. Warmer temperatures cause an earlier onset of spring, leading to earlier snowmelt and increased runoff. This can result in higher river flows during the spring season, increasing the risk of flooding.

Climate change can also disrupt seasonal patterns, including the timing and duration of specific weather events. Warming temperatures can cause earlier snowmelt in mountainous regions, resulting in increased spring runoff and a higher likelihood of flooding during that season (Beniston and Stoffel 2016). Additionally, shifts in the timing and duration of the rainy season can also impact flood frequencies in regions with a monsoonal climate. Changes in temperature can alter the timing and duration of seasons, affecting the overall hydrological regime. For instance, a warmer spring can cause snowmelt to occur earlier, resulting in an earlier peak in river flows and potentially increasing flood frequencies. Similarly, a prolonged or intensified rainy season can lead to more frequent and severe flooding events (Wasko et al. 2021). In colder regions, climate change-induced temperature rise can lead to the thawing of permafrost, which is permanently frozen ground. Thawing permafrost can alter the hydrological regime by increasing soil moisture content and reducing the ground's ability to absorb water. This can result in increased surface runoff and contribute to higher flood frequencies in these regions (Najibi and Devineni 2018).

Permafrost acts as a natural barrier that restricts the movement of water through the soil. As temperatures rise, permafrost thaws and the soil become saturated, reducing its ability to retain water. This leads to increased surface runoff and a higher likelihood of flooding. Additionally, thawing permafrost can also disrupt the stability of infrastructure, such as roads and buildings, further exacerbating flood risks.

Extreme heat events are occurring more often and with greater intensity because of climate change. Elevated temperatures may cause evapotranspiration or the sum of transpiration from plants and evaporation from the ground (Zeng et al. 2018). This, in turn, reduces soil moisture and increases the risk of flash flooding during intense rainfall events. During extreme heat events, the ground becomes dry and hardened, reducing its ability to absorb water. During periods of intense precipitation, the soil's ability to absorb water is hindered, leading to a higher volume of water flowing over the ground surface and an increased likelihood of flooding. This phenomenon is particularly significant in urban areas with a high percentage of impervious surfaces, such as concrete or asphalt, which further hinder water absorption (Mohajerani et al. 2017). In conclusion, the impact of temperature rise on flood frequencies is multifaceted. The increase in evaporation and atmospheric moisture content changes in snow accumulation and snowmelt dynamics, shifts in seasonal patterns, permafrost thaw, and changes in extreme heat events all contribute to the increased risk of flooding. Comprehending these connections and their geographical diversity is essential for efficient flood risk management and adaptation methods in response to climate change. Geospatial modeling techniques play a vital role in quantifying and predicting the impact of temperature rise on flood frequencies, enabling policymakers and stakeholders to make informed decisions to mitigate the risks associated with climate change-induced flooding.

2.3 *Geospatial Modeling in Flood Frequency Analysis*

2.3.1 Introduction to Geospatial Modeling

Geospatial modeling is a powerful tool used to analyze and understand the relationship between climate change and flood frequencies. It involves integrating geographic information systems (GIS), remote sensing data, and statistical models to assess the spatial distribution and temporal changes in flood events. Geospatial modeling provides a comprehensive framework for analyzing flood frequency patterns, identifying vulnerable areas, and predicting future flood scenarios under different climate change scenarios (Chakrabortty et al. 2021). This section introduces geospatial modeling and its relevance in flood frequency analysis.

GIS technology facilitates the gathering, retention, examination, and representation of geographical data. It allows researchers to integrate diverse datasets, such as topographic features, land-use/land cover information, hydrological networks, and climate data, into a single platform. GIS provides a spatial framework for organizing and analyzing flood-related data, facilitating the identification of flood-prone areas, and understanding the underlying factors contributing to flood frequencies. Remote sensing plays a crucial role in geospatial modeling by providing detailed and up-to-date information on land surface characteristics, such as elevation, vegetation cover, and impervious surfaces (Yin et al. 2021). In flood frequency analysis, remote sensing sources such as satellite images, aerial pictures, and light detection and ranging (LiDAR) data are often used. These data sources enable the extraction of essential variables, such as digital elevation models (DEMs), land cover classifications, and surface roughness parameters, which are vital inputs for flood modeling. Similarly, geospatial modeling incorporates statistical models to analyze flood frequency patterns and predict future flood scenarios. Various statistical techniques, such as frequency analysis, regression analysis, and time series analysis, can be applied within a geospatial modeling framework. These models help quantify the relationship between climate variables (e.g., precipitation and temperature) and flood frequencies, identify trends and patterns in historical flood data, and estimate the likelihood of future flood events.

2.3.2 Applications of Geospatial Modeling in Flood Frequency Analysis

Geospatial modeling has a wide range of applications in flood frequency analysis, providing valuable insights into the impacts of climate change on flooding. This section highlights the key applications of geospatial modeling in understanding flood frequencies and their relationship with climate change (Mishra et al. 2022). Geospatial modeling enables the creation of flood hazard maps that depict the spatial extent and severity of flood events. By integrating topographic information, hydrological data, and historical flood records, flood hazard maps provide a visual representation of areas prone to flooding. These maps help in identifying high-risk zones, informing land-use planning decisions, and implementing appropriate flood mitigation measures. Geospatial modeling is instrumental in assessing the impact of climate change on flood frequencies (Politi et al. 2016). By incorporating climate projections from climate models into hydrological models, researchers can simulate future flood scenarios under different climate change scenarios. This allows for the evaluation of potential changes in flood magnitudes, frequencies, and spatial extents, providing critical information for climate change adaptation planning.

Geospatial modeling facilitates the assessment of vulnerability and risk associated with flood events. By integrating socioeconomic data with flood frequency analysis, researchers can identify vulnerable populations, critical infrastructure, and areas with high potential for flood damage (Nofal and Van De Lindt 2022). This information helps prioritize resources for flood risk management, emergency response planning, and the development of resilient infrastructure. Geospatial modeling supports land-use planning decisions in flood-prone areas. By considering flood frequency analysis, hydrological modeling, and land cover information, policymakers can determine appropriate land-use regulations, such as zoning restrictions and buffer zones, to reduce flood risks. Geospatial modeling also aids in evaluating the effectiveness of floodplain management strategies, such as levees, flood control channels, and natural floodplain restoration.

2.3.3 Advantages and Limitations of Geospatial Modeling

Geospatial modeling offers several advantages in flood frequency analysis, but it is also important to consider its limitations. This section discusses the advantages and limitations of geospatial modeling in the context of assessing the impact of climate change on flood frequencies (Barnard et al. 2019). Geospatial modeling allows for a spatially explicit analysis of flood frequencies, providing detailed information on flood patterns at a local, regional, and global scale. This enables targeted flood risk management strategies and adaptation measures based on specific flood-prone areas. Geospatial modeling integrates diverse data sources, including topographic data, land cover information, climate data, and historical flood records. This multidisciplinary approach provides a comprehensive understanding of the factors influencing flood frequencies and enables the incorporation of multiple variables into flood models. Geospatial modeling enables the assessment of future flood scenarios under different climate change projections. By integrating climate models and hydrological models, researchers can simulate potential changes in flood frequencies, allowing policymakers to make informed decisions regarding flood risk management and adaptation strategies. Geospatial modeling serves as a decision support tool for policymakers, planners, and emergency managers (Barnard et al. 2019). By providing visual representations of flood hazards, vulnerability assessments, and floodplain maps, geospatial modeling helps stakeholders understand the complex relationships between climate change and flood frequencies, leading to more effective flood risk management and mitigation measures.

However, geospatial modeling relies heavily on the availability and quality of data. Data gaps, inconsistencies, and uncertainties can impact the accuracy and reliability of flood frequency analysis. Obtaining high-quality and up-to-date data, especially in data-scarce regions, can be a challenge. Geospatial modeling involves simplifications and assumptions to represent complex hydrological processes (Fenicia et al. 2016). These simplifications can introduce uncertainties in flood frequency analysis, as they may not fully capture the intricacies of local hydrological systems. Climate change projections and future climate scenarios are subject to uncertainties. Geospatial modeling relies on these projections to simulate future flood scenarios, and uncertainties in climate models can propagate into flood frequency analysis, affecting the accuracy of predictions. Geospatial modeling primarily focuses on the physical aspects of flood frequencies, often overlooking the influence of human factors and socioeconomic dynamics. Human activities, land-use changes, and socioeconomic conditions can significantly impact flood risk and vulnerability. Integrating these factors into geospatial models remains a challenge. Despite these limitations, geospatial modeling provides a robust framework for understanding the impact of climate change on flood frequencies (Kumar et al. 2023). With ongoing advancements in data collection techniques, modeling approaches, and climate projections, geospatial modeling will continue to play a critical role in flood frequency analysis, supporting evidence-based decision-making for flood risk management and adaptation strategies in the face of climate change.

In conclusion, geospatial modeling offers a powerful approach to analyze the impact of climate change on flood frequencies. By integrating GIS, remote sensing data, and statistical models, geospatial modeling enables the assessment of flood hazards, evaluation of climate change impacts, and identification of vulnerable areas. While it has advantages in providing spatially explicit analysis, integrating diverse data sources, and supporting decision-making, limitations related to data availability, simplifications, uncertainties, and human factors should be considered. Geospatial modeling continues to evolve, and advancements in data collection, modeling techniques, and climate projections will further enhance its effectiveness in flood frequency analysis and flood risk management.

2.4 Data Acquisition and Preprocessing

2.4.1 Collection of Climate and Hydrological Data

Accurate and reliable data are crucial for geospatial modeling to assess the impact of climate change on flood frequencies. This section focuses on the collection of climate and hydrological data, which serve as the foundation for geospatial modeling. Climate data, including precipitation and temperature, are essential for understanding the climatic conditions driving flood events. Climate data can be obtained from various sources, such as meteorological stations, weather satellites, and reanalysis datasets. Meteorological stations provide ground-based measurements of temperature and precipitation, but they are often limited in spatial coverage (Sun et al. 2018). Weather satellites, on the other hand, provide global coverage and can capture large-scale climatic patterns. Reanalysis datasets combine observations with climate models, providing gridded climate data with global coverage and long-term records. These diverse sources of climate data should be carefully selected and evaluated based on their spatial and temporal resolution, accuracy, and availability.

Hydrological data, such as river discharge measurements and water level records, are essential for characterizing the hydrological processes and understanding flood frequencies (Wilhelm et al. 2018). Hydrological data can be collected from gauging stations strategically located along rivers and streams. These stations measure water flow and stage levels, which are critical for calibrating and validating hydrological models. In addition to the gauging stations, other sources of hydrological data include remotely sensed data, such as altimetry data from satellites, which provide information on water levels in large rivers and lakes. Historical flood records and floodplain mapping data also contribute to understanding the occurrence and magnitude of past flood events.

2.4.2 Preprocessing of Data for Geospatial Modeling

Once climate and hydrological data are collected, they often require preprocessing to ensure their compatibility and suitability for geospatial modeling. This section discusses the key steps involved in preprocessing the data. Climate and hydrological data collected from different sources may have varying formats and units. Data formatting involves standardizing the data format, unit conversion, and ensuring consistency across different datasets (Gorgolewski et al. 2016). This step enables seamless integration and analysis of the data within a geospatial modeling framework. Climate and hydrological data are often collected at different spatial and temporal resolutions. Spatial aggregation involves interpolating or aggregating point data to a consistent spatial grid, allowing for uniform analysis across the study area. Temporal aggregation involves aggregating data at a consistent time step, such as daily, monthly, or annual, depending on the requirements of the geospatial modeling analysis (Turschwell et al. 2016). Missing data are common in climate and hydrological datasets. Missing data can arise due to instrument failures, data transmission issues, or gaps in long-term records. Missing data handling techniques, such as data imputation or interpolation, can be employed to fill the gaps and ensure a complete dataset. Care should be taken to select appropriate methods that preserve the statistical properties of the data and minimize potential biases. Calibration and validation of hydrological models are crucial steps in geospatial modeling. Calibration involves adjusting model parameters to match observed data, such as river discharge or water level records. Validation involves assessing the model's performance by comparing simulated outputs with independent observed data. The calibration and validation process ensures that the hydrological model accurately represents the real-world hydrological processes and provides reliable estimates of flood frequencies.

2.4.3 Data Quality Control and Uncertainty Assessment

Data quality control and uncertainty assessment are essential for geospatial modeling to ensure the reliability and robustness of the results. This section discusses the key aspects of data quality control and uncertainty assessment in the context of climate and hydrological data used for flood frequency analysis (Hu et al. 2020). Data quality control involves identifying and addressing errors, outliers, and inconsistencies in the collected data. Quality control procedures include data screening, outlier detection, and consistency checks. Data screening involves examining the data for any obvious errors or inconsistencies. Outlier detection techniques, such as statistical tests or visual inspection, can be used to identify extreme values that deviate significantly from the expected range. Consistency checks involve evaluating the relationships between different variables to identify any implausible or illogical values. Data quality control ensures that the data used for geospatial modeling are reliable and representative of the actual conditions. Uncertainty assessment is crucial in geospatial modeling to understand the reliability and limitations of the results. Uncertainties can arise from various sources, such as data errors, model parameterization, and climate projections (Alizadeh 2022). Uncertainty assessment techniques include sensitivity analysis, Monte Carlo simulations, and ensemble modeling. Sensitivity analysis helps identify the most influential variables or parameters in the geospatial modeling analysis. Monte Carlo simulations involve generating multiple realizations of the input data, incorporating their uncertainties, to quantify the uncertainty propagation in the modeling process. Ensemble modeling combines multiple climate models or hydrological models to capture the range of uncertainties associated with the modeling process (Her et al. 2016). Uncertainty assessment provides insights into the robustness of the geospatial modeling results and helps in decision-making processes.

Data acquisition and preprocessing are critical steps in geospatial modeling to assess the impact of climate change on flood frequencies. Collecting accurate and reliable climate and hydrological data from various sources is essential. Preprocessing the data involves formatting, aggregating, handling missing data, and calibrating and validating hydrological models. Data quality control and uncertainty assessment ensure the reliability and robustness of the results. By following these steps, researchers can obtain high-quality data and generate reliable geospatial modeling outputs that contribute to a better understanding of the impact of climate change on flood frequencies.

2.5 Geospatial and Geostatistical Modeling Techniques

Geospatial modeling techniques play a crucial role in assessing the impact of climate change on flood frequencies. These techniques enable the integration of climate and hydrological data within a spatial framework, allowing for a comprehensive understanding of flood dynamics. This section provides an overview of the geospatial modeling techniques commonly used in flood frequency analysis (Samanta et al. 2018). Deterministic modeling approaches are based on physical principles and mathematical equations that describe the hydrological processes governing flood generation. The hydrological system's water flow is simulated by these models, which take runoff, evapotranspiration, precipitation, and infiltration into account. Climate, land cover, and topography data are among the inputs used for deterministic models like the Soil and Water Assessment Tool (SWAT) and the Hydrological Simulation Program-Fortran (HSPF). These models provide insights into the hydrological response of a catchment to different climatic conditions and land-use scenarios.

Statistical modeling approaches focus on analyzing historical flood data to estimate the frequency and magnitude of future flood events. These models establish relationships between flood characteristics (e.g., peak discharge and volume) and relevant predictors (e.g., precipitation and catchment characteristics). To match the flood data and estimate flood frequencies, statistical models like the generalized extreme value (GEV) distribution or the Log-Pearson Type III distribution are often utilized (Rulfová et al. 2016). These models provide probabilistic estimates of flood magnitudes for a given return period. Hybrid modeling approaches combine deterministic and statistical models to leverage their respective strengths. These models integrate the physics-based representation of hydrological processes from deterministic models with the statistical analysis of historical flood data. Hybrid models, such as the Flood Estimation Handbook (FEH) in the United Kingdom, use a two-step approach. First, the deterministic component simulates the rainfall–runoff processes (Formetta et al. 2018). Then, the statistical component analyzes the simulated flood data to estimate flood frequencies. Hybrid models can provide more robust and reliable flood frequency estimates by incorporating both physical and statistical information (Paramaguru et al. 2022).

Statistical modeling approaches are widely used in flood frequency analysis due to their simplicity and ability to provide probabilistic estimates of flood magnitudes. This section discusses some commonly employed statistical modeling techniques in flood frequency analysis. Probability density functions describe the distribution of flood characteristics, such as peak discharge or volume. Commonly used PDFs include the Log-Pearson Type III distribution, which is suitable for annual maximum flood peaks, and the generalized extreme value (GEV) distribution, which can be used for both annual maximum and partial duration flood data (Kumar et al. 2022). These distributions are fitted to the observed data using methods like the method of moments or maximum likelihood estimation. Once the distribution parameters are estimated, they can be used to estimate flood magnitudes for different return periods. Index flood methods are based on the concept that the flood magnitude is related to a specific index, such as catchment area or mean annual precipitation. These methods assume that the relationship between the index and flood magnitude is stable over time. The most commonly used index flood method is the regional flood frequency analysis, which groups catchments with similar characteristics into homogeneous regions (Paramaguru et al. 2023). The flood frequency relationship is then established for each region, allowing for the estimation of flood magnitudes in ungauged catchments. L-moments are statistical moments used to estimate flood quantiles. They provide robust estimates of the distribution parameters, particularly in the presence of outliers or non-normal data. L-moments are calculated from order statistics of the observed flood data and used to derive the L-moment ratios, which can be related to the distribution parameters of various probability distributions. L-moment-based methods, such as the Hosking and Wallis method or the TL-moments method, are widely used in flood frequency analysis.

Machine learning techniques have gained popularity in recent years for their ability to capture complex relationships between flood characteristics and predictors, including climate variables and catchment characteristics. This section discusses some machine learning techniques commonly employed in flood frequency analysis. According to Anilan et al. (2016), artificial neural networks (ANN) are computational models that mimic the structure and functions of the human brain. Artificial neural networks (ANNs) consist of interconnected layers of nodes, often known as neurons. Every neuron takes in stimuli, applies activation functions to process them, and then generates an output. ANNs can be trained using historical flood data and relevant predictors to establish the relationship between the inputs and flood magnitudes. Once trained, ANNs can predict flood magnitudes for new inputs. ANNs are renowned for their capacity to manage large and intricate datasets and to identify nonlinear correlations. According to Swain et al. (2024), support vector machines (SVMs) are supervised learning models that categorize data into several categories after analysis. Using a high-dimensional feature space as a mapping tool, support vector machines (SVMs) choose the best hyperplane to divide the classes with the greatest margin. In the context of flood frequency analysis, SVMs can be used for regression tasks to predict flood magnitudes based on input variables. SVMs aim to find the hyperplane that maximizes the margin between flood data points, allowing for accurate predictions. Especially for small or high dimensionality datasets, SVMs work very well.

Several decision trees are combined into a single forecast using ensemble learning techniques called random forests (Desai and Ouarda 2021). A random subset of the predictors and a random subset of the data are used to form each decision tree. The sum of the forecasts made by each tree yields the final forecast. Measures of varying relevance and nonlinear correlations may be handled by random forests. They can also manage missing data. In the context of flood frequency analysis, random forests can be used to predict flood magnitudes based on a set of predictors such as climate variables, catchment characteristics, and historical flood data. Another ensemble learning method that builds a strong predictive model by combining many weak prediction models is called gradient boosting machines (GBM) (Saravanan et al. 2023). GBMs construct a series of decision trees, with each new tree being taught to fix the mistakes produced by its predecessors. This iterative process results in a powerful predictive model. GBMs can capture complex nonlinear relationships, handle missing data, and provide measures of variable importance. In flood frequency analysis, GBMs can be used to predict flood magnitudes based on relevant predictors and historical flood data.

Machine learning techniques offer the advantage of flexibility and adaptability to capture complex relationships between flood characteristics and predictors. Nevertheless, it is crucial to acknowledge that these methods are driven by data and significantly depend on the quality and representativeness of the training data. Thorough deliberation should be devoted to data preparation, feature selection, and model validation in order to guarantee dependable and precise forecasts. Geospatial modeling techniques, including statistical modeling approaches and machine learning techniques, play a vital role in flood frequency analysis. Statistical modeling approaches provide probabilistic estimates of flood magnitudes based on historical flood data and relevant predictors. Machine learning techniques offer the ability to capture complex relationships and make predictions based on a wide range of variables. Researchers may use geospatial modeling tools to provide significant insights into the influence of climate change on flood frequencies. This information can then be used to develop successful strategies for flood management and mitigation.

3 Results and Discussion

3.1 *Flood Runoff Estimation Using Advanced Geostatistical Approaches*

In this chapter, a meta-heuristic geostatistical technique for flood magnitude prediction developed by Singh et al. (2022a, b) was used as a case study. Four machine learning (ML) models, viz. support vector machine (SVM), random forest (RF), multi-linear regression (MLR), and multiple adaptive regression splines (MARS), were employed for flood magnitude estimation for the river basin.

3.2 *Study Area Description*

The study was conducted for the Gola basin (Uttarakhand) located in the foothills of the Himalayas (Fig. 1). The main branches of the river include Kanchi, Karkari, and Kharkai. The watershed is located between latitudes 29.16° to 29.27° N and longitudes 79.8° to 79.53° E in the northern part of India. The geographical area of the Gola basin is approximately 615 km^2. The climate in the Gola river basin is typically moderately warm. The highest and lowest elevations within the catchment are 2302 m and 252 m, respectively, above the mean sea level. The Gola watershed experiences a subtropical climate with a significant amount of rainfall occurring during specific seasons. The mean annual precipitation is 1700 mm, with the monsoon season being the primary contributor. Due to its location near the eastern border of the Himalayan Mountain range, the area receives substantial rainfall. The river is predominantly fed by springs and serves as a water source for Kathgodam and Haldwani. The monsoon period lasts from July to September and accounts for most of the annual rainfall.

3.3 *Data Acquisition and Methodology*

The rainfall and runoff data for a period of 12 years (2009–2020) were utilized to analyze the modeling of rainfall and runoff in the Gola watershed. Information on runoff was gathered from the Gola River monitoring station at Kathgodam Barrage. Data on rainfall were gathered from the irrigation departments of Nainital, Bhimtal, and Kathgodam using three rain-gauge stations. The average areal rainfall for the Gola watershed was determined using the Thiessen polygon technique. The models were developed and validated using the 12-year dataset. 80% of the whole dataset was allocated for testing, while the remaining 20% was allocated for training. It is essential to cross-validate the dataset during the division process to ensure statistical consistency. An overview of the proposed methodology is described as a flowchart in Fig. 2.

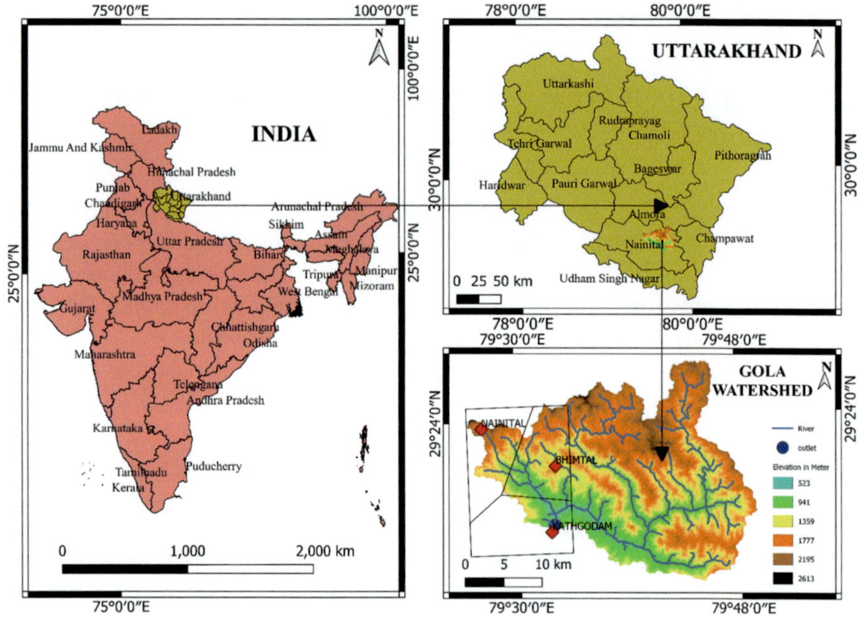

Fig. 1 Index map of Gola river basin (Singh et al. 2022a, b)

3.4 *Outcomes of Study*

The study aims to find out the most efficient machine learning approach for the prediction of flood runoff magnitude for the Gola river basin. Thus, a comparison of four ML models, viz. MLR, RF, SVM, and MARS, was carried out. For the development of the ML model, discharge lagged by three daily time steps ($Q_{(t-1)}$, $Q_{(t-2)}$, $Q_{(t-3)}$) and original rainfall data ($R_{(t)}$) along with its three daily step lagged variants ($R_{(t-1)}$, $R_{(t-2)}$, $R_{(t-3)}$) were used as predictors in the study.

When compared to the MLR model, the comparative findings showed that the four heuristic techniques were more accurate at predicting flood runoff. In terms of machine learning models, the RF model outperformed the MARS, SVM, and MLR models in predicting daily flood runoff for all scenarios examined. Specifically, during the training period (Figs. 3a and 4a), the RF model (RMSE (m^3/s), R^2, NSE, and PBIAS (%) = 6.31, 0.96, 0.94, and − 0.20 during the testing period (Figs. 3b and 4b), respectively. During the training and testing phases, the RF model continuously outperformed the other three models. Conclusively, the RF model emerges as the most promising model with considerable potential for precisely projecting the flood runoff in the Gola watershed.

Based on the quantitative analysis and indicators used, the models were ranked as follows: RF, MARS, SVM, and MLR for the training period and RF, MARS, MLR, and SVM for the testing period. These rankings may have been influenced by factors such as the division of data, uncertainties in inputs, and optimization of model

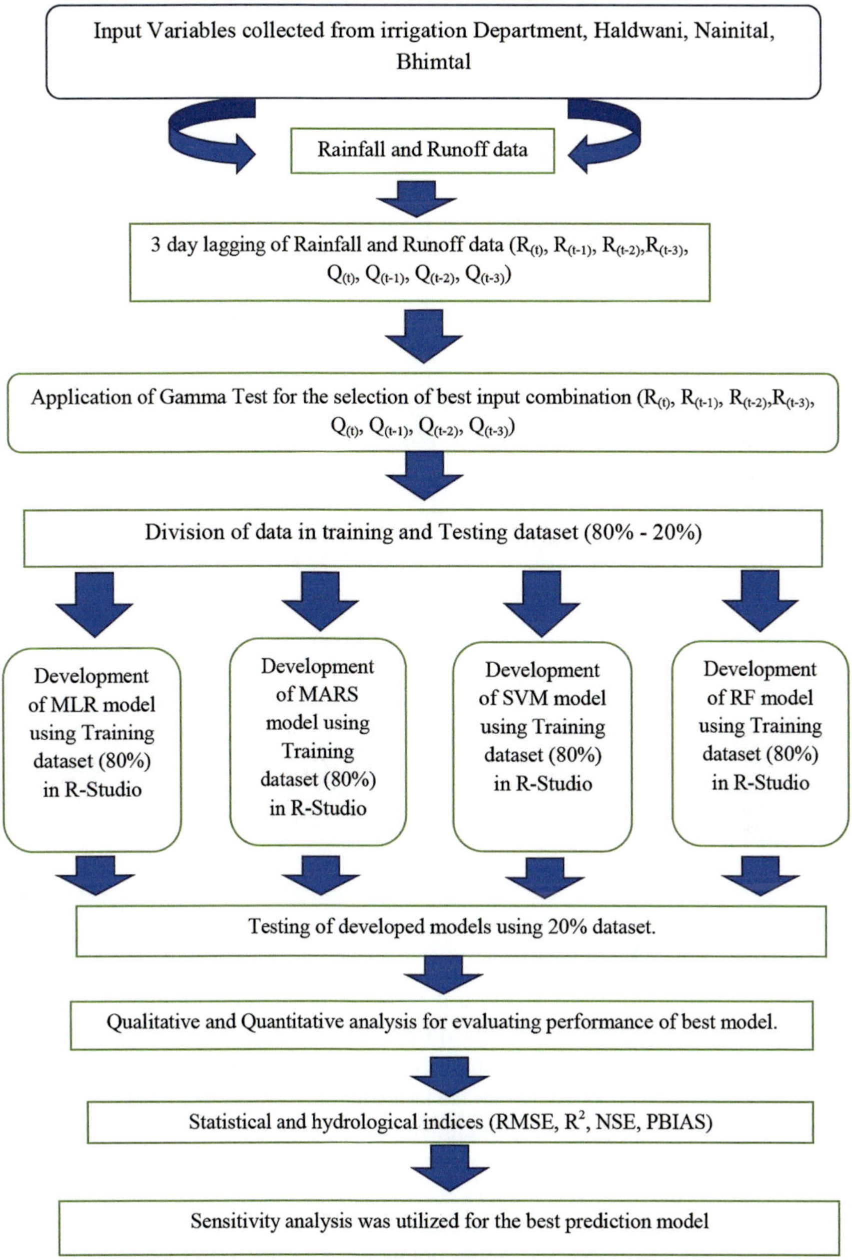

Fig. 2 Flowchart of the methodology used for flood prediction (Singh et al. 2022a, b)

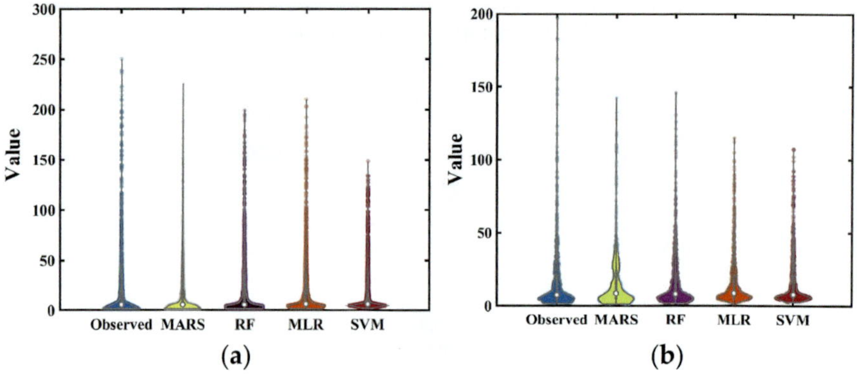

Fig. 3 Violin plot indicating the ability of models to capture the distribution and extreme values for different models for **a** training and **b** testing periods (Singh et al. 2022a, b)

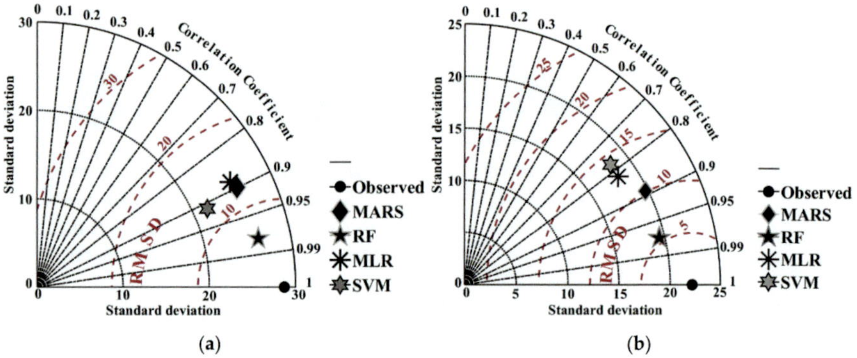

Fig. 4 Taylor diagram indicating the efficiencies of different model during **a** training and **b** testing periods (Singh et al. 2022a, b)

parameters. To assess the reliability of the models for flood magnitude prediction, it is recommended to test them using different data lengths and variations in the training–testing splits. The results obtained indicate that the MLR, MARS, SVM, and RF techniques exhibit satisfactory accuracy in modeling rainfall and flood runoff parameters. From the study, it was concluded that the RF model was the most efficient model for flood runoff prediction in tropical watersheds.

The study would help researchers for the development of advanced machine learning approaches for precise flood runoff predictions. These ML models could be integrated with land use and downscaling models for future predictions. The outcomes of the research could help policymakers for development of a robust plan to combat disasters such as flash floods and minimize their impacts.

4 Conclusion

Throughout this chapter, we have explored the impact of climate change on flood frequencies using geospatial modeling techniques. The findings presented here shed light on the significant implications of climate change for flood events and highlight the importance of incorporating geospatial modeling in understanding these impacts. Our analysis has demonstrated that climate change has a discernible effect on flood frequencies. The geospatial modeling approach allowed us to examine the spatial distribution of flood events and assess how they have changed over time. The results revealed a clear increase in the frequency and intensity of floods in many regions, indicating a direct link to climate change. This finding underscores the urgent need for effective adaptation and mitigation strategies to address the escalating risks posed by flooding. Geospatial modeling has proven to be a valuable tool for assessing the vulnerability of different areas to flood events. By integrating various data sources, such as topography, land cover, and rainfall patterns, we were able to identify areas that are particularly susceptible to increased flood frequencies. Such insights are crucial for the development of targeted measures to enhance resilience and reduce the potential impacts of flooding on communities and infrastructure.

The use of geospatial modeling techniques has been instrumental in enhancing our understanding of how climate change affects flood frequencies. By integrating spatial data and advanced modeling algorithms, geospatial modeling enables us to capture the complex interactions between climate variables and flood patterns. This approach provides a comprehensive and spatially explicit assessment of flood risks, allowing for informed decision-making and effective planning. Geospatial modeling enables us to explore the intricate relationships between climate drivers and flood events at various scales. It allows us to analyze the impacts of changes in temperature, precipitation patterns, and land use on flood frequencies. Additionally, by incorporating historical data and future climate projections, geospatial models offer valuable insights into the potential future scenarios of flood risk. This knowledge is essential for policymakers, urban planners, and other stakeholders to develop robust adaptation strategies and implement appropriate mitigation measures. Furthermore, geospatial modeling facilitates the identification of vulnerable regions and populations, supporting targeted interventions to reduce the impacts of floods. By integrating socioeconomic data and demographic information, these models enable us to assess the social and economic consequences of flood events. This knowledge can guide the allocation of resources and the implementation of measures that enhance preparedness, early warning systems, and emergency response capabilities.

Geospatial modeling has proven to be a valuable tool for understanding the impacts of climate change on flood frequencies. The findings presented in this chapter underscore the urgency of addressing the escalating risks posed by flooding and highlight the need for informed decision-making and proactive measures. By refining geospatial models, expanding data collection efforts, and fostering interdisciplinary collaborations, we can enhance our understanding of climate change impacts on flood frequencies and develop effective strategies to build resilience in the face of a changing climate.

References

Aceves EA, Plata-Rocha W, Monjardin-Armenta SA, Rangel-Peraza JG (2023) Geospatial modelling of floods: a literature review. Stoch Env Res Risk Assess 37(11):4109–4128

Alizadeh O (2022) Advances and challenges in climate modeling. Clim Change 170(1–2):18

Anilan T, Satilmis U, Kankal M, Yuksek O (2016) Application of Artificial Neural Networks and regression analysis to L-moments based regional frequency analysis in the Eastern Black Sea Basin, Turkey. KSCE J Civ Eng 20:2082–2092

Barnard PL, Erikson LH, Foxgrover AC, Hart JAF, Limber P, O'Neill AC, Van Ormondt M, Vitousek S, Wood N, Hayden MK, Jones JM (2019) Dynamic flood modeling essential to assess the coastal impacts of climate change. Sci Rep 9(1):4309

Beniston M, Stoffel M (2016) Rain-on-snow events, floods and climate change in the Alps: events may increase with warming up to 4 C and decrease thereafter. Sci Total Environ 571:228–236

Bill R, Blankenbach J, Breunig M, Haunert JH, Heipke C, Herle S, Maas HG, Mayer H, Meng L, Rottensteiner F, Werner M (2022) Geospatial information research: state of the art, case studies and future perspectives. PFG—J Photogram Remote Sens Geoinf Sci 90(4):349–389

Cai W, Santoso A, Collins M, Dewitte B, Karamperidou C, Kug JS, Lengaigne M, McPhaden MJ, Stuecker MF, Taschetto AS, Zhong W (2021) Changing El Niño-Southern oscillation in a warming climate. Nat Rev Earth Environ 2(9):628–644

Chakrabortty R, Pal SC, Janizadeh S, Santosh M, Roy P, Chowdhuri I, Saha A (2021) Impact of climate change on future flood susceptibility: an evaluation based on deep learning algorithms and GCM model. Water Resour Manag 35:4251–4274

Desai S, Ouarda TB (2021) Regional hydrological frequency analysis at ungauged sites with random forest regression. J Hydrol 594:125861

Fenicia F, Kavetski D, Savenije HH, Pfister L (2016) From spatially variable streamflow to distributed hydrological models: analysis of key modeling decisions. Water Resour Res 52(2):954–989

Fischer S, Schumann AH (2021) Multivariate flood frequency analysis in large river basins considering tributary impacts and flood types. Water Resour Res 57(8):e2020WR029029

Formetta G, Prosdocimi I, Stewart E, Bell V (2018) Estimating the index flood with continuous hydrological models: an application in Great Britain. Hydrol Res 49(1):123–133

Gorgolewski KJ, Auer T, Calhoun VD, Craddock RC, Das S, Duff EP, Flandin G, Ghosh SS, Glatard T, Halchenko YO, Poldrack RA (2016) The brain imaging data structure, is a format for organizing and describing outputs of neuroimaging experiments. Sci Data 3(1):1–9

Her Y, Yoo SH, Seong C, Jeong J, Cho J, Hwang S (2016) Comparison of uncertainty in multi-parameter and multi-model ensemble hydrologic analysis of climate change. Hydrol Earth Syst Sci Discuss 1–44

Hirabayashi Y, Alifu H, Yamazaki D, Imada Y, Shiogama H, Kimura Y (2021) Anthropogenic climate change has changed frequency of past flood during 2010–2013. Prog Earth Planet Sci 8(1):1–9

Hu L, Nikolopoulos EI, Marra F, Anagnostou EN (2020) Sensitivity of flood frequency analysis to data record, statistical model, and parameter estimation methods: an evaluation over the contiguous United States. J Flood Risk Manag 13(1):e12580

Kumar A, Nagar S, Anand S (2021) Climate change and existential threats. In: Global climate change. Elsevier, pp 1–31

Kumar A, Tripathi VK, Sachan P, Rakshit A, Singh RM, Shukla SK, Pandey R, Vishwakarma A, Panda KC (2022) Sources of ions in the river ecosystem. In: Ecological significance of river ecosystems. Elsevier, pp 187–202

Kumar V, Sharma KV, Caloiero T, Mehta DJ, Singh K (2023) Comprehensive overview of flood modeling approaches: a review of recent advances. Hydrology 10(7):141

Mishra A, Mukherjee S, Merz B, Singh VP, Wright DB, Villarini G, Paul S, Kumar DN, Khedun CP, Niyogi D, Stedinger JR (2022) An overview of flood concepts, challenges, and future directions. J Hydrol Eng 27(6):03122001

Mohajerani A, Bakaric J, Jeffrey-Bailey T (2017) The urban heat island effect, its causes, and mitigation, with reference to the thermal properties of asphalt concrete. J Environ Manag 197:522–538

Najibi N, Devineni N (2018) Recent trends in the frequency and duration of global floods. Earth Syst Dyn 9(2):757–783

Nofal OM, Van De Lindt JW (2022) Understanding flood risk in the context of community resilience modeling for the built environment: research needs and trends. Sustain Resilient Infrastruct 7(3):171–187

Panda KC, Kumar A, Pradhan SN, De N, Meena VS (2021) Impact of soil moisture stress on rice productivity in warming climate over Indian Mid-Indo-Gangetic Plain. Clim Change Environ Sustain 9(1):21–31

Panda KC, Singh RM, Singh SK (2024) Advanced CMD predictor screening approach coupled with cellular automata-artificial neural network algorithm for efficient land use-land cover change prediction. J Clean Prod 141822

Panda KC, Singh RM, Singh VK, Singla S, Paramaguru PK (2023) Impact of climate change induced future rainfall variation on dynamics of arid-humid zone transition in the western province of India. J Environ Manag 325:116646

Panda KC, Singh RM, Thakural LN, Sahoo DP (2022) Representative grid location-multivariate adaptive regression spline (RGL-MARS) algorithm for downscaling dry and wet season rainfall. J Hydrol 605:127381

Paramaguru PK, Panda KC, Suna T, Rajput J (2023) Assessment of spatiotemporal variability of rainfall and surface wind speed over the eastern coastal agro-ecological zones of India using advanced trend detection approaches. Theor Appl Climatol 154(1–2):311–335

Paramaguru PK, Paul JC, Panigrahi B, Panda KC (2022) Assessment of replenishable groundwater resource and integrated water resource planning for sustainable agriculture. Geospatial technology for landscape and environmental management: sustainable assessment and planning, pp 21–47

Politi E, Rowan JS, Cutler ME (2016) Assessing the utility of geospatial technologies to investigate environmental change within lake systems. Sci Total Environ 543:791–806

Qin Y, Abatzoglou JT, Siebert S, Huning LS, AghaKouchak A, Mankin JS, Hong C, Tong D, Davis SJ, Mueller ND (2020) Agricultural risks from changing snowmelt. Nat Clim Change 10(5):459–465

Rulfová Z, Buishand A, Roth M, Kyselý J (2016) A two-component generalized extreme value distribution for precipitation frequency analysis. J Hydrol 534:659–668

Sahoo SP, Panda KC (2020) Prediction of climate change using statistical downscaling techniques. New frontiers in stress management for durable agriculture, pp 311–328

Samanta RK, Bhunia GS, Shit PK, Pourghasemi HR (2018) Flood susceptibility mapping using geospatial frequency ratio technique: a case study of Subarnarekha River Basin, India. Model Earth Syst Environ 4:395–408

Saravanan S, Abijith D, Reddy NM, Parthasarathy KSS, Janardhanam N, Sathiyamurthi S, Sivakumar V (2023) Flood susceptibility mapping using machine learning boosting algorithms techniques in Idukki district of Kerala India. Urban Clim 49:101503

Saroughi M, Mirzania E, Vishwakarma DK, Nivesh S, Panda KC, Daneshvar FA (2023) A novel hybrid algorithms for groundwater level prediction. Iran J Sci Technol Trans Civ Eng 1–18

Singh AK, Kumar P, Ali R, Al-Ansari N, Vishwakarma DK, Kushwaha KS, Panda KC, Sagar A, Mirzania E, Elbeltagi A, Heddam S (2022a) An integrated statistical-machine learning approach for runoff prediction. Sustainability 14(13):8209

Singh VK, Panda KC, Sagar A, Al-Ansari N, Duan HF, Paramaguru PK, Vishwakarma DK, Kumar A, Kumar D, Kashyap PS, Elbeltagi A (2022b) Novel Genetic Algorithm (GA) based hybrid machine learning-pedotransfer Function (ML-PTF) for prediction of spatial pattern of saturated hydraulic conductivity. Eng Appl Comput Fluid Mech 16(1):1082–1099

Sun Q, Miao C, Duan Q, Ashouri H, Sorooshian S, Hsu KL (2018) A review of global precipitation data sets: data sources, estimation, and intercomparisons. Rev Geophys 56(1):79–107

Swain SS, Khura TK, Sahoo PK, Chobhe KA, Al-Ansari N, Kushwaha HL, Kushwaha NL, Panda KC, Lande SD, Singh C (2024) Proportional impact prediction model of coating material on nitrate leaching of slow-release Urea Super Granules (USG) using machine learning and RSM technique. Sci Rep 14(1):3053

Tulla PS, Kumar P, Vishwakarma DK, Kumar R, Kuriqi A, Kushwaha NL, Rajput J, Srivastava A, Pham QB, Panda KC, Kisi O (2024) Daily suspended sediment yield estimation using soft-computing algorithms for hilly watersheds in a data-scarce situation: a case study of Bino watershed, Uttarakhand. Theor Appl Climatol 1–25

Turschwell MP, Peterson EE, Balcombe SR, Sheldon F (2016) To aggregate or not? Capturing the spatio-temporal complexity of the thermal regime. Ecol Ind 67:39–48

Walsh JE, Ballinger TJ, Euskirchen ES, Hanna E, Mård J, Overland JE, Tangen H, Vihma T (2020) Extreme weather and climate events in northern areas: a review. Earth Sci Rev 209:103324

Wang W, Lee X, Xiao W, Liu S, Schultz N, Wang Y, Zhang M, Zhao L (2018) Global lake evaporation accelerated by changes in surface energy allocation in a warmer climate. Nat Geosci 11(6):410–414

Wasko C, Nathan R, Stein L, O'Shea D (2021) Evidence of shorter more extreme rainfalls and increased flood variability under climate change. J Hydrol 603:126994

Wasko C, Sharma A (2017) Global assessment of flood and storm extremes with increased temperatures. Sci Rep 7(1):7945

Wilhelm B, Canovas JAB, Aznar JPC, Kämpf L, Swierczynski T, Stoffel M, Støren E, Toonen W (2018) Recent advances in paleoflood hydrology: from new archives to data compilation and analysis. Water Secur 3:1–8

Wright DB, Yu G, England JF (2020) Six decades of rainfall and flood frequency analysis using stochastic storm transposition: review, progress, and prospects. J Hydrol 585:124816

Wyżga B, Kundzewicz ZW, Ruiz-Villanueva V, Zawiejska J (2016) Flood generation mechanisms and changes in principal drivers. Flood risk in the Upper Vistula Basin, pp 55–75

Yin J, Dong J, Hamm NA, Li Z, Wang J, Xing H, Fu P (2021) Integrating remote sensing and geospatial big data for urban land use mapping: a review. Int J Appl Earth Obs Geoinf 103:102514

Zeng Z, Peng L, Piao S (2018) Response of terrestrial evapotranspiration to Earth's greening. Curr Opin Environ Sustain 33:9–25

Management

Surface and Sub-surface Hydrological Management: A Systematic Review

Zahra Ebrahimzadeh, Akbar Norouzi-shokrlu, and Khodayar Abdollahi

Abstract Groundwater–surface water (GW-SW) interactions are essential for understanding the physical, chemical, and biological attributes of both sub-surface and surface ecosystems, as well as the functioning of riparian ecosystems. The interaction between GW-SW involves the exchange of water and chemicals between the land surface and the sub-surface. This interaction plays a crucial role in Integrated Watershed Studies, where methods such as field/experimental, analytical modeling, numerical modeling, and semi-analytical methods are used to analyze and predict GW-SW water interactions. Understanding these interactions is essential for managing water resources effectively and addressing issues like flooding and water scarcity. Factors such as hydrological conditions, geological structure, water–rock interaction, climate, land use, and the presence of substances like nutrients and toxic substances influence the interactions between GW-SW. Various methods, including field/experimental, analytical modeling, numerical modeling, and semi-analytical methods, are employed to study these interactions and assess their impact on water resources. These interactions play a crucial role in sustainable river basin management and environmental sustainability. In this chapter, we explore the definition and importance of GW-SW management. This chapter focuses on the methods measuring and modeling their interaction and the hydrological management techniques to optimize GW-SW interactions. Also, in this chapter, the capability of remote sensing in the management of GW-SW and their interaction, the impact of climate change on the interaction of GW-SW, and their management have been described.

Z. Ebrahimzadeh (✉)
Faculty of Natural Resources and Earth Sciences, Shahrekord University, Shahrekord, Iran
e-mail: z.brahimzadeh@stu.sku.ac.ir; zahraebrahimzadeh696@gmail.com

A. Norouzi-shokrlu
Water Resources Planning and Allocation at Regional Water Company of Ardabil, Ardabil, Iran
e-mail: norouzishokrlu65@gmail.com

K. Abdollahi
Department of Environmental Engineering, Faculty of Natural Resources and Earth Sciences, Shahrekord University, Shahrekord, Iran
e-mail: kabdolla@sku.ac.ir

S. C. Pal and U. Chatterjee (eds.), *Surface, Sub-Surface Hydrology and Management*, Springer Geography, https://doi.org/10.1007/978-3-031-62376-9_23

Keywords Groundwater · Surface water · Interaction · Remote sensing · Climate change

1 Introduction

Water is an invaluable natural resource that is extremely stressed in numerous parts of the world, especially in semi-arid regions. Water scarcity is evident in this region due to low rainfall and elevated evapotranspiration rates. Groundwater serves as a dependable water source in arid and semi-arid areas and is extensively utilized for irrigation, household, and industrial needs (Abdollahi et al. 2017). However, increasing depletion of groundwater has led to aquifer overexploitation and deterioration. Human and industrial wastes disposal, as well as the percolation of pesticides and herbicides, contribute to this degradation. Water plays an important role for human survival, socioeconomic development and ecological health. However, water resources are under threat due to overuse, pollution, and climate change (Cosgrove and Loucks 2015; Jakeman et al. 2016). Surface and groundwater management has emerged as a critical approach to address these challenges. Surface water resources are generally socially and fairly well managed, while groundwater resources are often hidden and more complicated to consider (Jakeman et al. 2016). Ensuring the sustainability of important ecosystems and meeting the increasing water demand heavily rely on the integrated management of surface and groundwater resources (Reilly et al. 2008). Understanding the interaction between surface and sub-surface hydrology is important for several reasons, including:

1. Water management: Surface and sub-surface water are interconnected, and changes in one can affect the other. Understanding this interaction is crucial for effective water management, including flood control, irrigation, and groundwater recharge.
2. Ecosystem health: Surface and sub-surface water play a vital role in supporting ecosystems, and its varying can have notable impacts on the other. For example, changes in groundwater levels can affect the health of wetlands and other surface water bodies.
3. Climate modeling: Accurate climate modeling requires an understanding of an intricate interactions between surface and sub-surface water. This includes the role of groundwater in regulating surface water flows and the impact of land surface heterogeneity on atmospheric boundary layer development (Simon et al. 2021).
4. Carbon storage: Peatlands are an important carbon sink, and their health depends on the interaction between surface and sub-surface water. Understanding this interaction is crucial for effective management of peatlands and other carbon storage systems (Qiu et al. 2019; Pandey et al. 2023).

Overall, understanding the interaction between surface and sub-surface hydrology is necessary for effective water management, ecosystem health, climate modeling, and carbon storage.

The aim of this chapter is to display and survey information about surface and sub-surface hydrological interactions, overview of the interactions between surface and sub-surface hydrology, factors affecting the interactions (such as soil properties, land use, and climate), methods for measuring and modeling the interactions, surface and sub-surface hydrological management techniques, techniques for managing both surface and sub-surface hydrology to optimize their interactions, examples of integrated management approaches, such as green infrastructure and floodplain restoration, remote sensing capabilities in surface and sub-surface hydrological management, geospatial and geostatistical techniques for surface and sub-surface hydrological management, overview of geospatial and geostatistical techniques for surface and sub-surface hydrological management, examples of how these techniques have been applied to manage the interactions between surface and sub-surface hydrology, discussion of the advantages and limitations of these techniques, impacts of climate change on surface and sub-surface hydrological interactions and management, overview of the potential impacts of climate change on surface and sub-surface hydrology, discussion of how these impacts may affect the interactions between surface and sub-surface hydrology, strategies for adapting surface and sub-surface hydrological management to climate change, systematic review of surface and sub-surface hydrological management literature, overview of the systematic review methodology used, summary of the literature reviewed on surface and sub-surface hydrological management, analysis of the findings and trends in the literature, discussion of the implications of the literature for surface and sub-surface hydrological management, case studies of successful surface and sub-surface hydrological management, overview of the case studies presented, analysis of the management techniques used in the case studies, discussion of the lessons learned from the case studies, future directions for surface and sub-surface hydrological management, emerging trends and challenges in surface and sub-surface hydrological management, opportunities for future research and innovation to improve the understanding and management of the interactions between surface and sub-surface hydrology, and at the end conclusion and recommendations for future surface and sub-surface hydrological management practices.

Surface hydrology represents the study of water on the Earth's surface, including rainfall, runoff, and the courses that surface water can be transported through rivers or being stored in reservoirs. It is requiring above earth water, in differentiate to groundwater hydrology, which bargains with water underneath the surface of the Earth. Surface water hydrology is utilized to forecast the effects of water infrastructures like dams and canals, taking into account the arrangement of the watershed, geology formation, soil types, vegetation distribution, nutrients content, energy sources, and wildlife population. Surface water storage incorporates water collected on the soil surface or underground, intercepted water, and water from soil (Gentine et al. 2012).

Sub-surface hydrology, on the other hand, presents the study of water movement in the sub-surface, including groundwater hydrology, which deals with the sub-surface water which lies beneath the water table in soils and geological formations which are completely saturated. Hydrology of groundwater includes the study of the sub-surface and the generally science of water movement therein. When surface water infiltrates soil up to bedrock, it shall be classified as groundwater, and its rate at which this happens specifies base flow requirements for instream flow, as well as sub-surface water levels in wells. Whereas groundwater is not portion of surface-water hydrology, it needs to be considered in order to fully understand the behavior of surface water (Sapriza-Azuri et al. 2018).

1.1 Overview of the Interactions Between Surface and Sub-surface Hydrology

All articles and researches demonstrate that surface and sub-surface hydrology are interlinkage, and changes in one can affect the other. Understanding the interactions between surface and sub-surface hydrology is crucial. Some of these researches are shown in Table 1.

The interaction of surface water and groundwater refers to the exchange of water and chemicals between the land surface and the sub-surface. Surface water infiltrates into the ground and recharges the underlying aquifer, while groundwater discharges to the surface and contributes the stream with base flow. These exchanges are evaluated in the Integrated Watershed Studies with regard to surface water and groundwater quality and quantity. There are several methods for analyzing groundwater–surface water interaction, including field/experimental methods, analytical modeling methods, numerical modeling methods, and semi-analytical methods (Lewandowski et al. 2020).

To assess the impact of groundwater pumping on surface waters, a model is necessary to predict the interaction between groundwater wells and water rights for management purposes (Valett and Reinhold 2022). Integrated surface water and groundwater management may assist decreasing the risk of flooding and water scarcity in arid and semi-arid areas. Effective policies and regulations are essential to support sustainable water management (Lewandowski et al. 2020).

Based on the research conducted, there are five different types of surface water-aquifer systems rely on the number of interactions, which are as follows:

(1) Gaining stream: Occurs when underground water enters the surface flow
(2) Losing stream: Water penetrates into the aquifer from surface flow
(3) Transition losing stream: In this case, both of the above interactions occur
(4) Parallel stream: The property is the groundwater level below or within the stream bed
(5) Flow through stream: It has different underground water levels on both sides of the stream bed (Valett and Reinhold 2022).

All of the above are shown in Table 2.

Table 1 Overview of the interactions between surface and sub-surface hydrology

Topic	Result	References
Surface water groundwater interaction in water-stressed semi-arid western India: insights from environmental isotopes	Hypersaline water from the salt pan is recharging groundwater in the coastal area near the Gulf of Kachchh	Pandey et al. (2023)
A review on the research advances in groundwater–surface water interaction with an overview of the phenomenon	This work provides a concise overview of the progress achieved in the field, identifies areas where further research is needed, and proposes a path forward. It offers valuable insights for hydrologists, policymakers, and practicing engineers involved in water resources development and management	Banerjee and Ganguly (2023)
Modeling groundwater and surface water interaction: an overview of current status and future challenges	The present research makes a valuable contribution in addressing typical shortcomings of model applications, fills in gaps in the current body of literature, and emphasizes the significance of the modeling process in effectively planning the sustainable management of water resources	Ntona et al. (2022)
Surface and groundwater interactions: a review of coupling strategies in detailed domain models	The review sets out strategies to be followed in developing models of groundwater flow that reflect the real-world conditions, between surface water and groundwater interactions at regional scales	Haque et al. (2021)
Interaction between surface water and groundwater in Yinchuan Plain	This research demonstrates the effectiveness of integrated approach that merges numerical simulation, hydro geochemistry, and isotope data analysis to determine GW-SW interactions in Yinchuan Plain	Cai et al. (2020)
Interaction between lakes' surface water and groundwater in the Pantanal wetland, Brazil	The result of this study shows that the groundwater recharge in the study area takes place in the topographical lows, through a process referred to as depression-focused recharge which contradicts with earlier investigations of the same area	Freitas et al. (2019)

Table 2 Various scenarios for interaction of streams and groundwater

Scenarios	From stream to aquifer	From aquifer to stream
Gaining stream		*
Transition losing stream	*	*
Losing disconnected stream	*	
Losing connected stream	*	
Parallel stream	*	*
Flow through stream	*	*

1.2 Factors Affecting the Interactions (Such as Soil Properties, Land Use, and Climate)

Several factors affect the interactions between surface water and groundwater. These factors include:

Hydrological conditions: River runoff and groundwater level were found to be key factors affecting the interaction between surface water and groundwater (Meng et al. 2021).

Geological structure: The geological structure of the territory can affect the interaction between surface water and groundwater. For example, disjunctive tectonic faults can be employed as conduits for the discharge of mine water and deeply mineralized solutions to the surface earth where they are transported into the river system (Sklyarenko et al. 2021).

Water–rock interaction: Hydro-chemical components of surface water and groundwater can be dominated by water–rock interaction, such as calcite dissolution and silicate weathering (Yu et al. 2022).

Climate: Recharge source of groundwater can be affected by climate, such as atmospheric precipitation (Yu et al. 2022).

Land use: Land use can affect the interaction between surface water and groundwater. For example, sand-gravel exposure in the alluvial fan was found to be more conducive to surface water and groundwater interaction.

Nutrients, bacteria, sediment, and toxic substances: These substances can assist water from both point and nonpoint sources, affecting surface and groundwater quality (DVandas et al. 2002).

The interactions between surface water and groundwater are complex and require a comprehensive framework to review the climate, geological, and hydrological elements.

2 Methods for Measuring and Modeling the Interactions

Generally, there are several approaches to analyze the surface water and groundwater interaction which according to the research are as follows:

1. Field/experimental methods: These methods involve direct measurements of water levels, water quality, and other parameters in the field. Examples of field/experimental methods include aquifer intensive methods, surface water intensive methods, and transition zone intensive method (Banerjee and Ganguly 2023).
2. Analytical modeling methods: These methods involve the use of mathematical equations to describe the physical processes that govern groundwater and surface water interaction. Analytical modeling methods may be exerted to evaluate the rates of water exchange between groundwater and surface water (Banerjee and Ganguly 2023).

Table 3 Method for analyzing groundwater and surface water interaction

Methods for GW-SW interaction	Field/experimental methods	
	Analytical modeling approach	
	Numerical modeling approach	
	Semi-analytical method	Aquifer intensive methods
		Transition zone intensive methods
		Surface water intensive methods

3. Numerical modeling methods: These methods involve the use of computer models to simulate the physical processes that govern groundwater and surface water interaction (Banerjee and Ganguly 2023). Numerical modeling methods can be applied to predict the effects of changes in groundwater pumping rates or land use on surface water and groundwater interaction.
4. Semi-analytical methods: These methods involve a combination of analytical and numerical modeling methods. Semi-analytical methods may be exerted to simulate groundwater and surface water interaction in complex hydrogeologic settings (Banerjee and Ganguly 2023).

All of the above are shown in Table 3.

2.1 *Surface and Sub-surface Hydrological Management Techniques*

Surface and sub-surface hydrological management techniques are necessary to the sustainable use of water resources. Some of the techniques used for surface and sub-surface hydrological management are as follows:

1. Low-impact development (LID): LID is a technique that manages stormwater runoff through the use of small-scale features such as rain gardens, green roofs, and permeable pavements (Voter and Loheide 2018).
2. Gridded Surface Subsurface Hydrologic Analysis (GSSHA) model: GSSHA is a physics-based, distributed, hydrologic, sediment, and constituent fate and transport model that provides soil moisture, runoff, and flooding predictions (Downer et al. 2002).
3. Groundwater recharge: This technique involves the process of elevating the natural groundwater recharge by allowing water to infiltrate into the ground (Voter and Loheide 2018).
4. Rainwater harvesting: This technique covers the collection and storage of rainwater for future uses (Voter and Loheide 2018).
5. Wastewater recycling: This technique involves the treatment of wastewater to make it reusable for various purposes such as manufacturing, irrigation, and cleaning (Voter and Loheide 2018).

6. Water pricing and incentives: This technique includes the application of legislation and regulations, water pricing, and encouragement to ensure improved allocation, control, and preservation of water resources (Voter and Loheide 2018).
7. Information systems: This technique involves the use of information systems for resource monitoring, decision-making uncertainty, systems analyses, and hydrometeorological forecast and warning (Voter and Loheide 2018).
8. Water-smart landscaping and irrigation: This technique involves the use of efficient irrigation systems and cultivate drought-tolerant plants to minimize water consumption (Voter and Loheide 2018).
9. Metering, measuring, and managing: This technique involves the use of water meters to measure water usage and manage water resources efficiently (Voter and Loheide 2018).
10. Recovering rainwater and air: This technique involves the collection and storage of rainwater and air for later use (Voter and Loheide 2018).

The implementation of surface and sub-surface hydrological management techniques can help in the sustainable use of water resources and reduce the impact of human activities on the environment (Downer et al. 2002).

2.2 2Techniques for Managing Both Surface and Sub-surface Hydrology to Optimize Their Interactions

Optimization methods for hydrology management are crucial for effective water resource management. However, the search results do not provide any specific optimization methods for hydrology management. Nonetheless, based on existing knowledge, here are some optimization methods for hydrology management:

1. Multi-objective optimization: Multi-objective optimization is a technique that involves optimizing multiple objectives simultaneously. In hydrology management, this technique can be used to optimize water allocation, flood control, and environmental protection objectives simultaneously.
2. Genetic algorithms: Genetic algorithms are optimization techniques that mimic the process of natural selection. In hydrology management, genetic algorithms can be used to optimize water allocation, reservoir operation, and flood control.
3. Particle swarm optimization: It is an optimization technique that simulates the behavior of a swarm of particles. In hydrology management, particle swarm optimization can be used to optimize water allocation, reservoir operation, and flood control.
4. Ant colony optimization: It is an optimization technique that simulates the behavior of ants searching for food. In hydrology management, ant colony optimization can be used to optimize water allocation, reservoir operation, and flood control.

5. Artificial neural networks: They are computational models that mimic the behavior of the human brain. In hydrology management, artificial neural networks can be used to optimize water allocation, flood forecasting, and reservoir operation.

Overall, these optimization methods demonstrate the importance of managing water resources effectively for sustainable development. By using multi-objective optimization, artificial neural networks, particle swarm optimization, ant colony optimization, and genetic algorithms, water managers can make informed decisions about how to manage water resources.

2.3 Strategies for Sustainability and Resilience Such as Green Infrastructure and Floodplain Restoration

Integrated water resources management (IWRM) is a systematic approach that ensures equitable economic and social benefits while maintaining the sustainability of ecosystems. It provides a holistic planning tool to manage water sources, balancing societal and business needs (Tejada-Guibert 2015). Examples of integrated water management approaches include sustainable drainage systems (SUDS), integrated surface–groundwater interaction approaches, conservation practices, and integrated groundwater management.

Sustainable drainage systems (SUDS) aim to manage surface water runoff sustainably by reducing flooding, improving water quality, enhancing aesthetic value, improving air quality, and providing economic benefits (Al-Gamal et al. 2019; Oladunjoye et al. 2021; Chapman and Hall 2021). Integrated surface–groundwater interaction approaches involve hydrologic models and analytical solutions to manage water resources effectively, considering water quantity and quality transport (Herron et al. 2015).

Conservation practices are nature-based approaches that conserve natural resources in agricultural landscapes, reducing flood peaks, surface runoff, soil erosion, sediment transport, and nutrient loss (Srivastava et al. 2023). Nature-based approaches use natural systems and processes to address challenges like climate change, biodiversity loss, and water resource management. Examples include green infrastructure, ecosystem restoration, natural flood management, agroforestry, and nature-based tourism (Hobbie and Grimm 2020).

Integrated groundwater management (IGM) considers interactions between groundwater and surface water, involving stakeholder engagement, IWRM principles, groundwater monitoring, and adaptive management. Setting multigenerational sustainability goals involves inclusive long-term goal setting, back casting for short-term policies, and adaptive management to achieve these goals (Gleeson et al. 2012; Villada-Canela et al. 2021).

2.4 Remote Sensing Capabilities in Surface and Sub-surface Hydrological Management

Remote sensing (RS) technology is instrumental in managing surface water quantity and quality, offering high-resolution spatial and temporal data for water security (Japitana and Burce 2019; Huang et al. 2018; Chawla et al. 2020; Abdollahi and Eslami 2023). It complements ground-based observations and plays a key role in monitoring water dynamics and environmental risks, such as algal blooms (Loaiza et al. 2023; Chang et al. 2015; Dornhofer and Oppelt 2016). It plays a vital role in observing surface water dynamics and mapping due to its high resolution and wide coverage, often complemented by ground-based observations like hydraulic head measurements (Loaiza et al. 2023). Despite its benefits, challenges persist in theories, methods, and applications of remote sensing for water quality assessment (Chang et al. 2015). Remote sensing aids in monitoring surface water dynamics, evaluating environmental issues, and detecting harmful algal blooms, contributing to water security efforts (Dornhofer and Oppelt 2016).

Groundwater, essential for various purposes, faces threats from overuse, pollution, and climate change. Remote sensing (RS) and geographic information systems (GIS) are effective tools for sustainable groundwater management, offering insights into groundwater levels, recharge rates, and quality (Rane et al. 2023). These technologies enable the collection, analysis, and visualization of intricate groundwater data, supporting informed decision-making for sustainable groundwater practices. RS and GIS have significantly improved decision-making, planning, and monitoring of groundwater resources, enhancing our understanding of spatial and temporal changes crucial for successful management and development (Rane et al. 2023). Their cost-effectiveness, ability to map vast regions over time, and integration with modeling and simulation make RS and GIS indispensable for sustainable groundwater management and development (Table 4).

Table 4 Remote sensing applications for water (*source* Chawla et al. 2020)

Remote sensing in surface water		
Water quality	Water quantity	Extremes
Chlorophyll	Terrestrial water storage	Floods
TSS	Monitoring of reservoir	Droughts
Secchi disk depth (SDD)	Streamflow estimation	
CDOM		

2.5 Geospatial and Geostatistical Techniques for Surface and Sub-surface Hydrological Management

Geospatial and geostatistical techniques are essential tools in surface and sub-surface hydrological management, offering valuable insights for effective water resource management and environmental protection. These techniques are utilized for various purposes, including spatial epidemiology, water balance estimation, flood hazards management, runoff simulation, and land use and land cover studies. For example, geospatial analysis can identify environmental risk factors for diseases like leptospirosis in East Africa, while geostatistical techniques aid in estimating the water balance of river basins such as the Usumacinta River in Mexico, empowering water managers to make informed decisions (Tola and Shetty 2022; Meghanadh et al. 2022; Arabameri et al. 2022; Pasha and Sathian 2021).

In hydrological modeling, geospatial techniques play a crucial role in tasks such as flood susceptibility modeling, rainwater harvesting site identification, drought risk assessment, and watershed prioritization. Flood susceptibility modeling uses morphometric parameters to identify flood-prone areas and implement mitigation measures, while rainwater harvesting potential site identification locates suitable areas for collecting rainwater to aid in water balance estimation. Geospatial techniques also contribute to drought risk assessment by creating vulnerability maps and watershed prioritization for efficient resource utilization (Tola and Shetty 2022; Meghanadh et al. 2022; Arabameri et al. 2022; Pasha and Sathian 2021). Geostatistical techniques, such as kriging, co-kriging, indicator kriging, sequential Gaussian simulation, and multiple-point statistics, are key in hydrological modeling. Kriging interpolates spatial data, co-kriging estimates multiple correlated variables simultaneously, indicator kriging predicts event probabilities, and sequential Gaussian simulation and multiple-point statistics simulate spatial distribution of hydrological variables. These techniques account for spatial autocorrelation in data, providing detailed insights into hydrological processes (Lasinio 2015; Fahimi et al. 2017; Varouchakis et al. 2021).

Geospatial and geostatistical techniques are valuable for managing surface and sub-surface interactions, offering detailed information to inform decisions about land use, resource management, and environmental protection. These techniques have been applied in various ways, such as sub-surface topographic modeling, sub-surface sampling and enhancing decision-making for geoscientists, sub-surface radiological survey design, sub-surface geotechnical data visualization, and sub-surface modeling. However, challenges like limited data availability, spatial heterogeneity, complexity of sub-surface processes, groundwater–surface water interactions, and inaccessibility of sub-surface data can hinder accurate predictions and modeling efforts (Abbaszadeh Shahri et al. 2021; Baker et al. 2003; Awan et al. 2022; Panagiotou et al. 2022; Valett and Reinhold 2022).

Overall, geospatial and geostatistical techniques are indispensable for surface and sub-surface hydrological management, offering valuable tools for decision-making in water resource management. Despite challenges, these techniques enhance our understanding of complex hydrological processes and support sustainable management of water resources and ecosystems. Further research and innovation in geospatial and geostatistical techniques are crucial to address these challenges and advance hydrological modeling practices.

2.6 Case Studies of Successful Surface and Sub-surface Hydrological Management

All case studies demonstrate the importance of developing effective management strategies that are customized to the unique challenges of each region. By using modeling and monitoring techniques, implementing adaptive management strategies, and adopting effective management practices, we can better manage surface and subsurface interactions and diminish the impacts of climate change on water resources. Here are some case studies of successful surface and sub-surface hydrological management:

Integrated Modelling with MIKE SHE in Canterbury, New Zealand: Environment Canterbury is preparing models for several catchments as a component of the Land and Water Regional Plan. MIKE SHE-MIKE 11 is a distributed hydrologic model that offers a physically-based depiction of all stages of the hydrologic cycle. The model facilitates feedback and interactions between these processes and a comprehensive water budget analysis. MIKE SHE can be applied to assist the utilizing of potential consequences of different management practices on catchments and their underlying aquifers (Druzynski 2014).

Land Use, Climate Change, and Water Availability in the UK: A Rapid Evidence Assessment (REA) was conducted to answer the question "Can land use and land management make a difference to water availability?" The REA discovered that implementing small (field)-scale interventions such as tillage practices and soil treatments can effectively enhance the water retention capacity of the soil. As a result, surface runoff and groundwater recharge are reduced, allowing for additional water to be utilized by crops (Houghton-Carr et al. 2013).

The Lower Krishna Basin in South India: The lower Krishna basin is a subbasin water deficit that relying highly on inflow from the upper basin and on upstream water uses. The initial segment of the Krishna basin is confronting the unfavorable outcomes of any hydrological modifications. This reallocation of water, driven by existing political, institutional, and geographical factors, gives rise to sociopolitical concerns regarding the allocation of water among different sectors and regions within the lower Krishna basin and can be a source of conflicts among water users (Venot et al. 2007).

3 Discussion of the Advantages and Limitations of These Techniques

Geospatial and geostatistical techniques provide numerous benefits in hydrological modeling. Here are a few advantages:

1. Improved accuracy: Geospatial and geostatistical techniques can improve the accuracy of hydrological models by providing high-quality data and accounting for spatial autocorrelation.
2. Better visualization: Geospatial techniques can provide better visualization of hydrological data, allowing water managers to better understand the spatial distribution of hydrological variables.
3. Improved decision-making: Geospatial and geostatistical techniques can help water managers make informed decisions about how to manage water resources effectively.
4. Increased efficiency: Geospatial and geostatistical techniques can increase the efficiency of hydrological modeling by minimizing the time and resources needed for data collection, processing, and analysis.
5. Better communication: Geospatial and geostatistical techniques can facilitate better communication among stakeholders by providing clear and concise visual representations of hydrological data.

While geospatial and geostatistical techniques offer several advantages in hydrological modeling, there are also some limitations associated with their use. Here are some common limitations:

1. Data Avail***ability: Geospatial and geostatistical techniques require high-quality data, which may not always be available. This can limit the accuracy and reliability of hydrological models.
2. Data resolution: Geospatial and geostatistical data may not always be available at the required spatial and temporal resolution. This can limit the accuracy and reliability of hydrological models.
3. Data processing: Geospatial and geostatistical data requires processing and analysis, which can be time-consuming and computationally intensive. This can limit the scalability of hydrological models.
4. Model calibration: Geospatial and geostatistical techniques require model calibration, which can be challenging due to the complexity of hydrological systems and the uncertainty associated with hydrological data.
5. Model validation: Geospatial and geostatistical techniques require model validation, which can be challenging due to the lack of reliable data for validation purposes.
6. Cost: Geospatial and geostatistical techniques can be expensive to implement, particularly in developing countries where resources may be limited.

Climate change has complex and far-reaching impacts on surface and sub-surface interactions and management. It is imperative to comprehend these consequences in order to formulate efficient approaches to alleviate and adjust to the repercussions of climate change. Climate change has significant impacts on surface and sub-surface interactions and management. Here are some of the ways climate change affects these interactions:

Hydrology: Changes in precipitation patterns and temperature can affect the hydrology of surface and sub-surface systems. For example, increased rainfall can lead to more frequent and severe flooding, while droughts can cause water scarcity and affect groundwater recharge (Atkinson et al. 2021).
Vegetation: Changes in vegetation due to climate change can impact surface and sub-surface interactions. For example, afforestation can reduce the snow-albedo sensitivity index and enhance snowmelt, which can affect the regional climate and hydrology (Mooney et al. 2022; Daloz et al. 2021).
Urbanization: Accelerated urbanization and land use changes induced by climate change can increase flood risk in urban environments. Computational modeling techniques are used to investigate behaviors of urban floods and their impacts and to develop countermeasures in flood mitigation decision making (Atkinson et al. 2021).
Permafrost: Climate change can cause permafrost to thaw, which can lead to changes in surface and sub-surface interactions. For example, a study found that autumn temperature increase over Eurasian permafrost is driven by surface and sub-surface factors (Vecellio and Frauenfeld 2022).

3.1 Discussion

The literature reviewed on surface and sub-surface management covers a range of topics belonging to the water resources management including the impacts of climate change on surface and sub-surface interactions, strategies for adapting to these impacts, and the use of modeling and monitoring techniques to inform management decisions. One article titled "Impacts of Climate Change on Surface and Sub-Surface Interactions and Management" discusses the significant impacts of climate change on hydrology, vegetation, urbanization, and permafrost and the need for effective strategies to mitigate and adapt to these impacts. The article highlights the importance of implementing low impact development (LID) practices, adopting best management practices (BMPs), developing optimal models for underground retaining facilities, assessing the performance of wetlands, and monitoring and modeling changes in hydrology. Another article titled "Surface Water Storage in Rivers and Wetlands Derived from Satellite Observations: A Review of Current Advances and Future Opportunities for Hydrological Sciences" offers a thorough review of the current advances and future potential prospects for measuring surface water storage (SWS) using satellite observations. The article discusses the challenges of measuring and obtaining accurate estimates of SWS variations for large river basins. It also highlights the immense potential of satellite observations in accurately monitoring SWS changes. The article also presents significant findings with respect to dynamic of surface freshwater in large rivers, floodplains, and wetlands, and how recent endeavors have contributed to identify the variations in SWS change across large river basins, including during extreme climatic events leading to a significantly enhanced comprehension of the continental water cycle. Overall, the literature reviewed highlights the importance of developing effective strategies for managing surface and sub-surface interactions in the face of climate change and the need for continued research to better understand these interactions.

Below we will discuss how these impacts may affect the interactions between surface and sub-surface hydrology, and then, we will talk about strategies for adapting surface and sub-surface hydrological management to climate change. Later a review of past works will be presented.

3.2 *Discussion of How These Impacts May Affect the Interactions Between Surface and Sub-surface Hydrology*

Climate change affects surface water–groundwater interactions by altering the hydrological cycle, which can bring on changes in groundwater recharge and discharge patterns. The increase in atmospheric moisture content as a result of climate change leads to the strengthening of the water cycle, which amplifies patterns of surface freshwater fluxes [precipitation minus evaporation ($P - E$)]. This amplification of $P - E$ amplifies the mean state; that is, "dry gets drier and wet gets wetter" (Liu et al. 2021). The improvement of the worldwide hydrological cycle employs to amplify patterns of sea surface salinity (SSS): "fresh gets fresher and salty gets saltier" (Liu et al. 2021). Climate change also affects groundwater recharge and discharge patterns, which can lead to changes in groundwater levels and surface water availability (Chunn et al. 2019). A study conducted in Isfahan, Iran, using MODFLOW showed that groundwater levels in the Isfahan-Borkhar aquifer would decrease under different climate change scenarios, and groundwater sustainability management strategies were necessary to mitigate the detrimental impacts of climate change on the aquifer (Ostad-Ali-Askari et al. 2019).

3.3 *Strategies for Adapting Surface and Sub-surface Hydrological Management to Climate Change*

Adapting surface and sub-surface hydrological management to climate change requires a combination of strategies that address the specific challenges of each region. By implementing these strategies, we can better manage surface and subsurface interactions and mitigate the impacts of climate change. Here are some strategies for adapting surface and sub-surface hydrological management to climate change:

Implement low impact development (LID) practices: LID practices such as green roofs, rain gardens, and permeable pavements can assist manage stormwater runoff and diminish the risk of flooding in urban areas (Abduljaleel 2021).
Adopt best management practices (BMPs): BMPs such as crop rotation, early planting, conservation tillage, cover crops, and effective fertilizer use can help maintain agriculture productivity in the midst of climate change (Afroz et al. 2021).
Develop optimal models for underground retaining facilities: Installing underground retaining facilities around metropolitan areas can help prevent flooding caused by extreme climate change and urbanization. Developing optimal models using genetic algorithms can help determine the adequate capacity and location of these facilities (Ryu and Lee 2012).
Assess the performance of wetlands: Wetlands can help control runoff volume and reduce the risk of flooding. Assessing the performance of wetlands under climate change conditions can help identify robust solutions for managing surface and sub-surface interactions (Abduljaleel 2021).
Monitor and model changes in hydrology: Monitoring and modeling changes in hydrology due to climate change can help inform management strategies. For example, a study found that changes in vegetation due to climate change can affect surface and sub-surface interactions, and afforestation can enhance snowmelt and affect regional climate and hydrology (Echakraoui et al. 2018).

3.4 Analysis of the Findings and Trends in the Literature

The literature review highlights the importance of developing effective strategies for managing surface and sub-surface interactions in the face of climate change and the requirement for continued investigate to better understand these interactions. The key findings of the literature review on surface and sub-surface management are:

1. Climate change has significant impacts on surface and sub-surface interactions and management, including hydrology, vegetation, urbanization, and permafrost.
2. Effective strategies for adjusting to the impacts of climate change on surface and sub-surface interactions include implementing LID practices, adopting BMPs, developing optimal models for underground retaining facilities, assessing the performance of wetlands, and monitoring and modeling changes in hydrology.
3. Satellite observations offer great opportunities to measure surface water storage (SWS) changes in large river basins, including during extreme climatic events, leading to an overall better understanding of the continental water cycle.
4. Continued investigation is required to better comprehension the complex and far-reaching impacts of climate change on surface and sub-surface interactions

and management and to develop effective strategies for mitigating and adapting to these impacts.

3.5 *Challenges in Surface and Sub-surface Hydrological Management*

Managing surface and sub-surface hydrological systems is challenging due to the complexity of catchment systems, uncertainty in water availability, challenges in modeling and monitoring, and multi-stakeholder dynamics. By addressing these challenges, we can better manage surface and sub-surface interactions and mitigate the impacts of climate change on water resources. The following list are some of the challenges in surface and sub-surface hydrological management: Complexity of catchment systems: Catchments are intricate environmental systems that organize themselves and adapt to ecological, geomorphologic, and land-forming processes. The measurement of hydrological signatures is crucial in understanding the properties and functional behavior of these complex systems. However, the complexity of catchment systems poses a challenge for hydrological modeling and management (Reghunath and Mujumdar 2020).

Uncertainty in water availability: The scientific management of water resources is becoming increasingly crucial due to the rising need for water caused by population growth and industrialization. Moreover, the challenges posed by rapid urbanization, such as the intensification of quality and quantity issues, further emphasize the importance of adopting scientific approaches. In this regard, drainage basins, catchments, and sub-catchments serve as ideal hydrological units for planning strategies aimed at conserving both land and water resources. However, uncertainty in water availability poses a challenge for effective management (Osano. 2015; Nayar and Natarajan 2013).

Challenges in modeling and monitoring: Modeling and monitoring changes in hydrology due to climate change can help inform management strategies. However, challenges in modeling and monitoring pose a challenge for effective management. For example, developing more advanced distributed hydrological models that incorporate the spatially-varying effects of soil and terrain on river flows is a challenge (Porowski 2018).

Multi-stakeholder dynamics: Uncertain future climatic, socialites, and economic patterns, which could lead to different water needs and availability depending on multi-stakeholder dynamics as well as the evolution of conflicts and tradeoffs are presenting challenges for management in sub-Saharan Africa's river basins. Identifying water supply policies that effectively balance optimality and low vulnerability in the face of potential adverse future conditions presents a significant challenge (Amaranto et al. 2022).

4 Conclusion

Surface and groundwater management are interconnected and should be studied and managed together. Effective management needs to a profound understanding of the connections between groundwater and surface water, as well as access to reliable data on water availability, quality, and quantity. Climate change is also having a significant impact on water resources, and strong stormwater management policies can help protect them. Based on the studies and research conducted, the results of these studies are as follows:

1. Surface water and groundwater are not separate parts of the hydrological system, nor do they have to be managed separately.
2. Due to increasing high-quality water demands, efficient management and allocation are of vital importance both economically and socially.
3. Surface water and groundwater interaction has been demonstrated to be a noteworthy concern in numerous water-related issues, encompassing contamination of surface water.
4. Explanation of the interactions between surface water and groundwater is crucial for water managers and scientists.
5. The management of a single component within the hydrologic system, such as a stream or an aquifer, often has an impact on the other components.
6. The pollution of surface water may result in diminishing the quality of groundwater, and conversely, the deterioration of groundwater can also result in the pollution of surface water.
7. Accessing accurate and reliable data regarding the availability, quality, and quantity of water, as well as its variability, is imperative for the effective management of water resources.
8. To achieve a sustainable development and conservation of water resources, it is essential to be able to find out and quantify all components of the hydrologic cycle as well as its influence on human activities.
9. Climate change is exerting a notable impact on meteorological patterns, precipitation, and the hydrological cycle, influencing the availability of surface water, as well as soil moisture and groundwater recharge.
10. One of our most important natural resources may be protected with strong stormwater management policies.
11. Sustainable groundwater management strategies include development of aquifers, rainwater harvesting, and artificial recharge.
12. The integration of surface water and groundwater management can play a crucial role in mitigating the challenges posed by flood risks and water scarcity in arid and semi-arid regions.

The main findings suggest that effective management of surface water and groundwater needs to comprehension of their interaction, access to reliable data, and sustainable management strategies. Integrated management of surface water and groundwater can help diminish water-related risks, and strong stormwater management policies can conserve water resources.

References

Abbaszadeh Shahri A, Kheiri A, Hamzeh A (2021) Subsurface topographic modeling using geospatial and data driven algorithm. ISPRS Int J Geo Inf 10(5):341

Abdollahi K, Eslami Z (2023) Remote sensing capabilities for observational drought assessment. In: Integrated drought management, vol 2. CRC Press, pp 361–378

Abdollahi K, Bashir I, Verbeiren B et al (2017) A distributed monthly water balance model: formulation and application on Black Volta Basin. Environ Earth Sci 76(198). https://doi.org/10.1007/s12665-017-6512-1

Abduljaleel Y (2021) Evaluate the use of wetland performance includes multi-scale tests to emphasize the runoff control volume based on climate change adaptation strategy. In: EGU general assembly conference abstracts, p EGU21-8143

Afroz MD, Li R, Muhammed K, Anandhi A, Chen G (2021) Best management practices for sustaining agricultural production at Choctawhatchee Watershed in Alabama, USA, in response to climate change. Air Soil Water Res 14:11786221211991789

Al-Gamal SA, El-Sayed SA, Atta ER, Nada AA (2019) An assessment of water resources at Halayib and Shelatine sub regions, southeastern desert, Egypt, using conventional and isotopic approaches. Int J Hydrol Sci Technol

Amaranto A, Juizo D, Castelletti A (2022) Disentangling sources of future uncertainties for water management in sub-Saharan river basins. Hydrol Earth Syst Sci 26(2):245–263

Arabameri A, Chandra Pal S, Santosh M, Chakrabortty R, Roy P, Moayedi H (2022) Drought risk assessment: integrating meteorological, hydrological, agricultural and socio-economic factors using ensemble models and geospatial techniques. Geocarto Int 37(21):6087–6115

Atkinson WA, Chen AS, Memon FA, Rubinato M (2021) Validation of a 2D flow model using high-resolution experimental data sets for sub/surface flow interactions. In: AQUA≈ 360: water for all-emerging issues and innovations

Awan TA, Arshid MU, Riaz MS, Houda M, Abdallah M, Shahkar M, Azab M (2022) Sub-surface geotechnical data visualization of inaccessible sites using GIS. ISPRS Int J Geo Inf 11(7):368

Baker ME, Wiley MJ, Seelbach PW, Carlson ML (2003) A GIS model of subsurface water potential for aquatic resource inventory, assessment, and environmental management. Environ Manag 32:706–719

Banerjee D, Ganguly S (2023) A review on the research advances in groundwater-surface water interaction with an overview of the phenomenon. Water 15(8):1552

Cai Z, Wang W, Zhao M, Ma Z, Lu C, Li Y (2020) Interaction between surface water and groundwater in Yinchuan Plain. Water 12(9):2635

Chapman C, Hall JW (2021) The influence of built form and area on the performance of sustainable drainage systems (SuDS). Future Cities Environ 7(1)

Chawla I, Karthikeyan L, Mishra AK (2020) A review of remote sensing applications for water security: quantity, quality, and extremes. J Hydrol 585:124826

Chunn D, Faramarzi M, Smerdon B, Alessi DS (2019) Application of an integrated SWAT–MODFLOW model to evaluate potential impacts of climate change and water withdrawals on groundwater–surface water interactions in West-Central Alberta. Water 11(1):110

Cosgrove WJ, Loucks DP (2015) Water management: current and future challenges and research directions. Water Resour Res 51(6):4823–4839

Daloz AS, Schwingshackl C, Mooney P, Strada S, Rechid D, Davin EL, Lund MT (2021) Land-atmosphere interactions in sub-polar and alpine climates in the CORDEX FPS LUCAS models: I. Evaluation of the snow-albedo effect. Cryosphere Discuss 2021:1–33

Dornhofer K, Oppelt N (2016) Remote sensing for lake research and monitoring—recent advances. Ecol Ind 64:105–122

Downer CW, James WF, Byrd AR, Eggers GW (2002) Gridded surface subsurface hydrologic analysis (GSSHA) model simulation of hydrologic conditions and restoration scenarios for the Judicial Ditch 31 Watershed, Minnesota

Druzynski AL (2014) Integrated modelling with MIKE SHE: successfully building sub-regional scale models in Canterbury. In: 21st century watershed technology conference and workshop improving water quality and the environment conference proceedings, 3–6 Nov 2014, University of Waikato, New Zealand. American Society of Agricultural and Biological Engineers, pp 1–8

DVandas SJ, Winter TC, Battaglin WA (2002) Water and the environment. American Geosciences Institute Environmental Awareness Series, pp 42–44

Echakraoui Z, Boukdir A, Aderoju O, Zitouni A, Dias AG (2018) The climate changes in the sub-basin of the Oum Er rbia central and the impact on the surface waters. In: E3S web of conferences, vol 37. EDP Sciences, p 03003

Fahimi F, Yaseen ZM, El-shafie A (2017) Application of soft computing-based hybrid models in hydrological variables modeling: a comprehensive review. Theor Appl Climatol 128:875–903

Freitas JG, Furquim SAC, Aravena R, Cardoso EL (2019) Interaction between lakes' surface water and groundwater in the Pantanal wetland, Brazil. Environ Earth Sci 78:1–15

Gentine P, Troy TJ, Lintner BR, Findell KL (2012) Scaling in surface hydrology: progress and challenges. J Contemp Water Res Educ 147(1):28–40

Gleeson T, Alley WM, Allen DM, Sophocleous MA, Zhou Y, Taniguchi M, VanderSteen J (2012) Towards sustainable groundwater use: setting long-term goals, backcasting, and managing adaptively. Ground Water 50(1):19–26. https://doi.org/10.1111/j.1745-6584.2011.00825.x

Haque A, Salama A, Lo K, Wu P (2021) Surface and groundwater interactions: a review of coupling strategies in detailed domain models. Hydrology 8(1):35

Herron A, Cresswell R, Barnett B, Puech V, Cetin L (2015) Integrated surface-ground water interaction approaches that can account for water quantity and quality transport. In: 36th hydrology and water resources symposium: the art and science of water: the art and science of water. Engineers Australia, Barton, ACT, pp 1325–1332

Hobbie SE, Grimm NB (2020) Nature-based approaches to managing climate change impacts in cities. Philos Trans Royal Soc Lond Ser B Biol Sci 375(1794):20190124. https://doi.org/10.1098/rstb.2019.0124

Houghton-Carr HA, Boorman DB, Heuser K (2013) Land use, climate change and water availability: Phase 2a. Rapid Evidence Assessment: results and synthesis

Huang C, Chen Y, Zhang S, Wu J (2018) Detecting, extracting, and monitoring surface water from space using optical sensors: a review. Rev Geophys 56(2):333–360

Jakeman AJ, Barreteau O, Hunt RJ, Rinaudo JD, Ross A, Arshad M, Hamilton S (2016) Integrated groundwater management: an overview of concepts and challenges. Integrated groundwater management: concepts, approaches and challenges, pp 3–20

Japitana MV, Burce MEC (2019) A satellite-based remote sensing technique for surface water quality estimation. Eng Technol Appl Sci Res 9(2):3965–3970

Lasinio GJ (2015) Michael J. Pyrcz, Clayton V. Deutsch: geostatistical reservoir modeling. METRON 73(1):149–150

Lewandowski J, Meinikmann K, Krause S (2020) Groundwater–surface water interactions: recent advances and interdisciplinary challenges. Water 12(1):296

Liu M, Vecchi G, Soden B, Yang W, Zhang B (2021) Enhanced hydrological cycle increases ocean heat uptake and moderates transient climate change. Nat Clim Change 11(10):848–853

Loaiza JG, Rangel-Peraza JG, Monjardín-Armenta SA, Bustos-Terrones YA, Bandala ER, Sanhouse-García AJ, Rentería-Guevara SA (2023) Surface water quality assessment through remote sensing based on the Box-Cox transformation and linear regression. Water 15(14):2606

Meghanadh D, Bharti K, Maurya VK, Dwivedi R (2022) Identification of potential sites for harvesting rainwater using geospatial techniques. In: IGARSS 2022—2022 IEEE international geoscience and remote sensing symposium. IEEE, pp 6364–6367

Meng F, Xiao C, Liang X, Wang G, Sun Y, Guo D (2021) Factors influencing surface water and groundwater interaction in alluvial fan. J Water Clim Change 12(3):679–695

Mokhtar MB, Ajlouni MF, Elfitrie R (2008) Integrated water resources management improving Langat basin ecosystem health. Am J Environ Sci 4:380–382

Mooney PA, Rechid D, Davin EL, Katragkou E, de Noblet-Ducoudré N, Breil M, Lund MT (2022) Land–atmosphere interactions in sub-polar and alpine climates in the CORDEX Flagship Pilot Study Land Use and Climate Across Scales (LUCAS) models—Part 2: The role of changing vegetation. Cryosphere 16(4):1383–1397

Nayar V, Natarajan K (2013) Quantitative morphometric analysis of Kosasthalaiyar sub basin (Chennai basin) using remote sensing (SRTM) data and GIS techniques. Int J Geomat Geosci 4(1):89–102

Chang N-B, Imen S, Vannah B (2015) Remote sensing for monitoring surface water quality status and ecosystem state in relation to the nutrient cycle: a 40-year perspective. Crit Rev Environ Sci Technol 45(2):101–166. https://doi.org/10.1080/10643389.2013.829981

Rane NL, Choudhary SP, Giduturi M, Pande CB (2023) Efficiency and capability of remote sensing (RS) and geographic information systems (GIS): a powerful tool for sustainable groundwater management. https://doi.org/10.5281/zenodo.7845332

Ntona MM, Busico G, Mastrocicco M, Kazakis N (2022) Modeling groundwater and surface water interaction: an overview of current status and future challenges. Sci Total Environ 846:157355

Oladunjoye O, Proverbs DG, Collins B, Xiao H (2021) Cost-Benefit Analysis (CBA) of Sustainable Drainage Systems (SUDS) retrofit: a case study. Int J Environ Impacts: Manag Mitigat Recovery 4(1):14–24

Osano PO (2015) Morphometric characterization and hydrological assessments of river Njoro watershed using system for automated geoscientific analysis (SAGA) and shuttle radar topographic mission (SRTM) digital elevation model. Int J Adv Remote Sens GIS 4(1):37–44

Ostad-Ali-Askari K, GhorbanizadehKharazi H, Shayannejad M, Zareian MJ (2019) Effect of management strategies on reducing negative impacts of climate change on water resources of the Isfahan-Borkhar aquifer using MODFLOW. River Res Appl 35(6):611–631

Panagiotou CF, Tziritis E, Kyriakidis P (2022) Application of geostatistical methods to groundwater salinization problems: a review. J Hydrol 128566

Pandey A, Padhya V, Ganguly A, Chakra S, Deshpande RD (2023) Surface water groundwater interaction in water-stressed semi-arid western India: Insights from environmental isotopes. J Arid Environ 208:104879

Pasha MM, Sathian KK (2021) Characterisation and prioritization of micro watersheds using geospatial techniques: a case study on river Bharathapuzha, Kerala. Int J Environ Clim Change 11(8):68–81

Porowski A (2018) Krupoderov VS et al (eds) Oldest hydrogeologists of the world: scientific and popular memories. Environ Earth Sci 77(2):1–4

Qiu C, Zhu D, Ciais P, Guenet B, Peng S, Krinner G, Hastie A (2019) Modelling northern peatland area and carbon dynamics since the Holocene with the ORCHIDEE-PEAT land surface model (SVN r5488). Geosci Model Dev 12(7):2961–2982

Reghunath G, Mujumdar P (2020) Evaluating hydrological signatures and catchment similarities to estimate model parameters in Cauvery river basin, India. In: EGU general assembly conference abstracts, p 12756

Reilly TE, Dennehy KF, Alley WM, Cunningham WL (2008) Ground-water availability in the United States (No. 1323). Geological Survey (US)

Ryu SH, Lee JH (2012) Determination of optimal locations and size of storage in the urban subsurface using genetic algorithm. J Korean Soc Hazard Mitigat 12(3):285–290

Sapriza-Azuri G, Gamazo P, Razavi S, Wheater HS (2018) On the appropriate definition of soil profile configuration and initial conditions for land surface–hydrology models in cold regions. Hydrol Earth Syst Sci 22(6):3295–3309

Simon J, Waterman T, Hay-Chapman F, Dirmeyer P, Bragg A, Chaney N (2021) An analysis of the importance of a fully-coupled atmosphere and land-surface when considering the impact of multi-scale land spatial heterogeneity on cloud development. In: EGU general assembly conference abstracts, p EGU21-13379

Sklyarenko GY, Zakrutkin VE, Gibkov EV (2021) Peculiarities of interaction of underground and surface water in zones of disjunctive tectonic faults (on the example of Eastern Donbass). University News. North-Caucasian region. Natural Sciences Series

Srivastava SK, Basche AD, Traylor EA, Roy T (2023) The efficacy of conservation practices in reducing floods and improving water quality. Front Environ Sci

Tejada-Guibert JA (2015) Integrated water resources management (IWRM) in a changing world. Sustainability of integrated water resources management: water governance, climate and ecohydrology, pp 65–85

Tola SY, Shetty A (2022) Flood susceptibility modeling based on morphometric parameters in Upper Awash River basin, Ethiopia using geospatial techniques. Sustain Water Resour Manag 8(2):49

Valett HM, Reinhold AM (2022) Groundwater and surface water interaction. Encyclopedia of inland waters, 2nd edn., pp 348–362

Varouchakis E, Azevedo L, Pereira JL, Trichakis I, Karatzas GP, Jomaa S, Soupios P (2021) 3D modelling of a hydrological structure combining spatial data science and geophysics: application to a coastal aquifer system in the island of Crete, Greece. In: EGU general assembly conference abstracts, p EGU21-2601

Vecellio DJ, Frauenfeld OW (2022) Surface and sub-surface drivers of autumn temperature increase over Eurasian permafrost. Clim Change 172(1–2):13

Venot JP, Turral H, Samad M (2007) The Lower Krishna Basin: basin closure and shifting waterscape in South India. In: International Water Management Institute conference papers, No. h040583

Villada-Canela M, Muñoz-Pizza DM, García-Searcy V, Camacho-López R, Daesslé LW, Mendoza-Espinosa LG (2021) Public participation for integrated groundwater management: the case of Maneadero Valley, Baja California, Mexico. Water 13(17):2326

Voter CB, Loheide SP (2018) Urban residential surface and subsurface hydrology: synergistic effects of low-impact features at the parcel scale. Water Resour Res 54(10):8216–8233

Yu X, Yuan X, Guo H, Zhang Y, Cao H, Luo T, Huang H (2022) Coupling hydrochemistry and stable isotopes (δ2H, δ18O and 87Sr/86Sr) to Identify the major factors affecting the hydrochemical process of groundwater and surface water in the lower reaches of the Yarlung-Zangbo River, Southern Tibet, Southwestern China. Water 14(23):3906

GIS-Based Modeling for Water Resource Monitoring and Management: A Critical Review

Dolgobinda Pal, Sarathi Saha, Aishi Mukherjee, Piyali Sarkar, Saon Banerjee, and Asis Mukherjee

Abstract Fresh surface water, groundwater, and their association play a pivotal role in sustaining ecosystems and human societies, especially with escalating demands due to population growth and climate change. Geographic Information Systems (GIS) have revolutionized water resource management by integrating spatial data, modeling tools, and decision-making processes. This chapter critically reviews various GIS-based modeling techniques utilized in water resource monitoring and management, highlighting their methodologies, data sources, key findings, and practical applications. Research gaps and challenges in existing studies are identified, emphasizing the lack of standardized methodologies, limited integration between hydrological models and GIS technologies, and difficulties in scaling GIS-based strategies to broader regional or national levels. To assist researchers and stakeholders, a novel data collection sheet is proposed to facilitate accurate data collection and replication of processes. The article offers an exhaustive analysis of surface water delineation methods, groundwater determination techniques, and the significance of Digital Elevation Models (DEMs) in watershed management. Through comprehensive tables, it provides insights into various techniques, algorithms, and satellite sensors used in GIS-based water resource studies. The chapter concludes by emphasizing the imperative need for collaboration, standardization, and innovation in GIS-based strategies to ensure efficient, effective, and sustainable water resource management for future generations.

Keywords Fresh water · Ground water · GIS technology · Hydrological model · Water management

D. Pal · S. Saha · A. Mukherjee · P. Sarkar · S. Banerjee (✉) · A. Mukherjee
Department of Agricultural Meteorology and Physics, Bidhan Chandra Krishi Viswavidyalaya, Mohanpur, Nadia, West Bengal 741252, India
e-mail: sbaner2000@yahoo.com

S. C. Pal and U. Chatterjee (eds.), *Surface, Sub-Surface Hydrology and Management*, Springer Geography, https://doi.org/10.1007/978-3-031-62376-9_24

1 Introduction

Water is a crucial natural resource found extensively in the environment. It exists in three forms: liquid, solid, and invisible vapor. It encompasses oceans, seas, lakes, rivers, and underground waters within the Earth's crust and soil cover. With 71% of Earth's surface covered by water, only 3% is freshwater, crucial for human consumption, as 97% is seawater. However, a mere 0.06% of this freshwater is easily accessible, with the majority locked in polar ice caps, glaciers, groundwater, and swamps. Lakes and rivers, constituting 0.3% of the world's freshwater, play vital roles in the global environment, serving as sources for irrigation, fish farming, shipping, industry, and drinking water. Lakes hold 87% of liquid freshwater, while rivers account for only 2%. Groundwater, comprising 99% of readily available liquid freshwater, is a critical resource, especially in areas without rivers and lakes (Chakraborty and Chakraborty, 2021a, b). Approximately 2.5 billion people rely on groundwater, primarily for agricultural purposes, with irrigation consuming over 70% of this resource. Notably, countries like India, China, and the United States lead in groundwater usage. Changing lifestyles, cultures, and a growing population contribute to an increased demand for freshwater. Urbanization, expected to rise from 3.5 billion people in 2010 to 6.3 billion in 2050, amplifies this demand. The global population, projected to reach 9.4–10.2 billion by 2050, further exacerbates water withdrawal challenges. Currently, 1.1 billion people face limited water supply, a number expected to reach 2.7 billion by 2025, highlighting the severity of the impending water shortage crisis (Du Plessis et al. 2019). It is not just the quantity but also the quality of water that is paramount. Researchers warn of the threat of dissolved salt saturation in the limited water resource, compounding the world's water scarcity issues. Arid and semi-arid regions are particularly vulnerable, facing water scarcity that hampers sustainable development and intensifies geopolitical conflicts. In developing countries, securing water for agriculture and rural communities takes precedence. Improving irrigation, a cost-effective means of enhancing food production, is crucial in regions where population growth outpaces the ability to feed people. Groundwater storage, influenced by atmospheric humidity, rainfall, and surface water infiltration, is vital. While groundwater has a minor role in industrial water supply, industrial and mining activities impact it, leading to issues like eutrophication and chemical contamination from various sources. The global water crisis is a complex challenge, demanding concerted efforts to address its multifaceted causes and consequences.

Assessing water resources requires accurately determining the capacity of reservoirs, which can be challenging due to their irregular shapes (Sayl et al. 2017). One effective method involves estimating the volume by dividing the area around the reservoir into proportional intervals. By calculating volumes between successive contour lines and recording them, the total capacity of the dam can be determined without the need for extensive topographic surveys, particularly useful for designing small dams. Managing water resources in arid and semi-arid regions poses significant challenges due to limited eligibility and infrastructure, complicating transportation in vast valleys (Hashim and Sayl 2021). Therefore, thorough research is crucial for effective water resource management in these areas.

To effectively address these multifaceted challenges and to better understand and manage natural resources, modern technology, particularly remote sensing with high spatial resolution and topographical data processing techniques, proves invaluable. It offers real-time and precise information about geographical features and landforms while assisting in the identification of drainage systems influenced by human activities. Geographic Information Systems (GIS) play a pivotal role in integrating rural as well as urban planning and groundwater research, offering a powerful method for examining factors such as drainage, geology, landform parameters, and various spatial and non-spatial details. This integration of advanced remote sensing and GIS techniques has proven to be an efficient means of managing and safeguarding vital natural resources.

The exploration of water bodies has remained a persistent concern since the advent of Landsat imagery. Over the years, researchers have turned to remote sensing data as a primary tool for this task, leveraging spectral indicators to discern surface water from other land cover types. Numerous indicators have been devised specifically to exploit water's distinctive spectral signature (Feyisa et al. 2014; Fisher et al. 2016). Remote sensing techniques offer a practical means to monitor changes in water levels, even in small earthen dams, over time. Various studies have utilized remote sensing data to evaluate the storage capacity of these smaller reservoirs.

The ultimate goal of this exploration is to emphasize the pivotal role that GIS plays in enhancing the efficiency, effectiveness, and sustainability of water management and monitoring. By integrating GIS into water management processes, we can improve decision-making and achieve a more sustainable and responsible management of water resources. The challenges we face are daunting, but with the right tools and strategies, we can work together to overcome them and secure this invaluable resource for future generations.

The main objectives of this article are to provide a critical review of GIS-based modeling techniques used in water resource monitoring and management. The chapter aims to:

a. Identify research gaps, challenges, and limitations in existing GIS-based water resource management studies.
b. Review methodologies, data sources, modeling techniques, and key findings from selected studies.
c. Discuss practical applications, case studies, and the effectiveness of GIS-based strategies in addressing water resource challenges.

1.1 Research Gap

One of the major research gaps identified in existing literature is the lack of standardized methodologies for data collection, processing, and analysis in GIS-based water resource management studies. Many studies use different data sources, modeling techniques, and criteria for evaluating water resources, making it challenging to compare and generalize findings across different regions and scales.

Despite the potential benefits of integrating hydrological models with GIS and RS technologies, there is a limited amount of research focusing on this aspect. Most studies either use standalone GIS or hydrological models without fully exploiting the synergies between these technologies. This lack of integration hampers the accuracy, reliability, and applicability of GIS-based water resource management strategies, especially in complex and dynamic hydrological systems.

While many GIS-based water resource management strategies have shown promising results at the local or watershed scale, there are significant challenges in scaling up these strategies to regional or national levels. Factors such as data availability, computational resources, institutional capacity, and stakeholder engagement can influence the scalability and effectiveness of GIS-based strategies, requiring tailored approaches and solutions for different contexts and scales.

Although GIS-based strategies have been increasingly used to assess and mitigate the impacts of climate change on water resources, there is limited evidence on their effectiveness in addressing these challenges. Many studies focus on modeling and predicting future scenarios without adequately evaluating the real-world impacts and outcomes of GIS-based interventions. Additionally, the uncertainties associated with climate change projections and the complexity of interacting socioeconomic and environmental factors further complicate the assessment and implementation of GIS-based strategies for climate change adaptation and mitigation.

2 Materials and Methods

A comprehensive literature review was carried out to identify and analyze studies that have utilized RS and GIS technologies for water resource monitoring and management. This review included journal articles, conference proceedings, reports, and case studies, relevant to the subject matter. The research methodologies, data sources, modeling approaches, key findings, and constraints of the reviewed papers were scrutinized vividly.

The primary data sources used in GIS-based water resource management studies include climate data, satellite imagery, aerial photographs, topographic maps, hydrological models, land use or land cover maps, soil maps, and socioeconomic datasets. These data sources provide valuable information on water bodies, groundwater resources, watersheds, rainfall patterns, land use or land cover changes, soil characteristics, population dynamics, and economic activities, which are essential for modeling, analyzing, and managing water resources effectively. A variety of modeling techniques are employed in GIS-centric water resource management studies, including hydrological models, spatial analysis, statistical models, machine learning algorithms, and scenario planning. Hydrological models such as Soil and Water Assessment Tool (SWAT), and Modular Groundwater Flow Model (MODFLOW) are commonly used for simulating water flow, groundwater recharge, evapotranspiration, and water quality parameters. Spatial analysis techniques such as interpolation, buffering, overlay analysis, and network analysis are used for mapping, analyzing, and visualizing spatial patterns, relationships, and processes related to water resources.

Case studies from different geographical regions were reviewed to understand the applicability, effectiveness, and constraints of GIS-based modeling techniques in diverse environmental settings. The selected case studies included research on surface water body delineation, groundwater mapping, watershed management, rainwater harvesting zone identification, flood and drought risk assessment, water quality monitoring, and climate change impact analysis. The case studies provided insights into the practical applications of GIS-based strategies, the challenges encountered, and the lessons learned from implementing these strategies in real-world contexts.

A data collection sheet (Fig. 1) has been suggested for the first time through this article to provide a comprehensive guide for researchers, practitioners, and stakeholders engaged in similar studies. This sheet aims to help them to understand and replicate the data collection and processes accurately.

3 Results and Discussions

3.1 Identification and Delineation of Surface Water Bodies

The use of GIS for water body identification is not new. In actuality, it was a longstanding practice in India and other nations. Aim of GIS involves the use of geospatial data, software, and analytical techniques to delineate and monitor bodies of water, such as rivers, lakes, ponds, and reservoirs. The process typically begins with acquiring various data sources, including satellite imagery, topographic maps, and hydrological data. Preprocessing of the data is essential to correct for factors such as atmospheric interference, geometric distortions, and calibration issues, ensuring that the data is accurate and suitable for analysis. Once the data is prepared, GIS software is used to classify or segment the imagery, distinguishing water pixels from non-water pixels based on their spectral characteristics. This classification can be achieved through supervised or unsupervised techniques, including machine learning algorithms, depending on the target-specific requirements. Post-processing steps involve noise reduction, gap filling, and the refinement of water body boundaries to improve the accuracy of the results. GIS allows for the integration of classified data with other geographical information, including such as land use and land cover maps as well as elevation data, providing context for the water bodies and aiding in their identification. Additional refinements may include the delineation of watersheds and the incorporation of hydrological models for more precise water body mapping. Temporal analysis using time-series imagery enables the detection of changes in water body extent, including seasonal variations and long-term trends. Validation and accuracy assessment are essential to ensure the reliability of the results, often

Data collection:

Data Types	Data Sources	Temporal and spatial coverage	Parameters
Satellite imagery data	Landsat, Sentinel, RADARSAT	Monthly or Seasonal or Annual	Land cover or land use, Surface water dynamics
Digital elevation models (DEMs)	National mapping agencies, Global repositories	Study area or watershed	Elevation, Slope, Aspect, Curvature
Hydrological data	Government agencies, Research institutions, Field surveys	-	Streamflow measurements, Groundwater Levels, Rainfall data, Soil moisture content
Land use or land cover map	Satellite imagery (supervised or unsupervised classification)	Annual	Forests, Agricultural lands. Urban areas, Water bodies
Spectral indices data	Satellite imagery data (Landsat, Sentinel)	Monthly or Seasonal or Annual	Normalized difference water index (NDWI), Enhanced vegetation index (EVI)
Rainfall data	Meteorological departments, Weather stations	Daily or weekly or Monthly or Annual	Rainfall Intensity, Rainfall Distribution
Terrain slope data	DEMs, Satellite Imagery	Study area or watershed	Slope Gradient, Slope Aspect, Slope Curvature
Ground truth data	Field surveys, Ground based observations	-	Surface water bodies, Ground water levels, Land use or land cover

Data processing and analysis:

❑ Software:
- GIS software (ArcGIS, QGIS, etc.)
- RS software such as ERDAS Imagine, Environment for visualizing image (ENVI), etc.
- Statistical software (R studio, Python, etc.)

❑ Methods:
- Object based Image Analysis (OBIA)
- Supervised or Unsupervised Classification
- Spectral Analysis
- Hydrological Modeling
- Suitability Analysis
- Multi Criteria Decision Analysis (MCDA)

Fig. 1 Data collection sheet for GIS-based modeling in water resource monitoring and management study

involving ground truthing through field surveys and the use of accuracy metrics. Ultimately, the combination of GIS and remote sensing results in accurate maps of water bodies that find applications in fields such as urban planning, environmental management, agriculture, and disaster response.

In recent years, significant progress has been made in employing digital image processing on remote sensing data for cost-effective, time-efficient surface water detection and mapping. Target detection methods using remotely sensed images can be broadly categorized into four main classes: (i) methods based on single bands (ii) methods based on spectral index, (iii) methods based on machine learning, and (iv) methods based on spectral unmixing. Single band-based methods rely on a single band for water body detection, while the other three methods utilize information from multiple bands in the image. Table 1 provides a summary of surface water mapping and the techniques or algorithms used, and Table 2 lists commonly used satellite sensors in surface water delineation research.

3.2 Groundwater Determination Through Remote Sensing and GIS

Groundwater modeling has difficulties due to a lack of large and scattered input data, especially in dry and semi-arid locations. The challenge is exacerbated by the fact that regional groundwater models demand distributed input data, while conventional hydrological measurements provide only point data, resulting in unreliable model outputs. However, recent advances in remote sensing and GIS offer new sources of spatial input data, aiding groundwater applications. Groundwater flow models are essential tools for assessing the potential impact of future human activities on groundwater dynamics. However, the reliability of these models is dependent on the availability of high-quality data about the physical and hydrogeological conditions (Singh 2014). Physical settings include terrain, land use, soils, canals, drainage ditches, climate, and agricultural water demand, whereas hydrogeological settings comprise the aquifer system, boundary conditions, hydraulic parameters, and groundwater level dynamics. These variables vary spatially and temporally, making GIS an invaluable tool for coupling with groundwater models, aiding in conceptualization and characterization of hydrogeological systems (Chakraborty and Chakraborty 2021a, b). Calibrating groundwater models is an important step in achieving coherence between actual and calculated state parameters. The technique of trial and error is widely used to alter critical model parameters during calibration. However, an excessive number of parameters might lead to over-parameterization, reducing the model's predictive value (Gaganis 2009).

Several studies demonstrate the integration of GIS with groundwater models in diverse applications. For instance, Chenini and Mammou (2010) integrated the GIS software ARCVIEW with MODFLOW in Tunisia, demonstrating the potential of GIS-MODFLOW applications. Akbar et al. (2011) introduced ArcPRZM-3, a GIS-based modeling system that can be used for either a single site-specific simulation or many regionally synchronized simulations using spatially distributed data. The model was used in Woodruff County, Arkansas, the United States to develop a health risk map based on dissolved bentazon concentrations in soil. Meanwhile, Gaur et al. (2011) developed an approach merging MODFLOW with GIS. The established methodology was applied on the Banganga River sub-basin in India to delineate groundwater potential zones. They found that the developed model is extremely useful for assessing groundwater resources and determining the location of rainfall harvesting structures in semi-arid areas. In a study by Kumari et al. (2022), an investigation has conducted a study to formulate a map indicating the potential zone for groundwater recharge and proposed suitable location for constructing artificial recharge structures within the Hamirpur district of northern India. This study used

Table 1 Surface water delineation methods and techniques or algorithms

Method	Techniques or algorithms	References
(i) Single band	Intensity thresholding	Pervez et al. (2017)
	Histogram based thresholding	Yang et al. (2017)
	Active contour-based segmentation	Guo et al. (2017)
	SLIC (simple linear iterative clustering) based segmentation	Behnamian et al. (2017)
(ii) Spectral index based	Normalized Difference Vegetation Index (NDVI)	Kaplan and Avdan (2017)
	NDWI (Normalized Difference Water Index)	Zeng et al. (2017)
	TCW (Tasseled Cap Wetness Index)	
	AWEI (Automated Water Extraction Index)	
	LSWI (Land Surface Water Index)	
	NDMI (Normalized Difference Moisture Index)	Boschetti et al. (2014)
	SIWSI (Shortwave Infrared Water Stress Index)	
	NDII (Normalized Difference Infrared Index)	
	NDSWIR (Normalised Difference Shortwave Infrared Index)	
	NDPI (Normalised Difference Pond Index)	
	NDFI (Normalised Difference Flood Index)	
	EWI (Enhanced Water Index)	Trochim et al. (2016)
	HDWI (Hyperspectral Difference Water Index)	Wang et al. (2015)
	DFI (Desert Flood Index)	Baig et al. (2013)
	WNDWI (Weighted NDWI)	Gopal et al. (2018)
	MOWI (Modified Optimization Water Index)	Meng et al. (2019)
	WRI (Water Ratio Index)	Moradi et al. (2017)
	MVI (Modified Vegetation Index)	Bochow et al. (2012)
	MBI (Modified Brightness Index)	Thouvenin et al. (2017)

(continued)

Table 1 (continued)

Method	Techniques or algorithms	References
	SDI (Shoreline Development Index)	
	NDFI (Normalized Difference Feature Index)	Mishra and Prasad (2015)
	Terrain Indices (MrVBF and HAND)	Huang et al. (2016)
(iii) Machine learning based	Constrained energy minimization (CEM) based classification	Isikdogan et al. (2017)
	Decision or regression tree-based classification	Huang et al. (2017)
	ICA algorithms (FastICA, BSS SVD, JADE, NG-OL, SONS and SIMBEC)	Thouvenin et al. (2016)
	ISODATA	Hassan and Mahmud-Ul-Islam (2014)
	Regularised Kernel Fisher's discriminant analysis (rkFDA)	Huang et al. (2015)
	K-means clustering	Gao et al. (2016)
	Dempster–Shafer theory-based classification	Li et al. (2018)
	Artificial neural network (ANN)	Abd Manaf et al. (2016)
	Support vector machine (SVM)	
	Maximum likelihood (ML)	
	Minimum distance	
	Parallelepiped	
	Mahanalobis	
	Fully convolutional neural networks	Irwin et al. (2017)
	Baye's law-based classification	Verpoorter et al. (2012)
	Fuzzy-based methods	Pekel et al. (2016)
	Spectral angle mapper (SAM) classification	Csendes and Mucsi (2016)
(iv) Spectral mixing based	Linear spectral unmixing	Mukherjee and Samuel (2016)
	Spectral indices based unmixing	Bangira et al. (2017)
	Perturbed linear mixing model (PLMM)	Sisodia et al. (2014)
	Low Albedo Fraction (LAF) method based on LSU	Sisay (2016)

Table 2 Commonly used satellite sensors in surface water delineation research (*source* modified after Bijeesh and Narasimhamurthy 2020)

Sensor	Electromagnetic bands count	Band details	Spatial detailing (m)
Landsat MSS	Four	Green, red, near infrared1 and 2	60
Landsat TM	Seven	Blue, green, red, near infrared, shortwave infrared1 and 2, thermal	30
Landsat ETM+	Eight	Blue, green, red, near infrared, shortwave infrared 1 and 2, thermal, panchromatic	30 (PAN-15)
Landsat 8 OLI	Eleven	ETM+ bands and ultra blue, cirrus	30 (PAN-15)
Suomi NPP-VIIRS	Twenty-two	640–12,010 nm	375 (1–5), 750 (6–22)
MODIS	Thirty-six	620–14,385 nm	250, 500, 1000
SPOT 4	Five	Green, red, near infrared, shortwave infrared, monospectral	20 (monospectral-10 m)
SPOT 5	Five	Green, red, near infrared, shortwave infrared, panchromatic	10 (PAN-5 or 2.5)
EO-1 ALI	Ten	48–2350 nm	30 (PAN-10)
GF01 and GF02	Five	450–890 nm	8 (PAN-2)
Hyperion data	Two hundred twenty	357–2576 nm	30
Sentinel-2 MSI	Thirteen	444–2204 nm	60, 10, 20
Quick-Bird	Five	450–900 nm	2.62
World View 2	Nine	400–1040 nm	2.07
LISS IV	Three	520–860 nm	5.8
MERIS	Fifteen	390–1040 nm	Ocean: (1040 × 1200), land: (260 × 300)
MUX/CBERS-4	Four	450–890 nm	20

Multi-Influencing Factor (MIF) and Weighted Overlay Index (WOI) techniques leveraging GIS. The researchers generated a map delineating groundwater potential zone by overlaying seven thematic layers, including soil texture, geology, geomorphology, slope, drainage density, land use or land cover, and lineament density. Weightings were assigned to each parameter based on its influence on groundwater dynamics. The outcomes of this research offer valuable insights for effective planning and management of groundwater resources. Abijith et al. (2020) conducted a research study to assess the effectiveness of employing Analytical Hierarchy Process (AHP), Multi influence factor (MIF), GIS, and RS (RS) techniques for mapping groundwater potential in the Ponnaniyaru watershed, located in Tamil Nadu, India. The study considered various parameters such as geomorphology, drainage density, soil characteristics, slope, lineament density, geology, land use or land cover, and rainfall. The results, expressed as percentages for AHP and MIF, indicated very low (9.9%, 10.54%), low (12.9%, 11.53%), moderate (27.3%, 28.76%), high (29.5%, 31.46%), and very high (20.10%, 17.68%) groundwater potential. The validation results demonstrated a relatively strong predictive accuracy of 75% for AHP and 71% for MIF. Saha (2017) used the AHP (Analytical Hierarchy Process) for the groundwater potential zone mapping in the Md. Bazar Block of Birbhum district, West Bengal. The authors incorporated multiple thematic layers such as rainfall, geology, geomorphology, slope, elevation, soil texture, land use or coverage, lineament density, stream junction frequency, pond density, drainage density, and topographic wetness index. Data on lithology, hydrogeomorphology, drainage, and lineaments were extracted from Landsat-8 imagery to generate an ASTER DEM. Weightings for these various factors were determined using expert knowledge in AHP methodology. The resulting map indicated areas with very low groundwater potential, particularly in regions characterized by steep slopes, high drainage density, and impermeable rock formations. The accuracy of the map was validated through statistical methods, specifically using the Receiver Operating Curve (ROC) and by comparing groundwater depth data obtained from multiple sources. In Table 3, a detailed summary is provided, presenting a comprehensive overview of various methods, software, tools, and applications of data in the field of groundwater modeling.

3.3 Digital Elevation Model (DEM) and Watershed Management

A DEM is a crucial geospatial tool that represents the Earth's surface as a grid of elevation values. It provides a three-dimensional representation of terrain, capturing variations in elevation and depicting landforms such as mountains, valleys, and plains (Wu et al. 2019). DEMs are often derived from satellite-based technologies like the Shuttle Radar Topography Mission (SRTM) and airborne surveys like Light Detection and Ranging (LiDAR). Notable globally freely available DEMs include the AW3D30 (Advanced Land Observing Satellite World 3D 30 m), Copernicus DEM, NASADEM (National Aeronautics and Space Administration DEM), HydroSHEDS (Hydrological data and maps based on Shuttle Elevation Derivatives at Multiple Scales), TanDEM-X (TerraSAR-X add on for DEMs), and MERIT (Multi Error Removed Improved Terrain DEM). Copernicus provides worldwide DEMs at both 30m (hence referred to as COP30) and 90m resolution (Moges et al. 2023). These elevation models serve diverse applications, including land use planning, environmental monitoring, and water resource management. In the context of watershed management, DEMs play a pivotal role in delineating and analyzing watersheds.

Table 3 Comprehensive overview of groundwater modeling methods, software, tools, and data applications (*source* modified after Pavani et al. 2022)

Sl. No.	Methods	Tools	Data used	Outcomes
1	Weighted overlay index	ArcGIS, QGIS, ERDAS, ENVI	Toposheets, DEM, LULC, soil data, thematic maps, online resources	Analysis of groundwater potential zone map
2	Analytical hierarchy process (AHP)	ArcGIS, QGIS, LSTM	LANDSAT data, SRTM or USGS (DEM), thematic layers, Catastrophe theory (CT), yield data	Identification of groundwater potential zone map
3	MODFLOW	Visual Modflow Flex	LISS, SRTM or USGS (DEM), aquifer parameters	Validation of groundwater model
4	SWAT-MODFLOW	QSWATMOD, SWAT and SWAT-MODFLOW	DEM, LULC, soil maps, river network, HRUs, Subbasins, aquifer parameters	Groundwater recharge rates, scenarios, and decadal trends
5	Spatial interpolation	Arc GIS, Spatial Analyst Tool, IDW Tool, Local and Global Polynomial Interpolation, TSPLINE	Location of wells as point data, well information about depth of water level	Water level fluctuations of annual, pre-monsoon, and post-monsoon seasons groundwater level modeling
6	Kriging method	Arc GIS, Ordinary Kriging, Empirical Bayesian Kriging, Simple Kriging, Universal Kriging, Density Sampling Network Tool and Priority Index	Samples of well points, its location and water levels, piezometer levels	Forecasted spatial level differences, to predict the standard errors, estimation of groundwater potential site
7	LISA	LISA, gradient and surface interpolation Depth to water level, electrical conductivity Availability of groundwater tools	Depth to water level, electrical conductivity	Depth to water level, electrical conductivity

(continued)

Table 3 (continued)

Sl. No.	Methods	Tools	Data used	Outcomes
8	GRACE	GRACE, CLSM, GLDAS, GTCH, EOF	Canopy water storage, soil moisture, vertical deformation of water levels, terrestrial water storage and precipitation anomalies	Mapping of groundwater storage and its variations
9	GS FLOW	GSFLOW, PRMS, MODFLOW, AHP and GSFLOW, pyGSFLOW, modified index overlay mode	Percolation, evaporation, and stream flow data	Impacts of surface and subsurface water interfaces, integrated hydrologic model
10	HYDRUS-1D	HYDRUS-1D	Well information, meteorological data, soil and aquifer data	Groundwater recharge and soil water fluxes

Figure 2 illustrates a sequential methodology for conducting morphometric analysis specifically tailored for watershed management purposes.

Several studies globally have scrutinized the effectiveness of DEMs in watershed modeling. For instance, a recent investigation by Tran et al. (2022) focused on the La Giang River basin in Central Vietnam, assessing six DEMs (MERIT, SRTM, NASA, ASTER GDEM2, TanDEM-X, and AW3D30) using the SWAT model. NASA and SRTM 30 DEM are recognized for their high accuracy, whereas ASTER GDEM2 exhibits increased uncertainty in delineating river networks. MERIT, SRTM 90m, and ASTER GDEM2 showed inaccuracies in basin delineation, especially at the intersection of subbasins. The proposed cross-validation approach reveals the positive impact of higher resolution DEM calibration on hydrological simulations and vice versa, particularly regarding flood peak observations. Another study conducted by del Rosario Gonzalez-Moradas et al. (2020) to evaluate five widely used DEMs such as ASTER GDEM, SRTM, TanDEM-X, AW3D30, and Copernicus and focused on their performance in flow routing. The results showed that Copernicus and AW3D30 effectively estimated the flow in major streams. However, Copernicus displayed discrepancies in catchments and generated a higher number of non-existent tributaries compared to AW3D30. The impact of DEM resolution on water flow simulation has been investigated by Roostaee and Deng (2020) in three U. S. watersheds with different terrains. They use the HSPF (Hydrological Simulation Program-Fortan) watershed model and DEMs ranging from 3.5 to 100 m. Lower DEM resolutions result in reduced simulated flow, especially in flatter areas. The resolution of the DEM had a minor impact on manual watershed delineation. In two of the three

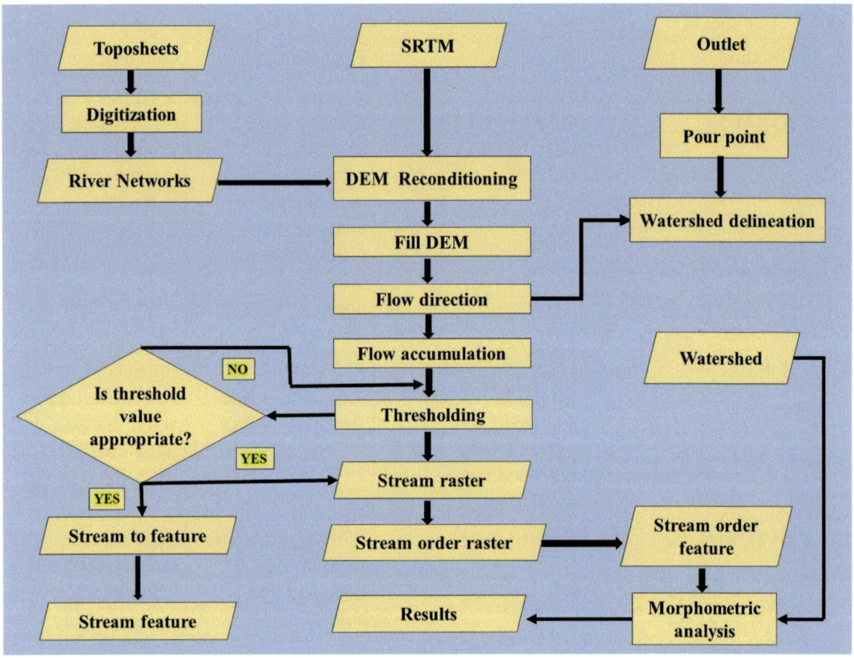

Fig. 2 Step-by-step methodology for conducting morphometric analysis for watershed management (adopted from Singh et al. 2021 and slightly modified)

watersheds studied, the uncertainty in parameters outweighed the uncertainty in resolution. This suggests that adjusting water balance variables can help alleviate the challenges posed by lower-resolution DEMs in watersheds with moderate slopes. The findings guide water resources management based on watershed modeling, emphasizing the importance of addressing uncertainty from DEM resolutions. In studies beyond hydrological modeling, DEMs had been applied in morphometric analysis and watershed prioritization studies to address specific environmental challenges. Singh et al. (2021) addressed soil erosion in the Dudhnai Watershed. They employed a combined approach using toposheets, remotely sensed DEMs, and the ArcGIS toolbox, highlighting the speed and cost-effectiveness of this methodology. Roy et al. (2020) investigated gully erosion susceptibility in the Gandheswari Watershed, West Bengal, India, focusing on the fringing area of the Chhotanagpur plateau. Researchers employed the MLPC (Multilayer Perception approach) and its ensembles (MLPC-Bagging, MLPC-Dagging, MLPC-Decorate). The study, considered 20 geoenvironmental factors such as elevation, distance from road, distance from river, distance from lineament, soil texture, soil moisture, soil erodibility, rainfall, drainage density, slope, slope aspect, geology, geomorphology, NDVI, Land use and land cover (LULC), topographical wetness index, stream power index, terrain ruggedness index, plan, and profile curvature. The findings revealed that the MLPC-Decorate model stands out with impressive accuracy, boasting Receiver Operating

Characteristic (ROC) curve AUC values of 0.924 and 0.906 for training and validation databases, respectively. Shifting focus to southern India, Kasi et al. (2020) explored the accuracy and parameter uncertainty of DEMs for drainage morphometric parameters within the Champavathi River sub-basin. Morphometric analysis revealed that the basin is elongated, with low relief, gentle slopes, and medium drainage density. Importantly, the results were consistent across multiple DEM sources, indicating minimal dependence on scale and source characteristics. Innovative approaches also surfaced in Wu et al. (2023) study, where deep learning models were developed to classify the images containing the locations of flow barriers. The study developed and compared various CNN (Convolutional Neural Network) models, utilizing High Resolution DEMs (HRDEMs) and aerial orthophotos. Results showed most models consistently achieved over 90% accuracy. Notably, the CNN model using only HRDEMs performed better than those with additional data. The study highlighted the minimal impact of aerial orthophotos on model accuracy, and the chosen model worked well in various geographic settings. Despite the advancements and widespread applications of DEMs, it is crucial to regularly assess their accuracy and performance. Ferreria and Carbal (2022) meticulously examined the vertical accuracy of SRTM, GMTED2010, ALOS PALSAR, and Topodata DEMs in line with the Brazilian Cartographic Accuracy Standard. To achieve this, the authors utilized the official high-accuracy network data from the Brazilian Geodetic System. All four DEMs evaluated showcased satisfactory accuracy for mapping purposes at scales up to 1:100,000. Utilizing the inverse distance weighting (IDW) method, the study found that more than 90% of the sampled points from these DEMs had altimetric errors of under 25 m when compared to the high-accuracy reference points of the Brazilian Geodetic System (Fig. 3). Furthermore, the research revealed that altimetric error did not demonstrate any significant correlation with either altitude or slope within the study area. This study underscores the importance of regularly assessing the accuracy of DEMs and highlights avenues for future research to further refine our understanding of the factors influencing DEM accuracy. Similarly, Tesema (2021) explored the impact of DEM resolution and sources on morphometric parameters in the Tena watershed, Ethiopia. Employing three DEMs ($ASTER_{30m}$, $ALOS_{30m}$, and $SRTM_{30m}$) in conjunction with ArcGIS software, the author uncovers substantial disparities in parameter outcomes depending on the selected DEM. However, $ASTER_{30m}$ emerges as the preferred option for tasks such as soil erosion analysis, sediment evaluation, flood risk assessment, and overall watershed modeling. Conversely, $ALOS_{30m}$ demonstrates superiority in peak discharge analysis. Despite these variations, all examined DEMs demonstrate capability in computing watershed shape parameters. The study emphasizes the pivotal role of informed DEM selection in accurately gauging hydrological and hydraulic characteristics, thus carrying significant implications for decision-making in watershed management and development. While providing valuable comparative insights, Tesema's research could benefit from further elucidation of methodology and validation procedures to bolster its credibility and applicability.

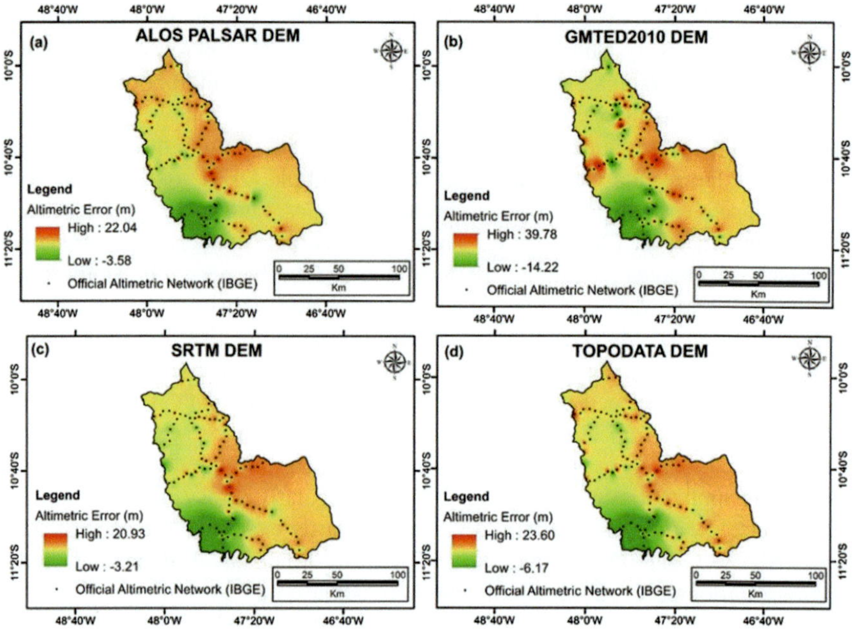

Fig. 3 Spatial variations of altimetric errors of **a** ALOS PALSAR DEM, **b** GMTED2010 DEM, **c** SRTM DEM, **d** TOPODATA DEM using IDW method

3.4 Determination of Potential Rain Water Harvesting Zone Through RS and GIS

Rainwater harvesting is a crucial strategy for dealing with water scarcity in agriculture, especially in arid and semi-arid regions. These areas mainly depend on rainfed farming, facing challenges due to factors like aridity and unpredictable rainfall. The varying rainfall amount and distribution create obstacles for farmers, affecting their ability to sustain crops and livestock production. In response to the need for more water, people in arid and semi-arid regions have developed various rainwater harvesting techniques like ponds, dams, terracing, percolation tanks, Nala bunds, etc. These methods help capture and reuse rainfall effectively. Choosing the right place for rainwater harvesting involves considering six important factors: hydrology, climate, soil, agronomic practices, topography, and socioeconomic conditions. Balancing these factors is a complex task, especially in large watersheds. Evaluating these elements carefully is crucial for the success and sustainability of rainwater harvesting projects.

Different techniques, both direct and indirect, are used to gather information about rainwater harvesting. Geospatial technology, including RS and GIS, is particularly useful. These tools create detailed maps with information on slope, soil types, drainage patterns, land use, and lineaments, helping in site selection. The integration of GIS and remote sensing has greatly improved the efficiency of selecting sites for rainwater harvesting. Researchers can identify optimal locations more precisely and reduce the number of proposed sites, enhancing the success and sustainability of projects. Geospatial data and GIS go beyond just identifying sites; they also assist in ongoing monitoring, management, and conservation efforts. This comprehensive approach addresses water scarcity in arid and semi-arid regions. Geospatial technology emerges as a promising tool in rainwater harvesting, contributing to the sustainable management of water resources in water-stressed environments. Figure 4 presents a conceptual methodology flowchart outlining the methodology for mapping potential rainwater harvesting zones. In Fig. 5, two maps are presented that delineate the potential zones for rainwater harvesting and identify suitable locations for rainwater harvesting structures in the districts of Mirzapur, Chandauli, and Sonbhadra in the state of Uttar Pradesh, India, adopted from Waghaye et al. (2023). Table 4 provides a comprehensive overview of the application of RS and GIS techniques in the investigation of rainwater harvesting.

4 Conclusions

The comprehensive research highlights the pivotal role of GIS and RS in water resource management, particularly in the identification and delineation of surface water bodies, groundwater determination, and digital elevation modeling for watershed management. The integration of GIS with various surface water delineation approaches, including single band methods, spectral index-based methods, machine learning-based methods, and spectral mixing-based methods, has resulted in more accurate and efficient water body mapping. Moreover, the application of GIS and RS in groundwater determination has substantially improved groundwater modeling by providing spatial input data, which helps with the conceptualization and characterization of hydrogeological systems. Various studies have demonstrated the successful integration of GIS with groundwater models in diverse applications such as groundwater potential assessment and the identification of suitable sites for artificial recharge structures. Additionally, DEMs derived from satellite-based technologies and airborne surveys have proven indispensable in watershed management. DEMs allow for the three-dimensional representation of terrain, capturing variations in elevation and assisting in the delineation and analysis of watersheds. Several studies have evaluated the accuracy and performance of DEMs in watershed modeling, emphasizing the importance of selecting appropriate DEMs for specific applications.

The future of GIS-based modeling for water resource monitoring and management appears promising. New technologies like artificial intelligence, internet of things, machine learning, and big data analytics offer opportunities to improve data integration, analysis, prediction, visualization, and decision-making. Future research endeavors should work on combining GIS with hydrological modeling, making RS data more accurate and reliable, and enhancing the performance and scalability of

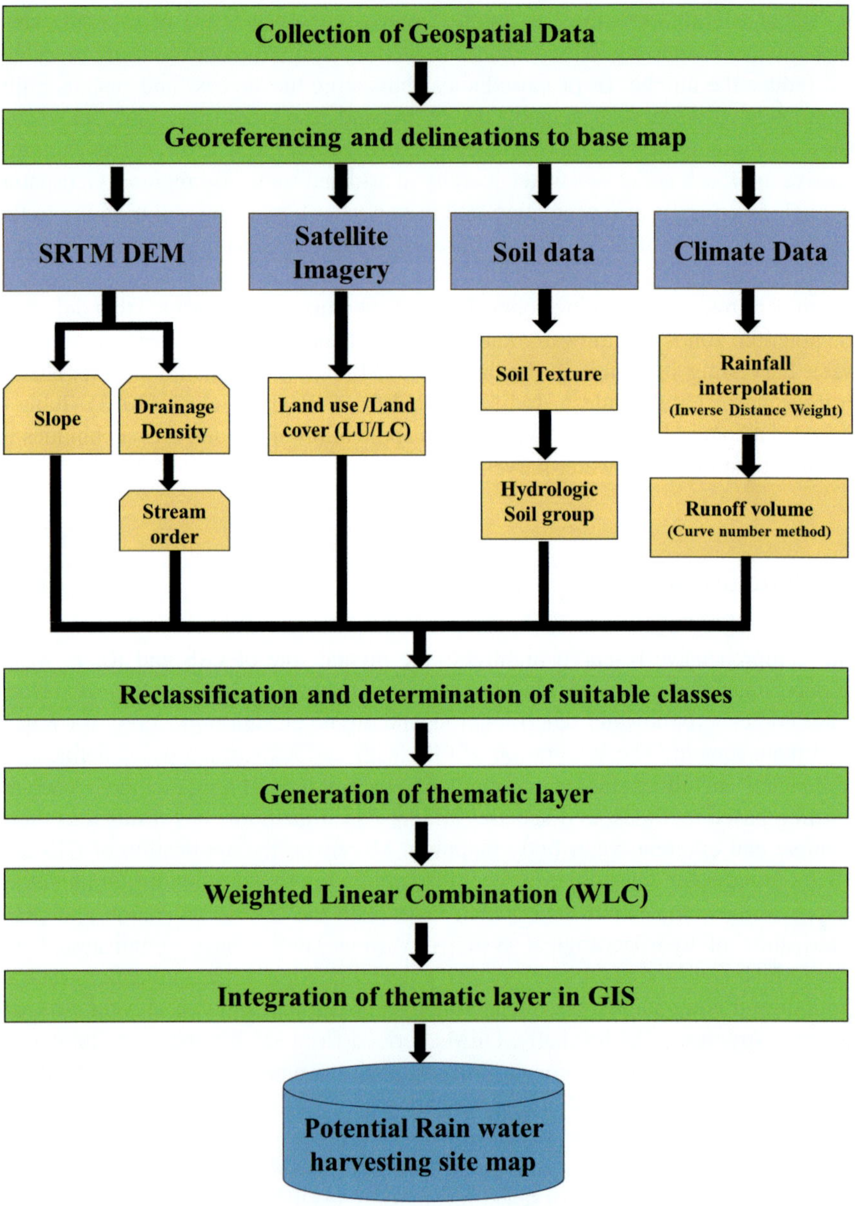

Fig. 4 Conceptual methodology for the mapping of potential rainwater harvesting zone

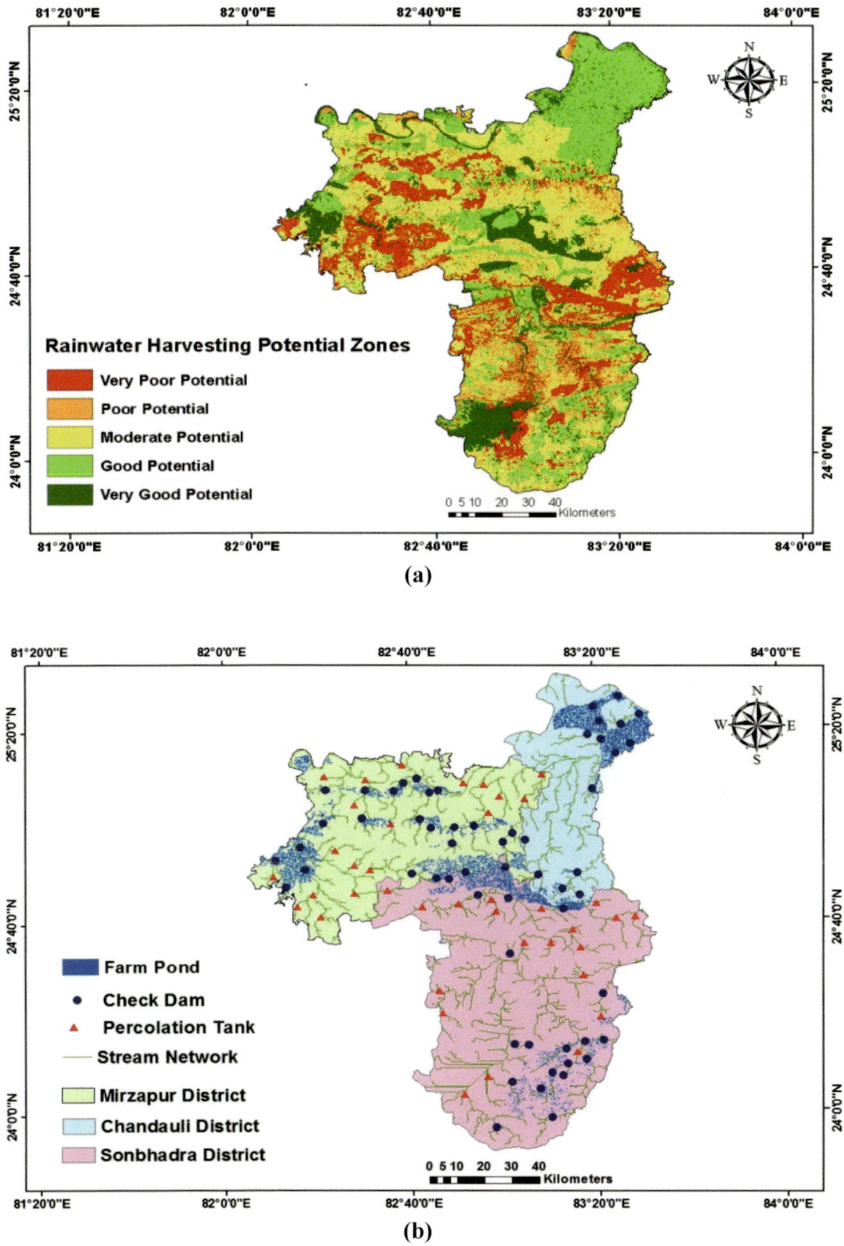

Fig. 5 Map delineates **a** rainwater harvesting potential zones and **b** suitable locations for rainwater harvesting structures

Table 4 Application of RS and GIS in rainwater harvesting study

Sl. No.	Study site	Brief description of the scientific research	References
1	Madhya Pradesh, India	The authors used geospatial and Multiple Criteria Decision Analysis (MCDA) techniques to evaluate rainwater harvesting potential and identify suitable locations for rainwater harvesting structures	Singhai et al. (2017)
2	Khushkera-Bhiwadi-Neemrana Investment Region (KBNIR), Rajasthan, India	The research paper proposes a GIS-based method for finding good locations for rainwater harvesting. By leveraging freely available imagery and diverse data layers, including surface elevation, land use, soil, drainage, and depression maps, the study applies the Soil Conservation Service—Curve Number (SCS-CN) method to assess depression volume and surface runoff	Tiwari et al. (2018)
3	Uttar Pradesh (Mirzapur, Chandauli, and Sonbhadra), India	The authors utilized GIS and MCDA techniques to identify potential zones and specific locations for rainwater harvesting, leveraging geospatial tools and AHP. Employ Boolean logic concepts to map suitable locations for RWH based on identified potential zones	Waghaye et al. (2023)
4	Damodar River Basin, West Bengal, India	Identification of rainwater harvesting potential and suitable sites using GIS-based MCDA	Singh et al. (2017)

(continued)

Table 4 (continued)

Sl. No.	Study site	Brief description of the scientific research	References
5	West Medinipur, West Bengal, India	Evaluate the potential for rainwater harvesting and pinpoint appropriate locations or zones for various rainwater harvesting structures through the integration of RS, GIS, and MCDA techniques	Jha et al. (2014)
6	Escambia County, Florida, the United States	GIS surface maps were used alongside local utility consumption data to assess the potential decrease in household potable water use. This analysis incorporated LiDAR and National Agriculture Imagery Program (NAIP) data	Grant et al. (2018)
7	Al-Qadisiyah Governorate, Iraq	Authors aimed to identify suitable water harvesting area through application of RS technique integrated with GIS	Al-Khuzaie et al. (2020)
8	Tekeze basin, Ethiopia	Identify sites with the potential for rainwater harvesting and determine suitable sites using a GIS-based MCDA	Alene et al. (2022)
9	Pernambuco State, Brazil	Identify roof areas and estimate rainwater harvesting (RWH) potential by employing a method that integrated DEM, digital terrain model (DTM), and a vegetation index	Gomes et al. (2021)
10	West Bank, Palestine	Evaluate the development of domestic water poverty and domestic rainwater harvesting suitability maps, utilizing the AHP and the GIS-based Weighted Overlay Summation Process (WOSP)	Shadeed et al. (2019)

hydrological models. It is also important to include socioeconomic and environmental data, address the impacts of climate change, and assess the cost-effectiveness and sustainability of GIS-based approaches. In conclusion, GIS and RS technologies enable more accurate, efficient, and long-term water resource planning and decision-making. These breakthroughs significantly contribute to the preservation and sustainable management of water resources, particularly in arid and semi-arid regions. Continuous innovations in GIS and RS, coupled with ongoing research efforts and interdisciplinary collaborations, are essential to ensuring the conservation and long-term use of water resources for the present and forthcoming generations.

References

Abd Manaf S, Mustapha N, Sulaiman MN, Husin NA, Hamid MRA (2016) Comparison of classification techniques on fused optical and SAR images for shoreline extraction: a case study at northeast coast of Peninsular Malaysia. J Comput Sci 12(8):399–411

Abijith D, Saravanan S, Singh L, Jennifer JJ, Saranya T, Parthasarathy KSS (2020) GIS-based multi-criteria analysis for identification of potential groundwater recharge zones-a case study from Ponnaniyaru watershed, Tamil Nadu, India. HydroResearch 3:1–14

Akbar TA, Lin H, DeGroote J (2011) Development and evaluation of GIS-based ArcPRZM-3 system for spatial modeling of groundwater vulnerability to pesticide contamination. Comput Geosci 37(7):822–830

Alene A, Yibeltal M, Abera A, Andualem TG, Lee SS (2022) Identifying rainwater harvesting sites using integrated GIS and a multi-criteria evaluation approach in semi-arid areas of Ethiopia. Appl Water Sci 12(10):238

Al-Khuzaie MM, Janna H, Al-Ansari N (2020) Assessment model of water harvesting and storage location using GIS and remote sensing in Al-Qadisiyah, Iraq. Arab J Geosci 13:1–9

Baig MHA, Zhang L, Wang S, Jiang G, Lu S, Tong Q (2013) Comparison of MNDWI and DFI for water mapping in flooding season. In: 2013 IEEE international geoscience and remote sensing symposium-IGARSS. IEEE, pp 2876–2879

Bangira T, Alfieri SM, Menenti M, Van Niekerk A, Vekerdy Z (2017) A spectral unmixing method with ensemble estimation of endmembers: application to flood mapping in the Caprivi floodplain. Remote Sens 9(10):1013

Behnamian A, Banks S, White L, Brisco B, Millard K, Pasher J, Chen Z, Duffe J, Bourgeau-Chavez L, Battaglia M (2017) Semi-automated surface water detection with synthetic aperture radar data: a wetland case study. Remote Sens 9(12):1209

Bijeesh TV, Narasimhamurthy KN (2020) Surface water detection and delineation using remote sensing images: a review of methods and algorithms. Sustain Water Resour Manag 6:1–23

Bochow M, Heim B, Küster T, Rogaß C, Bartsch I, Segl K, Kaufmann H (2012) Automatic detection and delineation of surface water bodies in airborne hyperspectral data. In: 2012 IEEE international geoscience and remote sensing symposium. IEEE, pp 5226–5229

Boschetti M, Nutini F, Manfron G, Brivio PA, Nelson A (2014) Comparative analysis of normalised difference spectral indices derived from MODIS for detecting surface water in flooded rice cropping systems. PLoS ONE 9(2):e88741

Chakraborty SK, Chakraborty SK (2021) Water: its properties, distribution, and significance. Riverine ecology, volume 1: eco-functionality of the physical environment of rivers, pp 23–55

Chakraborty SK, Chakraborty SK (2021) Geo-hydrological perspectives of riverine flows. Riverine ecology, volume 1: eco-functionality of the physical environment of rivers, pp 375–476

Chenini I, Mammou AB (2010) Groundwater recharge study in arid region: an approach using GIS techniques and numerical modeling. Comput Geosci 36(6):801–817

Csendes B, Mucsi L (2016) Inland excess water mapping using hyperspectral imagery. Geographica Pannonica 20(4):191–196
del Rosario Gonzalez-Moradas M, Viveen W (2020) Evaluation of ASTER GDEM2, SRTMv3. 0, ALOS AW3D30 and TanDEM-X DEMs for the Peruvian Andes against highly accurate GNSS ground control points and geomorphological-hydrological metrics. Remote Sens Environ 237:111509
Du Plessis A, du Plessis A, Schmuhl (2019) Water as an inescapable risk, pp 147–172
Ferreira ZA, Cabral P (2022) A Comparative study about vertical accuracy of four freely available digital elevation models: a case study in the Balsas River Watershed, Brazil. ISPRS Int J Geo-Inf 11(2):106
Feyisa GL, Meilby H, Fensholt R, Proud SR (2014) Automated Water Extraction Index: a new technique for surface water mapping using Landsat imagery. Remote Sens Environ 140:23–35
Fisher A, Flood N, Danaher T (2016) Comparing Landsat water index methods for automated water classification in eastern Australia. Remote Sens Environ 175:167–182
Gaganis P (2009) Model calibration/parameter estimation techniques and conceptual model error. In: Uncertainties in environmental modelling and consequences for policy making. Springer, pp 129–154
Gao H, Wang L, Jing L, Xu J (2016) An effective modified water extraction method for Landsat-8 OLI imagery of mountainous plateau regions. In: IOP conference series: earth and environmental science, vol 34, no 1. IOP Publishing, p 012010
Gaur S, Chahar BR, Graillot D (2011) Combined use of groundwater modeling and potential zone analysis for management of groundwater. Int J Appl Earth Obs Geoinf 13(1):127–139
Gomes YRM, Jucá MVQ, Batista LFDR, Ribeiro Neto A, dos Santos SM (2021) Rainwater harvesting potential in water abundant and scarce regions in northeast Brazil using remote sensing. Sustain Water Resour Manag 7:1–9
Grant ATJ, McKinney NL, Ries R (2018) An approach to quantifying rainwater harvesting potential using imagery, geographic information systems (GIS) and LiDAR data. Water Sci Technol: Water Supply 18(1):108–118
Guo Q, Pu R, Li J, Cheng J (2017) A weighted normalized difference water index for water extraction using Landsat imagery. Int J Remote Sens 38(19):5430–5445
Hashim HQ, Sayl KN (2021) Detection of suitable sites for rainwater harvesting planning in an arid region using geographic information system. Appl Geomat 13:235–248
Hassan MS, Mahmud-Ul-Islam S (2014) Detection of waterlogging areas based on passive remote sensing data in Jessore District of Khulna Division, Bangladesh. Int J Sci Res Publ (IJSRP) 4:12
Huang C, Chen Y, Zhang S, Li L, Shi K, Liu R (2016) Surface water mapping from Suomi NPP-VIIRS imagery at 30 m resolution via blending with Landsat data. Remote Sens 8(8):631
Huang C, Nguyen BD, Zhang S, Cao S, Wagner W (2017) A comparison of terrain indices toward their ability in assisting surface water mapping from Sentinel-1 data. ISPRS Int J Geo Inf 6(5):140
Huang X, Xie C, Fang X, Zhang L (2015) Combining pixel-and object-based machine learning for identification of water-body types from urban high-resolution remote-sensing imagery. IEEE J Sel Topics Appl Earth Observ Remote Sens 8(5):2097–2110
Irwin K, Beaulne D, Braun A, Fotopoulos G (2017) Fusion of SAR, optical imagery and airborne LiDAR for surface water detection. Remote Sens 9(9):890
Isikdogan F, Bovik AC, Passalacqua P (2017) Surface water mapping by deep learning. IEEE J Sel Topics Appl Earth Observ Remote Sens 10(11):4909–4918
Jha MK, Chowdary VM, Kulkarni Y, Mal BC (2014) Rainwater harvesting planning using geospatial techniques and multicriteria decision analysis. Resour Conserv Recycl 83:96–111
Kaplan G, Avdan U (2017) Object-based water body extraction model using Sentinel-2 satellite imagery. Eur J Remote Sens 50(1):137–143
Kasi V, Pinninti R, Landa SR, Rathinasamy M, Sangamreddi C, Kuppili RR, Dandu Radha PR (2020) Comparison of different digital elevation models for drainage morphometric parameters: a case study from South India. Arab J Geosci 13:1–17

Kumari S, Poddar A, Kumar N, Shankar V (2022) Delineation of groundwater recharge potential zones using the modeling based on remote sensing, GIS and MIF techniques: a study of Hamirpur District, Himachal Pradesh, India. Model Earth Syst Environ 8(2):1759–1770
Li L, Vrieling A, Skidmore A, Wang T, Turak E (2018) Monitoring the dynamics of surface water fraction from MODIS time series in a Mediterranean environment. Int J Appl Earth Obs Geoinf 66:135–145
Meng Q, Wen X, Yuan L, Xu H (2019) Factorization-based active contour for water-land SAR image segmentation via the fusion of features. IEEE Access 7:40347–40358
Mishra K, Prasad P (2015) Automatic extraction of water bodies from Landsat imagery using perceptron model. J Comput Environ Sci
Moges DM, Virro H, Kmoch A, Cibin R, Rohith AN, Martínez-Salvador A, Conesa-García C, Uuemaa E (2023) How does the choice of DEMs affect catchment hydrological modeling? Sci Total Environ 164627
Moradi M, Sahebi M, Shokri M (2017) Modified optimization water index (MOWI) for Landsat-8 OLI/TIRS. Int Arch Photogramm Remote Sens Spat Inf Sci 42:185–190
Mukherjee NR, Samuel C (2016) Assessment of the temporal variations of surface water bodies in and around Chennai using Landsat imagery. Indian J Sci Technol 9(18):1–7
Pavani CSL, Vital TR, Haleem S (2022) Geospatial tools and techniques for ground water management: a review. J Geointerface 1:57–66
Pekel JF, Cottam A, Gorelick N, Belward AS (2016) High-resolution mapping of global surface water and its long-term changes. Nature 540(7633):418–422
Pervez W, Khan SA, Hussain E, Amir F, Maud MA (2017) Evaluate the capability of Landsat8 operational land imager for shoreline change detection/inland water studies. Int Arch Photogramm Remote Sens Spat Inf Sci 42:145–152
Roostaee M, Deng Z (2020) Effects of digital elevation model resolution on watershed-based hydrologic simulation. Water Resour Manag 34:2433–2447
Roy P, Chakrabortty R, Chowdhuri I, Malik S, Das B, Pal SC (2020) Development of different machine learning ensemble classifier for gully erosion susceptibility in Gandheswari Watershed of West Bengal, India. Machine learning for intelligent decision science pp 1–26
Saha S (2017) Groundwater potential mapping using analytical hierarchical process: a study on Md. Bazar Block of Birbhum District, West Bengal. Spat Inf Res 25(4):615–626
Sayl KN, Muhammad NS, El-Shafie A (2017) Optimization of area–volume–elevation curve using GIS–SRTM method for rainwater harvesting in arid areas. Environ Earth Sci 76:1–10
Shadeed SM, Judeh TG, Almasri MN (2019) Developing GIS-based water poverty and rainwater harvesting suitability maps for domestic use in the Dead Sea region (West Bank, Palestine). Hydrol Earth Syst Sci 23(3):1581–1592
Singh A (2014) Groundwater resources management through the applications of simulation modeling: a review. Sci Total Environ 499:414–423
Singh LK, Jha MK, Chowdary VM (2017) Multi-criteria analysis and GIS modeling for identifying prospective water harvesting and artificial recharge sites for sustainable water supply. J Clean Prod 142:1436–1456
Singh WR, Barman S, Tirkey G (2021) Morphometric analysis and watershed prioritization in relation to soil erosion in Dudhnai Watershed. Appl Water Sci 11(9):151
Singhai A, Das S, Kadam AK, Shukla JP, Bundela DS, Kalashetty M (2017) GIS-based multi-criteria approach for identification of rainwater harvesting zones in upper Betwa sub-basin of Madhya Pradesh, India. Environ Dev Sustain 21:777–797
Sisay A (2016) Remote sensing based water surface extraction and change detection in the central rift valley region of Ethiopia. Am J Geogr Inf Syst 5(2):33–39
Sisodia PS, Tiwari V, Kumar A (2014) Analysis of supervised maximum likelihood classification for remote sensing image. In: International conference on recent advances and innovations in engineering (ICRAIE-2014). IEEE, pp 1–4
Tesema TA (2021) Impact of identical digital elevation model resolution and sources on morphometric parameters of Tena watershed, Ethiopia. Heliyon 7(11)

Thouvenin PA, Dobigeon N, Tourneret JY (2016) Online unmixing of multitemporal hyperspectral images accounting for spectral variability. IEEE Trans Image Process 25(9):3979–3990

Thouvenin PA, Dobigeon N, Tourneret JY (2017) A hierarchical Bayesian model accounting for endmember variability and abrupt spectral changes to unmix multitemporal hyperspectral images. IEEE Trans Comput Imaging 4(1):32–45

Tiwari K, Goyal R, Sarkar A (2018) GIS-based methodology for identification of suitable locations for rainwater harvesting structures. Water Resour Manag 32:1811–1825

Tran TND, Nguyen QB, Vo ND, Marshall R, Gourbesville P (2022) Assessment of terrain scenario impacts on hydrological simulation with SWAT model. Application to Lai Giang Catchment, Vietnam. In: Advances in hydroinformatics: models for complex and global water issues—practices and expectations, pp 1205–1222

Trochim ED, Prakash A, Kane DL, Romanovsky VE (2016) Remote sensing of water tracks. Earth Space Sci 3(3):106–122

Verpoorter C, Kutser T, Tranvik L (2012) Automated mapping of water bodies using Landsat multispectral data. Limnol Oceanogr Methods 10(12):1037–1050

Waghaye AM, Singh DK, Sarangi A, Sena DR, Sahoo RN, Sarkar SK (2023) Identification of suitable zones and sites for rainwater harvesting using GIS and multicriteria decision analysis. Environ Monit Assess 195(2):279

Wang S, Baig MHA, Zhang L, Jiang H, Ji Y, Zhao H, Tian J (2015) A simple enhanced water index (EWI) for percent surface water estimation using Landsat data. IEEE J Sel Topics Appl Earth Observ Remote Sens 8(1):90–97

Wu D, Li R, Rekabdar B, Talbert C, Edidem M, Wang G (2023) Classification of drainage crossings on high-resolution digital elevation models: a deep learning approach. Giscience Remote Sens 60(1):2230706

Wu T, Li J, Li T, Sivakumar B, Zhang G, Wang G (2019) High-efficient extraction of drainage networks from digital elevation models constrained by enhanced flow enforcement from known river maps. Geomorphology 340:184–201

Yang X, Zhao S, Qin X, Zhao N, Liang L (2017) Mapping of urban surface water bodies from Sentinel-2 MSI imagery at 10 m resolution via NDWI-based image sharpening. Remote Sens 9(6):596

Zeng L, Schmitt M, Li L, Zhu XX (2017) Analysing changes of the Poyang Lake water area using Sentinel-1 synthetic aperture radar imagery. Int J Remote Sens 38(23):7041–7069

Enhancing Watershed Management Through Advanced Geospatial and Morphometric Approaches for Wainganga River Basin, Central India

Jayesh Mukherjee, Anuva Chowdhury, and Surajit Ghosh

Abstract Effective river basin management with innovatory geospatial and morphometric techniques helps in comprehending flood mitigation, ecosystem conservation, and optimising land and water resource allocation. This chapter is a case study based on the Wainganga River basin, a northern sub-basin of the Godavari River in peninsular India. The present study deals with a newly developed erosivity risk index based on the combination of sub-basin areal morphometry and anthropogenic parameters like agricultural area and population density. The research utilises various open-access multi-resolution raster and vector datasets, which are integrated into the GIS platform to compute key sub-catchment morphometrics for 22 sub-basins. These parameters determine the basin's hydrological response, sediment transport, and erosion dynamics. Subsequently, prioritisation is obtained based on average compound (C_p) scoring considering the areal morphometric data. Geostatistical computations, including principal component analysis and hierarchical clustering, showed first two components explained 90.56% of the dataset verifying the C_p score findings. LULC for the years 1985, 2005, and 2021 registered a rising dominance of farmlands and settlements followed by a major decline in forests and conversion of fallow lands for agriculture. The formulated erosivity risk index determined low to moderate soil erosion for all the sub-basins except for sub-basin 16, which exhibits

J. Mukherjee (✉)
Centre for the Study of Regional Development, School of Social Sciences, Jawaharlal Nehru University, New Delhi, India
e-mail: jam169@aber.ac.uk

Earth Surface Processes Research Group, Department of Geography and Earth Sciences, Aberystwyth University, Aberystwyth, Wales, UK

A. Chowdhury
Department of Remote Sensing, Birla Institute of Technology, Mesra, Ranchi, Jharkhand, India
e-mail: phdrs10052.21@bitmesra.ac.in

S. Ghosh
International Water Management Institute (IWMI), Colombo, Sri Lanka
e-mail: s.ghosh@cgiar.org

S. C. Pal and U. Chatterjee (eds.), *Surface, Sub-Surface Hydrology and Management*, Springer Geography, https://doi.org/10.1007/978-3-031-62376-9_25

an optimum index value of 1.00 due to high population density and cropping intensity. Hence, the chapter provides a lucid index for soil erosivity easily adoptable by policymakers and water resource managers enabling focused resource allocation and conservation efforts towards the most critical areas within a watershed.

Keywords Drainage basin morphometry · Erosivity risk index · River basin management · Wainganga River basin · Sustainable development goals

1 Introduction

Effective management of river basins is essential for ensuring the sustainable utilisation of water resources (Sabale et al. 2023; Roestamy and Fulazzaky 2021; Loucks et al. 2000), promotion of conjunctive usage of surface and groundwater (Osorio Olivos et al. 2023; Sabale et al. 2023; Khan et al. 2014) and mitigating environmental challenges (Zeitoun et al. 2013; Palmer et al. 2008). In recent years, advanced geospatial (Rai et al. 2022; Lyon and Lyon 2022; Krishna 2005) and morphometric approaches (Khan and ElKashouty 2023; Rakesh et al. 2023; Meshram and Sharma 2017) have emerged as powerful tools for enhancing river basin management (RBM). This chapter explores the significance of these approaches and their applications in optimising water resource management (WRM), soil erosivity and potential land conservation practices, mitigating natural hazards like floods, and promoting environmental sustainability.

Geospatial technologies facilitate the integration of surface and sub-surface hydrological data within river basins (Srivastava et al. 2022; Papa and Frappart 2021). The advent of advanced geospatial tools and the evolution of Big Data have enabled precision mapping of land use (Zhang and Li 2022; Parece and Campbell 2015), aiding in devising better land use policies (Roy and Murthy 2009), decipher hydrological patterns (Sytsma et al. 2020; McColl and Aggett 2007), and hydrogeomorphic delineation through new GIS-based toolkit (David et al. 2023) for morphometric parameters (Mukherjee and Patel 2022; Sahu et al. 2018; Bhatt and Ahmed 2014), providing a holistic understanding of watershed dynamics. Moreover, applying geostatistical techniques enhances hydrological modelling accuracy (Ferreira et al. 2021; Meyer et al. 2016), allowing for a nuanced analysis of surface and sub-surface water interactions. The combination of these approaches empowers decision-makers for effective watershed planning to mitigate droughts (Drisya and Sathish Kumar 2023; Escobedo Garcia and Ulibarri 2023; Cai et al. 2015), flood risks (Hasanuzzaman et al. 2022; Asdak and Supian 2018; Yazdi et al. 2013), optimise water allocation (Zehtabian et al. 2023; Molinos-Senante et al. 2014), and address challenges induced by climate change (Tsakiris and Loucks 2023; Lemieux et al. 2014; Sivakumar 2011). By intertwining geospatial and geostatistical methodologies, a comprehensive framework emerges, fostering sustainable practices in RBM that balance the intricate interplay of surface and sub-surface hydrology. Geographic Information Systems (GIS) and remote sensing (RS) have transformed how we perceive and manage river basins. GIS allows for the integration of diverse socio-spatial data (Albaladejo-García et al. 2023; McLain et al. 2017), such as population density, topography, land use, and hydrological parameters, providing a comprehensive understanding of the basin's characteristics across varying scales (Asgari 2021). Remotely sensed satellite imagery and aerial photography enable real-time monitoring of land use and land cover (LULC) changes (Mashala et al. 2023; Mukherjee and Patel 2022; Weslati et al. 2020), fluctuations in water levels and quality of water (Kalyanapu et al. 2023; Younos and Heyer 2015), land degradation (Gabriele et al. 2023; Dubovyk 2017), desertification (Mashala et al. 2023; Kumar et al. 2022), and vegetation health (Kureel et al. 2022; Houborg et al. 2015), offering valuable insights for environmental planners. Understanding how land is utilised within a basin is crucial for assessing its impact on erosion and overall riverine ecosystem, riparian vegetation (Pace et al. 2022), and landscape ecological health (Du et al. 2023; Heggem et al. 2000). Geospatial modelling and tools facilitate the creation of accurate LULC maps, helping to identify areas vulnerable to soil erosion (Pal 2016; Biswas et al. 2015), deforestation based on multi-criteria parameters (Mekasha et al. 2020; Monjardin-Armenta et al. 2020), or urbanisation (Dhanaraj and Angadi 2022; Bharath et al. 2018). This information is invaluable for developing targeted strategies to address specific challenges and promote sustainable land use practices (Moghadam et al. 2023; Balasubramani 2018). Furthermore, geospatial techniques aid in watershed and sub-watershed delineation (Singh et al. 2023; Bharath et al. 2021) and hydrological modelling (Mishra et al. 2023), allowing for the precise mapping of the fluvial networks (Gautam et al. 2023; Sajadi et al. 2022), flow type (Mani et al. 2022), drainage pattern (Trong et al. 2023), stream density, and drainage texture based on open-access digital elevation models (DEM) (Oyedotun 2022; Das et al. 2016). This information is essential for predicting surface runoff and designing effective water management infrastructure through site suitability analysis of check dams within the watershed.

Other than geospatial solutions to quantify and manage watersheds, the advent of the latest technology, like the digital twin, is revolutionising RBM by creating virtual replicas of real-world river systems (Botai et al. 2023; Morlot et al. 2023). This innovative application enables stakeholders to monitor, analyse and optimise the entire river basin in a digital environment. By integrating real-time data from various sources such as sensors, satellites, and models, digital twins offer a comprehensive understanding of the river's dynamics, ecosystem services, and riparian health. Countries like China are currently using pilot projects to re-engineer watershed management and water projects under digital twin frameworks (Sheng et al. 2023). Digital twins facilitate effective decision-making by simulating different scenarios and predicting the impact of interventions. To assist decision-making for flood response and water resource management, an innovative dam and watershed management platform has been developed using a geospatial framework that makes use of real-time data and simulation models (Park and You 2023).

Drainage morphometric techniques are salient for a river basin's geometric characterisation and interpretation. It is a quantification of the altitude, volume, area, land profiles, and slope of the study area (Kudnar and Rajasekhar 2020). The term 'morphometry' is the measurement and mathematical analysis of the configuration of part or whole of the earth's surface, shape, and dimension of its landforms (Mukherjee and Patel 2022; Reddy et al. 2004a; Agarwal 1998). The quantitative delineation of various areal and linear parameters based on the extracted stream network (Erosemiah and Viji 2023) from a DEM helps in a comprehensive study of catchment dynamics. The surface drainage characteristics of various river basins worldwide have been quantified, mapped, and studied earlier using conventional methods (Horton 1945; Leopold and Miller 1964; Morisawa 1959; Strahler 1954, 1957, 1964). Such watershed-based morphometric analysis primarily aims to evaluate stream properties by measuring various channel network attributes. Evaluation of morphometric parameters thus necessitates analysis of different drainage parameters, i.e. ordering of various streams, measurement of area and perimeter of the watershed, length of drainage channels, drainage density, drainage frequency, bifurcation ratio, and texture ratio (Kumar et al. 2000)—to mention but a few. To aid this quantitative description of watershed morphometry, measurement of linear and areal features, the gradient of channel network, and contributing ground slopes are required (Vijith and Satheesh 2006; Nag and Chakraborty 2003; Reddy et al. 2004a, b; Kumar et al. 2000; Krishnamurthy et al. 1996). This analysis is vital for drainage morphometry and its interactive impacts on landforms and their characteristics (Reddy et al. 2004a). Sub-basin prioritisations and morphometric analysis have been the most widely used method by researchers to discern a range of environmental applications in watershed science, for instance, management of floods (Sutradhar and Mondal 2023; Ghasemlounia and Utlu 2021), flash floods (Akay and Baduna Koçyiğit 2020), soil (Ali et al. 2024; Verma et al. 2023), and water (Pathare and Pathare 2021) conservation and groundwater recharge potential (Sahoo et al. 2023; Choudhari et al. 2018).

Catchment morphometry plays a pivotal role in crafting a well-defined and effective RBM plan. Firstly, morphometry aids in identifying vulnerable areas prone to erosion (Verma et al. 2023), sedimentation (Panwar et al. 2017), and other geomorphic changes (Ali et al. 2024). This information is indispensable for implementing targeted interventions to mitigate soil erosion, safeguarding soil fertility, and preventing downstream impacts. Secondly, morphometric analysis contributes to flood risk assessment (Sutradhar and Mondal 2023; Ghasemlounia and Utlu 2021; Akay and Baduna Koçyiğit 2020) by delineating the drainage network and identifying areas susceptible to inundation. This knowledge is essential for developing flood management strategies, designing infrastructure, and implementing land use policies to minimise the impact of flooding on communities and ecosystems. Moreover, morphometric parameters serve as input data for hydrological models, enabling the simulation of various scenarios (Basahi et al. 2016). Morphometry-based hydrodynamic modelling (Ozdemir and Akbas 2023) helps predict water runoff, streamflow patterns, and the response of the river basin to different hydro-metrological scenarios. Such predictive capabilities are essential for crafting resilient management plans that account for seasonal variations, climate change impacts, and human activities. In essence, morphometry acts as the foundation for informed decision-making in RBM. It provides a roadmap for effective resource utilisation, hazard mitigation, and preserving biodiversity, contributing to long-term sustainable water resource management within the river basin.

Achieving the UN Sustainable Development Goals (SDGs) at the river basin level (He and James 2021) can be significantly enhanced through the integration of advanced geospatial and morphometric approaches. These technologies provide a robust framework for addressing multiple SDGs simultaneously. Morphometric analysis aids in understanding basin dynamics, supporting SDG 15 (Life on Land) (Zhou et al. 2022) through habitat preservation and erosion control. By employing these approaches, SDG 13 (Climate Action) can be addressed through the assessment of climate change impacts on river basins (Kørnøv et al. 2020), altering agriculture and water systems (Nhemachena et al. 2020; Campbell et al. 2018). Integrating geospatial data assists in achieving SDG 11 (Sustainable Cities and Communities) (Avtar et al. 2020; Im 2020) by optimising urban planning around water resources. Additionally, these technologies facilitate informed decision-making, promoting SDG 17 (Partnerships for the Goals) through collaboration among stakeholders (Sousa et al. 2021) on a sub-national to a global scale. Thus, the latest geospatial and computational technologies provide a holistic understanding of the basin's characteristics, dynamics, and vulnerabilities, enabling decision-makers to develop targeted and sustainable strategies to achieve the SDGs (Sajeva et al. 2020). As we confront the challenges of a changing climate, growing demand for water resources, and threats to food security, the continued advancement and application of these technologies are essential for ensuring the resilience and health of river basins.

Therefore, geospatial-based sub-basin prioritisation and catchment morphometric analysis is a powerful medium for natural resource management (Shelar et al. 2022) related to land, soil, surface and ground water, agriculture, and natural vegetation. The present chapter aims to delineate a new and lucid risk index based on erosivity assessment of the sub-basins of the Wainganga watershed considering a composite framework by combining sub-basin scale morphometric prioritisation along with present-day population density and agricultural area. Various case studies from the arid and semi-arid granitic hard rocky terrain of central India including those from Bundelkhand (Singh et al. 2023), Marathwada (Sahu et al. 2018), and Deccan Volcanic Province region (Manjare et al. 2019; Khanday and Javed 2016) on sub-basin scale mophometrics have highlighted the vitality of such research in deciphering the potential erosion risk and hydro-geomorphological alterations in different sub-basins for watershed management. The Wainganga River basin (WRB) is a vital source of sediment flux and water supply in the larger fluvial system of the Godavari river basin. Many scholars have studied the WRB based on different aspects of environmental alterations like precipitation (Taxak et al. 2014), anomalies in rainfall trends (Nair and Mirajkar 2022; Kudnar et al. 2022), spatio-temporal shifts in temperature (Thakur et al. 2020), streamflow (Das and Nanduri 2018), topography, hydrology, morphology (Kudnar 2020a, b), morphometric quantification and erosional cycles (Kudnar and Rajasekhar 2020), fluctuations in groundwater level (Chandan and Yashwant 2017), and estimation of flood frequency in data scarce environment (Vesuviano et al. 2022). Results from the recent meteorological and landscape research works on the WRB have raised curiosity about effective RBM planning based on the integration of watershed and environmental science. Hence, this article seeks to develop a new erosivity risk index based on the integration of multiple parameters, including drainage basin morphometry, LULC changes, and the dominant anthropogenic parameters like farming area and population density.

2 The Geographical Setting

The WRB is situated in the subcontinental interior part of peninsular India forming a major contributing sub-basin to the central Godavari River basin, the largest catchment of southern India (Kudnar 2020a, b; Kudnar and Rajasekhar 2020). The location extends from 78°00′ to 81°00′ E longitudes and 19°60′ to 22°07′ N latitudes, as shown in Fig. 1. The total basin area is 51,421 km^2 with a relative relief of 1064 m. The regional physiographic composition of WRB is comprised of mountains (10.56%), denuded plateau (33.92%), and extensive plain (55.51%) (Kudnar 2020b). The basin is bounded in the north by the extended basaltic hills of the Deccan trap region, the Eastern Ghats to the south and east, and the Maikala hill range in the west. The Chiroli hills form the watershed divide between the Wainganga and Narmada basins. The highest elevation is found in the north-western and along the eastern part of the river basin, represented with higher to lower shades of brown followed by the shade of yellow, representing the comparative lower elevation denuded by the initial stream orders/headwaters, followed by the entire central section of the catchment represented in shades of green eroded away by the Wainganga River along with its principal tributaries (Fig. 1). The WRB lies in the trijunction of three central Indian states, namely, Maharashtra (27,350 km^2), Madhya Pradesh (23,109 km^2), and Chhattisgarh (962 km^2), hence having a spatial spread across twelve districts from the mentioned states. The Wainganga River intially flows westwards and then turns southwards in Madhya Pradesh. It continues to flow southwards through Maharashtra until its confluence with the Godavari River near Ashti town (Taxak et al. 2014).

The climatic regime within the WRB is characterised by tropical wet and dry and semi-arid seasons after the Köppen-Geiger climate classification (Beck et al. 2018). Semi-arid summers are predominantly sustained from March to May, while the monsoon season is from June to September, with some local precipitation in the post-monsoon season as well. The mean daily maximum temperature ranges from 26 to 30 °C in July and 32 to 33 °C in October. The minimum temperatures observed vary within a range of 10 to 15 °C in the winter month of January. A rainfall of ~ 150 cm is received during the monsoon months, and there are an insignificant number of water conservation structures on the main river. Thus, only a few irrigation schemes exist in the watershed (Taxak et al. 2014).

The geological formations of the WRB (Fig. 2i) have a combination of igneous and metamorphic rock complexity in the northern and central sections, followed by a sedimentary environment towards the southern part of the basin. Basaltic Deccan trap covers the entire northern section of the WRB, followed by the gneissic complex from

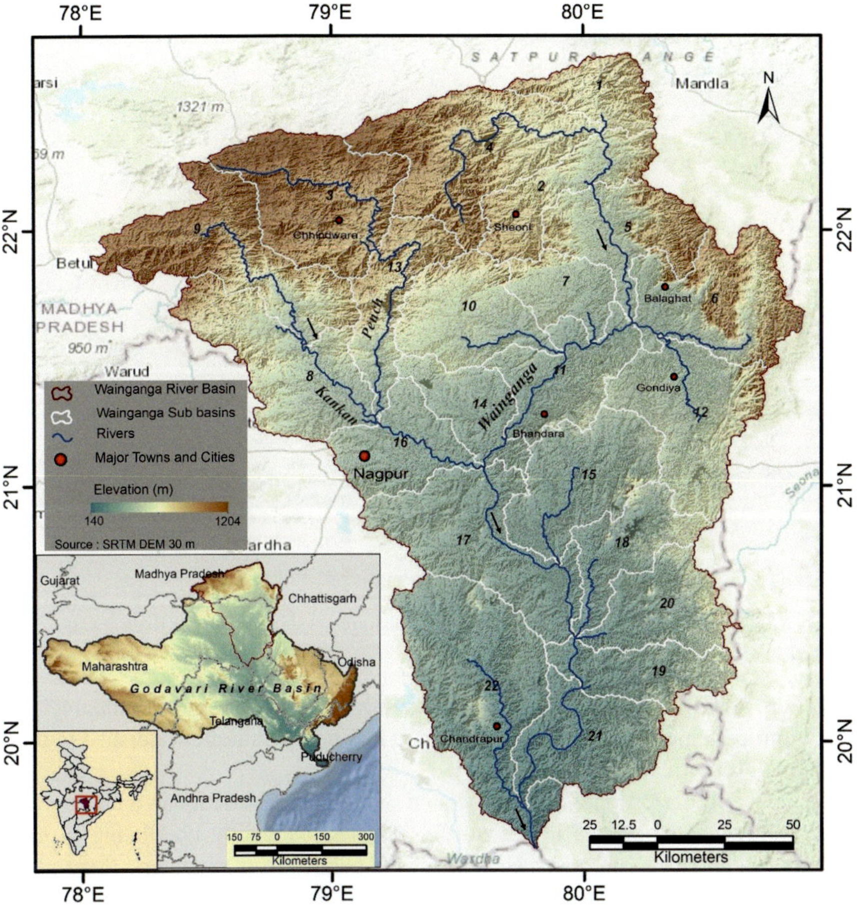

Fig. 1 Locational setting of the Wainganga River basin, central India

the Achaean era, covering 48.18% (Kudnar 2020a). The other prominent geological units observed are the granitic complex of Dongargarh, with a few patches of intrusive rock groups and undifferentiated fluvial and aeolian deposits throughout the WRB. The geomorphology of the WRB (Fig. 2ii) explains the dynamics of the earth's surface processes forming various types of landforms. Dissected structural and denudational hills, valleys, and plateaus are found in the northern, north-eastern, north-western, and southwestern sections of the watershed, which continuously progress towards the formation of an extensive pediment pediplain complex. Older alluvial plains and active and older flood plains are associated with the fluvial activities of the Wainganga River and its major tributaries. Water bodies like ponds, lakes, and some other unclassified water areas or storages indicate the vitality of water conservation in the WRB, which is important for watershed management strategies,

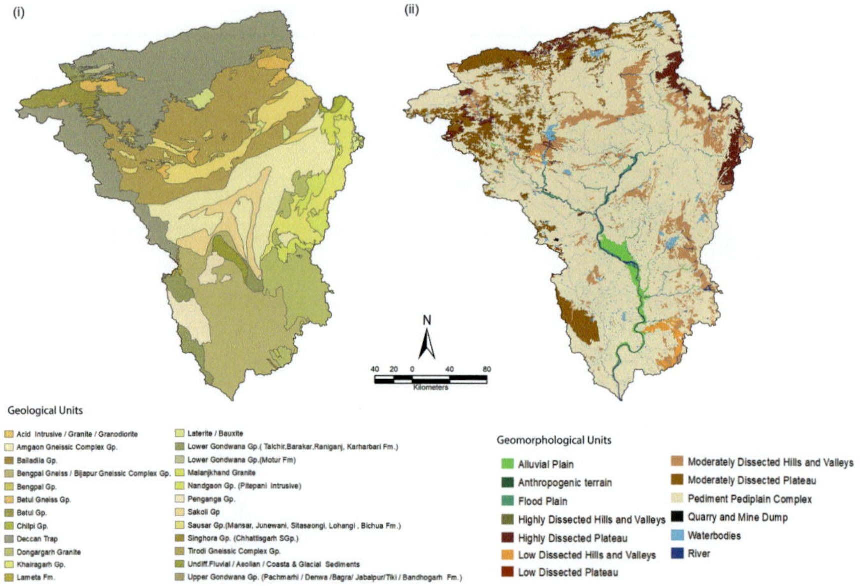

Fig. 2 **a** Geological units and **b** geomorphological features in the Wainganga River basin

irrigation, and drought mitigation projects. Active quarry mines and some unclassified anthropogenically altered landscapes are mapped within the watershed to mark the human footprints.

Degrees of high slopes are associated with the denudational and dissected hills, valleys, and plateaus in the north, followed by gentle undulating grounds with minimum slope are observed towards the central river valley and flood plains (Fig. 3i). The drainage density (Fig. 3ii) explains the presence of total length of the rivers from all stream orders per km^2 (Horton 1945) within the watershed. Initiation and confluence of major first to third-order streams in the northern and north-western parts of the river basin explain the high density of drainage, varying between 3.2 and 3.8 km/km^2. This progressively diminishes towards the central section of the basin, which is mainly drained by higher stream orders.

Soil types for the WRB (Fig. 3iii) were obtained from the Digital Soil Map of the World database to gain an overview of the soil genesis. The process and nature of soil depend on the underlying geology and the resultant geomorphological processes. Chromic vertisols and lithosols represent moderate to high soil erodibility across the entire central and adjoining areas of the WRB. The pellic vertisol group of soil found towards the mouth of the river basin has the highest erodibility marked by extensive riverine and human actions. The vertic cambisols and ferric luvisols have a low impact on soil erodibility because of the comparative higher elevation and rigid geological formation of the area that brings down its soil erodibility.

Therefore, the importance of underlying geology, exogenic and endogenic processes, and considerable changes in the climate influence the development of

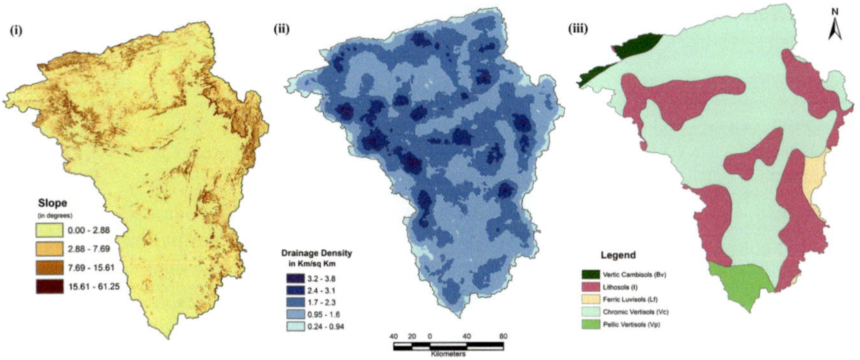

Fig. 3 **a** Distribution of slope, **b** drainage density, and **c** types of soil in the Wainganga River basin

drainage networks, the genesis and morphology of landforms, soil properties, and land degradation processes. It is a well-known fact that different landform features, in conjunction with certain types of parent material and along the local drainage, greatly influence soil type that forms in topographic sequence under specific geo-pedological conditions. Soil drainage reflects the local conditions such as surface drainage conditions, sub-surface geology, type of landform, slope, and aspects important for landscape planning and implementing blueprints of RBM.

3 Materials and Methods

Different digital datasets (raster and vector) obtained from various international and national agencies have been used for the present study (Table 1). The workflow and methods for delineation of the sub-basin scale erosivity risk index have been represented in Fig. 4. Importance and use of open-access datasets have been on the rise since the past decade and have been delivering accurate results. This case study is entirely based on various resolution datasets which have been integrated in GIS environments using ArcMap 10.3 and QGIS 3.28.8. The geostatistical analysis for the study has been evaluated in IBM SPSS 23. The rest calculations were all performed in MS Excel.

The SRTM DEM 30 m has been obtained for the study area to delineate the watershed, drainage, and sub-watershed properties. The hypsometric integral (HI) is evaluated based on the elevation-relief ratio method (Pike and Wilson 1971). Contour lines extracted from the DEM have been used to enumerate the HI value. The areal morphometry has been worked out in the GIS platforms, while, the histogram function in MS Excel was utilised to classify the shape-based morphometric characteristics for the sub-basin prioritisation and ranking. Statistical evaluation of the sub-basins is done using the principal component analysis (PCA) method to obtain the first factor-loading matrix and, thereafter, the rotated loading matrix using orthogonal

Table 1 List of different raster and vector datasets used in the study

Attributes	Data product	Resolution/scale	Source	Web link	Time period
Elevation, slope, drainage, catchments	SRTM DEM v.3.0	30 m	USGS Earth Explorer	https://earthexplorer.usgs.gov/	2014
Geology	Bhukosh	1:2 million	GSI	https://bhukosh.gsi.gov.in/Bhukosh/Public	2023
Geomorphology		1:250,000			
Soil	Digital Soil Map of the World v3.6	1:5 million	FAO-UNESCO	https://data.apps.fao.org/map/catalog/static/api/records/446ed430-8383-11db-b9b2-000d939bc5d8	2007
LULC	Sentinel-2 based LULC	10 m	ESRI	https://www.arcgis.com/apps/instant/media/index.html?appid=fc92d38533d440078f17678ebc20e8e2	2021
	IRS LISS-III LULC	1:50,000	NRSC, ISRO	https://www.nrsc.gov.in/EO_LULC_Objective?language_content_entity=en	2005
	Decadal LULC Classifications across India	100 m	ORNL DAAC, NASA	https://daac.ornl.gov/VEGETATION/guides/Decadal_LULC_India.html	1985
Population density	Population counts (population per pixel) v.2.0	100 m	WorldPop Hub	https://hub.worldpop.org/geodata/summary?id=92	2020
India Administrative Boundaries	National and state boundaries	–	Survey of India	https://onlinemaps.surveyofindia.gov.in/Home.aspx	2023

Note: All web links are accessible as on 27 March 2024

transformation. The scree plot and hierarchical clustering (dendrogram) have been computed in the statistical software to reflect the findings of compound scores and PCA. The LULC products were acquired from different sources for the analysis for years—1985 (Oak Ridge National Laboratory (ORNL) Distributed Active Archive Center (DAAC)) with a spatial resolution of 100 m; 2005 (IRS LISS-III LULC, NRSC, ISRO) with a spatial resolution of 30 m, and 2021 (Sentinel-2 based land

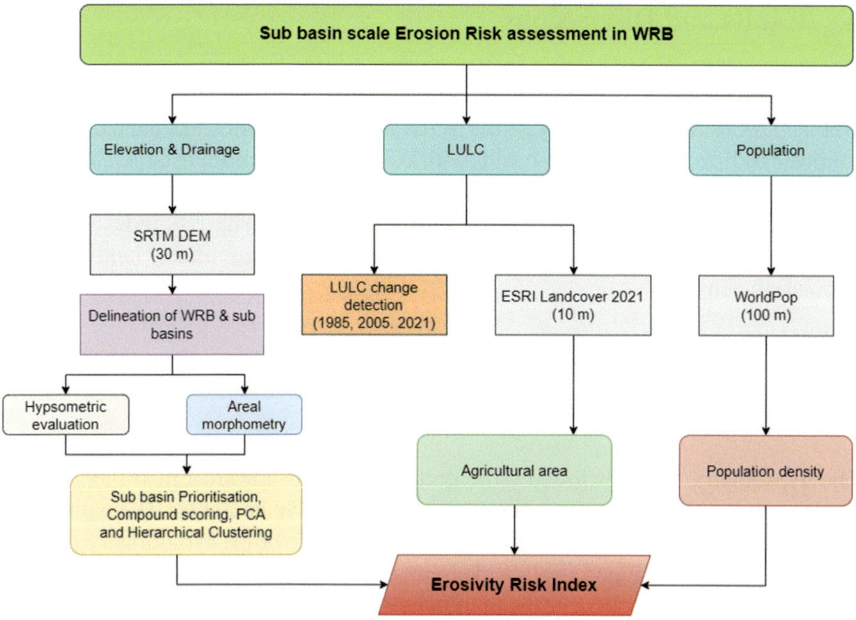

Fig. 4 Methodology diagram showcasing the workflow

use/land cover time series of the world, produced by Impact Observatory, Microsoft, and Esri) with a spatial resolution of 10 m. LULC maps were prepared based on resampling technique where the resolution was set to 100 m for maintaining uniformity during the analysis. The area under agriculture was extracted from the ESRI's Sentinel-2-based LULC 10 m using ArcMap 10.3's extract by attribute feature tool. For calculation of the erosivity risk index, agricultural area pixels were extracted for each sub-basin using zonal statistics and then computed by multiplying the pixel count (sensor resolution 10 m). Population density for evaluating the erosivity risk has been obtained from the population count (per pixel) dataset of the WorldPop Hub by the School of Geography and Environmental Science, University of Southampton, UK. Using the zonal statistics tool, the population density for each sub-watershed was determined. The estimation of erosivity risk is determined based on the product of normalised scores of C_p value, agricultural area, and population count for each sub-basin (refer to Table 7). Finally, the histogram tool in MS Excel was used to categorise the calculated risk for each sub-basin from high to low.

4 Results and Discussions

This section contains six sub-sections describing and comparing the outcomes of the article. The areal morphometry of the WRB and its sub-basins has been computed, followed by prioritising the sub-basins with a compound score. Geostatistical analysis of the sub-basins is done to cross-validate the findings of priority ranking. A temporal mapping of the LULC for the WRB, followed by an erosivity risk index, has been calculated based on the integration of compound scores, agricultural land area, and population density.

4.1 Basin Hypsometry

The hypsometric curve is found to be quite curvilinear upwards, near the X and Y axes, representing cumulative area (a/A) and cumulative height (h/H), respectively, indicating that the WRB is in its monadnock (old) stage of the evolutionary process. The value of the HI (Fig. 5) obtained signifies that only 24.56% of the land surface remains to be eroded to achieve the final stage of gradation (base level) where the river will neither have erosion nor deposition (Kesseli 1941). On contrary to HI, Strahler (1952) defined erosional integral (EI) as the total rate of erosion done by a fluvial system to a present-day landscape. For the WRB, the value of EI is 75.44%, which means that the river has already eroded more than 75% of the existing landscape within the watershed and is presently in its senile old stage. The flat shape of the WRB derived from morphometric calculations like circulatory ratio, form factor, and compactness coefficient also validate the values of HI and EI (Table 2).

4.2 Shape-Based Watershed Morphometry

Quantification of a landscape has been practised since the 1940s based on three main aspects—linear, areal/shape, and relief. Morphometry is the physical measurement of any watershed that correlates hydrology, geomorphology, and other consistent environmental parameters (Mahala 2020; Aher et al. 2014). This helps in providing a comprehensive overview for efficient planning to enroute RBM and conservation measures. Physical dimensions or shape-based morphometry of a catchment is vital in understanding the nature and character of the river basin at various spatial scales, varying from the entire watershed to the smallest sub-watershed. Table 2 exhibits and discusses eight areal morphometric parameters that help in understanding the watershed's nature and behaviour in terms of climatic influence shaping the hydro-morphological scenario.

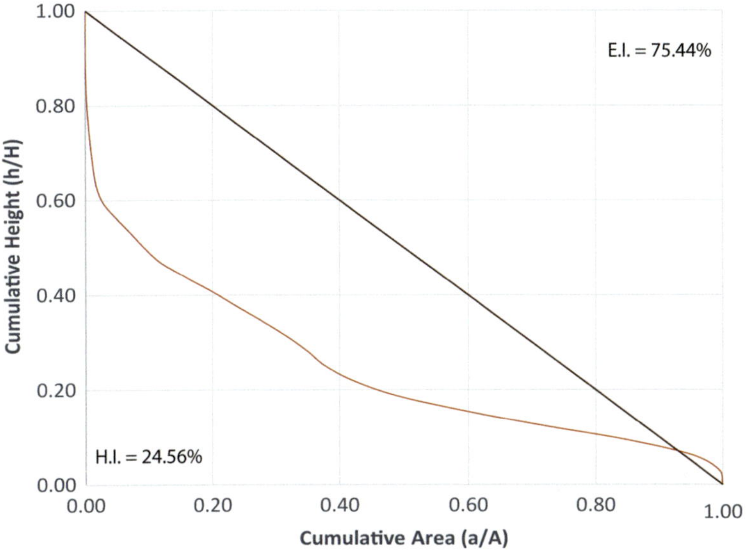

Fig. 5 Hypsometric curve of Wainganga River basin

4.3 Sub-basin Prioritisation

To decipher and evaluate the physical conditions of a larger watershed, it is essential to disintegrate it into further smaller units or sub-watersheds which can be ranked and prioritised (Ali et al. 2024; Bharath et al. 2021; Manjare et al. 2018). This is beneficial for carrying out RBM plans for estimation of runoff potential (Shaikh et al. 2022), water and soil conservation (Verma et al. 2023; Pathare and Pathare 2021), sustainable agricultural practices (Mokarram and Hojati 2017), sediment yield and site suitability for natural embankments or check dams (Nooka Ratnam et al. 2005) for better irrigation and afforestation policies. The linear parameters of geomorphometry, such as stream frequency, drainage density, drainage texture, bifurcation ratio, length of overland flow, and others, share a direct or positive relationship, i.e. higher the value, more the erodibility. While, the areal parameters such as elongation ratio, compactness coefficient, circularity ratio, basin shape, and form factor serve an inverse or negative relationship with erodibility, i.e. lower the value, the more the risk of erodibility and vice-versa. A compound score (C_p) is calculated as the average of all the values from the eight shape-based parameters for individual sub-basins. Based on the C_p value, sub-basins having the least rating value have been assigned with the highest priority (rank 1), the next value was assigned second priority (rank 2), and so on till the highest C_p value is given the least priority as rank 18 (Table 3). The spatial distribution of each morphometric parameter is mapped in Fig. 6. Based on priority and ranks assigned to each sub-basin, these are grouped into three categories: high, medium, and low priorities (Table 3). The prioritisation analysis results exhibit that sub-basins—1, 2, 3, 4, 6, 8, 14, 16, 17, and 21 are the zones having low composite

Table 2 Catchment morphometry based on eight areal parameters

Parameters	Formula	Reference	Significance	Value
Index of shape (I_s)	$I_S = L^2/A$	US Corps of Engineers	Shape-based index is defined as the ratio of the square of the basin length to the area of the basin This was one of the earliest indices to judge the shape of the basin based on the length and area of the basin	2.28
Elongation ratio (R_E)	$R_E = D/L$	Schumm (1956)	Elongation ratio is the ratio of the diameter of a circle of the same area as the basin to the maximum basin length. The varying slopes of a watershed can be classified with the help of the index of elongation ratio, i.e. circular (0.9–0.10), oval (0.8–0.9), less elongated (0.7–0.8), elongated (0.5–0.7), and more elongated (< 0.5)	0.75
Circularity ratio (C_R)	$C_R = 4\Pi A/P^2$	Miller (1953)	For the outline form of the watershed (Strahler 1964; Miller 1953) a dimensionless circularity ratio as a quantitative method. C_R is defined as the ratio of watershed area to the area of a circle having the same perimeter as the watershed and it is pretentious by the lithological character of the watershed. The circularity ratio value (0.27) indicates that the watershed is circular, with high discharge and runoff with low subsoil permeability	0.27
Form factor (F_F)	$F_F = A/L^2$	Horton (1945)	The form factor is the ratio of the basin area to the square of the basin length. $F_F < 0.754$ is meant for a perfect circular watershed. Smaller F_F value leads to more elongated watershed shapes. A higher value of form factors is related to high peak flows of shorter duration, whereas a low form factor ranging from 0.42 indicates the basin to be elongated in shape and discharge for a longer duration	0.44

(continued)

Table 2 (continued)

Parameters	Formula	Reference	Significance	Value
Compactness coefficient (C_c)	$C_C = P_B/(2(\Pi.AB)^{0.5})$	Gravelius (1914)	The compactness coefficient is the ratio of the perimeter of the watershed to the circumference of the circular area, which equals the area of the watershed. The Cc is independent of the size of the watershed and dependent only on the slope. So, the Cc value is suggestive of a round shape and flooding	1.93
Length–width ratio (LW_R)	$LW_R = D_L/W$	Horton (1932)	It is stated that the ratio of length and width explains the maximum length and width dimension of a basin. Longer length along with shorter width explains that the basin is elongated and is in its primary stage of evolution and or structurally controlled, while the larger width and shorter length explain the typical size of a flood prone circular basin with high discharge over a long period. The LW_R for the Wainganga watershed is 1.17 suggesting that the basin is circular and is hence flood prone	1.17
Lemniscate ratio (L_R)	$L_R = Lb^2/A$	Chorley et al. (1957)	The lemniscate's value is used to determine the slope of the basin. In the formula, Lb is the basin length (km) and A is the area of the basin (km^2). The lemniscate value of 0.57 exhibits that the watershed occupies the minimum area in its inception area with a large number of initial stream orders	0.57
Ellipticity ratio (E_R)	$E_R = \pi \times Vl^2/4A$	Stoddart (1965)	Ellipticity is the degree of deviation from circularity (or sphericity). The E_R explains how much the river basin deviates from being circular. Here the value of 1.79 shows its nearness to a circular shape	1.79

scores and were considered under high priority. These sub-basins require immediate attention and effective planning to consider lesser hydro-morphological alterations. Out of the remaining ten sub-basins—5, 7, 9, 10, 11, 12, 15, 19, 20, and 22 are categorised under medium priority, while low-prioritisation is represented by sub-basins 13 and 18. Those having high $\underline{C}_p$ scores are in the most stable form in terms

Table 3 Compound value, ranking, and prioritisation of shape-based morphometric parameters

Sub basin	I_S	C_R	R_E	F_F	L_R	C_C	E_R	LW_R	C_p score	Priority ranking	Final priority
1	1.10	0.33	1.08	0.91	0.27	1.73	0.86	0.62	0.86	1	High
2	0.99	0.23	1.13	1.01	0.25	2.08	0.78	1.00	0.93	4	High
3	1.59	0.44	0.89	0.63	0.40	1.51	1.25	1.17	0.99	7	High
4	1.50	0.32	0.92	0.67	0.37	1.77	1.18	0.70	0.93	4	High
5	1.82	0.28	0.84	0.55	0.45	1.90	1.43	0.91	1.02	9	Medium
6	0.73	0.27	1.32	1.37	0.18	1.92	0.57	1.00	0.92	3	High
7	1.80	0.37	0.84	0.56	0.45	1.65	1.41	0.97	1.01	8	Medium
8	1.02	0.28	1.12	0.98	0.25	1.88	0.80	1.56	0.99	7	High
9	2.74	0.22	0.68	0.37	0.68	2.12	2.15	1.47	1.30	15	Medium
10	2.37	0.35	0.73	0.42	0.59	1.68	1.86	1.30	1.16	12	Medium
11	0.24	0.22	2.30	4.16	0.06	2.12	0.19	0.69	1.25	14	Medium
12	2.46	0.31	0.72	0.41	0.62	1.81	1.94	1.20	1.18	13	Medium
13	4.89	0.21	0.51	0.20	1.22	2.16	3.84	1.84	1.86	18	Low
14	0.79	0.27	1.27	1.27	0.20	1.93	0.62	0.59	0.87	2	High
15	1.95	0.33	0.81	0.51	0.49	1.74	1.53	1.65	1.13	11	Medium
16	1.37	0.39	0.96	0.73	0.34	1.61	1.08	1.06	0.94	5	High
17	1.44	0.25	0.94	0.69	0.36	2.01	1.13	1.00	0.98	6	High
18	3.63	0.23	0.59	0.28	0.91	2.07	2.85	1.43	1.50	17	Low
19	0.31	0.25	2.03	3.25	0.08	1.98	0.24	0.46	1.08	10	Medium
20	2.32	0.30	0.74	0.43	0.58	1.82	1.83	1.29	1.16	12	Medium
21	1.72	0.36	0.86	0.58	0.43	1.68	1.35	0.98	0.99	7	High
22	2.78	0.29	0.68	0.36	0.70	1.85	2.19	1.70	1.32	16	Medium

of erodibility as evident from their shape-based morphometric values. Constricted shapes with high values of elongation ratio and low circularity ratio also validate the point that these sub-basins are less prone to erosion with high values of mean bifurcation ratio (R_b) and HI. The medium ranks are assigned to those sub-basins that lie within an average C_p score and determine the progressive stage of evolution, which can be further justified with the values of HI and mean R_b.

4.4 Statistical Analysis of Sub-basins Using PCA and Hierarchical Clustering

An intercorrelation among the eight shape-based morphometric parameters for the twenty-two sub-basins is obtained to explain their physio-environmental conditions corresponding C_p values from the earlier sub-section. The negative values show a

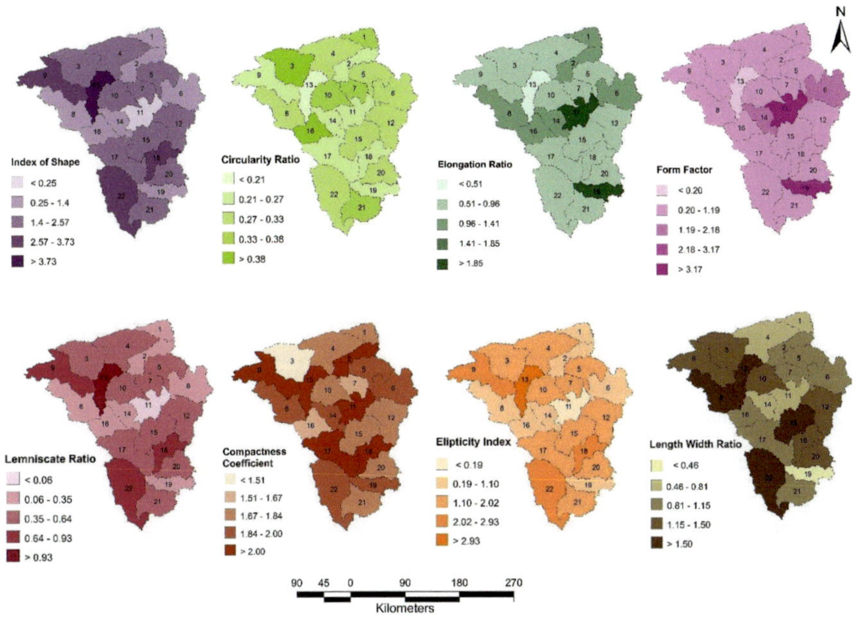

Fig. 6 Shape-based morphometry for the major sub-basins of the Wainganga River basin

declining trend of correlation among the parameters, while those having a value of 1 indicate the perfect correlation. From the correlation matrix (Table 4) of the eight shape-based parameters, the first unrotated factor-loading matrix is obtained. It can be observed from Table 5, that only the first two components, whose eigenvalues are greater than 1, together account for ~ 90.56% of the total variance in the WRB. Thus, two components are enough to explain the entire set of observations for the PCA, meaning that the dataset has a good intermixed bonding among the geomorphic variables. The first component (62.09%) explains more variance than the second component (28.47%). When it is assumed that there are two components, we can predict 90.56% of the information in all eight variables.

The scree plot (Fig. 7i) explains that the third component obtained after PCA shows and cross-verifies that the first two components are well enough to explain the dataset with eigen values of more than 2. The third component here shows the major break point in the scree plot that distinguishes the rest 6 components from the first two. The geometric, morphometric, and morphological attributes of the sub-basins have been analysed to identify classes and sub-classes of similar basins using the hierarchical clustering technique. This is based on the concept of proximity between points distributed in space, the distance between the points (representing areas) being a measure of the overall level of similarity between the areas concerned. A proto-cluster is initially formed by the most similar pair of areas, and the point halfway between them becomes the 'centroid' of this cluster. Other proto-clusters are formed similarly, and early proto-clusters may become amalgamated. This process is

Table 4 Pair correlation matrix between eight shape-based geomorphic variables

Correlation matrix[a,b]		I_S	C_R	R_E	F_F	L_R	C_C	E_R	LW_R
	I_S	1.000	-0.137	-0.782	-0.664	1.000	0.171	1.000	0.759
	C_R	– 0.137	1.000	– 0.241	– 0.302	– 0.137	– 0.988	– 0.137	– 0.072
	R_E	– 0.782	– 0.241	1.000	0.982	– 0.782	0.243	– 0.782	– 0.662
	F_F	– 0.664	– 0.302	0.982	1.000	– 0.664	0.313	– 0.664	– 0.582
	L_R	1.000	– 0.137	– 0.782	– 0.664	1.000	0.171	1.000	0.759
	C_C	0.171	– 0.988	0.243	0.313	0.171	1.000	0.171	0.096
	E_R	1.000	– 0.137	– 0.782	– 0.664	1.000	0.171	1.000	0.759
	LW_R	0.759	– 0.072	– 0.662	– 0.582	0.759	0.096	0.759	1.000

Note: [a] Determinant = 0.000
[b] This matrix is not positive definite

repeated until all points are included in a 'universal cluster', which may be achieved in several ways. The method used here is the average linkage method. The order of combination is known as the 'pairing sequence', and this may be represented by a linkage tree or dendrogram (Fig. 7ii). In the dendrogram, the *Y* axis displays the sub-basins from 1 to 22 and how they are related to the alternative sub-basin, which are marked beside the number of sub-basins. For instance, sub-basin 22 is related to sub-basin 7 in terms of areal morphometry; likewise, sub-basin 21 has no match, while sub-basin 20 is linked to sub-basin 5 and so on. Now, the second group of clustering shows that sub-basin pairs 18 and 16 are connected to pairs 12 and 1; similarly, 7 and 12 are linked to 4 and 18. This second group of clusters is related to the third order, i.e. (18, 16), (12, 1), which is linked with (7, 12) and (4, 18). These succeed on till the entire linkage is established between the groups of sub-basins. The differential basin geology and its effect on the resulting physiography is seen to significantly influence the stream and basin parameters, though often its effect is subtle and cannot be readily or explicitly quantified. Cluster analysis, therefore, provides a good way to create further sub-groups of basins, deriving a degree of homogeneity among them that could prove quite useful when planning and management measures need to be adopted, providing the pathway for each sub-class of basins to be looked into and treated uniquely.

4.5 Land Use and Land Cover Dynamics

Analysing the alterations of LULC is essential to envisage the continuous environmental and anthropogenic transformations. With the advancement of remote sensing technology and availability of various multi-resolution earth observation open-access datasets with global coverage, it has become possible to meticulously map and quantify LULC changes across desired temporal scales (Salhi et al. 2024). LULC is

Table 5 Total variance explained based on PCA method

Component	Initial eigen values			Extraction sum of square loadings			Rotation sums of squared loadings		
	Total	% of variance	Cumulative %	Total	% of variance	Cumulative %	Total	% of variance	Cumulative %
1	4.967	62.091	62.091	4.967	62.091	62.091	4.965	62.068	62.068
2	2.277	28.466	90.557	2.277	28.466	90.557	2.279	28.488	90.557
3	0.404	5.056	95.613						
4	0.339	4.235	99.847						
5	0.100	0.122	99.969						
6	0.002	0.031	100.000						
7	5.900E−16	7.375E−15	100.000						
8	− 2.128E−16	− 2.660E−15	100.000						

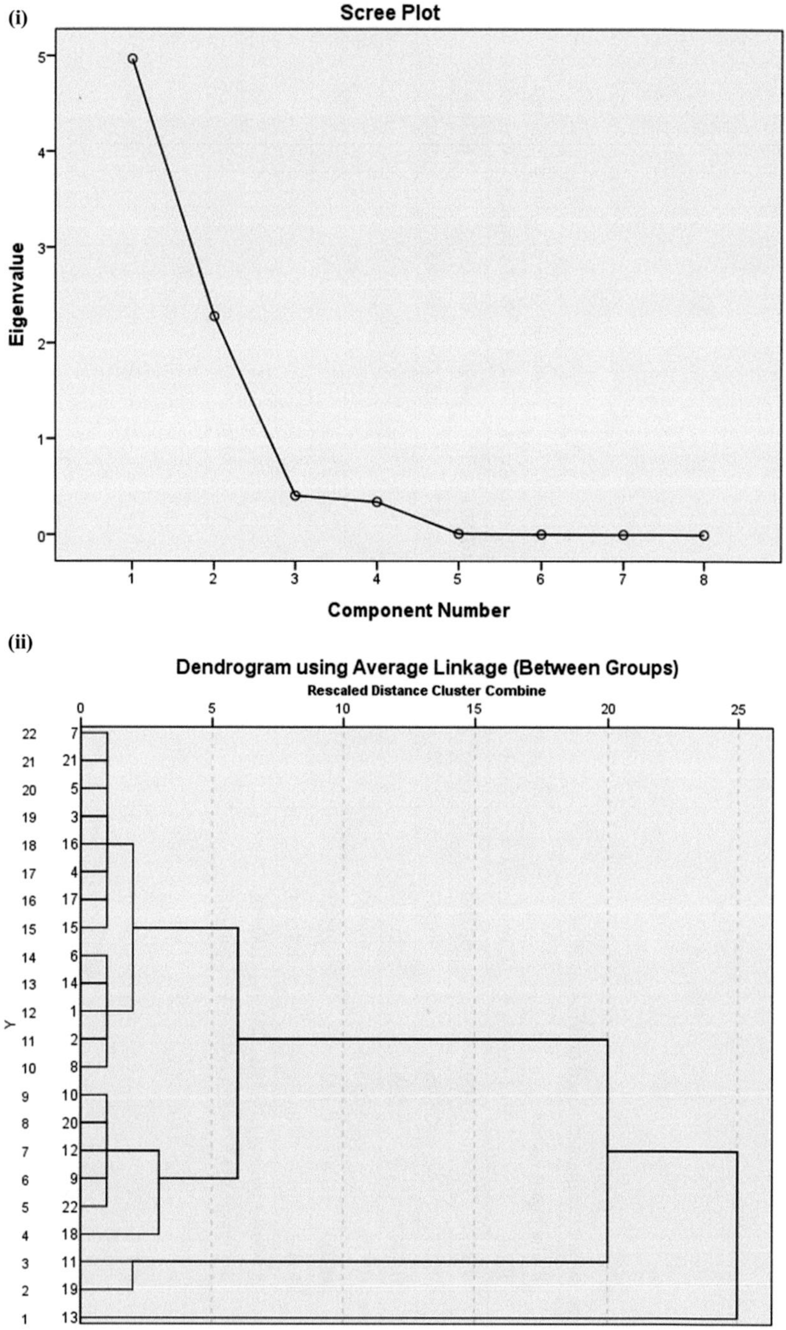

Fig. 7 **a** Scree plot explaining the result obtained from PCA. **b** Dendrogram exhibiting the hierarchical basin cluster analysis

crucial in watershed prioritisation as it directly influences hydrological processes, ecosystem services, water availability (Fan and Shibata 2014), and informed land management. Understanding LULC patterns helps assess the impact on basin health, identifying areas vulnerable to degradation or pollution (Shrestha et al. 2022). Prioritising basins based on LULC inform sustainable watershed management, guiding conservation efforts and land use planning. It enables the identification of critical zones for protection, restoration, or sustainable development, ensuring the resilience of ecosystems and the provision of essential water-related services for communities. Effective RBM, accompanied by LULC analysis, supports integrated and adaptive strategies for water resource management and environmental conservation.

The LULC alterations of the WRB are mapped in Fig. 8 and the change matrix is tabulated in Table 6. Agriculture is the dominant land use class in the WRB. This is followed by dense forest, which shows a decline in area between 1985 and 2005 and a further decrease between 2005 and 2021. So, the change in dense forest area decreased by − 0.38% between 1985 and 2005 and by − 10.31% between 2005 and 2021, suggesting possible deforestation and heightened susceptibility to high runoff, destabilised slopes, local landslides, higher sediment flux in the river, increased evapotranspiration and flooding. Over time, there is a discernible rise in both degraded vegetation and settlement and a drop in fallow land. The implications of these shifts extend to human activity, water supplies, and biodiversity. For sustainable development and informed environmental management, an understanding of LULC dynamics is necessary. Water bodies shows, a significant decline between 1985 and 2005 and a subsequent rise between 2005 and 2021. This may be attributed to the major drought event of 2005, as reported in the EM-DAT database. Dynamic shifts in land cover and land use throughout time explain man-made phenomena, including urbanisation, agricultural expansion, and deforestation. The alterations in LULC may impact runoff, soil stability, and water retention in a manner akin to how changing agriculture, settlement, and degraded vegetation may impact the region's resiliency. Monitoring and managing these LULC changes are essential for sustainable development and checking the risk scenarios.

4.6 Delineation of Erosivity Risk Index

The risk of erosivity is a critical factor in sub-basin prioritisation since it determines an area's susceptibility to soil erosion, sedimentation, and associated environmental issues. The term 'erosivity' describes the ability of rainfall to separate and move soil particles, resulting in soil erosion (Wang et al. 2024; Yin et al. 2017). Table 7 offers a thorough summary of the twenty-two sub-basins and evaluates the erosivity risk susceptibility (score) related to anthropogenic interventions. Normalising the values is a crucial component as it makes easier to evaluate the performance of each sub-basin with the targeted parameters like population count per pixel and agricultural area as mapped in Fig. 9. Here, the erosivity risk index for determining soil erosion is calculated based on the risk equation used in disaster studies, which

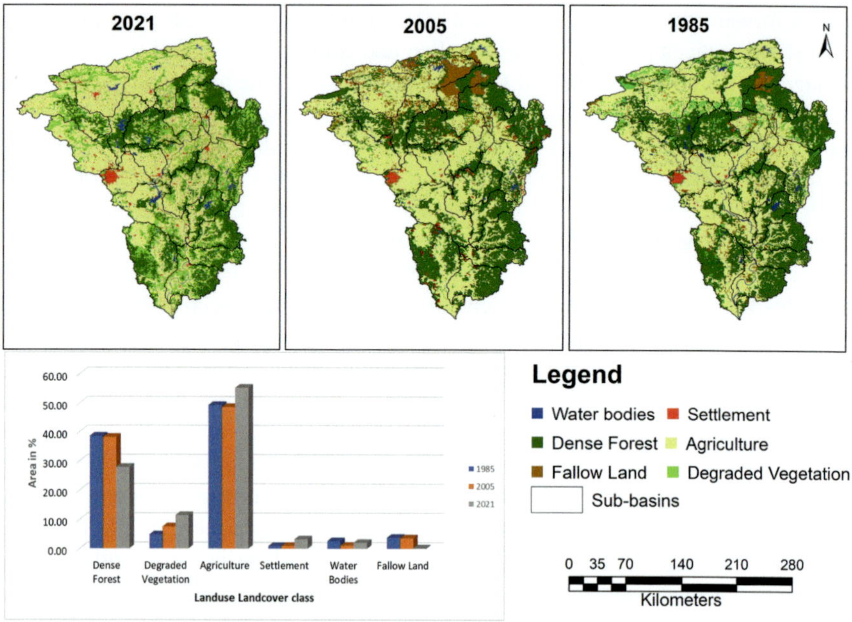

Fig. 8 Temporal mapping of LULC for the Wainganga River basin for the years 2021, 2005, and 1985

is expressed as a product of hazard (C_p score), exposure (population density) and vulnerability (agricultural area). The C_p score describes the physical property of a basin which helps in equating the hazard of the catchment, while, the population density is identified as the aggregate community exposed to a potential threat like soil erosion. The agricultural lands are considered vulnerable to soil loss with major runoff potential because of less informed land and water management factors. Hence, determining the erosivity risk stands on the three foundations of hazard, exposure, and vulnerability factors.

Lower values of the erosivity risk index indicate a more resilient and sustainable watershed, while the higher values are meant to showcase higher risk of soil erosion (Fig. 10). The normalised C_P values show the degree of vulnerability depending on the areal catchment parameters. Sub-basin 13 is particularly noteworthy since it has a normalised C_P value of 1.00, indicating that it is more vulnerable to erosion. Anthropogenic footprints within the sub-basins are clarified by the normalised agricultural area and population metrics. Sub-basin 16 is notable for having a normalised population value of 1.00, indicating a high population density due to the presence of Maharashtra's Nagpur city signifying higher erosivity risk generated from urban runoff posing a challenge for the watershed management. There are prospective places where human activities may considerably contribute to environmental stress,

Table 6 Change matrix of the LULC classes

LULC class	1985		2005		2021		1985–2005		2005–2021	
	Area (km^2)	Area (%)	Area (km^2)	Area (%)	Area (km^2)	Area (%)	Area (km^2)	Area (%)	Area (km^2)	Area (%)
Dense forest	19,855.17	38.54	19,660.99	38.16	14,346.44	27.84	– 194.18	– 0.38	– 5314.55	– 10.31
Agriculture	25,406.56	49.31	25,147.71	48.81	28,456.98	55.23	– 258.85	– 0.50	3309.27	6.42
Settlement	455.57	0.88	368.05	0.71	1663.89	3.23	– 87.52	– 0.17	1295.84	2.52
Degraded vegetation	2488.50	4.83	3914.26	7.60	5883.86	11.42	1425.76	2.77	1969.6	3.82
Fallow land	1959.28	3.80	1873.15	3.64	118.41	0.23	– 86.13	– 0.17	– 1754.74	– 3.41
Water bodies	1359.61	2.64	560.53	1.09	1055.35	2.05	– 799.08	– 1.55	494.82	0.96
	51,524.69	100.00	51,524.69	100.00	51,524.93	100.00				

Table 7 Calculation of erosivity risk index

Sub basin	C_P score	Normalised C_p	Agricultural area (km^2)	Normalised agricultural area	Population density (count/pixel)	Normalised population density	Erosivity risk	Normalised erosivity risk score
1	0.86	0.00	871.14	0.23	12,355.94	0.00	9,289,600.88	0.00
2	0.93	0.07	525.21	0.09	17,814.33	0.01	8,737,803.72	0.00
3	0.99	0.12	2300.38	0.79	225,720.21	0.09	511,920,464.38	0.17
4	0.93	0.07	2841.49	1.00	86,484.08	0.03	228,074,055.07	0.08
5	1.02	0.16	516.33	0.09	54,470.70	0.02	28,734,714.20	0.01
6	0.92	0.06	606.84	0.12	42,074.02	0.01	23,499,909.24	0.01
7	1.01	0.14	534.74	0.09	40,519.51	0.01	21,816,772.74	0.01
8	0.99	0.12	1898.07	0.63	136,687.41	0.05	256,145,041.58	0.09
9	1.30	0.44	1704.03	0.55	56,499.71	0.02	125,605,773.73	0.04
10	1.16	0.30	311.27	0.01	11,683.50	0.00	4,232,528.07	0.00
11	1.25	0.39	871.35	0.23	79,604.91	0.03	86,584,872.47	0.03
12	1.18	0.32	1122.34	0.33	198,965.54	0.08	264,025,538.41	0.09
13	1.86	1.00	776.39	0.19	6143.99	0.00	8,880,596.92	0.00
14	0.87	0.00	974.71	0.27	91,912.59	0.03	77,625,165.09	0.03
15	1.13	0.26	1180.33	0.35	57,834.80	0.02	76,951,798.89	0.03
16	0.94	0.08	1253.49	0.38	2,499,393.02	1.00	2,952,069,941.84	1.00
17	0.98	0.12	1917.81	0.64	77,049.53	0.03	144,569,754.69	0.05
18	1.50	0.64	422.71	0.05	13,088.14	0.00	8,297,744.42	0.00
19	1.08	0.21	293.58	0.00	6722.71	0.00	2,124,995.59	0.00
20	1.16	0.30	359.81	0.03	5268.50	0.00	2,206,292.79	0.00

(continued)

Table 7 (continued)

Sub basin	C_P score	Normalised C_p	Agricultural area (km^2)	Normalised agricultural area	Population density (count/pixel)	Normalised population density	Erosivity risk	Normalised erosivity risk score
21	0.99	0.13	635.07	0.13	17,061.09	0.00	10,766,727.78	0.00
22	1.32	0.46	1558.90	0.50	41,978.92	0.01	86,255,484.70	0.03

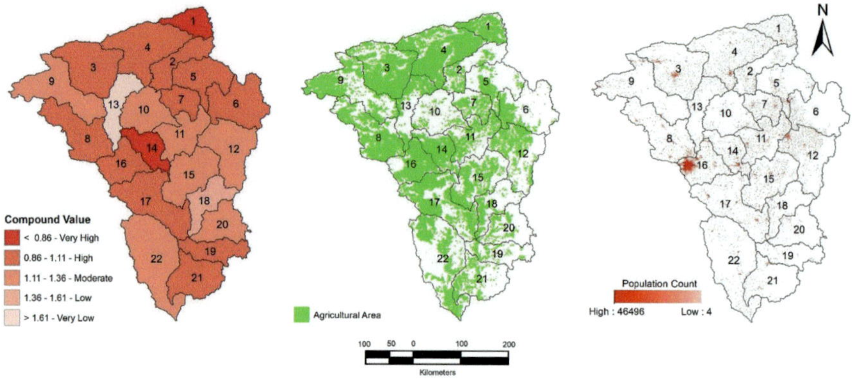

Fig. 9 Mapping of the compound scores, area under agriculture, and population per pixel (density) within the Wainganga River basin

for example, in sub-basins 3, 4, 8, 9, 12, and 22 due to their elevated values in agricultural areas and population density (Table 7). Sub-basin 4 comprises the optimum agricultural area where the normalised score is 1.00.

Therefore, the erosivity risk index compiles a comprehensive evaluation of the sub-basin's vulnerability to soil (splash) erosion through rainfall-induced high overland flow and increased runoff either from farm fields or urban centres. The highest risk region is observed to be sub-basin 16. This emphasises to put on strong conservation plans into place and environmental friendly watershed practices in these areas to reduce potential threats of erodibility. The study draws attention to the complex interactions that exist between human activity, physio-environmental factors, and the risks that arise in different sub-basins. Long-term environmental resilience and sustainable development in the area depend on the implementation of strategies to address the unique issues found in each sub-basin.

For sustainable RBM, erosivity risk integration into sub-basin prioritisation has proved essential. It permits the development of customised approaches to reduce the broader environmental impact of erosivity on ecosystems and communities within a particular watershed, prevent soil erosion, preserve agricultural productivity, and safeguard water security. There is a necessity to put sustainable watershed management techniques into action right away, especially in sub-basins with heavy agricultural demand and dense population. Understanding how physiographic and anthropogenic factors are intertwined is essential for reducing environmental hazards and building disaster resilience.

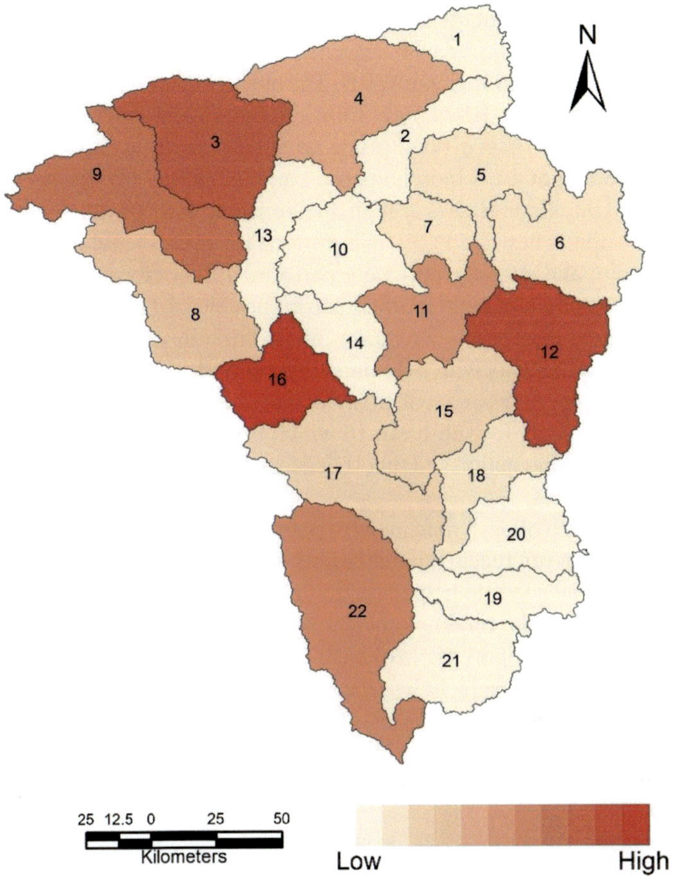

Fig. 10 Erosivity risk index for the Wainganga River basin

5 Conclusion

The physical quantification and anthropogenic factors within a river basin need to be intertwined with the delineation of erosivity zones. This will enhance better policy formulations to implement soil conservation and watershed management strategies so that the objective of prevention of soil loss and efficient water allocation can be fulfilled in reality benefiting the local population. The applications of advanced geospatial and geostatistical approaches for the WRB offer a promising avenue for enhanced RBM through the new erosivity risk index having a combination of sub-basin morphometry and land use parameters like proportion of farmlands and population density.

River basins like that of Wainganga are mainly composed of hard rock granitic terrain with extensive fluvial landforms (active and old floodplains) along the main river channel and its tributaries. The distribution of slope and drainage density conforms the local topography with presence of higher elevation and initial stream orders in the northern section of the WRB. The chapter discusses the old monadnock stage of WRB through the HI 24.56%, followed by, watershed's areal morphometry suggesting a flat oval shaped flood prone river basin with twenty-two major sub-basins. Based on eight areal morphometric parameters, the twenty-two sub-basins were prioritised into three classes as high, medium, and low. To validate the findings, geostatistical approaches like PCA (pair correlation matrix) and hierarchical clustering (scree plot and dendrogram) were performed to conform the priority scores and explain the strength of eight morphometric variables. LULC mapping and quantification explained the primary alterations in the different classes with major rise in agriculture and settlements with a diminishing trend of natural vegetation (forests). On the final stride, the assessment of erosivity risk index showcases a low to moderate normalised score except for sub-basin 16 which is considered to be mostly affected by higher risk of sedimentation and runoff due to the expansion of urban and farmland areas.

Thus, this research offers a road map for putting into practice sustainable methods like afforestation, river reach rehabilitation, ecological restoration drive, natural check dams, to name a few, to help in reducing the hazards associated with erosivity. This shall pave the way for implementation of efficient land use planning, and creating policies that support the resilience of this significant river basin in central India. Various stakeholders, including local communities, administrators, and developmental organisations, must work together going ahead to translate these results into practical policies and ground-level actions for the comprehensive and sustainable management to achieve the SDGs for the WRB.

Acknowledgements The authors would like to express their sincere gratitude to Prof Padmini Pani from CSRD, JNU for her valuable suggestions on the initial draft of the manuscript.

References

Agarwal CS (1998) Study of drainage pattern through aerial data in Naugarh Area of Varanasi District, U.P. J Indian Soc Remote Sens 24:169–175. https://doi.org/10.1007/BF02990795

Aher PD, Adinarayana J, Gorantiwar SD (2014) Quantification of morphometric characterization and prioritization for management planning in semi-arid tropics of India: a remote sensing and GIS approach. J Hydrol 511:850–860. https://doi.org/10.1016/j.jhydrol.2014.02.028

Akay H, Baduna Koçyiğit M (2020) Flash flood potential prioritization of sub-basins in an ungauged basin in Turkey using traditional multi-criteria decision-making methods. Soft Comput 24:14251–14263. https://doi.org/10.1007/s00500-020-04792-0

Albaladejo-García JA, Zabala JA, Alcon F, Dallimer M, Martínez-Paz JM (2023) Integrating socio-spatial preference heterogeneity into the assessment of the aesthetic quality of a Mediterranean agricultural landscape. Landsc Urban Plan 239:104846. https://doi.org/10.1016/j.landurbplan.2023.104846

Ali R, Sajjad H, Masroor M, Saha TK, Rahaman MH (2024) Morphometric parameters based prioritization of watersheds for soil erosion risk in Upper Jhelum Sub-catchment, India. Environ Monit Assess 196(1):1–20. https://doi.org/10.1007/s10661-023-12226-1

Asdak C, Supian S (2018) Watershed management strategies for flood mitigation: a case study of Jakarta's flooding. Weather Clim Extremes 21:117–122. https://doi.org/10.1016/j.wace.2018.08.002

Asgari M (2021) A critical review on scale concept in GIS-based watershed management studies. Spat Inf Res 29(3):417–425. https://doi.org/10.1007/s41324-020-00361-7

Avtar R, Aggarwal R, Kharrazi A, Kumar P, Kurniawan TA (2020) Utilizing geospatial information to implement SDGs and monitor their progress. Environ Monit Assess 192:1–21. https://doi.org/10.1007/s10661-019-7996-9

Balasubramani K (2018) Physical resources assessment in a semi-arid watershed: an integrated methodology for sustainable land use planning. ISPRS J Photogramm Remote Sens 142:358–379. https://doi.org/10.1016/j.isprsjprs.2018.03.008

Basahi J, Masoud M, Zaidi S (2016) Integration between morphometric parameters, hydrologic model, and geo-informatics techniques for estimating WADI runoff (case study WADI HALYAH—Saudi Arabia). Arab J Geosci 9:1–18. https://doi.org/10.1007/s12517-016-2649-6

Beck HE, Zimmermann NE, McVicar TR, Vergopolan N, Berg A, Wood EF (2018) Present and future Köppen-Geiger climate classification maps at 1-km resolution. Sci Data 5(1):1–12. https://doi.org/10.1038/sdata.2018.214

Bharath A, Kumar KK, Maddamsetty R, Manjunatha M, Tangadagi RB, Preethi S (2021) Drainage morphometry based sub-watershed nagitization of Kalinadi basin using geospatial technology. Environ Chall 5:100277. https://doi.org/10.1016/j.envc.2021.100277

Bharath HA, Chandan MC, Vinay S, Ramachandra TV (2018) Modelling urban dynamics in rapidly urbanising Indian cities. Egypt J Remote Sens Space Sci 21(3):201–210. https://doi.org/10.1016/j.ejrs.2017.08.002

Bhatt S, Ahmed SA (2014) Morphometric analysis to determine floods in the Upper Krishna basin using Cartosat DEM. Geocarto Int 29(8):878–894. https://doi.org/10.1080/10106049.2013.868042

Biswas H, Raizada A, Mandal D, Kumar S, Srinivas S, Mishra PK (2015) Identification of areas vulnerable to soil erosion risk in India using GIS methods. Solid Earth 6(4):1247–1257. https://se.copernicus.org/articles/6/1247/2015/se-6-1247-2015.pdf. Accessed 30 Dec 2023

Botai JO, Ghosh S, Matheswaran K, Dickens C, Langa N, Andarcia MG (2023) Options for digital twin application in developing country river basin management: a review. International Water Management Institute (IWMI). Colombo, Sri Lanka. CGIAR Initiative on Digital Innovation, 56p. https://www.researchgate.net/profile/Surajit-Ghosh-5/publication/376185615_Options_for_Digital_Twin_Application_in_Developing_Country_River_Basin_Management_A_Review/links/656cc9b63fa26f66f44d7c5d/Options-for-Digital-Twin-Application-in-Developing-Country-River-Basin-Management-A-Review.pdf. Accessed 30 Dec 2023

Cai X, Zeng R, Kang WH, Song J, Valocchi AJ (2015). Strategic planning for drought mitigation under climate change. J Water Resour Plan Manag 141(9):04015004. https://doi.org/10.1061/(ASCE)WR.1943-5452.0000510

Campbell BM, Hansen J, Rioux J, Stirling CM, Twomlow S (2018) Urgent action to combat climate change and its impacts (SDG 13): transforming agriculture and food systems. Curr Opin Environ Sustain 34:13–20. https://doi.org/10.1016/j.cosust.2018.06.005

Chandan KS, Yashwant BK (2017) Optimization of groundwater level monitoring network using GIS-based geostatistical method and multi-parameter analysis: a case study in Wainganga Sub-basin, India. Chin Geogra Sci 27:201–215. https://doi.org/10.1007/s11769-017-0859-9

Chorley RJ, Malm DE, Pogorzelski HA (1957) A new standard for estimating drainage basin shape. Am J Sci 255(2):138–141

Choudhari PP, Nigam GK, Singh SK, Thakur S (2018) Morphometric based prioritization of watershed for groundwater potential of Mula river basin, Maharashtra, India. Geol Ecol Landsc 2(4):256–267. https://doi.org/10.1080/24749508.2018.1452482

Das J, Nanduri UV (2018) Assessment and evaluation of potential climate change impact on monsoon flows using machine learning technique over Wainganga River basin, India. Hydrol Sci J 63(7):1020–1046. https://doi.org/10.1080/02626667.2018.1469757
Das S, Patel PP, Sengupta S (2016) Evaluation of different digital elevation models for analyzing drainage morphometric parameters in a mountainous terrain: a case study of the Supin-Upper Tons Basin, Indian Himalayas. Springerplus 5:1–38. https://doi.org/10.1186/s40064-016-3207-0
David SR, Murphy BP, Czuba JA, Ahammad M, Belmont P (2023) USUAL Watershed Tools: a new geospatial toolkit for hydro-geomorphic delineation. Environ Model Softw 159:105576. https://doi.org/10.1016/j.envsoft.2022.105576
Dhanaraj K, Angadi DP (2022) Geospatial analysis of contemporary urbanisation and rural–urban transition in Mangaluru, India. Asia-Pac J Reg Sci 6(2):515–539. https://doi.org/10.1007/s41685-022-00239-6
Drisya J, Sathish Kumar D (2023) Evaluation of the drought management measures in a semi-arid agricultural watershed. Environ Dev Sustain 25(1):811–833. https://doi.org/10.1007/s10668-021-02079-4
Du L, Dong C, Kang X, Qian X, Gu L (2023) Spatiotemporal evolution of land cover changes and landscape ecological risk assessment in the Yellow River Basin, 2015–2020. J Environ Manag 332:117149. https://doi.org/10.1016/j.jenvman.2022.117149
Dubovyk O (2017) The role of Remote Sensing in land degradation assessments: opportunities and challenges. Eur J Remote Sens 50(1):601–613. https://doi.org/10.1080/22797254.2017.1378926
Erosemiah D, Viji R (2023) Study on drainage network pattern and morphometric analysis by using hydrology support algorithm. J Geol Soc India 99(7):1006–1014. https://doi.org/10.1007/s12594-023-2421-y
Escobedo Garcia N, Ulibarri N (2023) Planning for effective water management: an evaluation of water management plans in California. J Environ Plan Manag 66(13):2739–2759. https://doi.org/10.1080/09640568.2022.2082930
Fan M, Shibata H (2014) Spatial and temporal analysis of hydrological provision ecosystem services for watershed conservation planning of water resources. Water Resour Manag 28:3619–3636. https://doi.org/10.1007/s11269-014-0691-2
Ferreira RG, da Silva DD, Elesbon AAA, dos Santos GR, Veloso GV, de Souza Fraga M, Fernandes-Filho EI (2021) Geostatistical modeling and traditional approaches for streamflow regionalization in a Brazilian Southeast watershed. J S Am Earth Sci 108:103355. https://doi.org/10.1016/j.jsames.2021.103355
Gabriele M, Brumana R, Previtali M, Cazzani A (2023) A combined GIS and remote sensing approach for monitoring climate change-related land degradation to support landscape preservation and planning tools: the Basilicata case study. Appl Geomat 15(3):497–532. https://doi.org/10.1007/s12518-022-00437-z
Gautam PK, Kumar D, Singh AK, Singh DS (2023) Drainage network analysis and tectonics interference of the Reth River, Central Ganga Plain (India) using geospatial technology. Geotectonics 1–16. https://doi.org/10.1134/S0016852123030032
Ghasemlounia R, Utlu M (2021) Flood prioritization of basins based on geomorphometric properties using principal component analysis, morphometric analysis and Redvan's priority methods: a case study of Harşit River basin. J Hydrol 603:127061. https://doi.org/10.1016/j.jhydrol.2021.127061
Gravelius H (1914) Flusskunde. Goschen'sche Verlagshandlung, Berlin
Hasanuzzaman M, Islam A, Bera B, Shit PK (2022) Flood susceptibility mapping using morphometric parameters and GIS. In: Spatial modelling of flood risk and flood hazards: societal implications. Springer, Cham, pp 15–31. https://doi.org/10.1007/978-3-030-94544-2_2
He C, James LA (2021) Watershed science: linking hydrological science with sustainable management of river basins. Sci China Earth Sci 64(5):677–690. https://doi.org/10.1007/s11430-020-9723-4

He C, Harden CP, Holden J, Mdee A (2023) Watershed science: coupling hydrological science and water resources management. Hydrol Processes e14889. https://doi.org/10.1002/hyp.14889

Heggem DT, Edmonds CM, Neale AC, Bice L, Jones KB (2000) A landscape ecology assessment of the Tensas River Basin. In: Monitoring ecological condition in the Western United States: proceedings of the fourth symposium on the Environmental Monitoring and Assessment Program (EMAP), San Francisco, CA, 6–8 Apr 1999. Springer, Netherlands, pp 41–54. https://doi.org/10.1007/978-94-011-4343-1_5

Horton RE (1932) Drainage-basin characteristics. Trans Am Geophys Union 13(1):350–361

Horton RE (1945) Erosional development of streams and their drainage basins; hydrophysical approach to quantitative morphology. Geol Soc Am Bull 56(3):275–370

Houborg R, Fisher JB, Skidmore AK (2015) Advances in remote sensing of vegetation function and traits. Int J Appl Earth Obs Geoinf 43:1–6. https://doi.org/10.1016/j.jag.2015.06.001

Im J (2020) Earth observations and geographic information science for sustainable development goals. Giscience Remote Sens 57(5):591–592. https://doi.org/10.1080/15481603.2020.1763041

Kalyanapu A, Owusu C, Wright T, Datta T (2023) Low-cost real-time water level monitoring network for falling water river watershed: a case study. Geosciences 13(3):65. https://doi.org/10.3390/geosciences13030065

Kesseli JE (1941) The concept of the graded river. J Geol 49(6):561–588

Khan MR, Voss CI, Yu W, Michael HA (2014) Water resources management in the Ganges Basin: a comparison of three strategies for conjunctive use of groundwater and surface water. Water Resour Manag 28:1235–1250. https://doi.org/10.1007/s11269-014-0537-y

Khan MYA, ElKashouty M (2023) Watershed prioritization and hydro-morphometric analysis for the potential development of Tabuk Basin, Saudi Arabia using multivariate statistical analysis and coupled RS-GIS approach. Ecol Ind 154:110766. https://doi.org/10.1016/j.ecolind.2023.110766

Khanday MY, Javed A (2016) Prioritization of sub-watersheds for conservation measures in a semi arid watershed using remote sensing and GIS. J Geol Soc India 88:185–196. https://doi.org/10.1007/s12594-016-0477-7

Kørnøv L, Lyhne I, Davila JG (2020) Linking the UN SDGs and environmental assessment: towards a conceptual framework. Environ Impact Assess Rev 85:106463. https://doi.org/10.1016/j.eiar.2020.106463

Krishna IM (2005) Geospatial information technology in watershed management. Advances in water science methodologies, p 171

Krishnamurthy J, Srinivas G, Jayaraman V, Chandrasekhar MG (1996) Influence of rock types and structures in the development of drainage networks in typical hardrock terrain. ITC J (3/4):252–259

Kudnar NS (2020a) GIS-based assessment of morphological and hydrological parameters of Wainganga River Basin, Central India. Model Earth Syst Environ 6(3):1933–1950. https://doi.org/10.1007/s40808-020-00804-y

Kudnar NS (2020b) GIS-based investigation of topography, watershed, and hydrological parameters of Wainganga river basin, Central India. Sustain Dev Pract Using Geoinform 301–318. https://doi.org/10.1002/9781119687160.ch19

Kudnar NS, Rajasekhar M (2020) A study of the morphometric analysis and cycle of erosion in Waingangā Basin, India. Model Earth Syst Environ 6(1):311–327. https://doi.org/10.1007/s40808-019-00680-1

Kudnar NS, Diwate P, Mishra VN, Srivastava PK, Kumar A, Pandey M (2022) Spatio-temporal variability and trend analysis of rainfall in Wainganga river basin, Central India, and forecasting using state-space models. Theor Appl Climatol 150(1–2):469–488. https://doi.org/10.1007/s00704-022-04168-4

Kumar R, Kumar S, Lohani AK, Nema RK, Singh RD (2000) Evaluation of geomorphological characteristics of a catchment using GIS. Gis India 9(3):13–17

Kumar S, Meena RS, Sheoran S, Jangir CK, Jhariya MK, Banerjee A, Raj A (2022) Remote sensing for agriculture and resource management. In: Natural resources conservation and advances for sustainability. Elsevier, pp 91–135. https://doi.org/10.1016/B978-0-12-822976-7.00012-0

Kureel N, Sarup J, Matin S, Goswami S, Kureel K (2022) Modelling vegetation health and stress using hypersepctral remote sensing data. Model Earth Syst Environ 1–16. https://doi.org/10.1007/s40808-021-01113-8

Lemieux CJ, Gray PA, Douglas AG, Nielsen G, Pearson D (2014) From science to policy: the making of a watershed-scale climate change adaptation strategy. Environ Sci Policy 42:123–137. https://doi.org/10.1016/j.envsci.2014.06.004

Leopold LB, Miller JP (1964) Fluvial processes in geomorphology. Freeman & Co., San Francisco, p 522

Loucks DP, Stakhiv EZ, Martin LR (2000) Sustainable water resources management. J Water Resour Plan Manag 126(2):43–47. https://doi.org/10.1061/(ASCE)0733-9496(2000)126:2(43)

Lyon JG, Lyon L (eds) (2022) Geospatial information handbook for water resources and watershed management, volume III: advanced applications and case studies. CRC Press

Mahala A (2020) The significance of morphometric analysis to understand the hydrological and morphological characteristics in two different morpho-climatic settings. Appl Water Sci 10(1):1–16. https://doi.org/10.1007/s13201-019-1118-2

Mani A, Kumari M, Badola R (2022) Morphometric analysis of Suswa River Basin using geospatial techniques. Eng Proc 27(1):65. https://doi.org/10.3390/ecsa-9-13225

Manjare BS, Khan S, Jawadand SA, Padhye MA (2018) Watershed prioritization of Wardha River Basin, Maharashtra, India using morphometric parameters: a remote sensing and GIS-based approach. In: Hydrologic modeling: select proceedings of ICWEES-2016. Springer, Singapore, pp 353–366. https://doi.org/10.1007/978-981-10-5801-1_25

Manjare B, Paunikar S, Shrivatra J (2019) Prioritization of sub-watersheds of Chandrabhaga river from Purna river basin, Maharashtra using geospatial techniques. J Geosci Res 2:111–120

Mashala MJ, Dube T, Mudereri BT, Ayisi KK, Ramudzuli MR (2023) A systematic review on advancements in remote sensing for assessing and monitoring land use and land cover changes impacts on surface water resources in semi-arid tropical environments. Remote Sens 15(16):3926. https://doi.org/10.3390/rs15163926

McColl C, Aggett G (2007) Land-use forecasting and hydrologic model integration for improved land-use decision support. J Environ Manag 84(4):494–512. https://doi.org/10.1016/j.jenvman.2006.06.023

McLain RJ, Banis D, Todd A, Cerveny LK (2017) Multiple methods of public engagement: disaggregating socio-spatial data for environmental planning in western Washington, USA. J Environ Manag 204:61–74. https://doi.org/10.1016/j.jenvman.2017.08.037

Mekasha ST, Suryabhagavan KV, Gebrehiwot M (2020) Geo-spatial approach for land-use and land-cover changes and deforestation mapping: a case study of Ankasha Guagusa, Northwestern, Ethiopia. Trop Ecol 61:550–569. https://doi.org/10.1007/s42965-020-00113-6

Meshram SG, Sharma SK (2017) Prioritization of watershed through morphometric parameters: a PCA-based approach. Appl Water Sci 7:1505–1519. https://doi.org/10.1007/s13201-015-0332-9

Meyer S, Blaschek M, Duttmann R, Ludwig R (2016) Improved hydrological model parametrization for climate change impact assessment under data scarcity—the potential of field monitoring techniques and geostatistics. Sci Total Environ 543:906–923. https://doi.org/10.1016/j.scitotenv.2015.07.116

Miller VC (1953) A quantitative geomorphic study of drainage basin characteristics in the Clinch Mountain area, Virginia and Tennessee, vol 3. Columbia University, New York

Mishra AK, Upadhyay A, Mishra PK, Srivastava A, & Rai SC (2023) Evaluating geo-hydrological environs through morphometric aspects using geospatial techniques: a case study of Kashang Khad watershed in the Middle Himalayas, India. Quatern Sci Adv 100096. https://doi.org/10.1016/j.qsa.2023.100096

Moghadam ES, Sadeghi SH, Zarghami M, Delavar M (2023) Developing sustainable land-use patterns at watershed scale using nexus of soil, water, energy, and food. Sci Total Environ 856:158935. https://doi.org/10.1016/j.scitotenv.2022.158935

Mokarram M, Hojati M (2017) Morphometric analysis of stream as one of resources for agricultural lands irrigation using high spatial resolution of digital elevation model (DEM). Comput Electron Agric 142:190–200. https://doi.org/10.1016/j.compag.2017.09.001

Molinos-Senante M, Hernández-Sancho F, Mocholí-Arce M, Sala-Garrido R (2014) A management and optimisation model for water supply planning in water deficit areas. J Hydrol 515:139–146. https://doi.org/10.1016/j.jhydrol.2014.04.054

Monjardin-Armenta SA, Plata-Rocha W, Pacheco-Angulo CE, Franco-Ochoa C, Rangel-Peraza JG (2020) Geospatial simulation model of deforestation and reforestation using multicriteria evaluation. Sustainability 12(24):10387. https://doi.org/10.3390/su122410387

Morisawa ME (1959) Relation of morphometric properties to runoff in the Little Mill Creek, Ohio. Drainage Basin, Tech. rep. 17. Columbia University, Department of Geology, ONR, New York

Morlot M, Rigon R, Formetta G (2023) Hydrological digital twin model of a large anthropized Italian alpine catchment: the Adige River basin. J Hydrol 130587. https://doi.org/10.1016/j.jhydrol.2023.130587

Mukherjee J, Patel PP (2022) Landscape characterization using geomorphometric parameters for a small sub-humid river basin of the Chota Nagpur Plateau, Eastern India. In: Drainage basin dynamics: an introduction to morphology, landscape and modelling. Springer, Cham, pp 127–152. https://doi.org/10.1007/978-3-030-79634-1_6

Nag SK, Chakraborty S (2003) Influence of rock types and structures in the development of drainage network in hard rock area. J Indian Soc Remote Sens 31:25–35. https://doi.org/10.1007/BF03030749

Nair SC, Mirajkar AB (2022) Rainfall trend anomalies over Wainganga Basin, Central India. In: Recent advancements in civil engineering: select proceedings of ACE 2020. Springer, Singapore, pp 917–928. https://doi.org/10.1007/978-981-16-4396-5_78

Nhemachena C, Nhamo L, Matchaya G, Nhemachena CR, Muchara B, Karuaihe ST, Mpandeli S (2020) Climate change impacts on water and agriculture sectors in Southern Africa: threats and opportunities for sustainable development. Water 12(10):2673. https://doi.org/10.3390/w12102673

Nooka Ratnam K, Srivastava YK, Venkateswara Rao V, Amminedu EKSR, Murthy KSR (2005) Check dam positioning by prioritization of micro-watersheds using SYI model and morphometric analysis—remote sensing and GIS perspective. J Indian Soc Remote Sens 33:25–38. https://doi.org/10.1007/BF02989988.pdf. Accessed 30 Dec 2023

Osorio Olivos LM, Méllo Junior AV, Soraes GADS (2023) An approach for water allocation with a couple surface and groundwater model. J Appl Water Eng Res 1–15. https://doi.org/10.1080/23249676.2023.2253151

Oyedotun TDT (2022) Quantitative assessment of the drainage morphometric characteristics of Chaohu Lake Basin from SRTM DEM data: a GIS-based approach. Geol Ecol Landsc 6(3):174–187. https://doi.org/10.1080/24749508.2020.1812147

Ozdemir H, Akbas A (2023) Is there a consistency in basin morphometry and hydrodynamic modelling results in terms of the flood generation potential of basins? A case study from the Ulus River Basin (Türkiye). J Hydrol 625:129926. https://doi.org/10.1016/j.jhydrol.2023.129926

Pace G, Gutiérrez-Cánovas C, Henriques R, Carvalho-Santos C, Cássio F, Pascoal C (2022) Remote sensing indicators to assess riparian vegetation and river ecosystem health. Ecol Ind 144:109519. https://doi.org/10.1016/j.ecolind.2022.109519

Pal S (2016) Identification of soil erosion vulnerable areas in Chandrabhaga river basin: a multi-criteria decision approach. Model Earth Syst Environ 2:1–11. https://doi.org/10.1007/s40808-015-0052-z

Palmer MA, Reidy Liermann CA, Nilsson C, Flörke M, Alcamo J, Lake PS, Bond N (2008) Climate change and the world's river basins: anticipating management options. Front Ecol Environ 6(2):81–89. https://doi.org/10.1890/060148

Panwar S, Agarwal V, Chakrapani GJ (2017) Morphometric and sediment source characterization of the Alaknanda river basin, headwaters of river Ganga, India. Nat Hazards 87:1649–1671. https://doi.org/10.1007/s11069-017-2838-y

Papa F, Frappart F (2021) Surface water storage in rivers and wetlands derived from satellite observations: a review of current advances and future opportunities for hydrological sciences. Remote Sens 13(20):4162. https://doi.org/10.3390/rs13204162

Parece TE, Campbell JB (2015) Land use/land cover monitoring and geospatial technologies: an overview. Advances in watershed science and assessment, pp 1–32. https://doi.org/10.1007/978-3-319-14212-8_1

Park D, You H (2023) A digital twin dam and watershed management platform. Water 15(11):2106. https://doi.org/10.3390/w15112106

Pathare JA, Pathare AR (2021) Watershed prioritization for soil and water conservation in Darna River basin: a PCA approach. Sustain Water Resour Manag 7(4):49. https://doi.org/10.1007/s40899-021-00531-x

Pike RJ, Wilson SE (1971) Elevation-relief ratio, hypsometric integral, and geomorphic area-altitude analysis. Geol Soc Am Bull 82(4):1079–1084. https://doi.org/10.1130/0016-7606(1971)82[1079:ERHIAG]2.0.CO;2

Rai PK, Mishra VN, Singh P (2022) Geospatial technology for landscape and environmental management. Springer, Singapore. https://doi.org/10.1007/978-981-16-7373-3

Rakesh CJ, Govindaraju Lokanath S, Kumar AK (2023) Prioritization of soil erosion prone sub-watersheds through morphometric analysis using geospatial and weighted sum approach: a case study of Boranakanive reservoir catchment in Tumkur district, Karnataka, India. Environ Earth Sci 82(12):306. https://doi.org/10.1016/j.wsee.2022.11.011

Reddy GPO, Maji AK, Gajbhiye KS (2004a) Drainage morphometry and its influence on landform characteristics in basaltic terrain—a remote sensing & GIS approach. Int J Appl Earth Observ Geoinform 6:1–16. https://doi.org/10.1016/j.jag.2004.06.003

Reddy GPO, Maji AK, Gajbhiye KS (2004b) GIS for morphometric analysis of drainage basin. GIS India 1:9–14

Roestamy M, Fulazzaky MA (2021) A review of the water resources management for the Brantas River basin: challenges in the transition to an integrated water resources management. Environ Dev Sustain 1–16. https://doi.org/10.1007/s10668-021-01933-9

Roy PS, Murthy MSR (2009) Efficient land use planning and policies using geospatial inputs: an Indian experience. Land Use Policy 31–72. https://www.researchgate.net/profile/Parth-Roy/publication/260706692_Efficient_Land_Use_Planning_and_Policies_using_Geospatila_Inputs_An_Indian_Experience/links/0a85e5320893b38f9d000000/Efficient-Land-Use-Planning-and-Policies-using-Geospatila-Inputs-An-Indian-Experience.pdf. Accessed 30 Dec 2023

Sabale R, Venkatesh B, Jose M (2023) Sustainable water resource management through conjunctive use of groundwater and surface water: a review. Innov Infrastruct Solut 8(1):17. https://doi.org/10.1007/s41062-022-00992-9

Sahoo S, Ramole MM, Dahiphale P, Awasthi S, Pateriya B (2023) Geospatial technology based morphometric analysis and watershed prioritization of lower Satluj basin in India for groundwater recharge potential. Trop Ecol 1–16. https://doi.org/10.1007/s42965-023-00307-8

Sahu U, Panaskar D, Wagh V, Mukate S (2018) An extraction, analysis, and prioritization of Asna river sub-basins, based on geomorphometric parameters using geospatial tools. Arab J Geosci 11:1–15. https://doi.org/10.1007/s12517-018-3870-2

Sajadi P, Singh A, Mukherjee S, Sang YF, Chapi K, Salari M (2022) Drainage network extraction and morphometric analysis in an Iranian basin using integrating factor analysis and geospatial techniques. Geocarto Int 37(3):896–925. https://doi.org/10.1080/10106049.2020.1750060

Sajeva M, Maidell M, Kotta J (2020) A participatory geospatial toolkit for science integration and knowledge transfer informing SDGs based governance and decision making. Sustainability 12(19):8088. https://doi.org/10.3390/su12198088

Salhi W, Heddoun O, Honnit B, Saidi MN, Kabbaj A (2024) Characterizing land use-land cover changes in N'fis watershed, Western High Atlas, Morocco (1984–2022). Appl Geomat 1–15. https://doi.org/10.1007/s12518-024-00549-8

Schumm SA (1956) Evolution of drainage systems and slopes in badlands at Perth Amboy, New Jersey. Geol Soc Am Bull 67(5):597–646

Shaikh M, Yadav S, Manekar V (2022) Application of the compound factor for runoff potential in sub-watersheds prioritisation based on quantitative morphometric analysis. J Geol Soc India 98(5):687–695. https://doi.org/10.1007/s12594-022-2045-7

Shelar RS, Shinde SP, Pande CB, Moharir KN, Orimoloye IR, Mishra AP, Varade AM (2022) Sub-watershed prioritization of Koyna river basin in India using multi criteria analytical hierarchical process, remote sensing and GIS techniques. Phys Chem Earth Parts a/b/c 128:103219. https://doi.org/10.1016/j.pce.2022.103219

Sheng D, Lou Y, Sun F, Xie J, Yu Y (2023) Reengineering and its reliability: an analysis of water projects and watershed management under a digital twin scheme in China. Water 15(18):3203. https://doi.org/10.3390/w15183203

Shrestha S, Poudyal KN, Bhattarai N, Dangi MB, Boland JJ (2022) An assessment of the impact of land use and land cover change on the degradation of ecosystem service values in Kathmandu Valley using remote sensing and GIS. Sustainability 14(23):15739. https://doi.org/10.3390/su142315739

Singh P, Prakash K, Kumar A, Kannaujiya AK, Deep A, Mohanty T (2023) GIS-based morphometric analysis with emphasis on hypsometric appraisal for sub-watershed prioritization: a case study of the Shyamari River Basin, Central India. Sustain Water Resour Manag 9(4):122. https://doi.org/10.1007/s40899-023-00900-8

Sivakumar B (2011) Global climate change and its impacts on water resources planning and management: assessment and challenges. Stoch Env Res Risk Assess 25:583–600. https://doi.org/10.1007/s00477-010-0423-y

Sousa M, Almeida MF, Calili R (2021) Multiple criteria decision making for the achievement of the UN sustainable development goals: a systematic literature review and a research agenda. Sustainability 13(8):4129. https://doi.org/10.3390/su13084129

Srivastava A, Singhal A, Jha PK (2022) Geospatial technology for sustainable management of water resources. In: Ecological significance of river ecosystems. Elsevier, pp 105–132. https://doi.org/10.1016/B978-0-323-85045-2.00008-X

Stoddart DR (1965) The shape of atolls. Mar Geol 3(5):369–383. https://doi.org/10.1016/0025-3227(65)90025-3

Strahler AN (1954) Quantitative geomorphology of erosional landscapes. In: Proceedings of the 19th international geological congress—Algiers, pp 341–354

Strahler AN (1957) Quantitative analysis of watershed geomorphology. Trans Am Geophys Union 6(8):913–920

Strahler AN (1964) Quantitative geomorphology of basins and channel networks. In: Chow VT (ed) Handbook of applied hydrology. McGraw Hill, New York

Strahler AN (1952) Hypsometric (area-altitude) analysis of erosional topography. Geol Soc Am Bull 63(11):1117–1142

Sutradhar S, Mondal P (2023) Prioritization of watersheds based on morphometric assessment in relation to flood management: a case study of Ajay river basin, Eastern India. Watershed Ecol Environ 5:1–11. https://doi.org/10.1016/j.wsee.2022.11.011

Sytsma A, Bell C, Eisenstein W, Hogue T, Kondolf GM (2020) A geospatial approach for estimating hydrological connectivity of impervious surfaces. J Hydrol 591:125545. https://doi.org/10.1016/j.jhydrol.2020.125545

Taxak AK, Murumkar AR, Arya DS (2014) Long term spatial and temporal rainfall trends and homogeneity analysis in Wainganga basin, Central India. Weather Clim Extremes 4:50–61. https://doi.org/10.1016/j.wace.2014.04.005

Thakur A, Mishra PK, Nema AK, Sahoo SP (2020) Spatio-temporal trends and shift analysis of temperature for Wainganga sub-basin, India. Int J Environ Stud 77(3):464–479

Trong NG, Quang PN, Cuong NV, Le HA, Nguyen HL, Tien Bui D (2023) Spatial prediction of fluvial flood in high-frequency tropical cyclone area using TensorFlow 1D-convolution neural networks and geospatial data. Remote Sens 15(22):5429. https://doi.org/10.3390/rs15225429

Tsakiris GP, Loucks DP (2023) Adaptive water resources management under climate change: an introduction. Water Resour Manag 1–13. https://doi.org/10.1007/s11269-023-03518-9

Verma N, Patel RK, Choudhari P (2023) Watershed prioritization for soil conservation in a drought prone watershed of Eastern India: Tel River Basin, Odisha. Geol Ecol Landsc 7(4):405–418. https://doi.org/10.1080/24749508.2021.2022830

Vesuviano G, Griffin A, Stewart E (2022) Flood frequency estimation in data-sparse Wainganga Basin, India, using continuous simulation. Water 14(18):2887. https://doi.org/10.3390/w14182887

Vijith H, Satheesh R (2006) GIS based morphometric analysis of two major upland sub-watersheds of Meenachil river in Kerala. J Indian Soc Remote Sens 34:181–185. https://doi.org/10.1007/BF02991823.pdf. Accessed 30 Dec 2023

Wang L, Li Y, Gan Y, Zhao L, Qin W, Ding L (2024) Rainfall erosivity index for monitoring global soil erosion. CATENA 234:107593. https://doi.org/10.1016/j.catena.2023.107593

Weslati O, Bouaziz S, Serbaji MM (2020) Mapping and monitoring land use and land cover changes in Mellegue watershed using remote sensing and GIS. Arab J Geosci 13:1–19. https://doi.org/10.1007/s12517-020-05664-5

Yazdi J, Salehi Neyshabouri SAA, Niksokhan MH, Sheshangosht S, Elmi M (2013) Optimal prioritisation of watershed management measures for flood risk mitigation on a watershed scale. J Flood Risk Manag 6(4):372–384. https://doi.org/10.1111/jfr3.12016

Yin S, Nearing MA, Borrelli P, Xue X (2017) Rainfall erosivity: an overview of methodologies and applications. Vadose Zone J 16(12):1–16. https://doi.org/10.2136/vzj2017.06.0131

Younos T, Heyer CJ (2015) Advances in water sensor technologies and real-time water monitoring. Adv Watershed Sci Assess 171–203. https://doi.org/10.1007/978-3-319-14212-8_7

Zehtabian E, Masoudi R, Yazdandoost F, Sedghi-Asl M, Loáiciga HA (2023) Investigation of water allocation using integrated water resource management approaches in the Zayandehroud River basin, Iran. J Clean Prod 395:136339. https://doi.org/10.1016/j.jclepro.2023.136339

Zeitoun M, Goulden M, Tickner D (2013) Current and future challenges facing transboundary river basin management. Wiley Interdiscip Rev: Clim Change 4(5):331–349. https://doi.org/10.1002/wcc.228

Zhang C, Li X (2022) Land use and land cover mapping in the era of big data. Land 11(10):1692. https://doi.org/10.3390/land11101692

Zhou X, Moinuddin M, Renaud F, Barrett B, Xu J, Liang Q, Zhao J, Xia X, Bosher L, Huang S, Hoey T (2022) Development of an SDG interlinkages analysis model at the river basin scale: a case study in the Luanhe River Basin, China. Sustain Sci 17(4):1405–1433. https://doi.org/10.1007/s11625-021-01065-z

Modelling and Mapping of Natural Resource Management Using RS & GIS: A Case Study of Monitoring Coastline Changes in Kerala

Soumili Dutta and Prasun Kumar Gupta

Abstract Coastal regions are rich in natural resources due to their unique hydrogeological and ecological characteristics. Changes in these regions, induced by shoreline shifts have been a major concern for coastal communities around the world. With increasing world population, global warming and associated climate change impacts influencing the global environment, the world's coastlines are under threat. Thus, their mapping and change determination are necessary for sustainable management and conservation measures. This study on the coastline of Kerala attempts to study its change over a 20-year time span, using remote sensing and GIS techniques. The delineation and extraction of shoreline have been performed in two ways—using the AI-based Segment Anything Model for Geospatial data (SAMGeo) for land–water segmentation and coastline delineation, and then index-based method for coastline extraction. While SAMGeo works well for coastline delineation in smaller areas, hardware limitations prevented us from testing it in larger areas. Thereafter, using the Digital Shoreline Analysis System (DSAS) tool, shoreline change rate has been estimated. The study reveals that the coastline of Kerala experiences varying rates of erosion and accretion, owing to several natural and anthropogenic factors, emphasising the dynamic nature of coastline. The maximum erosion rates are in the districts of Kasargod and Kollam at around 18m/year, while Kollam had the highest accretion rates at around 14 m/year. Overall, the findings emphasise the need for further investigation in areas of high erosion to come up with effective conservation and management strategies to protect the coastal communities and the ecosystem.

Keywords Artificial intelligence · NDWI · Shoreline change rate · End point rate · Kerala

S. Dutta (✉) · P. K. Gupta
Geoinformatics Department, Indian Institute of Remote Sensing, Indian Space Research Organisation (ISRO), Dehradun, Uttarakhand, India
e-mail: soumili@iirsddn.ac.in

P. K. Gupta
e-mail: prasun@iirs.gov.in

S. C. Pal and U. Chatterjee (eds.), *Surface, Sub-Surface Hydrology and Management*, Springer Geography, https://doi.org/10.1007/978-3-031-62376-9_26

1 Introduction

Shorelines mark the transition of land into water, like rivers, seas, and oceans. In the wake of climate change and its associated impacts on the ocean, shoreline change, and its prediction has become a topic of increasing importance, especially for the coastal communities. This comes in the light of vulnerability of coast due to erosional or accretional activities, owing to several influencing factors like rise in sea level and frequency of extreme weather events like cyclones, floods, storm surges, etc. (Tsiakos and Chalkias 2023). Such events lead to coastal erosion and shoreline shifting, that heavily impacts the livelihood of people residing in coastal areas (Nandi et al. 2016). The world's coastlines are home to billions of people and vital ecosystems, besides acting as economic hubs; and therefore, the assessment of their susceptibility to coastal geohazards has become an issue of paramount importance. As per UNESCO 2011 report, more than 40% of global population reside within a radius of 100 km from shoreline (Sekovski et al. 2020). In India, as census report of 2011 states, about 15.5% of the population resides in coastal districts, which can only be assumed to have increased due to migration patterns and growing population in such areas (Adger and Kelly 1999). Globally shoreline-based studies have been widely carried out, like by Palanisamy et al. (2024) who studied a 60km stretch of the southern coast of Rio de Janeiro. Gharnate et al. (2024) carried out a similar study in Morocco. As noted by Sahana et al. (2021), India is the 6th most vulnerable nation, susceptible to coastal hazards. Thus, studying the coastal vulnerability in terms of shifting shoreline in Indian context is necessary to implement mitigation strategies and improve livelihood of the coastal community.

India boasts a 7500 km long coastline, with around 23% of it being affected by different levels of erosion (Mukhopadhyay and Karisiddaiah 2014). While the eastern coast, lying next to the Bay of Bengal, is characterised by sandy beaches and coastal dunes, the west coast, between the Western Ghats and the Arabian Sea is overlain by rocky headlands and estuaries. Mukhopadhyay and Karisiddaiah (2014) note that the western coast including Lakshadweep islands witnesses more erosional activities than the eastern coast. According to reports, West Bengal, Puducherry, Tamil Nadu, and Kerala are among the most erosion-prone states of India. A recent study conducted by Mondal et al. (2024) identifies coastline dynamics and erosion levels in Ghoramara Island in West Bengal, India. On the western coast of the Indian peninsula, except Kerala, the shoreline appears to be stable. Identifying such areas that are under threat is important to conserve the coastline and the coastal community.

The state of Kerala has been selected for conducting this study to determine shoreline change rates. An article published on July 13, 2023, highlights how livelihood of fishermen are under threat as coastal erosion is washing away their homes (Rajwi 2023). A NASA-based survey projects a sea-level rise of 0.58m by 2100 in Kochi, a city in Kerala (Global Order 2023). There are also several interviews being conducted of the coastal population in Kerala which are available on YouTube giving a glimpse of their condition under the increasing threats of coastline erosion. However, when surveys are conducted only some portions where it is focused on get identified. In situ data collection is not only a tedious process, but also past data are not always available. Also, it is quite expensive to survey entire coastline of India. In this regard, the application of remote sensing and earth observation imagery for shoreline change and evolution studies has gained huge popularity because of its inexpensive, easy availability for identification of areas that need attention. After the areas have been identified detailed survey can then be conducted to analyse the dynamics and come up with conservation measures.

This study thus uses Landsat Earth observation satellites' data to study the shoreline change in Kerala state. The main objectives of this chapter are to test the AI-based Segment Anything Model for Geospatial data (SAMGeo) for land–water segmentation and coastline delineation, and estimation of rate of shoreline change, that is studying the shoreline dynamics, using remote sensing and GIS techniques.

2 Rationale of the Study

Shorelines have been defined by several researchers considering different water level factors and along-shore variations. Boak and Turner (2005) define shoreline as a land–water interface, the position of which keeps changing temporally, giving it a dynamic status. This change can be attributed to sediment movement across and along the littoral zone owing to changing water levels. To represent the "true" shoreline position, shoreline indicators are used which are physically visible (Boak and Turner 2005).

The dynamic nature of the shoreline in recent years has aggravated due to global warming and associated climate change impacts. There are two main types of shoreline movement—erosion, or the receding of land and accretion, or advancement of land (Nandi et al. 2016). Coastal erosion events can lead to submergence of coastal tracts, beaches, and low-lying land affecting the coastal communities. Since coastal regions also act as important economic hubs of the region hosting infrastructures such as harbours, ship docks, jetties, and activities such as aquaculture, mining, and tourism (Thakare and Shitole 2023), coastline change detection and prediction of future shoreline position is of paramount importance to implement hazard mitigation and Coastal Zone Management activities (CZM) (Kannan et al. 2015).

Shoreline change analysis study requires data of shoreline position at different time scales of the region. Traditionally, ground-based in situ observation was carried out, from 1807 to 1927, later followed by introduction of aerial photography, which was then succeeded by satellite-based imagery (e.g., Landsat) in the 80's (Kumar et al. 2020). With the advent of satellite imagery, the study and analysis of large coastlines, remote locations, became possible, while being easy and inexpensive (Thakare and Shitole 2023). Both optical and SAR images (Wei et al. 2021) as well as LiDAR and Unmanned Aerial Vehicles (UAV) generated digital surface maps (Kumar et al. 2020) are useful for this. Many scientists and researchers have thereafter carried out extensive research to study the coastline dynamics, coastal vulnerability (Narayana and Priju 2006) and made predictions regarding future trend of the same (Chrisben Sam and Gurugnanam 2022).

The first step towards shoreline change calculation is the detection of shorelines. After performing atmospheric and radiometric corrections, this can be done, by digitising the shoreline manually from standard false colour composite images (Narayana and Priju 2006; Basheer Ahammed and Pandey 2022; Thakare and Shitole 2023) or the shoreline can be automatically extracted using value thresholding after calculating indices (Salauddin et al. 2018).

For automatic shoreline extraction, different water-based indices have been used by different researchers. The NDWI or the Normalised Difference Water Index given by McFeeters (McFEETERS 1996) using the green and NIR band and the Modified NDWI using the SWIR and green bands given by Xu (2006) has been used by Salauddin et al. (2018) to differentiate land from water and further thresholding the image to extract shoreline for an island in Bangladesh. Feyisa et al. (2014) gave a new Automated Water Extraction Index (AWEI) used by Gandhi (2023) for automatic shoreline extraction of part of Mumbai coast using Google Earth Engine.

Nandi et al. (2016) applied Tasseled cap transformation technique on Landsat imagery to delineate land–water boundary of Sagar Island, West Bengal by first extracting pixels of the shore, then converting the image into binary raster and finally transforming it into vector format to extract the coastline. Another method of shoreline extraction is by using linear edge enhancement techniques namely Roberts, Canny, Sobel, etc. (Yasir et al. 2020; Hossain et al. 2021). Kannan et al. (2015) used non-linear edge enhancement techniques for detecting shoreline in Visakhapatnam coast. Histogram thresholding from optical satellite imagery and a new band ratioing technique have been used to delineate the land–water boundary by Kourosh Niya et al. (2013) for Bushehr province in Iran.

The shoreline changes calculation, prediction and analysis have most commonly been carried out using the Digital Shoreline Analysis System (DSAS), extension available in ArcGIS software in most of the reviewed literatures cited above (Dutta et al. 2022; Hossain et al. 2021; Nandi et al. 2016; Salauddin et al. 2018).

With the era of AI and machine learning coming in, researchers have made use of different machine learning algorithms for automatic shoreline extraction and change estimation. Kumar et al. (2020) have used K-nearest neighbour (KNN), Support Vector Machines (SVM) and Artificial Neural Network (ANN) techniques to model shoreline change of part of Orissa coast. Other researchers have also used other supervised and unsupervised machine learning algorithms for extracting and modelling shoreline change (Tsiakos and Chalkias 2023). Murray et al. (2023) uses the open-source software CoastSat that uses a combined approach of neural network algorithm with Otsu thresholding using MNDWI for shoreline position extraction automation.

However, the above-cited literature that used various techniques to conduct shoreline change analysis, have been primarily focused on small portions of states or districts, which is one of the research gaps being addressed in this chapter. Also, application of the new AI-based GeoSAM model for shoreline delineation has not yet been attempted when the study was conducted. This study thus aims to apply and test the Segment Geospatial model (Wu and Osco 2023), launched in May 2023 as a

Python package for segmenting geospatial data using the Segment Anything Model, developed by Meta in April 2023 (Kirillov et al. 2023), for automatic shoreline delineation. Thereafter index-based masking for delineation and extraction of shoreline for the entire coastline of Kerala to study the shift from multi-temporal satellite images of Landsat 7 (2002) and Landsat 8 (2014, 2022) has been performed.

3 Study Area

Located on the Southwest corner of the Indian Peninsula, Kerala hosts 10% of India's total coastline stretching about 593 km (Selvan et al. 2020), running through the nine coastal districts of Trivandrum (13%), Kollam (6%), Alappuzha (14%), Ernakulam (8%), Thrissur (9%), Malappuram (12%), Kozhikode (12%), Kannur (14%), Kasargod (12%). Extending from latitudes 8°18'N–12°76'N and longitudes 72°22' E–74°52' E (Fig. 1), Kerala has a unique physiographic setting with 41 west-flowing rivers and a rocky coast. With an elevation of 0m to 8m above msl (Narayana and Priju 2006), the coastal plain of Kerala overlooks the Arabian Sea to its west. The coastline of Kerala hosts various physical features such as sandy beaches, lagoons, estuaries, rocky cliffs, sand dunes, barriers, spits, etc. that are significant in shaping its structure.

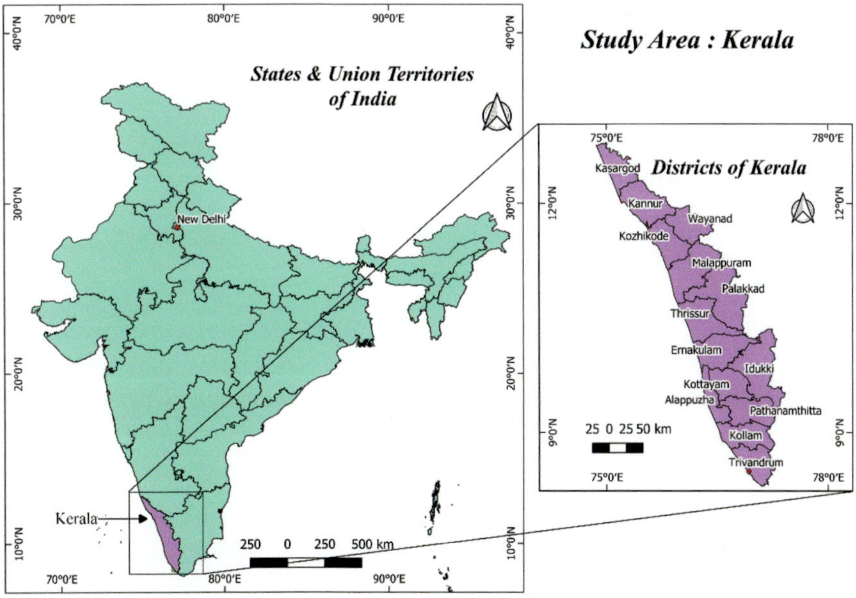

Fig. 1 Study Area: Kerala, India

Table 1 Satellite data used

Satellite image	Processing level	Date (year of acquisition)	Sensor and band used
Landsat 7	Level 2, collection 2, tier 1	Feb–March, 2002	ETM + (Band 2, band 3, band 4)
Landsat 8	Level 2, collection 2, tier 1	Feb–March, 2014	OLI (Band 3, band 4, band 5)
Landsat 8	Level 2, collection 2, tier 1	Feb–March, 2022	OLI (Band 3, band 4, band 5)

The coastal region of Kerala observes two monsoon seasons every year, namely south-west (summer) monsoon and north-east (winter) monsoon. The southwestern summer monsoon has a more severe impact as during this season, the intensity of waves increases, causing more devastation along the coast. Otherwise, the shoreline witnesses gentle waves (Selvan et al. 2020).

The coast of Kerala is lined with a major port, harbours, estuaries, lagoons, mud banks, etc. along with islands and islets. It is also famous for its fisheries and other resources such as mangroves, minerals, etc. This coastal zone is home to around 40% of Kerala's total population with a population density of 2500 person/sq.km (Sachin Pavithran et al. 2014). Owing to the over-exploitation of resources, excessive developmental activities, the coastal environment is getting degraded along with alteration of shoreline, which is a result of not only natural but anthropogenic activities as well. Thus, studying the shoreline, and its change over the years is important for understanding the importance of management strategies, to protect it.

4 Materials and Methods

4.1 Materials Used

This study uses Landsat Earth observation series satellite imageries to study coastline changes in the state of Kerala. The datasets used to conduct this study are listed in Table 1. Data processing and preprocessing is conducted mainly in python, google earth engine (javascript), QGIS and ArcGIS.

4.2 Methodology

The methodology followed for this study can be divided into two parts –

Shoreline delineation and extraction, and
Estimation of shoreline change

4.2.1 Shoreline Delineation and Extraction

The demarcation of land–water boundary and its extraction has been done first for a small portion of Kerala shoreline (using artificial intelligence-based method), and then for the entire shoreline of Kerala (using index-based method).

AI-Based:

Attempt was made to utilise the new Segment Anything Model and its geospatial python package developed by Kirillov et al. (2023); Wu and Osco (2023). A small part of shoreline (approx. 3 km) was selected as region of interest (ROI) as in Fig. 2, to demonstrate the extraction of shoreline using the Segment Anything Model for geospatial data. This was done in Google Colab in Python. Data for the region of interest was downloaded from Tile Map Services (TMS) and the Segment geospatial python package developed by Wu and Osco (2023) that uses Kirillov et al. (2023)'s Segment Anything Model, was imported. In the selected region of interest (Fig. 2), background and foreground were marked interactively to generate segment mask (Fig. 3). After segmentation, raster to vector conversion was carried out to convert the raster mask into shapefile. The polygon boundary was then exported as line geometry to device. This can be later opened in QGIS, and the shoreline- boundary can be snipped using GIS tools such as "split features" tool (Fig. 4). The background and foreground were marked interactively to generate segment mask as shown in Fig. 3. After segmentation raster to vector conversion was carried out to convert the raster mask into shapefile. The polygon boundary was then exported as line geometry to device. This can be later opened in QGIS, and the shoreline- boundary can be snipped using GIS tools such as "split features" tool (Fig. 4).The methodology followed for extracting shoreline using the Segment Anything Model is described in Fig. 5.

Fig. 2 Selecting ROI

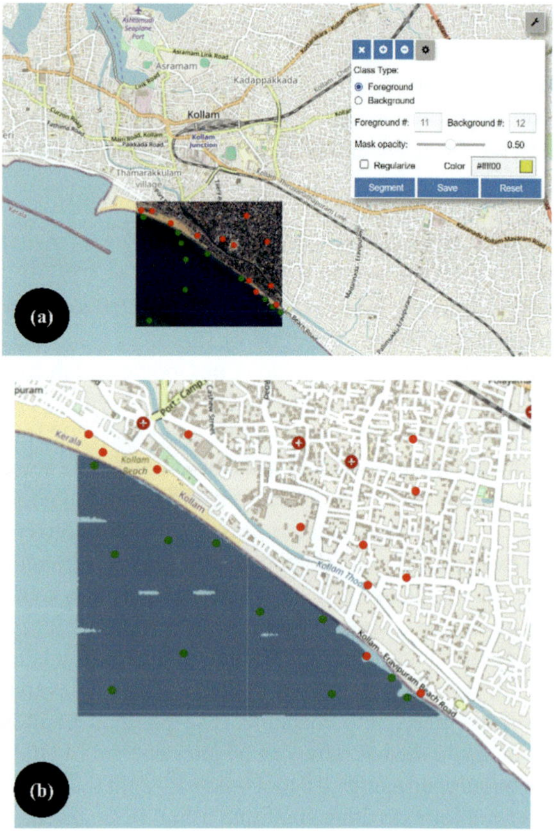

Fig. 3 (**a**) Background and Foreground Markers. (**b**) Generated segmented mask

Index Based:

To extract the shoreline of the study area, that is, the shoreline of the state of Kerala, NDWI-based thresholding was used. First, the western boundary of Kerala state was digitised from downloaded state level vector map of India from Survey of India website. Next a 20km buffer was made to create the region of interest for downloading the multitemporal satellite images. Standard FCC images containing three bands namely—NIR, Red and Green were downloaded and clipped for the region of interest and exported from Google Earth Engine. Composite satellite images for 3 years were downloaded to study the shift of shoreline from Landsat 7 (2002) and Landsat 8 (2014, 2022) collection 2, level 2, tier 1, surface reflectance data (Fig. 6).

Using McFEETERS (1996) Normalised Difference Water Index (NDWI), the composite images of the 3 years were classified into five classes. The NDWI equation as given by McFEETERS (1996) is:

$$\text{NDWI} = \frac{\text{Green} - \text{NIR}}{\text{Green} + \text{NIR}}$$

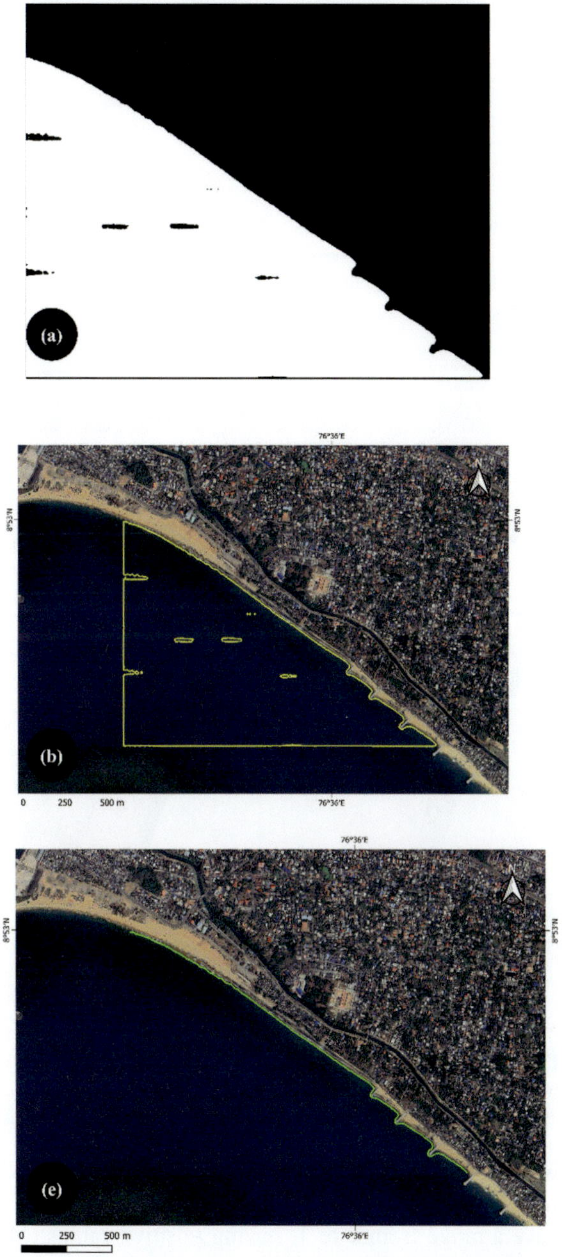

Fig. 4 (**a**) Segmented mask showing Land and water. (**b**) Line boundary converted from polygon. (**c**) Shoreline demarcated

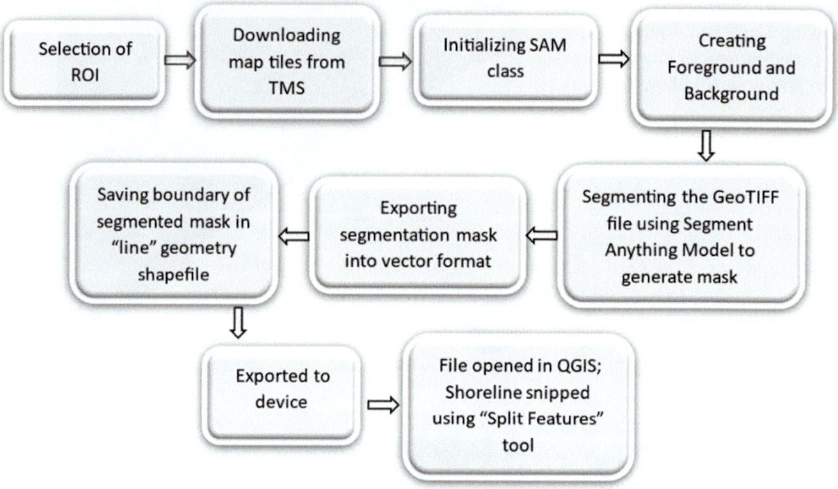

Fig. 5 Methodology Steps for AI-based shoreline delineation

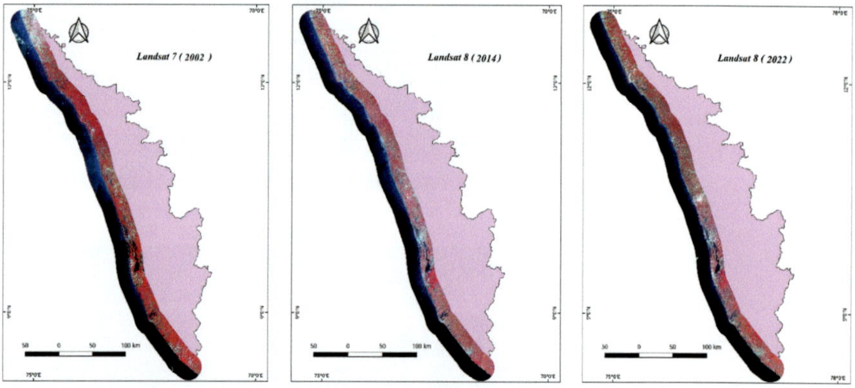

(a) (b) (c)

Fig. 6 Downloaded Landsat images of the selected years: (**a**) Landsat 7 for 2002; (**b**) Landsat 8 for 2014; (**c**) Landsat 8 for 2022

where Green refers to reflectance of green band, and NIR refers to reflectance of Near Infrared Band. The value of NDWI ranges from − 1 to + 1 where water bodies usually have a range from 0 to 1, having positive values (McFEETERS 1996). Using this concept, the NDWI images were reclassified into two classes (binary classification)—Land and Water (1 and 0).

Next, the raster reclassified was converted to vector polygon wherein two separate polygons were obtained. The land polygon was deleted, and the water polygon vector was then converted to line after fixing its geometry. The line was snipped using "split features" tool, to extract the shoreline.

However, for generating transects the lines cannot have abrupt bends. Thus, the line was then simplified using Douglas Peucker distance algorithm in "simplify" tool. The line was then manually modified to smoothen the sharp edges.

4.2.2 Estimation of Shoreline Change

Shoreline change analysis was performed with the help of the DSAS (v 5.1) tool in ArcGIS software (version 10.5). A personal geodatabase was created, and shoreline and baseline feature classes were formed. The three shorelines were merged into a single file and stored in "shoreline" class with required attributes. The baseline was generated by forming a buffer of 1500 m around the shoreline, then converting the merged buffer polygon to line and extracting the right boundary of the buffer polygon to get the baseline, which was then stored in "baseline" class with required attributes. To calculate the rate of shift of shoreline, orthogonal transects are casted from the baseline at an interval of 50 m, after the shoreline and baseline classes are ready. The necessary attribute including dates are entered in the attribute table. For the coastline of Kerala, a total number of 11,380 transects were generated at 50m interval. The intersection of the shorelines with the transects are used to calculate the rate of change statistics.

The DSAS tool can be used for different statistical change rate calculations such as End Point rate (EPR), Linear Regression Rate (LRR) and Weighted Linear Regression Rate (WLR). Different distance measurements performed by DSAS involve Net Shoreline Movement (NSM) and Shoreline Change Envelope (SCE) (Himmelstoss et al. 2021). In this study, NSM and EPR have been considered.

The NSM is a measure of the distance between the oldest and youngest shorelines for a given transect. It is expressed in meters. This in turn is used for the EPR calculation.

The EPR is a measure of shoreline change that is calculated by dividing the distance between the oldest and most recent shorelines, that is the NSM by the time elapsed between them.

$$\text{EPR} = \frac{\text{Net Shoreline Movement}}{\text{Time between oldest and recent shoreline}}$$

where Net Shoreline Movement = Distance between oldest and most recent shoreline in m

The values obtained were categorised and colour-coded for visualisation. Also, the EPRs were district wise divided and plotted for further analysis of nature of shoreline movement in terms of erosion (for negative values) and accretion (for positive values).

The methodology framework followed for the study is as in Fig. 7.

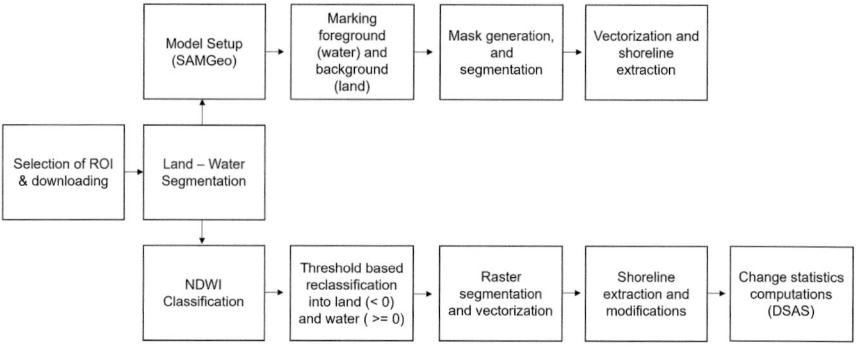

Fig. 7 Methodology Flowchart

5 Results and Discussion

5.1 Results

This study was carried out in two parts—first, the demarcation of the land–water boundary and its extraction and second, the estimation of change in the three shorelines, thus obtained for the years 2002, 2014 and 2022. The results obtained after carrying out the processes described above are as follows.

5.1.1 Shoreline Delineation and Extraction

AI-Based

Using artificial intelligence-based segment anything model that uses supervised machine learning algorithms to segment features, shoreline has been delineated. User-generated masking has been used in this case to create custom segmentation mask for differentiating land and water by providing the model with points defining the background and foreground for segment mask generation. The final extraction of shoreline has been done in QGIS, and the result is as in Fig. 8.

Index Based

To extract the entire shoreline of the state of Kerala NDWI based reclassification method was used. The results obtained after vectorising the reclassified layers for the 3 years are as shown in Fig. 9.

The shorelines were extracted after converting the vector "polygon" to "line" geometry. However, since it is converted from vector it showed abrupt pixel-boundary-like structure. It was then simplified and manually edited to obtain the shorelines to be used for change estimation.

Fig. 8 Result of SAM-generated shoreline delineation

This was required as the DSAS tool cannot process abrupt changes of slope of line and generated incorrect transects. The result obtained for 2002 is as in Fig. 10. Similarly, the other two shorelines for 2014 and 2022 were simplified and modified manually and extracted.

The shorelines of the three years after extraction were obtained as shown in Fig. 11.

5.1.2 Estimation of Change in the Shoreline from 2002 to 2022:

This part of the process was carried out in ArcGIS that mainly aimed at calculation of shoreline change using the DSAS tool extension. The NSM or the net horizontal movement is expressed in negative for erosion and positive to represent accretion. A maximum eroded distance of about 370 m is seen in Kasargod district whereas a maximum accretion value of about 286 m is seen in Kollam district since the year 2002–2022. The NSM values and EPR values have been categorised into seven classes as in Tables 2 and 3 respectively for visualisation (Fig. 12).

The obtained values for the transects were district wise divided to generate graph plots to visualise district wise erosion or accretion trends (Fig. 13). Since NSM showed similar trends with that of EPR, further analysis was carried on based on EPR only. The plotted graphs show the erosion and accretion rates indicated by brown showing erosion and green colour indicating accretion activities.

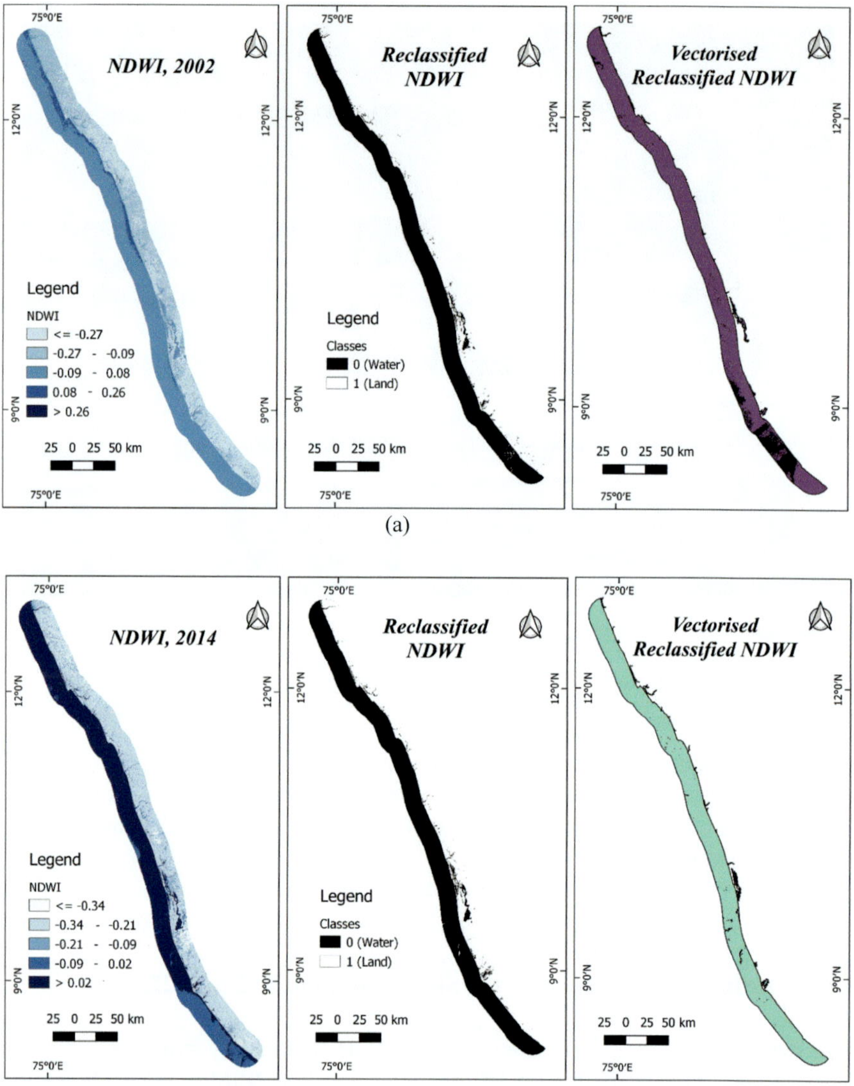

(b)

Fig. 9 Results of NDWI, Reclassification and vectorisation for 2002 (**a**), 2014 (**b**) and 2022 (**c**)

Kasargod shows maximum erosion rate of around 18 m/year, and an accretion rate of around 8 m/year. Kannur district shows a maximum erosion rate of around 12 m/year and a maximum accretion rate of about 7 m/year is seen.

The Kozhikode district shows a maximum erosion rate of around 6m/year and a maximum accretion rate of around 4 m/year. Malappuram district shows highest erosion as about 4 m/year and highest accretion as 5 m/year. Trissur district shows a

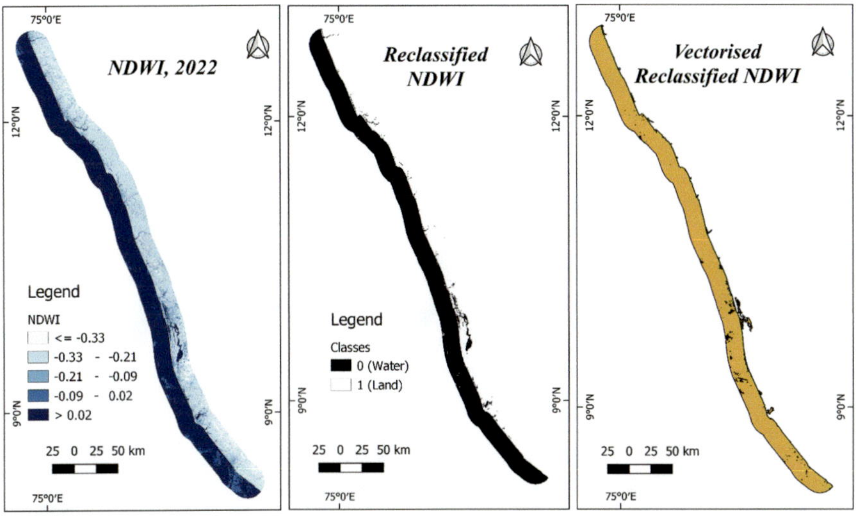

(c)

Fig. 9 (continued)

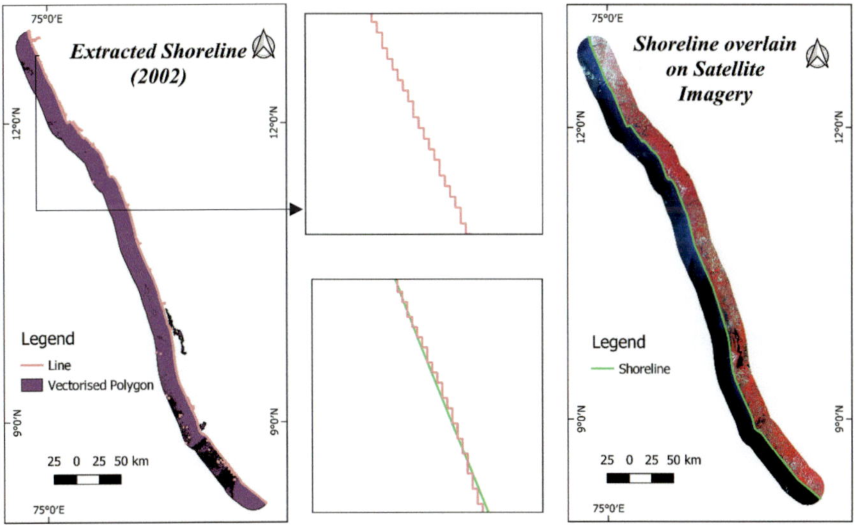

Fig. 10 Shoreline modified and overlain on Satellite imagery

trend of highest accretion as 7 m/year and highest erosion rate of 6 m/year in certain places. Ernakulam and Alappuzha show a maximum erosion rate of about 15 m/year and 8 m/year respectively, whereas Trivandrum shows area with highest erosion rate

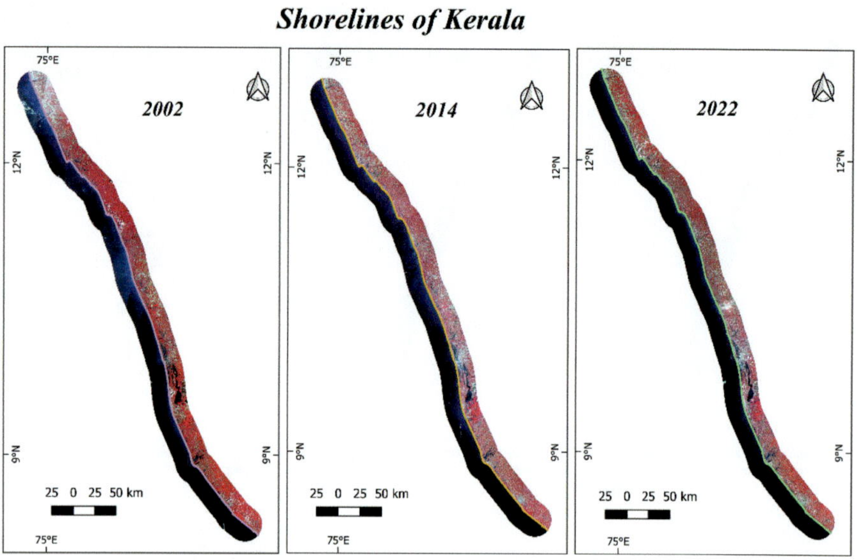

Fig. 11 Vectorised Line modified and overlain on Satellite imagery

Table 2 NSM categorisation

Category	NSM
High erosion	< − 100 m
Moderate erosion	− 100 m to − 50 m
Low erosion	− 50 m to − 5 m
Stable	− 5 m to 5 m
Low accretion	5 m to 50 m
Moderate accretion	50 m to 100 m
High accretion	> 100 m

Table 3 EPR categorisation

Category	EPR in m/year
High erosion	< − 8
Moderate erosion	− 8 to − 4
Low erosion	− 4 to − 1
Stable	− 1 to 1
Low accretion	1–4
Moderate accretion	4–8
High accretion	> 8

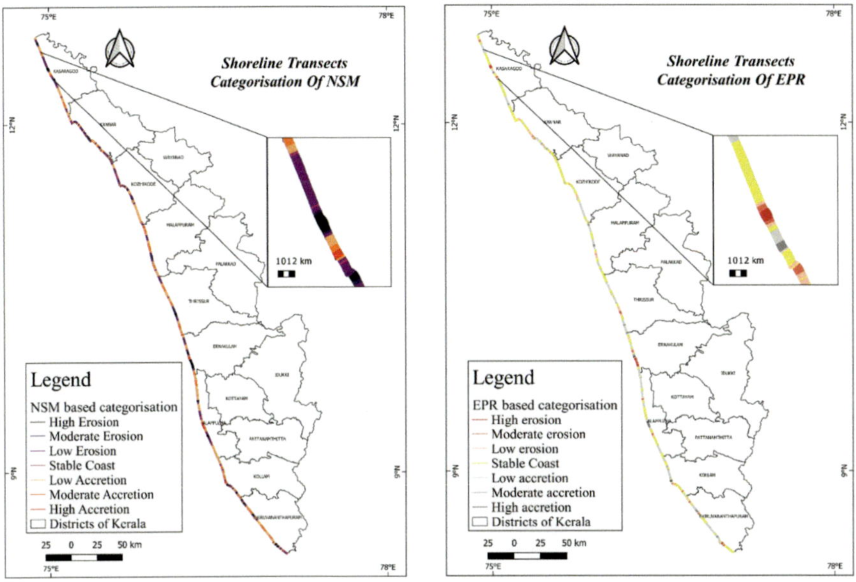

Fig. 12 Shoreline Transects of NSM (**a**) and EPR (**b**) categorisation

of 12 m/year. Kollam has area with highest erosion rate in the entire state of about 18m/year.

Box plots (Fig. 14) have been generated of the EPR rates to get idea about the district's shoreline change and get a comparative picture of erosion and accretion activities among the districts. However, since the data contains high range of variability in some cases, it has considered minimum and maximum as 1.5 times the interquartile range.

It is observable from the generated boxplots that while some districts show a lower interquartile range, like Kasargod, Kannur, Alappuzha, Kozhikode, Kollam, other districts namely Ernakulam, Thrissur and Malappuram show a larger interquartile range. Districts having outlier values indicate extreme levels of erosion and accretion. It is also observable that the median values for most of the districts lie above the zero line, suggesting majority of data points having positive values, indicating accretional activities.

Thus, it is observable that Kerala has a highly dynamic coastline with regions experiencing erosional and accretional activities. While some states show less variability exhibiting a more stable condition, some states show high amount of variability like the state of Kollam that has both the highest erosional and accretional rates in the state. The states having a greater number of outliers represent that some areas show extreme levels of erosion (outliers on negative side). Thus, surveys can be conducted

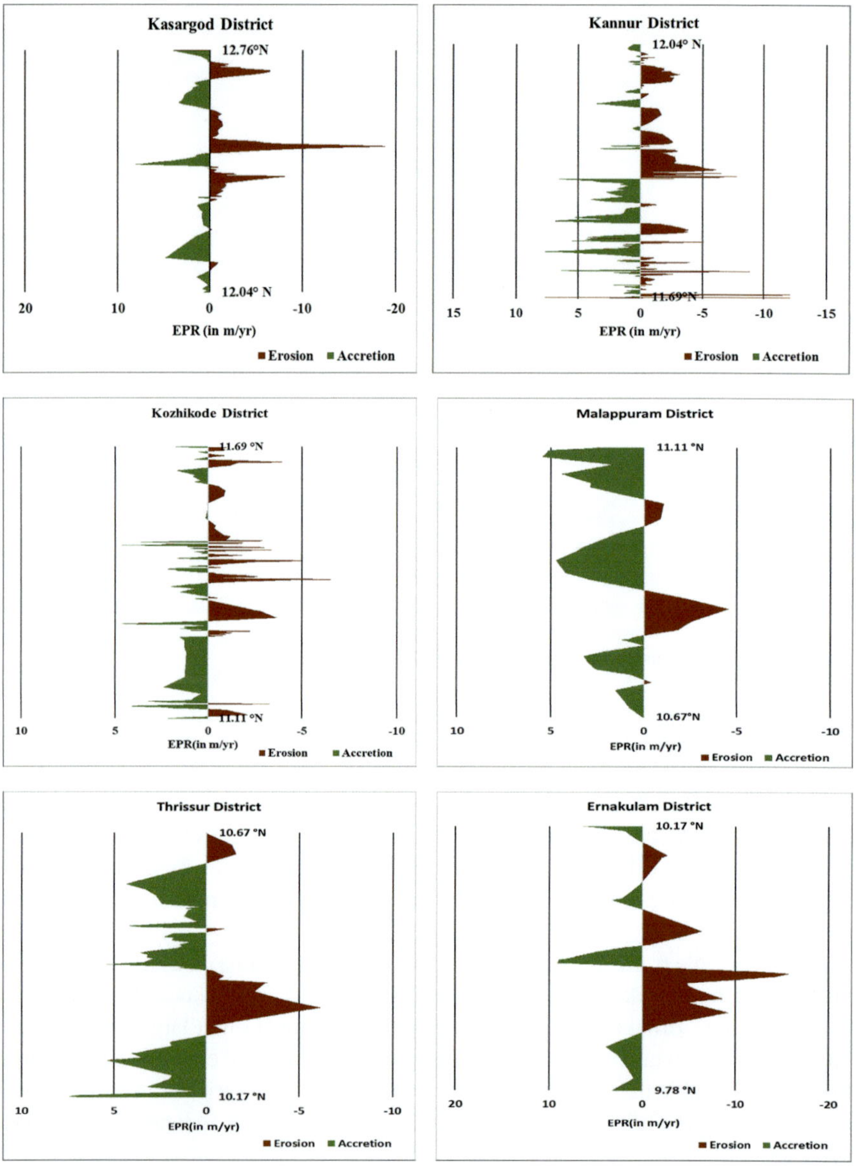

Fig. 13 EPR values plotted district wise, arranged geographically from North to South

in such regions to find out the exact cause-and-effect scenario so that coastal management plans can be devised and implemented. This will not only be cost-effective but also, accelerate the stabilisation of the threatened coast of Kerala.

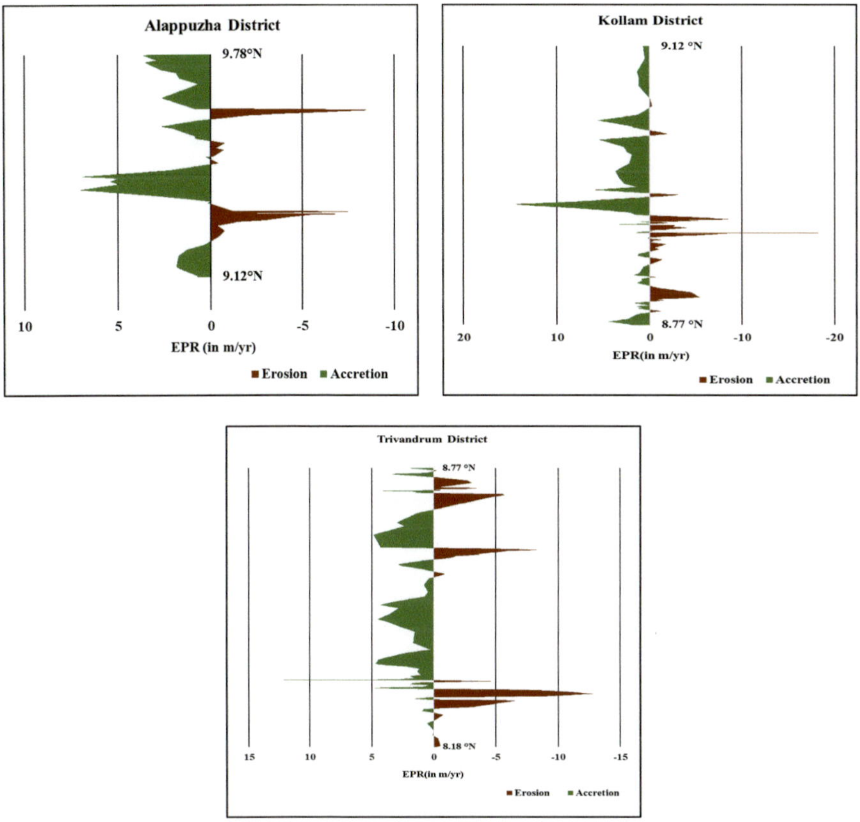

Fig. 13 (continued)

5.2 Discussion

In this study two main objectives were addressed that is shoreline extraction and rate of shift calculation. While initially attempt was made to segment the water from land, to extract the land–water boundary, that is, the shoreline of Kerala using the new segment geospatial package that makes use of the Segment Anything Model. However, due to hardware limitations, and the length of the entire shoreline (around 590kms) being very long, the data was too heavy to be loaded and processed in the available hardware system. Thus, a small portion of Kerala coastline (around 3 kms), was selected as region of interest to make sure smooth running of the model in the available computer system.

The model generated good results and demarcated the shoreline with decent accuracy based on visual analysis. However, how good the Segment Anything Model will perform to delineate a larger coastline is to be determined, provided fair hardware facilities, and ground data.

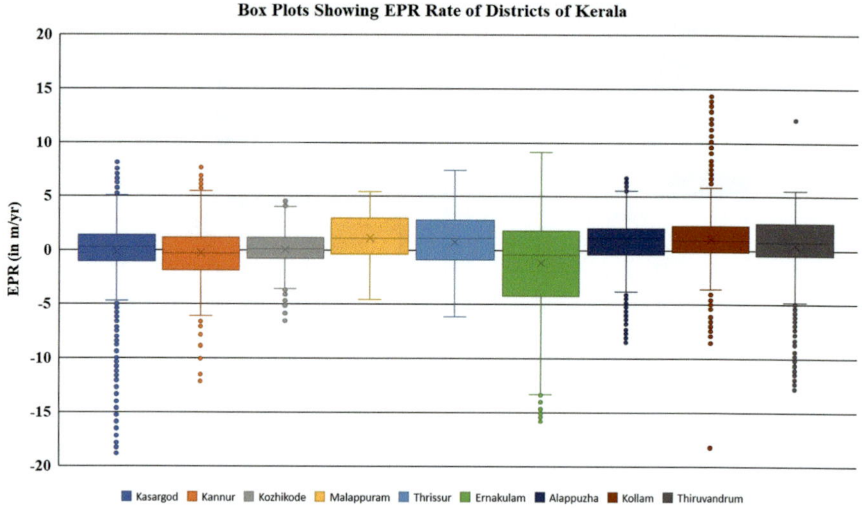

Fig. 14 Box plots showing district wise EPR values

NDWI based extraction was then performed on the downloaded satellite data from Google Earth Engine, to extract the entire coastline of Kerala. The extracted coastline however showed pixel-boundary like structure as it was vectorised from raster. A good coastline was obtained, and few errors were manually fixed. The shorelines for the three years are generated in this way, which was then used for shoreline shift calculation in DSAS tool extension.

The shoreline change rate can be estimated using the generated EPR values from the DSAS tool. The 20 years study generated results with about 40% of the total transects generated at 50m interval shows erosional rates. This confirms with the report by Ministry of Earth Sciences that 41% of Kerala's coastline is experiencing different rates of erosion (Kallungal 2021). The district wise generated graphs (Fig. 13) show a maximum erosion rate of around 18 m/year in Kasargod and Kollam districts, as well as highest accretion rates are also seen in Kollam district. This dynamic nature and high variability of Kollam district shoreline could be attributed to the fact that the district hosts major mining activities of companies such as India Rare Earth Ltd. (IREL) and Kerala Mineral and Metals Ltd. (KMML), which result in such high erosion rates in the district (Selvan et al. 2020). Artificial structures such as groynes, breakwaters, etc. though built to lower erosion rates, can sometimes lead to further erosion, if not built properly. The intensity of waves gets influenced due to such structures, and areas adjacent to such features face heavy erosion. Besides these, other activities influencing erosion in the coastline of Kerala are the mangrove forest destructions, done to accommodate the increasing population pressure along the coast

(Sachin Pavithran et al. 2014). The rates of erosion and accretion are heavily influenced by long-term sediment transportation. The Kasargod coast beach are dominated by sand, silt and clay which are prone to erosion (Das 2022). The geomorphology and shape and orientation of the shoreline also heavily influences its erosive and accretive nature. The severity of erosion in certain areas can also be attributed to the steepness of the near shore and the angle at which waves interact with the coastline (Rao et al. 1985). A natural factor that heavily influences erosion is monsoon showers and occurrences of cyclones such as cyclone Ockhi, occured in 2017 and Tauktae, struck in 2021. Though monsoon time data could not be obtained, many researchers including Selvan et al. (2020) notes the influence of monsoon showers on shoreline alteration. Data for this could not be taken as the images contained too much cloud content.

Accretion along the coast of Kerala, besides natural reasons of sediment deposition at river mouths, and orientation of coastline is being anthropogenically induced by construction of artificial beaches and structures such as breakwaters, seawall, groins, etc. Specially near the fishing harbours, breakwaters and protective structures like seawalls trap sediment flow and cause accretion of the coast, such as in the Muthalapozhi fishing harbour in Trivandrum, Manjeshwara in Kasargod district, Azhekkal fishing harbour in Kannur, Thangaserry in Kollam, etc. Formation of mud banks, a distinct feature in the coast of Kerala (Mallik 1987) also induces heavy accretion by trapping sediments like in the Thrissur district in Chavakkad region (Selvan et al. 2020). In the Vizhingjam region of the Trivandrum district, the presence of rocky outcrop traps the sediment causing accretion of coast.

The boxplots (Fig. 14) created show a comparative picture of variation in shoreline change rates in the districts. The districts showing presence of large number of outliers indicate excessive erosion and accretion rates, whereas districts not showing presence of outliers, like Thrissur and Malappuram show lower rates of erosion and accretion. The median values give an idea about the average trend of the district, and mostly all districts show a median value above zero, indicating on average higher accretion levels. This could be attributed to the artificial beach conservation measures being taken nowadays to protect the shoreline. However, despite protection measures, some locations show high to very high erosion rates due to mining and other such destructive activities along the coast.

6 Limitations and Recommendations

6.1 Limitations of the Study

The limitations of this study can be listed broadly into three points, the first being due to memory limitations in google colab, Landsat satellite images could not be uploaded for shoreline delineation using SAM, due to the file size. Thus, images were downloaded from map tiles in colab. However, due to RAM exhaust error, map tiles for only a small area (~3 kms) had to be considered. For processing the model over larger areas, better hardware facilities are needed. The second limitation of the study is that median composite of 2 months satellite images had to be used due to cloud cover over the coastal areas. Thus, the consideration of reference shoreline or datum line as the mean high tide line could not be taken into account. Also, as satellite passes over a fixed time over the area (10.30 a.m. local time) it is difficult to capture the mean high tide line in satellite images. The final major limitation was that post-monsoon period images, which reflect maximum changes in shoreline, could not be considered due to heavy cloud cover, which is an inherent problem over coastal areas.

6.2 Future Scope and Recommendations

The study can be improved greatly by including various aspects of coastline delineation and advanced computing. The Segment Anything Model and its Python package Segment Geospatial (samgeo) can be tested for long shoreline delineation and extraction. The accuracy assessment of NDWI-based delineated shoreline can be carried out using ground-based in situ data collection. Further research is required for validation of the obtained rates of change, using in situ data. The factors influencing the vulnerability of the coast need to be explored in detail. Satellite data fusion can be explored for improved spatio-temporal accuracy. Additionally, SAR data can be utilised to assess shoreline change during monsoon seasons when there is cloud cover.

7 Conclusions

In this chapter, an attempt has been made to utilise artificial intelligence-based model—Segment Anything Model for coastline delineation. However, due to hardware limitations, NDWI-based approach has been used for extracting the entire Kerala shoreline for three years covering a 20-year span from 2002 to 2022. The extracted shorelines are then processed and used as input in the DSAS tool, which estimates their shift and rate of shift.

The rates indicated a maximum of about 18 m/year erosion rate in the state while a maximum of about14 m/year accretion rate. The state of Kollam appears to have a very dynamic nature hosting regions experiencing highest erosion rate in the state as well as highest accretion rate. However, there can be errors as no ground truth data has been used, and since the shoreline is extracted based on index-based reclassification method from satellite-obtained imagery. Despite that, the fact that Kerala's coastline is facing a serious threat has been a topic of discussion since the last decade.

While natural equilibrium of erosion and accretion has been prevalent in all coasts around the world, anthropogenic intervention into the natural system, has started threatening this equilibrium. The National Centre for Coastal Research (NCCR) reported a multitude of reasons, both anthropogenic and natural for the change in shoreline of Kerala. Global warming and its consequent sea-level rise, increased intensity of precipitation, cyclones, are threatening the coastline. Besides, construction and mismanagement of artificial beaches, and structures and economical projects along the shoreline are affecting the coastline. Reclamation of wetlands is another factor that heavily affects the coast.

Efforts are being put from local to government level to save the coastline of Kerala. The government has already started the process of relocating vulnerable population residing within around 50m from the coast. As noted by the former director of ISRO's Vikram Sarabhai Space Cente (VSSC), along Kerala coast, the intensity and prevalence of coastal erosion is higher due to the higher energy stored in Arabian Sea compared to the Bay of Bengal. Implementing shore conservation measures such as construction of sea walls and diaphragm walls is already being carried out to mitigate the impact of high energy waves. There however, is a need for a more thorough CZM plan to be upheld at the government level. Such studies can help to identify the areas in urgent need of protection, and accordingly measures can be taken for protection and conservation of the coastline.

Acknowledgements This study was conducted as part of M. Tech course carried out by the first author under supervision and guidance of the second author. Thanks to the Director, Indian Institute of Remote Sensing, Indian Space Research Organization (ISRO), Dehradun, India for his support. We thank the anonymous reviewers for their useful comments, which greatly helped improve the manuscript.

References

Adger WN, Kelly PM (1999) Social vulnerability to climate change and the architecture of entitlements. Mitig Adapt Strateg Glob Chang 4:253–266. https://doi.org/10.1023/A:1009601904210

Basheer Ahammed KK, Pandey AC (2022) Assessment and prediction of shoreline change using multi-temporal satellite data and geostatistics: a case study on the eastern coast of India. J Water Clim Change 13:1477–1493. https://doi.org/10.2166/wcc.2022.270

Boak EH, Turner IL (2005) Shoreline definition and detection: a review. J Coast Res 2005:688–703. https://doi.org/10.2112/03-0071.1

Chrisben Sam S, Gurugnanam B (2022) Coastal transgression and regression from 1980 to 2020 and shoreline forecasting for 2030 and 2040, using DSAS along the southern coastal tip of Peninsular India. Geod Geodyn 13:585–594. https://doi.org/10.1016/j.geog.2022.04.004

Das GK (2022) Coastal Environment—Case Studies. Coastal environments of India: a coastal West Bengal perspective. Springer International Publishing, Cham, pp 43–58

Dutta D, Kumar T, Jayaram C, Akram W (2022) Shoreline change analysis of hooghly estuary using multi-temporal landsat data and digital shoreline analysis system. In: Zhang Y, Cheng Q (eds) Geographic information systems and applications in coastal studies. IntechOpen

Feyisa GL, Meilby H, Fensholt R, Proud SR (2014) Automated water extraction index: a new technique for surface water mapping using landsat imagery. Remote Sens Environ 140:23–35. https://doi.org/10.1016/j.rse.2013.08.029

Gandhi U (2023) Automated coastline extraction using google earth engine. In: spatial thoughts
Gharnate A, Taouali O, Mhammdi N (2024) Shoreline change assessment of the Moroccan atlantic coastline using DSAS techniques. J Coast Res 40:418–435. https://doi.org/10.2112/JCOASTRES-D-23-00013.1
Global Order (2023) How global warming threatens Kerala's unique coastline. https://www.globalorder.live/post/how-global-warming-threatens-kerala-s-uniquecoastline#:~:text=According%20to%20the%20NASA%20Sea,causes%20of%20erosion%20in%20Kerala
Himmelstoss EA, Henderson RE, Kratzmann MG, Farris AS (2021) Digital shoreline analysis system (DSAS) version 5.1 user guide open-file report 2021–1091
Hossain MS, Yasir M, Wang P et al (2021) Automatic shoreline extraction and change detection: a study on the southeast coast of Bangladesh. Mar Geol 441:106628. https://doi.org/10.1016/j.margeo.2021.106628
Kallungal D (2021) 41% of Kerala's coastline under erosion threat: Study. The New Indian Express. https://www.newindianexpress.com/states/kerala/2021/Dec/12/41-of-keralas-coastline-under-erosion-threat-study-2394576.html
Kannan R, Kanungo A, Murthy MVR (2015) Detection of shoreline changes Visakhapatnam coast, Andhra Pradesh from multi-temporal satellite images. J Remote Sens GIS. https://doi.org/10.4172/2469-4134.1000157
Kirillov A, Mintun E, Ravi N, Mao H, Rolland C, Gustafson L, Girshick R (2023) Segment anything. In Proceedings of the IEEE/CVF international conference on computer vision, pp 4015–4026
Kourosh Niya A, Alesheikh A, Soltanpour M, Kheirkhahzarkesh M (2013) Shoreline change mapping using remote sensing and GIS case study: Bushehr province. Int J Remote Sens Appl 3(3):102–107
Kumar L, Afzal MS, Afzal MM (2020) Mapping shoreline change using machine learning: a case study from the eastern Indian coast. Acta Geophys 68:1127–1143. https://doi.org/10.1007/s11600-020-00454-9
Mallik TK (1987) Coastal zone management programme in Kerala, India. Environ Geol Water Sci 10:95–102. https://doi.org/10.1007/BF02574666
McFEETERS SK (1996) The use of the normalized difference water index (NDWI) in the delineation of open water features. Int J Remote Sens 17:1425–1432. https://doi.org/10.1080/01431169608948714
Mondal BK, Mahata S, Das R et al (2024) Analysis of the shoreline changes using geoinformatics in Ghoramara Island of Hugli Estuary West Bengal in India. J King Saud Univ Sci 36:103014. https://doi.org/10.1016/j.jksus.2023.103014
Mukhopadhyay R, Karisiddaiah SM (2014) The Indian Coastline: Processes and Landforms. In: Kale VS (ed) Landscapes and landforms of India. Springer, Netherlands, Dordrecht, pp 91–101
Murray J, Adam E, Woodborne S et al (2023) Monitoring shoreline changes along the Southwestern Coast of South Africa from 1937 to 2020 using varied remote sensing data and approaches. Remote Sens Basel. https://doi.org/10.3390/rs15020317
Nandi S, Ghosh M, Kundu A et al (2016) Shoreline shifting and its prediction using remote sensing and GIS techniques: a case study of Sagar Island, West Bengal (India). J Coast Conserv 20:61–80. https://doi.org/10.1007/s11852-015-0418-4
Narayana AC, Priju CP (2006) Landform and shoreline changes inferred from satellite images along the Central Kerala coast, India. J Geol Soc India 68:35–49
Palanisamy P, Sivakumar V, Velusamy P, Natarajan L (2024) Spatio-temporal analysis of shoreline changes and future forecast using remote sensing, GIS and kalman filter model: A case study of Rio de Janeiro Brazil. J South Am Earth Sci 133:104701. https://doi.org/10.1016/j.jsames.2023.104701
Rajwi, T (2023). Fishermen in Kerala live on edge as coastal erosion destroys homes. The Hindu. https://www.thehindu.com/news/national/kerala/living-on-the-edge-kerala-coastal-erosion/article67068428.ece

Rao PP, Nair MM, Raju DV (1985) Assessment of the role of remote sensing techniques in monitoring shoreline changes: a case study of the Kerala coast. Int J Remote Sens 6:549–558. https://doi.org/10.1080/01431168508948477
Sachin Pavithran AP, Menon NR, Sankaranarayanan KC (2014) An analysis of various coastal issues in Kerala. Int J Sci Res Educ 2:1993–2001
Sahana M, Rehman S, Ahmed R, Sajjad H (2021) Analyzing climate variability and its effects in Sundarban Biosphere Reserve, India: reaffirmation from local communities. Environ Dev Sustain 23:2465–2492. https://doi.org/10.1007/s10668-020-00682-5
Salauddin M, Hossain KT, Tanim IA et al (2018) Modeling spatio-temporal shoreline shifting of a coastal island in bangladesh using geospatial techniques and DSAS extension. Ann Valahia Univ Targoviste Geogr Series 18:1–13. https://doi.org/10.2478/avutgs-2018-0001
Sekovski I, Del Río L, Armaroli C (2020) Development of a coastal vulnerability index using analytical hierarchy process and application to Ravenna province (Italy). Ocean Coast Manag 183:104982. https://doi.org/10.1016/j.ocecoaman.2019.104982
Selvan SC, Kankara RS, Prabhu K, Rajan B (2020) Shoreline change along Kerala, south-west coast of India, using geo-spatial techniques and field measurement. Nat Hazards 100:17–38. https://doi.org/10.1007/s11069-019-03790-2
Thakare LM, Shitole TA (2023) Shoreline change detection analysis along the Sindhudurg coast of India using DSAS technique. Ind J Sci Technol 16:1340–1348. https://doi.org/10.17485/IJST/v16i18.162
Tsiakos CAD, Chalkias C (2023) Use of machine learning and remote sensing techniques for shoreline monitoring: a review of recent literature. Appl Sci Switz. https://doi.org/10.3390/app13053268
Wei X, Zheng W, Xi C, Shang S (2021) Shoreline extraction in SAR image based on advanced geometric active contour model. Remote Sens Basel 13:1–19. https://doi.org/10.3390/rs13040642
Wu Q, Osco LP (2023) Samgeo: a python package for segmenting geospatial data with the segment anything model (SAM). https://doi.org/10.5281/ZENODO.7966658
Xu H (2006) Modification of normalised difference water index (NDWI) to enhance open water features in remotely sensed imagery. Int J Remote Sens 27:3025–3033. https://doi.org/10.1080/01431160600589179
Yasir M, Sheng H, Fan H et al (2020) Automatic coastline extraction and changes analysis using remote sensing and GIS technology. IEEE Access. https://doi.org/10.1109/ACCESS.2020.3034707

Evaluation of Surface Water Quality in Mahanadi River Basin, Odisha, for Drinking Purposes Based on GIS, MEREC and a Hybrid MACROS Approach

Abhijeet Das

Abstract Water is one of the essentials for life on Earth. Everyone has the right to access safe drinking water, but sadly, industrialization, poor waste management, and overpopulation are polluting water supplies. Thus, in order to ensure environmental sustainability, it is imperative to model and map surface water (SW) susceptible areas and their responses to runoff and spatiotemporal land use changes. The purpose of this investigation was to highlight the differences and similarities among the different tributaries of the Mahanadi River, Odisha, to determine the primary causes of pollution in the river basin, using Method based on the Removal Effects of Criteria (MEREC/M) and Measurement of Alternatives and Ranking according to Compromise Solution (MACROS) methodology. The study region was separated into 19 zones for ease of sample collection. A total of 21 water samples were collected during the 2020–2023 period and subjected to extensive physicochemical analysis. According to an analysis of the physicochemical data gathered, the water's TKN and EC were determined to be high when compared to WHO recommendations. The observation could be because there are a lot of possible anthropogenic sources in this watershed and it is primarily utilized for agricultural purposes. Additionally, the M-WQI value ranged from 34 to 381, suggesting excellent to unsuitable category. Around 57.89% of water locations ($n = 11$) exhibit poor-unsuitable class, thus illustrating overuse of inorganic fertilizers, contributes to pollution of water bodies. Also, it arises because of climatic changes along the agroecological zones. In order to meet increasing demand for clean water, sustainable use and conservation of available water resources, MACROS study was done to find out effects of agricultural nutrient pollution, along with different anthropogenic practices. As per its classification, Site Z-(9) represents the most polluted site, holding a highest score of $U_i = 0.0469$, followed by sites Z-(8), (19) and (2). The results demonstrate that WQ is affected by both point and non-point sources of pollution in rural and urban areas. Some of these sources includes use of fertilizers, sewage discharge, manures and

A. Das (✉)
Research Scholar, Department of Civil Engineering, C.V. Raman Global University (C.G.U), Bhubaneswar, Odisha 752054, India
e-mail: das.abhijeetlaltu1999@gmail.com; abhijeetlaltu1994@gmail.com

S. C. Pal and U. Chatterjee (eds.), *Surface, Sub-Surface Hydrology and Management*, Springer Geography, https://doi.org/10.1007/978-3-031-62376-9_27

pesticides, animal husbandry activities, inefficient irrigation practices, deforestation of woods, pollution due to industrial discharge, aquaculture, and agricultural runoff, domestic sewage, mining, and recreational activities. From the overall findings, it is concluded that samples from the upstream sites show better water quality (WQ) than downstream portions. The suggested contributions at the downstream side for WQ deterioration are industrial effluents from various industries such as nut-bolt industry, electroplating, and galvanizing and cycle industries residing in the nearby regions. Therefore, the implementation of MEREC and MACROS methodology in the present work, would offer a necessary step for proper management of the river channel, and it will be a very helpful tool for the public and decision makers to evaluate the WQ for sustainable management.

Keywords Mahanadi · MEREC · MACROS · Inorganic · Pollution · Anthropogenic · Sustainable

1 Introduction

Surface water is the predominant water source for consumption, domestic services, industry, manufacturing, agriculture and almost all aspects of human life. Notably, dynamic spatiotemporal evaluation of water quality can be used to analyse issues with water pollution, pinpoint possible sources of contamination, and offer data assistance and guidance for efficient water resource management (Kiprotich et al. 2021). However, pollutants can, however, infiltrate river systems through a number of different channels, including atmospheric deposition, seepage, leaching, stream and ditch discharge, and storm water runoff. In general, the amount of water used for residential use only makes up 4.86%, with 21% being employed by the manufacturing industry, leaving 74.89% for use in agriculture (Sulamo et al. 2021). The rate of pollution in the environment has increased due to rapid industrialization, urbanization, and human habitation development. Thus, scenarios of land use change and climate change brought about by humans have caused concern about the sustainability of water resources all over the world (Tumsa et al. 2022). However, runoff from agricultural activities and the careless discharge of untreated industrial and mining pollutants also significantly contaminate water systems. They are the ones who gather and carry organic and inorganic materials that have the tendency to change the water's natural composition and quality, including wastewater from homes, businesses, and farms (Berihun et al. 2020). Meanwhile, water pollution has a substantial negative influence on socioeconomic development, human health, aquatic ecological balance, and prosperity. The higher concentration of heavy metals in the samples is partly caused by the use of agrochemicals on agricultural land. Following that, there is always a chance for bioaccumulation and biomagnification due to the potentially hazardous metals (Town et al. 2021). In addition, changes in land use and variations in climate parameters can affect hydrologic processes including transpiration, evaporation, runoff, infiltration, and percolation. These fluctuations eventually change the region's water supply. Therefore, implementing a thorough program for the development of water resources that prioritizes the management and monitoring of water quality is essential (Gashaw et al. 2021). The Water Quality Index (WQI), a comprehensive indicator, provides an easy-to-read, scientifically grounded evaluation of current water quality conditions as well as potential future changes. This technique relies mostly on the physicochemical characteristics of water to monitor and regulate the quality of water meant for human consumption; however, it is important to select the water quality measures that can most accurately depict spatiotemporal variations and pinpoint the different sources of contamination (Yigez et al. 2021). Consequently, statistical methods like as visualisation and descriptive statistics have been used to evaluate the sources of nutrients and other contaminants in surface water. It is a helpful method for assessing how the quality of the aquatic environment has changed and trended as it combines several original factors into a single index (Belay and Mengistu 2021). Its fundamental idea is to evaluate the degree of excess indicator contamination in surface water and calculate the health effects on people. In order to describe the overall water quality condition and grade, the Water Quality Index (WQI) integrates several factors into a dimensionless variable and assigns a relative weight to each parameter depending on its significance for protecting the aquatic environment (Kidane et al. 2019). It involves quickly interpreting the quality of water by combining many physical, chemical, and microbiological criteria to produce a numerical number that permits a certain degree of contamination and to display the water's current condition in rivers and other bodies of water (Moisa 2021); however, the majority of these indices do not take into account the realities of each zone; rather, they are typically chosen, given a weight, and then converted to a scale based on subjective considerations (Duguma 2022). But calculating WQI is a time-consuming process that takes into account a lot of national and international norms when it comes to water consumption criteria. This complicates the procedure even if the computations are straightforward (Berihun et al. 2020). One of the best methods to categorize the various parameters' weights is mentioned to be **MEREC** (Method based on the Removal Effects of Criteria (M)). To deal with that (Keshavarz-Ghorabaee et al. 2021), this approach is employed as an objective way of weighting variables in order to determine their relative weights while avoiding the drawbacks of traditional information entropy. The basic idea is to calculate the objective weight of the indicators using two concepts—the evaluation indicators' conflict and contrast intensity—based on the indicators' correlation (Li et al. 2020). This strategy proves to be a helpful technique for identifying likely surface water zones because it represents efficient and economical techniques. To evaluate the state of the water quality in inland and coastal waters, this process could be utilized as a proxy indicator. Similar relationships between these characteristics have been found in a number of studies, which can improve the assessment of the trophic condition of inland waters (Trung, 2022). As a result, it functions as a useful tool for acquiring, organizing, storing, and retrieving a range of spatial and non-spatial data in addition to using these data for spatial analysis and integration to provide insightful modelling and outputs (Gurara et al. 2020). Within a certain geographic context, interpolating spatial parameters with instruments such as the

Geographical Information System (GIS), that promotes integrating field inventory, which has made scientific representation more flexible. This is widely employed in the spatial register sector to collect a range of spatial data and to utilize overlay analysis to characterize spatially variable occurrences (Skourtos et al. 2021). In relation to spatial modelling, Inverse Distance Weighted (IDW) interpolation was employed, and parameter values were categorized using the WHO's guidelines. This is essentially an algorithm that estimates values and shows the spatial distribution number of contaminants, which expresses the minimum number of points that have been set. The output cell size was set to 0.01 to obtain a smooth prediction of the values in the unsampled/unmeasured areas and provide a detailed explanation of the spatial distribution of each parameter in relation to the others (Huang 2022). Allocating unknown samples and extracting relevant information becomes a more challenging challenge as the number and dimension of measurement parameters in samples rise. Accordingly, earlier studies have demonstrated that simulations based on deterministic water quality models frequently produce false findings because these models are unable to handle the flaws and ambiguity in the underlying data or to adequately depict the complexity of water quality, which varies on numerous aspects (Xu et al. 2020b). To get past this, Decision-making methods have been applied universally to the evaluation of water quality based on intricate data (Tan et al. 2022). However, it is important to note that optimization tools can make the processes for building WQI for water quality assessment simpler by highlighting their strengths and weaknesses (Ougahi 2022). The development of novel approaches to data collection and statistical interpretation for preventive or management purposes is imperative, as the practical and trustworthy assessment of water quality during monitoring programs is intricate and challenging to interpret (Rezk et al. 2021). Therefore, Multi-Criteria Decision Analysis (MCDA) is employed to assess and contrast several, frequently contradictory factors in order to arrive at the optimal conclusion. When assessing the requirements, MCDA takes into account spatial data models, the evaluation criteria's spatial dimension, and decision qualities when determining how the criteria relate to one another (Wang et al. 2019). Due to its capacity to acknowledge and weigh the importance of intricate details, researchers usually make use of Measurement of Alternatives and Ranking according to Compromise Solution (MACROS) to assess susceptible areas in order to aid in decision-making during the environmental planning and natural resource management processes (Gong et al. 2021). It helps in ranking potential actions or alternatives in order of priority. The fundamental objective is to calculate the number of factors required to produce a set of variables, analyze and assess the parameter weight, specify the degree of correlation between the variables and relevant statistical factors, and investigate each of the factors found by means of each score (Devici et al. 2021). **MACROS** mechanism excels at handling complex decision scenarios by simulating the uncertainties and vagueness associated with the multiple criteria. Also, by lowering the relative error found by ignoring the artificial weights, it improves WQI. This methodology can be applied effectively to determine the value and utility degree of various areas and establish the priority order for their implementation. Using these methods, the problem of evaluating a discrete set of alternatives can be examined based on a set of decision criteria (Wagner et al.

2022). As a result, implementing a viable and cost-effective approach for reliable water quality evaluation is critical. Because of the benefits of these approaches, the WQIs may be calculated in real-time from a large amount of data (Burgan 2022). However, the Mahanadi River serves as a source of water for industry, agriculture, energy production, drinking, and aquaculture; nevertheless, the excessive urbanization of the area, intensive farming practices, and the discharge of domestic and industrial effluent have all had an impact on the waterways. As a result, there is still little geographic analysis of zoning susceptible risk and water appropriateness (Gavit et al. 2020). Because of this, MEREC and MACROS were used in this study to weigh the quality parameters and aid in understanding the current state of the water quality, determining whether surface waters are suitable for human consumption, offering scientific advice to managers of water resources, and ultimately identifying the most important parameters and those in charge of the spatial differentiation of the water quality. Additionally, to examine probable sources of contamination, such as natural processes or human activity, and analyze the factors that could affect the quality of the water in these places with varying levels of contamination, using Arc GIS 10.5. In the end, it is anticipated that this activity would produce scientific data regarding surface water quality, which will be helpful for managing and developing the state of Odisha's surface water resources sustainably.

2 Description of Study Area

The Mahanadi River analysis is located within 20°30′07″ N to 20°29′38.5″ N latitude and 84°52′54.2″ E and 85°46′01.2″ E longitude (Fig. 1). The river, Mahanadi, which also signifies 'mighty' and Great River and this river is the people's lifeblood in Chhattisgarh and Odisha (Das 2022a, 2023a). Approximately 38,606,665 people who depend on it for their basic needs and survival receive water from it. Hydrographically, its drainage area is described as 141, 589 Km^2, which denotes the vicinity of 4.3% of the nation's geographic area (Mishra et al. 2022). The river's catchment area is larger in some areas in Orissa State (65,600 Km^2) and Chhattisgarh (74,970 Km^2), while smaller components present within Jharkhand (650 Km^2), Maharashtra (250 Km^2), and Madhya Pradesh (130 Km^2). A majority of the district of Chhattisgarh and the state of Odisha are protected by this river basin, with very little aiding the Jharkhand states (0.45%) and the Maharashtra state (0.07%). It estimates an annual total load of sediment, including bed and suspended loads, of about 29.77 million tons. This research region includes some of the most productive croplands and contributes significantly to the state and national economies (Baghel et al. 2023). The mean annual precipitation in this watershed is approximately 140 cm, and the maximum summer temperature recorded was 45 °C. The minimum winter temperature is approximately 25 °C. This watershed has a hot tropical environment. Additionally, it demonstrates how difficult it is to accommodate the large influx in the restricted reservoir capacity because of competing demands for flood control and irrigation when there is considerable rain in the upstream watershed, especially in September close to the conclusion of the monsoon season (Pandey et al. 2022). About eighty percent of this comes from the south-west monsoon that lasts from July to September. Approximately 87% of the water used in the basin is used for irrigation. The primary crops that are grown in this river are oilseeds, sugarcane, and rice. Because of this, the basin's primary economic activity is agriculture, and when compared to other land uses, agricultural land use is the most significant feature (Saxena et al. 2022; Das 2024a). The local farmers who grow crops for sale in and around cities or towns as their primary source of income are better off as a result of this exchange. Additionally, the river receives municipal sewage and effluents from numerous urban and industrial areas, as well as man-made problems such as the construction of groins, jetties, ports, seawalls, industries, aquaculture, and farming, all of which frequently cause extensive erosion by altering the movement of sediment along the Mahanadi River, changing the WQ.

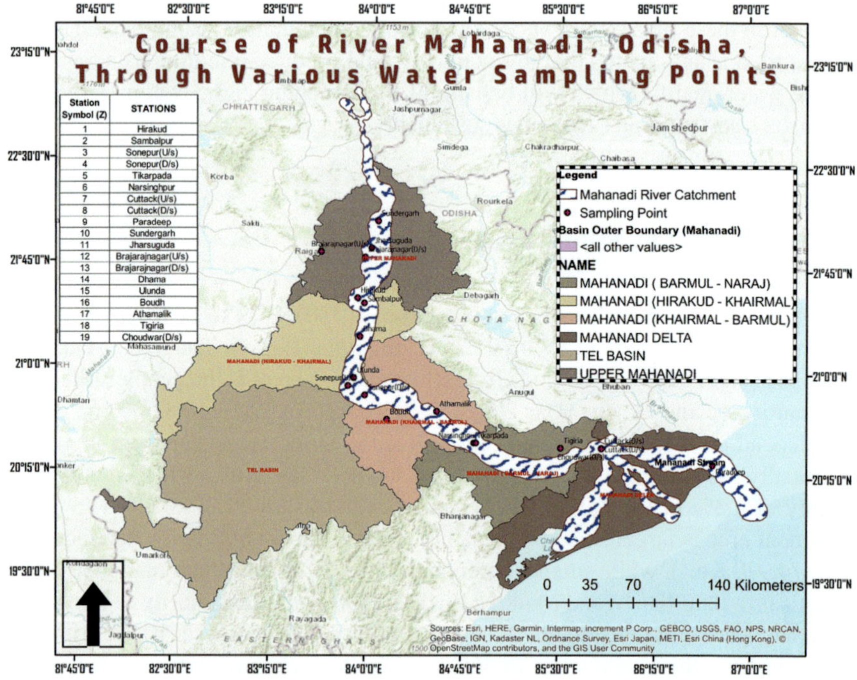

Fig. 1 Study area with monitoring stations

3 Sampling and Analysis

The Odisha State Pollution Control Board (OSPCB) is in charge of keeping an eye on the water quality along the Mahanadi River in the state of Odisha. Water samples were gathered at 19 sampling locations, and during the monsoon season, the OSPCB database provided the physicochemical monitoring data for the years i.e., 2020–2023. The river's sampling locations were selected in close proximity to river ghats, where a variety of human activities are concentrated near the water body; sewage disposal sites; encroachment areas where human activities such as bathing and laundry contribute to pollution; and seasonal nallahs, where surface runoff from the rainy season drains into the river (Li et al. 2021). When selecting the parameters to assess the water quality, it is crucial to carefully take into account a number of factors. First, reviewing earlier studies on assessments of water quality in similar rivers or locations may provide information on features that are often considered. Understanding the terrain hydrology, human activities, and local conditions is crucial. Consulting with specialists, scientists, and stakeholders in water resource management can yield valuable insights based on their expertise and experience (Meng et al. 2020). Hence, 20 water quality parameters were chosen in compliance with World Health regulations (WHO 2008). The base map and the stream network of river basin were prepared, based on Survey of India toposheets on 1:50,000 scale and Arc GIS 10.5 software was used to digitize the above maps. The study region was delineated using the GIS, and these toposheets were first georeferenced using known geographic coordinates. But particular attention was paid to the methods used in each experiment, including sample collection, preservation, and handling. The location of sampling points was recorded by GPS interfacing and all the information pertaining to the samples were saved on a cloud server. The temperature, pH, TDS (total dissolved solids) and EC (electrical conductivity) were gauged in a field by utilizing digital thermometer, pH/TDS/EC meter. While TH (total hardness) is established by the titration method (EDTA). The Chloride (Cl^-) ion was calculated utilizing the titration method with silver nitrate. The sulphate ($SO_4{}^{2-}$), fluoride (F^-), free ammonia (free—NH_3) and ammoniacal nitrogen (NH_3-N) ions were computed utilizing a spectrophotometer. The soluble Na^+ in sodium adsorption ratio (SAR), iron (Fe^{2+}) and boron (B^+) were tested by using Flame Photometer. Alkalinity was examined by using titration of H_2SO_4, while COD was ascertained by employing digestion, followed by titration and microbial analysis is employed to measure the TC (total coliform) and FC (faecal coliform). Nitrate ($NO_3{}^-$) is measured by ion-selective electrodes while BOD (biochemical oxygen demand) is obtained using standard methods. Rest parameters were analyzed using standard procedures (APHA 1995). Considering, Quality assurance and quality control (QA/QC), before each in-situ assessment, all electrodes were thoroughly cleaned with deionized water. The equipment used for all analyses was pre-calibrated using standard solutions or as indicated by their company guidelines to preserve accuracy and precision in observations. The chemical analysis's accuracy has been confirmed by Charge Balance Errors (CBE %) assessment, calculated using Eq. 1.

$$\text{CBE } (\%) = [(\Sigma \text{ cations} - \Sigma \text{ anions}) / (\Sigma \text{ cations} + \Sigma \text{ anions})] * 100 \quad (1)$$

Values of cations and anions were taken in mg/l. Ionic errors the samples fell within the acceptable limit of $\pm$ 5%. The weights of the parameters for each month were assigned to each of the sampling points that were taken into consideration for the spatial variation of the water quality parameters. Then, "Radial Basis Functions" and the IDW interpolation technique were applied to the imported data, resulting in the generation of maps with smooth curves that showed the variation of all indicators along the entire stretch of the river for the three-year period.

4 Methodology

The construction of a water quality index allowed for the trustworthy evaluation of the dataset comprising several variables examined at several sample locations. Thus, MEREC (M) weighing method was included in this novel study to examine the similarity of variance in water quality indicators between the sampling sites. In addition, it is a rating system that shows the combined impact of all the chosen water quality parameters on the overall water quality (Aredehey et al. 2020). This method illustrates a family of correlation-based techniques that rely on analytical testing of the decision matrix to extract the data from the criteria that are used to assess the weights of the criterion (Das 2023a). These are an upgrade over traditional WQIs, which are often based on the expert survey method, the analytical hierarchy process, and the Delphi method, which assigns weights to parameters based on expert and subjective judgments. These methods entail giving various parameters weights and ranking a subset of parameters according to expert opinion within a particular thematic context. They can also be used to detect spatiotemporal patterns in water quality and analyze potential causes of variations in water quality over time that could have an impact on the wellbeing of aquatic ecosystems (Subramani et al. 2005). For example, these serve as superb research instruments that facilitate the interpretation of intricate data sets, pinpoint the variables influencing variations in water quality, and enable the creation of indexes that impartially assess the water resource (Cunha et al. 2022). The primary goal of this review was to develop a straightforward WQI calculation process that can be used to determine the quality of surface and subsurface water using MCDM techniques, thereby evaluating the spatiotemporal variability of the water quality of a given water source with less effort and better accuracy (Pan et al. 2021). The stages involved in computing the results include identifying and prioritizing the criteria for making decisions, estimating the coefficients for each criterion, determining the relative relevance of each criterion, and recalculating the weight and finally, computing the relative weight (W_j) are addressed by Li et al. (2020). The details of each step and their procedures can be found in Fig. 2.

Lastly, a quality rating scale and the summation of weights $(K)_i$, resulted in the M-WQI, which is articulated as follows:

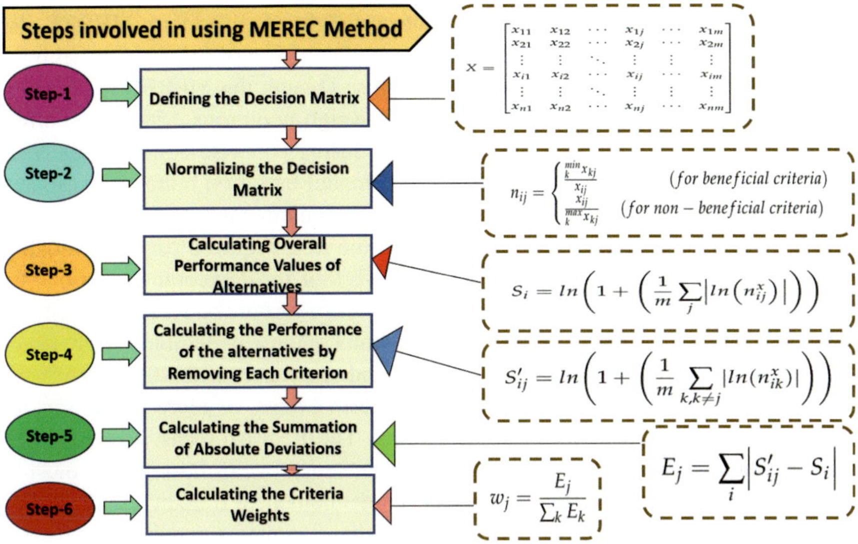

Fig. 2 A methodological framework of MEREC technique

$$M - \text{WQI} = \sum (W)_j * (K)_i \tag{2}$$

where $(K)_i$ for each parameter, represented in Eq. (3), that exhibits the ratio of the monitored value (Y_i) to its standard value (S_i). The polynomial expression results as

$$(K)_i = (Y_i/S_i) * 100 \tag{3}$$

The grading system for quality of the suggested index is being reviewed as per Dash & Kalamdhad, (2021): values between 0 and 50 are regarded as excellent, while values fall from 50 to 100 are implied as good; if the reported values varied in a range of 100–150, it implies about medium water. However, water with a range of 150–200, it indicates as poor, while if value > 200, it represents about worst/extremely poor water quality.

Several authors have used the MCDM method in the past few decades for a variety of reasons, including managing water resources, evaluating the quality of the water, and solving environmental, energy, and sustainability-related, risk and safety, technology, and information management issues (Zhang et al. 2022). This makes it necessary for MACROS, which is illustrated as a multi-criteria decision-making method (MCDM) for placing the alternative order. According to a number of scientists, MACROS is a useful technique that can reduce model uncertainty and improve the weighting procedure's accuracy. It is a method for making decisions in complicated situations where a wide range of factors or criteria are taken into account while ranking and choosing options (He et al. 2022). It enables one to ascertain the best weightings for certain characteristics that are indicative of their influence on overall water quality in the context of WQI parameter weightings. The term MEREC also finds its essence in the quantification, storage, and sharing of information with the development of information to generate weights. Though the MEREC approach is frequently utilized in MCDM, it is utilized to weigh numerous criteria and effectively handle discrepancies that could surface during pairwise comparisons (Pant et al. 2021). The MACROS forecasts a single model based on numerous response variables or indicators, which is necessary to close the information gap between the data obtained by the monitoring systems and the data required for water quality. This technique is also proven to be an effective tool for developing solutions for challenges related to the management of groundwater resources, such as identifying groundwater potential zones and assessing the water quality in relation to the research site's intrinsic properties, such as geology and land use (Sun et al. 2020). It also serves as a gauge for the unpredictable nature of random events, or more accurately, the average knowledge gained from their occurrence. The precise forecast is greatly influenced by the quantity and quality of input components, but all information must be affordable and readily accessible. This offers a mathematical framework for dealing with ambiguity and uncertainty by recognizing that statements may be partially true or incorrect and enabling the mathematical description of varying degrees of truth. This ultimately results in a decrease in the total cost by lowering the amount of elements utilized in the chemical analysis to compute WQIs (Kabiri et al. 2021). This mechanism was applied to all water quality criteria to create overall ranks in order to gain understanding knowledge about the overall pollution levels. It incorporates the concepts of hierarchical structure analysis, systematic weighting, and alternative strategy selection (Han et al. 2020). As a result, by taking into account all significant factors and allocating weight based on their variability, it is a scientific method of evaluating the water quality. However, because of the novel way in which it assigns weight based on the observed data sets, this approach is not region-specific and may be applied elsewhere (Qin et al. 2020). Details of the MACROS steps are presented in Fig. 3.

5 Results and Discussion

In the present review, surface water quality maps have been created using ArcGIS software 10.3 and are compared to WHO standards based on the parameters that have been chosen as listed below. In order to effectively manage water resources, these maps were shown to be useful in identifying the study area's good, marginal, and vulnerable water quality locations. However, in order to assess the drinking acceptability of surface water samples taken from the river basin, a newly developed WQI technique has been applied to the physicochemical study.

In the current topic, Temperature is a significant physical factor that regulates the water's taste, viscosity, solubility, odours, and chemical reactions. In addition, temperature has an impact on various other biochemical processes and characteristics of water, including the processes of sedimentation and chlorination, heavy metal

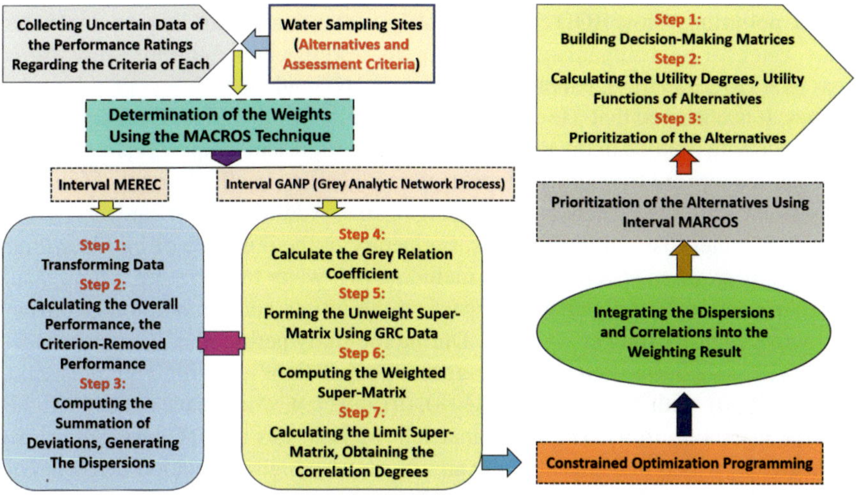

Fig. 3 Methodological framework of MACROS for selecting robust combination of water quality indicators in order to develop a ranking for WQ model

biosorption, and biological oxygen demand (BOD). In the ongoing research, the water temperature ranges from 25.72 to 30.36 °C over the different sampling stations on the river. This variation is because of fluctuation in temperature of river water usually depends on the season, geographic location, sampling time and temperature of effluents entering the stream. One essential aspect of water quality for figuring out if the environment is acidic or alkaline is the pH scale, which calculates the volume of dissolved hydrogen ions in water. A pH number of less than 7, 7, or greater than 7, suggests a base solution, while a pH value between 0 and 14 indicates an acidic water solution. In the present work, the value ranged from 7.741 to 7.913 mg/l, signifying the slightly alkaline nature. It is noteworthy that water's pH is impacted when carbon dioxide and water combine to generate carbonic acid (Zhao et al. 2020). As a result, high temperatures can occasionally produce a pH rise. Additionally, the photosynthetic action causes the pH levels to rise. However, DO is regarded as the most important index among all the water quality criteria for assessing the environmental quality. The reason behind that is all aquatic life relies on it to digest food and produce energy for survival and growth. Conversely, DO has no direct impact on public health; however, some people find it unpleasant to consume water that contains little to no oxygen. On the other hand, fish and other living things that use DO to digest organic matter include bacteria and microorganisms. Moreover, the microbial biodegradation of phytoplankton biomass and wet materials may contribute to the anoxic conditions in the reservoir by significantly lowering DO concentrations (Zhang et al. 2020b). In this work, its value spanned from 7.257 to 7.812 mg/l. In almost all stations, DO is above the permissible limit (6.0 mg/l) for drinking water. Based on BOD, the amount of oxygen required to stabilize commercial and residential waste is calculated. A higher amount in a water sample indicates a higher level of organic pollution. A low BOD concentration indicates that there is less organic matter in the water for microorganisms to oxidize. Its readings in the current study, varied from 1.095 to 2.389 mg/l, which, according to WHO guidelines, is less than 5 mg/l at all sites. It is observed that a low BOD level suggests that there is less organic matter for microbes to oxidize in the water sample. Thus, if present in appropriate amounts, these parameters are frequently advantageous for human health. Further, TC is used today to indicate the presence of microorganisms. Additionally, it can survive and flourish in water distribution systems, frequently in the presence of biofilms. Inhibition happens when specific factors including pH, water temperature, salinity, and substrate content have a negative impact on the nitrate and ammonium oxidizers' ability to function (Yang et al. 2016). During the study period, the TC count ranged from 1219 to 42,530 MPN/100 ml, suggesting all stations are within the prescribed limits (> 5000), with an exception of Z-(8), (9) and (19), which have readings higher than the threshold limits. The graph that follows indicates that TC levels are rising downstream, which indicates that the area has a very high population density. This might be the result of local industries' effluent seeping into the water and polluting it. Moreover, this indication frequently appears in water as a result of inappropriate sediment and hazardous waste management, industrial discharges, and sedimentation. In later time, Greater TSS also suggests the existence of microscopic particles that are released into rivers from several sources, such as industrial effluents and city sewage. In addition to urban anthropogenic causes, intensive local agricultural practises that cause variance in surface runoff also have a considerable impact on TSS levels. At greater concentrations, it irritates human gastrointestinal tracts and may have laxative effects. The concentration varies from 28.62 to 74.88 mg/l, in the current investigation. As a result, it was observed from the data that the reported TSS levels in the river water were within the ideal range of 100 mg/l. It is depicted that good water is seen in the upstream section due to an increase in water volume that increased the dilution activity. The alkalinity of water is mostly brought about by the existence of hydroxide ions (OH^-), bicarbonate ions (HCO_3^-), and carbonate ions (CO_3^{2-}), or a combination in the water of these two ions. While high alkalinity may be a sign of chemical or industrial contamination, low alkaline water may have corrosive qualities that indicate poor water quality and make it dangerous for use (Wu et al. 2020). It was observed in a range of 70.398–100.88 mg/l, in the current research, which is detected in the 200 mg/l range. Because of this, the taste of water gets harsher above this threshold (200 mg/l). It can be observed that the water at Z-(9) was comparatively more alkaline than the water at other stations. This could be because there were more salts present. Later on, COD is helpful in locating hazardous areas and organic compounds that are resistant to biological processes. It determines how much oxygen is needed to oxidize organic materials and oxidizable inorganic molecules chemically. This indicator signals the beginning of organic acid generation as a result of anaerobic circumstances brought on by a high concentration of dissolved organic matter, which lowers pH. The value obtained in the current work exhibits a range between 6.75 and 21.87 mg/l. As directed by WHO, the highest amount that is allowed is 30 mg/l in drinking water. The following farming indicators namely, NH_3-N and free NH_3 are the primary supply of nutrients for aquatic life, including fish and microorganisms. The four types of nitrogen that can be found in water and wastewater are organic nitrogen, ammonia nitrogen, nitrite nitrogen, and nitrate nitrogen. These are the most prevalent pollutants in water and can occasionally be utilized to track the flow of water through the soil. Furthermore, elevated levels of organic and ammonia nitrogen suggest sewage-related water contamination, while elevated nitrate concentrations cause fast-growing algae that deteriorate the quality of the water (Samsudin et al. 2017). The reported findings of NH_3-N and Free-NH_3 ranged from 0.511 to 1.928 and 0.021 to 0.059 mg/l respectively. Hence, the recommended limit for drinking water should be 2 mg/l. In general, synthetic fertilizers, residential and commercial wastewater, inflammatory processes that remove natural deposits, and microbial activation of nitrogen-containing organic molecules are the initial sources of TKN in surface water. A surplus of nutrients, mainly nitrogen, in the water body can be detrimental since it initiates the eutrophication process. The algal bloom that turns the water green and blocks out the sun is surpassed by this process. Dead zones are created for fish and other aquatic life when bacteria consume the DO left behind by the breakdown of dead algae. Furthermore, newborns are immediately and seriously at risk for serious health problems when they drink water with TKN concentrations greater than 5 mg/L. The levels in the current studied region, ranged between 3.279 and 11.791 mg/l. Nevertheless, according to WHO guidelines, its readings should not go beyond the 5 mg/l cutoff point. The results demonstrate that, in most cases, values are greater. This results from the mingling of human and animal waste, runoff from agricultural land, and anthropogenic contamination. In regards to EC, it shows the concentration of ionized substances in water and also, is a good measure of salinity. Elevated EC scores results because of saline source mixing, aquifer material leaching or dissolving, or a combination of these processes (Pei et al. 2019). Moreover, elevated mineralization levels suggest the existence of inorganic pollutants, and EC provides crucial data on the mineralization levels in water. The outcomes shown in a frame of 138.10–7779.35 μS/cm, demonstrated that water samples were determined to be within the advised limits of 2250 mg/l, and can be used without further treatment to make drinking water, with the exception of site i.e., Z-(9). Similarly, the addition of anions such as Cl^-, PO_4^{3-} and NO_3^- to sewage discharge or agricultural runoff would likewise increase conductivity. Moreover, salinity specifies the amount of dissolved salt in a body of water that may be measured using a variety of unique factors. It has a direct impact on soil erosion, flood danger, terrestrial biodiversity, agricultural output, water quality, and stream ecology. Na + reacts with the soil by decreasing its permeability and structure, hence measuring the concentration is crucial for irrigation. Thus, SAR is produced when sodium in water is exchanged for Ca^{2+} and Mg^{2+} in soils. This process lowers permeability and eventually produces soil with inadequate internal drainage. The faster the rate of water infiltration, which makes the soil more suited for irrigation, the higher the SAR caused by sodium absorption. In the period of concerned study, the surface water shows a score of 0.412–16.589, which is comfortably accord a value of 10 mg/l, except at place, Z-(9). The land that is irrigated with water that has a high SAR value at one location also has a high TKN, and this decomposition of $NaHCO_3$ also, renders the land unusable. In case of cationic analysis, Leaching or the use of washing detergents containing borate compounds combined with waste water input may be a legitimate cause of boron (B^+) in the studied area. Greater concentration in the current study region might have resulted from sea water intrusion during the tides. Reduced oxygen in river water due to higher value impacts the material and nutrient cycle processes' uniformity (Hua et al. 2020). Due to the increased quantity of this signal, it is likely caused by industrial waste, which is linked to the handling of garbage that has been improperly or mistreated. This could be brought on by direct effluent discharge, excessive surface water use, efficient ion leaching, and agricultural effects. Nonetheless, the concentration in the river was at a level that did not pose a health risk, and the current readings in the research area were lower than the WHO standard of 2 mg/l. Another indicator of water quality namely, TDS are employed to evaluate dissolved particles in water samples directly. A body of water's TDS is determined by the amount of organic matter and inorganic salts present; the lower the TDS, the greater the water quality. From the investigated work, it has been estimated that TDS concentration is found to be high at Z-(9) & (19). Additionally, it emphasizes that all sampling stations' TDS readings fell under the allowable level (WHO, 100 mg/l). This pattern lowers the TDS concentration in the river and supports the diluting effect. Additionally, the discharge of industrial effluent, agricultural runoff, and household sewage into the river causes increased readings. The data points in the study region point to the chemical weathering of rock-forming minerals as the primary component that effectively regulates the chemistry of surface water. TDS is increased by higher evaporation brought on by chemical weathering and/or human activity, and samples typically migrate from rock dominance to the evaporation zones. Water TH is computed using the cation concentrations like Ca^{2+} and Mg^{2+} and anions such as HCO_3^-, Cl^- and SO_4^{2-}. This indicator is related with Ca^{2+} and Mg^{2+}, that is produced when carbonate minerals dissolve in geological formations. Elevated total hydrostatic head (TH) can lead to scaling of pots and boilers, blockage of irrigation systems, and even health issues in humans. The observed values are seen to vary within the range of 51.18–2194.90. The WHO (2017) states that 300 mg/l is the highest permissible limit. Based on these results, it is determined that the area is under the TH limit and is suitable for drinking, apart from Z-(9). High hardness at Z-(9) is typically prohibited since it can lead to the accumulation of lime, commonly known as scaling, in pipes and water heaters. This can eventually reduce the efficiency of the water heater and the amount of soap that lathers. Meanwhile, chloride ions occur naturally in surface water and, unless present in high concentrations, do not normally pose a health risk to humans. Instead, they give food a disagreeable, saline flavour. Cl^- concentrations of 250 mg/L or more in fresh water may also be a sign of pollution from wastewater, runoff from farms, and rocks that contain chloride (Gao et al. 2021). Additionally, it gives the water a saltier, more corrosive flavour, and when people are exposed to it, it may have a laxative effect. The permissible limit of Cl^- is 250 mg/l as per WHO standards. Its concentration in the current study varied from 9.648 to 4904.88 and at every sampling location, its results stayed below the upper permissible limit, except at Z-(9). The reason for greater concentration at Z-(9) site is due to the region's geology, agricultural runoff, industrial and residential wastes, and other factors. However, sulphur minerals, heavy metal sulphides that are frequently found in igneous and metamorphic rocks, biological processes, and human activities are some of the potential sources. This parameter may get into water streams as a result of surface runoff from contaminated areas and sewage discharge (Cheng et al. 2020). Due to industrial and domestic waste, the concentration of sulphate in water increases. In this regard, the value was taken of a value that represents about 4.97–376.07 mg/l, in the present study. Higher values in Z-(9), as compared to other locations, also be caused by anthropogenic activities such as leaching and the emission of sulphur gases into the environment by manufacturers and municipal utilities. Given the research area's high level of agriculture-driven activity, it may also be connected to runoff from agriculture. The element fluoride (F^-) is commonly found as soluble compounds in the crust of the planet (Das 2024a). Excessive water intake has been connected to diseases like UTIs, paediatric mental retardation, reduced ovulation, and harm to the neurological system. The parameter F^-, migrates with moving water, much like a lot of other ions, and is eventually strengthened by evaporation. On top of that, common minerals like gypsum naturally occur in water as a result of leaching from surrounding rock formations. Fertilizers can also be added to water to enhance its natural mineral content (Ghaffari et al. 2021). The quantify of F^- concentration in the present study was found to be 0.258–1.0 mg/l. Following a disinfection procedure, the water could be used for drinking at all sites since the concentrations were within the WHO acceptable range. Further, NO_3^- primarily enters the river environment from fertilizer-containing agricultural runoff and residential wastewater. Although surface washing, phytoplankton absorption, and bacterial nitrate denitrification can cause its levels to rise quickly, changes in ground water nitrate levels usually happen gradually. While low levels of this indication are necessary for plant metabolism and human nutrition, they can also promote the growth of undesirable bacteria within a waterworks and supply system, which can lead to the deposition of a slushy coating on the pipes (Guan et al. 2019). They frequently result in unpleasant tastes and odours in drinkable water, as well as major consequences for the environment and public health. The results in the current region, revealed that its readings were in the range of 1.289–2.689 mg/l. In all sites, levels are within the criteria limit of 45 mg/l. The biggest problem with water supply in remote regions is Fe^{2+}. However, a small amount of Fe^{2+} is required in the food of humans and for plant metabolism. An excess caused by a high Fe^{2+} concentration can cause hemochromatosis, hyperglycaemia, nausea, and gastrointestinal problems (Hong et al. 2016; Das 2024a). It also explains how rainwater that came into touch with the soil raised the river's Fe^{2+} content. Moreover, highly concentrated iron water can cause turbidity and turn a reddish-brown colour. Within the permitted limit of 1 mg/l, the iron fluctuates from 0.59 to 2.609 mg/l during the study period.

In the present discussion, from the results of physicochemical parameters, it is observed that Fe^{2+} and Cl^- are the most prevalent anion and cation, respectively, in samples of surface water. The study area's relative cation and anion abundances were arranged in the following order: $Fe^{2+} > B^+$ and $Cl^- > SO_4^{2-} > NO_3^- > F^-$, respectively. As, from the findings, most of the parameters in this case indicate a stronger trend to the downstream side. When the river flows downstream, the discharges from the different cities along its length mix with the river water, raising the pollution levels. As a result, some of the downstream variables show a declining trend.

In summary, it is not possible to determine with precision how much the river water quality deviates from the drinking standard only on the values of these parameters. As a result, this study was carried out to evaluate the current state of the river water quality using MEREC and GIS. The contamination sources were found using MCDM techniques, specifically MACROS, and the proper treatment steps needed to make the water fit for consumption were suggested. Here, understanding the general state of surface water and the health of an ecosystem can be done effectively and affordably by using the MEREC technique to water quality profiling. This highlights the necessity of using it as a crucial tool to compare surface water quality and management in a given area (Das 2023a). It also aids in the selection of an appropriate, practically priced treatment technique to address the quality challenges at hand. The pattern of water use, the sources of the water, the geolocation, and

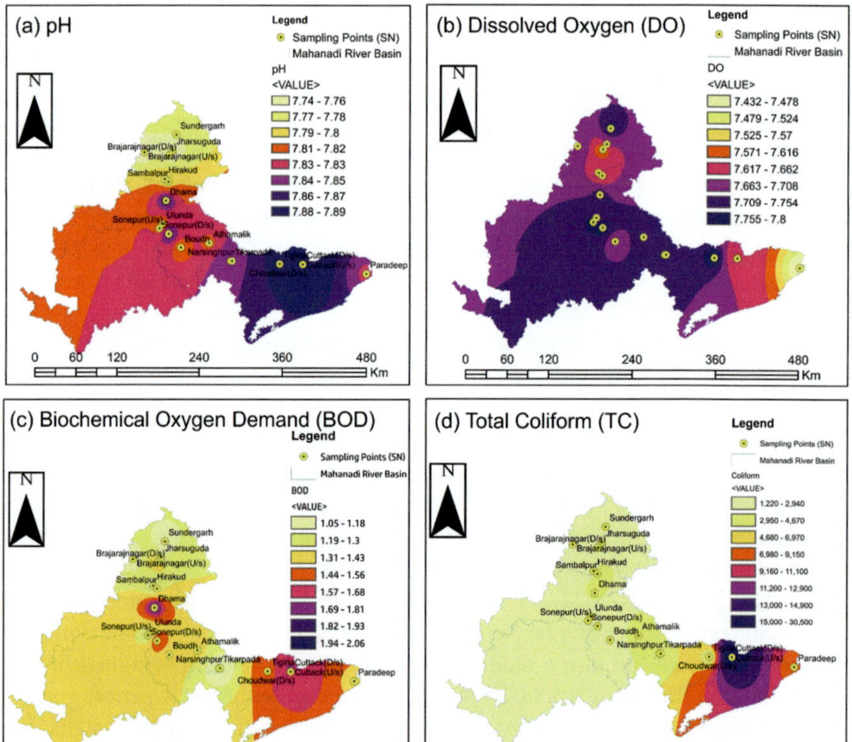

Fig. 4 Spatial interpolation of parameters by IDW, such as **a** pH, **b** DO, **c** BOD, **d** TC, **e** TSS, **f** Alkalinity, **g** COD, **h** NH_3-N, **i** Free NH_3, **j** TKN, **k** EC, **l** SAR, **m** B^+, **n** TDS, **o** TH, **p** Cl^-, **q** SO_4^{2-}, **r** F^-, **s** NO_3^- and **t** Fe^{2+}

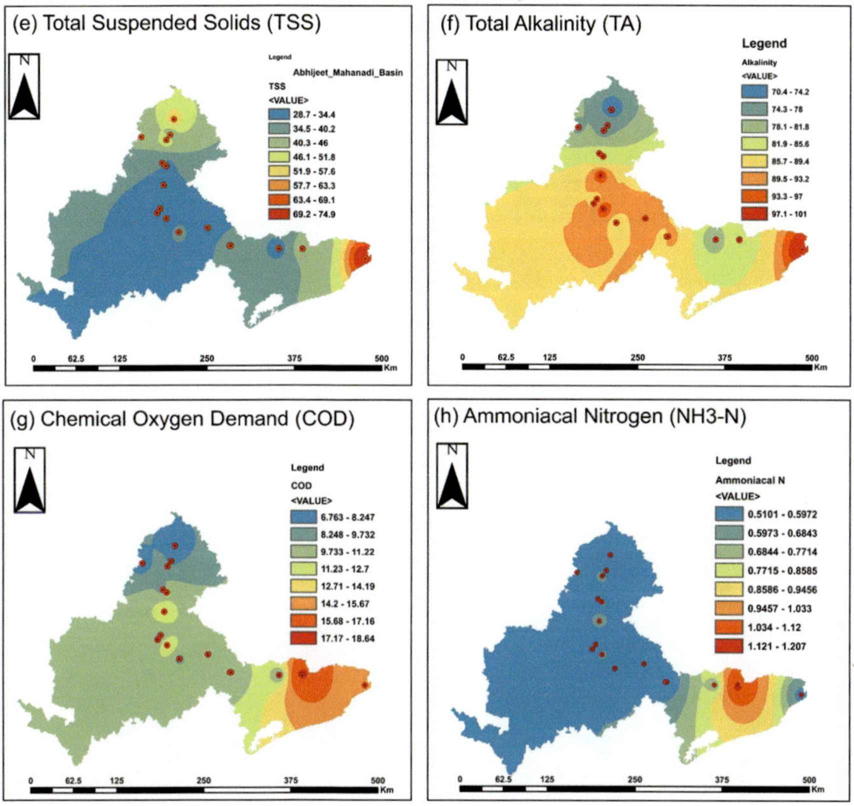

Fig. 4 (continued)

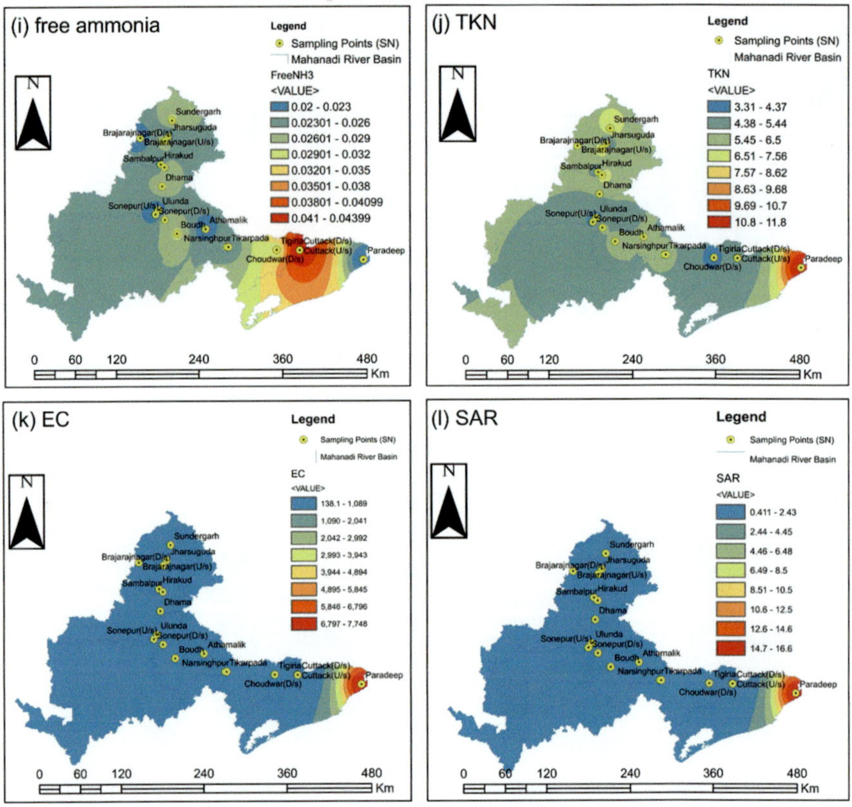

Fig. 4 (continued)

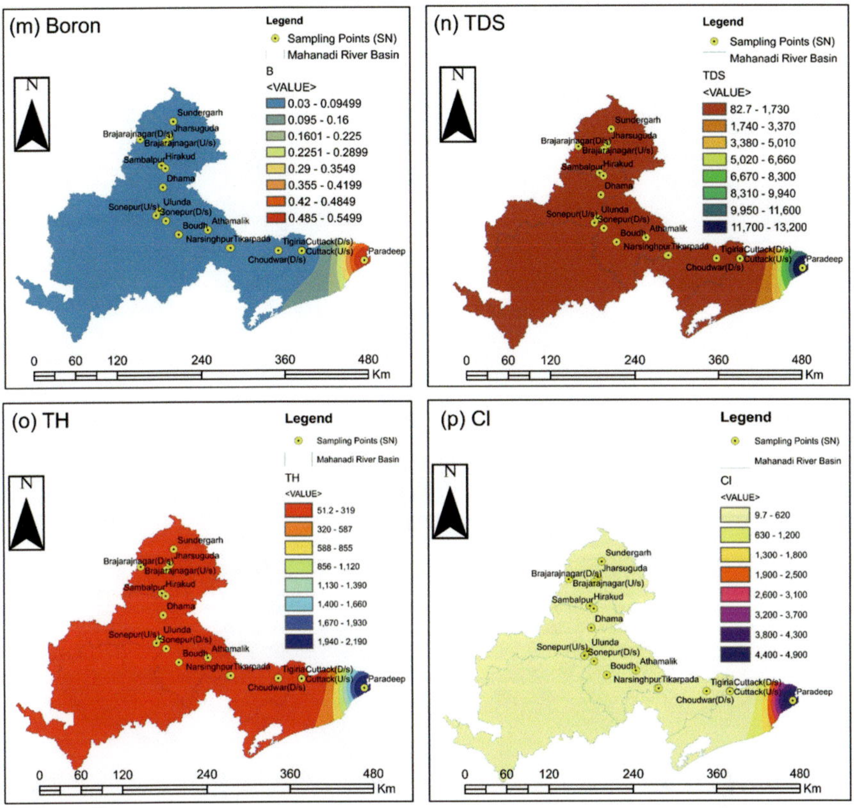

Fig. 4 (continued)

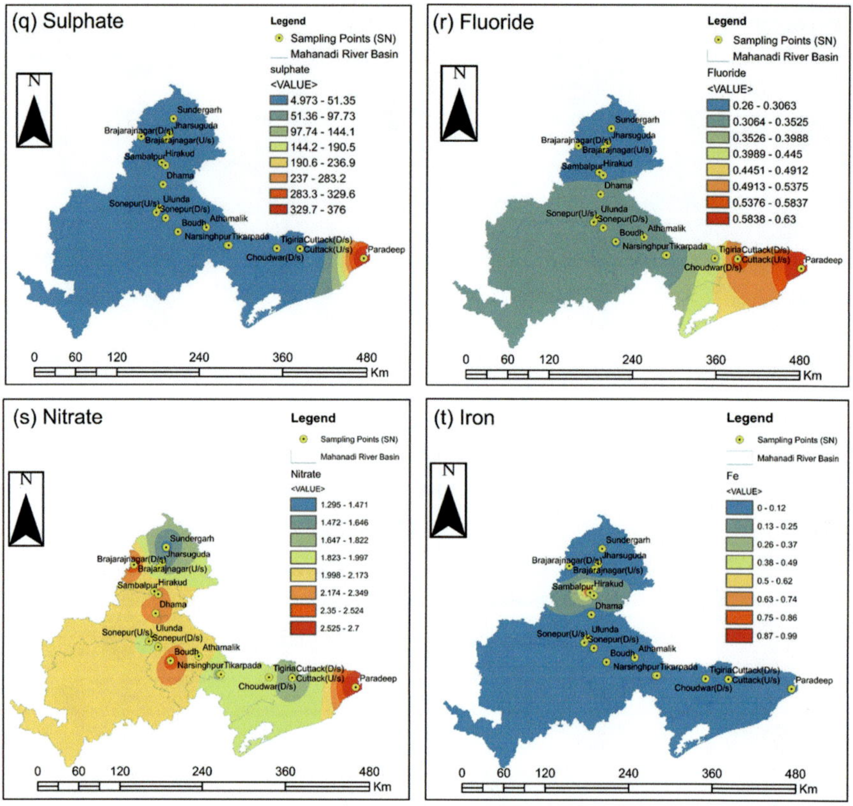

Fig. 4 (continued)

the degree of quality and purity measurement all support this method. This method involves conducting surveys or interviews with experts who, based on their knowledge, provide the proper set of standards for evaluating a particular body of water. It also involves a methodical approach to standardize and formalize the parameter selection procedure for widespread use. This approach's main idea is to combine the significant factors affecting water quality into a single index that can be used as a simple scale to assess a target stream's water quality. Water quality was often categorized into five groups using the M-WQI computed values: excellent, good, poor, extremely poor, and inappropriate. Table 1 shows the calculations for every station. The findings of the present work suggests that the observed M-WQI range is 34 to 381, illustrating excellent to unsuitable category. The M-WQI values in this study show that the surface water samples were classified as good water in each of the corresponding 31.57%, respectively. The fluctuations in M-WQI across the sites are seen in Fig. 5a. Six samples fall in the excellent class, 2 fall under good, 2 samples belong to poor, six fall in very poor zone, and finally, 3 samples fall in the unsuitable category. However, 57.89% displays inadequate/poor water quality and is not fit for human consumption (Fig. 5b). The primary cause of the elevated WQI values in surface water samples at these locations was the presence of higher concentrations of Cl^-, Ca^{2+}, SO_4^{2-}, TDS and EC. The majority of the upstream section has good water due to an increase in water volume that increased the dilution activity. Nevertheless, there are significant variations in the water quality at upstream and midstream areas, including Z-(2), (5), (6), (7), (10), (11), and (12), because of the influence of pollution sources as well as geographical regions. The samples under poor/very poor class have been connected to unkempt ghats where people dispose of trash, plastics, animal carcasses, and engage in open defecation. At the downstream side, there is a significant pressure at places Z-(8), (9), and (19). Thus, it is the consequence of industrial and agricultural activity, as well as the buildup of contaminants brought on by downstream water flow. Based on the data, it can be concluded that the river's pollution level increases from its upper to lower regions (Das 2023c). The study region's polluted zones were classified and identified using the results of the spatial analysis. Approximately 58% of the study region has to be monitored before use in order to draw the boundary by pollutants. The IDW tool in ArcGIS was used to interpolate the spatial data with the aid of reported WQI readings. Figure 6 displays the map interpolation for the research area. Furthermore, the aforementioned indicators' spatial distribution was not evenly distributed across the research region. Due to being a coastal area at Z-(9), one of the main causes of the decline in the quality of the water that is fit for human consumption in some areas is the influx of saltwater. It is possible to draw the conclusion that, in comparison to other study area regions, the drinking water quality was higher in the northeastern regions (Das 2023d).

Given this method's results, experts have determined that the model eclipsing problem is the source of uncertainty in this strategy. For example, a model eclipsing problem would result from the adoption of absolute values of comparable severity regarding their physical and biological repercussions. It therefore results from the fact that the water quality parameters' genuine nature is not reflected in the WQ model's output (Gurara et al. 2023). Besides, the findings indicate that when parameters surpass the standard mean, the quality of drinking water declines, which consequently impacts the well's quality grade. These findings stem from the relative weights allocated to each of the qualitative criteria. Therefore, each concentration should be taken into account based on its expected toxicity and the overall pollution load that an area is experiencing through index weighting in order to produce a conclusive evaluation of the pollution status when implementing a MACROS model.

However, using normalized value of each parameter in MCDM calculation could give rise to an increase its impact in M-WQI ranking. So, MACROS method also ranked these locations as the most polluted sites and identifies the relative pollution level, by giving them an overall rank. Its influence on the water quality is determined by the weight of its water quality criteria. The parameters that have the highest weight and lowest value have the most effect on the overall quality. In certain cases, decision analysis makes extensive use of this theory, especially when dealing with MCDM issues (Berihun et al. 2020). It performs exceptionally well in managing intricate decision-making situations by mimicking the ambiguity and uncertainty brought

Table 1 Classification of water type based on M-WQI and its ranking for each site by MACROS approach

Location No (Z)	M-WQI	Type of water	MACROS score (U_i)	Rank of MACROS
1	34	Excellent	0.0169	18
2	241	Very poor	0.0174	4
3	67	Good	0.0170	17
4	92	Good	0.0173	7
5	176	Poor	0.0171	13
6	182	Poor	0.0171	12
7	265	Very poor	0.0172	9
8	317	Unsuitable	0.0194	2
9	381	Unsuitable	0.0469	1
10	281	Very poor	0.0174	6
11	277	Very poor	0.0174	5
12	282	Very poor	0.0170	15
13	290	Very poor	0.0170	14
14	41	Excellent	0.0161	19
15	44	Excellent	0.0170	16
16	39	Excellent	0.0172	10
17	35	Excellent	0.0171	11
18	38	Excellent	0.0172	8
19	341	Unsuitable	0.0185	3

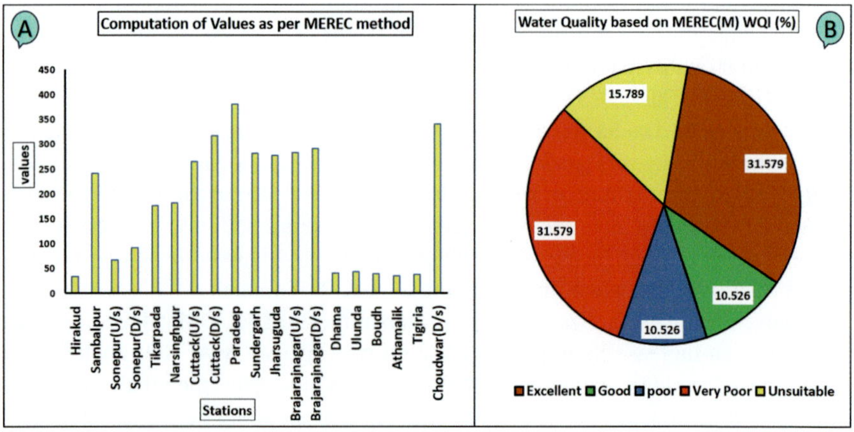

Fig. 5 Distribution of values in terms of % for each parameter, across all sampling sites using MEREC Method

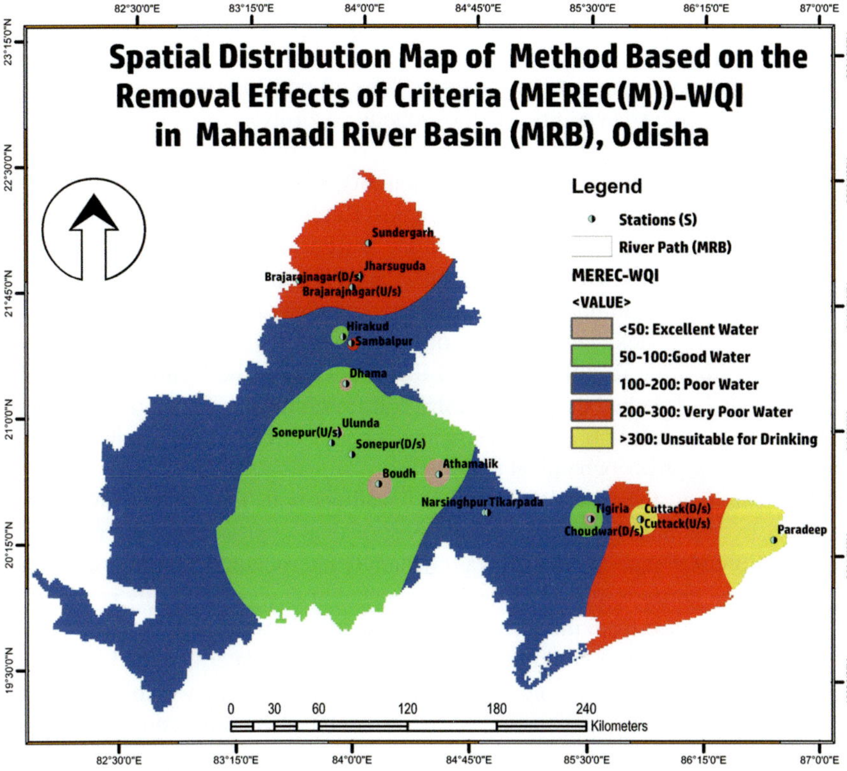

Fig. 6 A Spatial distribution map of MEREC-WQI, across all sampling locations

forth by the numerous criteria. By lowering the relative error found by ignoring the artificial weights, it improves WQI.

From analysis, the current investigation found that the first five parameter weights were in order of TC > TKN > EC > TDS > Cl^-. Because of this, employing this procedure yields a dependable analysis from various parameter weights and is based on MEREC weights and rough set theory. The MACROS flow chart, which is displayed in Table 1, illustrates how mathematical computations are used to determine the values at each site. The value of MACROS, in the current work, ranged between 0.016 and 0.047. Figure 7(a, b) displays the distribution and ranking of samples. When compared to other places, the sampling site Z-(9) had the highest level of pollution during the sampling period. This conclusion was amply demonstrated by the greatest WQI results in this location. While, it holds also high values of TC, TKN, EC, Sulphate, Chloride, and TDS, which were also higher than their desirable concentration. Here, EC parameter points caused by sewage, high dissolved solids concentrations, high evaporation rates, and high-water flow rates, while TDS may be linked to the discharge from catchment regions and the influx of domestic and industrial sewages into river systems, which encourage the leaching of soluble metals from surface soil and show that they are coming from interactions between soil and dust (Das 2022b). Also, TKN, TC and Cl^- are connected to human activity and urban discharges. Its dual origins can be traced to home effluent, primarily detergents, and agricultural operations, in addition to the reductive hydrolysis of minerals containing phosphate.

It is also observed that there is a high alkalinity, which could serve as a buffer to lower the pH and hardness of the water samples. But SO42- is related in the research area because of the inflow of agricultural runoff and domestic sewage, which together create major alterations in the water chemistry and bioavailability in the tropical river system (Das 2022c). Similarly, site Z-(8), (19), and (2) grouped under 2nd, 3rd, and 4th most polluting sites, having scores of (0.019), (0.016), and (0.0174). This was also evident from the results of M-WQI that reflects that the locations Z-8 (317), 19 (341), and 2 (241) had the second, third and fourth highest M-WQI indicating it as poor/very poor/unsuitable water class, and a high TC-TKN value, indicating it as contaminated by physicochemical parameters.

After obtaining the MACROS ratings, locations with low water quality were identified by contouring the data using geostatistical approaches. Nevertheless, a water quality index map in ArcGIS was created by interpolating the provided values throughout the entire research region using a spatial analyst tool; this is shown in Fig. 8. The sequence from the interpolation map also showed that the river's water quality was good prior to entering the urbanized area and later declined as a result of the addition of untreated industrial and domestic effluent, which flowed past the municipal boundary and further in the direction of the downstream sampling sites (Das 2023b).

Since the capacity of treatment plants is less than the amount of sewage produced in the cities, the poor zone thus exhibits approximately residential sewage that discharged from these sites into the river. According to reports, untreated sewage wastes are dumped into the river, which could be the main source of pollution in

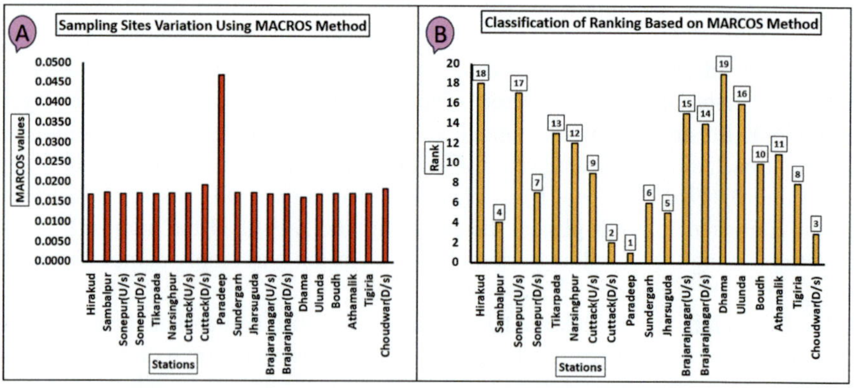

Fig. 7 Distribution of values in terms of ranking for each parameter, across all sampling sites using MACROS Method

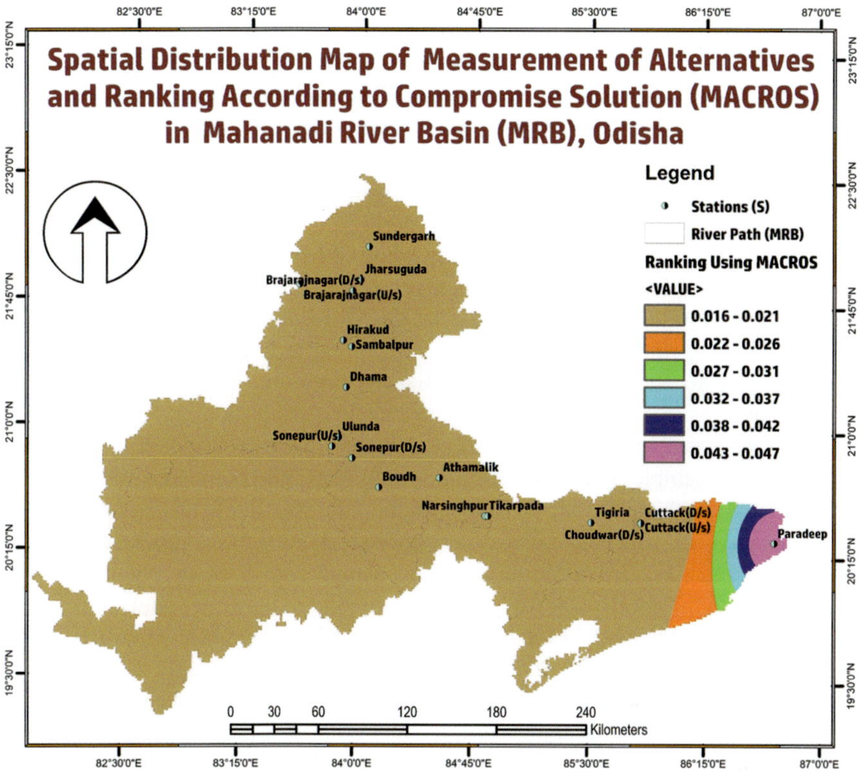

Fig. 8 A Spatial distribution map of ranking categorization by MACROS Method, across all sampling locations

the area around it (Das 2024b). Furthermore, because of the excessive extraction of surface water brought on by rapid population increase, commercial activity, and other factors, salt water incursion is a regular occurrence in coastal aquifers. In order to implement treatment procedures to improve the water quality and ultimately increase water consumption efficiency, this interpolation map provides information on zones with lower surface quality water. Since river water serves as the main supply of water for homes, farms, and industries, its degradation can have a negative influence on aquatic habitat, agriculture, human health, and even microbial diversity, which can create ecological imbalances (Zhang et al. 2020).

Consequently, one of the most important tasks in monitoring and conserving water resources is identifying possible sources of contamination and conducting periodic surface water quality monitoring in the river basin. Thus, by taking into account all significant factors and allocating weight based on their variability, the MACROS approach provides a scientific method for evaluating the quality of the water. However, because of the novel way in which it assigns weight based on the observed data sets, this approach is not region-specific and may be applied elsewhere.

6 Conclusions

Nowadays, the primary danger to human health is caused by declining water quality (WQ). The public is alerted to a new sign for waterborne disease prevention due to the quality of the water. This research makes use of sensitive water quality parameters, that is acquired from the Mahanadi River basin, Odisha during monsoon season, 2020–2023, at 19 different locations, considering 21 WQ parameters. Nonetheless, the current work effectively illustrates the value of MEREC (M) methodologies in conjunction with MACROS and graphical representation techniques. As the criteria were compared to the WHO recommendations, according to the data set's results, TC and TKN together address the main influencing indicators that affect the SW chemistry due to anthropogenic factors like farming and household waste water runoff, as well as variables like ion exchange and material leaching. The M-WQI of the measured water in the current work, together with the corresponding categories for each water sample, is displayed on a spatial map. The computed values varied from 34 to 381 with an average of 180.16. In this case, smaller WQI values (less than 50) were excellent, whereas higher values (more than 300) should not be consumed. It is observed that 31.57% of the surface water samples are in the excellent class, and 10.526% are in the good water category. Further, WQ of the river water samples has M-WQI > 100, containing 11 locations (57.89%), which represents poor-unsuitable WQ of analyzed river water samples. It is observed that around eight sites (42.10%) meet the requirement of excellent to good water quality. In order to clearly visualize the areas that are classified under various WQ criteria, a geographic diagram was created utilizing IDW and the WQI values of 19 sampling locations. It illustrates how using contaminated water side by side and continuously in 11 different areas has a negative impact on socioeconomic growth and public health. In addition, site (9) encompasses a high score of 381, is also having a difficult time meeting the growing demand for fresh water because of the incursion of extra saline water, which is causing problems for the locals and obstructing the provision of clean drinking water. Further, to investigate the combined impacts of different physicochemical factors, which regulates the entire WQ, which is represented holistically by the MACROS ranking technique. The results were obtained in a range of 0.016–0.047, and it shows that station Z-(9), (8), (19) and (2) falls into the poor to inappropriate category and is not fit for drinking. Thus, water pollution is mostly caused by anthropogenic activity, although seasonal variations in precipitation, surface runoff, interflow, groundwater flow, and bacterial contamination are contributing factors. It is determined that the remaining site area WQ is passably excellent and suitable for drinking. The variance in the WQ parameters at these four sites is displayed via the spatial distribution analysis approach using the IDW interpolation method. Additionally, it shows that the concentrations of sulphate and chloride, TC, TKN, EC, TDS, and other factors are excessively high. However, the site 9 is located on the river basin's coast as a result of the population's constant increase, the modern infrastructure, and the small-to-large industry, the WQ of the area may be impacted by runoff from agriculture, then waste water from homes and businesses. Due to urbanization, industry, agriculture, and exponential population increase, these four sites, which are ranked from 1 to 4, are under extreme stress. As a result, poor waste management, particularly along the bank, and the discharge of partially or untreated sewage into rivers are degrading the quality of the river, which in turn is impacting the surface water quality (SWQ) in nearby communities. Further, the river water contains a significant concentration of organic contaminants. Therefore, routine SW quality monitoring is a very helpful step to restrict the use of contaminated water, especially for drinking purposes at the poor or inappropriate sites. In the end, specific water conservation and water body preservation techniques ought to be implemented for the health of people and the water body.

Acknowledgements The author extends their appreciation to State Pollution Control Board, Odisha. The author would like to acknowledge anonymous reviewers for their contribution to this manuscript. Author would also like to thank Rani Prativa Das, Ruchika Square, Bhubaneswar, for providing financial support for carrying out this innovative work.

Authors Contributions Abhijeet Das: Conceptualization, Data Curation, Methodology, Software, Validation, Formal analysis, Investigation, Resources, Project Administration, Funding Acquisition, Writing – review and editing, Visualization.

References

APHA (1995) Standard methods for the examination of water and wastewater, 19th edn. American Public Health Association, New York

Aredehey G, Mezgebu A, Girma A (2020) The effects of land use land cover change on hydrological flow in Giba catchment, Tigray, Ethiopia. Cogent Environ Sci. 6 (1). https://doi.org/10.1080/23311843.2020.1785780

Baghel S, Tripathi MP, Khalkho D, Al-Ansari N, Kumar A, Elbeltagi A (2023) Delineation of suitable sites for groundwater recharge based on groundwater potential with RS, GIS, and AHP approach for Mand catchment of Mahanadi Basin. Sci Rep 13(1):9860

Belay T, Mengistu DA (2021) Impacts of land use/land cover and climate changes on soil erosion in Muga watershed, Upper Blue Nile basin (Abay), Ethiopia. Ecol Process. https://doi.org/10.1186/s13717-021-00339-9

Berihun ML, Tsunekawa A, Haregeweyn N, Dile YT, Tsubo M, Fenta AA, Meshesha DT, Ebabu K, Sultan D, Srinivasan R (2020) Evaluating runoff and sediment responses to soil and water conservation practices by employing alternative modeling approaches. Sci Total Environ 747:141118. https://doi.org/10.1016/j.scitotenv.2020.141118

Burgan HI (2022) The short-term and seasonal trend detection of sediment discharges in Turkish rivers. Rocznik Ochrona Środowiska 24:214–230

Cheng S, Dong Y, Liu K (2020) Characteristics of organic pollution in the shallow groundwater, Jinan pilot area Shandong Province. Miner Explor 11(12):2749–2756 (in Chinese)

Cunha DPS, Pontes KV (2022) Desalination plant integrated with solar thermal energy: a case study for the Brazilian semi-arid. J Clean Prod 331:129943

Das A (2022a) Multivariate statistical approach for the assessment of water quality of Mahanadi basin, Odisha. Mater Today Proceed 65:A1–A11

Das A (2022b) Water criteria evaluation for drinking purposes in Mahanadi river Basin, Odisha. In: International conference on trends and recent advances in civil engineering. Singapore: Springer Nature Singapore. pp 237–260

Das A (2022c) Surface water quality modelling using water quality index (WQI) and geographic information system (GIS) on the Mahanadi Basin, Odisha. In: International conference on water technologies. Singapore: Springer Nature Singapore. pp 21–46

Das A (2023a) Anthropogenic effects on surface water quality assessment in Baitarani River Basin, Odisha using GIS and MCDM techniques. Eng Res Trans 5:37–64

Das A (2023b) Assessment of potability of surface water and its health implication in Mahanadi Basin, Odisha. Mater Today Proceed

Das A (2023c) Assessment of surface water quality by means of MCDM approach in Mahanadi River Basin (MRB) Odisha. Resilience 360:51

Das A (2023d) Identification of surface water contamination zones and its sources on Mahanadi River, Odisha using entropy-based WQI and MCDM techniques. Eng Res Trans 4:67–92

Das A (2024a) Assessment of surface water quality using entropy-WQI, Fuzzy-TOPSIS analysis, irrigation indices and spatial interpolation approaches in Mahanadi River Basin, Odisha India. J Environ Nanotechnol 13(1):182–212

Das A (2024b) Water quality assessment using water quality index (WQI) under GIS framework in Brahmani Basin, Odisha. In: International conference innovation in smart and sustainable infrastructure. Springer, Singapore. pp 131–150

Dash S, Kalamdhad AS (2021) Discussion on the existing methodology of entropy-weights in water quality indexing and proposal for a modification of the expected conflicts. Environ Sci Pollut Res 28(38):53983–54001

Deveci M, Özcan E, John R, Pamucar D, Karaman H (2021) Offshore wind farm site selection using interval rough numbers based best—worst method and MARCOS. Appl Soft Comput 109:107532

Duguma TA (2022) Soil erosion risk assessment and treatment priority classification: a case study on Guder watersheds, Abay river basin, Oromia Ethiopia. Heliyon 8:e10183. https://doi.org/10.1016/j.heliyon.2022.e10183

Gao Z, Han C, Xu Y, Zhao Z, Luo Z, Liu J (2021) Assessment of the water quality of groundwater in Bohai Rim and the controlling factors—a case study of northern Shandong Peninsula, north China. Environ Pollut 285:117482. https://doi.org/10.1016/J.ENVPOL.2021.117482

Gashaw T et al (2021) Evaluating the effectiveness of best management practices on soil erosion reduction using the SWAT model: for the case of Gumara Watershed, Abbay (Upper Blue Nile) Basin. Environ Manage 68(2):240–261. https://doi.org/10.1007/s00267-021-01492-9

Gavit BK et al (2020) Hydrological modelling using SWAT. September

Ghaffari M, Chavoshbashi AA, Eslami A, Hatami H, Pourakbar M, Hashemi M (2021) Spatial and temporal variation of groundwater quality around a volcanic mountain in northwest of Iran. Groundw Sustain Dev 14:100627. https://doi.org/10.1016/J.GSD.2021.100627

Gong XM, Yang M, Du PL (2021) Renewable energy accommodation potential evaluation of distribution network: a hybrid decision—making framework under interval type-2 fuzzy environment. J Clean Prod 286:124918

Guan QH, Li F, Wang A, Feng P, Tian C, Chen X, Liu D (2019) Hydrochemical characteristics and evolution of karst spring groundwater system in Jinan. Carsol Sinica 38(05):653–662. https://doi.org/10.11932/karst20190501

Gurara MA, Jilo NB, Tolche AD (2023) Modelling climate change impact on the streamflow in the upper wabe bridge watershed in wabe shebele river basin, Ethiopia. Int J River Basin Manag 21(2):181–193

Han Q, Tong R, Sun W, Zhao Y, Yu J, Wang G, Shrestha S, Jin Y (2020) Anthropogenic influences on the water quality of the Baiyangdian Lake in North China over the last decade. Sci Total Environ 701:134929

He W, Lu Y, An HH, Zhou X, Su PF, Han D (2022) Parametric analysis of humidification dehumidification desalination driven by photovoltaic/thermal (PV/T) system. Energy Convers Manage 259:115520

Hong T, Xie Y, Yu Q, Zhao Y, Zhao G, Yang L (2016) Hydrochemical charateristics study and genetic analysis of groundwater in a key region of the Wumeng Mountain Southwestern Chian. Earth Environ 44(1):11–18. https://doi.org/10.14050/j.cnki.1672-9250.2016.01.002. (in Chinese)
Hua K, Xiao J, Li S, Li Z (2020) Analysis of hydrochemical characteristics and their controlling factors in the Fen River of China. Sustain Cities Soc 52:101827. https://doi.org/10.1016/j.scs.2019.101827
Huang QJ (2022) Selecting sustainable renewable energy-powered desalination: an MCDM framework under uncertain and incomplete information. Clean Technol Environ Policy 24(5):1581–1598
Kabiri S, Manesh MHK, Yazdi M, Amidpour M (2021) New procedure for optimal solar repowering of thermal power plants and integration with MSF desalination based on environmental friendliness and economic benefit. Energy Convers Manage 240:114247
Keshavarz-Ghorabaee M, Amiri M, Zavadskas EK, Turskis Z, Antucheviciene J (2021) Determination of objective weights using a new method based on the removal effects of criteria (MEREC). Symmetry 13(4):525
Kidane M et al (2019) The impact of land use and land cover (LULC) dynamics on soil erosion and sediment yield in Ethiopia. Heliyon 5(12):e02981. https://doi.org/10.1016/j.heliyon.2019.e02981
Kiprotich P et al (2021) Assessing the impact of land use and climate change on surface runoff response using gridded observations and SWATþ.1–29
Li WC, Ren XS, Ding SM, Dong LC (2020) A multi-criterion decision making for sustainability assessment of hydrogen production technologies based on objective grey relational analysis. Int J Hydrogen Energy 45(59):34385–34395
Li J, Gao Z, Li C (2021) Study on distribution characteristics of Sr element of karst water and its indication in Jinan. Groundwater 43(04):5–8. https://doi.org/10.19807/j.cnki.DXS.2021-04-002. (in Chinese)
Meng Q, Wang X, Xing L, Dong Y, Zhu H, Wu Z, Li C, Yu M, Hou Y (2020) A study of the difference in supply sources of the four groups of springs in Jinan. Hydrogeol Eng Geol 47(01):37–45. https://doi.org/10.16030/j.cnki.issn
Mishra RL, Sahoo S, Dash C, Biswal TK (2022) Preliminary observations on a paleo-fluviatile deposition and uplift-driven migration of the Mahanadi River in the Gondwana basin of Odisha, eastern India. Arab J Geosci 15(9):808
Moisa MB (2021) Impact of land-use and land-cover change on soil erosion using the RUSLE model and the geographic information system: a case of Temeji watershed. Western Ethiopia 12(7):3404–3420. https://doi.org/10.2166/wcc.2021.131
Ougahi JH (2022) Application of the SWAT model to assess climate and land use/cover change impacts on water balance components of the Kabul River Basin Afghanistan. Uncorrected Proof. https://doi.org/10.2166/wcc.2022.261
Pan W, Xing L, Yu M, Deng X (2021) Review on the pollution situation and behavior of antibiotics in the karst groundwater system. Groundwater 43(02):5–10. https://doi.org/10.19807/j.cnki.DXS.2021-02-002. (in Chinese)
Pandey L, Sain K (2022) Porosity mapping of shallow subsurface sediments: a case study in Mahanadi basin, eastern margin of India. Explor Geophys 53(5):517–531
Pant N, Rai SP, Singh R, Kumar S, Saini RK, Purushothaman P, Nijesh P, Rawat YS, Sharma M, Pratap K (2021) Impact of geology and anthropogenic activities over the water quality with emphasis on fluoride in water scarce Lalitpur district of Bundelkhand region. India. Chemosphere 279:130496. https://doi.org/10.1016/J.CHEMOSPHERE.2021.130496
Pei W, Fu Q, Li T, Wang P, Hou R (2019) Study of the spatiotemporal variability in agricultural drought vulnerability based on a dynamic classification projection pursuit model. Arab J Geosci 12(2):426–442. https://doi.org/10.1007/s12517-019-4949-0
Qin G, Liu J, Xu S, Wang T (2020) Water quality assessment and pollution source apportionment in a highly regulated river of Northeast China. Environ Monit Assess 192(7):446. https://doi.org/10.1007/s10661-020-08404-0

Rezk H, Mukhametzyanov IZ, Al-Dhaifallah M, Ziedan HA (2021) Optimal selection of hybrid renewable energy system using multi- criteria decision-making algorithms. CMC Comput Mater Con 68(2):2001–2027

Samsudin MS, Azid A, Khalit SI, Shakir A, Zaudi MA (2017) River water quality assessment using APCS-MLR and statistical process control in Johor River Basin, Malaysia. Int J Adv Appl Sci 4(8):84–97. https://doi.org/10.21833/ijaas.2017.08.013

Saxena A, Khan MM, Raychowdhury N, Singh KJ (2022) Early Permian macrofloral diversity in Indian Gondwana: evidence from Talchir formation of Singrauli coalfield, Son-Mahanadi valley basin, central India. J Earth Syst Sci 131(2):70

Skourtos M, Damigos D, Kontogianni A, Tourkolias C, Marafie A, Zainal M (2021) A combined probabilistic framework to support investment appraisal under uncertainty in desalination projects: an application to Kuwait's water/energy nexus. Water Supply 21(1):276–288

Subramani T, Elango L, Damodarasamy SR (2005) Groundwater quality and its suitability for drinking and agricultural use in Chithar River Basin, Tamil Nadu India. Environ Geol 47:1099–1110

Sulamo MA, Kassa AK, Roba NT (2021) Evaluation of the impacts of land use/cover changes on water balance of Bilate watershed, Rift valley basin Ethiopia. Water Pract Technol 16(4):1108–1127. https://doi.org/10.2166/wpt.2021.063

Sun K, Li T, Liu X (2020) The influences of sponge city construction on spring discharge in Jinan city of China. Hydrol Res 51(5):959–975. https://doi.org/10.2166/nh.2020.008

Tan X et al (2022) Impacts of climate change and land use/cover change on regional hydrological processes: case of the Guangdong-Hong Kong-Macao greater bay area. Front Environ Sci 9:1–16. https://doi.org/10.3389/fenvs.2021.783324

Town SN et al (2021) Impacts of land use and land cover changes on peakdischarge and flow volume in Kakia and Esamburmbur sub-catchments of Narok Town, Kenya

Trung DD (2022) Development of data normalization methods for multi-criteria decision making: applying for MARCOS method. Manuf Rev 9:22

Tumsa BC, Kenea G, Tola B (2022) The application of SWATþ model to quantify the impacts of sensitive LULC changes on water balance in Guder catchment, Oromia Ethiopia. Heliyon 8:e12569. https://doi.org/10.1016/j.heliyon.2022.e12569

Wagner PD et al (2022) Representation of hydrological processes in a rural lowland catchment in Northern Germany using SWAT and SWATþ. Hydrol Process. 36 (5). https://doi.org/10.1002/hyp.14589

Wang ZF, Wang YJ, Xu GY, Ren JZ (2019) Sustainable desalination process selection: decision support framework under hybrid information. Desalination 465:44–57

WHO (2008) Guidelines for drinking-water quality: incorporating first and second addenda, Recommendations, 3rd (ed). WHO Press, Geneva, Switzerland, 1, 668

Wu T, Yuan L, Han S, Li F (2020) Hydrochemical characteristics and water quality evaluation of groundwater in Angulinao Basin. Environ Sci Technol 43(3):198–205. https://doi.org/10.19672/j.cnki.1003-6504.2020.03.028. (in Chinese)

Xu D, Ren JZ, Dong LC, Yang YK (2020) Portfolio selection of renewable energy-powered desalination systems with sustainability perspective: a novel MADM-based framework under data uncertainties. J Clean Prod 275:124114

Yang L, Liu C, Qi X (2016) Study on characteristic variation of hydro-chemistry of Jinan spring. J Water Resour Water Eng 27(01):59–64. https://doi.org/10.11705/j.issn.1672-643X.2016.01.10. (in Chinese)

Yigez B et al (2021) Spatial distribution of soil erosion and sediment yield in the Koshi River Basin, Nepal: a case study of Triyuga watershed. J Soils Sediments 21(12):3888–3905. https://doi.org/10.1007/s11368-021-03023-9

Zhang H, Li H, Yu H, Cheng S (2020) Water quality assessment and pollution source apportionment using multi-statistic and APCS-MLR modeling techniques in Min River Basin, China. Environ Sci Pollut Res Int 27:1–14. https://doi.org/10.1007/s11356-020-10219-y

Zhang ZM, Zhang F, Du JL, Chen DC (2022) Surface water quality assessment and contamination source identification using multi—variate statistical techniques: a case study of the Nanxi River in the Taihu Watershed. China Water 14(5):778

Zhao X, Fang Y, Liu L, Xu M, Zhang P (2020) Ameliorated moth-flame algorithm and its application for modeling of silicon content in liquid iron of blast furnace based fast learning network. Appl Soft Comput J 94:106418. https://doi.org/10.1016/j.asoc.2020.106418

Modeling Agricultural Land Suitability Using MCDM-AHP Techniques in Semi-arid Region of West Bengal, India

Prosenjit Kayal, Susmita Das, and Indrajit Roy Chowdhury

Abstract Agriculture serves as the backbone of Bankura District, West Bengal, contributing substantially to its economic sustenance. This study endeavors to ascertain optimal land utilization for agriculture, crucial for sustainable development in the semi-arid region. Employing a comprehensive approach, it integrates data from diverse sources, encompassing parameters like slope, elevation, land use land cover (LULC), normalized difference water index (NDWI), soil type, proximity to river and road, geology, rainfall, and aspect. Through weighted overlay analysis, based on the Analytic Hierarchy Process (AHP), the study identifies and models potential agricultural zones. The resultant classification delineates five categories: very highly suitable, highly suitable, moderately suitable, marginally suitable, and currently not suitable. Cross-validation with district reports, Google Earth imagery, and field surveys bolster the credibility of findings, yielding an overall accuracy of 92.14%. Despite the substantial areas identified as agriculturally suitable, the study emphasizes the imperative of adopting sustainable management practices and resilient farming techniques to augment agricultural output. The insights garnered from this research hold significant implications for local farmers, regional planners, and governmental bodies. They can inform judicious decision-making processes, facilitate the identification of prime agricultural sites, foster agricultural expansion, and nurture self-sufficient local economies.

Keywords Agricultural land suitability · Analytic hierarchy process · Pair-wise comparison matrix · Weighted overlay analysis · Bankura

---

P. Kayal (✉) · I. R. Chowdhury
Department of Geography and Applied Geography, University of North Bengal, Siliguri, West Bengal, India
e-mail: rs_prosenjit@nbu.ac.in

I. R. Chowdhury
e-mail: irchowdhury_geo@nbu.ac.in

S. Das
School of Oceanographic Studies, Jadavpur University, Kolkata, West Bengal, India

S. C. Pal and U. Chatterjee (eds.), *Surface, Sub-Surface Hydrology and Management*, Springer Geography, https://doi.org/10.1007/978-3-031-62376-9_28

1 Introduction

It is commonly recognized that one of the key indicators of economic growth lies in the prudent and sustainable utilization of non-renewable natural resources, notably land (Akıncı 2013). The World Commission on Environment and Development (WCED) linked sustainable development to land appropriateness and described sustainable development as satisfying present demands without jeopardizing the potential of future generations to satisfy their own needs (Feizizadeh and Blaschke 2012). As efficient land-use planning is a key component of agricultural productivity, land is important to the agricultural sector (Sarkar et al. 2021). Since the early 1900s, the world's population has continually increased, and it now numbers 7.6 billion people (PRB 2018). Food insecurity brought on by overcrowding has given rise to resource-depleting agriculture and irreversible environmental harm. In the next decades, there will be substantial but unpredictable effects on food security from a changing climate, a growing world population, rising food prices, and environmental stressors (FAO 1976). Developing countries account for 60% of the world's population (Mustafa et al. 2011). Additionally, the pressure on natural and agricultural resources has increased due to population growth, particularly in developing countries (Zolekar and Bhagat 2014; Roy et al. 2022). In many developing countries, like India population pressures continue to tip the scales against effective land and water management. High population rates in these countries, along with decreasing resources and inadequate agricultural land management practices, have resulted in a majority of food and environmental concerns. As the world's population grows and there is a greater need for food, the only dependable economic sector that can respond to the situation by making food available is agriculture. Sustainable agricultural development is a primary goal in all countries around the world, whether developed or developing (Mustafa et al. 2011).

The land suitability study found that organizing available resources and assessing how well they can produce crops can increase agricultural security (Roy et al. 2022). Using an integrated strategy to strengthen nations' capacity to plan and monitor better exploitation and management of their land resources is important to raise agricultural productivity without undermining land and environmental security (Molla 2022). Organizing available resources and figuring out how well they can produce crops may increase agricultural security (Estrada et al. 2017). The approaches used for evaluating land look at how well it fits the needs of land use and sustainable production (Zolekar and Bhagat 2014; Abdel Rahman et al. 2016). If certain management and planning measures are followed after a study on the suitability of the land, it might be possible to utilize the land effectively (Mandal et al. 2021; Roy et al. 2022). One of the most crucial stages in land-use planning for growing a particular crop in light of the regional circumstances is evaluating the suitability of the land (FAO 1993). For decision-makers to come up with the optimal management strategies for achieving sustainable land productivity, studies of land suitability evaluation could be taken into account (Pramanik 2016). The development of agricultural productivity can be assisted by land suitability analysis. A site's suitability for agricultural development is determined by a number of factors, including physiographic data, climatic characteristics (temperature and rainfall), internal soil conditions (depth, moisture, texture, salinity, and natural fertility), and external soil conditions (slope, accessibility, and flooding) (Wang et al. 2012).

Since the 1990s, academics have widely embraced multi-criteria decision analysis (MCDM) techniques regarding the process of making spatial decisions. (Bhattacharya et al. 2021; Romano et al. 2015; Roy et al. 2022). Additionally, it is possible to resolve significant issues with spatial distribution and location selection by combining GIS and remotely sensed data (Kazemi et al. 2016; Hussein et al. 2018; Alkaradaghi et al. 2019). Information and communication technology advancements have led to the development of computer-based decision-support models for site suitability analysis of various activities (De la Rosa et al. 1992; Shim et al. 2002). Geospatial techniques can be used to identify suitable agricultural land sites based on criteria such as geology, topography, soil characteristics, drainage, and transport patterns in the study area. The potential locations for the agricultural land use pattern can also be identified and prioritized using this technique (Pramanik 2016). Additionally, the Analytic Hierarchy Process (AHP) is widely used to make multi-criteria decisions about the suitability of a piece of land for a particular field (Thapa and Murayama 2008; Feizizadeh and Blaschke 2012; Akıncı et al. 2013; Yalew et al. 2016a; 2016b; Hussein et al. 2018; Banerjee et al. 2020; Tashayo et al. 2020a; 2020b; Everest et al. 2021; Bandyopadhyay et al. 2009 ; Sarkar et al. 2021; Mangan et al. 2022; Sonia et al. 2023). By comparing different parameters in pairs according to their relative significance, it establishes the weight of importance for various land uses.

Many investigations have explored the viability of agriculture in West Bengal and across India, yet none have specifically addressed the semi-arid climatic zone of Bankura District in West Bengal. Despite the district's propensity for drought, it can increase food production when there is adequate rainfall (Census of India 2011). Rice is the primary crop of the district besides, it also produces potatoes, wheat, Vegetables, Mustard, Summer Til, etc. The suitability of the land is not taken into consideration by the farmers in this district as they engage in a variety of agricultural activities. It is therefore critical to determine which regions in the area are suitable for agricultural activity. Due to a large database limitation, a thorough evaluation of the region's suitability for agriculture has not yet been done. However, to prepare the agricultural land suitability for the area, multi-source data from about ten parameters were used for the first time. To determine the various factors influencing the suitability of agriculture in the semi-arid region of Bankura District, the current study used weighted overlay analysis based on the Analytic Hierarchy Process (AHP) technique (Mandal et al. 2021; Roy et al. 2022). To put it another way, this weighting method, when used in conjunction with spatial mapping to analyze agricultural suitability, is extremely sophisticated because it takes into account how each factor affects the suitability of agricultural land (Nganga et al. 2020; Roy et al. 2022). Therefore, the main goal of the current study is to assess suitable agricultural land sites in order to satisfy future generations' basic needs for food. Therefore, from a regional perspective, the research also helps to improve the living conditions for rural communities in the inspected area and the development of livelihood strategies in the study area.

There are four main sections to the chapter. An introduction and a description of the study site are included in the first section. The second section contains the parameters that were chosen, the data source, and a thorough methodology description. The third section contains the results and discussions that elaborate on the status of particular study parameters, identification of spatial distributions and associations of agricultural site suitability pockets', and validate the findings. Concluding observations make up the final section.

2 Study Area

The Bankura district, located in the state of West Bengal's southwestern part, is semi-arid and drought-prone due to the Tropic of Cancer passing through it (Das et al. 2013; Nag and Das 2017). Moreover, this region constitutes a portion of the tropical plateau fringe area situated on the eastern Chhotanagpur plateau. Low-lying alluvial plains can be found in the east and northeast. The land gradually rises to the west, giving place to rolling terrain peppered with rocky outcrops. The district extends longitudinally from 86° 36 to 87° 44 E and latitudinally from 22° 37 to 23° 38 N. ~~Due to its granite gneiss geology, undulating plateau upland with inter-fluvial lateritic upland, undulating river system, low to medium precipitation (100–140 cm), high temperature (35–40 °C), and tropical dry deciduous forest cover, the Bankura district is a true representation of the tropical plateau fringe (Nandi and Sarkar~~ 2021~~).~~ Geologically, the eastern part of the district is composed of alluvium, and the western half is made up of an undulating plateau area (Sutradhar and Chattaraj 2019). This district has a total land area of 6882 km^2. Three subdivisions—Bankura Sadar, Khatra, and Bishnupur, as well as 22 Community Development Blocks, comprise this district (Patra et al. 2021). According to the Census of India (2011), there are 766,902 households in the Bankura district, with 3,596,674 people living there, of which 1,838,095 are male and 1,758,579 are female. The population density is 523 persons per km^2; 8.33% of people live in urban areas, compared to 91.67% of people who live in rural areas. Despite being one of the most drought-prone regions of West Bengal, the economy in Bankura district is underdeveloped and heavily reliant on agriculture. Agriculture provides nearly 70% of the district's income. Therefore, the study area's varied physical and socioeconomic conditions pose a challenge to agricultural practices. This study aims to identify the best agricultural zones that will aid in the development of new sustainable agricultural management techniques for the semi-arid region of West Bengal. The location map of the study area is depicted in Fig. 1.

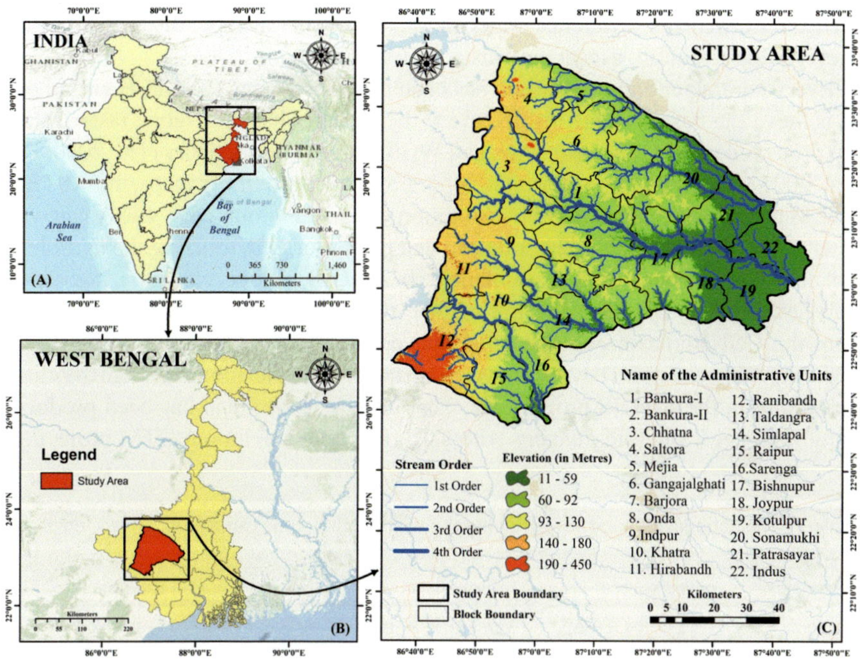

Fig. 1 Location and geographical settings of the study area **a** the location of West Bengal state in India **b** the location of Bankura District in West Bengal **c** Study area showing administrative units along with important drainage network and surface elevation map

3 Materials and Methodology

3.1 Data Source and Application of Thematic Layers

To evaluate and model the potential location of agricultural development in the semi-arid area of Bankura District, a significant amount of multi-source spatial and non-spatial data has been considered for the present study. Table 1 and Fig. 2 provide a summary of the methodology and information used to assess the suitability of the agricultural land in the study area. However, ten criteria were chosen to assess land suitability based on their applicability in agriculture, and these criteria were selected following expert advice and in-depth literature reviews (Bandyopadhyay et al. 2009; Akıncı et al. 2013; Everest et al. 2021; Zolekar and Bhagat 2015; Tashayo et al. 2020a; Kumar et al. 2021; Rajasekhar et al. 2021; Molla 2022; Roy et al. 2022) Table 2. To identify viable and suitable sites for agricultural development by applying a set of limitations and standards, multi-criterion site suitability modeling was developed in the present study. Ten different constraints and criteria were chosen, and they were ranked according to their significance and importance in agriculture. The parameters which were considered for the study include slope, elevation, land use land cover

(LULC), normalized difference water index (NDWI), soil type, distance from the river, geology, rainfall, aspect, and distance from the road.

Every parameter that was chosen has been processed using the GIS environment. To achieve the best possible results, every theme layer has undergone rasterization and projected using UTM Projection and WGS Datum Zone 45 (Roy et al. 2020; Das et al. 2021). To create slope elevation, aspect, and river proximity maps, the ASTER-GDEM data has been used 30 m in resolution, obtained through the NASA gateway. After the GDEM data had been collected and pre-processed, utilizing the hydrology tool, the river proximity map was obtained. In ArcGIS, a slope, elevation, and aspect map have been prepared using the surface tool. Geology information was taken from the GSI Bhukosh geo-scientific data portal at https://bukosh.gsi.gov.in/Bhukosh/MapViewer.aspx. They are converted to raster data after going through the initial processing in the ArcGIS programme. The land-use/cover map has been produced using Sentinel-2 satellite imagery with a 10 m spatial resolution, UTM projection,

Table 1 Thematic layers used for agricultural site suitability zone estimation, their sources and details

Parameters	Data source	Information	References
Slope	ASTER GDEM	Version 3, 30 m × 30 m	https://search.earthdata.nasa.gov
Elevation	ASTER GDEM	Version 3, 30 m × 30 m	https://search.earthdata.nasa.gov
LULC	United States Geological Survey (USGS)	Sentinel-2 Satellite Imagery, 10 m Spatial resolution, Projection: UTM, Zone: 45N, Datum: WGS84, Date of acquisition: 29th January 2022	https://earthexplorer.usgs.gov/
NDWI	Landsat 8 OLI/TIRS	USGS, 30 m × 30-m, 2019	https://earthexplorer.usgs.gov/
Soil type	Food and Agriculture Organization (FAO)	Digital Soil Map of the World—ESRI shapefle	http://www.fao.org
Distance from river	ASTER GDEM	Version 3, 30 m × 30 m	https://search.earthdata.nasa.gov
Geology	Geological survey of India (GSI) data	Vector layer, 1:1,091,958	https://www.gsi.gov.in
Rainfall	CRU TS v. 4.04 (Climatic research unit gridded time series version 4 data)	High-resolution gridded data, 0.5° × 0.5°	https://sites.uea.ac.uk/cru/data
Aspect	ASTER GDEM	Version 3, 30 m × 30 m	https://search.earthdata.nasa.gov
Distance from road	Road network shape file	Open Street Map Data	https://www.openstreetmap.org/

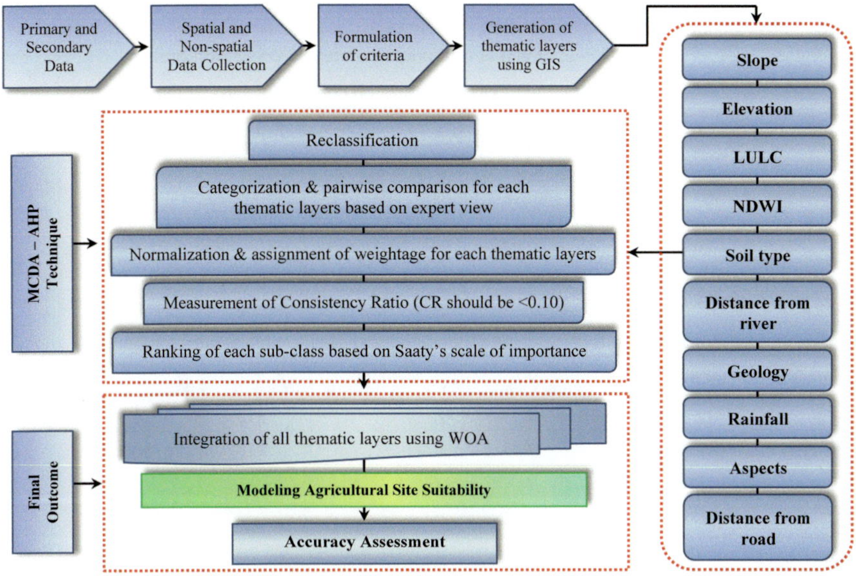

Fig. 2 Methodological flow diagram of the study

zone 45N, and WGS 84 datum. After that, the collected images were mosaiced, and techniques for edge augmentation, band arrangement, and atmospheric modification were used. Finally, the classification under supervision using a maximum likelihood algorithm has been used to create the LULC map and the corresponding classes (Patil et al. 2012; Abbas and Jaber 2020; Roy et al. 2021a; Kayal et al. 2022). Using Kappa statistics, the accuracy assessment was carried out during the post-classification phase (Kraemer 2014). With a kappa coefficient of 0.887, the LULC is classified with a satisfactory level of accuracy. Then, a high-resolution rainfall map for the study area with a $0.5^\circ \times 0.5^\circ$ grid was downloaded from a Climatic Research Unit (CRU) Web satellite. The final annual rainfall map was produced using the Kriging interpolation technique. Previous studies (Kim et al. 2011; Sarkar et al. 2021; Roy et al. 2022) have shown that there seems to be greater accuracy and dependability with the Kriging interpolation method than any other method. Data was then gathered from the open street map server to prepare a road proximity map. The soil type map was created using the Food and Agriculture Organization (FAO) Digital Soil Map of the World (DSMW). The Normalized Difference Water Index (NDWI), is often used to determine the moisture content of the soil (Zolekar and Bhagat 2015; Roy et al. 2022). Landsat 8 OLI (30-m resolution) data was used for the NDWI map and was obtained from the US Geological Survey's (USGS) Earth Explorer portal. In a GIS environment, the following equation was used to create the moisture map:

$$\text{NDWI} = \frac{G - \text{NIR}}{G + \text{NIR}} \tag{1}$$

Table 2 Review of the literature on the theme layers

Literature	Factors influencing agriculture site suitability									
	SL	EL	LULC	NDWI	ST	DRI	GL	RF	AS	DRO
Bandyopadhyay et al. (2009)	√		√		√					
Akıncı et al. (2013)	√	√	√		√				√	
Everest et al. (2021)	√	√	√						√	
Zolekar and Bhagat (2015)	√		√	√				√		
Pramanik (2016)	√	√	√	√	√	√	√		√	√
Yalew et al. (2016)	√	√	√	√	√	√				√
Yohannes and Soromessa (2018)	√	√				√		√	√	√
Biswas et al. (2020)	√		√	√	√		√	√		
Saha et al. (2021)	√	√	√	√		√	√	√	√	√
Tashayo et al. (2020)	√	√						√		
Kumar et al. (2021)	√	√	√		√	√	√		√	√
Mandal et al. (2021)	√		√					√		
Rajasekhar et al. (2021)			√	√	√			√		
Sarkar et al. (2021)	√	√	√	√				√		
Mangan et al. (2022)	√						√	√		√
Molla (2022)							√	√		
Roy et al. (2022)	√	√	√			√		√		√

Note SL Slope, *EL* Elevation, *LULC* Land use land cover, *NDWI* Normalized difference water index, *ST* Soil Type, *DRI* Distance from river, *GL* Geology, *RF* Rainfall, *AS* Aspect, *DRO* Distance from roads

In this case, NIR stands for near-infrared bands, and G stands for green.

Thus, it is evident that numerous data sources were consulted to produce the intended outcome in the current assessment of agriculture land suitability. To prevent any output errors, the data is rasterized, reclassified, and transformed into identical cells with a resolution of 30 m, UTM projection, Zone 45 N after all thematic layers have been prepared. After that, ten thematic layers were integrated to create the ultimate model of agricultural land suitability for the Bankura district using weighted overlay analysis (WOA) in a GIS environment.

3.2 Analytic Hierarchy Process (AHP)

There are many approaches, but the most appreciated is the multi-criteria decision-making (MCDM) based analytic hierarchy process (AHP), a productive method of decision-making developed by Saaty (1980, 1990) for choices made in a hierarchical structure using a range of criteria. The AHP method is advantageous because it makes it possible to organized parameters hierarchically. In addition, this approach is effective, saves time, yields high accuracy, and integrates qualitative and quantitative approaches (Biswas et al. 2020; Orhan 2021; Senapati and Das 2021; Roy et al. 2021a, b; Sarkar et al. 2021; Mangan et al. 2022). A multi-criteria approach is necessary to identify the suitability zone containing agriculture in the semi-arid region because numerous factors affect agricultural productivity (Yalew et al. 2016b; Shailaja et al. 2018). First of all, the weights of each factor have been re-assigned in accordance with their relative significance for fostering agriculture (Table 3). The results of a comprehensive literature analysis and expert guidelines were used to create the pair-wise comparison matrix or PCM (Table 4; Fig. 3). Additionally, the PCM weightage values have been normalized and standardized using the principal vector technique (Tables 5 and 6) for each of the thematic layers. The process then needs to be run through two different measurements, namely consistency index (CI) and consistency ratio (CR), to assess the uniformity of the PCM.

These two measurements are as follows:

Step: 1 PCM computed using Eq. 2

$$\text{PCM} = \begin{bmatrix} X_{11} & X_{12} & \ldots X_{1n} \\ X_{21} & X_{22} & \ldots X_{2n} \\ \vdots & \vdots & \vdots\vdots \\ X_{n1} & X_{n1} & \ldots X_{nm} \end{bmatrix} \tag{2}$$

where X is the set of parameters used to construct PCM, and numerical values signify the relative weights assigned to the criteria.

Step: 2 Utilizing Eq. 3, the weight has been normalized (Roy et al. 2021a, b; Kayal et al. 2022):

Table 3 Description of scales for pair-wise comparison with AHP based on Saaty (1980)

Scales	Degree of preferences	Description
1	Equal importance	Two criteria contribute equally to the objective
3	Moderate importance of one over another	Experience and judgment slightly favor one criterion over another
5	Essential or strong importance	Experience and judgment strongly favor one criterion over another
7	Very strong or demonstrated importance	A criterion is favored very strongly over another; its dominance demonstrated in practice
9	Extreme importance	The evidence favoring one criterion over another is of the highest possible order of affirmation
2, 4, 6, 8	Intermediate values between the adjacent scale values (when compromise is needed)	
Reciprocals	If criteria i has one of the above numbers assigned to it when compared with criteria j, then j has the reciprocal value when compared with i	

Table 4 Relative importance scales (1–9) and random consistency index (RCI) values up to 11 parameters based on Saaty (1980, 1990)

Intensity of importance	Definition	*n* (No. of parameters selected)	RCI values
1	Equal	3	0.58
2	Week	4	0.90
3	Moderate	5	1.12
4	Moderate plus	6	1.24
5	Strong	7	1.32
6	Strong plus	8	1.41
7	Very strong	9	1.45
8	Very very strong	10	1.49
9	Extreme	11	1.51

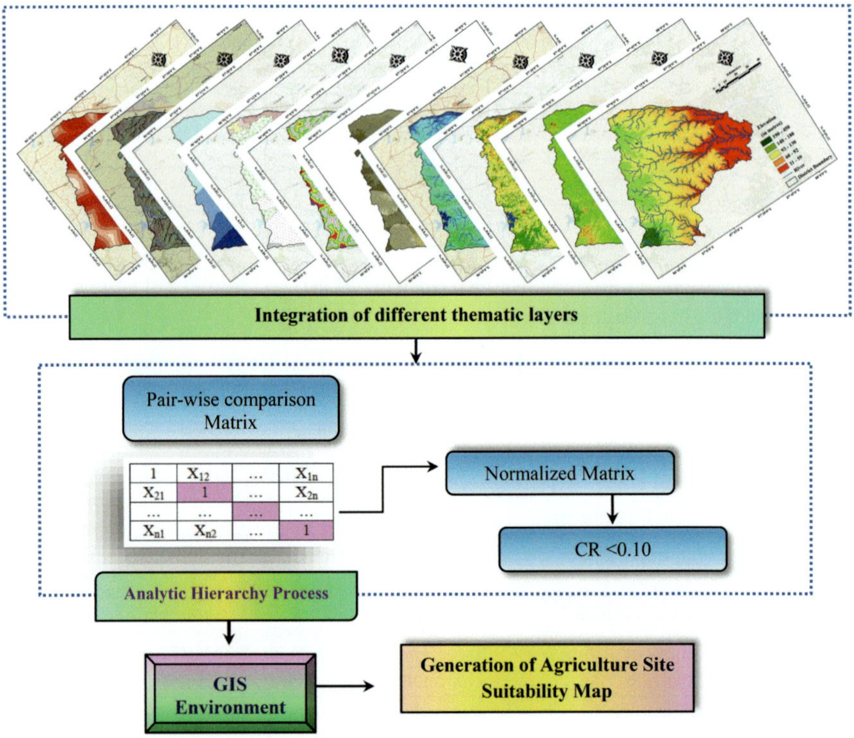

Fig. 3 Generation of the final agriculture suitability map through the integration of diverse layers using the MCDA-AHP methodology and GIS techniques

Table 5 PCM of the ten selected thematic layers

Parameters	SL	EL	LULC	NDWI	ST	DRI	GL	RF	AS	DRO
SL	1.000	2.000	2.000	3.000	6.000	4.000	8.000	9.000	9.000	9.000
EL	0.500	1.000	2.000	3.000	6.000	4.000	7.000	8.000	9.000	9.000
LULC	0.500	0.500	1.000	2.000	4.000	3.000	5.000	5.000	7.000	8.000
NDWI	0.330	0.330	0.500	1.000	3.000	2.000	5.000	5.000	6.000	7.000
ST	0.166	0.166	0.250	0.330	1.000	3.000	1.000	4.000	2.000	3.000
DRI	0.250	0.250	0.330	0.500	0.330	1.000	5.000	3.000	6.000	7.000
GL	0.125	0.143	0.200	0.125	1.000	0.200	1.000	2.000	1.000	3.000
RF	0.111	0.125	0.200	0.125	0.250	0.333	0.500	1.000	1.000	2.000
AS	0.111	0.111	0.143	0.167	0.500	0.167	1.000	1.000	1.000	2.000
DRO	0.111	0.111	0.125	0.143	0.330	0.143	0.330	0.500	0.500	1.000
SUM	3.204	4.736	6.748	10.390	22.410	17.843	33.830	38.500	42.500	51.000

Note SL Slope, *EL* Elevation, *LULC* Land use land cover, *NDWI* Normalized difference water index, *ST* Soil Type, *DRI* Distance from river, *GL* Geology, *RF* Rainfall, *AS* Aspect, *DRO* Distance from roads

$$\mathrm{NW} = \left(\frac{\mathrm{AM}}{\sum_{n-1}^{N_f} \mathrm{AM}_n}\right) \tag{3}$$

Here the arithmetic mean for the nth row of the PCM is denoted by AM_n, and NW stands for normalized weights.

Additionally, AM can be represented by Eq. 4.

$$\mathrm{AM} = \frac{1}{n}\sum_{n=1}^{1} a_1 \tag{4}$$

where a_1 stands for the values in the data set, n for the number of attributes, and AM for the arithmetic mean.

Step: 3 Building the consistency ratio (CR) with Eq. 5

$$\mathrm{CR} = \frac{\mathrm{CI}}{\mathrm{RI}} \tag{5}$$

where the consistency ratio index is denoted by RI and the consistency index by CI.

Based on Saaty's RI scale and the ten parameters that the current study examined, the consistency ratio (CR) value is 1.49 (Table 4).

Maximum eigenvalue ($\lambda_{\max}$) = 10.746, Number of Parameters (n) = 10, Ratio Index (RI) = 1.49,

$$\mathbf{CI} = \frac{(10.746 - 10)}{(10 - 1)} = 0.08294$$

Table 6 Normalized pairwise matrix and standardized weight for each thematic layers based on AHP

Parameters	SL	EL	LULC	NDWI	ST	DRI	GL	RF	AS	DRO	Normalized weight
SL	0.312	0.422	0.296	0.289	0.268	0.224	0.236	0.234	0.212	0.176	0.267
EL	0.156	0.211	0.296	0.289	0.268	0.224	0.207	0.208	0.212	0.176	0.225
LULC	0.156	0.106	0.148	0.192	0.178	0.168	0.148	0.130	0.165	0.157	0.155
NDWI	0.103	0.070	0.074	0.096	0.134	0.112	0.148	0.130	0.141	0.137	0.115
ST	0.052	0.035	0.037	0.032	0.045	0.168	0.030	0.104	0.047	0.059	0.061
DRI	0.078	0.053	0.049	0.048	0.015	0.056	0.148	0.078	0.141	0.137	0.080
GL	0.039	0.030	0.030	0.012	0.045	0.011	0.030	0.052	0.024	0.059	0.033
RF	0.035	0.026	0.030	0.012	0.011	0.019	0.015	0.026	0.024	0.039	0.024
AS	0.035	0.023	0.021	0.016	0.022	0.009	0.030	0.026	0.024	0.039	0.025
DRO	0.035	0.023	0.019	0.014	0.015	0.008	0.010	0.013	0.012	0.020	0.017

Note SL Slope, *EL* Elevation, *LULC* Land use land cover, *NDWI* Normalized difference water index, *ST* Soil Type, *DRI* Distance from river, *GL* Geology, *RF* Rainfall, *AS* Aspect, *DRO* Distance from roads

$$\mathbf{CR} = \frac{0.08294}{1.49} = 0.05567$$

With 'n' equal to 10, the Consistency Ratio (CR) is calculated as 0.05567, indicating a value less than 0.10. This suggests a reasonable level of consistency within the PCM.

Step: 4 For the Construction of CI, Eq. 6 has been taken into consideration.

$$\text{CI} = \frac{(\lambda \max - n)}{(n-1)} \tag{6}$$

In this context, CI represents the Consistency Index, λ max denotes the average of the Eigenvalue, and 'n' signifies the total number of criteria. The Saaty scale requires that the value of CR be less than 0.10 in this case; otherwise, the results will not be consistent (Table 7).

3.3 Utilization of Weighted Overlay Technique (WOA) for Modeling Agricultural Land Suitability Map

In this study, weighted overlay analysis was used to identify the agricultural site suitability zone in the semi-arid region of Bankura district. The application of the chosen raster layers and corresponding weights results in a model of cumulative likelihood that agricultural suitability zones will occur, which is a useful tool for researchers (Ibrahim-Bathis and Ahmed 2016; Roy et al. 2020; Biswas et al. 2020). Additionally, this overlay study offers a quick, easy, and appropriate dataset for analytical operations (Mandal et al. 2021). In this study, the WOA was carried out using a GIS platform that compiles the output's ten thematic layers. Before moving on to WOA, equal cell sizes and projection systems must be used in all relevant layers. WOA can be expressed using Eq. 7. (Roy et al. 2020; 2021a, b; Kayal et al. 2022)

$$\text{WOA} = \sum_{i=1}^{n} Wi \times Ri \tag{7}$$

Here 'Wi' represents a specific decision criterion, 'Ri' denotes the raster layer linked with that criterion, and 'n' signifies the number of the decision matrix.

However, because the present study used the method of influencing factors to assess the suitability of the land for agriculture based on ten factors, it is essential to determine the relationships between each of the factors that were chosen (Datye and Gupte 1984; Zolekar and Bhagat 2014). To comprehend how these factors are connected and how they influence one another, Roy et al. (2022) suggest that it is essential to evaluate spatial suitability using a variety of influential factors. Because

Table 7 Weights of selected parameters and their subclasses according to AHP method and rank assignment

Parameters	Sub-parameters	AHP normalized weights	Influence (%)	Rank of sub parameters (after saaty's scale)
Slope (in degrees)	72.00–89.99	0.267	26.7	2
	54.00–71.99			4
	36.00–53.99			6
	18.00–35.99			8
	0.00–17.99			10
Elevation (in Metres)	190–450	0.225	22.5	2
	140–180			4
	93–130			7
	60–92			9
	11–59			10
LULC	Water body	0.155	15.5	Constraint
	Vegetation cover			6
	Plantation			7
	Agricultural land			10
	Built up area			4
	Sand bar			3
	Bare ground			2
NDWI	– 0.060 to 0.236	0.115	11.5	9
	– 0.154 to – 0.061			7
	– 0.199 to – 0.155			5
	– 0.253 to – 0.200			3
	– 0.451 to – 0.254			1
Soil type	Loam	0.061	6.1	10
	Sandy clay loam			7
	Sandy loam			5
Distance from River (in Km)	4.1–5	0.080	8.0	2
	3.1–4			3
	2.1–3			4
	1.1–2			5
	0–1			6
Geology	Unclassified metamorphic	0.033	3.3	5
	Talchir, Barakar, Raniganj, Karharbari formation			3

(continued)

Table 7 (continued)

Parameters	Sub-parameters	AHP normalized weights	Influence (%)	Rank of sub parameters (after saaty's scale)
	Kuilapal granite			3
	Singhbhum			3
	Chhotanagpur gneissic complex			3
	Undiff. fluvial/ Aeolian/coastal & glacial sediments			7
	Panchet, Sura Panchet, Parsora formation			2
Rainfall (in mm)	1230–1250	0.024	2.4	9
	1200–1220			7
	1170–1190			5
	1140–1160			3
	1100–1130			1
Aspect	Flat	0.025	2.5	0
	S–SW–W			9
	SE–E			8
	NW			7
	NE			6
	N			5
Distance from Roads (Km)	0–3	0.017	1.7	5
	3.1 − 6			3
	6.1–9			2
	9.1–12			1
	12.1–16			1

of this, the study used correlation study to uncover the connections between each factor and assess the stability among the factors influencing suitability of the land for agriculture. Using a multivariate tool and the PCA technique, the ultimate correlation matrix was acquired (Fig. 4). The results of the correlation matrix demonstrate both favorable and unfavorable interactions between all variables impacting the appropriateness of agriculture in the study area.

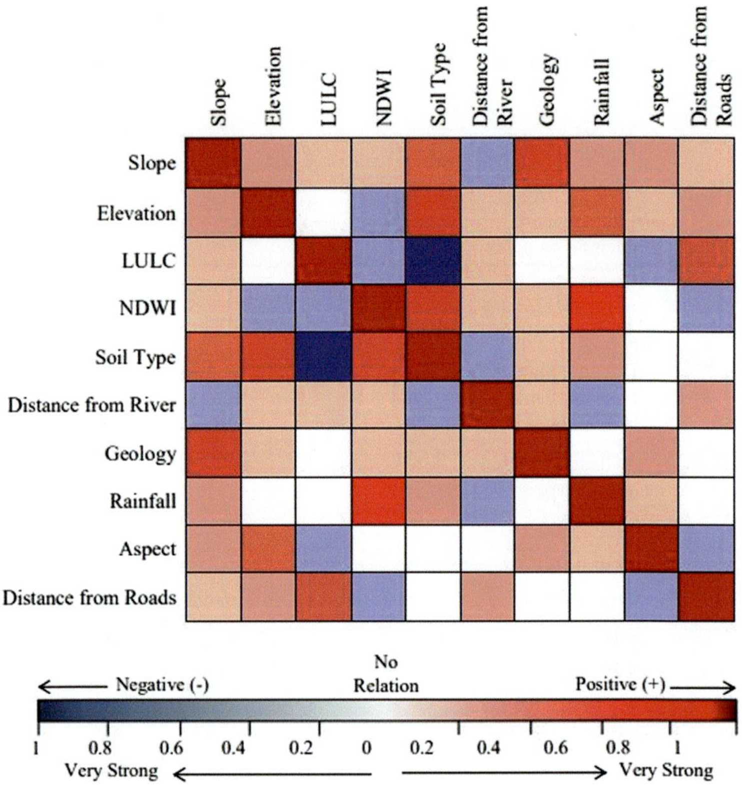

Fig. 4 Correlation matrix of ten factors influencing agricultural suitability using Pearson's correlation coefficient (*r*)

4 Result and Discussion

4.1 Variables Affecting Agriculture Suitability

4.1.1 Slope

The slope is a crucial element in agriculture. The degree of slope directly affects a variety of soil properties, including soil depth, susceptibility to erosion, soil tillage, irrigation, and plant adaptation. The amount of soil and water lost in a region is directly impacted by the slope's steepness, length, and sharpness (Everest et al. 2021). The slope map was created using ASTER-GDEM data with a 30-m spatial resolution that was gathered from Earth data by NASA. The study data is derived from a digital elevation model that uses a grid format with pixels that measure the value of each pixel from every angle and is made up of cells with a height value (Nganga et al. 2020). The extracted data shows that the slope of the Bankura district

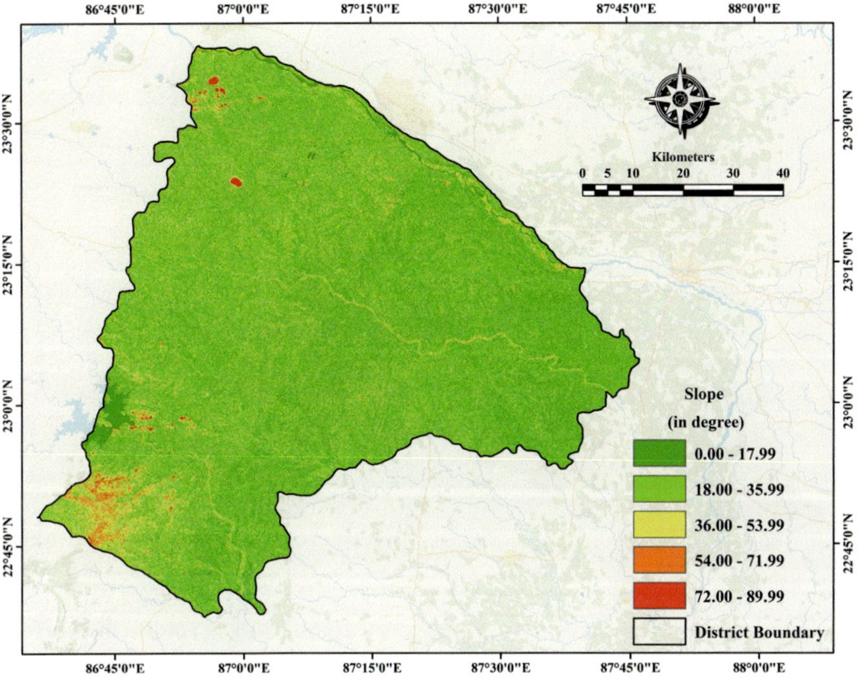

Fig. 5 Slope map of the study area

varies from 0 to 88.99 degrees (Fig. 5) and the region with a typical gradient running from west to east. Flood plains and interfluves characterize the eastern portion, while the elevation gradually climbs in the west, giving birth to isolated hillocks, broken low ridges, and valleys.

4.1.2 Elevation

Elevation is one of the most important factors in agricultural production and diversity (Everest et al. 2021). Several studies agreed that elevation significantly affects both agricultural diversity and output (Atalay 2006; Roy et al. 2022). According to Kapluhan (2013), elevation rises from the coast to the interior, influencing climatic conditions. Similar crops differ psychologically and morphologically as a result of elevation changes, influencing yield production. However, the topographic characteristics of the study area can be identified using the map below (Fig. 6). Data extracted from DEM has revealed that the Bankura district's elevation ranges from 11 to 450 m. Compared to the eastern part, the southwestern portion of the study area has the highest topography. Due to its undulating topography, the southwest region mostly consists of rocky surfaces, while the eastern part has a gentle flat plain that offers ample opportunity for agricultural production.

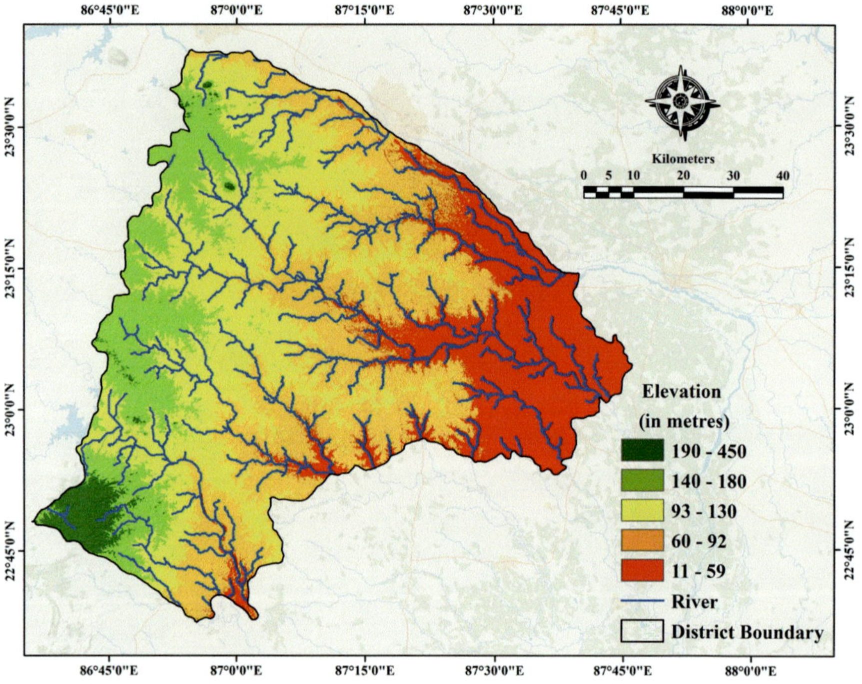

Fig. 6 Elevation map of the study area

4.1.3 Land Use Land Cover (LULC)

Data on land use are useful in determining an area's productivity for irrigation. The supervised classification method has been used to create land cover thematic maps. The inputs for the classes created from training sites are calculated using statistics (mean, variance, and covariance) to create a Bayesian probability function. The class to which each pixel most likely belongs is subsequently determined. Utilizing maximum probability techniques, the watershed was classified in the Sentinel-2 satellite imagery (10-m spatial resolution) collected from the United States Geological Survey (USGS) website.

The ERDAS-Imagine software was used to identify the training sets of the seven major classes (Fig. 7), including the water body (2.48%), vegetation cover (18.87%), plantation (0.05%), agriculture (64.15%), built-up areas (7.22%), shrubs (0.74%), and bare ground (6.50%). Agriculture is dominant among seven different LULC classes in the district, containing the highest percentage of land area. Paddy is the main crop produced in terms of agriculture, but the district also produces other crops like potato, wheat, vegetables, mustard, summer til, etc. Similar to paddy, the district produces an abundance of vegetables and potatoes. The district's economy is therefore based primarily on agriculture, and as a result, the majority of its residents work in the sector (DCHB 2011). Hence, meticulous planning is imperative for ensuring

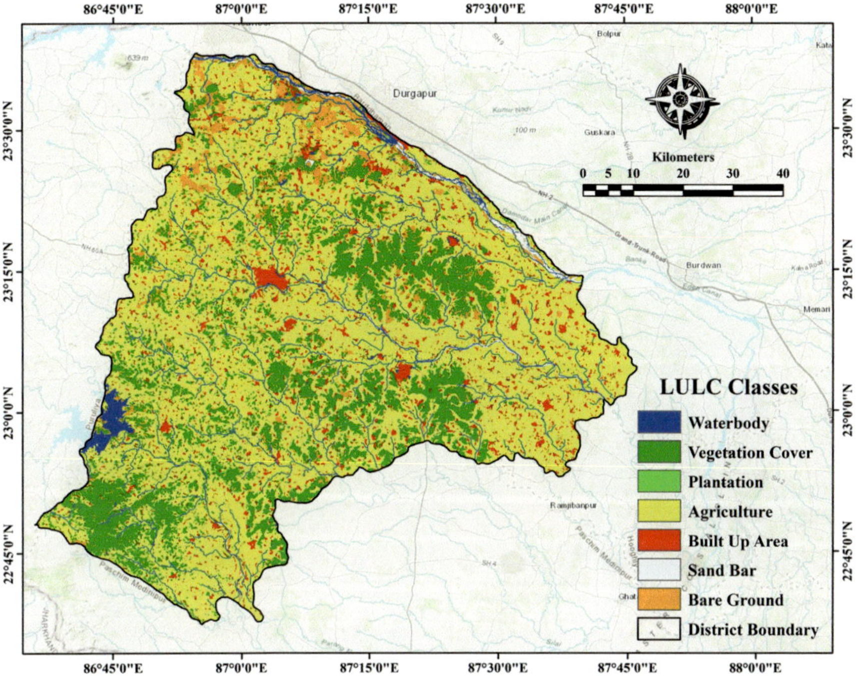

Fig. 7 Land use land cover map of the study area

the sustainable utilization of land, underscoring the significance of conducting a comprehensive land suitability assessment for the region.

4.1.4 Normalized Difference Water Index (NDWI)

According to Zolekar and Bhagat (2015), spectral reflectance varies depending on the terrain's features, such as soils, water bodies, rocks, plant cover, and built-up areas. Water has a lower spectral reflectance than other surfaces as a result of significant absorption (Hui et al. 2008). Consequently, moist soils appear darker than dry soils in remote sensing images (Xu et al. 2006). NDWI is a useful technique for determining the soil's moisture content that is necessary for agricultural production (Sow et al. 2013; Zolekar and Bhagat 2014; Roy et al. 2022). Additionally, NDWI is a good indicator of soil quality and positively affects crop productivity (Wang et al. 2012). Thus in this present study moisture map was calculated by NDWI which is measured using Infra-Red (IR) and Green (G) bands of Landsat 8(OLI/TIRS) satellite images. The NDWI value usually ranges from ' + 1' to '− 1' where the built-up area and bare surface are represented by "− 1" and the moist area by " + 1". Five categories, including very high (− 0.060 to + 0.236), high (− 0.154 to − 0.061), medium (− 0.199 to − 0.155), low (− 0.253 to − 0.200), and very low (− 0.451 to − 0.254) have

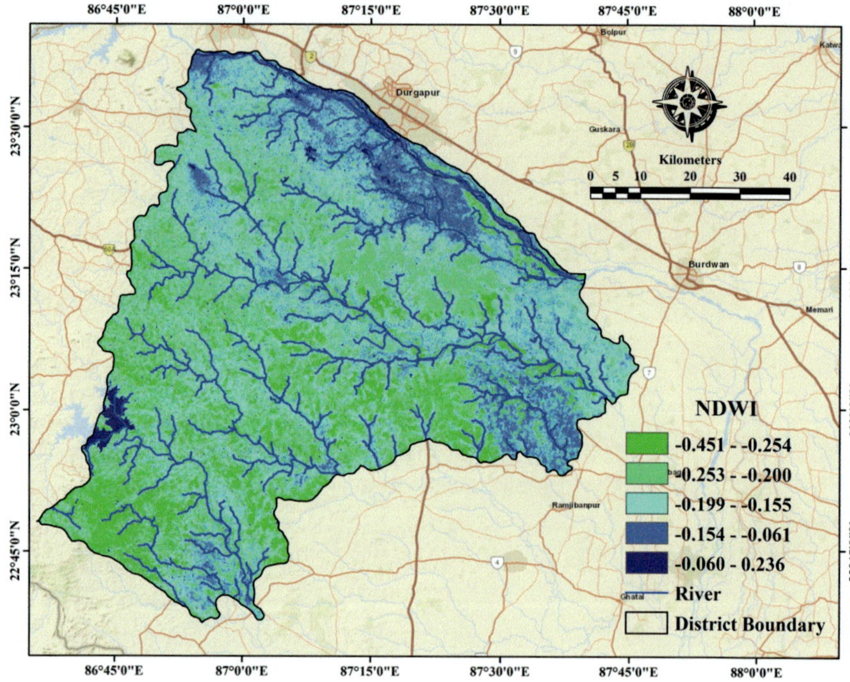

Fig. 8 NDWI map of the study area

been assigned to the soil moisture maps (Fig. 8). In the northern, eastern, and south-western regions of the study area, high NDWI values in blue indicate high moisture with maximum plant moisture content, which is suitable for agriculture. Conversely, negative values in green indicate a dry area with low vegetation moisture content, which is less suitable for agriculture and more susceptible to drought. However, 62 percent of the region exhibits negative NDWI values, indicating that this area is more vulnerable to drought. Therefore, the implementation of site suitability modeling for agriculture in this area is crucial for the management of sustainable agriculture.

4.1.5 Soil Type

Among the essential elements required to cultivate any kind of agricultural product is soil. Soil biology expertise and land suitability are both crucial (Roy et al. 2022). Because soil texture affects the soil's capacity to retain water and nutrients for plant uptake, it has an impact on soil fertility. The availability of nutrients like nitrogen (N), phosphorus (P), and potassium (K) as well as the biochemical components of the soil are all directly influenced by the texture of the soil, including depth, wetness, and water holding capacity (Pramanik 2016). The health of the soil and agricultural systems must be maintained through agricultural practices that encourage

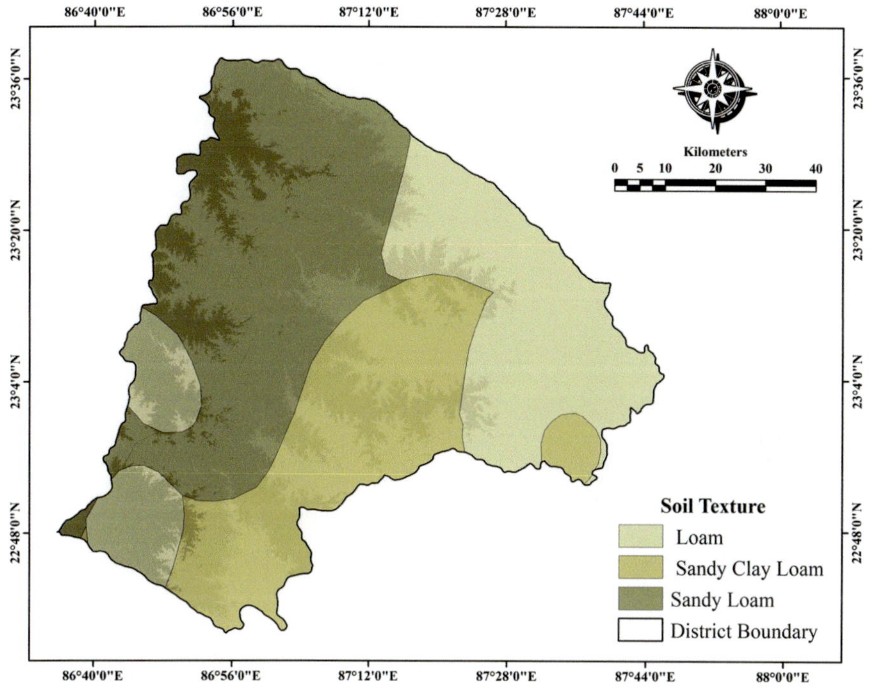

Fig. 9 Soil type map of the study area

maximum yield with minimal water and fertilizer use (Chakraborty and Mistri 2015; Roy et al. 2022). Soil information for the study area has been collected from the FAO Digital Soil Map of the World-ESRI shapefile. According to Fig. 9, the study area is comprised of three major soil groups, with loam being the most prevalent in the eastern part, sandy clay loam being found in the middle and southern portions of the study area, and sandy loam being present in the northern and western portions of the study area. With a ratio of sand, silt, and clay that is ideal for crops, loam soil is among the types of soil that has the highest agricultural productivity.

4.1.6 Distance from Rivers

Water is necessary for all life forms, for all facets of socioeconomic development, and for the preservation of thriving ecosystems (FAO 1976). Being close to a river enhances agricultural output because it supplies water for irrigation (Cook et al. 2009). Furthermore, the moisture content of soils decreases as one gets farther away from rivers and streams. The ideal location for agricultural production is approximately a one kilometer distance from the river (Saha et al. 2021; Roy et al. 2022). To account for the river's influence on the suitability of the land for agriculture, we have prepared multi-buffer maps for the study. To analyze the impact of distance

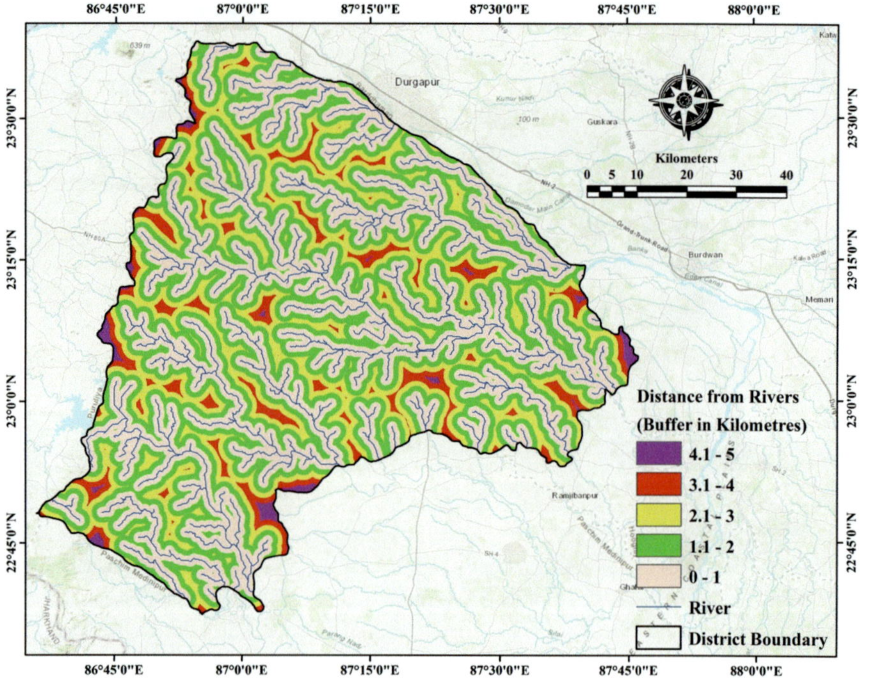

Fig. 10 Distance from river map in the study area

from the river on agriculture, five classes: very high (4.1–5 km), high (3.1–4 km), moderate (2.1–3 km), low (1.1–2 km), and very low (0–1 km) has been prepared (Fig. 10). However, in this study, very close proximity to a river is thought to have a significant impact on agriculture, whereas very high distance from a river has a negligible impact on agriculture.

4.1.7 Geology

Geology affects soil formation and defines the type of minerals found in soil. The geologic conditions that produce the minerals are crucial for growing our food because they make up about 45 percent of the soil composition. By releasing and reacting with plant nutrients, the mineral content of soil can impact its fertility and production. Also, geology controls the flow of water (river, stream), which is also required for agricultural productivity (Chowdhury et al. 2010; Yeh et al. 2016; Abijith et al. 2020). ~~The geological data of the study area has been extracted from the Geological Survey of India (GSI) website.~~ Geologically the study area is divided into seven geological age groups (Fig. 11; Table 8). The middle-eastern and eastern parts of the district have undefined fluvial, aeolian, and coastal sediment of quaternary age with the largest area coverage. The second-largest geological formation is

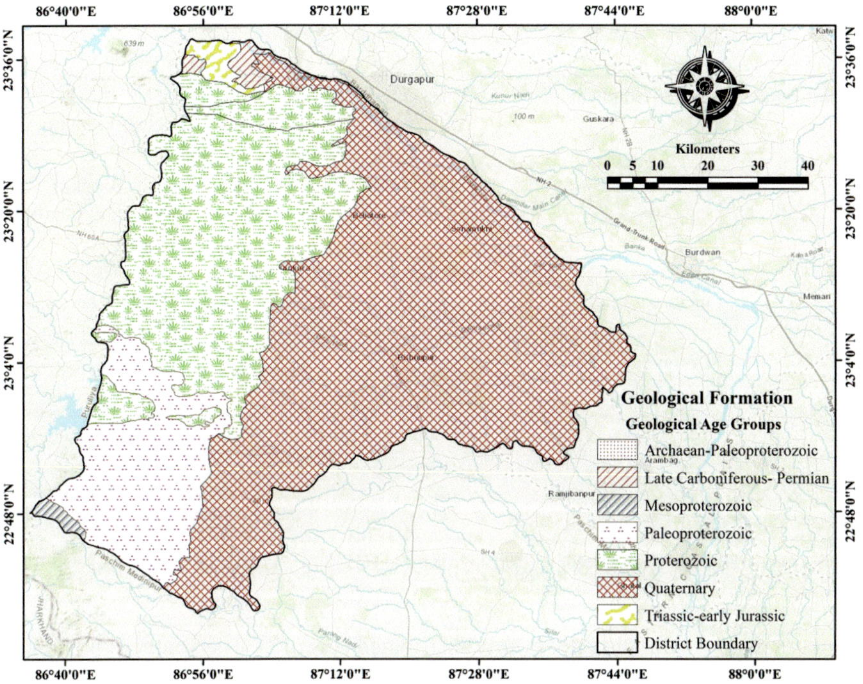

Fig. 11 Geological formation of the study area

the Chhotanagpur gneissic complex of Proterozoic age. The third largest geological formation in terms of area coverage is the Paleoproterozoic age group, which mainly occupied the western and northern parts of the district. The other four formations viz. are scattered throughout the southern and northern parts of the study area. In our study, the highest priority has been given to the sediments that were deposited during the most recent quaternary period and are undifferentiated Fluvial-Aeolian and Coastal sediments.

4.1.8 Rainfall

Climatic factors like the amount of rainfall affect whether a piece of land is suitable for agriculture (Saha et al. 2021; Roy et al. 2022). Water plays a crucial role in plant growth and agricultural production in all of its forms because it facilitates the flow of nutrients and food throughout the plant (Arulbalaji et al. 2019). Additionally, the climatic factor has a positive impact on every facet of land productivity and yield, such as crop productivity, grazing intensity, and so on (Iizumi and Ramankutty 2015). But it has been demonstrated that some climatic elements are very closely related to agricultural output (Yohannes and Soromessa 2018; Akpoti et al. 2019; Worqlul et al. 2019; Alemayehu et al. 2020; Tashayo et al. 2020b). However, in the

Table 8 Overview of stratigraphic sequence for rock units in the study area

Super group	Group	Age	Formation
Gondwana	Upper Gondwana	Triassic-early Jurassic	Panchet, Sura Panchet, Parsora formation
	Lower Gondwana	Late Carboniferous-Permian	Talchir, Barakar, Raniganj, Karharbari formation
	Unclassified metamorphic	Archaean-Paleoproterozoic	
	Chhotanagpur gneissic Complex	Proterozoic	
	Singhbhum	Paleoproterozoic	
		Mesoproterozoic	Kuilapal Granite
		Quaternary	Undiff. Fluvial/ Aeolian/ Coastal & Glacial Sediments

Source ~~Data extracted from~~ geological survey of India (GSI)

Bankura district rainfall is not evenly distributed (Fig. 12). The southern portion of the study area has the highest annual rainfall (1230–1250 mm), while the northern and northwestern portions of the district have the lowest (1100–1130 mm). Due to the direct relationship between rainfall and agriculture, regions with high rainfall are thought to have a greater possibility of having agricultural sites.

4.1.9 Aspect

Plants require periodic exposure to the sun to maintain their physiological processes; though it varies depending on the type of plant. However, most cultigens grow at their best on the southern and western sides that are exposed to sunlight for a significant portion of the day (Akıncı et al. 2013). Because of this, the aspect is taken into account as a measure of assessment for choosing the land to be used for agriculture. Generally, aspect-slope maps typically show the slope's aspect (direction) and steepness (degree) simultaneously. Red, orange, yellow, and other hues are used to represent aspect categories. Saturation, or brilliance of color, is used to map the degree of slope classes so that steeper slopes appear brighter (Everest et al. 2021). Aspect is connected to plant variety, species dispersion, vegetation time, and crop yield thus it has been considered as one of the key components in crop productivity. Sunlight is necessary for plants to function physically and metabolically (Bale et al. 1998). Figure 13 displays an aspect map of the study area.

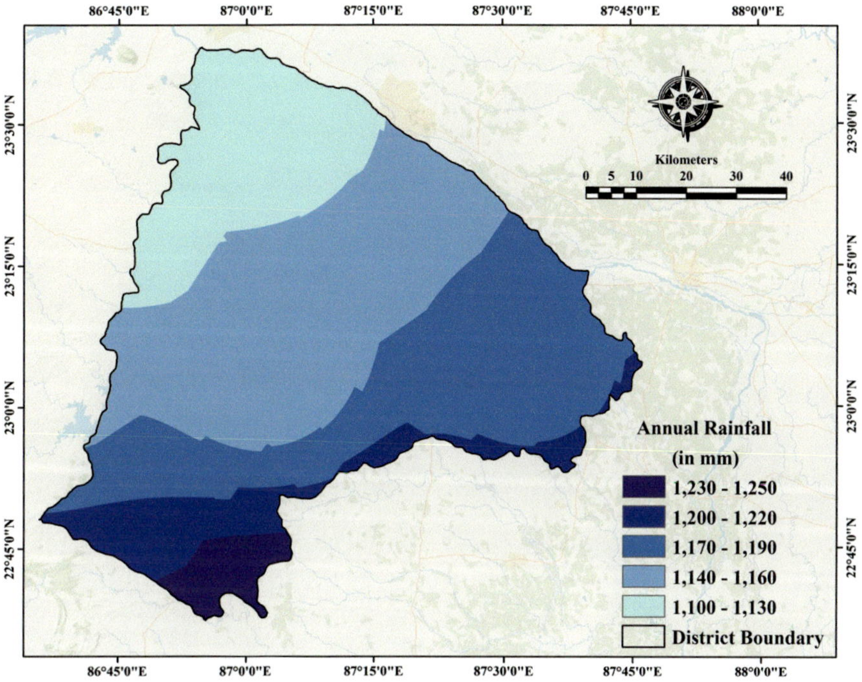

Fig. 12 Rainfall map of the study area

4.1.10 Distance from Roads

FAO (2007) asserts that to manage and develop agriculture, the suitability of a piece of land depends on its accessibility, such as the distance between it and the availability of a transport network. The availability of roads makes it easier to supply the crucial raw materials needed for agricultural development (Pramanick 2016; Yalew et al. 2016a). Previous studies proved that the ideal location for agricultural production is approximately one kilometer from the road (Saha et al. 2021; Roy et al. 2022). Thus, we created multiple buffers for the study to take into account the impact of transport networks on the suitability of the land for agriculture. Five classes have been created to analyze the effects of road proximity on agriculture: very high (12.1–16 km), high (9.1–12 km), moderate (6.1–9 km), low (3.1–6 km), and very low (0–3 km) (Fig. 14). However, it has been assumed that very close proximity to a road has a significant impact on agriculture, whereas very high distance from roads has negligible effect on agricultural land suitability. Thus in this present study, greater priority has been given to the nearness to accessibility has a higher importance.

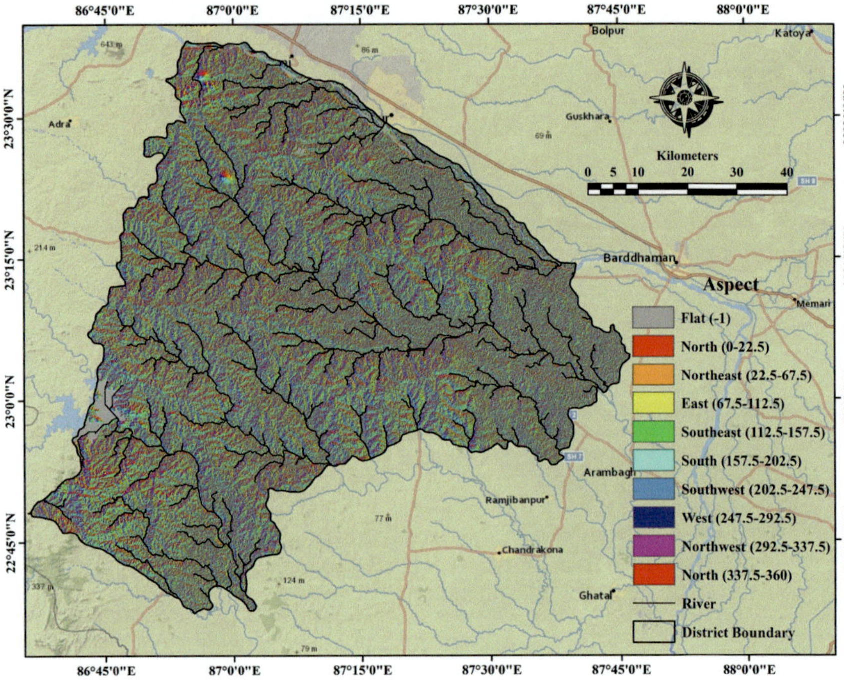

Fig. 13 Aspect map of the study area

4.2 *Estimation of Agriculture Land Suitability Zone and its Geospatial Distribution*

The agriculture land suitability zone has been mapped using the weights of selected criteria which were determined through AHP analyses and assigned scores of sub-criteria. The prepared map has finally been divided into five categories (Fig. 15; Table 9) which are examined by the following heads:

4.2.1 Very Highly Suitable

In the study area, this is the agriculturally most advantageous area. A total of 642.18 sq. km. of the district's land area (9.33%) has been designated as "Very highly suitable" for agriculture. It includes the central and eastern parts of the Khatra subdivision, the southern part of the Bishnupur subdivision, and a very low concentration in the Bankura Sadar subdivision; the southern and northern border sides have very few agriculturally suitable areas. This area is distinguished by very low to low elevated areas (less than 100 m above sea level), a very gentle slope (less than 17 degrees), extremely high soil moisture, loamy soil texture, ideal annual rainfall

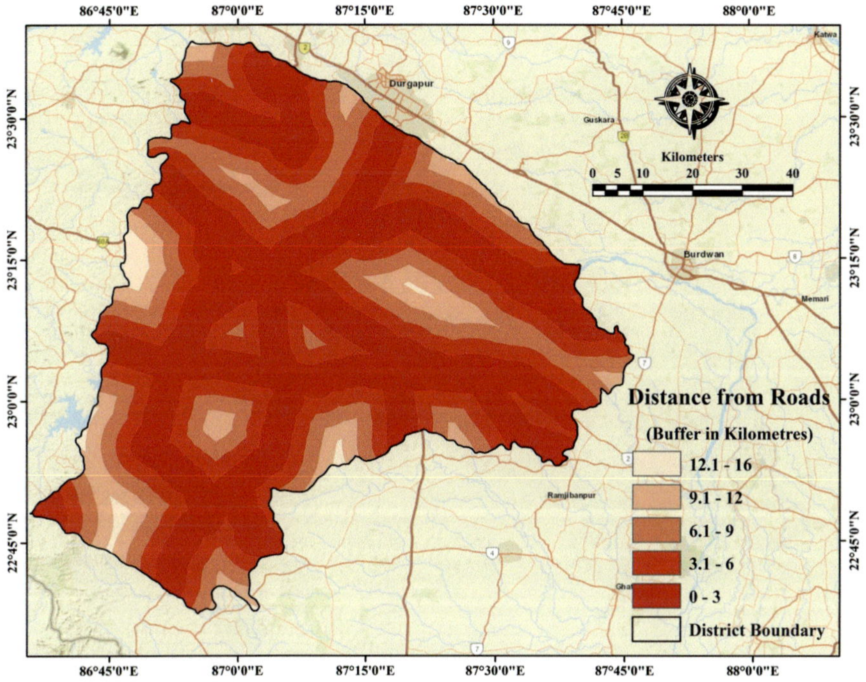

Fig. 14 Distance from road map of the study area

(1170–1250 mm), and proximity to rivers and a transportation network. Due to all of these favorable factors, agriculture is the most appropriate here.

4.2.2 Highly Suitable

A total of 987.37 sq. km area (14.35%) out of the district's total area is classified as highly suitable. It is the study area's second most important agricultural site suitability zone. The subdivision-wise distribution of highly suitable zones is very similar to that of the preceding zones. This class of agriculture suitability includes the southern part of the Bishnupur subdivision, the central and eastern parts of the Khatra subdivision, and the southern and northern border sides of the Bankura Sadar subdivision. This region has flat terrain (less than 100 m), a gentle to steep slope, high to moderate soil moisture (− 0.154 to + 0.236), sandy clay loamy soil texture, ideal rainfall, and is economically accessible due to its proximity to the road network. This region is ideal for cultivation if the proper management techniques are used. However, for better results, soil erosion prevention measures must be implemented.

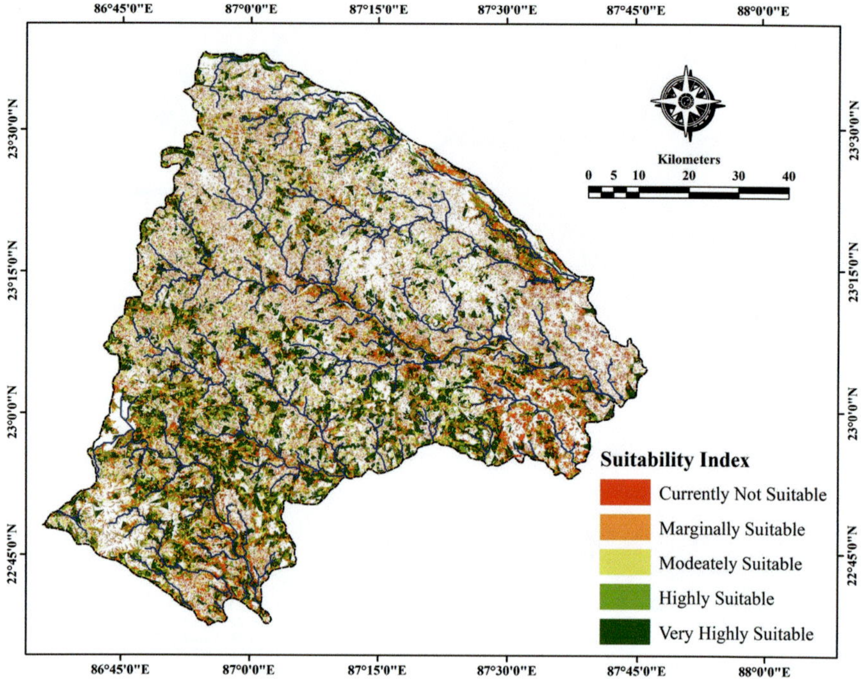

Fig. 15 Agriculture site suitability map of ~~semi-arid region of~~ Bankura district ~~prepared using AHP technique~~

4.2.3 Moderately Suitable

Moderately suitable agriculture zone covers 1345.35 sq. km areas (19.55%) out of the total area of the district. It covers the central, western, southern, and south-eastern parts of the Khatra subdivision; southern, southwestern and little patches in the northern part of the Bishnupur subdivision; and little patches parallel to the river bank flowing from the western to the eastern part of the Bankura Sadar subdivision. This zone has been characterized by flat to undulating terrain (60–130 m), gentle to steep slope, moderate to low soil moisture, sandy clay loamy to sandy loam soil texture, moderate to low rainfall, proximity of forest and bushes, moderate to high distance from river (2.1–4 km) and transport network (6.1–12 km). Because of these characteristics, the third zone can be considered moderately suitable for agriculture; however, proper management with costly inputs is an urgent requirement.

Table 9 Subdivision-wise distribution and characteristics of agricultural land suitability classes

Agricultural suitability class	Area (in sq. km)	Area (in %)	Subdivision wise distribution	Land characteristics and qualities	Remarks
Very highly suitable	642.18	9.33	Central and eastern part of the Khatra subdivision; Southern part of the Bishnupur subdivision; and the concentration is very low in Bankura Sadar subdivision; the southern and northern border sides have very few areas that are highly suitable for agriculture	It includes very low to low elevated areas (< 100 m altitude), very gentle slope (< 17°), very high soil moisture, loamy soil texture, receives ideal annual rainfall (1170–1250 mm), and proximity to rivers and transport network	Due to all of these favorable factors, agriculture is the most appropriate. Intensive agriculture is possible with the help of efficient irrigation
Highly suitable	987.37	14.35	The subdivision wise distribution of highly suitable zones is quite similar like the previous zones. Southern part of the Bishnupur subdivision; central and eastern part of the Khatra subdivision; and the southern and northern border sides of Bankura Sadar subdivision	This region contains flat terrain (< 100 m), gentle to steep slope, high to moderate soil moisture (– 0.154 to + 0.236), sandy clay loamy soil texture, ideal rainfall, and very close proximity to river and economically accessible due to proximity to road network	With the right management techniques, ideal for cultivation. Soil erosion prevention measures must be implemented

(continued)

Table 9 (continued)

Agricultural suitability class	Area (in sq. km)	Area (in %)	Subdivision wise distribution	Land characteristics and qualities	Remarks
Moderately suitable	1345.35	19.55	Central, western, southern and southeastern part of Khatra subdivision; southern, southwestern and little patches in northern part of the Bishnupur subdivision; and little patches parallel to the river bank flowing from western to eastern part of the Bankura Sadar subdivision	It comprises flat to undulating terrain (60-130 m), gentle to steep slope, moderate to low soil moisture, sandy clay loamy to sandy loam soil texture, moderate to low rainfall, proximity of forest and bushes, moderate to high distance from river (2.1–4 km) and transport network (6.1–12 km)	Moderately suitable for agriculture when managed properly and with expensive inputs
Marginally suitable	2277.27	33.09	Southern, eastern and central part of the Khatra subdivision; eastern and northeastern part of the Bishnupur subdivision; and southeastern part of the Bankura Sadar subdivision	It is the part of western hill region and northwestern dry region. This region having steep slope, low moisture content, receives low to very low precipitation annually, High to very high distance from river and road network	This area is not suitable for agriculture. The situation can be improved by improvising necessary safeguards against drainage and severe erosion
Currently not suitable	1629.84	23.68	North eastern, south eastern, and north western part of the Bankura Sadar; Northern, south eastern and southern part of the Bishnupur subdivision and throughout the entire part of the Khatra subdivision in a disperse manner	Open rock outcrop, vegetation cover, high concentration of builtup area, water body, bare ground, sand bar, low to very low annual perception, and lack of irrigation	These lands are not suitable for farming. Agriculture is not permitted in areas with habitation, open rocks, roads, or forest areas
Total	**6882.00**	**100.00**			

4.2.4 Marginally Suitable

A total of 2277.27sq. km area (33.09%) out of the total area has been classified as marginally suitable for agriculture. It is the largest zone among five classified agriculture site suitability zones. It comprises the southern, eastern, and central parts of the Khatra subdivision; the eastern and north-eastern parts of the Bishnupur subdivision; and the south-eastern part of the Bankura Sadar subdivision of the Bankura district. It is the part of western hill region and the north-western dry region. This region has a steep slope, low moisture content, receive low to very low precipitation annually, and is a high to very high distance from the river and road network. Thus, this area is not suitable for agriculture. However, the situation can be improved by improvising necessary safeguards against drainage and severe erosion.

4.2.5 Currently Not Suitable

It covers 1629.84 sq. km (23.68% of the total area of the study area), making it the second-largest zone among five classified agriculture site suitability zones. This is the agriculturally least advantageous area. This zone comprises the northeastern, southeastern, and northwestern parts of the Bankura Sadar; the Northern, southeastern, and southern part of the Bishnupur sub-division, and the entire part of the Khatra subdivision in a dispersed manner. Open rock outcrop, vegetation cover, high concentration of built-up area, water body, bare ground, sand bar, low to very low annual perception, and lack of irrigation are the major features of this region. Therefore, agricultural activities cannot be carried out.

4.3 Validation

Assessing the accuracy of any model is primarily dependent on comparing the outcomes with the real ground truth. The study has incorporated three validation techniques, including:

(a) Firstly, images from Google Earth were compared to the prepared agriculture suitability map (Fig. 16). To conduct this procedure, areas identified as highly suitable were extracted from Google Earth images and compared with the results obtained. The comparison revealed that the agricultural suitability maps closely mirrored the locations identified on Google Earth, validating the accuracy of the findings.
(b) Besides, according to the district agricultural report, approximately 1,62,507 hectares, or 1625.07 sq. km, are estimated to be areas under non-agricultural use and unsuitable for agriculture combined (BDP Report 2011; DSHB 2014). This estimate is nearly identical to our result because adding the marginally and currently unsuitable zones yields a total area of 1629.84 sq. km.

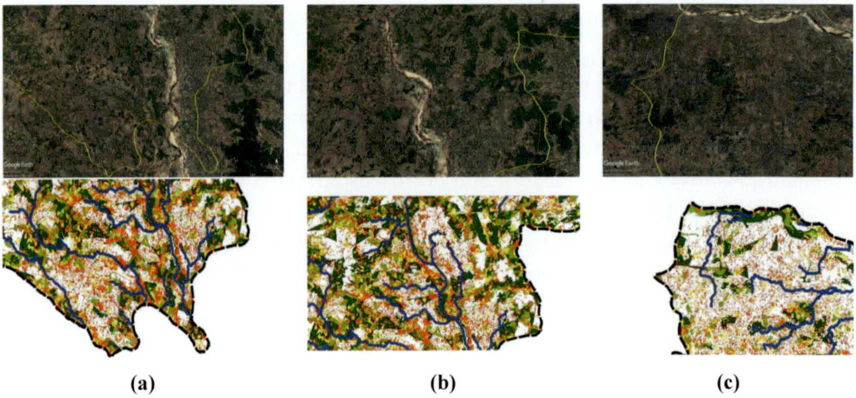

Fig. 16 Validation of agriculture suitability zones through comparison with google earth images

(c) Using an error matrix, the accuracy was estimated for users, producers, and overall accuracy (Congalton 1991; Banko 1998; Zolekar and Bhagat 2015). In our study, a total number of 233 ground reference points were collected from the field using GPS-equipped devices and verified using classified data. The outcome reveals that, with variation in producer and user accuracy, the classified map's overall accuracy is estimated to be about 92.14%. The accuracy estimates for the very highly suitable lands and the currently unsuitable agriculture zones are 94.71% and 92.22% for producers and 90.84% and 95.45% for users, respectively (Table 10).

5 Conclusion and Recommendations

The imperative task of identifying suitable land and assessing its quality holds paramount importance in delineating an appropriate land utilization framework and its efficacy in conserving natural resources for posterity. Accordingly, the principal objective of this research endeavor was to comprehensively consider all pertinent criteria, encompassing slope, elevation, land use/land cover (LULC), Normalized Difference Water Index (NDWI), soil composition, proximity to rivers, geological attributes, precipitation patterns, aspect orientation, and proximity to road networks, with the aim of delineating potential agricultural zones. Furthermore, the economic valuation of land, predicated on its distance from road infrastructure, was undertaken to discern optimal agricultural locales within the semi-arid expanse of Bankura District. Moreover, the MCDM-AHP methodology was employed in the present study to adjudicate the relative significance of diverse criteria in shaping agricultural productivity. The present analysis has undergone rigorous validation procedures, instilling confidence in its applicability for planning and management directives. Owing to its favorable physical and climatic attributes, a substantial swath of Bankura

Table 10 Error matrix: accuracy assessment of the classified agriculture suitability classes

	Reference classes					Total sample	User's accuracy (%)
	Very highly suitable	Highly suitable	Moderately suitable	Marginally suitable	Currently not suitable		
Classified classes							
Very highly suitable	47	03	01	00	00	51	**90.84**
Highly suitable	01	36	02	00	0	39	**90.05**
Moderately suitable	00	02	48	03	0	53	**85.71**
Marginally suitable	00	00	03	49	3	55	**89.07**
Currently not suitable	00	00	00	01	34	35	**95.45**
Total sample	**48**	**41**	**54**	**53**	**37**	**233**	
Producer's accuracy (%)	**94.71**	**85.41**	**86.27**	**89.33**	**92.22**		
Overall accuracy	**92.14**						

district exhibits suitability for agricultural production, notably concentrated in the central and eastern sectors of the Khatra subdivision, the southern reaches of the Bishnupur subdivision, and the northern and southern peripheries. Conversely, areas characterized by open rock outcrops, dense vegetation cover, high concentrations of built-up infrastructure, presence of water bodies, extensive bare ground, sandbars, low to very low annual precipitation levels, and inadequate irrigation infrastructure, predominantly evident in the northeastern, southeastern, and northwestern sectors of Bankura Sadar; the northern, southeastern, and southern fringes of the Bishnupur subdivision; and throughout the entirety of the Khatra subdivision, pose limitations to agricultural suitability. Despite the district's ample land resources conducive to agricultural activities, unsustainable farming methodologies and population density contribute to land degradation and soil fertility depletion. Furthermore, the region's fragmentation of arable land into small, dispersed parcels renders it incapable of meeting the escalating demand for agricultural produce. Consequently, urgent measures are warranted to ameliorate the deteriorating land quality and foster alternative livelihood avenues through localized resource management initiatives. In areas deemed marginal or presently suitable for agricultural pursuits, the adoption of alternative methodologies such as agroforestry and precision agriculture emerges as imperative strategies to safeguard soil integrity. Nevertheless, the authors suggest that for optimal results in future research, it is advisable for researchers to consider additional factors that are pivotal for modeling agricultural land suitability in Bankura district. In conclusion, this current evaluation holds significant importance as a decision-support mechanism for guiding the expansion of agricultural land utilization strategies in the semi-arid region, ultimately addressing the crucial need to enhance food security.

Acknowledgements First and foremost, the authors would like to thank the Department of Geography and Applied Geography at the University of North Bengal for giving them the chance to conduct the study. Although the study wasn't specifically funded by the authors, it was finished during the UGC-JRF period. Moreover, the authors extend their heartfelt gratitude to the editor and two anonymous reviewers for their valuable comments and suggestions, which significantly contributed to enhancing the quality of the manuscript.

Author Contributions **PK** conceptualization, data curation, formal analysis, Investigation, methodology, software, validation, visualization, writing-original draft, writing-review & editing; **SD** formal analysis, writing-review & editing; **IRC** supervision, writing-review & editing.

Funding This research did not receive financial support from public, commercial, or non-profit organizations.

Data Availability All data utilized in this study are accessible to any interested party through appropriate request for further research purposes.

Conflict of Interest The authors affirm that there are no conflicts of interest to disclose.

Consent for Publication All authors have thoroughly reviewed the manuscript and have provided their consent for its submission.

References

Abbas Z, Jaber HS (2020) Accuracy assessment of supervised classification methods for extraction land use maps using remote sensing and GIS techniques. In: IOP Conference series: materials science and engineering, IOP Publishing, Vol 745, No 1, p 012166

Abdel Rahman MA, Natarajan A, Hegde R (2016) Assessment of land suitability and capability by integrating remote sensing and GIS for agriculture in Chamarajanagar district, Karnataka, India. Egypt J Remote Sens Space Sci 19(1):125–141. https://doi.org/10.1016/j.ejrs.2016.02.001

Abijith D, Saravanan S, Singh L, Jennifer JJ, Saranya T, Parthasarathy KSS (2020) GIS-based multi-criteria analysis for identifcation of potential groundwater recharge zones—a case study from Ponnaniyaru watershed, Tamil Nadu, India. HydroResearch 3:1–14. https://doi.org/10.1016/j.hydres.2020.02.002

Akıncı H, Özalp AY, Turgut B (2013) Agricultural land use suitability analysis using GIS and AHP technique. Comput Electron Agric 97:71–82. https://doi.org/10.1016/j.compag.2013.07.006

Akpoti K, Kabo-bah AT, Zwart SJ (2019) Agricultural land suitability analysis: state-of-the-art and outlooks for integration of climate change analysis. Agric Syst 173:172–208. https://doi.org/10.1016/j.agsy.2019.02.013

Alemayehu S, Ayana EK, Dile YT, Demissie T, Yimam Y, Girvetz E, Worqlul AW (2020) Evaluating land suitability and potential climate change impacts on alfalfa (Medicago sativa) production in ethiopia. Atmosphere 11(10):1124. https://doi.org/10.3390/atmos11101124

Alkaradaghi K, Ali SS, Al-Ansari N, Laue J, Chabuk A (2019) Landfll site selection using MCDM methods and GIS in the Sulaimaniyah Governorate. Iraq. Sustainability Ayoade, J.O. (2004). Introduction to climatology for the tropics. Spectrum Book Limited, Ibadan. 11(17), 4530
Arulbalaji P, Padmalal D, Sreelash K (2019) GIS and AHP techniques based delineation of groundwater potential zones: a case study from southern western Ghats, India. Sci Rep 9:1–17. https://doi.org/10.1038/s41598-019-38567-x
Atalay I (2006) Toprak Oluşumu, Sınıfandırılması Ve Coğrafyası, Meta Basım, Baskı
Bale C, Williams J, Charley J (1998) The impact of aspect on forest structure and floristics in some Eastern Australian sites. For Ecol Manage 110:363–377. https://doi.org/10.1016/S0378-1127(98)00300-4
Bandyopadhyay S, Jaiswal RK, Hegde VS, Jayaraman V (2009) Assessment of land suitability potentials for agriculture using a remote sensing and GIS based approach. Int J Remote Sens 30(4):879–895. https://doi.org/10.1080/01431160802395235
Banerjee P, Ghose MK, Pradhan R (2020) Analytic hierarchy process based spatial biodiversity impact assessment model of highway broadening in Sikkim Himalaya. Geocarto Int 35(5):470–493. https://doi.org/10.1080/10106049.2018.1520924
Banko G (1998) A review of assessing the accuracy of classifications of remotely sensed data and of methods including remote sensing data in forest inventory
BDP Report (2011) Bankura district profle, Govt. of West Bengal, Retrieved on 25th June 2023, from https://bankura.gov.in/document-category/district-profile/
Bhattacharya S, Das S, Das S, Kalashetty M, Warghat SR (2021) An integrated approach for mapping groundwater potential applying geospatial and MIF techniques in the semiarid region. Environ Dev Sustain 23(1):495–510. https://doi.org/10.1016/j.envc.2021.100396
Biswas S, Mukhopadhyay BP, Bera A (2020) Delineating groundwater potential zones of agriculture dominated landscapes using GIS based AHP techniques: a case study from Uttar Dinajpur district, West Bengal. Environ Earth Sci. https://doi.org/10.1007/s12665-020-09053-9
Census of India (2011). District census handbook: Bankura Village and town directory (Part XII-A). Directorate of Census Operations West Bengal. Government of India, Retrieved from https://censusindia.gov.in/nada/index.php/catalog/1364
Chakraborty K, Mistri B (2015) Importance of soil texture in sustenance of agriculture: a study in Burdwan-I CD Block, Burdwan, West Bengal. Eastern Geographer 21:475–482
Chowdhury A, Jha MK, Chowdary VM (2010) Delineation of groundwater recharge zones and identifcation of artifcial recharge sites in West Medinipur district, West Bengal, using RS, GIS and MCDM techniques. Environ Earth Sci 59(6):1209–1222. https://doi.org/10.1007/s12665-009-0110-9
Congalton RG (1991) A review of assessing the accuracy of classifications of remotely sensed data. Remote Sens Environ 37(1):35–46. https://doi.org/10.1016/0034-4257(91)90048-B
Cook SE, Fisher MJ, Andersson MS, Rubiano J, Giordano M (2009) Water, food and livelihoods in river basins. Water Int 34(1):13–29. https://doi.org/10.1080/02508060802673860
Das S, Choudhury MR, Nanda S (2013) Geospatial assessment of agricultural drought (a case study of Bankura District, West Bengal). Int J Agric Sci Res 3(2):1–27
Das M, Parveen T, Ghosh D, Alam J (2021) Assessing groundwater status and human perception in drought-prone areas: a case of Bankura-I and Bankura-II blocks, West Bengal (India). Environ Earth Sci. https://doi.org/10.1007/s12665-021-09909-8
Datye VS, Gupte SC (1984) Association between agricultural land use and physico- socio-economic phenomena: a multivariate approach. Trans Inst Ind Geogr 6(2):61–72
DCHB (2011) District census handbook, Bankura District. Retrieved on 15th June 2023, from https://censusindia.gov.in/nada/index.php/catalog/1364
De La Rosa D, Moreno JA, Garcia LV, Almorza J (1992) MicroLEIS: a microcomputer-based Mediterranean land evaluation information system. Soil Use Manag 8:89–96
DSHB (2014) District statistical handbook, Bankura District, West Bengal, Government of India. Retrieved from http://wbpspm.gov.in/publications/District%20Statistical%20Handbook

Estrada LL, Rasche L, Schneider UA (2017) Modeling land suitability for Cofea arabica L. in Central America. Environ Model Softw 95:196–209. https://doi.org/10.1016/j.envsoft.2017.06.028

Everest T, Sungur A, Özcan H (2021) Determination of agricultural land suitability with a multiple-criteria decision-making method in Northwestern Turkey. Int J Environ Sci Technol 18(5):1073–1088. https://doi.org/10.1007/s13762-020-02869-9

FAO (1976) A framework for land evaluation. Italy

FAO (1993) Guidelines for land-use planning. FAO development series I, FAO, Rome

FAO (2007) Land evaluation, towards a revised framework. Land and Water Discussion Chapter 6. Rome: FAO Electronic publishing division

Feizizadeh B, Blaschke T (2012) Land suitability analysis for Tabriz County, Iran: a multi-criteria evaluation approach using GIS. J Environ Plann Manage 2012:1–23. https://doi.org/10.1080/09640568.2011.646964

Hui F, Xu B, Huang H, Yu Q, Gong P (2008) Modelling spatial-temporal change of Poyang Lake using multitemporal Landsat imagery. Int J Remote Sens 29(20):5767–5784. https://doi.org/10.1080/01431160802060912

Hussein AF, Jameel BI, Abd KK (2018) Comparative analysis of Fuzzy MCDM methods for material selection in biomedical application. Assoc Arab Univ J Eng Sci 25(2):137–148

Ibrahim-Bathis K, Ahmed SA (2016) Geospatial technology for delineating groundwater potential zones in Doddahalla watershed of Chitradurga district, India. Egypt J Remote Sens Space Sci 19(2):223–234. https://doi.org/10.1016/j.ejrs.2016.06.002

Iizumi T, Ramankutty N (2015) How do weather and climate infuence cropping area and intensity. Glob Food Sec 4:46–50. https://doi.org/10.1016/j.gfs.2014.11.003

Kapluhan E (2013) Drought in Turkey and efect of drought on agriculture. Marmara Geogr Mag ISSN, 1303–2429

Kayal P, Majumder S, Chowdhury IR (2022) Modeling the spatial pattern of potential groundwater zone using MCDM-AHP and geospatial technique in sub-tropical plain region: a case study of Islampur sub-division, West Bengal India. Sustain Water Resour Manag 8(6):185. https://doi.org/10.1007/s40899-022-00759-1

Kazemi A, Attari MYN, Khorasani M (2016) Evaluating service quality of airports with integrating TOPSIS and VIKOR under fuzzy environment. Int J Serv Econ Manag 7(2–4):154–166

Kim S, Lee W, Shin K, Kafatos M, Seo D, Kwak H (2011) Comparison of spatial interpolation techniques for predicting climate factors in Korea. For Sci Technol 6:97–109. https://doi.org/10.1080/21580103.2010.9671977

Kraemer HC (2014) Kappa coefcient. Wiley StatsRef: statistics reference online. Wiley, pp 1–4. https://doi.org/10.1002/9781118445112.stat05301.pub2

Kumar A, Pramanik M, Chaudhary S, Negi MS (2021) Land evaluation for sustainable development of Himalayan agriculture using RS-GIS in conjunction with analytic hierarchy process and frequency ratio. J Saudi Soc Agric Sci 20(1):1–17. https://doi.org/10.1016/j.jssas.2020.10.001

Mandal P, Mandal S, Halder S, Paul S (2021) Assessing and mapping cropland suitability applying geospatial and MIF techniques in the semiarid region with an integrated approach. Arab J Geosci 14(18):1–19. https://doi.org/10.1007/s12517-021-08171-3

Mangan P, Pandi D, Haq MA, Sinha A, Nagarajan R, Dasani T, Keshta I, Alshehri M (2022) Analytic hierarchy process based land suitability for organic farming in the arid region. Sustainability 14:4542. https://doi.org/10.3390/su14084542

Molla SH (2022) Investigating the suitability for rice cultivation using multi-criteria land evaluation in the Sundarban region of south 24 Parganas District, West Bengal, India. J Indian Soc Remote Sens 50(2):359–372. https://doi.org/10.1007/s12524-021-01441-3

Mustafa AA, Singh M, Sahoo RN, Ahmed N, Khanna M, Sarangi A (2011) Land suitability analysis for different crops: A multi-criteria decision-making approach using remote sensing and GIS. Researcher 3(12):61–84

Nag SK, Das S (2017) Assessment of groundwater quality from Bankura I and II Blocks, Bankura District, West Bengal, India. Appl Water Sci 7:2787–2802. https://doi.org/10.1007/s13201-017-0530-8

Nganga WB, Ng'etich KO, Macharia MJ, Kiboi NM, Adamtey N, Ngetich KF (2020) Multi-infuencing-factors' evaluation for organic-based soil fertility technologies out-scaling in Upper Tana Catchment in Kenya. Sci Afr 7:e00231. https://doi.org/10.1016/j.sciaf.2019.e00231
Orhan O (2021) Land suitability determination for citrus cultivation using a GIS—based multi-criteria analysis in Mersin Turkey. Comput Electron Agric 190:106433. https://doi.org/10.1016/j.compag.2021.106433
Patil MB, Desai CG, Umrikar BN (2012) Image classifcation tool for land use/land cover analysis: a comparative study of maximum likelihood and minimum distance method. Int J Geol Earth Environ Sci 2(3):189–196
Patra U, Gayak J, Khan K, Karim SK, Halder S, Sen A, Paul G (2021) A comparative study of tribal education development in India: with special references to Purulia District, West Bengal. J Critical Rev, 110–121
Pramanik MK (2016) Site suitability analysis for agricultural land use of Darjeeling district using AHP and GIS techniques. Model Earth Syst Environ 2(2):56. https://doi.org/10.1007/s40808-016-0116-8
PRB (2018) Population reference bureau (PRB), retrieved on 24th July 2023, from https://www.prb.org/2018-world-population-datasheet-with-focus-on-changing1age-structures
Rajasekhar M, Sudarsana RG, Nanabhau K, Ramachandra M, Pradeep KB (2021) Multi-criteria land suitability analysis for agriculture in semi-arid region of Kadapa District, Southern India: geospatial approaches. Remote Sens. Land 5:59–72. https://doi.org/10.21523/gcj1.2021050201
Romano G, Dal Sasso P, Liuzzi GT, Gentile F (2015) Multi-criteria decision analysis for land suitability mapping in a rural area of Southern Italy. Land Use Policy 48:131–143. https://doi.org/10.1016/j.landusepol.2015.05.013
Roy S, Hazra S, Chanda A, Das S (2020) Assessment of groundwater potential zones using multi-criteria decision-making technique: a micro-level case study from red and lateritic zone (RLZ) of West Bengal India. Sustain Water Resour Manag 6(4):1–14. https://doi.org/10.1007/s40808-021-01174-9
Roy S, Bose A, Chowdhury IR (2021a) Flood risk assessment using geospatial data and multicriteria decision approach: a study from historically active food-prone region of Himalayan foothill India. Arab J Geosci 14(11):1–25. https://doi.org/10.1007/s12517-021-07324-8
Roy S, Bose A, Mandal G (2021b) Modeling and mapping geospatial distribution of groundwater potential zones in Darjeeling Himalayan region of India using analytical hierarchy process and GIS technique. Model Earth Syst Environ. https://doi.org/10.1007/s40808-021-01174-9
Roy S, Singha N, Bose A, Basak D, Chowdhury IR (2022) Multi-influencing factor (MIF) and RS–GIS-based determination of agriculture site suitability for achieving sustainable development of Sub-Himalayan region, India. Environ Dev Sustain. https://doi.org/10.1007/s10668-022-02360-0
Saaty TL (1980) The analytic hierarchy process: planning, priority setting, resource allocation. McGraw-Hill International Book Co., New York
Saaty TL (1990) How to make a decision: the analytic hierarchy process. Eur J Oper Res 48(1):9–26. https://doi.org/10.1016/0377-2217(90)90057-I
Saha S, Sarkar D, Mondal P, Goswami S (2021) GIS and multi-criteria decision-making assessment of sites suitability for agriculture in an anabranching site of sooin river India. Model Earth Syst Environ 7(1):571–588. https://doi.org/10.1007/s40808-020-00936-1
Sarkar B, Das P, Islam N, Basak A, Debnath M, Roy R (2021) Land suitability analysis for paddy crop using GIS-based Fuzzy-AHP (F-AHP) method in Koch Bihar district West Bengal. Geocarto Int. https://doi.org/10.1080/10106049.2021.2007299
Senapati U, Das TK (2021) Assessment of basin-scale groundwater potentiality mapping in drought-prone upper Dwarakeshwar, River basin West Bengal, India, using GIS-based AHP techniques. Arab J Geosci 14(1):960. https://doi.org/10.1007/s12517-021-07316-8
Shailaja G, Kadam AK, Gupta G, Umraikar BN, Pawar NJ (2018) Integrated geophysical, geospatial and multiple-criteria decision analysis techniques for delineation of groundwater potential zones

in a semi-arid hard-rock aquifer in Maharashtra India. Hydrogeol J 27(2):639–654. https://doi.org/10.1007/s10040-018-1883-2

Shim JP, Warkentin M, Courtney JF, Power DJ, Sharda R, Carlsson C (2002) Past, present and future of decision support technology. Decis Support Syst 33:111–126

Sonia S, Verma RK, Ghosh T, Gaur RK (2023) Agricultural land suitability categorization and evaluation using GIS assisted AHP in the Arid Western Plain Zone of Rajasthan, India. Indian J Agric Res, 1, 11. https://doi.org/10.18805/IJARe.A-6085

Sow M, Mbow C, Hély C, Fensholt R, Sambou B (2013) Estimation of herbaceous fuel moisture content using vegetation indices and land surface temperature from MODIS data. Remote Sens 5(6):2617–2638. https://doi.org/10.3390/rs5062617

Sutradhar A, Chattaraj M (2019) Identification of agricultural productivity regions and its major determinants: a case study from Bankura district, West Bengal. Int J Res Soc Sci 9(5):661–678

Tashayo B, Honarbakhsh A, Akbari M, Eftekhari M (2020a) Land suitability assessment for maize farming using a GIS-AHP method for a semi- arid region Iran. J Saudi Soc Agric Sci. https://doi.org/10.1016/j.jssas.2020.03.003

Tashayo B, Honarbakhsh A, Azma A, Akbari M (2020b) Combined fuzzy AHP–GIS for agricultural land suitability modeling for a watershed in southern Iran. Environ Manage 66(3):364–376. https://doi.org/10.1007/s00267-020-01310-8

Thapa RB, Murayama Y (2008) Land evaluation for peri-urban agriculture using analytical hierarchical process and geographic information system techniques: a case study of Hanoi. Land Use Policy 25(2):225–239. https://doi.org/10.1016/j.landusepol.2007.06.004

Wang S, Fu BJ, Gao GY, Yao XL, Zhou J (2012) Soil moisture and evapotranspiration of different land cover types in the Loess Plateau China. Hydrol Earth Syst Sci 16(8):2883–2892. https://doi.org/10.5194/hess-16-2883-2012

Worqlul AW, Dile YT, Jeong J, Adimassu Z, Lefore N, Gerik T, Clarke N (2019) Efect of climate change on land suitability for surface irrigation and irrigation potential of the shallow groundwater in Ghana. Comput Electron Agric 157:110–125. https://doi.org/10.1016/j.compag.2018.12.040

Xu H (2006) Modifcation of normalised diference water index (NDWI) to enhance open water features in remotely sensed imagery. Int J Remote Sens 27(14):3025–3033. https://doi.org/10.1080/01431160600589179

Yalew SG, Van Griensven A, Mul ML, van der Zaag P (2016a) Land suitability analysis for agriculture in the Abbay basin using remote sensing, GIS and AHP techniques. Model Earth Syst Environ 2(2):1–14. https://doi.org/10.1007/s40808-016-0167-x

Yalew SG, van Griensven A, van der Zaag P (2016b) AgriSuit: a web—based GIS MCDA framework for agricultural land suitability assessment. Comput Electron Agric 128:1–8. https://doi.org/10.1016/j.compag.2016.08.008

Yeh H, Cheng Y, Lin H, Lee C (2016) Mapping groundwater recharge potential zone using a GIS approach in Hualian River, Taiwan. Sustain Environ Res. https://doi.org/10.1016/j.serj.2015.09.005

Yohannes H, Soromessa T (2018) Land suitability assessment for major crops by using GIS-based multi-criteria approach in Andit Tid watershed Ethiopia. Cogent Food Agric 4(1):1470481. https://doi.org/10.1080/23311932.2018.1470481

Zolekar RB, Bhagat VS (2014) Use of IRS P6 LISS-IV data for land suitability analysis for cashew plantation in hilly zone. Asian J Geoinform 14(3):23–35

Zolekar RB, Bhagat VS (2015) Multi-criteria land suitability analysis for agriculture in hilly zone: remote sensing and GIS approach. Comput Electron Agric 118:300–321. https://doi.org/10.1016/j.compag.2015.09.016

Assessment of Water Risk and Feasibility of Rooftop Rainwater Harvesting: A Case Study

Satabdi Biswas, Satiprasad Sahoo, and Anupam Debsarkar

Abstract A significant share of rural households in Bengal has long been suffering from arsenic and iron contamination of drinking water producing adverse effects on health as well as socio-economic environment. The present study is focused on the assessment of water risk using Remote Sensing and GIS platforms and socio-economic primary data. The surface water deficit zoning map shows a very high deficit (46.07–51.99 mm) in the three study blocks. All the wells have a decline in groundwater (ranging from 1.40 to 4.25 mbgl), and in a few cases, negative recharge is also observed. The levels of As and Fe vary from 13 to 69.5 ppb and 0.3 to 6.77 ppm, respectively. 99.2% of surveyed rural households face multiple water-related risks. However, high rainfall (1719.9 mm) in the monsoon season has substantial potential for groundwater recharge. Households are willing to adopt RRWH if adequate support from the government is provided. Two indices, viz., water risk index (WRI) and capacity index (CI), have been developed based on 712 rural socio-economic household survey data. A binary logistic regression of WRI was carried out with a set of socio-economic variables, which were found to have a significant effect on WRI. The logistic regression showed 76.1% correct predictability of the supplied water risk. However, CI was not found to be dependent on WRI. The feasibility of rooftop rainwater harvesting (RRWH) was explored as an alternative to ensure the sustainable well-being of the affected population.

Keywords Surface water deficit zoning map · Socio-economic study · Binary logistic regression · Rooftop rainwater harvesting

S. Biswas (✉)
Department of Geography, Mrinalini Datta Mahavidyapith, North Dumdum, West Bengal, India
e-mail: satabdibiswas2009@gmail.com

S. Sahoo
International Center for Agricultural Research in the Dry Areas (ICARDA), Cairo, Egypt

A. Debsarkar
Department of Civil Engineering, Jadavpur University, Jadavpur, West Bengal, India

S. C. Pal and U. Chatterjee (eds.), *Surface, Sub-Surface Hydrology and Management*, Springer Geography, https://doi.org/10.1007/978-3-031-62376-9_29

1 Introduction

Finite freshwater resources have become the most critical component for man-environment security especially when water demands exceed the supply. Half of India's food is produced using groundwater abstracted by STWs, which is almost 230 km^3 of groundwater (85%) every year for agriculture, implying the largest use of the world's groundwater (UNESCO 2023). Free access to groundwater for high-yielding crop cultivation involving STW (30 m^3/hr) did not provide any cost-effective solution. The ease of access to the shallow aquifer, launching of the electrified pump, cheaper boring cost, availability of loans from a bank, and subsidized electricity, all are responsible for the unethical drafting of the groundwater by STWs (Biswas et al. 2020). While drafting groundwater, particularly for irrigational usage, many contaminants, e.g., arsenic, iron, and fluoride leach out from the subsurface to the surface causing a threat to public health (UNESCO 2023). The unethical and excessive drafting of groundwater and its multiple compounding effects further aggravate qualitative and quantitative water stress in the form of declining groundwater table, shortage of potable water, deterioration of surface water quality, increased contamination in groundwater, and loss of agricultural-economic-piscicultural productivity and all these ultimately lead to jeopardizing the ecosystem (Chowdhury el al. 2016; Jones et al. 2023). Half of the total withdrawal of groundwater is supplied for domestic uses. The inadequacy of fresh water is also reflected by the socio-economic profile of a region characterized by the vulnerability of human lives and livelihoods ending up with poverty (Bandyopadhyay and Basu 2017). To overcome these water risks the United Nations targeted to ensure supply of 20 lpcd of safe water as a Millennium Development Goal within 2030. Presently, a safe water supply is treated as a human right (UNETAD 2022). According to the Census of India (2011), more than 68.8% of people still live in rural areas, and most of them are engaged in agricultural activities. Rural India depends on groundwater to the extent of more than 80%. In West Bengal, only 33.69% of households have access to improved sources of drinking water (Roy et al. 2023). The entire southern part of West Bengal including the district of North 24 Parganas dominates calcium, magnesium, arsenic, and iron in groundwater with relatively higher pH values (7.42–9.07) (Kumar et al. 2021; Soren et al. 2023). In North 24 Parganas dependence on pond-based surface water and groundwater is quite common for all sectors. The main source of drinking water is hand tube wells which are usually 100–200 m away from the residences. Thus, high morbidity has happened by the outbreak of various water-borne diseases (Bandyopadhyay and Basu 2017). However, the toxicity of arsenic has caused arsenicosis and the problem has certainly been worsened under poor socio-economic conditions, illiteracy, and food habits (ADB 2018; Bhattacharya et al. 2019). North 24 Parganas has a huge rainfall, but wastage of rainwater is a common phenomenon due to lack of awareness. Artificial recharge by collected rainwater via a recharge pit into a shallow high-arsenic tube well (16 m deep) in Ashoknagar, Habra I block, North 24 Parganas, successfully reduced the concentrations of arsenic, in 2009 from an initial concentration of 128 pg/L in the tube well was reduced to 80 pg/L after

1 month (ADB 2020). Though a considerable amount of runoff happens in North 24 Parganas, which can be used for harvesting rainwater to overcome water scarcity, the scheme has not yet been implemented (Chowdhury et al. 2016). Understanding the severity of the problem the government of West Bengal undertook a drinking water sector improvement project to provide safe and reliable drinking water to nearly 1.65 million people in the selective arsenic-fluoride and salinity-affected selected areas. North 24 Parganas district was also selected as the district is one of the state's most arsenic-affected districts (ADB 2020). However, the state government installed a few arsenic mitigation plants which failed to perform satisfactorily in the long run. The reasons behind this failure were identified as (i) lack of participation of the affected community in the process of policymaking and implementation. (ii) Weak governance over groundwater control and conservation due to rampant use of shallow tube well (STW). (iii) Inadequate effort to popularize rainwater harvesting (RWH) as an alternative water supply. It is thus a need of the hour to implement the outcome of new scientific researches after prioritizing social structure, cultural behaviors, political limitations, and environmental changes in water management plans to protect fragile water resources (UNCTAD 2022; Abbas et al. 2023). There have been numerous researches recommending the widespread application of RWH coupled with a cost-effective purification system, model-based GIS technique, and suitable site and structure of RWH to overcome the crisis (Mahmoud et al. 2015; Maina and Raude 2016; Roy 2019; Sayl et al. 2020; Kumar et al. 2021; Karimi and Zeinivand 2021).

As such, no significant research has been made on the combination of surface water deficit zones, rainwater quality, and the socio-economic aspect of rooftop RWH (RRWH). In the present study, water hotspot areas have been identified over the recent period (2010–2020) along with different socio-economic factors influencing the extent of water scarcity considering rural households of different arsenic-iron contaminated areas, i.e., Amdanga, Barasat II, and Basirhat I of North 24 Parganas District of West Bengal.

However, RWH can play a key role as a viable alternative for groundwater across North 24 Parganas. The RRWH does not only depend on bio-physical phenomena but also a lot on social attitude and acceptability for its success (Adham et al. 2018).

Thus, the objectives of this chapter can be summarized as (i) to assess the feasibility of RRWH concerning water deficit zone analysis and (ii) to evaluate the socio-economic feasibility of RRWH among rural households.

2 Study Area

Nearly 2.19 million people (42.4% of the district population) of North 24 Parganas are at high risk of arsenic poisoning (PHED 2018). Declining groundwater with seasonal water logging is a common phenomenon in this district (Table 1.1). The Ministry of Drinking Water and Sanitation, Government of India, tested 47,062 groundwater samples collected from 23 blocks of North 24 Parganas throughout (2014–2017) and found that 8,609 (18.3%) exceeded the safe arsenic limit of 10 μg/L. Samples mostly collected from Baduria, Basirhat I, Gaighata, Habra I, and Swarupnagar blocks had arsenic concentrations greater than 10 μg/L. Thus, the ministry planned to implement an arsenic-free piped water supply scheme (groundwater- or surface water-based) for a large number of districts, including North 24 Parganas (ADB 2020).

The selection of the study area was based on the extent of arsenic contamination in groundwater, i.e., Barasat II, Basirhat I, and Amdanga, as per data provided by Central Ground Water Board (CGWB), CGWB Kolkata (2021). Villages belonging to these blocks were selected based on 15 years of tube well data provided by the State Water Investigation Directorate (SWID) (Fig. 1.1). Twelve selected wells had a highly declining trend of groundwater with arsenic and iron[1] contamination. The household survey was carried out using snowball sampling, where the surveyor received the answer for 'experience of harvested water' or 'suffering from arsenicosis' wherever possible. Seven hundred and twelve rural households (6.29% of current census households) scattered over 21 villages were surveyed. The details of the study area are given in Table 1.2.

The study area covering 372.12 km^2 of three blocks had two types of water distribution systems, i.e., predominantly groundwater-based and a few surface water-based piped water supply schemes (PWSS) introduced recently for arsenic-contaminated areas. More than 95% of the population under Amdanga and Barasat II relied on STWs for drinking purposes (Banerjee 2016). It was often the case that the governmental department (PHED) monitored the tube wells, and put a red cross to make the people aware of arsenic contamination, acknowledging the fact that the rural households had been consuming contaminated water for the last 40 years without having any alternative.

3 Materials and Methodology

The methodology consists of four phases: (i) data collection; (ii) identification of water deficit zones; (ii) assessment of water vulnerability/risk (i.e., calculation of indices based on household-based primary data to find out the water risk of rural households); (iii) binary logistic regression and implementation of RRWH (Fig. 1.2). Census data (2011) were used for the analysis of the pressure exerted by population growth over groundwater.

[1] The permissible limit of arsenic, iron, and pH are 0.01 ppm, 0.3 ppm, and 6.5–8.5 according to the BIS 10500 (2012).

Table 1.1 About the study block

No. of blocks of North 24 parganas	No of habitats	Population	Supplied piped water (%)	Arsenic affected habitations	Arsenic affected population	% Of arsenic-affected population	Annual rainfall (bcm)	Annual water resource from upstream runoff (bcm)	Annual replenishable groundwater with discharge (bcm)
All 23 blocks	7334	5,184,365	61.4	2699	2,196,158	42.4	1.91	50.33	1.93

Source Government of West Bengal, Public Health Engineering Department, Integrated Management Information System (IMIS), 2016; (ADB 2020)

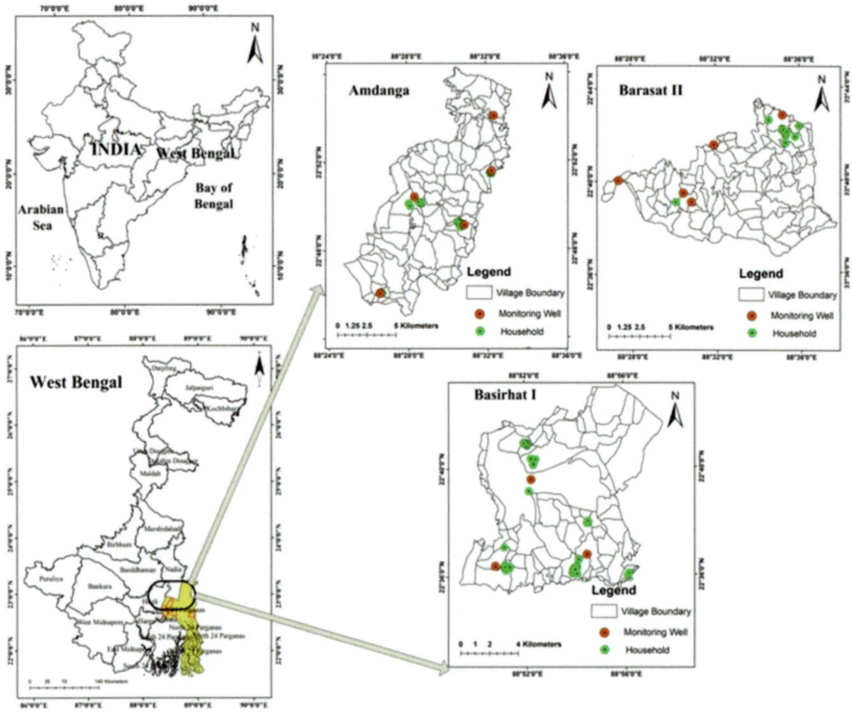

Fig. 1.1 Study area

Table 1.2 Study sites

Block	No. of surveyed villages	Households (census 2011) (for selected villages only)	No. of surveyed households (%)
Amdanga	6	4079	179 (25.1)
Barasat II	7	4439	247 (34.69)
Basirhat I	8	4001	286 (40.98)
Total	21	12,519	712

Source SWID (2001–2015), Census of India (2011), PHED (2001–2015), CGWB Kolkata (2021), Primary data (2019)

3.1 Identification of Water Scarcity

The extent of water scarcity has been identified following three steps mentioned underneath:

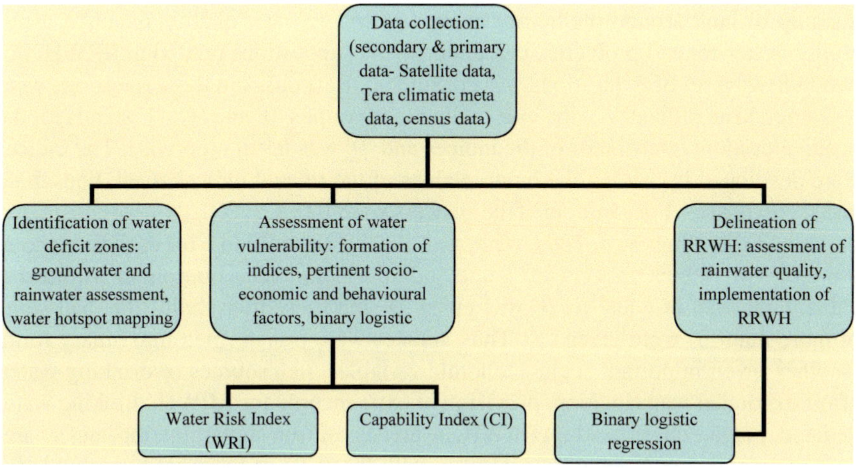

Fig. 1.2 Graphical representation of the methodology followed

3.2 Rainwater and Groundwater Assessment

Rainfall data were collected from the Indian Meteorological Department (IMD, Pune) for 45 years (1971–2015) and analyzed. Qualitative and quantitative assessments of the groundwater scenario were done based on the groundwater data collected from (SWID 2001–2015), PHED, and CGWB.

3.3 Identification of Hotspot Zones Using Terra Climatic Meta Data

A block-wise surface water deficit zoning map was developed for the identification of water deficit areas from the terra climatic monthly metadata ($0.5° \times 0.5°$ grid-wise resolution) (https://climatedataguide.ucar.edu). The terra climatic data included temperature, precipitation, evaporation, vapor pressure, runoff, soil moisture, etc.

3.4 Selection of Variables for Indices and Statistical Analysis

To find out water hazards and feasibility of RRWH among rural households, two indices, namely 'water risk index' (WRI) and 'capability index' (CI), were developed based on a perception-based questionnaire survey carried out in study blocks during 2019–2020. The questionnaire was about the respondents' background including the quantity of land around the home, sources of water, water quality, quantity, availability, water-related problems, perceptions, and suggestions related to RRWH. For two indices, a set of 6 questions was considered. Responses to these questions were converted to an indicator of the index with binary values ‘0’ and ‘1’. ‘1’ stands for the positive or more contributor of the indices and ‘0’ was taken otherwise. The indices were developed by adding the binary values of the related indices. Each household had six responses. For example, if the answers were 1, 0, 1, 1, 0, 0, then we considered the sum of the values as 3. Thus, we had a value between 0 and 6 for each household. We took two values as a cut-off value and considered each household within that value. Less than or equal to ‘3’ was given ‘0’ and those households that had some of more than ‘3’ were given ‘1’. Thus, indices were transformed into binary form. Another seven important socio-economic variables, i.e., sources of drinking water, effort to combat waterlogging, provision of extra income for RRWH, drinking water demand, preparedness to adopt RRWH, awareness of iron contamination, and awareness of arsenic contamination were used with the WRI of the rural households for binary logistic regression. The interrelation between two indices and other variables was explored, and the results showed significant dependency. All the independent variables were strongly associated with dependent variables in the chi-square test (χ^2). We considered logistic regression to assess the combined effect of some socio-economic variables on the ‘water risk index’ (WRI). In the WRI, different sources of drinking water were considered, viz., PHE tube well, personal tube well, groundwater withdrawal using a personal pump, dug well. Variables like availability of water in summer, occurrences of water-borne diseases, occurrences of arsenicosis, outbreaks of mosquito-related diseases, provision of a concrete roof, availability of spaces for a storage facility, ownership of a house, possession of water tank, and experiences of rainwater use were clubbed into CI. The questionnaire was coded using SPSS software version 1.5.

4 Result and Discussion

4.1 Rainfall

The distribution of 45 years of historical rainfall data indicates that rainfall in the study area had been high during monsoon and significantly low in pre- and post-monsoon. The ranges of the departure and cumulative departure from yearly average rainfall were (1058.40–2108.31) mm and (2012.91–3635.56) mm, respectively, while the average rainfall was 1719.91 mm. The average monsoon and post-monsoon rainfall were obtained as 1289.01 $\pm$ 252.12 mm and 205.62 $\pm$ 94.52 mm, respectively. Average rainfall during monsoon and post-monsoon seasons varied from (921.03–1772.14) mm and (32.9–466.97) mm. The trend line of the average monsoonal rainfall has a positive slope, implying an increasing trend.

4.2 Groundwater Profile

The quantity of groundwater drafted by big diameter STW (discharge 20 m^3/hr) in Amdanga, Barasat II, and Baisrhat I blocks was 1964, 2318, and 1991 ha.m, respectively, out of the district total of 54,095 ha.m, which was mainly utilized for winter rice (Boro paddy) cultivation purpose (PHED 2018). The depth of the groundwater varied from (5–11) m below ground level (mbgl). The quantity of the net groundwater available was comparatively low for Barasat II. However, the quantity of groundwater drafting was found to be the maximum for agricultural pursuit (3813.80 ha.m) for this block. The net available groundwater for future usage was estimated as 27.14% indicating an alarming scenario. In the cases of Amdanga and Basirhat I blocks, 41.05% and 45.27% of groundwater have been estimated to be available, respectively, implying safe access for future drafting.

4.3 Pre-Monsoon and Post-Monsoon Water Level

Well-wise yearly SWID and CGWB data (2001–2015) are used for quantitative analysis of groundwater over the study area. In Fig. 1.3, well numbers (W1–W5), (W6–W9), and (W10–W12) are located in Amdanga, Barasat II, and Basirhat I block, respectively.

All the wells had a sharp decline in groundwater levels during the post-monsoon season with the average value ranging from 4.25 to 1.40 mbgl. The highest yearly fluctuation of well water levels was found to be more than 4.25 m bgl in W1. The yearly average fluctuation was observed to be 2.75 mbgl for four wells (W6 to W9) of the Barasat II block. A sharp decline was observed during the post-monsoon season, which happened after recharge during the monsoon. A negative declining trend was observed during the post-monsoon season which varied from 3.60 to 1.00 mbgl in seven wells (W2, W3, W4, W5, W6, W7, and W12) for the all-study blocks. However, W2 and W4 remained non-functional throughout 2004–2008 and 2013–2014, respectively, and incidentally were revived again. All wells were observed to have a gradually widened gap between pre- and post-monsoon well water levels throughout 2011–2015. Another set of groundwater monitoring data is shown in Table 1.3, indicating the same decline trend of water level.

4.4 Groundwater Quality

The quality of well water is largely degraded due to the presence of arsenic in the shallow depth. Arsenic, iron, and pH were the matters of real concern for the study area out of the listed standard water-quality parameters (Table 1.4).

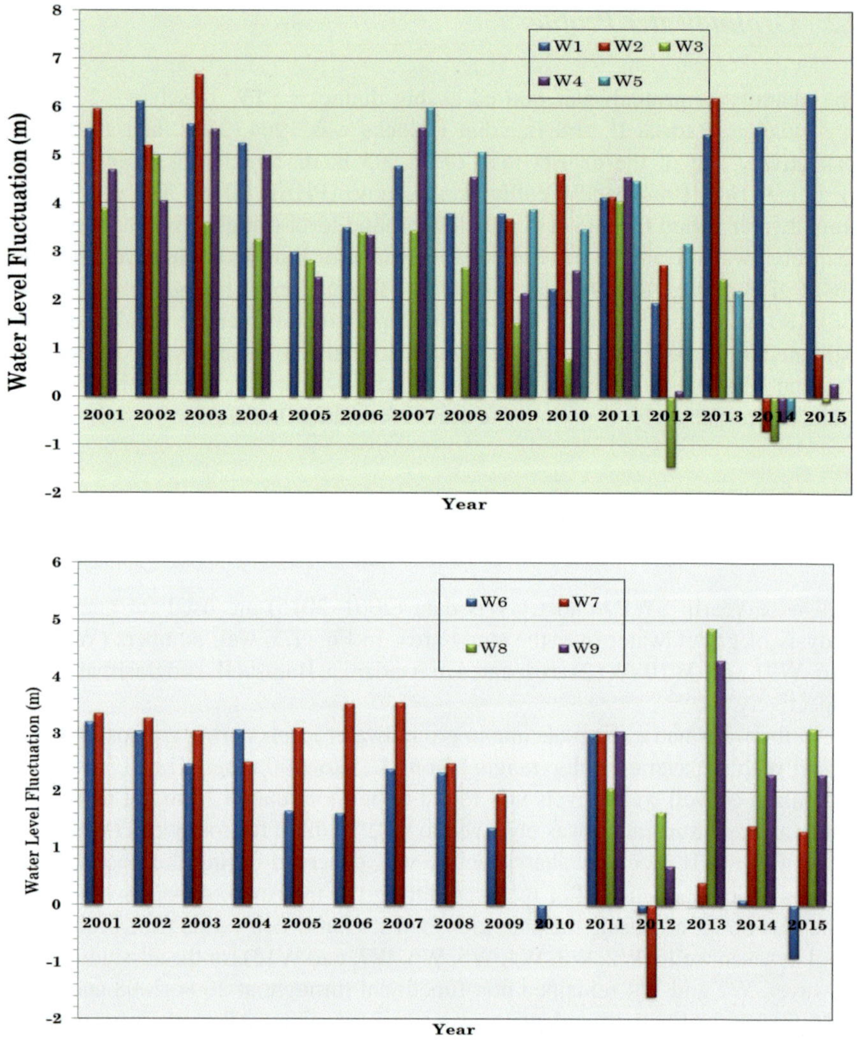

Fig. 1.3 Well water level fluctuation in Amdanga, Barasat II, and Basirhat I (2001–2015) (Source: SWID)

The pH level of all the wells of the three blocks fluctuated marginally over the period 2001–2015 during both pre-monsoon and post-monsoon seasons. It had been within the range specified by IS 10500–2012 (i.e., between 6.5 and 8.5) but on the higher side especially in post-monsoon samples as observed for W11 and W12. Iron concentration for all the wells was found above the Indian standards (0.3 mg/l). Iron levels increased during post-monsoon compared to pre-monsoon for wells W1, W5, W8, W9, and W12, which is a little bit unusual. In the case of arsenic, all wells had a concentration over the safe limit (i.e., 0.01 ppm) as specified in IS: 10,500.

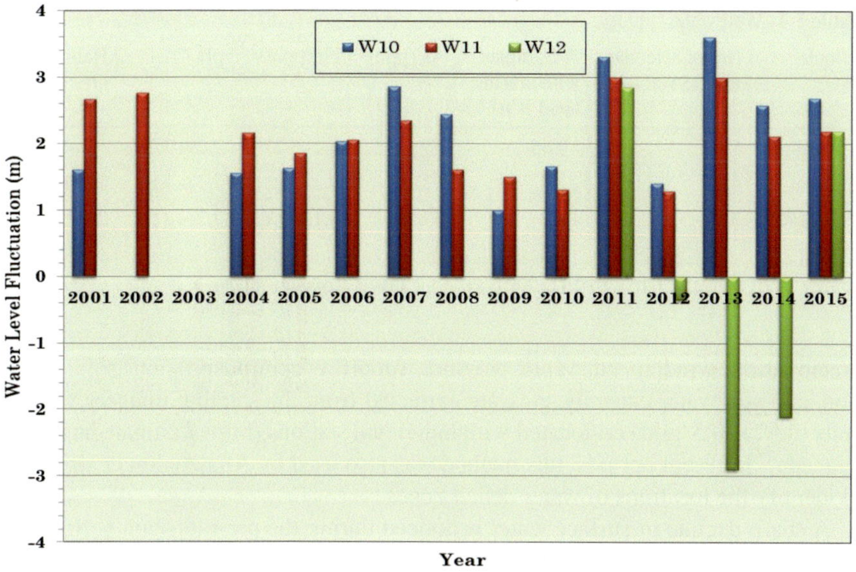

Fig. 1.3 (continued)

Table 1.3 Water level depth of groundwater monitoring wells of CGWB

Block	Location	Type of monitoring well	Post monsoon (mbgl)	Pre monsoon (mbgl)
Barasat II	Barasat	Pz	14.45	16.55
	Digberia	TW	9.27	10.80
	Chowmoha	Pz	**1.36**	**5.67**
	Bagaband	TW	5.15	6.93
	Baidyapur	Tw	4.9	7.31
	Ganganagar	Pz	17.04	18.90
Basirhat I	Basirhat	Pz	5.8	4.5
Amdanga	Rafipur	Pz	**7.55**	**11.82**
	Sirasini	Pz	**3.20**	**12.40**
	Moriche	TW	4.92	6.97
	Sadhanpur	TW	**7.90**	**12.70**

Source CGWB, Kolkata (2017), Pz = Piezometric well, TW = Tube well)

4.5 Water Hotspot

The block-wise surface water deficit zoning map shows the locations of water hotspot areas (Fig. 1.4). Terra climatic data, viz., maximum and minimum temperature,

Table 1.4 Well water quality

Block	No. of selected well (SWID)	% Samples with arsenic level > 10	As (ppb)	Fe (ppm)	pH	Habitation affected
Amdanga	5	7.66	13	0.2–6.77	7.29–7.78	58 (< 50%)
Barasat II	4	22.7	125	0.3–2.0	7.38–7.81	135 (75%)
Basirhat I	3	41.9	69.5	0.13–0.71	7.25–8.51	234 (> 75%)

Source SWID (2001–2015), PHED (2001–2015), CGWB Kolkata (2021)

precipitation, wind speed, vapor pressure, runoff, evapotranspiration, solar radiation, and soil water capacity, etc., are extracted from the satellite imagery for grid cells (0.5° × 0.5°grid) co-located with individual stations *(*https://climatedataguide.ucar.edu*)*. In recent periods, this method was also used by Abatzoglou et al. (2017) to identify the locations of water deficit zones.

A sharp decline in surface water is noticed during the pre-monsoon period over the north to north-eastern portion of the Amdanga block. More than 50% area comes under high (45.94 mm–46.07 mm) to very high (46.07 mm–46.32 mm) surface water deficit categories. While the middle portion of the block falls under the medium (45.76 mm–45.94 mm) surface water deficit zone. In the Barasat II block, nearly 30% of the south-eastern part of the block belongs to the high (48.14 mm–48.63 mm) to very high (48.63 mm–49.68 mm) surface water deficit zone. Almost 60% area of the Basirhat I block falls under the medium (51.00–51.18 mm) to high (51.18 mm–51.38 mm) and very high (51.38 mm–51.99 mm) surface water deficit zone.

4.6 Findings of Socio-Economic and Binary Logistic Regression Analysis

The adequacy of water availability has various positive socio-economic impacts (UNCTAD 2022). Present society faces morbidity or mortality and deaths caused by various water risks (UNU-INWEH 2023).

4.7 Population Trend with Water Demand

Water demand varies directly with the population. The gap between the demand and the availability of water is expected to be challenging in the coming days. The real growth in water demand is reported to be location-specific (UNESCO 2023). In India, the domestic water demand in rural areas is considered as 55 lpcd (NRDWP 2013). The population and water demand for study blocks are shown in Fig. 1.5.

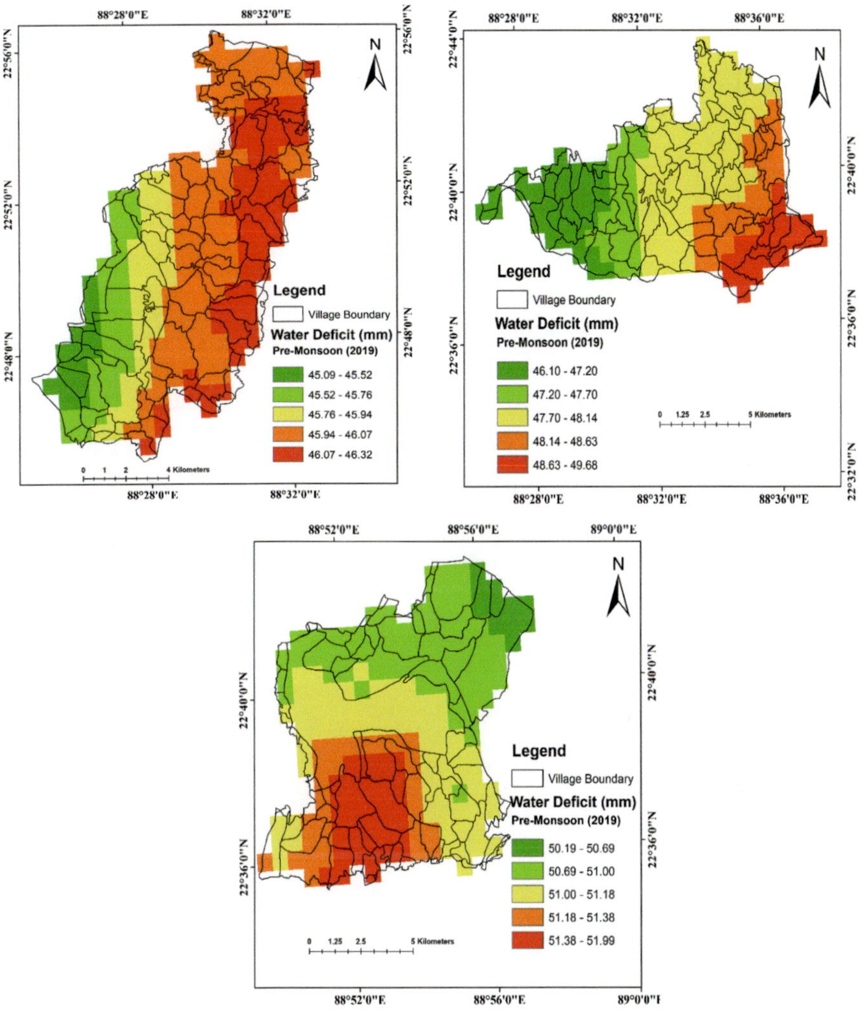

Fig. 1.4 Water hotspot maps for Amdanga, Barasat II, and Basirsat I blocks (2010–2020)

In terms of population, Barasat II block had the highest negative growth (–29.50%) during the period 1991–2001 as the major portion of the population shifted toward Madhyamgram municipal area (DHDR 2010). A similar negative growth was noticed for the Basirhat I block from 1981 to 1991 due to the establishment of the Basirhat Municipality (DHDR 2010). The maximum population density in Barasat II is expected to be 2487.27 persons/sq km with a water demand of 15,600,635 lpd in the year 2031. The recent decadal variation was recorded as 31,820 for the same block (Census 2011). The steep water demand would be undoubtedly very challenging for the study area. Thus, there was an urgent need for comprehensive synchronization of the socio-economic factors with RRWH.

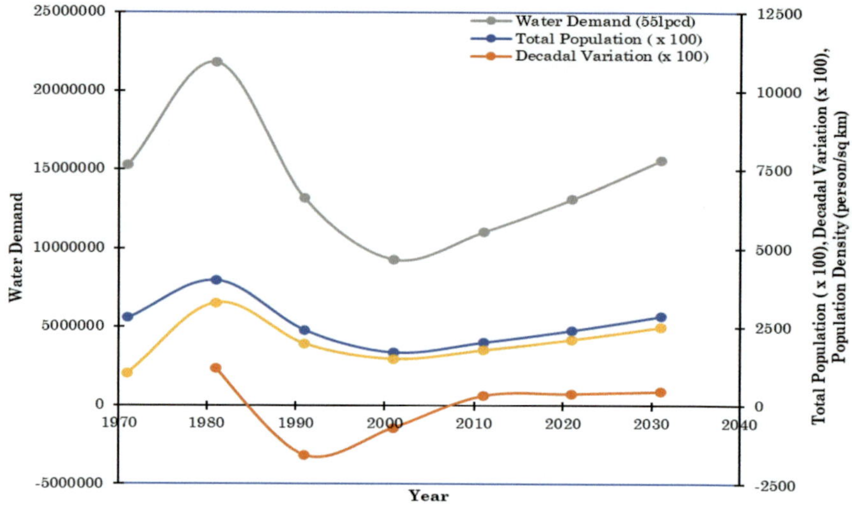

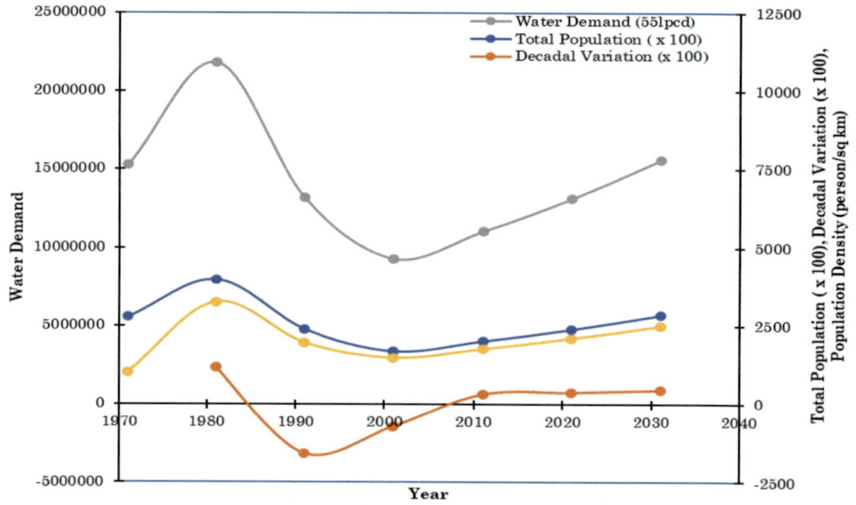

Fig. 1.5 Population, population density, decadal variation, and water demand of Amdanga, Barasat II, and Basirhat I Blocks, 1971–2031 (from the left-hand top corner in a clockwise direction) (Source: PHED and Census 1971, 1981, 1991, 2001 and 2011)

4.8 Socio-Economic Characteristics of the Households

The binary logistic regression was carried out to assess how rural households suffer from water risk and whether the rural households are willing to adopt RRWH or not (Table 1.5). The response variable WRI (y) was assumed to have binary characteristics, having values '0' and '1' representing low and acute WRI. We regressed the

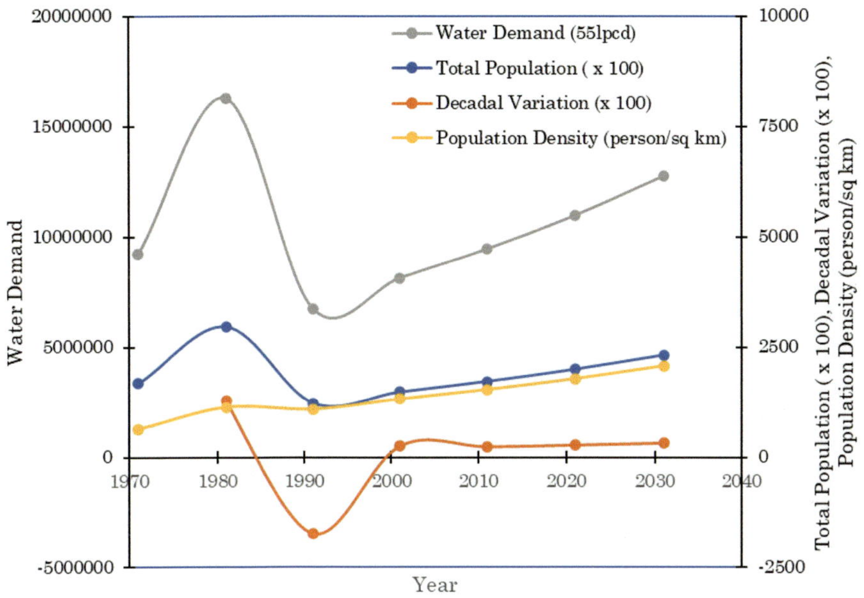

Fig. 1.5 (continued)

WRI on the seven important socio-economic variables along with CI. Each independent variable was taken as binary with '0' for low and '1' for high. Data were analyzed with binary logistic regression as judged by rural households.

Table 1.5 Results of logistic regression of 'water risk index' on the related independent variables

	B	S.E	Sig	Exp (B)
Sources of drinking water	2.024	0.221	0.000***	7.572
Effort to combat waterlogging	0.400	0.368	0.277	1.492
Provision of extra income for RRWH	– 0.736	0.370	0.047**	0.479
Drinking water demand	2.021	0.221	0.000***	7.549
Preparedness to adopt RRWH	0.523	0.281	0.063*	1.687
Capability index (CI)	– 0.080	0.194	0.680	0.923
Awareness of iron contamination	1.059	0.275	0.000***	2.884
Awareness of arsenic contamination	2.406	0.277	0.000***	11.095

Note -2log likelihood = 670.110, percentage of correct predictions: 76.1%
Dependent variable (Y) = 'Water risk index (WRI)',
*Significant at 10%, **Significant at 5% & ***Significant at 1%
Sources of drinking water: Safe = 1, Unsafe = 0; Effort to combat waterlogging: Yes = 1, No = 0; Provision of extra income for RRWH: Yes = 1, No = 0; Drinking water demand: Upto 20 l = 0, Above = 1, Preparedness to adopt RRWH: Yes = 1, No = 0; CI: Less = 0, More = 1, Awareness of arsenic contamination: Yes = 1, No = 0; Awareness of iron contamination: Yes = 1, No = 0

It is obvious from Table 1.5 that most of the variables expectedly affected the WRI with variables like sources of drinking water, provision of extra income for RRWH, drinking water demand, preparedness to adopt RRWH, awareness of iron contamination, awareness of arsenic contamination being significant at least at 10 percent level of confidence. Here, we consider water from personal shallow hand tube well, groundwater withdrawn using a motor pump, PHE-marked tube well water, un-standardized bottled water, groundwater withdrawn using a personal motor pump, dug well water as a source of unsafe drinking water (52.7%), and groundwater withdrawn using deep tube well of PHE as safe drinking water. Ownership of house, ownership of water tank, ownership of pond, provision of a concrete roof, any experience of using rainwater, availability of spaces for the development of storage facility, and exercising any effort to combat water-logging—these seven variables are clubbed together into CI. However, the CI is not found to be significantly correlated with WRI while the rest of the variables have a positive correlation with WRI. The reason behind this result can be explained as most households had their own house (95.6%), out of which 80.8% of residents had more than 101 sq. m land, 55.8% of surveyed households had concrete roofs, 9.1% had water tanks, 16.7% had ponds, and 8.7% gave the effort to combat waterlogging. However, none of them were aware of the proper process of RRWH. Regarding arsenic and iron contamination, the majority of surveyed population were found to be aware of the contamination but surprisingly they were found to be indifferent about the issue of water safety. Nearly, 99.2% of households expressed their concern about the prevalence of multiple water-related health issues. Thus, these four variables, i.e., sources of drinking water, water demand, awareness of arsenic contamination, and awareness of iron contamination, have a positive significant influence with a 1% level of significance. 'Provision of extra income for adopting RRWH' harmed WRI as they could not understand the importance of the indirect cost of supplied water. Thus, the provision of extra income for adopting RRWH and preparedness to adopt RRWH, these two variables were statistically significant with a 5% and 10% level, respectively. Overall the model had a 76.1% ability to predict the water risk of rural households.

4.9 Quality of Rainwater

Rainwater samples were collected from each study block in July 2021 (Fig. 1.6). The results of rainwater quality (Table 1.6) show that except pH all the parameters, e.g., TDS, chloride, and ammoniacal nitrogen, lay well within the safe limit prescribed under IS: 10,500 (2012).

The pH value of the collected rainwater samples appeared to be slightly acidic except for Basirhat I.

It can be concluded that rainwater available in the study area, i.e., Amdanga, Barasat II, and Basirhat I blocks of the district, can be utilized for various designated purposes.

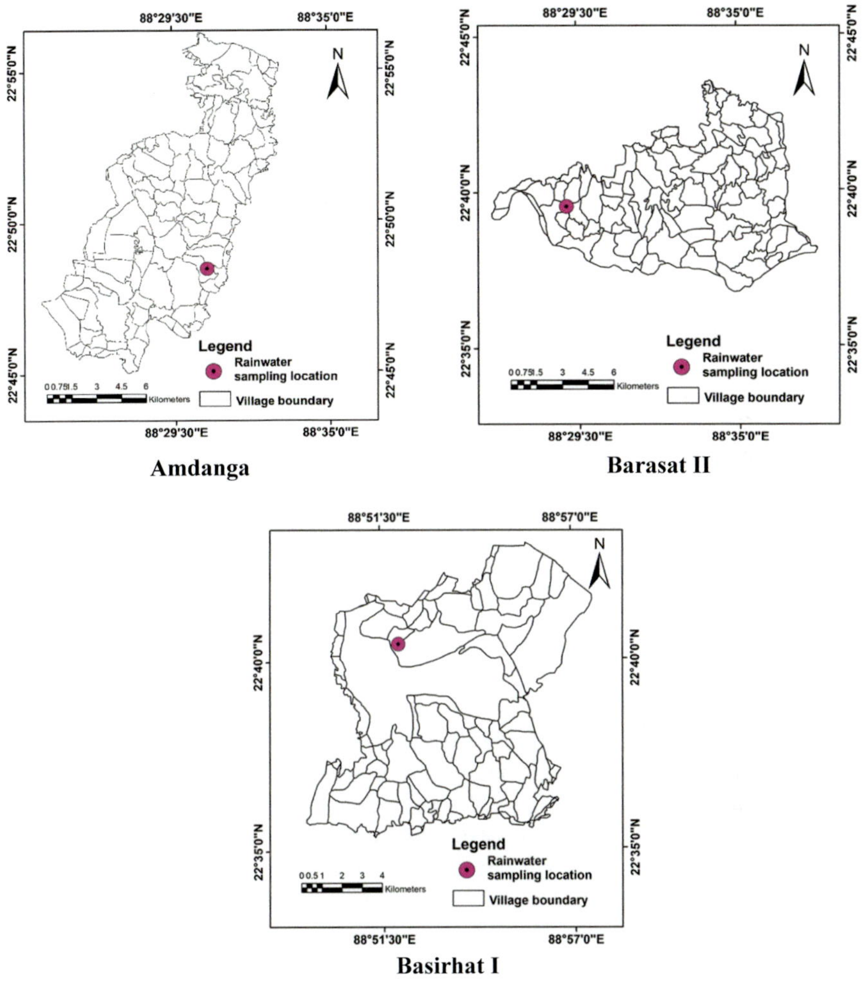

Fig. 1.6 Location maps for rainwater sampling points

5 Discussion

Bandyopadhyay and Basu (2017) reported the average annual rainfall of South 24 Parganas, adjacent to our study area, as 1762 mm. In this district, the monsoon usually spans from June to September and the rest of the year faces a water crisis. Chowdhury et al. (2016) used two empirical equations to calculate the amount of harvested water at an institute building based on normal rainfall of 1960 mm, Gosaba, South 24 Parganas. The harvested water could satisfy a supply of 70 lpcd. In our study, we found rainfall trends of six years, i.e., 1971, 1973, 1975, 1986, 1993, and 1999, showing the occurrence of greater rainfall in the monsoon period indicating favorable

Table 1.6 Rainwater quality

Parameters	WHO guideline (2008)	IS 10500 (2012)	Rainwater quality of the study area (July 2021)		
			Barasat II	Basirhat I	Amdanga
pH	6.5–8.5	6.5–8.5	6.01	6.59	6.06
TDS (mg/l)	–	500	BDL	22.0	16.0
Chloride (mg/l)	> (0.2–0.5) and < 5 mg/l	250	4.0	9.0	4.0
Sulfate (mg/l)	–	200	–	–	–
Ammonia-N (mg/l)	< 1.5 mg/l	0.5	0.15	< 0.1	0.24

Source primary data (2021)

recharge conditions. Few high values of rainfall in the post-monsoon season can also be pointed out corresponding to the years 1971, 1973, 1975, 1989, 1991, 1995, and 1998, The average recharge by the monsoonal rain was recorded at 4410.07 ha.m for 3 selected blocks. The rainfall distribution pattern implied a huge potential for RRWH in the study area. Administration of the South 24 Parganas has prioritized RWH all over the district through pond excavation, waterbody renovation, and water conservation through community participation as a part of the Mahatma Gandhi National Rural Employment Guarantee Scheme (Bandyopadhyay and Basu 2017). However, in our study, we found a major portion of the rainwater flowing down as runoff in the absence of an appropriate recharge system in the study areas.

The situation worsened from March to June when most of the hand tube wells failed to cater to water due to a fall in the well water level. Women and children were the worst victims, as they had to collect drinking water for their families (Bandyopadhyay and Basu 2017). Jasrotia et al. (2019) developed a groundwater model for the identification of suitable sites of recharge. Various thematic maps were developed using field data, pre and post-monsoon well data, satellite data, etc. The water table fluctuation was found (5–29 m) and (4–25 m) during the pre-monsoon and post-monsoon periods. However, the groundwater level was found declining due to easy excess to the shallow aquifer after having very high recharge (154–204 mm/year) by monsoon rain (Paul and Das 2021). In our present study, we found few wells got choked and incidentally were revived again, indicating an over-abstraction of groundwater with time. All well water levels were observed declining throughout (2011–2015). The highest fluctuation of well water level between pre- and post-monsoon was found more than 4 mbgl. This may be attributed to an increase in the agricultural area covered under boro and other winter crops along with pisciculture by an amount of 33.15% and 17.29% respectively during the 5th MI Census data compared to the 4th MI Census in the district (PHED 2018). All tube wells (TWs) except DTWs are supposed to run 6–10 h per day. However, they run more than this period and at the same time illegally rented drafting also happens. The average drafting depth of the tube wells up to 130 mbgl indicates drafting up to both 1st and 2nd aquifers, thus becoming responsible for decreasing the drawdown cone along with arsenic-iron contamination (Biswas et al. 2020). In our study area supply of safe drinking water is a critical issue. Generally, shallow aquifers are the reservoirs of harvested rainwater. Installations of cheap (< 50 USD) hand tube wells by a large number of households with three pipes and one filter (about 20 m) without maintaining the minimum safe distance (between the STWs or DTWs) have resulted in a sharp decline in water level. This over-drafting of groundwater has jeopardized the rural water supply (WaterAid 2011). In our study area, the situation was found more alarming in the case of W6, W7 of Barasat II, and W12 of Basirhat I block, where the water level declined significantly during the post-monsoon compared to the pre-monsoon level. During our survey, we noticed that rural households use groundwater for agriculture and pisciculture purposes year round, and illegal drafting of groundwater had become responsible for choking the wells. In this way, the Barasat II block has gradually attained the semi-critical stage in terms of 'groundwater development'.

Soren et al. (2023) calculated the spatio-temporal groundwater quality index (WQI) of drinking and irrigation water throughout (2011–2019) across the entire South Bengal. The results of WQI showed that 54.9% of the study area had 'poor', 'very poor', and 'not suitable for drinking' categories of water. According to the 'water suitability index' (WSI) results, they found 32.39% of samples were 'unsuitable' and 'doubtful to unsuitable' for the use of drinking and irrigation purposes. Moderately alkaline pH (7.42–9.07) was observed for the groundwater in the districts of North 24 Parganas, Purba Medinipur, Bankura, South 24 Parganas, and parts of Purulia, Burdwan, and Howrah exceeding the WHO and BIS standards. The inhabitants of these areas use two types of water resources, viz., groundwater and surface (pond) water even in the dry season for drinking purposes. The quality of the water was not found to satisfy the potable water quality (Bandyopadhyay and Basu 2017). The PHED reported arsenic concentration in Jayanagar, Budge Budge, and Mathurapur-II blocks of South 24 Parganas above the safe limit of 0.05 mg/l. The iron concentration in groundwater was found as 6.15 ppm and 3.96 ppm at Magrahat II and Jayanagar blocks, respectively, which is beyond the acceptable limit according to IS 10500:2012. Bhattacharya (2019) stated that the number of affected villages with more than 300 ppb arsenic concentration in tube well samples was found to increase in every consecutive survey in North 24 Parganas. Basu et al. (2003) found high arsenic concentration at 15 m depth and considered high risk for drinking water in Barasat II and Amdanga blocks. In our study, we found a direct relationship between arsenic and iron during the post-monsoon season (Table 1.4). Nearly 13.2% of the surveyed population suffered from arsenicosis, while 64.9% of households were still using shallow tube well for drinking purposes. Though a large number of respondents used to buy bottled water for drinking purposes which they considered reasonably safe, it hardly complied with the BIS standards. Thus, supplied or available water became a burden for rural households. As a whole, 75% of habitation was found to be affected by the high arsenic contamination in both the Barasat II and Basirhat I blocks while for Amdanga block it was comparatively low (50%). During the dry season and monsoon season, people use pond water, and due to lack of awareness water-borne diseases especially diarrhea increased in South 24 Parganas (Bandyopadhyay and Basu 2017). In our study, 16.7% of households

STW giving rent

Rich Households used a submersible pump

Poor Households used a submersible pump

Photo. 1.1 STWs are given on rent to farmers and households with submersible pumps fitted to STWs

had used ponds for domestic purposes. The percentage of households suffering from other major water-borne diseases like dysentery, diarrhea, jaundice, typhoid, and gastrointestinal problems was 65.7% and more than 8%, respectively.

Bandyopadhyay and Basu (2017) reported that irrigation water was not available free of cost. Thus, most of the cultivators could not afford irrigation due to a financial constraint. On the contrary, in our survey farmers of almost all-study blocks were found to use STW, LDTW, MDTW, HDTW, and DTW for boro and other winter cultivation. STWs were given rent on an hourly basis throughout the study blocks at a cheaper rate (Photo. 1.1). Wastage of groundwater is also an important issue in the agricultural sector.

In North 24 Parganas, the top layer from the surface is the clay layer (thickness 15–70 m), below that numerous alternative layers exist which are composed of fine to coarse sand-gravels-silt, and clay. Groundwater occurs in shallow depths under confined conditions (Bhu-Jal News 2009). Paul and Das (2021) found that the groundwater is very near to the surface (2–5 mbgl) and covers almost 56.33% of the North 24 Parganas. The southern, and south-western part of the district has a continuous patch of silt with a clay medium layer having homogeneous capillary permeability. Thus, there is a huge potential for groundwater recharge by RWH.

Karimi and Zeinivand (2021) developed a runoff coefficient map and runoff productivity map GIS-based model for identifying RWH sites and found almost entire areas had the potentiality of RWH. After storage of harvested rainwater, sufficient dosage of chlorine was provided to keep the pH range between 6.5 and 8.5 in Gosaba, South 24 Parganas (Chowdhury et al. 2016). Paul and Das (2021) developed a DRASTIC model, based on geological, topographical, and hydrological components on the GIS platform to find out the aquifer vulnerability of the North 24 Parganas district. The result showed that the maximum portion of the (85.27%) district such as Bagdah, Bongaon, Sandeshkhali-II, part of Minakhan and Hasnabad, Hingalganj, Barrackpore I, and II blocks were under moderately high to very high vulnerable zone. The land use and land cover change map showed the key factors for determining the subsurface water deficit in this area. They further reported that all arable land with STWs had been falling into a danger zone. The percolation tank played a vital role in the recharge of groundwater with a relatively flat slope (< 2%) (WaterAid 2011; Rana and Suryanarayana 2020). Bera and Chatterjee (2019), based on over 27 years of study (1990–2017) on LULC, identified the hotspot developed due to an increase in the built-up area by 22.59% in the same district. They identified the same problem for Rajarhat, Barasat, and Barrackpore. However, the increased proportion of water bodies happened due to the conversion of low-lying agricultural land into a shallow pond (Bheri) for profitable aquacultural activities. A huge number of brick kilns were constructed over unproductive agricultural land. In our study, the south-eastern part of Basirhat I block was found more vulnerable to surface water deficit. The entire eastern part of Amdanga block and the southern part of Barasat II block fell under a high-water deficit zone indicating a steady decline in water level and consequent urgency for artificial recharge after studying 'water hotspot maps'. Maina and Raude (2016) argued and found that 50% of the catchment area failed to find potential sites for rainwater harvesting after developing a digital elevation model (DEM) for the potential catchment of RWH. Mahmoud et al. (2015) developed a decision support system (DSS) model in a GIS platform, based on remote-sensing data, and filed a survey for an in-situ RWH (IWH) site. The result showed that 1.5 and 27.8% of the study area had excellent to good suitability for IWH. Karimi and Zeinivand (2021) suggested the exact location of RWH and these could be used for potable purposes, and other emergencies. They suggested that the final decision of making RWH structures depends on field research, property land issues, and socio-economic conditions, especially regional needs and capacity. Another study showed a 50% potential catchment area for RWH after developing a digital elevation model and thematic maps under a GIS platform. The study concluded that socio-economic aspects such as public willingness, needs, and preferences of local people should be considered before implementing RWH (Maina and Raude 2016). In our socio-economic study, we found near about 5% of respondents were found to be suffering from malaria, dengue, and chikungunya. Interestingly, we found that 67.3% and 52.4% of the respondents had been facing water logging and water insufficiency during the monsoon and summer seasons, respectively. However, 46.5% of the households admitted that they collect rain occasionally in any type of container. Thus, the collection of rainwater implied that households had already accepted rain as an alternative water source but were unaware of its utility. It was an important indicator of RRWH. While none of them knew whether rainwater was free from arsenic or iron or not. None of the surveyed households were aptly aware of the proper technique of RWH. Nearly 17.4% of the respondents agreed to initiate RRWH provided proper training is imparted free of cost. 19.9% of the female respondents mentioned the adoption of RRWH depends on the willingness of the male members of the family though a female member shoulders the sole responsibility of collection and use of water for the family. However, the test results of collected rainwater samples were found quite excellent for the use of rainwater for drinking purposes after adding disinfectants. The West Bengal Water Sector Improvement Project suggested a few pond-based water supply schemes for Barrackpore I (which covers 15 villages with 61,814 people according to 2011 census), River Hooghly-based water supply schemes for large number of blocks, viz., Habra I & II, Gaighata, part of Amdanga and Barrakpore I & II, Barasat I, Deganga and one groundwater-based water supply scheme for Barrackpore I & Barasat I but surprisingly no scheme for Barasat II (ADB 2020). How far the river Hooghly will supply safe water to numerous villages and municipalities without imposing any meterage system? Water is not always available to support competing demands of economic and social welfare. It is a necessary prerequisite to identify the patterns and trends of emerging water challenges (Tzanakakis et al. 2020; UNCTAD 2022). The focus should be subsequently given to the water governance and implementation of analytical research knowledge before the application of integrated water management to combat the issue of uncertainty in the quality of supplied drinking water (UNCTAD 2022; UNU-INWEH 2023). To address these threats, certain steps could be explored e.g., arranging for initial investments for the integrated water project, coordination and alignment of different policies, generation of public awareness, ensuring local participation and cooperation, strengthening institutional-professional capacity, regular monitoring of water quality, enforcement of laws, implementation of cross-sector programs, sharing data and information with the mass to achieve the full potential of RWH systems (UNCTAD 2022). Finally, emphasis is to be given to the social outcomes of the integrated water management plan (UNESCO 2023).

However, people should be made aware of the importance of RRWH after the eradication of non-technical barriers (UNCTAD 2022). The availability of safe water quality is only ensured when it is continuously monitored (UNCTAD 2022).

Rahman et al. (2014) reported the overall quality of harvested rainwater was quite satisfactory and encouraging for the households to adopt RRWH to achieve self-sufficiency in water availability.

In our study, the households, as a whole, expressed their willingness to adopt the RRWH, which can be established very well from the results of logistic regression analysis. Though there is a need for granting incentives from the Government side, the beneficiaries are supposed to contribute a minimum for the long-term management of the RRWH system. By introducing a rebate, the capital cost would be excluded and households are expected to adopt RRWH provided they receive proper training and maintenance (Bizoza and Umutoni 2012). As a recommendation, the government should offer a 10% waiver in property tax on the adoption of RRWH and vice versa for non-adoption (Barenhoff 2011). In our study, 33.3% of surveyed households are cultivators, daily laborers, or are engaged in any other low-wage occupations, who expressed their inability to bear the cost of RRWH. Though any kind of roof can be used for RWH, concrete roofs are considered, as two governmental schemes are presently running for the construction of concrete roofs for low-income households. More precisely, availability of spaces for a storage facility, possession of a water tank, provision of a concrete roof, previous experience of using rainwater, an effort to combat waterlogging, large water demand, water-borne diseases, and mosquito-related risks, water insufficiency in lean season all these factors are indirectly influencing the choice for RRWH. A significant portion of households (56.2%) were found to be suffering from acute WRI making it a key determinant for the adoption of RRWH. In the study area, there are no such provisions for water meters. It is noteworthy that the installation of smart meters and automated leakage prevention systems in 40,000 households in Tamil Nadu, India reduced water consumption by nearly 35% (UNCTAD 2022). In the present study, two other variables, viz., 'having male members as the decision-making authority for the implementation of RRWH' and 'provision of extra income for adopting RRWH', were found to be 19.9% and 7.4%, respectively, which can be considered as an important observation.

6 Recommendations

To supply safe water for the well-being of the affected community the following strategies are likely to be adopted.

- Promote job opportunities and alternative livelihoods to reduce over-dependency on "Boro" cultivation.
- Increase the number of parks with water bodies and injection wells for all built-up areas, percolation, and farming ponds to avoid land acquisition in rural areas.
- Generate a sense of ownership by giving 'pond patta' to the households in a rural area, i.e., one kind of pond audit every 5 years and reward households by giving some kinds of incentives.
- Under the RWH scheme, the government should include the affected local community, and make them aware through training and workshops involving medical personnel, NGOs, and local clubs at the gram panchayat level.
- It is necessary to introduce appropriate land use for the exact purpose, water meter, restrict and penalize illegal construction, and landfilling.
- RRWH should be made mandatory even for small rooftops in rural areas with low-cost water treatment technology.
- Subsidize the underprivileged section of the population to install the RRWH.
- Need to coordinate with the concerned Government department to overcome barriers like social, legal, and political issues toward the adoption of RWH.
- Combining state and central governmental concrete housing schemes introduced for low-income families with RRWH into one scheme.

7 Conclusions

In any area, the magnitude of the water problem is aggravated by the complexity of socio-economic aspects. It is pertinent to mention that all the wells not only suffered from the decline of groundwater (ranging from 1.40 to 4.25 mbgl), but also cases of negative recharge were also observed. Furthermore, (30%–60%) vulnerability was observed in the surface water deficit map. As population density is high, water demand was steep due to obvious reasons and there had been an issue of the quality of the supplied water. The levels of As (13 to 69.5 ppb) and Fe (0.3 to 6.77 ppm) observed were higher than the respective safe limits, which further implied that 99.2%

of the rural households faced multiple water-related risks. In the logistic regression model, four variables were found to have a 1% level of significance, whereas another variable had a 5% level of significance. This implies that the model indicates a 76.1% probability of water risk. However, high rainfall (1719.9 mm) in the monsoon season has substantial potential for groundwater recharge. Despite this fact, there has been a decline in groundwater levels in the post-monsoon period. Barasat II block has an alarming scenario as available net groundwater for future use is only 27.14%, with the situation heading toward a semi-critical condition. Unfortunately, people were forced to drink shallow groundwater and the excess rainwater was wasted through runoff due to sheer unawareness. Households are conditionally willing to adopt RRWH provided adequate governmental support is available. This part of the present research, so far unexplored would be useful for the local administration in formulating an action plan for contaminated areas to introduce RRWH in the study area. Though this study was successful in achieving its key objectives, there are some issues yet to be resolved, e.g., well log assessment for in-situ recharge coupled with cost–benefit analysis.

8 Conflicts of Interest

The authors declare no conflict of interest.

9 Ethical Statement

Not applicable.

Acknowledgements The authors thank the SWID, PHED, CGWB Kolkata, Government of West Bengal, and IMD India for providing necessary support for this research work.

Funding This research received no specific grant from any funding agency in the public, commercial, or not-for-profit sectors.

References

Abatzoglou JT, Dobrowski SZ, Parks SA, Hegewisch KC (2017) Data descriptor: TerraClimate, a high-resolution global dataset of monthly climate and climatic water balance from 1958 to 2015. Sci Data 5:1–12. https://doi.org/10.1038/sdata.2017.191

Abbas F, Al-Naemi S, Farooque AA, Phillips M (2023) A review on the water dimensions, security, and governance for two distinct regions. Water Switz 15(1):1–20. https://doi.org/10.3390/w15010208

Adham A, Sayl KN, Abed R, Abdeladhim MA, Wesseling JG, Riksen M, Fleskens L, Karim U, Ritsema CJ (2018) A GIS-based approach for identifying potential sites for harvesting rainwater

in the Western Desert of Iraq. Int Soil Water Conserv Res 6(4):297–304. https://doi.org/10.1016/j.iswcr.2018.07.003
Asian Development Bank (ADB) (2018) West Bengal drinking water sector improvement project: arsenic and fluoride in drinking water in West Bengal—Characteristics, Implications, and Mitigation. https://www.adb.org/projects/documents/ind-49107-006-rrp
Asian Development Bank (ADB) (2020) West Bengal drinking water sector improvement project: arsenic and fluoride in drinking water in West Bengal—Characteristics, Implications, and Mitigation. 0–26.
Bandyopadhyay M, Basu R (2017) Crisis of fresh water in South 24 Parganas District, West Bengal: causes and consequences. IOSR J Humanit Soc Sci 22(06):04–15. https://doi.org/10.9790/0837-2206040415
Banerjee P (2016) Report on case study selection (Working paper). South Asia Consortium for Interdisciplinary Water Resources Studies (SaciWATERs). http://saciwaters.org/shiftinggrounds/pdfs/Reportoncasestudyselection-poulomi.pdf
Barenhoff L (2011) Urban rainwater harvesting in Hyderabad an option to overcome the water crisis of a megacity of tomorrow ? In Lutz Barenhoff Prof. (Issue May)
Basu B, Sil S (2003) Environment planning arsenic mapping for North 24- Pargana District of West Bengal—using GIS and Remote Sensing technology. Map India
Bera S, Das Chatterjee N (2019) Mapping and monitoring of land use dynamics with their change hotspot in North 24-Parganas district, India: a geospatial—and statistical-based approach. Model Earth Syst Environ 5(4):1529–1551. https://doi.org/10.1007/s40808-019-00601-2
Bhattacharya A, Sarani C, Kumar Roy A, Karthik D, Singh A, Kumari S, Kumari V, Daksh K, Kumar P, Kumar Mishra A (2019) An analysis of arsenic contamination in the groundwater of India, Bangladesh and Nepal with a special focus on the stabilisation of arsenic-laden sludge from arsenic filters. Electron J Geotech Eng 24(01):1–34
Bhu-Jal N (2009) Central ground water board, 24, 1
Biswas S, Debsarkar A, Pal M (2020) Water insufficiency, health hazards and rainwater harvesting in North 24 Parganas, West Bengal, India: results of a socio-economic survey. Arthaniti J Econ Theory Pract, 097674792096339. https://doi.org/10.1177/0976747920963399
Burean of Indian Standard (BIS) (2012). Indian standards drinking water specifications IS 10500:2012. Bureau of Indian Standard, Indian standards drinking water specifications, 2, 11. http://cgwb.gov.in/Documents/WQ-standards.pdf
Census of India (2011) District Census Handbook, North 24 Parganas, Village and Town Directory. https://censusindia.gov.in/2011census/dchb/DCHB_A/19/1911_PART_A_DCHB_NORTHTWENTYFOURPARGANAS.pdf
Chowdhury S, Chakraborty S, Lodh R (2016) Analysis of rainwater harvesting method for supply of potable water: a case study of Gosaba, South 24 Parganas, India. Intech, I (tourism), 13. https://doi.org/10.5772/57353
District human Development Report (DHDR) (2010). District human development report north 24 parganas. http://wbplan.gov.in/HumanDev/DHDR/24pgsnorth.PDF
Jasrotia AS, Kumar R, Taloor AK, Saraf AK (2019) Artificial recharge to groundwater using geospatial and groundwater modelling techniques in North Western Himalaya, India. Arab J Geosci, 12(24). https://doi.org/10.1007/s12517-019-4855-5
Jones ER, Bierkens MFP, Wanders N, Sutanudjaja EH, Van Beek LPH, Van Vliet MTH (2023) DynQual v1.0: a high-resolution global surface water quality model. Geosci Model Dev 16(15):4481–4500. https://doi.org/10.5194/gmd-16-4481-2023
Karimi H, Zeinivand H (2021) Integrating runoff map of a spatially distributed model and thematic layers for identifying potential rainwater harvesting suitability sites using GIS techniques. Geocarto Int 36(3):320–339. https://doi.org/10.1080/10106049.2019.1608590
Kumar R, Kumar S, Ghosh A (2021) Current scenario of groundwater arsenic contamination in West Bengal and its mitigation approach. Arsenic Tox Challenges Solutions. https://doi.org/10.1007/978-981-33-6068-6

Mahmoud SH, Mohammad FS, Alazba AA (2015) Delineation of potential sites for rainwater harvesting structures using a geographic information system-based decision support system. Hydrol Res 46(4):591–606. https://doi.org/10.2166/nh.2014.054

Maina CW, Raude JM (2016) Assessing land suitability for rainwater harvesting using geospatial techniques: a case study of Njoro catchment Kenya. Appl Environ Soil Sci. https://doi.org/10.1155/2016/4676435

National Rural Drinking Water Programme (NRDWP) (2013). National rural drinking water programme: movements towards ensuring people's drinking water security in Rural India. In: Ministry for drinking water & sanitation. http://indiawater.gov.in/imisreports/nrdwpmain.aspx

Paul S, Das CS (2021) An investigation of groundwater vulnerability in the North 24 Parganas District using DRASTIC and hybrid-DRASTIC models: a case study. Environ Adv 5:100093. https://doi.org/10.1016/j.envadv.2021.100093

Public Health Engineering Department or PHED (2018) IND : West Bengal Drinking water sector improvement project—Bulk water supply for two blocks in Bankura (Issue July)

Rahman S, Khan MTR, Akib S, Din NBC, Biswas SK, Shirazi SM (2014) Sustainability of rainwater harvesting system in terms of water quality. Sci World J. https://doi.org/10.1155/2014/721357

Rana VK, Suryanarayana TMV (2020) GIS-based multi criteria decision making method to identify potential runoff storage zones within watershed. Annals of GIS 26(2):149–168. https://doi.org/10.1080/19475683.2020.1733083

Roy C, Sati VP, Biswas A, Kumar S (2023) Status of drinking water, sanitation facilities, and hygiene in West Bengal: evidence from the National Family Health Survey of India (NFHS), 2019–2021. J Water Sanit Hygiene Dev 13(1):50–62. https://doi.org/10.2166/washdev.2023.228

Roy S (2019) Arsenic contamination of groundwater and its socio-economic impacts in West Bengal. 14(6), 59–62. https://doi.org/10.9790/2402-1406045962

Sayl KN, Mohammed AS, Ahmed AD (2020) GIS-based approach for rainwater harvesting site selection. IOP Conf Series Mater Sci Eng, 737(1). https://doi.org/10.1088/1757-899X/737/1/012246

Soren DDL, Barman J, Roy KC, Naskar S, Biswas B (2023) Evaluation of groundwater quality of South Bengal, India. J Earth Syst Sci, 132(3). https://doi.org/10.1007/s12040-023-02152-8

Tzanakakis VA, Paranychianakis NV, Angelakis AN (2020) Water supply and water scarcity. Water Switz 12(9):1–16. https://doi.org/10.3390/w12092347

UNCTAD (2022) Ensuring safe water and sanitation for all: a solution through science, Technology and Innovation. https://shop.un.org/

UNESCO (2023) Partnerships and co-operation for water

UNU-INWEH (2023) Global water security 2023 assessment

WaterAid (2011) Rainwater harvesting for recharging shallow groundwater

Integrating RS and GIS Techniques for Enhanced Natural Resource Management: A Comprehensive SWAT Model Approach

T. M. Sharannya

Abstract The integration of remote sensing (RS) and geographic information systems (GIS) technologies in hydrology has become increasingly important, offering solutions to understand the complexities of surface water and groundwater dynamics and their environmental interactions. This integration serves as a valuable tool for improving natural resource management by facilitating the acquisition of spatially explicit data, crucial for assessing environmental changes and predicting future trends. Focusing on the Valapattanam River Basin, this study utilizes the Soil and Water Assessment Tool (SWAT) to investigate the spatio-temporal variability of streamflow. The efficiency of the model is confirmed through calibration (R^2: 0.92, NSE: 0.87, PBIAS: 9.6) and subsequent validation (R^2: 0.89, NSE: 0.85, PBIAS: 3.9) covering the period from 1997 to 2022. Spatio-temporal analyses of surface runoff and groundwater contribution to streamflow reveal distinct patterns within the Valapattanam River Basin. Downstream streamflow is notably influenced by groundwater, peaking between August and October, while surface runoff predominantly shapes upstream regions, especially from June to August. Additionally, the analysis highlights a temporal lag, indicating a one-month delay in groundwater contribution to streamflow. This study emphasizes the crucial role of spatio-temporal streamflow analysis in formulating effective water resource management strategies and strategic planning in the study area. Future research endeavors may explore further integration of advanced modeling techniques and additional data sources to address emerging challenges in hydrological management and environmental conservation.

Keywords Remote sensing · GIS · SWAT · Spatio-temporal · Hydrological management

T. M. Sharannya (✉)
Hydrology and Climatology Research Group, Centre for Water Resources Development and Management (CWRDM), Kunnamangalam, Kozhikode, Kerala 673571, India
e-mail: sharannyatm@gmail.com; sharannyatm@cwrdm.org

S. C. Pal and U. Chatterjee (eds.), *Surface, Sub-Surface Hydrology and Management*, Springer Geography, https://doi.org/10.1007/978-3-031-62376-9_30

1 Introduction

In recent years, there has been significant attention given to the integration of remote sensing (RS) and geographic information systems (GIS) technologies in the field of hydrology. Hydrology, dealing with the study of surface water, groundwater, and their interactions with various elements of the environment, presents a multifaceted domain. RS, a widely utilized technique conducted from diverse platforms such as satellites and aircraft, has emerged as a valuable tool for enhancing our understanding of surface and subsurface water conditions (Thakur et al. 2016). This approach encompasses geophysical surveys involving gravity, magnetics, and electromagnetics, with geophysical surveys uniquely capable of providing insights into subsurface information (Brunner et al. 2006).

RS techniques offer a means of collecting data over large geographical areas without direct physical contact, utilizing various sensors mounted on satellites or aircraft (Jafarbiglu and Pourreza 2022). These sensors capture data across different spectral bands, enabling the assessment of land cover, land use, vegetation indices, and surface temperature, among other parameters. GIS, on the other hand, provide the platform for organizing, storing, analyzing, and visualizing spatial data within a geographical context (John et al. 2020). Through GIS, spatial relationships and patterns can be identified, facilitating the generation of valuable insights into environmental processes and phenomena. The integration of RS and GIS techniques enhances the capabilities of natural resource management by enabling the acquisition of spatially explicit data at various scales, from regional to local levels (Trivedi et al. 2022). This spatially explicit information is invaluable for understanding the dynamics of natural systems, assessing environmental changes, and predicting future trends. By leveraging RS and GIS, decision-makers can gain insights into land cover changes, deforestation, urbanization, soil erosion, and water quality, among other critical environmental parameters (Jafarbiglu and Pourreza 2022).

Effective management of natural resources is crucial, especially in the face of changing climate conditions. With climate change impacting ecosystems and water availability, it becomes imperative to employ advanced tools and techniques for informed decision-making (Dwivedi et al. 2022). The application and integration of RS and GIS in hydrological models play a vital role in enhancing natural resource management practices (Krivoguz 2024). These technologies provide valuable insights into spatial patterns, land cover changes, and hydrological processes, allowing for more accurate assessments and predictions. Among the various hydrological models available, the Soil and Water Assessment Tool (SWAT) stands out as one of the most widely used tools for simulating watershed processes. SWAT is a process-based model developed to simulate the impact of land-management practices on water, sediment, and agricultural chemical yields in large, complex watersheds (Abate et al. 2024; Asadi et al. 2024; Sharannya et al. 2021). SWAT integrates various hydrological and land-management processes, including land use, soil properties, climate, and topography, to predict the quantity and quality of water resources within a watershed (Arnold et al. 1998). The model's ability to simulate the hydrological cycle and quantify the impacts of land-management practices makes it a valuable tool for assessing the efficacy of conservation strategies and guiding land use planning decisions. Additionally, SWAT offers a range of components and features that are valuable for simulating various aspects of hydrology, including water balance, sediment depletion, climatic change, agricultural development, and land-management methods (Guiamel and Lee 2020).

The SWAT model has capability to integrate complex interactions among land use, soil properties, and climate variables within a watershed, facilitating the analysis of spatio-temporal variability in streamflow (Swain et al. 2020). By adopting this modeling approach, a holistic understanding of hydrological processes is achieved, allowing for accurate simulation and prediction of streamflow under changing conditions. In the present study, we focus on the spatio-temporal analysis of streamflow in the Valapattanam River Basin using the SWAT model. This chapter aim is to provide valuable insights that can inform sustainable water resource management practices in the basin. This research holds significance not only for the local communities reliant on the basin but also for the broader scientific community engaged in understanding and mitigating the impacts of climate change on water resources globally. Hence, our study aims to explore the integration of RS and GIS techniques with the SWAT model to enhance natural resource management practices.

2 Materials and Methods

2.1 Study Area

The Western Ghats (WG) region, stretching over approximately 2300 km along India's western coastline, holds significant importance in influencing the Indian tropical monsoon systems and harbors distinctive forest ecosystems (Mudbhatkal and Amai 2018). Encompassing Gujarat, Maharashtra, and Karnataka, the Western Ghats span 100–200 km inland from the seacoast, exhibiting remarkable ecological diversity. Among its features, the Valapattanam River Basin (VRB) stands out, situated within the coordinates of 11°47"–12°26"N and 75°14"–76°09"E, characterized by elevations ranging from sea level to 1566 m, with a total length extending approximately 117 km from source to outlet and covering a drainage area of 2517.22 square kilometers. The VRB is primarily characterized by agricultural and plantation areas, interspersed with forest cover, and is supplemented by the presence of the Pazhassi Barrage upstream, facilitating irrigation activities.

2.2 The Hydrological Model—SWAT

The Soil and Water Assessment Tool (SWAT) is a comprehensive hydrological model utilized in environmental and water resource management. The SWAT model is a collaborative effort between the United States Department of Agriculture, Agricultural Research Services (USDA-ARS), and the Agricultural Experiment Station in Temple, Texas (Lu et al. 2015). As outlined by Michael and Jain (2013), the computational elements of the model are categorized into eight primary divisions: hydrology, weather, soil temperature, crop growth, sedimentation, pesticides, nutrients, and agricultural management. It relies on various physical inputs such as weather variables, topography, soil properties, land-management practices, and vegetation within the catchment area. The calibration and validation process of the model utilized the sequential uncertainty fitting version 2 (SUFI2) method, as implemented in the SWAT calibration and uncertainty program (SWAT-CUP) tool specifically designed for SWAT. SUFI2 accounts for various sources of uncertainties including driving variables like rainfall, conceptual model, parameters, and measured data. Through multiple iterations, parameter ranges were adjusted based on the suggestions from the SUFI2 program and their physical constraints. Prior to calibration, sensitivity analysis was conducted to streamline parameters and reduce computational time. Each iteration involved updating parameter ranges based on sensitivity matrix calculations, ensuring that ranges became progressively narrower and centered around optimal simulation results. This iterative approach allowed for the refinement of parameter values until an optimized solution was achieved. Successful calibration and validation pave the way for model development and subsequent analysis. The execution of the SWAT model is visualized in the flowchart depicted in Fig. 1.

SWAT explicitly simulates the physical processes associated with water flow, sediment transport, crop growth, nutrient cycling, and other pertinent phenomena.

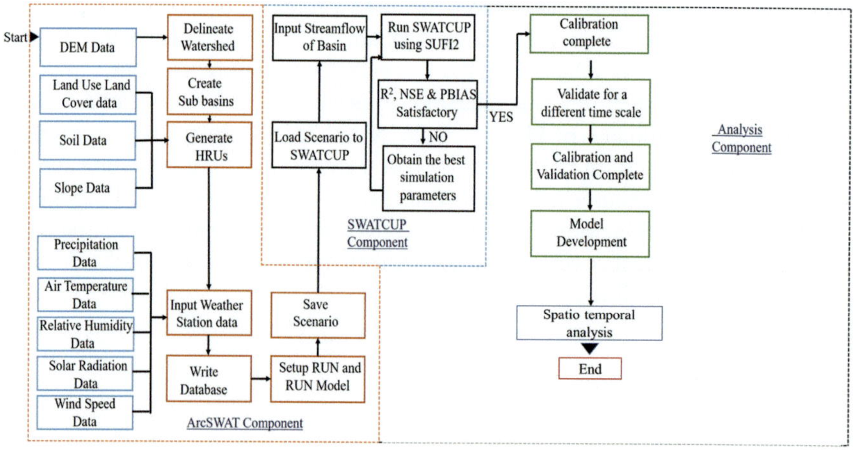

Fig. 1 Flowchart of the methodology

The water balance equation utilized to simulate the hydrologic cycle is represented by the following equation:

$$SW_t = SW_o + \left[R_{day} + Q_{surf} + E_a + W_{seep} + Q_{gw}\right]$$

$$SW_t = \text{Final soil water content in mm } H_2O$$

$$SW_o = \text{Initial soil water content in mm } H_2O,\ \text{t time in days},$$

$$R_{day} = \text{Amount of precipitation on day } i \text{ in mm } H_2O,$$

$$Q_{surf} = \text{Amount of surface runoff on day } i \text{ in mm } H_2O$$

$$E_a = \text{Amount of evapotranspiration on day } i \text{ in mm } H_2O$$

W_{seep} = Amount of percolation and bypass exiting the soil profile bottom on day i in mm H_2O

Q_{gw} = Amount of return flow on day i in mm H_2O.

The input data for SWAT model to generate runoff are digital elevation model (DEM), soil map, land use map, weather data, and outlet data which is shown in Table 1.

Table 1 Information on input data for SWAT model

SN	Data type	Source	Data parameter	Spatial/temporal resolution
1	Meteorological data	Indian meteorological department (IMD) https://mausam.imd.gov.in/	Precipitation, T-max and T-min	1995–2022
2	Topography map	Unites State geological survey (earth explorer)/shutter radar topographic mission (SRTM) http://srtm.csi.cgiar.org/	Digital elevation model (DEM)	30.48 m
3	Soil map	FAO (Digital soil map of the world)	Soil classification	916.72 m
4	Land use/land cover map	BHUVAN land cover database/ LANDSAT images USGS Archive; http://glovis.usgs.gov/	Land use classification	100 m
5	Hydrological data	India water resources information system, central water commission	Daily river discharge data (Perumannu)	1997–2022

3 Results and Discussions

3.1 Assessment of Runoff Using SWAT Model

The SWAT study facilitates adjustments to parameters within the river basin, encompassing crucial climate and soil data essential for hydrological modeling. To streamline database management, the SWAT offers the capability to refine SWAT input tables into a preferred format suitable for incorporation into model input files. This process involves condensing the SWAT databases to enhance accessibility, ultimately contributing to the development of the SWAT2012.mdb database. Through this initiative, the SWAT2012.mdb tables undergo editing and organization to render them compatible with various SWAT projects. It is advisable to generate a backup copy of the SWAT2012.mdb prior to embarking on any SWAT project to mitigate risks associated with data manipulation. The SWAT model was executed for the specified time period to establish the hydrological model for the river basin region. Subsequently, average monthly runoff values and other hydrological components, including sediment and evapotranspiration, were obtained. As shown in Fig. 2, the river basin experiences an annual average precipitation of 3697.1 mm, with 1331.06 mm contributing to surface runoff, 65.44 mm as lateral flow. The calculated weighted curve number for the entire river basin is 67, indicating a reasonable runoff potential in the study area.

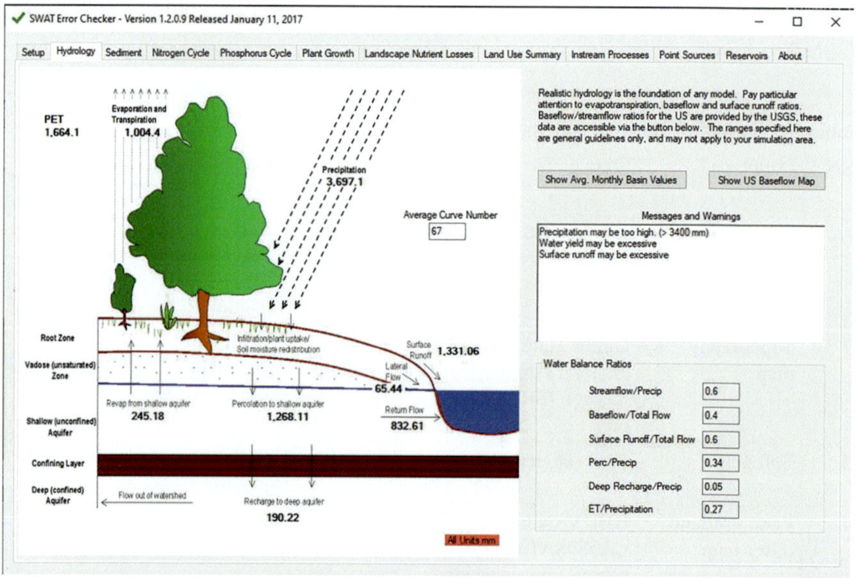

Fig. 2 Hydrology of the Valapattanam River Basin

Table 2 Water balance ratios for the Valapattanam River Basin

S N	Water balance ratio	Value
1	Streamflow/Precipitation	0.60
	Baseflow/total flow	0.40
	Surface runoff/total flow	0.60
2	Percolation/Precipitation	0.34
3	Deep recharge/Precipitation	0.05
4	ET/Precipitation	0.27

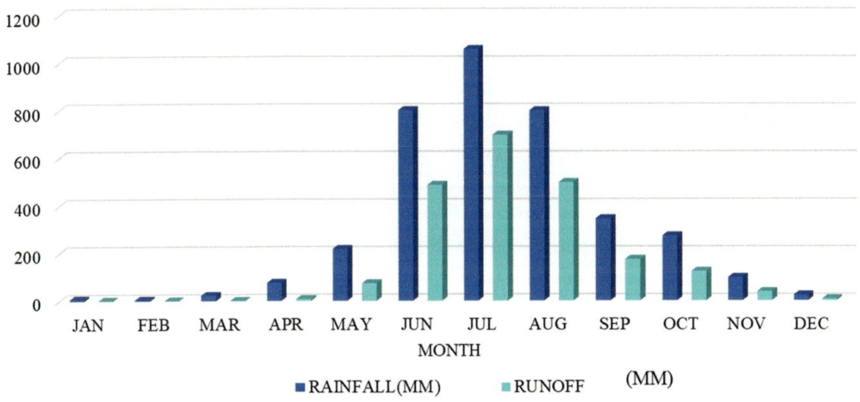

Fig. 3 Evaluation between average monthly rainfall and average monthly SWAT simulated runoff of the Valapattanam River Basin

Based on Table 2, 60% of the total precipitation contributes to the entire runoff, indicating a relatively high runoff efficiency. Furthermore, 60% of the total runoff contributes to the surface runoff.

The SWAT model simulation for the Valapattanam River Basin revealed that the highest average monthly surface runoff occurred in July (Fig. 3), whereas the lowest monthly average surface runoff was observed in February. Furthermore, the lateral flow was found to be at its maximum in July, while the lowest lateral flow was projected for January, February, and March.

3.2 Calibration and Validation of SWAT Model Using SUFI2 Algorithm for Runoff Simulation

SWAT Simulation

The SWAT simulation was conducted for the period from January 1, 1997, to December 31, 2022. The SWAT model was run at regular and monthly time intervals, with data divided into two, for calibration and validation. The calibration period used data from 1997 to 2013, while the validation period utilized data from 2014 to 2022, and the first two years (1995, 1996) were considered as a warm-up period.

The model setup was completed using DEM, soil, land use, and slope data. After setting up the model, it was run for the period 1997–2013 to assess the trend of model-simulated surface runoff and adjust the calibration parameters.

Model Calibration

For model calibration, monthly data of surface runoff during the monsoon season (June to October) of 1995, 1996 were used as a warming period. Data from 1997 to 2013 were then used for model calibration, and data from 2014 to 2022 were used for validation to compare the simulation results with observed values. Performance indices and the obtained values are tabulated in the Table 3.

Out of several calibration parameters, six parameters were selected for calibration and validation to achieve close agreement between observed and simulated surface runoff. These parameters were calibrated and validated using the SWAT-CUP model with the SUFI2 algorithm technique. The selected calibration parameters and the fitted values for these parameters during the monthly simulation period are presented in Table 4

The trend of observed and simulated runoff exhibits a significant similarity. During the calibration process, an excellent Coefficient of Determination (R^2) value of 0.92 was achieved. The Nash–Sutcliffe Efficiency (NSE) coefficient value of 0.87 and a Percent Bias (PBIAS) of 9.6 were observed for the period of 1997–2013. As a result, the trends between observed and simulated runoff in the entire river basin demonstrate a close agreement.

Table 3 Performance indices and the values obtained

Sl. No	Performance indices	Value during calibration	Value during validation
1	Coefficient of determination (R^2)	0.92	0.89
2	Nash–Sutcliffe coefficient efficiency (NSE)	0.87	0.85
3	Percent bias (PBIAS)	9.6	3.9

Table 4 Calibration parameters

SN	Parameters	Calibration fitted value
1	*r*__CN2.mgt	0.032082
2	*r*__SOL_BD().sol	0.298492
3	*r*__SOL_Z().sol	0.828393
4	*r*__RCHRG_DP.gw	0.616907
5	*r*__REVAPMN.gw	152.485153
6	*r*__ESCO.bsn	0.338580

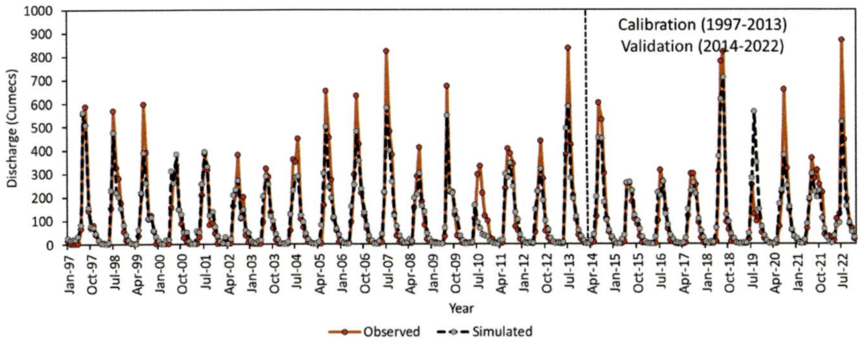

Fig. 4 Observed and simulated values after calibration and validation

Model Validation

Once calibration is done then the calibrated values is used for validation purpose on a monthly time step for a year of 2014–2022. In the case of validation (2014–2022) the model had given a Coefficient of Determination (R^2) value of 0.89. The Nash–Sutcliffe Efficiency (NSE) coefficient value of 0.85 and a Percent Bias (PBIAS) of 3.9 were observed. The graph after validation and calibration is shown below which shows a close relationship between observed and simulated results (Fig. 4).

3.3 *Sensitivity Analysis of Swat Model Parameters for River Basin*

Sensitivity Analysis of the Parameters

Sensitivity analysis of the SWAT model parameters for the Valapattanam River Basin was conducted to assess the impact of each parameter on the model output and to understand their individual influences. Six parameters were selected for sensitivity analysis as shown in Table 5. The input parameter ranges were determined using the SWAT-CUP user manual and the built-in parameter ranges provided in the Absolute_SWAT_Values.txt file of the SWAT-CUP model.

All the six parameters which are considered were mainly significantly sensitive parameters based on t-stat greater than 0.5 and *P*-value less than 0.7. These parameters include r__CN2.mgt, r__SOL_BD().sol, r__SOL_Z().sol, r__RCHRG_DP.gw, r__REVAPMN.gw, and r__ESCO.bsn. The global sensitivity analysis was performed to obtain the t-stat value and *P*-value of various SWAT model parameters for the river basin. These parameters were standardized at a 95% confidence level or 5% level of uncertainty to be considered acceptable in their calibrated states.

It is clear that the SCS runoff curve number is the most sensitive parameter of all. The higher sensitivity of groundwater parameters can be used to determine the irregular nature of the streams in the area. As most of the streams are isolated of

Table 5 Sensitivity analysis of parameters

SN	Rank	Parameters	*t*-stat	*p*-stat
1	1	*r*__CN2.mgt	59.142	0.000
2	7	*r*__SOL_BD().sol	15.5267	0.000
3	6	*r*__SOL_Z().sol	7.6603	0.000
4	8	*r*__RCHRG_DP.gw	1.1445	0.2529
5	10	*r*__REVAPMN.gw	0.8123	0.4169
6	5	*r*__ESCO.bsn	0.7094	0.4784

irregular flow, existence, the groundwater parameters play a significant role in influencing the hydrological response. On the other hand, the higher sensitivity of soil parameters can be explained by the fact that the soils in the river basin allow for a considerable amount of lateral flow, especially during the monsoon season. This lateral flow contributes significantly to the overall runoff in the basin.

3.4 Spatio-Temporal Analyses

A spatial map series consists of multiple pages which are generated from a single layout with each of the pages displaying a distinct map area. The map is created to distinguish and understand the differences and changes that occurred in the study area. The map is created by using the data which are to be selected from the SWAT model which is executed after the values of the parameters are added manually.

In this process of creating a map, several steps are involved which consists of the collection of data. The data used in the creation of the map is obtained from the SWAT model which is run after the values of the parameters are added manually. The spatial map created here consists of surface runoff and ground water. As spatial maps are created to observe the changes and differences in the study area, the years are divided into 3 decades. The 1st decade consists of the years from 1997 to 2005 (D1). The second decade includes 2006–2014 (D2) and the third decade from 2015 to 2022 (D3).

The spatio-temporal variations in surface runoff and groundwater contribution to streamflow are depicted in Figs. 5 and 6, respectively. A consistent observation is the elevated streamflow rates near to the reservoir. This is further delineated by distinct influences: while groundwater exhibits a dominant role in downstream areas, surface runoff emerges as a primary contributor in the upstream regions. The temporal dynamics, as illustrated in Figs. 5 and 6, surface runoff exhibits pronounced peaks from June through August. In contrast, the influence of groundwater on streamflow becomes notably prominent in the subsequent months from August to October. Such temporal patterns suggest a lagged response of over a month between groundwater dynamics and its consequential impact on streamflow. This intricate understanding underscores the importance of spatio-temporal streamflow analysis for informed water resource management, ecological sustainability, and strategic planning in the study area.

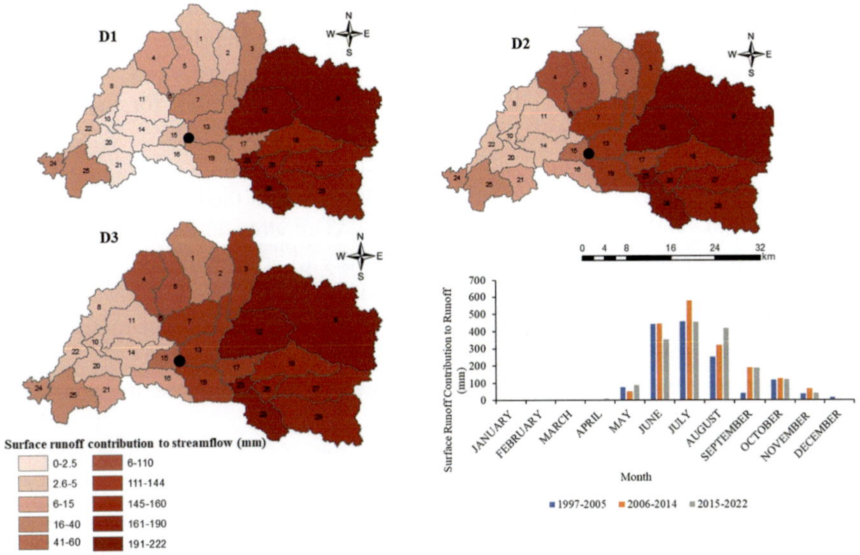

Fig. 5 Spatio-temporal variation in surface runoff

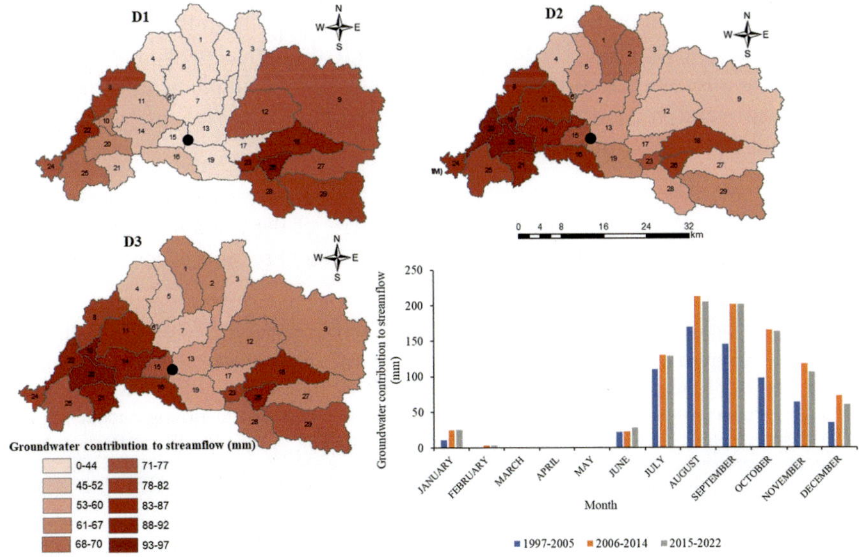

Fig. 6 Spatio-temporal variation in groundwater contribution to streamflow

4 Conclusion

The current study underscores the pivotal role of integrating RS, GIS, and hydrological modeling techniques in addressing the complexities of water resource management amidst the challenges posed by climate change. By using these advanced technologies, we can enhance our understanding of hydrological processes, assess environmental changes, and inform decision-making processes crucial for ensuring the sustainability of water resources for present and future generations.

In the study focusing on the Valapattanam River Basin, we employed the SWAT for hydrological analysis to identify the water dynamics of the study area. Through calibration and validation exercises using the SWAT model coupled with the SUFI2 algorithm, we demonstrated the model's capability to accurately simulate surface runoff, thus enhancing its predictive abilities. The model's efficiency was evident during both calibration (R^2: 0.92, NSE: 0.87, PBIAS: 9.6) and subsequent validation (R^2: 0.89, NSE: 0.85, PBIAS: 3.9) for the period spanning 1997–2022.

Furthermore, sensitivity analysis unveiled the significant influence of key parameters on model outputs, emphasizing the criticality of parameter calibration for reliable predictions. Spatio-temporal analyses of surface runoff and groundwater contribution to streamflow revealed intricate patterns within the Valapattanam River Basin. These analyses delineated the interplay between surface runoff and groundwater dynamics, with distinct temporal variations observed across different regions of the basin. Notably, groundwater significantly influenced downstream streamflow, with peak contributions observed between August and October, while surface runoff played a dominant role upstream, particularly during June to August. The analysis also indicated a temporal lag, signifying a one-month delay in the contribution of groundwater to streamflow. The insights from our hydrological model hold immense potential for informing decision-makers and stakeholders in implementing the necessary measures and precautions for water management to address the current and future water challenges of the study area.

References

Abate BZ, Alaminie AA, Assefa TT, Tigabu TB, He L (2024) Modeling climate change impacts on blue and green water of the Kobo-Golina River in data-scarce upper Danakil basin Ethiopia. J Hydrol Reg Stud. https://doi.org/10.1016/j.ejrh.2024.101756

Arnold JG, Srinivasan R, Muttiah RS, Williams JR (1998) Large area hydrologic modeling and assessment part I: model development. J Am Water Resour Assoc Am Water Resour Assoc 34(1):73–89

Asadi S, Mousavi SJ, López-Ballesteros A, Senent-Aparicio J (2024) Analyzing hydrological alteration and environmental flows in a highly anthropized agricultural river basin system using SWAT+, WEAP and IAHRIS. J Hydrol Reg Stud 52:101738. https://doi.org/10.1016/j.ejrh.2024.101738

Brunner P, Franssen HJH, Kgothang L, Gottwein BP, Kinzelbach W (2006) How can remote sensing contribute in groundwater modeling. Hydrogeol J 15:5–18

Dwivedi YK, Hughes L, Kar AK, Baabdullah AM, Grover P, Abbas R, Andreini D, Abumoghli I, Barlette Y, Bunker D, Chandra Kruse L, Constantiou I, Davison RM, De R, Dubey R, Fenby-Taylor H, Gupta B, He W, Kodama M, Wade M (2022) Climate change and COP26: Are digital technologies and information management part of the problem or the solution? An editorial reflection and call to action. Int J Inf Manage. https://doi.org/10.1016/j.ijinfomgt.2021.102456

Guiamel IA, Lee HS (2020) Watershed modelling of the Mindanao River Basin in the Philippines using the SWAT for water resource management. Civil Eng J 6:626–648

Jafarbiglu H, Pourreza A (2022) A comprehensive review of remote sensing platforms, sensors, and applications in nut crops. In: Computers and electronics in agriculture. Elsevier B.V. Vol 197 https://doi.org/10.1016/j.compag.2022.106844

John J, Bindu G, Srimuruganandam B, Wadhwa A, Rajan P (2020) Land use/land cover and land surface temperature analysis in Wayanad district, India, using satellite imagery. Ann GIS 26(4):343–360. https://doi.org/10.1080/19475683.2020.1733662

Krivoguz D (2024) Geo-spatial analysis of urbanization and environmental changes with deep neural networks: Insights from a three-decade study in Kerch peninsula. Eco Inform. https://doi.org/10.1016/j.ecoinf.2024.102513

Lu Z, Zou S, Qin Z, Yang Y, Xiao H, Wei Y, Zhang K, Xie J (2015) Hydrologic responses to land use change in the loess plateau: case study in the upper Fenhe river Watershed. Adv Meteorol, https://doi.org/10.1155/2015/676030

Mamo KH, Jain MK (2013) Runoff and sediment modeling using SWAT in Gumera catchment Ethiopia. Open J Mod Hydrol 03(04):196–205. https://doi.org/10.4236/ojmh.2013.34024

Mudbhatkal A, Amai M (2018) Regional climate trends and topographic influence over the Western Ghat catchments of India. Int J Climatol 38(5):2265–2279. https://doi.org/10.1002/joc.5333

Sharannya TM, Venkatesh K, Mudbhatkal A, Dineshkumar M, Mahesha A (2021). Effects of land use and climate change on water scarcity in rivers of the Western Ghats of India. Environ Monit Assess, 193(12). https://doi.org/10.1007/s10661-021-09598-7

Swain SS, Mishra A, Sahoo B, Chatterjee C (2020) Water scarcity-risk assessment in data-scarce river basins under decadal climate change using a hydrological modelling approach. J Hydrol 590:125260

Thakur JK, Singh SK, Ekanthalu VS (2016) Integrating remote sensing, geographic information systems and global positioning system techniques with hydrological modeling. Appl Water Sci 7(4):1595–1608. https://doi.org/10.1007/s13201-016-0384-5

Trivedi A, Rao KVR, Rajwade Y, Yadav D, Verma NS (2022) Remote sensing and geographic information system applications for precision farming and natural resource management. Indian Journal of Ecology. https://doi.org/10.55362/IJE/2022/3707

GPSR Compliance

The European Union's (EU) General Product Safety Regulation (GPSR) is a set of rules that requires consumer products to be safe and our obligations to ensure this.

If you have any concerns about our products, you can contact us on ProductSafety@springernature.com

In case Publisher is established outside the EU, the EU authorized representative is:

Springer Nature Customer Service Center GmbH
Europaplatz 3
69115 Heidelberg, Germany

Batch number: 08321895

Printed by Printforce, the Netherlands